# Lecture Notes in Networks and Systems

Volume 234

The series “Lecture Notes in Networks and Systems” publishes the latest developments in Networks and Systems—quickly, informally and with high quality. Original research reported in proceedings and post-proceedings represents the core of LNNS.

Volumes published in LNNS embrace all aspects and subfields of, as well as new challenges in, Networks and Systems.

The series contains proceedings and edited volumes in systems and networks, spanning the areas of Cyber-Physical Systems, Autonomous Systems, Sensor Networks, Control Systems, Energy Systems, Automotive Systems, Biological Systems, Vehicular Networking and Connected Vehicles, Aerospace Systems, Automation, Manufacturing, Smart Grids, Nonlinear Systems, Power Systems, Robotics, Social Systems, Economic Systems and other. Of particular value to both the contributors and the readership are the short publication timeframe and the world-wide distribution and exposure which enable both a wide and rapid dissemination of research output.

The series covers the theory, applications, and perspectives on the state of the art and future developments relevant to systems and networks, decision making, control, complex processes and related areas, as embedded in the fields of interdisciplinary and applied sciences, engineering, computer science, physics, economics, social, and life sciences, as well as the paradigms and methodologies behind them.

Indexed by SCOPUS, INSPEC, WTI Frankfurt eG, zbMATH, SCImago.

All books published in the series are submitted for consideration in Web of Science.

For proposals from Asia please contact Aninda Bose (aninda.bose@springer.com).

Svetlana G. Maximova · Roman I. Raikin ·
Alexander A. Chibilev · Marina M. Silantyeva
Editors

# Advances in Natural, Human-Made, and Coupled Human-Natural Systems Research

Volume 1

Springer

*Editors*
Svetlana G. Maximova
Altai State University
Barnaul, Russia

Alexander A. Chibilev
Steppe Institute of the Ural Branch
Russian Academy of Sciences
Orenburg, Russia

Roman I. Raikin
Department of Radiophysics
and Theoretical Physics
Altai State University
Barnaul, Russia

Marina M. Silantyeva
Biological Sciences, Faculty of Biology
Altai State University
Barnaul, Russia

ISSN 2367-3370 ISSN 2367-3389 (electronic)
Lecture Notes in Networks and Systems
ISBN 978-3-030-75482-2 ISBN 978-3-030-75483-9 (eBook)
https://doi.org/10.1007/978-3-030-75483-9



This Springer imprint is published by the registered company Springer Nature Switzerland AG
The registered company address is: Gewerbestrasse 11, 6330 Cham, Switzerland

# Contents

**Systemic Challenges and Changes in Education Systems in Russia and Around the Globe**

# Advances in Natural and Coupled Human-Natural Systems Research

# The Development of Scientific Approaches to Using Land Resources and Agro-Technologies within the Steppe Zone of the Orenburg Pre-urals Region

**Yury A. Gulyanov, Sergey V. Levykin, and Alexander A. Chibilyov (jr.)**

**Abstract** In this study, we developed the methods of increasing the productivity of spring and winter wheat in the steppe zone of the Orenburg Pre-Urals region, Russia. We used conventional methods for field tests, laboratory tests, attendant observations, and calculations. Statistical analysis was performed via Microsoft Excel Analysis ToolPak. We determined that the steppe zone of the Orenburg Pre-Urals has significant unused resources that could be utilized in increasing the gross yields of grain. The practice of water-conservation is widespread in this area. The climate and soil conditions allow growing special kinds of grain on high-fertile agricultural lands. However, this could only be achieved by introducing specialized methods of agricultural science, promoting rational use of land resources, nature conservation, and preservation of steppe biodiversity.

**Keywords** Stable development · Environment · Land resources · Efficient use · Winter and spring wheat

## 1 Introduction

We can estimate the food security of countries by examining its production of grain. In the changing environmental and climatic conditions, the steady production of high-quality wheat is a priority objective of the Russian Federation. In the steppe zones of Russia, the optimization of a land-use structure will inevitably reduce the usage of low-fertility land and increase the productivity of the high-fertile ones. The Orenburg

Y. A. Gulyanov (✉) · S. V. Levykin · A. A. Chibilyov (jr.)
Orenburg Federal Research Center of the Ural Branch of the Russian Academy of Sciences
Institute of Steppe of Ural Branch of the Russian Academy of Sciences, Orenburg, Russia
e-mail: iury.gulynov@rambler.ru

S. V. Levykin
e-mail: stepevedy@yandex.ru

A. A. Chibilyov (jr.)
e-mail: a.a.ml@mail.ru

 
S. G. Maximova et al. (eds.), *Advances in Natural, Human-Made, and Coupled Human-Natural Systems Research*, Lecture Notes in Networks and Systems 234,
https://doi.org/10.1007/978-3-030-75483-9_1

Region, located in the steppe zone in the European part of Russia, has considerable plots of annual crops and leguminous plants (around 2600–2700 thousand ha). This region is an important grain producer in Russia. It has large areas of fertile chernozem soils, long growth periods, and high amount of sunshine. However, the production of crops in the Orenburg Region is not stable. In the current decade, high crop yields (3.2 to 4.2 million tons) were harvested only in 2016 and 2017. According to the government report, high productivity (1.25–1.65 t/ha) in those years was caused by favorable weather conditions and competent technological policies. However, in the following years, around 2.0 million tons of crops (1.0 t/ha on average) were harvested from almost 2.4 million ha of crop plantations (EMISS n.d.).

This means that, despite the positive organizational and technological actions, the production of crops still majorly depends on other factors. The situation is worsened by global climate change. It causes unstable weather conditions during the growth period, increasing the frequency of droughts and hot, dry winds. In 2018, more than 350 thousand ha of cropland was lost to droughts and rapid temperature changes. Wheat is a staple food of the Russian population. It is used for baked goods, pasta, pastries, and other cereal products. The statistical analysis of agricultural indexes in the Orenburg.

Region of Russia for the last decade did not reveal a statistically significant dependence of the gross yield of spring wheat (r = 0.110) on the crop area (Table 1). However, the gross yield depended on productivity (r = 0.991), which accounted for 98.2% of the variation. The dependence can be described by a regression equation:

$$y = 1376x + 4.921, \qquad (1)$$

where $x$—productivity, t/ha; $y$—gross yield, thousand tons.

Winter wheat plantations displayed similar patterns. However, the gross yield of the winter wheat mainly depended on the sown area (r = 0.511) (Table 2). The yield instability is mainly caused by the extreme summer aridity and inability to accumulate enough moisture using existing agricultural technologies.

The relatively low productivity of spring wheat and winter wheat (0.72 t/ha and 1.50 t/ha, respectively) can be increased considerably to 1.38 t/ha and 2.49 t/ha (spring wheat and winter wheat, respectively) in different years. This confirms the existence of considerable reserves for increasing crop production in the Orenburg Region. Therefore, we must examine the reasons for low productivity and scientifically substantiate the methods of adapting to environmental changes.

Several researchers confirmed the significance of water-saving practices in the landscape-adaptive technologies of crop growing (Gostev 2017; Kiryushin 2018; Maksyutov et al. 2015). Such practices can serve as a foundation for the stabilization of crop production. In the steppe zone of the Orenburg Pre-Urals region, precipitations serve as a primary source of soil moisture. According to our observations, the rainfalls from the September of a preceding year to July of a current year are crucial for spring wheat. The correlation coefficient of precipitation and productivity (r) is stable and reaches 0.652 (Table 1). The winter wheat productivity is determined by precipitation

**Table 1** Dynamics of the productivity and gross yield of spring wheat in Orenburg Region, 2010–2019 data

| Indexes | Gross yield, thousand t | Sown area, thousand ha | Productivity, t/ha | Precipitation, mm |
|---|---|---|---|---|
| 2010 | 242.7 | 1400.62 | 0.17 | 260 |
| 2011 | 1540.2 | 1516.09 | 1.02 | 326 |
| 2012 | 528.3 | 1393.82 | 0.38 | 281 |
| 2013 | 884.6 | 1436.86 | 0.62 | 355 |
| 2014 | 862.0 | 1397.65 | 0.62 | 291 |
| 2015 | 1102.0 | 1397.76 | 0.79 | 267 |
| 2016 | 1321.5 | 1357.68 | 0.97 | 428 |
| 2017 | 1830.9 | 1328.92 | 1.38 | 369 |
| 2018 | 714.5 | 1312.71 | 0.54 | 247 |
| 2019 | 851.8 | 1224.89 | 0.7 | 338 |
| Average | 987.9 | 1376.7 | 0.72 | 316 |
| Coefficient of variation, % | 45.6 | 5.4 | 48.0 | 18.0 |
| Correlation (*r*) with productivity | 0.991 | | | |
| Correlation (*r*) with a sown area | 0.110 | | | |
| Correlation (*r*) with precipitation | 0.631 | | 0.652 | |

*Source* The Federal State Statistic Service (EMISS n.d.) and the author's calculations

from August of a preceding year to June of a current year (Table 2). The correlation coefficient is also high ($r = 0.641$).

To use natural resources efficiently, field crops must be flexible and adaptable. The stability of crop yields largely depends on developing drought-resistant varieties of crops (Alabushev 2013; Glukhovtsev and Rumyantsev 2014). The development of smart agricultural technologies accentuates the need for innovative crop varieties.

Smart digital technologies in agriculture include: (1) mapping the soil heterogeneity of a plot by portable optical sensors, (2) surveying by air drones, (3) using the data of remote sensors, (4) using informational technologies, software, and robotized agricultural machinery to differentiate technological activities (Costa et al. 2018; Guo et al. 2020; Mc Bratney et al. 2005).

This study aims to develop the methods of increasing the productivity of spring and winter wheat in the steppe zone of the Orenburg Pre-Urals region. These methods must be scientifically substantiated and adapted to the zonal soil and climate conditions.

To achieve the aim of the study, we solved several tasks:

Analyzing the current state of crop production and revealing the connection between the crop yields, the sown area, and crop productivity;

**Table 2** Dynamics of the productivity and the gross yield of winter wheat with different precipitation, 2010–2019 data

| Indicators | Gross yield, thousand ton | Sown area, thousand ha | Productivity, t/ha | Precipitation, mm |
|---|---|---|---|---|
| 2010 | 268.4 | 358.56 | 0.75 | 309 |
| 2011 | 239.4 | 142.91 | 1.67 | 332 |
| 2012 | 338.9 | 326.50 | 1.03 | 281 |
| 2013 | 222.7 | 210.5 | 1,06 | 289 |
| 2014 | 534.1 | 304.87 | 1.75 | 393 |
| 2015 | 232.7 | 189.64 | 1.23 | 248 |
| 2016 | 688.2 | 293.3 | 2.35 | 432 |
| 2017 | 860.4 | 344.97 | 2.49 | 338 |
| 2018 | 592.6 | 368.4 | 1.61 | 231 |
| 2019 | 435.4 | 413.2 | 1.05 | 240 |
| Average | 441.3 | 295.2 | 1.50 | 309 |
| Coefficient of variation, % | 50.0 | 29.5 | 39.8 | 21.0 |
| Correlation ($r$) with productivity | 0.829 | | | |
| Correlation ($r$) with a sown area | 0.511 | | | |
| Correlation ($r$) with precipitation | 0.412 | | 0.641 | |

*Source* The Federal State Statistic Service (EMISS n.d.) and the authors' calculations

Establishing the ways of increasing the productivity of spring wheat and winter wheat in the current environmental conditions;

Studying the specific features of different spring wheat varieties, determining the most adaptive varieties that have high productivity quality of grain;

Researching the influence of minimal tillage and no-till cultivation, and green manure on the productivity of winter wheat;

Revealing the links between the Normalized Difference Vegetation Index [NDVI] and the spring wheat crop areas, providing the basis for the integration of the index into the system of "smart" farming of landscape-adaptive agriculture;

Studying the phytometric structural parameters of crops and the spring wheat productivity in heterogeneously fertile soils.

## 2 Materials and Methods

In 2016–2019, we conducted a series of field experiments in the Orenburg Region. The soils in the region are dominated by loamy chernozem with a humus horizon of 50–52 cm, humus content of 4.1%, soil acidity of 7.6 (pH), labile phosphorus ($P_2O_5$) content of 2.7 mg/100 g, and the exchange potassium ($K_2O$) content of 30.0 mg/100 g in the arable layer of soil. The annual sum of effective temperatures was 2,627 °C. The average temperature of the warmest month (July) was 21.9 °C. The average temperature of the coldest month (January) was 14.8 °C. The depth of soil freezing fluctuated between 100 and 120 cm, with a snow cover of about 28 cm. The climate continentality index is 215. Annual precipitation was 360–370 mm, about 130 mm (35.0%) of which fell in the warm period of the year (May–August). Overall, this area has insufficient and unstable atmospheric humidity; long, frosty winters with little snow; short rapid springs; hot, dry summers; and long warm and dry autumns. The temperature regime and solar radiation in this area do not affect the yields.

The experiment was conducted on several varieties of soft spring and soft winter wheat:

Spring wheat:

*Favorit* (lutescens);

L-503 (lutescens);

*Yugo-Vostochnaya 2* (lutescens);

*Saratovskaya 74* (albidum);

*Kineliskaya Niva* (erythrospermum);

Winter wheat:

*Povolzhskaya 86* (Triticum aestivum L.).

These varieties were selected by the Research Institute of Agriculture of the South-East Regions (Saratov, Russia) and the Volga Research Institute of Selection and Seed Production by P. N. Konstantinov (Samara, Russia). The spring wheat varieties are recommended for growing in the central zone of the Orenburg Region. Spring wheat varieties were sown on winter and spring wheat predecessors in physically mature soil (10–15 May) at a standard rate of 4.0 mln seeds per 1 ha.

The winter wheat was grown in five-crop rotation with different methods of soil preparation methods: moldboard plowing at 23–25 cm, non-moldboard loosening at 23–25 cm, and shallow tilling at 5–7 cm. The crop rotation plan was: fallow → winter wheat → spring → wheat → chickpea → spring wheat. Green manure (preceding crop residue) was distributed on a field surface simultaneously with harvesting during crop rotation.

Weather conditions during the experiment were unstable—the temperature regime of soil and air, the atmospheric precipitation varied considerably during the experiment. The most humid years were 2016 and 2017, since around 350 mm of rain fell from August to June. Precipitation was evenly distributed in time, which promoted even sprouting, good wheat wintering, optimal productive density, and faster grain ripening.

The following years had less precipitation—109–120 mm (31.1–34.3%). It was unevenly distributed, which caused drought periods. The most arid month was August (4.3–9.8 mm); few precipitations fell in winter, especially in 2017–2018 (50 mm or 36.6%). The sample maximum of average air temperature was observed in 2019 (17.8 °C in May, 21.5 °C in June). The temperature was lower in the preceding years (1.2–3.5 °C in May, 2.7–3.3 °C in June).

We used conventional methods for field tests, laboratory tests, attendant observations, and recording (Dospehov 1985). The NDVI in spring wheat was calculated using Green Seeker Handheld Crop Sensor, Model HCS-100 with an active optical sensor. The plant assimilation was measured using digital laboratory scales (CAS CBL-2200H) with a measurement precision of 0.01 g. To estimate the phytometric parameters of crops, we chose plant samples from 0.25 $m^2$ plots. The experiment was conducted in four replications. Phenological observations, recording of a density of plant standing, and other accompanying studies were carried out based on a methodology of governmental variety testing (Dospehov et al. 1987).

The statistical analysis of data was performed via Microsoft Excel Analysis ToolPak.

## 3 Results

The good yield of spring wheat is key to ensuring stable harvests of high-quality crops in the Orenburg Region. The preceding crops in the crop rotation system significantly affected the yields of spring wheat. If spring wheat followed spring wheat, the amount of well-sprouted plants varied from 332.8 to 354.8 units/$m^2$. If the predecessor was winter wheat, the number of sprouts increased and reached 337.2–361.2 units/$m^2$. The *Kineliskaya Niva* variety formed the densest patterns of sprouting (354.8–361.2 units/$m^2$), while the pattern of *L-503* was the least dense (332.8–337.2 units/$m^2$). The number of fully mature plants was the greatest if spring wheat was sown after winter wheat (244.2–275.6 units/$m^2$). Repeated sowings of spring wheat reduced the mature plant amount by 11.2 units/$m^2$.

One of the main problems in modern agriculture is an increase in weed infestation. The increase of weed infestation serves as the main reason for the slowed adoption of no-till crop growing. Therefore, modern agricultural production requires nature-like and cheap technologies that are, at the same time, more efficient than chemical herbicides.

The results of our experiment demonstrated that the crop rotation system significantly affects weed infestations. Using winter wheat as a predecessor reduced the infestation with *Amaranthus retroflexus* (Green amaranth) by a factor of 1.09, and the infestation with *Echinochloa crus galli* (Barnyard grass) by a factor of 5.85. Repeated sowings of spring wheat increased the infestation by *Convolvulus arvensis* (Trailing bindweed) and Euphorbia virgata (Spurge) by a factor of 2.5 to 3.4.

The *L-503* variety plots with the spring wheat predecessor demonstrated the sample minimum of productivity—1.31 t/ha. The second lowest productivity was observed in the *Yugo-Vostochnaya 2* variety with the winter wheat predecessor—1.37 t/ha. The sample maximum of productivity (1.70 t/ha) was observed in the *Favorit* plot with the winter wheat predecessor. The productivity of other varieties was lower by 0.06–0.39 t/ha or 3.5–22.9%.

Moisture is extremely important for plant growth and development. Modern minimal tillage and no-till technologies help accumulate moisture and organic substances in the root-inhabited soil layer (Solonkin et al. 2019; Dridiger 2018). These technologies presuppose leaving green manure and distributing it simultaneously with harvesting.

Green manure forms a protective cover that traps moisture, precludes soil, inhibits the spread of weeds, assists soil microflora, and restores the fertile layer (Nascente et al. 2013; Quiroga et al. 2009). Green manure creates more favorable conditions for seed germination.

The principal task of steppe agriculture always was the improvement of the water regime. In this experiment, the most favorable conditions for growing winter wheat were created by the shallow tilling at a depth of 5–7 cm (Table 3).

The sample maximum of crop productivity (2.83 t/ha) was observed in shallow-tilled plots. It confirmed the high efficiency of nature-like water-saving practices in the landscape-adaptive systems of agriculture.

Since the mid-twentieth century, novel management techniques are being introduced into agriculture. These techniques led to increased productivity and the development of agrocenoces (Wall et al. 2008; Becker-Reshef, Vermote et al. 2010; Prasad et al. 2006). Currently, the "smart" land use technologies provide excellent opportunities for increasing the productivity of steppe agriculture in Russia.

**Table 3** Winter wheat productivity with various soil preparation methods, average of 2017–2019

| Method of soil preparation | Shoot pattern density, units/$m^2$ | Density of mature plants, units/$m^2$ | Mass of a grain from an ear, g | Grain productivity, t/ha |
|---|---|---|---|---|
| Plowing at 23–25 cm | 408 | 284.5 | 0.68 | 1.95 |
| Non-moldboard loosening at 23–25 cm | 446 | 352.2 | 0.69 | 2.46 |
| Shallow tilling at 5–7 cm | 466 | 399.7 | 0.71 | 2.83 |
| *Coefficient of variations, %* | 6.7 | *16.7* | 2.2 | *18.3* |
| *HCP* $_{05}$ | | | | *0.13* |

*Source* The authors' experimental data

**Table 4** Correlation of NDVI and the future productivity of winter wheat, average of 2018–2019

| Indicators | Phase of development | | | | |
|---|---|---|---|---|---|
| | Spring renewal | Spring tilling | Stem elongation | Earing | Ripening |
| NDVI | 0.26 | 0.37 | 0.61 | 0.66 | 0.44 |
| Correlation (*r*) with crop productivity | 0.63 | 0.72 | 0.94 | 0.95 | 0.75 |

*Source* The authors' experimental data

In this study, we used portable optical sensors, which helped us define the phytometric parameters of crops by calculating the NDVI. We concluded that NDVI is closely connected with the assimilation surface of crops and could be used for its efficient assessment. The NDVI was also connected with productivity of grain. However, the correlation varied substantially depending on the growth stage. The maximum correlation was observed in the stages of stem elongation and earing (0.94 and 0.95, respectively) (Table 4).

These equations represent the dependence of grain productivity on NDVI in stem elongation and earing:

$$y = 2.385x + 0.770, \tag{2}$$

$$y = 3.377x + 0.059, \tag{3}$$

where: $y$—grain productivity, t/ha; $x$—NDVI.

The correlation coefficient (r) indicates the dependence of grain productivity on the NDVI in 89.1 or 91.7% cases.

Therefore, we conclude that the yield of winter wheat may be increased by optimizing NDVI. This can be achieved by several means, such as non-root fertilizing with water solutions of nitrogen fertilizers in the earing stage.

The crop yield is affected by many natural and anthropogenic factors. The yield is spatially heterogeneous and depends on the variability of plant growth and development (Gulyanov 2019). In our opinion, climatic factors such as the frequency and amount of precipitation, insolation intensity, and temperatures of soil and air, are one of the main factors causing spatial variability. Soil factors, including texture, structure, fertility, accessibility of mineral nutrition components, pollution, and erosion processes, are also important. Geographical factors such as field slope and the aspect of the slope identify its microclimate. Biological factors can include pest infestations, weeds, and diseases.

However, agricultural techniques are the most extensive group of factors. They influence timeframes and the quality of agricultural activities, and may include unevenness of seeding, violation of crop rotation, usage of low-quality seeds, usage of unproductive variety, and usage of outdated agricultural machinery.

**Table 5** Dynamics of phytometric and structural indicators of spring wheat, average of 2017–2019

| Indicators | NDVI | Density of productive plants, units/m$^2$ | Mass of grain in an ear, g | Grain productivity, t/ha |
|---|---|---|---|---|
| Min value | 0.54 | 207.1 | 0.56 | 1.16 |
| Max value | 0.71 | 254.2 | 0.69 | 1.77 |
| Average value | 0.63 | 236.1 | 0.62 | 1.47 |
| Variation coefficient, % | 8.25 | 7.72 | 6.52 | 13.1 |
| Correlation with productivity (r) | 0.96 | 0.93 | 0.91 | |
| Coefficient of determination ($R^2$) | 0.92 | 0.86 | 0.82 | |

*Source* The authors' experimental data

Our study with spring wheat revealed a considerable heterogeneity of the vegetation cover in the experimental plot. It was observed in the dynamics of phytometric, structural parameters, and grain productivity on the experimental areas (Table 5).

The NDVI of crops in the earing phase varied from 0.54 to 0.71 units, with a variation coefficient of 8.25%. The structural components of the harvest varied significantly. The variability of productive plants was relatively stable, varying between 207.1 units/m$^2$ and 254.2 units/m$^2$ (a difference of 47.1 units/m$^2$). The mass of a grain varied from 0.56 to 0.69 g, with a variation coefficient of 6.52%. However, the sum variability of these two factors led to higher variability of productivity. The average productivity of grain plots was 1.47 t/ha, with a variation coefficient of 13%. The difference of 0.61 t/ha between the minimum and the maximum productivity signifies an additional reserve for increasing productivity. This increase can be achieved by leveling the impact of heterogeneous soil fertility via smart digital technologies.

However, the application of smart technologies to compensate for intra-plot soil heterogeneity usually presupposes the use of mineral fertilizers and pesticides (which is ecologically unacceptable).

We believe that the areas with high heterogeneity of fertility, and frequent infestations and diseases demand complex agrotechnical measures. In our opinion, digital technologies in agriculture should promote soil restoration, improve the environment, prevent further erosion, and reduce the impact of anthropogenic stress. These technologies include: (1) optimization of land structure, (2) anti-erosion arrangement of plots, (2) transition to landscape-adaptive systems, (3) introducing soil-protective crop rotations, and (4) developing new nature-like techniques and technologies.

## 4 Conclusions

The steppe zone of the Orenburg Pre-Urals region has a high potential for an increase in gross yield and productivity of crops. The increase is best achieved by landscape-adaptive systems of agriculture, rational use of land resources, protection of the environment, and conservation of natural diversity of steppes (Gulyanov et al. 2019; Levykin et al. 2019).

Crop rotation system can provide better soil humidity, higher germinability, denser plant patterns, and lower frequency of infestations by weeds and pests. The best results were demonstrated by sowing spring wheat after winter wheat. The varieties Favorit (lutescens, productivity), Saratovskaya 74 (albidum), and Kineliskaya Niva (erythrospermum) had the highest productivity in the experimental environment.

Rational water use is extremely important for the stability and productivity of winter wheat. Soil moisture depends on soil preparation methods and crop rotation. It defines the density of mature plant patterns and the mass of a grain. The best results were demonstrated by shallow tillage at 5–7 cm. This method provided the germinability of 93.3% and the shoot density of 466 units/$m^2$. The density of fully mature plants of 400 units/$m^2$ and the productivity of 2.83 t/ha indicate that shallow tillage is more efficient at conserving soil moisture. The deep non-moldboard loosening and plowing lowered productivity by 0.37 and 0.88 t/ha, respectively (13.1 and 31.1%, respectively). In the current condition, the introduction of shallow tilling can guarantee the stable harvests of winter wheat in the steppe zones of the Orenburg Pre-Urals region.

Digital technologies and the methods of "smart" agriculture allow controlling the fields more efficiently. Using portable optical sensors allows one to estimate the NDVI of crop fields. Low NDVI is caused by sparse seeding. Its impact can be quickly corrected by increasing the productivity of plants via automatic foliar nutrition in the stages of ripening and earing.

The application of "smart" agricultural technologies and innovative techniques of tillage and crop rotation can increase the crop productivity and stability of harvests not only in the Orenburg Pre-Urals region, but also in other steppe zones of the Russian Federation. The use of landscape-adaptive agriculture in the steppe zone might become a foundation for the environmentally safe production of organic products. We expect that the studies in this direction will continue.

**Acknowledgements** This study was conducted within the framework of the project "Steppes of Russia: Landscape and Ecological Grounds of Sustainable Development, Nature-like Technologies in the Context of Natural and Anthropogenic Changes in the Environment" (Project number GR AAAA-A17-117012610022-5).

## References

Alabushev AV (2013) An adaptive potential of crops varieties. Leguminous Plants and Cereals 2:47–51

Becker-Reshef I, Vermote E, Lindeman M, Justice C (2010) A generalized regression-based model for forecasting winter wheat yields in Kansas and Ukraine using MODIS data. Remote Sens Envir 114:1312–1323

Costa EM, Dos Anjos LYC, Samuel-Rosa A (2018) Digital elevation model quality on digital soil mapping prediction accuracy. Cienc Agrotecnol 42:608–622

Dospehov BA (1985) Methodology of field experience (with bases of statistical processing of results of the study), 5th edn. Agropromizdat, Moscow, USSR

Dospehov BA, Vasiliev IP, Tulikov AM (1987) Practical work on agriculture, 2nd edn. Agropromizdat, Moscow, USSR

Dridiger VK (2018) No-till technology and errors made in its learning. Agricult J 1:14–23

Glukhovtsev VV, Rumyantsev AV (2014) Selection of high drought-resisting adaptive sorts of crops. Her Russ Acad Agric Sci 6:30–32

Gostev AV (2017) Aspects of a resource-saving in agro-technologies of crops cultivation. Agrosnabforum 6:48–52

Gulyanov YuA (2019) Monitoring of phytometric parameters by use of innovative methods of crops scanning. Taurida Her Agrar Sci 3:64–76. https://doi.org/10.33952/2542-0720-2019-3-19-64-76

Gulyanov YuA, Chibilev AA, Levykin SV, Silantieva MM, Kazachkov GV, Sokolova LV (2019) Ecological-based adaptation of agriculture to the soil and climatic conditions in Russian steppe. Ukr J Ecol 9:393–398

Guo L, Fu P, Shi T, Chen Y, Zhang H, Meng R, … Wang S (2020) Mapping field-scale soil organic carbon with unmanned aircraft system-acquired time series multispectral images. Soil Tillage Res 196, 104477

Kiryushin VI (2018) Tasks of scientific-innovative securing of agriculture in Russia. Agriculture 3:3–12

Levykin SV, Chibilev AA, Gulyanov YuA, Silantieva MM, Kazachkov GV, Yakovlev IG (2019) Environmental and landscape significance of steppe megaprojects. Ukr J Ecol 9:371–375

Maksyutov NA, Zhdanov VM, Skorokhodov VY, Kaftan YV, Mitrofanov DV, Zenkova NA, … Zhizhin VN (2015) Moisture-saving methods and technologies in agriculture in the Orenburg Region. Grain Farming Russ 6:67–72

Mc Bratney A, Whelan B, Ancev T, Bouma J (2005) Future directions of precision agriculture. PRAG 6:7–23

Nascente AS, Li YC, Crusciol CAC (2013) Cover crops and no-till effects on physical fractions of soil organic matter. Soil Tillage Res 130:52–57. https://doi.org/10.1016/j.still.2013.02.008

Prasad AK, Chai L, Singh RP, Kafatos M (2006) Crop yield estimation model for Iowa using remote sensing and surface parameters. Int J Appl Earth Obs Geoinf 8:26–33

Quiroga A, Fernández R, Noellemeyer E (2009) Grazing effect on soil properties in conventional and no-till systems. Soil Tillage Res 105:164–170. https://doi.org/10.1016/j.still.2009.07.003

Solonkin AV, Boldyr' DA, Selivanova VY (2019) No-till—a beginning of development in the arid conditions of Volgogradskaya Region. Agricult J 3:31–37

The Federal State Statistic Service – [EMISS] (n.d.) Gross yield of agricultural crops. Retrieved from https://www.fedstat.ru/indicator/30950. Accessed 2 Oct 2020

Wall L, Larocgue D, Leger P-M (2008) The early explanatory power of NDVI in crop yield modeling. Int J Remote Sens 29:2211–2225

# Sustainability of Agriculture: The Vaccine Against Horse Strangles (*Adenitis Equorum*)

**Mikhail P. Neustroev, Evdokiya I. Elbyadova, and Sargylana G. Petrova**

**Abstract** On of the most common infectious disease in herd horse breeding in almost all countries is horse strangles. In the Republic of Sakha (Yakutia, Russia), the incidence of strangles in foals is 57.8–62.7%; the mortality rate is 4.0–22.0% . Vaccination is an effective method of combating this disease. Therefore, we aimed to develop a new effective and environmentally-safe strangles vaccine. In this study, we employed bacteriological and biochemical research methods. The vaccine was developed and tested according to the requirements of the Ministry of Agriculture of Kazakhstan. The performance trials were carried out in the farms of the Namsky and Megino-Kangalassky regions of the Republic of Sakha (Yakutia). Strangles vaccine is made of *Str. equi "H-34"* strain with the addition of an immunomodulator—a culture liquid of *Bacillus subtilis TNP-3*. The lethal dose of the strain and the immunogenicity of the vaccine were tested on laboratory white mice. The average lethal dose was 20,000 microbial bodies; the average immunogenic dose was 400,000 microbial bodies. During the vaccine tests, we gradually increased its dosage up to a threefold immunogenic dose. The vaccine was determined to be harmless, aside from the swelling that disappeared within 5–7 days. Foals of 7–8 months of age were vaccinated once in October–November. After the vaccination, their blood had increased bactericidal and lysozyme activity, a higher number of leukocytes, and, therefore, better phagocytosis. The precipitating antibody titer remained the same for 117 days after vaccination. The vaccine immunogenicity was 90% in white mice and up to 100% in foals. The developed strangles vaccine has sufficiently high immunogenicity and is environmentally safe.

**Keywords** Strangles · Streptococcus equi · Vaccine · Toxicity · Immunogenicity

---

M. P. Neustroev (✉) · E. I. Elbyadova · S. G. Petrova
M. G. Safronov Yakut Scientific Research Institute of Agriculture, Yakutsk, Russia
e-mail: mneyc@rambler.ru



S. G. Maximova et al. (eds.), *Advances in Natural, Human-Made, and Coupled Human-Natural Systems Research*, Lecture Notes in Networks and Systems 234,
https://doi.org/10.1007/978-3-030-75483-9_2

# 1 Introduction

Herd horse breeding is one of the main branches of animal husbandry in many countries. Restoring the livestock, increasing the productivity of herd horse breeding, obtaining high-quality products, and other factors depend on the effectiveness of veterinary measures, including infectious disease prevention. The most common equine infectious disease, especially in foals, is strangles—caused by *Streptococcus equi*. This disease has been known for a long time since it causes upper respiratory tract infections in horses (Sweeney et al. 1987, 2005). The symptoms of the disease feature: fever, mucopurulent nasal discharge, and head lymph node swelling with consequent abscesses. Horse strangles widespread and causes significant economic damage to horse breeding. Strangles is the most common in Novosibirsk and Irkutsk regions, the Altai Republic, Krasnoyarsk, and Altai Krais (Gustokashin 2013), the Republics of Tuva (Chysyma 1989), Khakassia, Sakha (Yakutia) (Neustroev 2000), as well as in Kazakhstan (Sansyzbaev 1993), Mongolia (Bayanzhargal et al. 2014), Kyrgyzstan (Raimbekov et al. 2016), the Netherlands (Hartford et al. 2002), Egypt (Neamat-Allah and Damaty 2016), Korea (Kim et al. 2017), and Brazil (Libardoni et al. 2016).

In the Republic of Sakha (Yakutia), the incidence of strangles in foals is 57.8–62.7%. Depending on the features of an epizootic, the mortality rate fluctuates between 4.0 and 22.0% (Neustroev 2000). In Kazakhstan, the incidence in foals is 30.1–46.7%; mortality reaches 16.0–28.3% (Sansyzbaev 1993). In Mongolia and Yakutia, the increase in the incidence of strangles is caused by the extreme environments, which decrease immunobiological reactivity, and the preproduction of kumis, which decreases the amount of milk that foals get from their mothers (Neustroev 2000).

*Streptococcus equi zooepidemicus*, the causative pathogen of strangles, can also cause arthritis in goat yeanlings (Boyle et al. 2018, 2017) and infectious abortions in mares (Neustroev 2000). Even clinically healthy horses may have *Streptococcus equi* (Libardoni et al. 2016; Neustroev 2000). The epidemiological significance of *Streptococcus equi* among the population has not yet been studied. Although white mice are used as a model for bacteriological diagnosis and study of the virulence of *Streptococcus equi*, the role of murine rodents in the spread of strangles has not been established. Since *Streptococcus equi* easily survives in soil and manure (Neustroev 2000), there is a need for general and specific control measures.

Vaccine prophylaxis is the most effective and low-cost measure in the fight against horse strangles. Different types of vaccines (inactivated, attenuated, live) are tested, manufactured, and used. Earlier (1943–1966) vaccines from living, weakened, and killed strains of *Streptococcus equi* did not find practical application (Bazelley 1942). Nevertheless, scientists worldwide continue to develop vaccines against horse strangles, which also have not found wide practical use. A live horse strangles vaccine from the deposited strain *Streptococcus equi TW 928* was developed in the Netherlands. The creators claim that it can be applied to animals from birth in various forms (parenteral, intramuscular, subcutaneous, intradermal, oral, and intranasal)

(Gustokashin 2013). However, this vaccine is not registered in Russia. Currently, in the United States, a live vaccine from an attenuated strain is used. It induces the production of antibodies in the blood after 7–10 days. A modified intranasal double-dose "Pinnacle IN" vaccine is also used. These vaccines are not used in Russia. They also require two and three doses with an interval of several weeks, which is practically inconvenient (Boyle et al. 2018, 2017).

In Kazakhstan, an inactivated subunit vaccine from the *Str. equi "YUS-15"*, and the "KazNIVI" and "Akyntai" vaccines, which contain antibiotics, were developed (Bizhanov et al. 2016; Bizhanov et al. 2009). In 1996–2000, during strangles outbreak at the Yakutsk Research Institute of Agriculture, an effective inactivated vaccine with the "*Poliribonat*" immunomodulator was developed and produced (Neustroev 2000). However, the registration of this vaccine expired. Consequently, there is a demand for an effective and environmentally friendly vaccine against horse strangles.

According to the results of our previous studies, the strangles vaccine should contain an immunomodulator that enhances the immunogenicity of inactivated vaccines, stimulates immunobiological reactivity, and helps the antibacterial component. The use of antibiotics in the prevention and treatment of horse strangles one of the most common methods in fighting this infection. However, antibiotics suppress gut flora, which is essential for the gastrointestinal tract, immune system, and metabolic processes. They may also produce antibiotic-resistant strains of bacteria (Neustroev 2000; Neustroev and Tarabukina 2003). Therefore, the use of antibiotics is impractical in organic animal husbandry.

We developed a new strangles vaccine and conducted preclinical tests on white mice. According to animal tests, the vaccine is non-toxic (Elbyadova and Neustroev 2019).

## 2 Materials and Methods

This study was conducted in the veterinary biotechnology laboratory of the Yakutsk Scientific Research Institute of Agriculture, the Scientific Production Center "Hotu-Bakt" LLC, and horse breeding farms in Yakutia.

The vaccine was named "Tabyn". For its production, we used the bacterial strain *Str. equi H-34*, which was deposited at the All-Russian State Research Institute for Control, Standardization, and Certification of Veterinary Medicines. The strain was harvested from a submandibular lymph node abscess of a strangles-infected foal. This strain meets the vaccine strain requirements on the tinctorial, morphological, cultural, biochemical, antigenic properties, and the ability to accumulate bacterial mass.

To accumulate the bacterial mass, we used meat peptone broth [MPB] with 1% glucose and horse blood serum. The bacterial base was inactivated with a 0.04% formalin solution, where active formaldehyde content was at least 36%. Aluminum hydroxide was used as an adjuvant.

We added an immunomodulator—a culture liquid from *Bacillus subtilis* TNP-3 strain—to the vaccine. The strain was cultivated for five days in MPB at +37 °C. To separate the culture liquid, the bacterial mass, containing 1 billion microbial cells, was centrifuged at 7,000 rpm for 15 min. The culture liquid was filtered through membrane filters into sterile vials, heated in a water bath at 95 °C for 15 min. Culture liquid was added to the finished vaccine in a 2:1 ratio. The vaccine is a clear light-yellow liquid with a white residue, which, when shaken, uniformly suspends in the liquid.

The toxicity and immunogenicity of the vaccine were tested in foals, using conventional methods.

The vaccine tests and the writing of technical documentation were carried out following:

- "Rules for conducting state registration (re-registration) and maintaining the State register of veterinary drugs in the Republic of Kazakhstan" (October 31, 2002 No. 349) (Ministry of Agriculture of the Republic of Kazakhstan 2002),
- "Instructions for the approval of scientific and technical documentation submitted by individuals and legal entities for new, improved veterinary drugs" (March 3, 2003 No. 105) (Ministry of Agriculture of the Republic of Kazakhstan 2003),
- Order of the Ministry of Agriculture of the Republic of Kazakhstan "On approval of the rules for the approval of regulatory and technical documentation for new, improved veterinary drugs and feed additives" (November 28, 2014 No. 7–1/625) (Ministry of Agriculture of the Republic of Kazakhstan 2014).

The performance trials were carried out in the farms of the Namsky and Megino-Kangalassky regions of Yakutia in the winter of 2016–2017. Foals of 7–8 months were preliminarily dewormed. The immunogenic dosage was 3 ml intramuscularly.

## 3 Results

To study the toxicity, we injected the foals of 7–8 months of age intramuscularly into the thigh area in increasing doses up to 3 times the immunogenic dosage. The previous studies established an immunogenic dose of 2 ml per head. We used three dosages: 2 ml, 3 ml, and 6 ml. For each dosage, four animals were injected. The animals were then placed under clinical monitoring for one month.

The monitoring has shown, that sometimes, within 5–7 days of vaccination, the injection site had doughy swelling. However, swelling passed after 2–4 days. In the entire observation period, foals displayed no signs of depression or refusal to feed.

To assess the immunogenicity, we tested the virulence of Streptococcus equi in vaccinated white mice. A pure culture of bacteria at a concentration of 2 billion microbial bodies in 1 ml was serially diluted from 200 million to 200 microbial cells. The dosage of bacterial suspension was 0.2 ml. The results are presented in Table 1.

In our calculations, we rely on the Kerber formula:

**Table 1** The virulence of *Streptococcus equi* strain

| No | Dilution | Microbal bodies | Died/total | L·i | Infection, % |
|---|---|---|---|---|---|
| 1 | $10^{-1}$ | $2 \times 10^8$ | 4/4 | 1.0 | 100 |
| 2 | $10^{-2}$ | $2 \times 10^6$ | 4/4 | 1.0 | 100 |
| 3 | $10^{-3}$ | $2 \times 10^7$ | 4/4 | 1.0 | 100 |
| 4 | $10^{-4}$ | $2 \times 10^5$ | 4/4 | 1.0 | 100 |
| 5 | $10^{-5}$ | 20,000 | 2/4 | 0.5 | 50 |
| 6 | $10^{-6}$ | 2,000 | 1/3 | 0.33 | 25 |
| 7 | $10^{-7}$ | 200 | 0/4 | 0.0 | 0 |
| | | | | 4.83 | 100 |
| | Control | | 0/4 | 0.0 | 0 |

$$\mathrm{lgLD}_{50} = \mathrm{lg}_{1.0} + \mathrm{lg}_{\mathrm{int}}\left(\sum - 0.5\right)$$

where $\mathrm{lg}_{1.0}$—the logarithm of dilution with reaction constant 1.0;

$\mathrm{lg}_{\mathrm{int}}$—the interval logarithm;

$\sum$—the sum of reaction constants.

$$\sum -0.5 = 4.83 - 0.5 = 4.33;\ 4.33 \times 1.0 = 4.33$$

$$4.33 + 1 = 5.33; \quad \mathrm{LD}_{50} = 10^{-5.33}$$

We calculated the standard error using the formula:

$$\log_{\text{standard error}}(\mathrm{LD}_{50}) = \pm \mathrm{L}_{\mathrm{int}} \frac{\sqrt{\sum \text{reaction rates} \times (1 - \text{reaction rates})}}{(\mathrm{n} - 1)}$$

$$\mathrm{L}_{\mathrm{int}} = \log \mathrm{f}.$$

With the use of the aforementioned formula, we obtain the following:

$$\log_{\text{standard error}}(\mathrm{LD}_{50}) = 1\frac{\sqrt{0.5 \times 0.5}}{6} + \frac{\sqrt{0.33 \times 0.67}}{6}$$

$$= \pm\sqrt{0.0416666 + 0.03685} = \sqrt{0.07851} = \pm 0.2802$$

Add or subtract to $\log LD_{50} = -5.050$ then $\log LD_{50} = -5.33$

$$\log LD_{50} = -5.610$$

Or $10^{-5.05}$ and $10^{-5.610}$ with statistical significance $1\text{-}\alpha = 0.72$ or 72%.

**Table 2** Immunity stress test

| Group | Quantity, ml | Quantity, ml | Microbial bodies, thousand | Died/total | L·i | % of infected |
|---|---|---|---|---|---|---|
| 1 | 0.5 | $10^{-0,2}$ | 1000 | 2/10 | 0.2 | 20 |
| 2 | 0.4 | $10^{-0,4}$ | 800 | 1/10 | 0.1 | 10 |
| 3 | 0.3 | $10^{-0,6}$ | 600 | 1/10 | 0.1 | 10 |
| 4 | 0.2 | $10^{-0,8}$ | 400 | 1/10 | 0.1 | 10 |
| 5 | 0.1 | $10^{-1}$ | 200 | 5/10 | 0.5 | 50 |
| | | | | | 1 | |
| Control | | | | 9/10 | 0.9 | 9 |

Therefore, the average lethal dose was $LD_{50} = 10^{-5.33}$ or 20 thousand microbial bodies (with the introduction of 0.2 ml of a diluted suspension).

We preliminarily established an immunogenic dosage. One ml of the vaccine contained 2 billion dead cells of *Streptococcus equi*.

The vaccine was tested on 60 white mice weighing 16–18 g. We formed six groups with ten animals in each. The control group was injected with saline.

We administered the vaccine doses of 0.5, 0.4, 0.3, 0.2, and 0.1 ml. Within ten days, the vaccinated animals displayed no signs of reactogenicity and vaccine toxicity.

Ten days after the immunization, the mice in all groups were injected subcutaneously in the back region with a titrated dose of a pathogenic strain of *Streptococcus equi* ($LD_{50}$). After ten days of observation, we obtained the following results:

- Group 1: 2 animals died;
- Group 2: 1 animal died;
- Group 3: 1 animal died;
- Group 4: 1 animal died;
- Group 5: 5 animals died;
- Group 6 (control): 9 animals died (Table 2).

According to the Kerber formula:

$$IgLD_{50} = Ig_{1.0} + Ig_{\text{int.}} \cdot \left(\sum - 0.5\right),$$

$$\sum = 1 - 0.5 = 0.5;\ 0.5 \cdot 0.2 = 0.1$$

$$0.1 + 1 = 1.1,\ \text{or}\ LD_{50} = 10^{-1.1}$$

The dose $LD_{50} = 10^{-1.1}$ induced immunity in 50% of the immunized mice after infection with a virulent strain. The immunogenic dose is $10^{-0.8}$ or 0.2 ml, with a content of 400 thousand microbial cells induced immunity in 90% of the vaccinated.

The immunizing ability of the "Tabyn" vaccine was compared to the previously developed strangles vaccine with the "Poliribonat" immunomodulator. Each vaccine was tested on 20 mice. After immunization, the mice were infected with a $5LD_{50}$

dose of *Streptococcus equi*. Within ten days of observation, two mice died in the first and second groups each. In the control group, 17 mice died.

The initial culture of streptococcus was isolated by a bacteriological examination of the blood of the heart, spleen, and liver of dead mice.

We compared the effects of the inactivated "Tabyn" vaccine and the "Poliribonat" vaccine in the horse breeding units suffering from horse strangles.

After weaning from mares (first half of November), we selected 60 foals of 6–7 months. In the first group, 20 animals were injected with the "Tabyn" vaccine. In the second group, 20 animals were injected with the "Poliribonat" vaccine. The 20 animals in the control group were not immunized.

The vaccines were injected intramuscularly at a dose of 3 ml per head. The animals were preliminarily dewormed.

The animals were then clinically monitored for three months.

Blood tests were taken from 5 animals of each group immediately after immunization, 24 days later, and 72 days later. On the 24th day after the "Tabyn" injection, blood serum displayed an increase in the immunobiological reactivity (bactericidal and lysozyme activity). By the end of the observation, we noted an increase in leukocytes and their phagocytic activity.

When using the strangles vaccine with the culture liquid of *Bac. subtilis TNP-3*, we noted the stimulation of the phagocytic ability of leukocytes. On the 30th day, the blood of immunized foals showed the highest titer of precipitating antibodies (1:35). In foals injected with the "Tabyn" vaccine, the average titer was higher (1:35) than in foals injected with the "Poliribonat" vaccine (1:20). The precipitating antibodies in the vaccinated animals remained at the same level for 117 days after vaccination.

Thirty days after re-vaccination, five animals from each group were infected with a 4 ml dose of the virulent *Str. equi H-34* culture. The virulent culture was applied to the nasal mucosa. For 30 days after the infection, we performed daily clinical observations.

The immunized foals of both groups displayed no signs of depression and refusal to feed. Body temperature was within the physiological norm for all 30 days of observation.

All non-immunized foals had the clinical signs of strangles. Their nasal cavity contained *Streptococcus equi*. There were no cases of strangles in vaccinated foals of the first and second groups.

The control infection demonstrated showed that the "Tabyn" vaccine induced immunity in 100% of vaccinated foals (not inferior to the "Poliribonat" strangles vaccine).

For the performance test, we immunized 161 foals of 6–7 months with the strangles vaccine. In "Khorobut" LLC, all 81 vaccinated foals displayed no symptoms of strangles. Before the vaccination, four foals had a mild form of strangles. At the "Partizan" site, two animals were sick with strangles before the immunization. During the observation period, the 19 vaccinated animals had no signs of strangles. Among 30 heads of foals immunized in the agricultural production complex "Baarys," there were no cases of strangles.

## 4 Discussion

Thus, we confirmed the harmlessness of the "Tabyn" vaccine in foals. These results correspond to the results of preclinical tests on mice (Elbyadova and Neustroev, 2019). The vaccine protected 90% of mice from Streptococcus equi infection. The control infection experiment in vaccinated foals confirmed the high immunogenicity of the inactivated vaccine. The performance tests proved that the vaccine is highly effective.

The calculation of the economic efficiency of vaccines has shown that the "Tabyn" vaccine is the most efficient. Its economic benefit is 22.47 RUB per RUB of costs. The economic benefit of "Poliribonat" was 8.2 rubles per ruble of cost.

Thus, we confirmed the viability of using a culture liquid of *Bac. subtilis TNP-3* in inactivated strangles vaccine from *Str. equi H-34* strain. The antiepizootic efficacy of the "Tabyn" vaccine (vaccine from *the Str. Equi H-34* strain + culture liquid of the bacterial strain *Bacillus subtilis*) is not inferior to the "Poliribonat" strangles vaccine. The economic efficiency of the "Tabyn" vaccine is 2–3 times higher than its analog.

The application of the "Tabyn" vaccine is ecologically, economically, and epizootologically justified. The technical documentation on this research was written and submitted for consideration to the Ministry of Agriculture and the Committee for Veterinary Control and Supervision of the Republic of Kazakhstan. Currently, the "Manual on the use of the vaccine" was approved and the "Marketing Authorization" was issued (RK-VP-1-3750-18 of November 27, 2018).

The high efficiency of the inactivated vaccine, in our opinion, can be explained by the antigenic activity of the vaccine strain and the immunomodulatory culture liquid of *Bacillus subtilis TNP-3* strain. According to the results of our previous studies, the bacterial strain *Bacillus subtilis TNP-3* can induce interferon synthesis, stimulate the immunobiological reactivity of the body, and enhance the immunogenicity of inactivated bacterial and viral vaccines (Neustroev et al. 2019; Osmaev et al. 2007). By immunogenicity, this vaccine is on par with those developed in the Netherlands, the USA, and Kazakhstan (Hartford et al. 2002; Boyle et al. 2017, 2018; Bizhanov et al. 2009, 2016). It surpasses them in terms of environmental friendliness and harmlessness since it does not contain antibiotics.

The non-toxicity of the vaccine is consistent with the previously obtained results of preclinical tests, and the data showing the harmlessness of the drug *"Sakhabactisubtil,"* which consists of *B. subtilis bacterial TNP-3* and *B. subtilis TNP-5*strains, in murine rodent tests (Neustroev et al. 2017).

## 5 Conclusion

We developed a new inactivated vaccine against horse strangles. We confirmed its non-toxicity, environmental safety, and sufficiently high immunogenicity. The developed vaccine, called "Tabyn", can be used in Kazakhstan and the EU countries to prevent horse strangles.

## References

Bayanzhargal B, Badmaeva OB, Tsydypov VT (2014) Epizootological aspects of horse infectious diseases in Mongolia. Bulletin of KrasGAU 3:156–159

Bazelley PL (1942) Studies with equine streptococci 3: Vaccination against strangles. Aust Vet J 18:141–155

Bizhanov AB, Karataev BS, Namet AM (2009) Vaccine against horse strangles (November 24, 2009 Patent KZ 23106). Astana, Kazakhstan

Bizhanov AB, Namet AM, Karataev BS (2016) Vaccine for horse strangles "Akyntai" (April 15, 2016 Patent KZ 31032). Astana, Kazakhstan

Boyle AG, Stefanovski D, Rankin SC (2017) Comparison of nasopharyngeal and guttural pouch specimens to determine the optimal sampling site to detect Streptococcus equi subsp equi carriers by DNA amplification. BMC Vet Res 13, 75:1–75:8. https://doi.org/10.1186/s12917-017-0989-4

Boyle AG, Timoney JF, Newton JR, Hines MT, Waller AS, Buchanan BR (2018) Streptococcus equi infections in horses: Guidelines for treatment, control, and prevention of strangles-revised consensus statement. J Vet Intern Med 32:633–647

Chysyma RB (1989) Epizootic features and age-related immunoreactivity of foals with horse strangles (Dis. of Cand. of Veterinary Sci.). All-Russian Scientific Research Institute of Experimental Veterinary Medicin, Moscow, Russia

Elbyadova EI, Neustroev MP (2019) The results of preclinical trials of the "Tabyn" vaccine prevent horse strangles. In: Parmon VN (ed) Proceedings of the conference Agrarian science—agricultural production in Siberia, Kazakhstan, Mongolia, Belarus, and Bulgaria (August 14–15, 2019) (pp. 275–276). Siberian Federal Scientific Center of Agrobiotechnology of the Russian Academy of Sciences, Novosibirsk, Russia

Gustokashin KA (2013) The distribution model of horse strangles in the Altai Territory from 1964 to 2011, based on epizootological monitoring. Bull Altai State Agrar Univ 11:79–80

Hartford OM, Foster TJ, Jacobs AA (2002) Strain and culture of *Streptococcus* equi tw 928 strain for the vaccination of horses (December 12, 2002 Patent RU2194752C2)

Kim JW, Jung JY, Lee K, Lee H, Kim HY, Yoon SS, ... Choi E (2017) A case of streptococcus equi zooepidemicus infection in a thoroughbred horse. J Comp Pathol 158:137.https://doi.org/10.1016/j.jcpa.2017.10.133

Libardoni F, Machado G, Gressler LT, Kowalski AP, Diehl GH, Santos LC, ... Vargas AC (2016) Prevalence of Streptococcus equi subsp. in horses and associated risk factors in the State of Rio Grande do Sul, Brazil. Res Vet Sci 104, 53–57.https://doi.org/10.1016/j.rvsc.2015.11.009

Ministry of Agriculture of the Republic of Kazakhstan (2002) Rules for state registration (re-registration) and maintenance of the State register of veterinary drugs in the Republic of Kazakhstan (October 31, 2002 No. 349). Astana, Kazakhstan

Ministry of Agriculture of the Republic of Kazakhstan (2003) Instructions for the approval of scientific and technical documentation submitted by individuals and legal entities for new, improved veterinary drugs (March 3, 2003 No. 105). Astana, Kazakhstan

Ministry of Agriculture of the Republic of Kazakhstan (2014) The order "On approval of the rules for the approval of normative and technical documentation for new, improved veterinary drugs and feed additives" (November 28, 2014 No. 7–1/625). Astana, Kazakhstan

Neamat-Allah ANF, El Damaty HM (2016) Strangles in Arabian horses in Egypt: Clinical, epidemiological, hematological, and biochemical aspects. Veterinary World 9:820–826

Neustroev MP (2000) Strangles in horses in Yakutia (etiology, epizootiology, control measures, and prevention), 1st edn. Siberian Branch of the Russian Academy of Agricultural Sciences, Novosibirsk, Russia

Neustroev MP, Tarabukina NP (2003) A method of treating horse strangles. Horse Breed Equest Sport 12:18–19

Neustroev MP, Murashev AN, Bondarenko DA, Stepanova AV, Tarabukina NP (2017) The study of the toxicity of the drug Sakhabactisubtil in rats. J Microbiol 5:59–64

Neustroev MP, Tarabukina NP, Petrova SG (2019) A way to improve the effectiveness of vaccination against infectious abortion in horse herd. Russ Agric Sci 1:55–57

Osmaev IA, Yurov KP, Neustroev MP (2007) Immunomodulatory properties of endogenous interferon in calves. Veterinary 1:11–12

Raimbekov DR, Jetigenov EA, Karypov KA, Arziev KA (2016) Epizootic features of horse strangles in the Chuisky region. Bull Kyrg Natl Agrar Univ Named after K I Scryabin 2:48–52

Sansyzbaev AR (1993) Horse strangles in Kazakhstan (distribution, properties of the pathogen, development of means of specific prevention and treatment) (Dis. of Doc. of Veterinary Sci.). Russian Academy of Agricultural Sciences, Moscow, Russia

Sweeney CR, Timoney JF, Newton JR, Hines MT (2005) Streptococcus equi infections in horses: Guidelines for treatment, control, and prevention of strangles. J Vet Intern Med 19:123–134

Sweeney CR, Whitlock RH, Meirs DA, Whitehead SC, Barningham SO (1987) Complications associated with Streptococcus equi infection on a horse farm. J Am Vet Med Assoc 191:1446–1448

# Economic Systems in the Age of Digital Changes and Unstable Environments

# Strategic Directions for Using the Potential of the Border Position for Sustainable Development of the Region

**Viktoria V. Lazareva, Inna V. Mishchenko, and Elina V. Gubernatorova**

**Abstract** This study focuses on the strategic directions for using the border position of regions in their sustainable development. In this study, we evaluated the status and ongoing changes in the use and development of the border position potential in the development of Russian-Chinese cross-border cooperation. In the evaluation, we accounted for the interests of various group entities: government, businesses, and society. We chose the Amur Region as the subject of this study. Amur Region has the longest border with China among the regions of Russia. Moreover, the territorial differences in the Amur Region are complemented by the preservation of different border regimes and the existing border infrastructure. We concluded that the ongoing changes in international economic cooperation and the transformation of the border position potential contribution to the regional competitiveness. However, the border potential is significantly under-utilized in achieving sustainable socio-economic development and improving the quality of life. Through our analysis, we confirmed the need to develop and implement regional and federal measures aimed at more effective use of the border potential. We examined the most promising areas of cross-border development, including the pressing and significant problems within the competence of regional governments.

**Keywords** Border position potential · Municipality · Sustainable development · Transformations · Strategic approach

## 1 Introduction

An important goal of spatial development is the maximum use of territorial potential. Sustainable development of several regions can be improved by using their border

V. V. Lazareva
Amur State University, Blagoveshchensk, Russia

I. V. Mishchenko (✉) · E. V. Gubernatorova
Altai State University, Barnaul, Russia
e-mail: mis.iv@mail.ru

 
S. G. Maximova et al. (eds.), *Advances in Natural, Human-Made, and Coupled Human-Natural Systems Research*, Lecture Notes in Networks and Systems 234,
https://doi.org/10.1007/978-3-030-75483-9_3

position, facilitating international economic cooperation. The use of existing border potential is important for the Far Eastern border regions. Their territorial remoteness hinders economic cooperation with the other regions of the country. However, their geographic position can contribute to external economic relations with the Asia–Pacific countries. Among these countries, the most prospective one for the development of cross-border cooperation is the People's Republic of China.

Currently, regional and municipal governments accumulated rich experience in using and developing cross-border potential. However, we determined that the potential is both fully unrealized and does not meet the current and future needs. Therefore, not only the practice of using the existing potential must be improved, but also its formation (e.g., changing the legal regime of borders and the status of border territories, granting special preferences for business entities in border areas). As the experience of foreign countries shows, the effective use of the border position potential can contribute to the sustainable development of even the least advantaged territories.

## *1.1 Potential of Border Position*

The impact of the border potential on the development of territories near country borders was only studied for the last hundred years. The increasing attention to socio-economic processes in the border areas was caused by several factors. The most important of these are the manifestations of spatial polarization effect: (1) economic degradation, (2) formation of economically-depressed enclaves, (3) narrowing of the economic space, (4) increasing social inequality, (5) rise of unemployment, and (6) decreasing quality of life (Dyachenko and Lazareva 2019).

The development of territorial imbalances stimulated theoretical and empirical research. The classical theories of production location are of significant interest for this study. These theories examine the specific features of development through the influence of the border on the location of enterprises (Losh 2007; Christaller 1966; Giersch, 1949–1950; Hoover 1948). Additionally, we used the theories on international and interregional trade, which examine the spatial effects of economic integration (Ohlin 1967; Rauch 1989); the growth pole theory of F. Perroux (Perroux 2007); and the core-periphery theory of Friedmann (Friedmann and Alonso 1964). These theories allowed us to analyze border territories through the relations between the regional center and the periphery. NEG models and concepts allowed us to evaluate the potential impact of international economic integration on the spatial redistribution of economic entities (Elizondo and Krugman 1996; Fujita and Mori 1996; Hanson 1998; Krugman 1993; Venables 1996).

The growth pole theory of F. Perroux and the core-periphery theory of J. Friedmann are especially important in the analysis of border regions. The remoteness of many border territories allowed us to examine them in the context of core-periphery relations. The P. Krugman core-periphery model and the theory "new economic geography" [NEG] significantly expanded the tools for analyzing regional development features (Krugman 1991). European countries widely use NEG models to analyze border territories. Before the start of integration processes in Europe, the near-border position of some regions was considered a hindrance. However, with the development of integration processes, "border regions became potentially equivalent to the central, coastal, and resource-rich territories, and possibly even better than them in some parameters" (Artyomenko 2009).

The topic of the influence of border position on regional development is still relevant. In this study, the research objects are the border territories at the macro-, meso- and micro-levels of the border area. Reviewing secondary literature allowed us to classify the studies on the topic into groups:

(1) Studies showing the constructive nature of border position influence on the development of territories (Artobolevskij and Vardomskij 2010; D'yachenko 2016; Baklanov and Tulokhonov 2010; Ryzhova 2008);
(2) Studies showing the ambiguous effects of border position on the development of territories (Novikov and Novikova 2008; Fadeeva 2014; Antonyuk and Kornienko 2016; Vardomskij 2017).

Even though scholarly opinions differ on the impact of border position on regional development, the scholars agree that the near-border position gives territories specific potential.

For example, P. Ya. Baklanov and S. S. Ganzey have the same opinion on the additional potential of development and international cooperation of border areas, calling it "the specific potential of the border area" (Artobolevskij and Vardomskij 2010). I. I. Dragileva establishes the term "contact potential," the use of which "can and should become the basis for improving the depressive state of Russian border regions" (Dragileva 2006). The work of G. A. Hmeleva and S. K. Umerbaeva dwells on the economic potential of the border location (Hmeleva and Umerbaeva, 2015).

Although the amount of secondary literature on border position influence on regional development is considerable, the concept of "border position potential of a region" has not yet received a clear unambiguous interpretation. Therefore, a meaningful interpretation of this concept.

Currently, the term "potential" in economic studies is interpreted as:

(1) A set of resources (resource approach);
(2) A set of opportunities (resultative approach);
(3) A set of "abilities" (targeted or strategic approach).

In interpreting the term "border position potential of a region," we employed the conceptual framework of these approaches. We defined the border position potential as the ability of the border territories to ensure long-term functioning, achieve strategic development goals, and improve the quality of life via the use of cross-border interaction opportunities.

The development and use of the cross-border potential is influence by several factors:

- Nature and status of international relations;
- Applied preferences that ensure cross-border cooperation;

- Potential of bordering territories;
- Complementarity of regional economies;
- State of the border infrastructure;
- Actions of regional authorities.

These factors, along with spatial factors (climatic conditions, economic and geographical position, agglomeration effects, human capital, the efficiency of institutions, etc.) define the characteristics and conditions of sustainable development in border regions.

The sheer amount of these factors determines the uniqueness of each border region and requires a differentiated approach to managing their sustainable development.

## 2 Materials and Methods

In assessing the status of the border position potential (i.e., specific opportunities associated with border proximity and the state of border infrastructure), one should keep in mind that the potential can differ significantly in terms of realizing the interests of various groups of factors (Fig. 1).

International economic relations treat border territories as transit zones. Therefore, the potential of these zones should be determined by assessing the transport and logistics capabilities and the development of border crossing infrastructure. The indicators for assessment are presented in Table 1.

Favorable placement of export-oriented production facilities is the most important factor for large businesses. The functioning of special regimes may indicate a high border position potential.

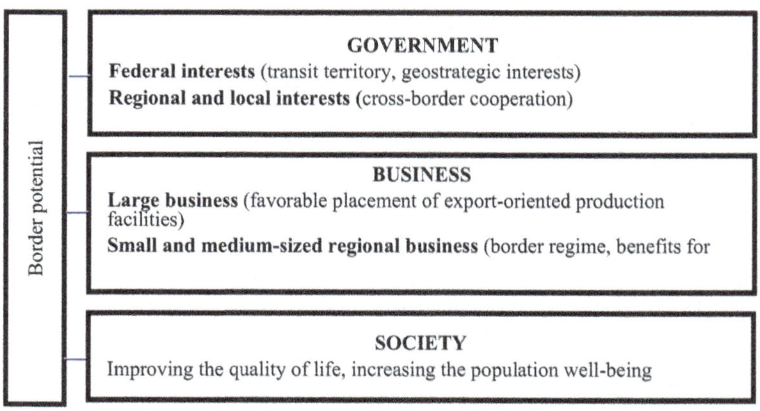


**Fig. 1** The combination of interests of various groups concerning the goals and results of the development of border territories. *Source* Compiled by the authors

**Table 1** Indicators characterizing the transit potential of the region

| Indicator | Units of measurement | Characteristics |
|---|---|---|
| Density of public railways | Km per 1,000 $km^2$ of territory | The ratio of the length of railways to the total area of the region |
| Density of paved public roads | Km per 1,000 $km^2$ of territory | The ratio of the length of roads to the total area of the region |
| Crossing points density (all) | Units per 1 km of border | The number of all established crossing points per 1 km of the border with the adjacent territory |
| Density of operating border crossing points | Units per 1 km of border | The number of operating crossing points per 1 km of the border with the adjacent territory |

*Source* Compiled by the authors

Small and medium-sized regional businesses are mostly interested in the border regime, and benefits for export–import operations and cross-border tourism.

The realization of regional interests is related to the development of cross-border cooperation. In assessing the use of the border position potential, one can analyze the following indicators: (1) the specific weight of the region in trade with a neighboring country; (2) export–import commodity structure; (3) the use of foreign labor in the regional economy.

In relation to societal interests, the border position potential is mainly connected to improving the standard of living. The standard of living can be described by such indicators as:

- GRP per capita, compared to the average in Russia;
- Per capita investment, compared to the average in Russia;
- Per capita income, compared to the average in Russia;
- Unemployment rate, compared to the average in Russia;
- Foreign trade turnover per capita, compared to the average in Russia.

The border intensity coefficient serving as an informative indicator can be calculated by Formula 1.

$$K = \frac{VTO}{K\_PAR'} \quad (1)$$

where K—*border intensity factor*; VTO—*foreign trade turnover per capita*; K_PAR—*gross regional product* (taking into account the level of purchasing power) per capita.

According to L. L. Bozhko, "the border intensity coefficient shows that the regional foreign trade turnover does not represent the importance of cross-border cooperation mechanisms, if taken separately, without considering the basic characteristics of the regional economy. Only by comparing the gross regional product per capita (taking into account the purchasing power) the volume of foreign trade per capita, one can calculate the intensity of cross-border economic cooperation and its importance for regional economy. Studies demonstrated that the border intensity coefficient strongly depends on the gross regional product. In highly developed regions, the border intensity coefficient is greater than or equal to 1" (Bozhko 2013).

## 3 Results

Recently, positive trends were observed in the transit potential of the Amur Region. The existing transport infrastructure (i.e., Trans-Siberian Railway, Baikal-Amur Mainline, the Amur highway, river transport on the Amur river, Zeya and Bureya Rivers, oil and gas pipelines) was supplemented by the bridge crossing over the Amur River in the Blagoveshchensk-Heihe area. Its commissioning opened a promising corridor for international transport. Currently, the Amur Region is turning into one of the largest centers of cargo transportation.

Table 2 presents the indicators of the transit potential in the Amur Region, compares them with the indicators of other regions of the Far Eastern Federal District sharing a common border with the People's Republic of China (PRC).

Currently, only two border crossings (out of eight total) in the Amur Region are operational: one in Blagoveshchensk and one in Poyarkovo, Mikhailovsky District. The remaining crossings have never been operational. More than 300 investment projects were implemented in the framework of export-oriented enterprises aimed at the Asia–Pacific region. Therefore, cross-border infrastructure (i.e., crossing points, speed and quality of work of customs authorities, formed logistics infrastructure) becomes a deciding factor in the transit potential of the Amur Region.

Large corporations have shown interest in placing their production facilities in the Far East, including the Amur Region, mainly due to the existence of *Territories of Advanced Social and Economic Development* [TAD], which provide tax benefits for its residents. In the Amur Region, three such territories were created: "Priamurskaya," "Belogorsk," and "Svobodny."

Moreover, TAD residents enjoy such benefits as: (1) free customs zone; (2) duty-free import of equipment and components; (3) non-quota foreign labor force; (4) easier procedures for conducting state and municipal control; (5) coordination of unscheduled inspections with the Ministry for the Development of the Russian Far East; (5) reduced time for obtaining permits; (6) simplified provision of land. However, other economic entities in the region have to rely solely on their own means. Their economic weakness and the failing state of municipal infrastructure prohibit most regional economic entities from using the internal transformation potential. Therefore, we consider the impact of TADs on the real sector of the economy to be negative since they draw labor resources from the already under-provided sectors. However, this is not a valid reason for abandoning the strategy of creating growth

**Table 2** Indicators of the transit potential of the Russian federation constituent entities sharing a common border with PRC, 2018

| Indicator | Amur region | Khabarovsk region | Primorsky Krai | Trans-Baikal region | Jewish autonomous region |
|---|---|---|---|---|---|
| Density of public railways | 81 | 27 | 95 | 56 | 141 |
| Density of paved public roads | 34 | 12 | 91 | 34 | 67 |
| Length of the border with PRC | 1,254 | 260 | 1,031.5 | 1,064 | 550 |
| Number of crossing points (not including air and sea ones) | 8 | 2 | 7 | 8 | 4 |
| Density of operating border crossing points (not including air and sea ones) | 0.0064 | 0.0077 | 0.0068 | 0.0075 | 0.0072 |
| Number of operating border crossing points (not including air and sea ones) | 2 | 2 | 6 | 6 | 2 |
| Density of operating border crossing points | 0.0016 | 0.0077 | 0.0058 | 0.0056 | 0.0036 |

*Source* Calculated by the authors (Federal Statistical Office 2018)

points. It only indicates its insufficiency for the development of the Far East Federal District (which includes the Amur Region).

The border regime, and the benefits for export–import operations and cross-border tourism are of special interest for small and medium-sized regional enterprises. Currently, many Amur Region entrepreneurs actively use the border crossings located in other Far East regions due to the high costs of export–import operations. The cost is increased by all structures involved in the activities of customs and border crossings. Blagoveshchensk is increasingly losing its leading role in developing cross-border ties with China. But in other regions, steady progress is being made (e.g., wholesale trade of Chinese goods is currently concentrated in Ussuriysk instead of Blagoveshchensk).

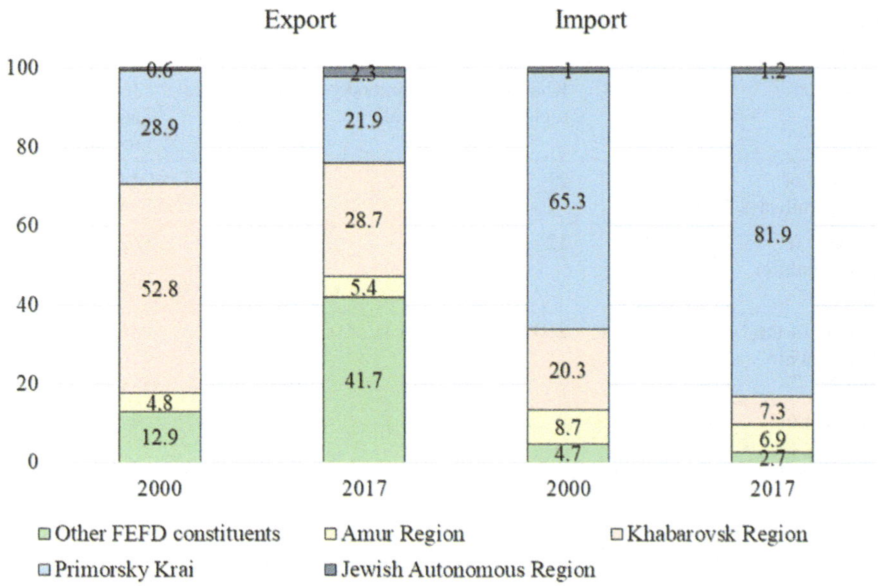


**Fig. 2** The proportion of the Far Eastern Federal District constituents in external trade with China in 2000 and 2017. *Source* (Federal Statistical Office 2018)

Amur Region plays a modest role in the development of cross-border cooperation with China among other regions of the Far East. Therefore, the border position of the Amur Region is under-utilized in foreign trade (Fig. 2).

The commodity structure of exports remained unchanged: the regions mainly export raw materials. The predominance of raw exports leads to the exhaustion of natural resources and limits their use in the future economic development of the region.

The use of foreign labor in the Amur Region is extremely underwhelming. The current policy limits the attraction of workers from China, including for seasonal work in agriculture, which is clearly not in the interests of the socio-economic development of the region. Moreover, it is a serious deterrent using the potential of foreign labor. Therefore, increasing to act on the potential, the government must abandon the far-fetched restrictions and attract foreign employees to work in the region.

Currently, Russia and China coordinate their international interests in developing cross-border relations in the Far East. However, on the regional level, the actions of Russian authorities limit the use of border position potential in the programs of socio-economic development (mainly due to inertia and the recent prevalence of restrictive practices) (Table 3).

The data indicates a significantly low socio-economic development of the Amur Region, compared to the average in Russia. The only exception is the per capita fixed investments the average in Amur Region is 1.66 times higher than in Russia.

**Table 3** Indicators for assessing the socio-economic development of the Amur Region in comparison with the average Russian values

| Indicator | 2010 | 2015 | 2016 | 2017 |
|---|---|---|---|---|
| GRP per capita, compared to the average in Russia, mln. RUB | 0.66 | 0.60 | 0.61 | 0.58 |
| Per capita investment, compared to the average in Russia, mln. RUB | 1.57 | 1.33 | 1.60 | 2.13 |
| Per capita income, compared to the average in Russia, RUB | 0.76 | 0.99 | 0.96 | 0.96 |
| Unemployment rate, compared to the average in Russia, % | 0.94 | 1.04 | 1.07 | 1.13 |
| Foreign trade turnover per capita, compared to the average in Russia, mln. USD | 0.20 | 0.35 | 0.33 | 0.27 |
| Border intensity coefficient | 0.07 | 0.15 | 0.11 | 0.10 |

*Source* Federal Statistical Office (2018)

The border intensity coefficient remains at a rather low level indicating low socio-economic development and inefficient use of the border position potential.

We conclude that the existing border position potential in the Amur Region is severely under-utilized. Moreover, in recent years, lucrative opportunities for developing cross-border ties were missed, and much of the earlier progress was lost. This mainly refers to:

- Servicing trade turnover between the Russian Federation and the People's Republic of China;
- Playing an intermediary role in the development of international trade;
- Using the potential of industrial cooperation to enter the Chinese market;
- Import-substitution;
- Using foreign labor;
- Attracting foreign capital;
- Creating tourist infrastructure.

## 4 Discussion

In this study, we considered only some promising areas of cross-border potential since they can contribute the most to sustainable development. These areas include the most pressing and significant problems within the competence of regional governments.

One of the largest underutilized resources in the region is arable land. A significant part of arable land is currently not in economic circulation (Table 4).

Currently, the development of agricultural production is limited by personnel shortage caused by the demographic decline in the districts, mainly the border ones. Labor supply can be provided by attracting seasonal workers from China.

However, the current migration policy limits the attraction of Chinese workers. This policy severely decreases the potential agricultural output. In 2017, the quota for foreign workers was 5,056 people, including 2,959 (approx. 60%) workers from the

**Table 4** Arable land in 2017 compared to 1989, %

| District | 2017 year to 1989 | District | 2017 year to 1989 |
|---|---|---|---|
| Tambovsky | 95.6 | Arkharinsky | 52.2 |
| Konstantinovsky | 92.0 | Selemdzhinsky | 40.2 |
| Ivanovsky | 81.4 | Bureysky | 39.2 |
| Belogorsky | 78.4 | Tyndinsky | 34.1 |
| Mikhailovsky | 78.2 | Skovorodinsky | 30.5 |
| Blagoveshchensky | 65.2 | Magdagachinsky | 28.6 |
| Seryshevsky | 63.0 | Mazanovsky | 25.3 |
| Oktyabrsky | 609 | Shimanovsky | 18.6 |
| Romnensky | 60.3 | Svobodnensky | 17.1 |
| Zavitinsky | 55.8 | Zeysky | 10.9 |

*Source* Federal Statistical Office (2018)

PRC. Implementing import-substitution requires significantly increasing the quota and adjusting the regional migration policy.

To determine the quotas, we calculated the need for labor resources based on the forecasting values of agricultural production indicators and labor productivity in the industry (Table 5).

These indicators show that by 2022 the Amur Region may severely lack labor resources in agriculture, meaning that the regional migration policy must be improved (especially the quotas for foreign workers). Interdepartmental commissions can use these calculations for determining quotas on foreign labor.

**Table 5** Forecasting the need for labor resources in agriculture

| Forecasting indicators | 2018 | 2019 | 2020 | 2021 | 2022 |
|---|---|---|---|---|---|
| Forecasting volume of agricultural production, mln. RUB | 59,299.7 | 64,056.0 | 68,812.3 | 73,568.6 | 78,324.9 |
| Forecasting labor productivity in agriculture, mln. RUB per thousand people | 1,996.5 | 2,234.3 | 2,472.1 | 2,709.9 | 2,947.6 |
| Forecasting average annual number of people employed in agriculture, thousand people | 24.3 | 19.6 | 14.9 | 10.2 | 5.6 |
| The industry's demand for labor resources, taking into account forecasting values, thousand people | 29.7 | 28.7 | 27.8 | 27.1 | 26.6 |

*Source* Calculated by the authors (Federal Statistical Office 2018)

# 5 Conclusion

In this study, we confirmed the need to develop and implement measures of more efficient use of the border position potential. At the federal level, the system of preferences in border regions should be uniform. At the regional level, the government efforts should aim to maintain and improve the border position potential, taking into account the goals of sustainable development. These tasks can be achieved via the use of a strategic approach in managing the development of border position potential.

The strategic documents developed by regional governments should include measures aimed at using the border position potential. These presuppose: (1) creating a system of incentives for business entities that cooperate in the important sectors of regional development, (2) modernizing the existing production facilities, and (3) efficiently using land resources to ensure food security via import-substitution.

# References


Antonyuk VS, Kornienko EL (2016) Theoretical aspects of sustainable socio-economic development of border areas. Innov Invest 12:181–183

Artobolevskij SS, Vardomskij LB (2010) Cross-border areas and regional policy in the Russian Federation. In: Podgrushnyj GP (ed) Problems of public geography. Institute of Geography of the National Academy of Sciences of Ukraine, Kyiv, Ukraine, pp 14–26

Artyomenko SV (2009) Border and cross-borderness as characteristics of a region. Sci Newsl Volinsky Natl Univ Ukr Ukr 4:180–184

Baklanov PY, Tulokhonov AK (2010) Cross-border and cross-border territories of Asian Russia and neighboring countries (problems and prerequisites for sustainable development). SB RAS, Novosibirsk, Russia

Bozhko LL (2013) The methodology for assessing the socio-economic development of the border area in the context of integration transformations. Reg Econ Manag 1. Retrieved from https://eee-region.ru/article/3311/9

Christaller W (1966) Central places in Southern Germany (C. W. Baskin Trans.). Prentice-Hall, Englewood Cliffs, NJ (Original work published 1933)

D'yachenko VN (2016) Features of programming the development of the Russian-Chinese border area. Econ Entrep 3-1(68):262–265

Dragileva II (2006) Cross-border cooperation in the Baltic region. In: Gutnika VP, Klemesheva AP (eds) The Baltic region as a pole of the economic integration of the North-West of the Russian Federation and the EU. Russian University, Kaliningrad, Russia, pp 77–100

Dyachenko V and Lazareva V (2019) The practice of use and development of border situation potential (based on the materials of the Amur region). Reg Econ Manag 2. DOI: https://doi.org/10.1088/1755-1315/272/3/032204

Elizondo PL, Krugman P (1996) Trade policy and third world metropolis. J Dev Econ 49:137–150

Fadeeva OP (2014) Transboundary rural territories of Transbaikal Region: development or degradation? (Notes of a sociologist). ECO 11:36–54

Friedmann J, Alonso W (1964) Regional development and planning. MIT Press, Cambridge, MA

Fujita M, Mori T (1996) The role of ports in the making of major cities: self-agglomeration and hub-effect. J Dev Econ 49:93–120

Giersch H (1949–1950) Economic union between nations and the location of industries. Rev Econ Stud 17(43):87–97

Hanson GH (1998) Regional adjustment to trade liberalisation. Reg Sci Urban Econ 28:419–444

Hmeleva GA, Umerbaeva SK (2015) The economic potential of the border location as a factor in increasing the level of socio-economic development of the region. Bull Samara State University Ser "Econ Manag" 9-1(131):48–54

Hoover EM (1948) The location of economic activity. McGraw Hill Book Company Inc., New York, NY

Krugman P (1991) Increasing returns and economic geography. J Polit Econ 3:483–499

Krugman P (1993) On the relationship between trade theory and location theory. Rev Int Econ 1:110–122

Losh A (2007) Spatial organization of the economy. In: Granberg AG (ed) Nauka, Moscow, Russia

Novikov AN, Novikova MS (2008) Regional features of the border position of southeastern Transbaikalia. Sci Notes Kazan State University Ser: Nat Sci 3:229–240

Ohlin B (1967) Interregional and international trade (Revised Edition). In: Harvard economic studies, XXXIX. Harvard University Press, Cambridge, MA

Perroux F (2007) Economic space: theory and applications. Spat Econ 2:77–93

Rauch JE (1989) Increasing returns to scale and the pattern of trade. J Int Econ 26:359–369

Ryzhova NP (2008) The development of the border regions of China: the role of the "frontier belt of openness." Spat Econ 4:36–59

Vardomskij LB (2017) New border regions of Russia in a changing post-Soviet space. In: Vardomskij LB, Popkova LI, Mądry C (eds) Strategy for the development of border territories: Traditions and innovations. Kursk State University, Kursk, Russia, pp 209–221

Venables AJ (1996) Equilibrium locations of vertically linked industries. Int Econ Rev 37:341–359

# A City-Forming Enterprise as a Subject of Sustainable Social and Economic Development: The Magnitogorsk Metallurgical Complex

**Svetlana N. Ispulova, Elena V. Oleynik, and Olga N. Kolesnikova**

**Abstract** The article concentrates on the consideration of the activity of the city-forming enterprise on the formation of conditions for the sustainable social and economic development of the mono industrial city, and also the definition of its most actual problems. The methodological basis of the research is based on the scientific statements of such scientists as O. Dinius, O. P. Ivanova, I. N. Ilyina, L. Karlson, I. D. Turgel, T. V. Uskova, K. Ehsani, etc. The results of a sociological study conducted in 2015–2019 in the city of Magnitogorsk (Chelyabinsk region), on the one hand, testifying to the significant measures of social support to the city-forming enterprise of its employees, and, on the other hand, the need, first of all, to invest in the socio-economic development of the city, to increase its attractiveness. The results of the survey correlate with the Report of the city's forward-looking enterprise in 2018—"Magnitogorsk Iron and Steel Works" PJSC, data from Chelyabinskstat, which allows the authors to make more accurate and balanced conclusions about the real situation in regards to this issue. The results of the study allowed to determine the critical positions that require special attention of the main subjects of urban management: self-government bodies, owners, administration of the city-forming enterprise, the urban community. These are the existence of double standards for the population of the city depending on the membership in the city-forming enterprise, the reduction of life expectancy of citizens, the dependence of all spheres of life of the city on the decisions of owners, representatives of the administrative body of the city-forming enterprise.

**Keywords** City-forming Enterprise · Mono-industrial City · Social and Economic Development · Social Partnership

S. N. Ispulova (✉) · E. V. Oleynik (✉)
Nosov Magnitogorsk State Technical University, Magnitogorsk, Russia
e-mail: oleynik.ev@mail.ru

O. N. Kolesnikova
Altai State University, Barnaul, Russia
e-mail: kolesnikova@socio.asu.ru



S. G. Maximova et al. (eds.), *Advances in Natural, Human-Made, and Coupled Human-Natural Systems Research*, Lecture Notes in Networks and Systems 234,
https://doi.org/10.1007/978-3-030-75483-9_4

## 1 Introduction

Under the current conditions of globalization, the modern world has lost stability and balance, and has become discrete and unpredictable. Residents of mono-industrial cities have found themselves in a particularly vulnerable situation. Dependence on the city-forming enterprise causes socio-economic problems for the urban community, such as problems of social and economic development, employment, monoprofilicity of the labor market and demographic problems (outflow of young and able-bodied people from the city, low birth rate). In quantitative terms, there are about 466 single-industry cities and more than 300 settlements of the same type in Russia (Ilyina et al. 2013). There is also a systemic crisis of mono profile cities, which makes the social fabric of cities of this type of development, a zone of risk with possible adverse consequences, which makes it necessary to pay more attention to this problem of the scientific community. Obviously, each mono-industrial city has its own issues, this article is devoted to the issues of sustainable development of the city of Magnitogorsk, Chelyabinsk region. Manitogorsk, a city with a non-government city-forming enterprise of the metallurgical industry, belonging to a large population, is isolated geographically. At the same time, it has significant resources for sustainable social and economic development, because the city-forming production has the potential for development and growth, that is almost resistant to economic crises, and it also has the potential of self-recovery.

"Magnitogorsk Metallurgical Kombinat" PJSC (hereinafter referred to as MMK) is a city-forming enterprise. The plant was built and launched, and then the plant began to plan and build the city. Thus, genetically, Magnitogorsk arose from the mill and lives at the mill, and this significantly affects the social and economic fabric of the city. Thus, MMK is one of the main subjects determining the vector of social and economic development of the city.

This state of affairs has led to the following contradictions: first of all, the city depends on the enterprise, not vice versa; second of all, the enterprise, pursuing commercial interests, uses the resources of the city (human, vocational education systems and natural resources), while reluctantly investing in the development of urban economy.

Contradictions served as the basis for determining the purpose of this work: to define the role of the city-forming enterprise in the sustainable social and economic development of the city, to prove the thesis on the mutual alignment of the success of the enterprise and the city based on the principle of social partnership.

## 2 Materials and Methods

The problems of development of cities with the mono-industrial type of development are devoted to the works of M. Androniceanu (Androniceanu et al. 2019), D. Gallardo-Vazquez (Gallardo-Vazquez et al. 2019), J. Green (2011), O. Dinius

(Dinius and Vergara 2011), O. P. Ivanova (2018), I. N. Ilyina (Ilyina et al. 2013), L. Carlson (2003), N. A. Simchenko (Simchenko and Piskun 2019), I. D. Turgel (2010), T. V. Uskova (Uskova et al. 2012), and K. Ehsani (2003). The analysis of modern scientific literature on social and economic problems of the development of mono industrial cities shows, that social partnership and cohesion are practically not considered as a necessary condition for the development of these territories. At the same time, European experience of regional development (Germany in particular) shows that it is social cohesion based on the principle of social partnership that has become a decisive organizing factor for development. Nevertheless, this fact has not been properly reflected in the scientific literature.

Taking into account the experience of the Czech Republic, the Netherlands, Japan, Great Britain, and the USA, in solving the problems of development of mono industrial cities, we have concluded that mono industrial cities can develop under the following scenarios: the transition from the production of obsolete products to promising ones, diversification of the economy, development of small and medium business, implementation of social activities. Any of these scenarios requires a long-term plan to attract investments, to unite the population, to actively participate in the development of the city of the city-forming enterprise.

The novelty of the work lies in the substantiation of the provision, that effective management of sustainable social and economic development of a mono industrial city is possible with a multisubject approach to management, which includes coordinated activities on the principles of the social partnership of local government, the owner, the administration of the city-forming enterprise and the local community.

In this regard, the results of a sociological study that we conducted in 2015–2019 in Magnitogorsk to study the opinion of locals on the current issues and prospects of the city and the role of the MMK.

The main method of our study was a handout survey. It was attended by 960 residents of the city, a third of whom were employees of the city-forming enterprise. It was a targeted analytical approach, the main criterias were: place of work, gender and age structure, social and professional characteristics. The main blocks of questions were related to the activities of the city-forming enterprise, working conditions, social partnership, objects of the city economy.

## 3 Results

The results of the survey showed that despite the fact that 90.3% of respondents see positive changes in the development of the city (especially noted the repair of roads and the development of sports), however, significant problems are noted in a number of sectors of social and economic development. Thus, according to 56.4% of respondents, the schedule of public transport routes is clearly oriented to MMK: the bulk runs in the direction of gatehouses, the number increases during shift hours at the mill.

72.1% of citizens believe that there are insufficient pre-school institutions in the city, which entails the inability of one of the spouses to get a job, and, therefore, a decrease in the income of the family. At the same time, MMK employees did not note such a problem, perhaps this is due to the fact that part of kindergartens, as in Soviet times, is sponsored by the mill. This reduces tension, on the one hand, and on the other hand, there is a division into children of MMK employees and others whose parents are engaged in health care, education, services, in ensuring the functioning of the urban economy.

Problems in the field of health care development, namely, unavailability of high quality medical care, lack of specialists, lack of modern diagnostic equipment were noted by 63.6% of respondents.

At the same time, the mill employees (96.2% of MMK employees) are satisfied with the volume and quality of medical care. It should be noted that the mill employees have the opportunity to receive all types of medical services on the territory of the mill, as there are modern workshop medical stations, one of the best hospitals in the city (with a modern medical base, doctors with high qualifications) belongs to the medical unit of MMK. The company provides its employees and their family members with sanatorium and resort treatments in country sanatoriums, health resorts, recreation bases.

Thus, health care is another area of urban life, where double standards are observed, which causes irritation in the deprived communities on the one hand, on the other hand, the mill employees feel the care of the employer, as some kind of exclusivity. At the same time, according to Chelyabinskstat, the average life expectancy in Magnitogorsk over the past three years has decreased from 64 years to 62.1 years (by 1.9 years): including among men—from 58.9 to 57.1 year (by 1.8 years), among women—from 69.8 to 68.1 years (by 1.7 years) (Ispulova, Oleynik, Potrikeeva & Koueva). Among the main reasons are unsatisfactory environmental conditions, which are largely influenced by the mill (Fig. 1).

Almost every family in the city is connected to the mill, which explains the fact that many respondents are aware of the mill's activities in terms of supporting their employees. For example, 93.5% of respondents noted the development of sports (as a director of the mill's activity), 91.3%—cultural and leisure direction, 61% care for their veterans, 80.1% support for working youth, respectively (Fig. 2).

Analysis of respondents' answers allows us to conclude the following: MMK invests in the creation of favorable conditions for work, rest, development of its employees. However, for development of the city, increase of its attractiveness, for

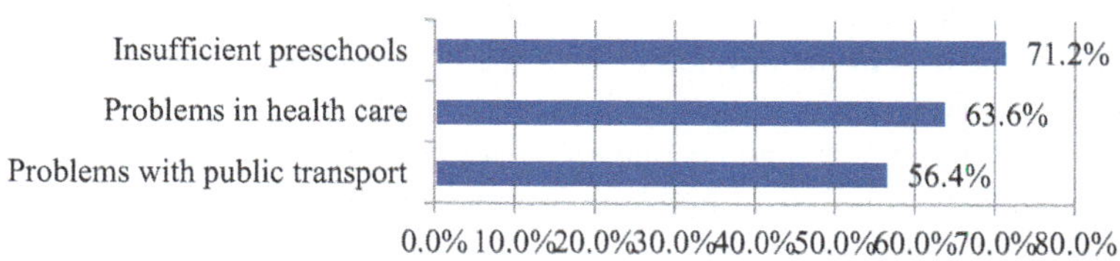


**Fig. 1** Respondents' opinion about problems in the city. *Source* Compiled by the authors

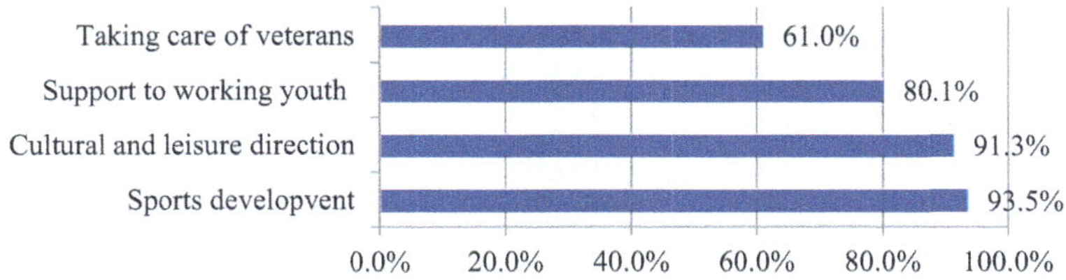


**Fig. 2** Respondents' opinion about the preferential directions of MMK social policy. *Source* Compiled by the authors

both residents of the city and visitors is not enough. Perhaps this is why in recent years there has been a significant outflow of population from the city (in 2018 more than 30 thousand people left the city), first of all, these are young people and people of working age, and these include potential employees of the mill.

## 4 Discussion

The opinion of respondents presented in the previous section is certainly important and valuable, but at the same time contradictory. Thus, based on the principles of social partnership and social responsibility, understanding that the population and development of the city depend on the city-forming enterprise, the following areas of social policy are implemented at MMK: improvement of working conditions; medical programs; performing mass sports; support for maternity; affordable housing; support for non-working pensioners; investments in the territory of presence, etc. In each of these areas, social programs for employees and their families are developed and successfully implemented.

Health improvement of families of employees was organized in sanatoriums of the Ural region, Metallurg (Essentuki), Ai-Petri (Black Sea coast). Children of employees and residents of the city during summer vacations have the opportunity to rest in modern children's health and educational centers "Ural Zori" and "Mountain Gorge."

Within the framework of the Employee Health Improvement Program, a set of medical and preventive measures are provided, a wide range of programs for diagnostics and early detection of diseases, personalized programs, medical examinations, dental services, hospital treatments.

The implementation of social order for the development of mass physical culture and sports is carried out by private institution of additional education "Sports Club "Metallurg-Magnitogorsk", ski centers "Metallurg-Magnitogorsk" and "Abzakovo", water park "Waterfall of Miracles". There are family sports holidays, competitions for employees during the summer and winter sports and athletic contests.

Support for high achievements in sports is provided by annual financing of competitions: I. H. Romazan Hockey Tournament, stages of the Russian Snowboard and Mountain Skiing Championship, International Boxing Tournament, stages of the

Russian Motocross Championship and Cup, Championship of the Ural Federal District in Table Tennis, Russian Kickboxing Championship, MMK Cups (judo, sport ballroom dancing).

Since 2004, the Maternity Support and Birth Promotion Program are implemented. Women participating in the program, future mothers, starting from week 13 of pregnancy, are released from work with average wages. According to the program, future mothers are obliged to visit the "Maternity" center, where physical and psychological training of future mothers is provided, free visits to the water park "Waterfall of Miracles," medical and preventive procedures in the city sanatorium-preventive center "Yuzhny."

The program "Support for large families" has been implemented since 2008. In 2018, 1,169 families of employees (3,688 children) with 3 or more children under the age of 18 received social support. In 2019, a program was implemented to assist large families with 5 or more children in purchasing a family car.

On behalf of the President of the Russian Federation V. V. Putin, during his visit to the Metallurgist's Day in 2019 and Chairman of the Board of Directors of MMK V. F. Rashnikov, expensive equipments were purchased for all children with cystic fibrosis living in Magnitogorsk.

The Corporate Housing and Investment Fund "Key" implements programs "Affordable Housing–Magnitka Metallurgists" and "Assistance to Young Families in the Purchase of Housing," for employees, installments of payments, and "social" mortgages are provided.

The Corporate Housing and Investment Fund "Klyuch" implements the programs "Affordable Housing–Metallurgists of Magnitogorsk" and "Assistance to Young Families in Purchasing Housing", for employees there are installment payments and "social" mortgage.

A charity fund "Metallurgist" has been created, where over 20,000 people in need are registered. The fund implements the Care program including (Ispulova et al. 2017):

- organization and provision of living for single people in the specialized house "Veteran";
- targeted cash and in-kind assistance, including for operations, purchase of expensive medicines, means of rehabilitation;
- provision of free vouchers to sanatoriums, health centers, geriatric center.

Implementation of these programs applies to all residents of the city, as in Magnitogorsk there is not a single family whose members would not work for MMK. This shows the mutual dependence of the enterprise and the city, as well as the leading role of the city-forming enterprise in the sustainable social and economic development of the city.

The polysubject approach in the social and economic development of the mono-industrial city is carried out by large charity projects aimed at the development of the urban environment and the creation of favorable living conditions. Thus, at the initiative of MMK and municipal authorities in Magnitogorsk, the implementation of a project unique for Russia for the comprehensive development of the territory

"Prityajenie" began (on the area of 400 hectares) building modern facilities for sports, medical facilities, museum, educational and cultural facilities, the core of the project will be the museum and educational cluster. According to the plan, construction works will be completed by 2025, and citizens will have new opportunities for self-development, sports, art, training and recreation. The project for improvement of the urban environment "Prityajenie" is included in the Strategy for the Development of Magnitogorsk until 2035 and is aimed at improving the infrastructure and improving the quality of life in the city.

## 5 Conclusion

MMK PJSC considers social responsibility projects as long-term investments that form the basis for the Company's sustainable development. There was a strong understanding that the sustainable social and economic development of the mono-industrial city needs to continue the implementation of coordinated activities with various subjects of management, the urban community, based on the principles of social partnership of cohesion and responsibility.

At the same time, the data presented in the official report of MMK for 2017 is as follows: net profit amounted to 30.678 billion rubles. MMK expenses for charity and implementation of social programs amounted to 860 million rubles, which is only 2.8% of the enterprise's profit, respectively.

It should be noted that in 2019 MMK allocated more than 2.5 million rubles for the implementation of social programs, 885 million rubles for the financing of the "social package" (benefits and guarantees for employees). According to preliminary estimates, the cost of the project "Prityajenie" will exceed 10 billion rubles.

The 2018 MMK Annual Report presents a social strategy: finding growth points for the development of small and medium-sized businesses, diversifying the city's economy, attracting investments (including foreign ones), creating new jobs, improving the living and recreation conditions of citizens as part of the implementation of the strategic initiative "Solving the Problem of Mono-City" to be implemented jointly with the Government of the Chelyabinsk Region and the administration of the city of Magnitogorsk.

In implementing this initiative, it is planned to work in the following areas:

(1) creating conditions for the development of small and medium-size businesses in the city, the main tool for its solution is "MMK-Industrial Park" LLC. The park has all the necessary resources to start any production, a decision has been made on possible co-financing of projects implemented by residents;
(2) based on Magnitogorsk State Technical University named after. G. I. Nosov, within the framework of the FTP "Development of Education," with the support of the Governor of Chelyabinsk Region and MMK (industrial partner), the children's techno park "Quantorium" was opened;

(3) industrial tourism was launched at MMK; the project corresponds to the developed Strategy of Social and Economic Development of the City of Magnitogorsk for the period up to 2035;
(4) planning long-term and sustainable development, ensuring environmental safety and improvement of the environmental protection management system, as in 2018 the implementation of the program of measures on the strategic initiative "Clean City" was continued, the goal of which is to achieve by 2025 a low level of pollution of the city atmosphere less than 5 units;
(5) improvement of the environmental situation, reduction of atmospheric air, and water bodies pollution remain a strategic priority.

The presented figures give the basis for analysis and formulation of conclusions: MMK management, realizing its responsibility not only to its employees but also to all residents of the city, expands the range of social programs, increases its investment in the development and improvement of the quality of life of the population of the city.

## References

Androniceanu M, Dräfguläfnescu I-V, Ciobanu A (2019) Corporate social responsibility and the quality of life. International business information management association conference: education excellence and innovation management through vision 2020. IBIMA, Granada, Spain, pp 1708–1718

Carlson L (2003) Company towns of the Pacific Northwest. University of Washington press, Washington, D.C.

Dinius OJ, Vergara A (2011) Company towns in the Americas: landscape, power, and working-class communities. University of Georgia Press, Athens, GA

Ehsani K (2003) Social engineering and the contradictions of modernization in Khuzestan's company tows: a look at Abadan and Masjed-Soleyman. Int Rev Soc Hist 3:361–399

Gallardo-Vázquez D, Valdez-Juárez LE, Lizcano-Alvarez JL (2019) Corporate social responsibility and intellectual capital: sources of competitiveness and legitimacy in organizations' management practices. Sustainability 11(20):5843

Green H (2011) The company town: the industrial edens and satanic mills that shaped the American economy. Basic Books, New York, NY

Ilyina IN, Lepsky VE, Molodtsov AV, Popov VG, Reznikova EV, Saveliev AM, ... Plisetskiy EE (2013) The development of monocities of Russia. Financial University, Moscow, Russia

Ispulova SN, Oleynik EV, Potrikeeva OL, Suprunenko GA (2017) Development of state and non-governmental gerontological social practices: the experience of Magnitogorsk city. Man in India 97(21):287–300

Ivanova OP (2018) Assessment of sustainable development of clusters in the territory of a monocity's leading social and economic development. Natl Interes: Priorities Secur 14(6):988–1004

Simchenko N, Piskun E (2019) Contribution of corporate social responsibility and social capital management to quality of life: models of evaluation. Int J Qual Ressss 13(3):689–704

Turgel ID (2010) Monofunctional cities of Russia: From survival to sustainable development. UralAGS, Moscow, Russia

Uskova TV, Iogman LG, Tkachuk SN, Nesterov AN, Litvinova NY (2012) Monogorod: development management. ISERT RAS, Vologda, Russia

# The Strategy for Sustainable Development of Territorial Entities: Territorial Marketing

**Victor I. Belyaev and Oksana N. Pyatkova**

**Abstract** The purpose of the study is to formulate fundamental methodological approaches to the use of means and methods of classical marketing in the development and substantiation of strategies for the sustainable development of large territorial entities. A separate district of the Altai Krai was taken as the research object. The research subject is the methodology for developing and marketing strategies for the sustainable development of districts of Russian constituent entities. The authors applied the methods of content analysis of scientific literature, analysis, synthesis, induction, and deduction. The analysis and synthesis of the contradictory principles of the concept of sustainable development use the methods of the systems approach and dialectics. The research results are the development of a methodology for the development and marketing substantiation of strategies for the sustainable development of Russian regions. Its application in the development of strategies for the sustainable development of particular districts of the constituent entities of the Russian Federation will allow focusing on the territory's unique natural features, to translate them into consumer properties of locally produced goods, which will provide them with higher demand and will make a particular contribution in increasing the sustainability of the region's development. This approach to formulate sustainable development strategies for particular territorial entities allows for more complete and efficient use of natural resources in the region's economic activity, which corresponds to the principles of sustainable development supported by the United Nations and ensures the sustainable development of the region and the country.

**Keywords** Sustainable development · Territorial marketing · Unique natural characteristics · Sustainable development strategy

V. I. Belyaev · O. N. Pyatkova (✉)
Altai State University, Barnaul, Russia
e-mail: pyatkovaon@mail.ru



S. G. Maximova et al. (eds.), *Advances in Natural, Human-Made, and Coupled Human-Natural Systems Research*, Lecture Notes in Networks and Systems 234,
https://doi.org/10.1007/978-3-030-75483-9_5

# 1 Introduction

Sustainable development is a continuous process of natural, economic, and social changes, carried out during balanced human activities to improve the quality of life, provided natural conditions are preserved, and the Earth's resources are conserved for future generations. Other objects, subjects, conditions, and factors of development are also included in the system of changes in sustainable development. These are, namely, technology, the organization of business conditioned by introducing new technologies, and the corresponding branches of natural and social science. Since sustainable economic development is closely related to nature, it is not advisable to consider it in relation to only individual enterprises. Enterprises using natural resources cannot be excluded from the concept of sustainable economic development, but their activities need to be organized in the context of the development of large territorial entities.

# 2 Materials and Methods

The study used official data (Federal State Statistics Service 2019; Government of the Russian Federation 2014; Altai Regional Legislative Assembly 2015; Administration of the Altaysky District of the Altai Krai 2016a, b) reflecting the content and features of natural resources of one of the districts of the Altai Krai. The authors carried out an expert survey of specialists with specific knowledge in the state of natural resources of the analyzed area and their possible use in solving problems of sustainable development of the area. Analysis, synthesis, deduction, induction, system approach, and dialectic method were used as cognition methods. The subject of the research is the methodology for the development and substantiation of the strategy for the sustainable development of such territorial entities as the districts of Russian constituent entities. As a methodological basis, classical marketing was adopted in its applied application to territorial entities, namely, territorial marketing.

# 3 Results

Sustainable development can only be ensured over a long-term perspective. Consequently, a clear and consistent methodology is needed to develop and substantiate strategies for the sustainable development of large territorial entities—countries and their associations (unions). The authors proposal consider the administrative-territorial structural units of countries to be the initial link in strategic planning for sustainable development. In the Russian Federation, such structural components (administrative-territorial units of management) are districts, regions, and autonomous formations.

Adopting territorial marketing as a methodological basis for developing sustainable development strategies for such territorial entities means the specification of its methodology (a more in-depth study of particular elements of this function of managing territorial entities and the introduction of new, previously unresolved tasks into it) and the inclusion of the proposed marketing principles for substantiating sustainable development strategies in the current methodology. We are talking about developing a methodology for the development and substantiation of a strategy for sustainable development of individual territorial entities, namely, the country's structural units that have clearly expressed economic, socio-economic, administrative, and territorial features.

According to the criteria of modern marketing, any strategy must be justified since sales volumes determine economic development and economic growth. In solving this problem, modern marketing proposes to strive to ensure the uniqueness of the marketing proposal (Kotler 2005). The uniqueness is necessary to distinguish a product on the market from among similar ones. The principle of uniqueness is quite applicable in territorial marketing (Belyaev et al. 2015). Territorial marketing is the corresponding marketing actions carried out "in the interests of the territory, its internal subjects, and external subjects, in whom the territory is interested" (Pankrukhin 2006). In this regard, F. Kotler writes that "many European countries, regions, and municipalities are in a frantic search for a unique role that they will play in the information society" (Kotler, Asplund, Rhine and Hyder 2005). Thus, the principle of uniqueness can be called universal, applicable not only in classical marketing of goods but also in its applied areas, including territorial marketing.

In the scientific literature, territorial marketing is most often associated with promoting the territory's name and information about its natural, cultural, historical, and other features to the potential consumers (Kotler et al. 2005; Pankrukhin 2006). The scholars mean the uniqueness of these features, although they do not directly stipulate this. In essence, these marketing approaches and activities aim to attract tourists, investors, and new residents to the territory. The theory and methodology of territorial marketing do not pay attention to the use of unique natural resources of territories in producing unique goods. This fact is a flaw in the concept of territorial marketing. Based on this, the authors previously proposed to supplement this concept with provisions on using the area's unique features in the creation of goods that embody these features in their consumer properties (Belyaev et al. 2015). There are many examples of such use of natural features. However, most of the goods that materialize the unique natural properties of the area are created on an intuitive basis, on the accidental discovery of its unique features, and not based on scientific research. Nature keeps many unobvious unique characteristics in secret. As a result, society does not use them in solving problems of sustainable development.

For example, the blue-mold Roquefort cheese was created, made from sheep's milk in the Count of Rouergue of France. The primary and unique feature of this cheese comes from its maturation. After its preliminary processing, the cheese mass is laid out for maturation in limestone grottoes, where spores of the *Penicillium roqueforti* fungi live. Theses spores do not live anywhere else. The presence of these spores in limestone grottoes is a unique natural feature of the Count of Rouergue.

This feature is not apparent, and it is impossible to find it without special devices. It was discovered by accident many centuries ago. Thus, the presence of spores of *Penicillium roqueforti* fungi in the limestone grottoes of the Count of Rouergue, as a unique natural feature of this area, is embodied (materialized) in the consumer properties of the product (Roquefort cheese), making it unique. There should be no doubt that this fact is incorporated into the sustainable development strategy of Rouergue.

Another example is Vologda oil. Its consumer properties are primarily determined by the specificity of the herbs growing in the Mologo-Sheksninsky interfluve of the Vologda region. Unlike Roquefort cheese, Vologda oil was not found by chance. Its unique features were found as a result of long and painstaking research by N. V. Vereshchagin, the author of the technology for producing this oil.

There are other examples, like these, showing how the unique territorial features are embodied in the consumer properties of goods (Belyaev et al. 2015). In most cases, they are based on accidental discoveries of the area's unique properties since the latter are of a non-obvious and secretive nature. On this basis, the authors propose to include in the current methodology for the development and substantiation of strategies for sustainable development of territorial entities, reflected in special legislative acts and other official regulations, a mandatory scientific search for non-obvious unique territorial features to use them in creating goods with unique consumer properties. This approach meets the fundamental provisions of modern marketing and does not contradict the postulates of J. Schumpeter's theory of economic development on the combination of resources and forces in entrepreneurial activity (Schumpeter 2007). Only such an approach can ensure an increase in the economic feasibility of using the unique natural resources of territorial entities provided by the UN Sustainable Development Agenda.

The content and essence of strategies, their development, and implementation are detailed in numerous scientific and educational literature. The strategies are presented in a variety of guises. That is, the question of the strategy's essence is largely debatable. However, in Russia's practice, strategies are perceived and used only as a long-term plan. The authors believe that this is not enough, and strategy is not only a plan. To perceive a strategy as a plan means to impoverish its essence, not to use the potential of strategic management in full, which will not contribute to the sustainable development of an economic object.

The essence of strategies can be understood from their definitions. All strategy definitions can be divided into two types. The first type includes definitions presented in the classical form, i.e., in the form of short, concise formulations reflecting the essence of the determined object's content. The second type includes definitions drawn up in the form of enumeration of procedures and circumstances necessary for explaining the phenomenon's essence. A. Chandler was one of the first to formulate a strategy concerning the management of objects of a socio-economic nature. According to his definition, the strategy is a system of "long-term goals and objectives of the enterprise, the approval of the course of action, the allocation of resources necessary to achieve these goals" (Petrov 2012). This is the first-type definition. The other definitions of this type are very similar in form and content to the definition given by A. Chandler.

The second type includes the definitions formulated by I. Ansoff and G. Mintzberg. Thus, I. Ansoff proposed to define strategies as "a set of rules for decision-making, which the organization is guided by in its activities." He offered four groups of rules (Ansoff 1989):

1. Rules for assessing the enterprise's results;
2. Rules for organizing relations with the external environment;
3. Rules for building relations within the enterprise;
4. Rules for the formation of internal regulations.

According to I. Ansoff, the development of rules using quantitative and qualitative indicators represents the organization's strategy, and its implementation in practice is strategic management.

The definition of G. Mintzberg's strategy also belongs to the second type. In his view, the organization strategy is a complex of private definitions consisting of one keyword. According to G. Mintzberg, the strategy is (Mintzberg et al. 2000):

- the plan;
- principles of behavior;
- the position of the organization in the external environment;
- the perspective;
- the method (dexterous maneuver).

These five definitions and the rules of I. Ansoff should be considered and applied in practice in a systematic unity. This means that when developing a strategy, all the components of the above definitions must be involved.

In Russia, the organization's strategy, as defined by I. Ansoff and G. Mintzberg, is perceived in theory but not used in practice. In Russia, a strategy is a long-term plan or the basis for such a plan.

Based on the preceding, the authors propose to supplement the current methodology for developing strategies for the sustainable development of territorial entities with two fundamental provisions. Firstly, it is necessary to substantiate the strategy from the marketing position. Secondly, it is necessary to substantiate the strategy from the standpoint of the content of strategic tasks, according to I. Ansoff and G. Mintzberg.

The essence of the principled marketing position is to conduct exploratory research on the territory to identify hidden non-obvious unique features of the area that can be used to create unique goods.

The principle's essence is that the strategy should not be guided only by a long-term plan, as it is currently observed. According to the definitions of the strategy of I. Ansoff and G. Mintzberg, it should represent a long-term plan, a principle (or rules) of behavior, and a position. Additionally, the strategy should have backup options of behavior to use them if favorable opportunities for development arise in the external environment or avoid the impact of threats (if such appear in the future).

This is the essence of proposals for the development of a methodology for the development and marketing substantiation of strategies for the sustainable development of territorial entities. Its fundamental provisions by no means contradict the laws and other official regulations of the Russian Federation, which are recommended to be applied when developing strategies. The proposed methodological principles were tested on real data on natural resources and spheres of activity of the Altaysky district of the Altai Krai.

When developing strategies for the development of Russian regions, it is recommended to comply with the requirements of the relevant laws and regulations. The laws and regulations set out the methodological principles of their development. Thus, the development of the strategy of the Altaysky district should be based on the following regulations, which determine the methodology for developing the strategy:

- Federal Law "On Strategic Planning in the Russian Federation" (June 28, 2014 No. 172-FZ) (Government of the Russian Federation 2014);
- Law of the Altai Krai "On strategic planning in the Altai Krai" (April 3, 2015 No. 30-ZS) (Altai Regional Legislative Assembly 2015);
- Regulations on strategic planning in the Altaysky municipal district, approved by the decision of the Altaysky District Assembly of Deputies (October 3, 2016 No. 3) (Administration of the Altaysky District of the Altai Krai 2016a, b);
- Decree of the Administration of the Altaysky District "On approval of the procedure for the development, adjustment, monitoring, and control of the implementation of the strategy of socio-economic development of the Altaysky municipal district and the action plan for the implementation of the strategy of socio-economic development of the Altaysky municipal district" (March 25, 2016 No. 356).

Without going into a detailed analysis of the content of the strategy development methodology reflected in these documents, we argue that the authors' proposals to supplement the methodology with the principles of marketing and the content of strategic objectives do not contradict the requirements of these official regulations but supplement them.

As a strategic goal of the project, the authors propose the sustainable development of the region based on the creation of a balanced competitive economy that employs the local population, is aimed at improving the quality of life and is based on a combination of regional, interregional, and national interests, which together create an attractive investment climate in the district. The strategy should be designed, taking into account the unique features of the area to develop programs for sustainable development of municipalities in the region and an action plan for developing the area itself. When substantiating the draft strategy, it is necessary to analyze the district's economic activity, its production capabilities, features of natural resources, and an assessment of their use. Additionally, in the draft strategy, it is necessary to present the population dynamics, an assessment of the quality of life, the achievements of the real sector of the economy, etc. These are the requirements of the official regulations indicated above. However, in the official regulations, there is no mention of the presence in the region's natural environment of unique features that could be used in the creation of goods with unique consumer properties, which will give the sustainable development of the region an additional impetus through the fuller use of its natural resources.

Additionally, since the future is unpredictable, the strategy must be considered a principle of behavior in the market, a position, a system of rules, and an opportunity to correctly respond to unexpectedly emerging favorable circumstances in the external environment. These facts are not mentioned in the draft and the official regulations. The authors believe that the addition of the official regulations for developing the district strategy with the two above-mentioned fundamental provisions will enrich the strategy development methodology and make it more effective in solving the problems of the district's sustainable development. It should be emphasized once again that the proposed methodological principles do not contradict the official requirements for the development strategies for the districts of the Altai Krai, the concept of sustainable development adopted by the United Nations (UN), and J. Schumpeter theory of economic development. At the same time, according to J. Schumpeter, they can make a certain contribution to the economic development of the region due to a unique combination of natural, economic, cultural, historical, and other factors inherent in the territory without causing damage to nature.

It is recommended to start implementing the first principle with an analysis of the natural features of the area to identify unique ones for use in sustainable development strategies through the creation of unique goods. The Altaysky district is a separate territorial unit of the Altai Krai. It is located in the southeastern part of the region and has absorbed all its natural resources. The region has a unique microclimate and features of the foothills, which favors the development of horticulture. This is facilitated by the unique location of the village of Altayskoye, the municipal center of the Altaysky district. This village is located in the Altai Mountains' foothill in a gorge, which led to the emergence of a unique ecosystem. The village is remote from industrial facilities, has a mild climate, warm summers, moderately frosty winters, sufficient moisture, calm weather, and a number of clear sunny days comparable with the Caucasus and Crimea.

The district produces about 90% of the total volume of fruits grown in the Altai Krai for industrial processing. Apple, pear, plum, cherry, currant, and strawberry bear fruit abundantly in the gardens available to all district residents. Moreover, almost every household plot very successfully cultivates thermophilic crops not typical for Siberia (grapes, peaches, apricots, and cherries) (Altai Regional Universal Scientific Library named after V. Shishkov n.d.). Polls of residents and gardening experts show that this area's features can be called unique due to its location in the relief of the foothills. Moreover, residents' experience in using the unique features of the area in the cultivation of fruits and berries not typical for the Altai Krai (grapes, peaches, and apricots) is also unique. In retrospect, there have been several attempts to cultivate grapes in the Altaysky district on an industrial scale. Even before the revolution, I. V. Michurin spoke and wrote about the possibility of cultivating grapes in Siberia. He drew attention to such unique natural features of the south of Siberia as the early appearance of snow falling on the unfrozen ground and the amount of summer heat sufficient for the maturation of thermophilic crops. In 1934, a strong point of the Research Institute of Horticulture named after I. V. Michurin was organized in the Altai Krai. Ivan Vladimirovich gave this institute about 50 varieties of grape cuttings (including French); this is how the first experienced vineyard appeared in Altai (Nedin 1954). Initially, the grape yield was low. However, it began to grow due to the accumulation of experience and the attraction of winegrowers and botanical scientists to cultivate this culture. Since 1950, the industrial production of grape wine began in the Altaysky district.

Nevertheless, in 1973, by a decision of the country's government, the cultivation of grapes in the Altai Krai on an industrial scale was stopped. Since then, viticulture has become an amateur activity in the Altai Krai (Tyryshkin 2017). Nevertheless, the local population retained the skill of cultivating grapes. This skill can be used when resuming the production of this crop on an industrial scale. A study of the opinions of the Altaysky district's inhabitants shows that they are ready to resume work on the cultivation of grapes and revive the vineyards.

The economic benefit from the renewal of viticulture in the Altaysky district will be observed if grapes are used in winemaking. According to expert estimates, it is known that around 90% of the grapes grown on plantations are used for wine production. Simultaneously, the level of cultivation of this culture in a particular area is judged solely by the quality of the wine. The unique feature of the Altaysky district in the strategy of its sustainable development is proposed to be used in the organization of industrial viticulture for the production of Altai grape wines. The beginning can be laid in the production of artisan wine. Using J. Schumpeter's terminology, new combinations of favorable climatic conditions with the existing attitudes of residents towards viticulture will contribute to the sustainable development of the Altaysky district's economy. This is confirmed by expert assessments, according to which the cultivation of grapes is one of the most attractive investment sectors of the AIC. Thus, according to experts of "RBC," the largest Russian media holding, with reference to the research of the "Deloitte" analytical agency, at the end of 2018, the profitability in this sector increased by 20% compared to 2014 and amounted to 27% (Sukhorukova 2019). This fact is also evidenced by data on the increase in the area of vineyards in Russia. Therefore, according to the Federal State Statistics Service, in 2018, the vineyard area in Russia increased by 5.3 thousand hectares compared to 2014. In 2019, it was increased by 5.7 thousand hectares (Federal State Statistics Service 2019).

This is the case with the implementation of the marketing methodological principle in developing a strategy for the sustainable development of territorial entities in a specific district of the Russian subject. As for the content of strategic tasks, its implementation is expressed in the inclusion with the strategy of sustainable development of the Altaysky district of the fundamental position (rule) of the constant search for unique features of the area to create new goods with unique features and develop the old ones, giving them unique consumer properties. The analysis of the natural environment of the Altaysky district shows that it has many unique natural features (resources). In particular, more than two hundred species of medicinal plants grow on its territory, many of which are listed in the Red Book of the

Russian Federation (golden root, maral root, echinacea, rose radiola, valerian officinalis, and some others). Russian companies for processing plant raw materials can choose the Altaysky districts as a supplier of unique raw materials grown in an ecologically clean environment marked with unique natural and climatic features capable of imparting unique properties to medicinal plants. Suppose this unique feature of the area is included in the strategy for sustainable development of the area and is creatively presented on the market of raw materials for medicine (that is, the principle of territorial marketing is applied). In that case, it can become an additional impetus in the sustainable development of the district. Another unique feature of the nature of the Altaysky district is its favorable conditions for fish farming. A retrospective analysis of the economic activities of the region shows that in the 1960s, in the Altaysky district, rich in a large number of rivers and lakes, feeding on the cleanest streams and springs, having a sufficient level of depth, and well warmed up by the sun's rays, a network of ponds was created to breed delicious fish, called the Altai mirror carp (Litvinova 1998). By 1970, over a hundred ponds, with a total area of about two thousand hectares, were created in the region. The productivity for the growth of carp was one kg per year (Gavrilov 2015). Currently, as the analysis of the region's natural resources shows, there are only three cascade ponds on its territory, in which one of the remaining fish farms breeds such fish species as silver carp and crucian carp in a small number mainly to provide paid fishing services. This is also one of the unique natural features that deserve to be included in the strategy.

The strategic management of territories is ambiguous. Much is determined by the tasks of the sustainable development concept. Around this concept, there is a constant discussion on the limited natural resources.

## 4 Discussion

Thomas Malthus, an English priest, demographer, and economist, should be named the first researcher to pay attention to limited resources. In 1898, he published the book "*Essay on the Principle of Population*," in which, relying on demographic and economic statistics of the eighteenth century, he formulated the law on population. The law's essence is in "a constant desire, manifested in all living things, to multiply faster than is allowed by the amount of food at their disposal" (Malthus 1993). The growth of population occurs exponentially every 25 years, while "the means of subsistence ... cannot grow faster than in arithmetic progression" (Malthus 1993). According to the assumptions of T. Malthus, in the absence of deterrent factors (wars, epidemics), the time will inevitably come when society will have problems with life support resources. Therefore, he concluded that society should restrain fertility by "preventing the conclusion of many early marriages" and "resisting ... natural drives" (Malthus 1993). Such proposals, focused exclusively on "obstacles to population reproduction" (Malthus 1993), were hardly acceptable due to their dubious humanity. However, one cannot fail to note his merit, which lies in the fact that he was one of the first to draw society's attention to limited space and resources. His statement became the origin of an extensive and multifaceted discussion about the need for environmental protection and natural resources' rational use. This discussion has acquired particular relevance in the last three to four decades as part of the formation of sustainable development.

In the second half of the 1950s, began the rapid growth of the world's population, which led to a sharp increase in consumption and production. The pressure on nature had increased, and a contradiction between economics and nature arose. Corresponding Member of the Academy of Sciences of the Russian Federation, the director of the Kurchatov Institute, M. V. Kovalchuk, speaking at a meeting of the Federation Council on September 30, 2015 ("M. Kovalchuk, the director of…," 2015), called this state a contradiction between nature and the technosphere that humanity has created over the past two centuries. Before the industrial revolution, a man who used only the energy of wind, water, and the strength of animals and his own was a part of nature. The waste from the products of his activity returned to nature and was processed by nature. There was no pollution of the environment from human activities. With the development of technology, the situation dramatically changed. This is how a contradiction arose between nature and the technosphere, or rather between nature and the economy, the goals of which are expressed in improving the quality of life. This contradiction is objective—a person, on the one hand, will always strive to improve the standard of living; on the other hand, a market economy, which can provide such an increase is impossible without economic growth. Thus, economic growth determines and ensures an increase in the quality of life and provides capital with the coveted profit directed towards further development, which will make it sustainable and ensure growth. Such a contradictory interweaving of factors of social life forms the dialectic of modern life. These circumstances determine the content of the discussion about the concept of sustainable development. In this concept, development and growth are interconnected. In the aggregate interaction, society's desire to improve the quality of life and the impossibility of a market economy to function normally without growth exacerbate the contradiction between the economy (technosphere) and nature. The issue caused by this contradiction's growth was noted at two UN conferences held in Rio de Janeiro (June 14, 1992, dedicated to the environment, and June 20–22, 2012, dedicated to sustainable economic development). The conferences were held because at the turn of 1970–1980 the aggravation of the contradictions mentioned above in the form of environmental pollution had reached such limits, that it could not but cause concern in the international community about the state of affairs in environmental management. In 1987, the UN created the World Commission on Environment and Development [WCED] to solve the problem. The task of the WCED was to develop recommendations for resolving the current contradiction between economics and environmental management.

Since sustainable economic development problems appeared and were realized by humanity quite recently, there is no clear, unambiguous, and consistent theory and methodology for creating conditions for sustainable development and sustainable development management. There are many controversial questions, hypotheses, assumptions, provisions, and disputable proposals that await their research and, ultimately, the formation of a theory in which fundamental positions would be proposed to resolve the contradiction between economics and environmental management in achieving goals of improving living standards. The authors' proposals on developing a methodology for developing strategies for sustainable development of territorial entities are one of the attempts to solve this problem. Ultimately, they are expressed in the more rational use of the area's unique resources without harming the natural environment and ensuring the growth of the region's economy.

It is impossible to reflect all the diversity of aspects of the debate on sustainable development in a short article. We can only pay attention to some particular aspects. One of the main ones related to sustainable development theory is the ratio of the terms "sustainable" and "development" used in this concept. Thus, sometimes, when discussing sustainable development problems, some participants in the discussion draw attention to the fact that "stability" presupposes the invariability of the equilibrium state of the system, and "development" is impossible without bringing the system out of equilibrium. Then how do we define the concept of "sustainable development" if the two terms used in its designation conflict? This question is still open.

Another general controversial issue is due to the perception of the concepts of "growth" and "development." Quite often, when talking about raising the standard of living, people mean the growth (increase) in the production of consumer goods, food, housing, services, forms of leisure, and cultural objects. Thus, the concepts of "growth" and "development" are sometimes unwittingly identified. Simultaneously, growth is understood exclusively as an increase in any goods' production and measured in aggregated and price terms. This leads to an understanding of growth in the terminology of T. Malthus. That is, it is assumed that economic growth without increasing the use of resources is impossible and is limited by the available resources. This is one more disputable problem. For example, Vaclav Havel, the first president of the modern Czech Republic, wrote in the introduction to Tomáš Sedlacek's book: "Why are we so dependent on ever-growing growth and the growth of growing growth?" (Sejlachek 2016). By this, he wanted to say that the economy should not be oriented only towards growth, namely, the growth driven by increased resource use.

Richard Heinberg sees the problem in a completely different light. He writes about the end of economic growth, designating the following reasons (factors) that determine the onset of this period: depletion of natural resources, strengthening the negative impact on the environment by increasing the extraction and use of natural resources, and the inability of the current financial system to adapt to the growing shortage of specific natural resources (Heinberg 2013). In the first paragraph, his hypothesis has something in common with T. Malthus's population law. Indeed, there is such a problem. Moreover, it can also be expressed in the underutilization of some resources, especially those that are unique and the uniqueness of which is not obvious. This problem is solved by introducing the principle of territorial marketing proposed by the authors into the methodology for developing a strategy for the sustainable development of territorial entities.

Joseph Stiglitz, Nobel Prize Laureate in Economics, sees the problems of economic growth and development in the inequality of income distribution between labor and capital: "...An increasing number of them [rich people] realize the impossibility of continuing the established economic growth on which their prosperity depends..." (Stiglitz 2016). He sees the onset of problems associated with economic growth and development from a completely different angle. Namely, he understands it from the point of view of the representatives of the so-called "golden billion," who, despite their myriad fortunes, nevertheless, is also concerned about the problems of economic growth and sustainable development.

We can cite many other examples showing various assumptions about the prospects, methods, and means of sustainable economic development. It seems that many of them can be answered from the standpoint of the theory of economic development by J. Schumpeter, who logically linked the concepts of growth and development in a single economic cycle. In his theory, development means qualitative changes in the economic cycle in the form of introducing new technologies and new combinations of available resources. This kind of development causes serious economic growth that can significantly improve the quality of life (Schumpeter 2007). As it was, J. Schumpeter's theory reconciles many, if not all, controversial positions of contemporary participants in the discussion. The authors' proposal to use the territorial marketing methodology in substantiating the development strategies of individual territorial entities is based on his theory. The search for unique features of the area and their embodiment in the consumer properties of a product is a new combination of natural resources, achievements of natural and social sciences, socio-economic conditions, and opportunities in a single integral strategic direction of sustainable development of a particular territorial entity.

## 5 Conclusion

Thus, the proposed fundamental principles of the methodology for developing sustainable development strategies are based on the theory of J. Schumpeter and the works of I. Ansoff and G. Mintzberg on the strategic management of organizations. They do not contradict the basic postulates of strategic development reflected in the works of these scholars, the provisions of sustainable development of the UN concept, and the official regulations of strategic planning of the Russian Federation.

## References

Administration of the Altaysky District of the Altai Krai (2016a) Regulation on strategic planning in the Altaysky municipal district (March 10, 2016a No. 3). Altayskoye village, Russia

Administration of the Altaysky District of the Altai Krai (2016b) Resolution "On approval of the development, adjustment, monitoring, and control of the implementation of the strategy of socio-economic development of the Altaysky municipal district and an action plan for the implementation of the strategy of socio-economic development of the Altaysky municipal district" (March 25, 2016b No. 356). Altayskoye village, Russia

Altai Regional Legislative Assembly (2015) Law of the Altai Krai "On strategic planning in the Altai Krai" (April 3, 2015 No. 30-LS). Barnaul, Russia

Altai Regional Universal Scientific Library named after V. Shishkov. (n.d.). *All Altay*. Retrieved from http://altlib.ru

Ansoff I (1989) Strategic management. Economics, Moscow, Russia

Belyaev VI, Bocharov SN, Goryaninskaya ON, Malakhov RG (2015) Territorial marketing: methodology and methods for substantiating strategic decisions for regional development. Publishing house of the Altay State University, Barnaul, Russia

Federal State Statistics Service (2019) Official statistics. Retrieved fromhttp://old.gks.ru/wps/wcm/connect/rosstat_main/rosstat/ru/statistics/publications/catalog/doc_1265196018516.

Gavrilov V (2015) Korshunov ponds. Altai tourist. Retrieved from https://www.vtourisme.com/otdykh-na-altae/rybalka-na-altae/427-nasledie-geroev-korshunovskie-prudy

Government of the Russian Federation (2014) Federal Law "On Strategic Planning in the Russian Federation" (June 28, 2014 No. 172-FL). Moscow, Russia

Heinberg R (2013) The end of growth. Book Club of Bookstores, Moscow, Russia

Kotler F (2005) Marketing management, 11th edn. Peter, St. Petersburg, Russia

Kotler F, Asplund K, Rhine I, Hyder D (2005) Place marketing. Attracting investments, enterprises, residents, and tourists to cities, communes, regions, and European countries. Stockholm School of Economics, St. Petersburg, Russia

Kovalchuk M (2015) The director of the Kurchatov Institute, spoke at a meeting of the Federation Council in the framework of the "Time of Experts. Retrieved fromhttp://council.gov.ru/events/news/59290

Litvinova MV (1998) History of fish farming. In: Korshunov VP (ed) In the foothills of Altai: essays on history and culture. Altayskoye village, Russia, Barnaul, Russia. pp 276–277

Malthus T (1993) Experience in the law of population. Anthology of economic thought, vol 2. Econov, Moscow, Russia

Mintzberg G, Alstrand B, Lampel J (2000) School of strategy. Peter, St. Petersburg, Russia

Nedin VK (1954) Growing grapes in Altai. Altai Book Publishing House, Barnaul, USSR

Pankrukhin AP (2006) Territory marketing, 2nd edn. Peter, St. Petersburg, Russia

Petrov A (ed) (2012) Strategic management, 3rd edn. Peter, St. Petersburg, Russia

Schumpeter J (2007) Theory of economic growth. In: Capitalism, socialism, and democracy. Eksmo, Moscow, Russia

Sejlachek T (2016) Economics of good and evil. In search of the meaning of economics from Gilgamesh to Wall Street. Ad Marginem Press, Moscow, Russia

Stiglitz J (2016) The Great Separation. Inequality in society, or What do the remaining 99% of the population do? Eksmo, Moscow, Russia

Sukhorukova E (2019) Analysts called grape growing the most cost-effective sector of the agricultural sector. RBC. Retrieved from https://www.rbc.ru/business/17/07/2019/5d2d94ee9a7947f18a0c0852

Tyryshkin A (2017) Valery Nedin: the legend of Altai viticulture. Biysk district. Retrieved from https://biysk-ok.ru/2017/10/16/valerij-nedin-legenda-altajskogo-vin

# The Legal Nature of Capital Shares in the Strategy of Corporate Sustainable Development

**Tatyana A. Filippova, Tatyana L. Platunova, and Pavel P. Travkin**

**Abstract** The purpose of the study was to analyze the concept of a share in the authorized capital of a limited liability company, to clarify its legal nature and to identify the peculiarities of application of legislation governing the legal regime of the share, and the use of existing legislation to build a sustainable trajectory and development of the corporation. The main methods of research are comparative-legal, historical, structural-systemic, diachronic analysis method. The main results of the study are to determine, based on the analysis of domestic and foreign legislation, the doctrine of sustainable development of the corporation, the concept and meaning of the legal regime of the share in the authorized capital of the corporation in the domestic legal system, as well as to assess the adequate reflection of these categories in the current Russian legislation. Conclusions: A share in the authorized capital of a limited liability company is defined as a legal instrument, a method of fixing a certain amount of property and corporate rights and obligations to a corporation member, identifying it among other entities not related to the corporation. The positions of the courts in the article allowed to look at the content and legal essence of the share in a different way from the doctrinal one. The link between the financial stability of the corporation and the authorized capital of the enterprise and its composition is obvious. Appealing to foreign sources for the purpose of consideration of social purpose and function of the share in the authorized capital allowed to understand their multidimensional essence and differences in normative fixation in different legal systems and general tendency of influence on financial stability of the enterprise of the size and composition of the authorized capital and its share. Justification of novelty: Absence of normative definition of the concept of share in the authorized capital of a corporation, presence of diametrically opposite views in the doctrine on its legal nature, weak study of problems of financial stability of a corporation in connection with the concept and composition of the authorized capital and its share, updates the authors' appeal to the stated topic. In addition to scientific interest, this problem is also of applied nature, as it causes difficulties in law enforcement and practice of specific organizations.

T. A. Filippova · T. L. Platunova (✉) · P. P. Travkin
Altai State University, Barnaul, Russia
e-mail: taplatunova@yandex.ru

 
S. G. Maximova et al. (eds.), *Advances in Natural, Human-Made, and Coupled Human-Natural Systems Research*, Lecture Notes in Networks and Systems 234,
https://doi.org/10.1007/978-3-030-75483-9_6

**Keywords** Sustainable development of the corporation · Limited liability company · Legal regime of the share · Share in the authorized capital · Financial stability of the corporation · Transactions with shares · Legal nature of the share · Civil legislation · Alienation of shares · Corporate rights · Mandatory rights

## 1 Introduction

Due to the overwhelming presence of limited liability companies (LLCs) in the Russian economic sector, they gain a leading role in the economic growth of the country. Therefore, the issue of sustainable development of these enterprises is urgently raised. In the face of various objective and subjective risks, companies are forced, among other things, to restructure their corporate strategies. Sustainable development of enterprises implies quantitative and qualitative characteristics. Precisely qualitative characteristics, which include corporate models of building the authorized capital of enterprises and the combination of their shares, form the initial matrix of further development of corporations. Some foreign scholars have raised questions on the designated theme since the 90th years of the twentieth century (Porter et al. 2007; Schumpater, 2007; Elkington 1994). However, we have not found any special works to assess the impact of the legal nature of a share in the authorized capital of enterprises on the sustainable development of corporations. In the paper, we have considered understanding the legal nature of a share in the authorized capital of corporations in connection with the provision of sustainable development of enterprises. Particularly, it is the formation of the first contributions to corporations. Their qualitative composition is reflected in the market opportunities of the activities of corporations in civil turnover and is a guarantee of stable interaction with the counteragents of corporations. According to the website of the Federal Tax Service (namely, Report No. 1-Yur *Number of Legal Entities, Information About which is Contained in the Unified State Register of Legal Entities*), 2.990.936 LLCs were included in the Unified State Register of Legal Entities as of March 1, 2020, and the total number of commercial organizations was 3.092.060 (Report No. 1-Yur, 2020).

The legal regime of a share in the authorized capital of a corporation (namely, in an LLC) is an interesting research subject, especially when it is considered from the scientific and practical points of view and in the context of the relationship between this institution and sustainable development of a corporation.

Not a legal company, but social realities and the applied nature, significant in the professional activity of a corporate lawyer, generate the problem of the concept, meaning, and legal nature of a share. Despite the fact that LLC is the most popular legal form of commercial legal entity in Russia, shares have many theoretical and practical barriers. Business societies are the most common form of business now, and their sustainability determines the economic well-being of the country.

We have designated the following as the research objectives:

- To analyze the legal nature of a share in the authorized capital of a corporation (LLC) from the positions formulated by scientists at different stages of development of legal science and presented in Russian and foreign sources.
- To identify the impact of forming the authorized capital of a corporation on the aspects of financial security and sustainable development of an enterprise.

In order to solve the tasks set, we have considered the problem of sustainable development of an enterprise in the works of foreign and Russian authors. Furthermore, we have designated the financial stability of an enterprise as an essential type of stability of a corporation, where the relationship between the authorized capital and its composition is reflected in the financial security of a corporation. We have also addressed the historical aspect of understanding a share in the authorized capital of a corporation and the analysis of modern approaches to revealing the legal essence of a share in the authorized capital of an LLC, which allowed us to draw certain conclusions, including the efficiency of the current legislation in the conditions of the risky activities of corporations in civil turnover.

The research novelty consists of discovering and substantiating the connection between understanding the legal nature of a share in the capital of a corporation and making balanced corporate and other decisions in a corporation that ensure stable development of an enterprise.

## 2 Materials and Methods

For the topic disclosure and the decision of research problems, we have used the theoretical and methodological approach to the problem of sustainable development developed and reflected in the works of Schumpater (2007), Porter et al. (2007), and Elkington (1994). The research basis is private and general methods of scientific research. Thus, we have asked ourselves what the legal meaning of a share in the authorized capital of a corporation is. Besides, we have consistently achieved the results of understanding the legal nature of a share by the method of legal and historical analysis. Using the method of diachronic analysis, we have considered the development of a share in the authorized capital of a corporation in science and legislation. When studying the relationship between the authorized capitals of a corporation and its structure with ensuring the financial stability of an enterprise, we have resorted to the system and structural method. The logical and legal method has also been significant for our research. Moreover, we have critically evaluated the positions stated in the available scientific, legal literature on understanding the legal nature of a share in the authorized capital of a commercial corporation and the works of scholars on the problems of sustainable development of an enterprise (Kucherova 2011) to solve the tasks set. A share in the authorized capital can be characterized as a key category used in regulations governing LLC activities. It is mentioned in the regulations of the Civil Code of the Russian Federation (Russian Federation 1994) and the Federal Law *On Limited Liability Companies* (Russian

Federation 1998). Thus, analyzing the provisions of Article 66 of the Civil Code of the Russian Federation, one can conclude that the split of the authorized capital into shares of participants is typical for all financial companies and partnerships. It is noteworthy that along with the term *share in the authorized capital*, the legislator uses the concept *share of a participant*. From the analysis of the provisions of Articles 66, 79, and 85 of the Civil Code of the Russian Federation, it becomes clear that the legislator uses them as synonymous. Therefore, the further material presentation will proceed from this conclusion.

For legislation and legal science, the term *share in the authorized capital* is not a novelty. Its earliest mention dates to the first century of the new era and belongs to Roman law (Muromtsev 1883). Addressing the historical aspect of the topic, we should note that there is a mention of *puppets* (i.e., shares in mining associations) in medieval Europe after the reception of the Roman law (Shtoph 1896). There is also a mention of shares in medieval maritime associations (Dolinskaya 2006). Such shares were of the non-documentary form and were equal to each other.

A share in the authorized capital of an LLC is subject to the regulatory approval by the current Civil Code of the Russian Federation and special Federal Law No. 14-FZ *On Limited Liability Companies* of February 8, 1998. Nevertheless, the Code and the Law mentioned above does not give a normative definition to this concept and does not define a share as an object of civil turnover (Russian Federation 1998).

The absence of a normative definition of a share in the authorized capital and the practical importance of understanding this category have led to a lively discussion in the legal literature.

Several authors have covered the legal nature of a share in the authorized capital of an LLC (Sukhanov 2009; Lapach 2002a, b; Lomakin 2004; Novoselova 2007; Mogilevsky 2010). In their works, the authors draw their attention to the diversity of theoretical approaches and views on the legal nature of a share and the paradigm of civil rights objects in general. Novoselova (2007) writes, "The paradoxicality of modern Russian turnover surprises since the objects, the legal nature of which is not conclusively defined, and theoretical approaches, concerning their essence, become the subjects of transactions. A share in the authorized capital of a limited liability company is one of such objects."

Many scholars have previously discussed the legal regime of a share in law and corporation (Bevzenko 2006; Platunova 2016; Timokhina 2000; Shevchenko 2004). More than that, scholars have paid their attention to the question of the negotiability of a share in a limited liability company (Imadaeva 2012; Bobkov 2002).

Foreign researchers have also considered various aspects related to corporate shares. For example, holding stakes in the capital of a legal entity in American society was considered as an opportunity to increase the income and well-being of employees of a company and, in general, as a tool to overcome economic inequalities. According to the researchers, corporate assets and equity interests also contribute to the development of a company (Blasi et al. 2010).

Applied aspects related to the financial security of a corporation are updated by modern legal practice, particularly by transactions with equity stakes.

## 3 Results

When examining the historical aspect of shareholding, we have established that this concept is mentioned in the laws of all countries, where corporations similar to a new LLC appear. This fact demonstrates the attention paid to the formation of the original property base of an organization as a basis for its development.

The current civil legislation does not contain a legal definition of the concept of a share in the authorized capital of an LLC. Consequently, it does not allow determining the legal nature of this concept. Meanwhile, such a definition is the first and necessary task for studying the legal regime of a share and its purpose.

Therefore, we can conclude from the provisions of the Civil Code of the Russian Federation and the Federal Law *On Limited Liability Companies* that a share is a part of the property, a percentage of the authorized capital of the company. According to Item 1 of Article 25 of the Law *On Limited Liability Companies* (Russian Federation 1998), At the request of creditors, levying on a share or part of a share of a participant in the authorized capital of a company for the debts of a participant is allowed only on the basis of a court decision in case of insufficiency of other property of a participant for covering the debts. However, this approach is formal and not logical enough since Article 128 of the Civil Code defines property as a broad concept, including several categories such as things, property rights, and other property. Thus, elucidating the legal nature of a share in the authorized capital of an LLC necessitates deciding which of the listed objects in the concept of the property a share should belong to (Russian Federation 1994).

Attempts to solve this issue are made in the doctrine of civil law. Lapach (2002a, b) adheres to the position that a share in the capital is a thing and writes, "A share in the authorized capital is an ideal quota (part) of the right of ownership of a company or a property of a partnership, a kind of analogue of a share in common ownership." Later, his understanding of a share has acquired several other shades. Zalessky (1998) believes that a share is a property right that determines the size of a mandatory claim owned by a participant in relation to a company. Mogilevsky (2010) adheres to a similar point of view and writes, "A share of a participant in the authorized capital is a type of object of property rights." This position seems disputable. According to the Civil Code of the Russian Federation (1994), the property owner is a company. Thus, a society is a particular obstacle to the conclusion that a share gives its owner the right to a part in the assets of the society. Article 26 of the Federal Law *On Limited Liability Companies* (Russian Federation 1998) testifies that it is possible to object that the owner of a share can leave the society at any time and demand payment. A similar right formalizes a claim of the participant of society on the part of the property, but this right is already obligatory, and defining it as a property is incorrect. Some representatives of the German Civil Law tried to prove the opposite, but they have not found support among domestic or European scholars.

Analysis of the literature on sustainable development of enterprises (Rotary and Niazyan 2015) allows us to distinguish a fundamental criterion of enterprise sustainability in the context of the declared topic, namely financial sustainability and financial security as its element. We understand the sustainable development of an enterprise as a set of production management, as well as financial and socio-economic indicators of sustainability. As for financial sustainability, we consider it as the creditworthiness and solvency of an enterprise. The depth of these categories is the research subject of economic sciences.

The absence of the uniform approach to understanding a share in the authorized capital of an LLC leads to the problem of law enforcement practice since it influences the choice of corresponding ways for protecting participants and society.

Thus, the problem of understanding the legal nature of a share is of applied nature. It is expressed in the influence on the financial stability and security of a corporation. The ways of its solution affect the activities of an LLC and its members, as well as the sustainable development of a corporation in civil turnover. In the literature on the issue of evaluation of financial and economic activities of corporations, it is noted that the key parameter in evaluating the financial stability of a company is the difference between the real equity capital and the authorized capital of a company (Sheremet 2016).

In recent literature, there are also attempts to consider a share in the authorized capital of an LLC through the prism of its economic and legal essence.

Thus, Shapkina (2001) considers the authorized capital of a company as a monetary expression of the value of the property that a legal entity should have at its establishment. However, such an opinion seems incomplete since there are specific objects included in this property. It is believed that such a position allows a relatively narrow understanding of a share in the authorized capital. In this sense, a share expresses a percentage of the value of the property of a company and gives a person the right, although limited, to dispose of the assets of a company. In this study, we find it much more interesting to clarify the legal essence of a share in the authorized capital and its connection with the financial stability of a corporation.

## 4 Discussion

When considering the historical aspect of the topic, we have found that in 1892, with the adoption of the Limited Liability Partnerships Act, a legal form of an LLC (GmbHG 1892) appeared in Germany, with which the concept of a share was formed. Later, in the twentieth century, neighboring states (particularly Austria and France) adopted similar laws (Klunzinger 2009).

At about the same time, the concept of a share was formed in the countries of the continental legal system and the Anglo-Saxon one (Law 1890 Partnerships Act 1907) (Flock n.d.; Klausner n.d.).

Similar trends were characteristic of the domestic legal system. Thus, in Kievan Rus, according to the charter of the Ivanovo Merchant Brotherhood, the first economic societies (merchant associations) began to appear from 1135 (Lyubavsky 1978).

The Charter of Veliky Novgorod for 1417 contains the first historical information about a share in partnership (Valka 1949). It informs about the folding associations of northwestern Russian cities. The partnership is a traditional legal form of legal entity in Russia. The partnership on shares, developed in the first quarter of the twentieth century, with its authorized capital divided into identical parts, is known to be an LLC prototype.

With the beginning of codification of the Russian legislation (namely, with the adoption of the Civil Code of the RSFSR in 1922), the limited liability partnership got the right to exist, but there were no articles in the Code devoted to the legal regulation of shares in partnerships (All-Russian Central Executive Committee, 1922). During the power of the councils, the issue of regulation of shares in limited liability partnerships was not solved. However, the detailed regulation of shares occurred in the last decade of the twentieth century when the Council of Ministers adopted Resolution No. 590 *Regulations on Joint Stock Companies and Limited Liability Companies*. In this resolution, the concept of share in the authorized capital gained its legal basis (The Council of Ministers of the USSR 1990).

As for the legal nature of a share in the authorized capital of an LLC, several scientific points of view exist.

Foreign literature presents the material and legal concept of intercompany relations. Judge (1995) considers a share as an expression of proprietary relationships.

According to Stepanov (2000), a share in the authorized capital does not represent a monetary expression in the form of an amount. More than that, it is not an independent right and not a set of rights, as well as not a security and not its type. A share is a particular legal instrument, the essence and purpose of which is disclosed in its role throughout the life of a company. The author notes that a share in the authorized capital is determined in numerical, percentage, and fractional terms and makes nominal value, which gives the right to speak about the complexity of the legal nature of a share.

Murzin (1998) defines a share as a form of the legal fiction that formally expresses a contract to which society and its participants are parties. Nevertheless, the question of the relationship of such fiction to civil legal objects relations remains open.

There is an opinion that a share in the authorized capital of an LLC is, first of all, a property. In favor of this point of view, its representatives (Fatkhutdinov 2009; Lapach 2002a, b) pay attention to Item 2 of Article 94 of the Civil Code of the Russian Federation, according to which when a member of an LLC submits an application for withdrawal from a company or makes a claim on the acquisition of a share by a company in cases stipulated by Item 1 of this article, a share passes to a company from the moment a company receives the respective application (claim) (Russian Federation 1994). The real value of a share in the authorized capital should be paid to this participant, or the property of the same value should be given with the consent of a participant in order and terms provided by the law on societies with limited liability and the social charter. Besides, according to Subitem 1 of Item 6 of Article 23 of the Law *On Limited Liability Companies*, the society is obliged to pay to the participant who has applied an exit from society the real cost of a share in the authorized capital of a society that is defined based on the accounting reporting data of a society for the last accounting period, previous to the day of submission of an application for an exit from society, or with the consent of this participant of a society to give a property of the same cost (Russian Federation 1998).

Special attention is paid to the concept of *nature*. In the explanatory dictionary by Dahl (1863–1866), nature is the naturalization or equivalence of something.

When considering Article 128 of the Civil Code of the Russian Federation, which defines the diversity of objects of civil rights of things, including money, securities, and other property, such as cashless monetary funds, non-documentary securities, digital rights, property rights, results of work and services protected by the results of intellectual activities and equivalent means of individualization (intellectual property intangible goods), we can conclude that the concept of property is broad. The scope of its understanding is in the civil aspect of the term *naturalization*.

From the philological point of view, only a thing as an object of the material world, possessing qualitative and quantitative characteristics, form, content, and physicality, can act as a property issued. However, in reality, it is not so. This fact is more a problem of word formation and semantics rather than rights. In practice, the said share is paid to the outgoing participant by movable or immovable property, including shares. In foreign law, capital shares can be materialized into securities. Therefore, they are sufficiently mobile in the civil circulation (Noack n.d.).

From the analysis of the provisions of the Russian Law *On Limited Liability Companies*, the Civil Code sometimes considers a share in the authorized capital of an LLC as a part of the property, and the percentage of capital (assets) of a company is an external expression of a share, characterizing its economic essence. A share is also a right of participation. It represents a claim to conduct by a participant of a company (subjective right), namely, participation and attitude to the organization, management, and activities of a company, etc. This is its internal legal side characterizing its legal essence. The share in the authorized capital of an LLC is a subjective right (right of participation), consisting of a complex of powers by a participant of a company. The basic understanding of a share is the competence to manage the affairs of a company (Serdyuk 2005).

During the discussion of the real nature of a share, the judgment about it as a subjective right was considered. Fatkhutdinov (2009), a representative of such judgment, believes that a share in the authorized capital of an LLC is an object of corporate relations, which reflects the activities of a company, its participants, and the interrelation between a company and its participants. In addition, he identifies a share that has a complex of corporate powers as an opportunity for a participant of a company to perform own actions, while own actions are a part of the activities of someone else.

The truth of this position requires clarification regarding each competence of a member of society. On this basis, the obligation of making contributions is not part of a share despite the fact that it arises with the advent of a particular legal fact (namely, from the moment of determining its size in decisions confirming the creation of society).

Defining a share of the authorized capital of an LLC as a right necessitates the consideration of such a right through the prism of legal certainty and its place in civil legal relations from a traditional position of its division into the property and obligatory ones. Fatkhutdinov (2009), when considering a complex of corporate powers, recognized their indivisibility.

According to the Civil Code of the Russian Federation, members of an LLC have mandatory rights with respect to it: when forming the authorized capital of a company, its founders delegate the right of ownership of this capital to the company, and in exchange, they acquire mandatory rights concerning it.

Lomakin (2004) holds such a position and denies the property nature of the rights granted by a share of its owner. At the same time, for example, shareholders of business companies and partnerships have a right to participate in the distribution of profits and receive part of the property remaining after settlements with creditors or its value in the event of liquidation of a partnership or company. These rights are undoubtedly like the rights of obligations.

Lapach (2006) shares this view but with an amendment that only one claim by a participant to a company may be binding. In particular, it is a claim to pay a part of the profit attributable to a share, but only to the extent that it may refer to the part of the profit distributed but not paid, while claims to distribute the property of a company being liquidated among the participants of the latter are property rather than binding.

Sukhanov takes a different position and believes that the relations between the members of a corporation are of a property nature and based on making a particular property contribution to its capital (Sukhanov 2009). The content of such relations is to provide members of an organization they have created by transferring a part of their property to it, an opportunity to manage its affairs in one form or another (vote at the general meeting when making relevant decisions, participate in its management bodies, and receive information about the state of its affairs), and participate in the property results of its activities (in the distribution of profits and losses, the balance of property in the liquidation of an organization) (Lomakin 2004). The property right is a set of claims (powers) to the behavior of a subject, namely, to the possession, use, and disposal of property, that is, a civil law construction created only by law. Therefore, the nature of this category differs significantly from the previous one.

As for a share in the authorized capital of an LLC, it means that its owner is entitled to a proportional (percentage) share of the total property of an organization. It is possible to detail this share only through its allocation when the number of objects of the material world (things) that define the structure of a share is established.

There are no special publications on the connection between understanding the legal nature of a share in the authorized capital of a company and the sustainable development of an enterprise. This topic is interdisciplinary and relatively new for research. However, there are some works with the economic content on the sustainable development of enterprises in general. The research emphasizes the organizational and management indicators of sustainability, production indicators, and economic, social, and environmental indicators. Mokhov and Kulmukhametova (2017) devoted their work to the main aspects of sustainable development of an enterprise. In their work, the authors state the multidimensional nature of sustainable development of corporations and pay special attention to the financial stability of a corporation as an indicator. Kucherova (2011) reflects the disclosure of the problem of ensuring sustainable growth of an enterprise.

We suppose that interdisciplinary research is required in the area of sustainable corporate development and the area of knowledge about the nature of the share capital and its share, as well as in the application of legislation that regulates these relationships.

## 5 Conclusion

We can conclude that a share in the authorized capital cannot be reduced to one of the civil rights objects specified in Article 128 of the Civil Code of the Russian Federation. More than that, it cannot be understood as a thing. This approach is very narrow and does not reflect the true nature of this category. However, it can be a subject of civil circulation, that is, it can act as a thing. A share is a multidimensional concept, which includes an economic element, legal content, and even a certain arithmetic index since the size of a share of a participant in the authorized capital is determined as a percentage or as a fraction (Article 14 of the Law *On Limited Liability Companies*). To a certain extent, a share performs an official function since it confirms that the participant has a certain set of rights that can be transferred to other persons. Characteristically, it can transfer individual rights and the entire set of rights owned by a particular participant. Upon withdrawal from an LLC, a participant has a right to receive a part of the property, which corresponds to a share in the authorized capital, and loses the status of a participant and the rights and obligations that bound it to this corporation. In this context, the provision of the law becomes clear that upon withdrawal of a share of the participant is alienated to the society, which means that the participant can no longer influence the decision-making in a corporation and distribute profits from its activities. Thus, the legislator seeks to protect an organization from persons who are ready to leave it and lose the status of the owner of a share in the authorized capital.

The concept of a share in the authorized capital can be defined as a way of fixing the rights of an LLC participant. We can find such an approach in court practice. For instance, when considering specific cases, the courts determine a share in the authorized capital as a method of acceptance and fixation of a specific volume of property and non-property rights and obligations for a person (participant) (Supreme Arbitration Court of the Russian Federation 2014). Moreover, when determining the place of a share among the civil rights objects, they indicate it as other property (not a thing).

This position is shared by R. S. Bevzenko, paying attention to the fact that "Bevzenko (2006)" shares this position:

> A share in the authorized capital is only a symbol or a sign of those rights, which belong to a participant of a company. It is only a percentage or a fraction showing the ratio of a part of the authorized capital paid by a person to the amount of the capital. Simultaneously, the object of a legal relationship is not a share in the authorized capital but the totality of the rights of a participant (Bevzenko 2006).

It is believed that a share in the authorized capital of an LLC gives a special legal status to its owner: when acquiring a share, a person immediately receives some abstract in the form of a series of claims for specific actions, which allows characterizing the legal nature of a share as a formal legal tool arising from the right.

A share has a special legal regime, which, among other things, takes into account the need to ensure the stability of a corporation and the interests of its members (i.e., to maintain a balance between the interests of a legal entity and its members, such as the right of a member to leave an LLC, the right to alienate the stake, etc.).

Authentic and doctrinal understanding of a share should be considered since science and legal practice impact each other, often consistent and progressive. Thus, the courts define a share as a fixation of rights, thereby pointing to its purely legal formation. While identifying it among the objects of civil rights, the court prejudges some difficulties arising in the process of civil circulation and refers a share to the formal legal tool, which does not reflect its full essence, but brings clarity to its technical and legal purpose.

Based on the identified types of sustainability of a modern enterprise, we focused on financial sustainability and its indicators. Financial security is an element of financial stability. First, this element reflects a number of complex indicators, including the authorized capital, its composition in shares contributed by the founders, and the content of each share (considering the range of rights of the founders to form a share in the capital of a corporation allowed by law).

Thus, multidimensionality in understanding a share in the authorized capital of an LLC allows us to conclude about its complex nature. Since the authorized capital performs several functions, it is composed of the nominal value of the shares of its participants. According to Item 1 of Article 14 of the Federal Law *On Limited Liability Companies*, the category share in the authorized capital is the key one in the analysis of the potential and realization of the rights of the participants of a corporation. A share in a corporation is not a natural form, as well as it is not only a legal abstraction and fortuity, but it is a complex legal and economic essence with a property, mandatory, and corporate component. Such understanding of a share predetermines the necessity to enshrine this notion in the form of legal definition in a concrete norm of civil legislation. With the proposed understanding of a share in the authorized capital, it becomes evident that the absence of a normative definition of a share in civil legislation is not a gap that needs to be filled, but rather the impossibility of a traditional technique to put such a multidimensional category into a single norm. Besides, we should note that such characterization of a share is not exhaustive, as for its full analysis, it is necessary to determine the ratio of its property component to the obligatory one, as well as to refer to the court practice and to determine the opinion of the law enforcer.

The two-component essential characteristic of a share in the authorized capital of an LLC is a feature considered in characterizing the research subject, including in the determination of the share negotiability and its place among the objects of civil rights. Besides, its legal essence is not limited to the technical fixation of various rights. It has the broadest applied legal meaning. In the legal understanding of a share in the authorized capital of a corporation, we can emphasize several necessary semantic accents. The first one is related to the applied understanding of a share as an indicator of the financial stability of a corporation. It is an understanding of the financial component of the capital that gives an understanding of a corporation as a stable subject of civil legal relations. The second accent is related to an understanding of a share as a part of the property of a corporation, which also affects the stability of a corporation in civil turnover.

In the topic under study, we can see some promising research areas related primarily to the analysis of risks and difficulties in the activities of Russian corporations (LLCs), which do not allow us to overcome the barriers to designate clear leadership in sustainable development. Further analysis requires a legal and regulatory rationale for the problem. The issue of LLC equity transactions also requires in-depth analysis in the context of ensuring the financial security of a corporation and sustainable development of enterprises.

## References

All-Russian Central Executive Committee (1922) The civil code of the RSFSR (November 11, 1922). RSFSR, Moscow

Bevzenko RS (2006) Withdrawal of a participant from a limited liability company: prospects for cancellation. Law 9:50–55

Blasi J, Kruse D, Freeman R (2010)Having a Stake: Evidence and implications for broad-based employee stock ownership and profit sharing. Retrieved from https://www.nber.org/chapters/c8093/

Bobkov SA (2002) Assignment of a share in the authorized capital of a limited liability company. J Russ Law 7:23–26

*company*. Moscow, Russia: INFRA-M.

Dal VI (1863–1866) Explanatory dictionary of living Great Russian language. M.O. Fellowship, Wolf, St.-Petersburg, Russian Empire

des Fachsenats für Unternehmensrecht und Revision der Kammer der Wirtschaftstreuhände (2014) DarstellungdesStammkapitalsim Jahresabschlussbei der gründungsprivilegierten Gesellschaft mitbeschränkter Haftung (GmbH) und der GmbH light (16. Juni 2014 als Stellungnahme KFS/RL 27). Wien, Österreich

Dolinskaya V (2006) Joint stock law: main provisions and trends. Wolters Kluwer, Moscow, Russia

Elkington J (1994) Towards the sustainable corporation: win-win-win business strategies for sustainable development. Calif Manage Rev 36(2):90–100

Fatkhutdinov RS (2009) Legal problems of conceding a share in the authorized capital of a limited liability company (Dis. of Cand. of Sciences). Academy of National Economy under the Government of the Russian Federation, Moscow, Russia

Federal Tax Service of Russia (2020) Report on Form No. 1-YUR. Retrieved fromhttps://www.nalog.ru/rn77/related_activities/statistics_and_analytics/forms/9558929/norms/9558929/

Flock C (n.d.) Praktischer Leitfadenzur Gründungeiner GmbH. Retrieved from http://www.wolff-flock.de/uploads/media/Gruendung_einer_GmbH.pdf
Fritz- G, Kanduth S (2014) Bilanzielle Abbildung des Stammkapitalseiner GmbH mit Gründungsprivilegierung. SWK-Heft 4:591–595
Gesetz betreffend die Gesellschaften mit beschränkter Haftung (GmbHG) (1892) Retrieved from https://www.gesetze-im-internet.de/gmbhg/GmbHG.pdf
Ilyushina MN (2012) Transactions with shares in the authorized capital of a limited liability company. RPA, Ministry of Justice of Russia, Moscow, Russia
Imadaeva MS (2012) The legal nature of the share in the authorized capital of a limited liability company. RPA of the Ministry of Justice of Russia, Moscow, Russia. Retrieved from http://www.law.edu.ru/doc/document.asp?docID=1251561/
Judge S (1995) Business law. Macmillan Press Ltd., London, UK
Klausner C (n.d.) Sonderprobleme der Bilanzierung. Retrieved from https://extrw.univie.ac.at/fileadmin/user_upload/ord_extrw/Lehrveranstaltungen_WS2009/Sonderprobleme_der_Bilanzierung.pdf/
Klunzinger E (2009) Grundzuge des Gesellschaftensrecht. BRD, Munchen, Germany
Kobras M, Schweizer N (2010) Reform of German law on limited liability companies. Schweizer Kobras Pty Ltd, Sydney, New South Wales
Kucherova EN (2011) Modern approaches to sustainable development of the enterprise. Bull OGU 9:76–81
Lapach LV (2006) Share in the authorized capital as “other property” in the system of civil rights objects. Bus Law 12:58–68
Lapach VA (2002a) System of civil rights objects: theory and judicial practice. Legal Center Press, St. Petersburg, Russia
Lapach VA (2002b) System of civil rights objects: theory and judicial practice. Legal Center Press, St.-Petersburg, Russia
Lomakin DV (2004) Corporate relations and the subject of civil law regulation. Legislation 5:32–37
Lyubavsky MK (1978) Big soviet encyclopedia, vol. 3. Soviet Encyclopedia, Moscow, USSR
Mochov VG, Kulmukhametova EZ (2017) Key aspects of sustainable development of modern corporation and approaches to their definition. In: Materials of III scientific and practical conference of young scientists “Joldasbekov Readings”. LCD “Design OT”, Taldykorgan, Kazakhstan, pp. 186–182
Mogilevsky SD (2010) Limited liability company: legislation and practice of its application. Statute, Moscow, Russia
Muromtsev SA (1883) Civil law of Ancient Rome. Moscow. Typography of A. I. Mamontov and Co., Moscow, Russia
Murzin DV (1998) Securities—wireless things. In: Legal problems of modern securities theory. Publishing House “Statute”, Moscow, Russia
Noack U (n.d.) The future of shares—a legal perspective (Die Zukunft der Aktieaus-rechtlicher Sicht). Retrieved fromhttps://www.researchgate.net/scientific-contributions/2163360413_Ulrich_Noack/
Novoselova LA (2007) The turnover of the share in the authorized capital of a limited liability company. In: The objects of civil turnover. Statute, Moscow, Russia, pp. 197–230
Nürnberg M (2008). Die GmbH und UG (haftungsbeschränkt): Ein Merkblattzum Gesellschaftsrecht. IHK Berlin, Berlin, BRD
Platunova TL (2016) Transactions on alienation of shares in the authorized capital of a limited liability company. Izvestia Altai State University 3:137–140
Porter M, Ketels K, Delgado M, Briden R (2007) Competitiveness at a crossroads: trends in development of Russian economy. Retrieved fromhttp://sp-ved.nalog.ru-MATERS/PORTER_RFstrategy.pdf
Presidium of the Supreme Arbitration Court of the Russian Federation (2014) Resolution of the Presidium of the Supreme Arbitration Court of the Russian Federation (June 10, 2014 No. 1999/14). Herald of the Supreme Arbitration Court of the Russian Federation, Moscow, Russia

Rotary TS, Niazyan VG (2015) Sustainable development of enterprise. Statistics and economics. Moscow, Russia
Russian Federation (1994) Civil Code of the Russian Federation (October 21, 1994 No. 51-FZ). Moscow, Russia
Russian Federation (1998) Federal Law "On Limited Liability Companies" (February 8, 1998 No. 14-FZ). Moscow, Russia
Schumpater Y (2007) Economic development theory. In: Capitalism, socialism and democracy. Eksmo, Moscow. Russia
Serdyuk EB (2005) Joint stock companies and shareholders: corporate and mandatory legal relations. Russia, Moscow
Shapkina GS (2001) New in joint stock legislation (amendments and supplements to the Federal Law "On Joint Stock Companies"). Authorized capital of the company. Vestnik VAS 12:102–103
Sheremet AD (2016) Analysis and diagnostics of financial and economic activity of the
Shevchenko SA (2004) Transfer of shares of a limited liability company. Legality 6:27–32
Shtoph A (1896) Mining law. Stasyulyavich Typography, St. Petersburg, Russian Empire
Stepanov DI (2000) Limited liability companies: legislation and practice. Econ Law 12:54–66
Sukhanov EA (2009) Civil law. Walters Clover, Moscow, Russia
Supreme Arbitration Court of the Russian Federation (2014) Determination of supreme arbitration Court (June 27, 2014 No. BAC-3640/14). Moscow, Russia
The Council of Ministers of the USSR (1990) Resolution "regulations on joint-stock companies and limited liability companies" (lost force) (June 19, 1990 No. 590). USSR, Moscow
The Parliament of the United Kingdom (1907) Law "On Limited Partnership" (Act.12/2000). London, UK
Timokhina E (2000) Sale of share in the authorized capital of LLC. Econ Life 7:34–39
Valka SN (1949) Diplomas of Veliky Novgorod and Pskov. Academy of Sciences of the USSR, Leningrad, USSR
Wendler M, Tremml B, Buecker B (2006) The key aspects of German business law. Springer, Berlin, Germany

# Talent Management for Endogenous Economic Growth

**Alla Yu. Gorbunova, Iurii V. Gorbunov, and Julia V. Ugarova**

**Abstract** To ensure the sustainability of economic growth and reduce dependence on the external environment, organizations are increasingly paying attention to internal (endogenous) factors, one of which is highly intelligent, talented employees. Managing such people requires unique methods and approaches. Therefore, management has developed a particular method of personnel management and employee development called *talent management*. The research aims to study the possibility of using the talent management method of personnel management in ensuring the economic growth of an endogenous organization during the transition of the economy to the post-industrial development stage. The research methodology is based on a systematic approach to the studied objects and processes. In addition, we applied the methods of theoretical generalization, systematization, analysis, synthesis, tabular, and graphical methods. The research results are the study of implementing the talent management method in an organization. The research novelty is determined by the possibility of not comprehensive application of this method due to insufficient resources and organizational competencies. We concluded that it is possible to obtain positive results when using the talent management method in a reduced, truncated form and to manage only one subsystem of the organization, which, in our opinion, makes this management method more accessible to Russian organizations.

**Keywords** Talent management · Endogenous economic growth · Highly intelligent talented employees

A. Yu. Gorbunova (✉) · I. V. Gorbunov
Altai State University, Barnaul, Russia
e-mail: gorbunova-alla@mail.ru

J. V. Ugarova
Polzunov Altai State Technical University, Barnaul, Russia

 
S. G. Maximova et al. (eds.), *Advances in Natural, Human-Made, and Coupled Human-Natural Systems Research*, Lecture Notes in Networks and Systems 234,
https://doi.org/10.1007/978-3-030-75483-9_7

## 1 Introduction

Society has always set scientists and practitioners the task of ensuring economic growth necessary to improve people's living standards, their well-being and solve social, cultural, environmental, and other problems. We need to increase the real volume of production of material goods, which is expressed at the government level, primarily in the growth of the (1) gross national product, (2) gross domestic product, and (3) national income. Economic growth is expressed in the resulting performance indicators at the organization level, such as output, revenue, and profit.

Society, wishing to ensure its economic and social development, strives to achieve high economic growth rates. There is a contradiction between the number of material goods received at present and in the future. Overuse of resources now (especially non-reproducible ones) can lead to economic, environmental, and other severe future problems. To resolve this contradiction, to find a balance between current and future consumption, environmental pressure, the concept of sustainable development arose in the 70s of the twentieth century (Meadows et al. 1972). This concept calls for economic growth that "satisfies the needs of the present, but does not jeopardize the ability of future generations to meet their own needs" (World Commission on Environment and Development 1987).

From the point of view of economic growth, the concept of "sustainability" implies a reasonable burden on the natural environment, ensuring its conservation and a happy future for generations to come and a long-term, continuous increase in material goods production. The ways of ensuring this growth vary depending on the development stage of society and the characteristics of the productive forces.

W. Rostow proposed an approach to identify the long phases of economic development of society, which is currently used in modern scholarship. This approach is defined based on such criteria as (1) technological innovation, (2) the rate of economic growth, and (3) changes in the structure of production. According to this approach, economically developed countries are currently moving from the stage of "high mass consumption" to the stage of "search for the quality of life." Moreover, the industrial type of economic growth is being replaced by the post-industrial one. (Rostow 1960). The transition to the post-industrial stage of economic development involves finding new methods of ensuring economic growth. At the same time, this growth should not be temporary or wave-like. Instead, it should be (1) stable, (2) systemic, (3) self-sustaining, and (4) irreversible. Therefore, economists and politicians pay more attention to the possibility of ensuring an endogenous type of economic growth, which occurs due to factors located inside the economic system and not outside it. In contrast to the exogenous one, such growth is more reliable since it does not depend on the external environment, which can turn from favorable to unfavorable.

Russian scientists determine the most important factors of economic growth of the endogenous type throughout the country: (1) increased investment in fixed assets, (2) the development of innovations that allow the introduction of the latest technologies of the sixth technological order – nano industry, medical technology, photonics, structural change in the economy to replace inefficient, non-innovative organizations with advanced and highly profitable ones. American economists believe that the macroeconomic factors of endogenous economic growth are crucial for increasing government spending on defense and other needs and stimulating consumption growth (Grinin et al. 2017; Ilyukhin et al. 2019; Kollias 2015; Mallick and Mohsin 2016).

At the organizational level, there are the following sources of endogenous economic growth: (1) an increase in production factors (capital, employees), (2) a rise in the efficiency of their use, (3) improvement of the planning system, (4) crisis management, (5) the use of modern mathematical models in managerial decision-making, (6) innovations, and (7) digital transformation of enterprises (Evdokimov and Suraeva 2019; Konnikov et al. 2019; Tsybareva 2019; Zub 2017).

Furthermore, we believe that the use of the talent management method and the development of employees are powerful factors of the endogenous economic growth during the transition of the economy to the post-industrial development stage.

The research aims to study the possibility of using the method of personnel talent management in ensuring the economic growth of an endogenous organization during the transition of the economy to the post-industrial development stage.

The research object is the talent management processes of the Chinese subsidiary Hengtong Group, located in Heilongjiang Province.

## 2 Materials and Methods

The research methodology is based on a systematic approach to the studied objects and processes. We used theoretical generalization, systematization, analysis, synthesis, tabular, and graphical methods.

## 3 Results

People have always been and remain the primary productive force. Nevertheless, the requirements for employees change over time. Science is a direct productive force of the post-industrial stage of economic development. During this stage, mass informatization and automation of all activities and constant changes require a person to invest in their work acquired knowledge and skills and their spiritual and intellectual strength. The presence of talented people in the organization makes it more flexible, innovative, and adaptive to changing conditions. Such an organization will actively use the existing growth points and create new ones, seeing the opportunities and threats to business development and withstand competition. First of all, the economic growth of the organization is determined by an internal factor—the presence of talented people who are interested in their work. Therefore, organizations can be less exposed to external negative factors and become more resilient.

The term *talent management* was first coined by an employee of the American company Softscape D. Watkins in 1998 (Watkins 1998). Subsequently, the term became widely used by foreign and Russian scientists and managers.

There are two approaches to determining the source of talent. According to the first approach, "talents are born," that is, people (not all) initially have a specific data set that they can use for themselves and society for good. According to the second approach, "they become talents." Each person has a talent from birth. However, its manifestation and development depend on the environment in which the person lives. The talent management method is based on the latter assumption.

American psychologist Howard Gardner created a classification of talent, identifying the following types (Gardner 2011):

1. Verbal-linguistic—the ability to express one's thoughts on a paper, widespread among writers, journalists, or linguists;
2. Digital—typical for mathematicians or programmers;
3. Auditory—inherent in musicians or linguists;
4. Spatial—possessed by artists, architects, and designers;
5. Physical talent is acquired in practice, found in dancers or athletes;
6. Talent for environmental perception—interaction with the outside world (e.g., farmers);
7. Entrepreneurial—having the gift of generating ideas, the ability to work in risky conditions.

Talent management is understood as a set of personnel management methods that attract and retain highly qualified, talented specialists and ensure their loyalty (Lyutikova and Nechushkina 2016).

Talent management is a system that includes the following elements (Asaul 2016):

1. Search and evaluation of talented people;
2. Adaptation of talent;
3. Talent development, training;
4. Effective use of the potential of employees;
5. Knowledge and competence management;
6. Career development plan;
7. Motivation and benefits.

It is not easy to implement a talent management system in a company. The difficulties are due to the fact that, when introducing this method of personnel management, it is necessary to implement an integrated approach, changing (1) the personnel management system, (2) the organizational structure, (3) communications, (4) organization culture, and many other factors. The second obstacle to implementing talent management is the high cost. High-class talented specialists do not work to feed themselves and their families; they quickly find high-paying jobs. In addition to high incomes, talented people need the quality of work, the possibility of further intellectual growth, and self-realization. It is quite expensive to create such conditions.

Russia lags far behind America and Europe in terms of talent management. However, we think that for domestic organizations, the experience of China, rather than the best North American and European countries in this regard, is more relevant for the following reasons:

1. This country has recently made a breakthrough in innovation and economic development. The level of innovation in some industries is comparable to the Russian one.
2. The PRC focuses on endogenous economic growth at both the macro and micro levels. Therefore, global economic crises only indirectly affect its economy. Reliance on internal sources of growth has always been relevant for Russia, and due to the ongoing international sanctions, its significance has increased.
3. Despite high achievements, not all Chinese companies have enough funds to implement a comprehensive talent management system, which makes them similar to Russian enterprises. Simultaneously, Chinese companies, focusing on the need for innovative development, have accumulated some experience of partial, not complex implementation of the talent management system, which gives good results. The study and use of this experience are essential for Russian organizations.

Highly innovative Chinese company Hengtong Group includes more than 50 subsidiaries and branches, specializing in power and fiber optic cables. The company is one of the world leaders in the cable industry. The organization is widely differentiated, with some subsidiaries engaged in other types of business.

The Hengtong company we are researching is a subsidiary with a registered capital of ten million, wholly owned by the Hengtong Group. The company is located in Heilongjiang, northeast China. The company specializes in the design, manufacture, delivery, installation, and maintenance of drainage and water purification systems. Consumers of products are both organizations and households. Moreover, some of the equipment is exported. Like all Hengtong Group organizations, the company is high-tech and innovative with a center for research and development [R&D]. More than thirty patents protect the company's inventions.

The total number of Hengtong employees in 2019 was 373 people (see Table 1).

Therefore, 200 people are directly involved in the production of equipment: (1) engineering and technical employees—50 people; (2) sales employees—61 people; (3) functional employees—56 people; and (4) 6 people—top management. The company's heart is the engineering and technical employees, who must ensure the leading competencies in the industry. A good designer and engineer should have a sound knowledge of technology, think creatively and extraordinary, use their imagination, and put all their intellect and soul into work. A person who offers a technical solution to a problem that is superior in quality to the current world level must be talented.

Hengtong executives faced the challenge of increased staff turnover, including among engineering and technical employees. In the three years since 2016, 21 people left this sector of the company, including 18 people who left voluntarily. Staff turnover increased from 13% in 2016 to 17% in 2018. New engineers replaced the departed.

**Table 1** The distribution of company personnel by the department

| Department | | Number of employees | Proportion (%) |
|---|---|---|---|
| Top management | | 6 | 1.61 |
| Production employees | Quality assurance department | 26 | 6.97 |
| | Production department | 143 | 38.34 |
| | Equipment department | 31 | 8.31 |
| Engineering employees | R&D Center | 11 | 2.95 |
| | Technical department | 23 | 6.17 |
| | Engineering department | 16 | 4.29 |
| Sales employees | Sales department | 38 | 10.19 |
| | Customer service department | 23 | 6.17 |
| Functional employees | Financial department | 11 | 2.95 |
| | HR department | 14 | 3.75 |
| | Purchasing department | 19 | 5.09 |
| | Department of capital construction | 12 | 3.22 |
| Total | | 373 | 100 |

*Source* Compiled by the authors

However, the loss of highly qualified specialists with extensive experience and developed competencies is quite dangerous for the company. Even a small flaw in the work, a slight lag behind the industry leaders can lead to the loss of both national and foreign markets by the company. The competition level between manufacturers of drainage and water purification systems is high both within China and abroad.

The management of the company decided to implement the talent management system. Nonetheless, due to the lack of experience in this area and insufficient resources, they chose to start with a pilot project, limiting the employees to engineering and technical staff. The talent management system was also curtailed. The executives chose only two components vital for them from the entire set of measures: (1) improving the remuneration system and training personnel.

In 2019, a new remuneration system was introduced for engineering and technical employees (see Table 2).

In developing such a remuneration system, the company management adhered to the recommendations of talent management. The wage level should be decent. Therefore, monthly surcharges are introduced. Also, it is necessary to make a reasonable differentiation of wages so that talented employees see the attention and care on the part of the company. To consider the individual contribution of each employee to the organization, a stimulating (variable) part is provided. However, the proportion of the variable part should not be too high. Talented people should be liberated in

**Table 2** Remuneration system (monthly) according to the talent management system for engineering and technical employees of the company

| Position | Base salary (yuan) | Incentive portion (% of base salary) | Fringe benefits | | Total (with maximum stimulating part) (yuan) | Share of incentives (variable) part of the total) (%) |
|---|---|---|---|---|---|---|
| | | | Surcharge (yuan) | Transport grant (yuan) | | |
| Head of the R&D center | 11,000 | Up to 60% | 3,500 | 400 | 21,500 | 30.6 |
| Deputy manager | 10,000 | Up to 60% | 3,500 | 400 | 19,900 | 30.2 |
| Multi-project manager | 8,000 | Up to 30% | 3,500 | 400 | 14,300 | 16.8 |
| Project manager | 6,000 | Up to 30% | 3,000 | 400 | 11,200 | 16.1 |
| 1st category engineer | 5,000 | Up to 30% | – | 400 | 6,900 | 21.7 |
| 2nd category engineer | 4,600 | Up to 30% | – | 400 | 6,380 | 21.6 |
| Technician | 3,200 | Up to 30% | – | 400 | 4,560 | 21.1 |

*Source* Compiled by the authors

their actions and have the right to make mistakes. Otherwise, the talent *squeezed* in hard blinders will not manifest itself, will not give the expected results. Therefore, the incentive part is slightly more than 30% only for the head of the R&D Center and their deputies—people who mainly solve managerial rather than technical tasks. The rest of the workers have a fixed salary of 80% or more, which determines the stability of income and in case of failure.

The second component of talent management applied in the company is a change in the system of professional development of highly qualified specialists. Until 2019, the organization practiced a uniform training system for all employees, including (1) theoretical courses, (2) training, and (3) master classes in various areas (team building, financial literacy, marketing skills, corporate culture training, etc.). The testing system revealed that highly qualified workers have a negative attitude to such obligations and tests. The information obtained during the training was not new for them. Thus, they consider the time spent on attending classes to be wasted. In this regard, the company management has changed the advanced training system for engineering and technical employees. Now, the head of the R&D Center, together with the employees, makes an individual retraining program for them, considering both the requirements of the company and the wishes of the employees. This approach allows one to show respect for talented people and create conditions for their further development.

**Table 3** Personnel dynamics of the engineering and technical sector of the company

| | 2016 | 2017 | 2018 | 2019 |
|---|---|---|---|---|
| Number of dismissed employees (person) | 6 | 7 | 8 | 4 |
| Voluntarily leaving (person) | 5 | 6 | 7 | 1 |
| Total number of employees in the engineering sector (people) | 46 | 47 | 47 | 50 |
| Staff turnover (%) | 13 | 14 | 17 | 8 |
| Coefficient of voluntary leaving (%) | 11 | 12 | 14 | 2 |

*Source* Compiled by the authors

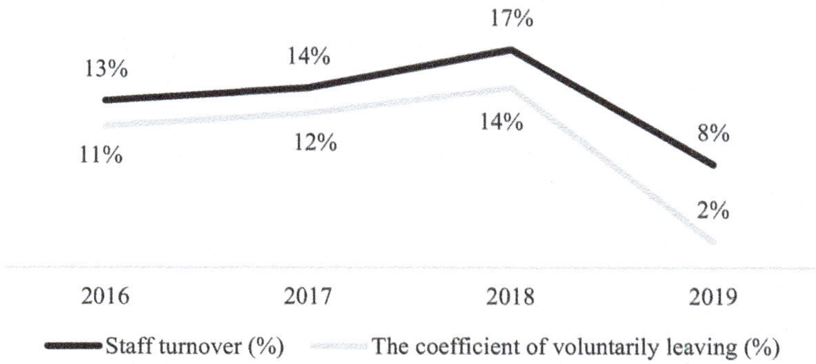


**Fig. 1** Dynamics of staff turnover rates of engineering and technical employees of the company. *Source* Compiled by the authors

By partially implementing the talent management system, Hengtong has mostly solved the problem of staff turnover among highly qualified, talented experts (see Table 3).

The staff turnover in 2019 decreased to 8%, with a voluntary leaving rate of 2% (see Fig. 1).

Therefore, even by partially introducing the talent management system, Hengtong significantly reduced staff turnover among highly qualified professionals. At the same time, the organization (1) strengthened the factors determining its endogenous economic growth, (2) reduced dependence on negative environmental impacts, and (3) increased its innovation, adaptability, and competitiveness.

## 4 Discussion

The research results confirmed the conclusions made by previous researchers about the particular relevance of modern organizations in ensuring economic growth of the endogenous type, which sources are internal factors. This kind of economic growth is sustainable, systemic, self-supporting, and irreversible. In the context of the transition of the economy to the post-industrial stage, its high innovation, rapid variability, the most critical factor in the sustainable growth of the organization are brilliant, talented employees. In this regard, the talent management method is becoming particularly relevant.

The peculiarity of our study is the identification of a positive result when using this method, not on the scale of the entire organization, but concerning only one of its subsystems. Furthermore, one should remember that the talent management method was introduced in an incomplete, truncated form in the given organization. We believe that such experience in applying talent management is relevant for Russian enterprises, which often do not have sufficient experience in implementing talent management or many resources.

## 5 Conclusion

The study showed the high relevance of providing organizations with endogenous economic growth. Besides, we revealed that currently, the main factor in the economic growth of organizations is highly qualified, talented employees. Moreover, the research indicated the high importance of the talent management method to attract and retain a high-quality workforce. As a result, the talent management method proved to be useful in obtaining positive results when applied in a reduced, truncated form to manage only one subsystem of the organization.

## References

Asaul AN (2016) "Talent management"—an innovative direction in personnel management. In: Asaul AN (ed) Proceedings of the 18th Scientific and Practical Conference: the global innovation competition. The role and place of Russia in it. Saint Petersburg, Russia: Autonomous Non-profit Organization "Institute of Problems of Economic Revival", pp. 383–391

Evdokimov AN, Suraeva MO (2019) Innovation and strategy—key elements of digital transformation of business. Econ Sci 5(174):89–92

Gardner H (2011) Truth, beauty, and goodness reframed: educating for the virtues in the 21st century. Basic Books, New York, NY

Grinin LE, Korotaev AV, Bondarenko VM (eds) (2017) N. D. Kondratev: crises and forecasts in the light of the theory of long waves. Look from the present. Uchitel Publishing House, Moscow, Russia

Ilyukhin AA, Ponomareva SI, Ilyukhina SV (2019) Economic growth and the factor of creative destruction. Russ J Econ Theory 16(4):630–639

Kollias C (2015) Defence and non-defence spending in the USA: stimuli to economic growth? Comparative findings from a semiparametric approach. Bull Econ Res 67(4):359–370

Konnikov EA, Konnikova OA, Lukashevich NS (2019) Vector model of assessing the potential performance of solutions aimed to improve the sustainability of development of industrial enterprises. Economic Sciences 1(170):47–50

Lyutikova MA, Nechushkina VS (2016) The usage of talent management systems in human resources. Nauka-Rastudent.ru 10(034)

Mallick SK, Mohsin M (2016) Macroeconomic effects of inflationary shocks with durable and non-durable consumption. Open Econ Rev 27(5):895–921

Meadows DL, Donella H, Randers J, Behrens WW III (1972) The limits to growth: a report for the Club of Rome's project on the predicament of mankind. Universe Books, New York, NY

Rostow WW (1960) The stages of economic growth: a non-communist manifesto. Cambridge University Press, New York, NY

Tsybareva ME (2019) The basics of economic development of Russia. Econ Sci 3(172):85–89

Watkins D (1998) Lightyear—an application framework for talent management that acts as a central feedback center for all organizational functions! Retrieved from https://ru.scribd.com/document/255608672/Talent-Managment

World Commission on Environment and Development (1987) Report: Our Common Future. Oxford. World Commission on Environment and Development, UK. Retrieved from http://www.un-documents.net/our-common-future

Zub AT (2017) Situational planning in crisis management: A recipe for success. *Public Administration. E-Journal, 64.* Retrieved from https://cyberleninka.ru/article/n/situatsionnoe-planirovanie-v-antikrizisnom-upravlenii-retsept-uspeha

# Legal Forms of Real Estate Development in Sustainable Development of Urban Environment

**Natalya A. Pronina**

**Abstract** The work is devoted to current problems of legislative and contractual regulation of real estate development business in Russian Federation. Relevance of the study was caused by the importance and demand for real estate development business in civil society and in the state as a whole, because the development projects are an integral element of the concept of sustainable development of urban environment. The main goal of sustainable development and its openness, safety, resilience and environmental sustainability of cities and towns, is to expand the scope of inclusive and environmentally sustainable urbanization and the opportunities for integrated and sustainable planning and management of human settlements by 2030. Partnerships between the authorities, the private sector and civil society are the main tools that will fully address economic, social, political and environmental issues of sustainable development. The urban right, as declared by the Aarhus Convention, equally recognizes the importance of integrated and sustainable human settlements planning, with a commitment to open decision-making and increased public support for environmental and urban decisions. A universal development logic can achieve the consensus in disputes between the civil society and government organizations, business and the private sector. Real estate development could be a professional activity aimed at qualitative transformation of a real estate object in order to increase its market value. The developer's task is to adequately fit the object of construction or reconstruction into the context of the urban environment, to obtain the necessary initial permissive documentation for this purpose, to achieve the reconstruction of the planned appearance and architectural attractiveness of the project without damaging the surrounding objects of planning and development of the territory. The application of analytical methods of research allowed the author to conclude that there are a significant number of studies on the concept of real estate development business activity, history of origin and real estate development business of the Institute. At the same time, the tasks of sustainable real estate development business predetermine the necessity of modernization of legal norms, systems of regulators, which promote effective and productive work in this direction. As a result, the focus of our research is

N. A. Pronina (✉)
Altai State University, Barnaul, Russia
e-mail: tashapronina@gmail.com



S. G. Maximova et al. (eds.), *Advances in Natural, Human-Made, and Coupled Human-Natural Systems Research*, Lecture Notes in Networks and Systems 234,
https://doi.org/10.1007/978-3-030-75483-9_8

the actual contractual regulation of the real estate development business activity. The author investigated the main goal of the developer's activity—investment, revealed its concept, which allows to define the mechanism of risk distribution. This risk is characteristic for the contractual model of the real estate development business project and should be considered as legal qualification of obligatory relations (internal and external) with the participation of the developer.

**Keywords** Real estate development business · Sustainable development · Investment activity · Investment contract · Risk allocation

## 1 Introduction

Real estate is an element of the dynamics of the development business in society, while one should consider the undeniable importance of it as a material existence form. As a long-term asset, real estate at a certain point in time reflects the vector of changes in this area, the needs of society in general, and its members in particular. The unprofessional approach to the construction and operation of buildings is fraught with serious mistakes that can lead to irreversible consequences, which must be considered with a number of the following factors:

- Durability of the object;
- Strong connection of the object with the land, land plot;
- Capital intensity;
- High cost of its transformation of the object.

One cannot exclude the economic component of any real estate construction and management project. Due to the limited material resources of future real estate owners and developers, there is a basic need to apply effective process organization methods of designing real estate and managing the financing of construction to minimize costs and maximize the land plot use. From the management point of view, the whole chain of realization of such a project—from land plot selection to sale to the final consumer—can be well-adjusted, especially if an experienced developer is involved in the project, as well as financial institutions, authorities, and local governments that provide obtaining permits (Kotlyarov 2020).

At the same time, strategic importance began to be attached to the paradigm of sustainable real estate development business of the territory, which was established as a priority direction of scientific and technological real estate development business of the Russian Federation.

According to Item 3 of Article 1 of the Town-planning Code of the Russian Federation (Russian Federation 2004a), "sustainable real estate development of territories is provided when implementing town-planning activities of safety and favorable conditions for life activity of the person, restriction of the negative impact of economic and other activity for the environment and ensuring protection and rational use of natural resources for the benefit of these and future generation."

Sustainable development of real estate business is necessary to ensure the openness, safety, environmental sustainability of cities and towns. The goal is to scale up inclusive and sustainable urbanization and opportunities for integrated and sustainable planning and human settlements management by 2030.

A successful sustainable real estate development business agenda cannot be achieved without partnership between authorities, the private sector, and civil society. This tool resolves economic, social, political, and environmental disputes during the specific investment project implementation. According to the Convention on Access to Information, Public Participation in Decision-making and Access to Justice in Environmental Matters (declared by the Aarhus Convention), the right to the city equally recognizes the importance of integrated and sustainable planning of human settlements, with a commitment to open decision-making and increased public support for environmental and urban real estate development business decisions (UN Economic Commission for Europe 1998). Russia still has not ratified it. A universal real estate development business logic can achieve consensus in disputes between civil society and government organizations, business and the private sector, balancing public and private interests without resorting to force. This logic is to reconcile resources, particularly land plots, and the intentions of their rights holders and other participants in public relations. We suppose that some land plots owned by private individuals are required for a complex real estate development business. The mechanism of withdrawal for municipal needs cannot help resolving the conflict since it is more of a forceful way of solving the problem. Implementation of the real estate development business project model, which includes the formation of "the concept of territory transformation, the implementation of which promises to bring economic benefits to all rights holders of real estate without exception and consolidation of all beneficiaries of the project through a variety of methods" (Trutnev 2017), will enable the project to be implemented effectively and painlessly for all parties involved.

Real estate development business opportunities make it possible to solve the task of attracting long-term investments in the sectors of the economy that are priority in terms of this concept. The goal of establishing partnerships at the global, regional, and local levels immanently includes the task of ensuring adequate financing of key sustainable urban real estate development business projects.

The importance of real estate development business as an integral part of sustainable urban real estate development business is generally recognized by the authors (Bychkov 2018; Popondopulo and Sheveleva 2015). Efficient application and adequate implementation scheme of the project requires detailed elaboration of legal instruments, system of incentives, contractual models, ways to minimize risks, and other legal issues of purely practical orientation that facilitate the implementation of such activities for the benefit of society and individual beneficiaries.

The city is the result of the activity. The activity is determined by the regulators: public-law and private-contract. The result of the activity depends on its organization and the content and quality of the activity (on the quality of its regulators—law and contract). Therefore, the quality of regulators inevitably predetermines both the efficiency of activity and the quality of the result—the quality of the urban environment (Trutnev 2019).

By the following circumstance, in a series of prerequisites for the real estate development business and increase in demand of the institute of real estate development business in the market is the state of Russian legal regulation of real estate and urban real estate development business, although, it is not ideal.

Natural consequence of legal uncertainty, namely, the construction and erection of real estate as types of economic activities acquire an unreasonable risk and minimum guarantee of planned results.

As an example of legal uncertainty can be cited case No. A56-48,136/2017 (the Supreme Court of the Russian Federation 2019). The Economic Board of the Supreme Court of the Russian Federation determined that the local government has the right to cancel the issued building permit in the order of self-control in case of its inconsistency with the law at the time of issuance (when issuing permit did not consider that the land is located in the protective zone of the monument of architecture). At the further real estate development business of law enforcement practice in this direction, it is highly probable that in the order of self-control may be revoked, building permits as objects of incomplete construction, and already built and commissioned objects, which translates them into the status of unauthorized construction, which can be demolished at the request of the interested body.

Real estate development business, as activity on the qualitative transformation of the real estate object for the increase in its market value, fairly and lawfully is carried to an independent kind of business activity. The companies rendering real estate development business services have seriously declared themselves in the market of construction of residential real estate and objects of non-residential purpose. There are well-known representatives of this industry who took an active part in projects significant for the country, for example, in preparation for the 2014 Olympics in Sochi. Large projects are being successfully implemented in the Siberian region by such developers as Yolka Development Group (Novosibirsk), Advantage Consulting Group (Barnaul), Project Development (Krasnoyarsk).

The first studies considering the peculiarities of the Russian real estate development business (Livshits 2003; Sheremet et al. 1998) are considered from the early 2000s. However, they were mainly devoted to the most general issues of real estate development business. The present study aims to review the contractual forms used in the real estate development business to formalize internal relations (client, project owner-developer) and external relations (developer-purchaser of the object). The novelty of our research has a practical orientation and sets the vector of the new research direction of the legal nature of the real estate development business contract.

The research focuses on the actual contractual regulation in the context of sustainable real estate development business of territories and urban environment, as the sphere of construction, financing of construction activities, especially housing construction is the subject of constant attention and, consequently, the constant regulation of the industry by the state. These changes cannot but affect the construction of economic relations in the real estate development business sphere. The following questions are the subject of the research:

- How these relations are formed today;
- What practical and theoretical questions arise within their connection;
- What are the real estate development business tendencies.

## 2 Materials and Methods

The current Russian legislation assumes a wide enough list of civil-law forms which can formalize the partner relations between the government and the private sector in sphere of erection of the real estate, objects of infrastructure, and other elements of sustainable real estate development business of the city environment. Consolidation of interests and investments can be formalized by:

- Agreement on public–private partnership;
- Investment agreement;
- Project financing agreement;
- Production sharing agreement;
- Concession agreement (Bychkov 2018).

Meanwhile, a classic real estate development business agreement cannot be attributed to the abovementioned obligations by a number of criteria.

American scientists suppose that, throughout history, every human society has had mechanisms for declaring, changing, administering, and enforcing the rules and definitions of social relations through which people live (Vago and Barkan 2018).

The concept of real estate development has not been enshrined in Russian legislation. In the literature, it is defined in different ways, for example, initiation, development and promotion of projects related to the creation of new real estate objects and reconstruction of existing ones for making profit (Soboleva 2014).

Another variant, with an emphasis on substantive characteristics "...qualitative material transformations in an object of real estate, which ensure its transformation into another, new object with a market value higher than that of the original object" are real estate development business (Masur and Shapiro 2010).

Maximov S. N. defines real estate development business as a way to resolve the contradiction between the changing and growing needs of the society for services provided with the use of real estate, and the available qualitative and quantitative characteristics of real estate (Maximov 2015).

Using the methodological basis of research activities, namely, a comprehensive approach to the provisions contained in regulations and scientific literature concerning the legal qualification of real estate development business, we derive a classical understanding of real estate development business. In turn, this classical understanding comes to the fact that it is a multi-stage entrepreneurial project aimed at making a profit by creating an added value of the object, consists in the transformation of a land plot or real estate object located on it (build a land plot or real estate object).

The emphasis on multistage real estate development business is not accidental. The object value can increase or decrease at each stage of the real estate development business project. The investment stage of the project is the stage at which the possibilities of own financing are assessed and analyzed, the issues of borrowed, attracted capital investments are resolved. During the stages of functional and urban planning, the function and purpose of the object, its architectural appearance, and the model of integration into the urban environment are determined. These stages are necessarily accompanied by the adoption of investment decisions, allowing to determine the appropriateness of a particular model of ownership of the object and the project effectiveness at a particular moment.

We used various research methods to ensure the implementation of an integrated approach to studying the investment purpose of development activities and the distribution of the identified value to the legal qualification of contracts carrying out economic activities. Using universal general scientific methods of observation, analysis, comparison, and modeling, we have studied the investment activity of a separate organization of the development business in Barnaul. In this regard, we assumed that the content of investment activities, their implementation through various contractual structures are far from controversial issues. Historical retrospective and comparative legal studies served as the basis for the conclusion that investment activity differs from the intersectoral, diverse nature, constantly imitating the need for the economy and economic turnover.

The real estate development business of a market economy originated back in the RSFSR. The concept definition of investment activity was given, which remained relevant and unchanged. The real estate development business is an investment and a set of practical actions for implementing investments (Supreme Soviet of the RSFSR 1991).

Literary sources do not so unambiguously define investment activity, especially if we turn to the authors of economic works. W. Sharp identified a common understanding of investment, the main substantive paradigm of "parting with money today to get more money in the future" (Sharpe et al. 1999).

In a particular study of legal problems of investment, it is noted that "an investment is any instrument in which money can be invested in the hope of preserving or increasing its value or providing a positive amount of income" (Igonina 2002). Time and risk are inherent characteristics of an investment.

## 3 Results

A real estate development business is a developer's professional competence, which closes the organization of activities of an architect, builder, contractor, designer, and marketer. At the same time, it satisfies the interests of the investor, creditor, and project owner in increasing the investment attractiveness and profitability of the project.

The developer is a professional participant of the business process, neither investor nor a customer within the meaning of the Law of the RSFSR of June 26, 1991 No. 14898-1 *On Investment Activity in the RSFSR* (Supreme Soviet of the RSFSR 1991). Besides, the investor is an independent person in the chain of the real estate development business project.

The developer increases the profitability of the project and consolidates the investment risks of the project to achieve or not achieve the planned level of profitability. These risks the investor would bear under the investment agreement. Moreover, in the real estate development business, these risks are transferred by the investor (owner, builder) to the developer, transforming them into mandatory claims under the real estate development business agreement, secured by the right to claim losses in case of failure to achieve specific investment horizons.

On the other hand, by selling the object as part of a real estate development business project, for example, at the initial stages of its construction, the developer does not give the buyer investment risks.

In other words, entering into contractual relations with contractors (purchasers of surplus-value of the project), there is no need to complicate the contract investment cause, the purpose and intentions of the buyer are covered by the acquisition of real estate and can be determined by contracts of sale, loan, perhaps, hybrid or convertible, or contract.

## 4 Discussion

We established that the investment objective is quite legal as a causal agreement with the developer on the real estate construction. Difficulties arise when the parties put their agreements in the form of an unnamed or mixed contract.

The Supreme Arbitration Court directly demonstrated the unwillingness to recognize investment agreements as sui generis contracts in its Resolution of July 11, 2011 No. 54 *On Certain Issues of Settlement of Disputes Arising out of Contracts Concerning Real Estate to be Created or Acquired in the Future* (the Supreme Arbitration Court of the Russian Federation, 2011) stated that "when considering disputes arising out of contracts related to investment activities in the sphere of financing construction or reconstruction of real estate objects, courts should determine the legal nature of the respective contracts and settle the dispute according to the rules of Chapters 30 ("Sale and Purchase"), 37 ("Contract"), 55 ("Sale and Purchase of Real Estate"), and 55 ("Sale and Purchase of Real Estate"). Unless stated otherwise, courts should assess agreements related to investment activities in the field of financing construction or reconstruction of real estate as agreements for the sale and purchase of future real estate" (clause 4 of Resolution No. 54).

The above explanation, intended to guide the interpretation of investment contracts by the courts, has become a guide for the tax authorities to actively requalify investment contracts into sales contracts, contracts, simple partnerships, followed by additional taxes and fees. This practice is far from the principle of turnover stability and cannot be considered favorable for developing real estate business in the country.

At the same time, we agree with the opinion of A.V. Maifat. The scientist argues that the contracts specified in paragraph 4 of Plenary Resolution No. 54 should not be considered as a basis for investments as a legal structure since they have a different legal nature. Only with a certain degree of convention, they can be called investments (Mayfat 2012).

In the Russian legal system, a real estate business contract is classified as untitled. According to the model of an assignment contract, the French Civil Code qualifies the development business contract, the actual conditions of which are the object, the price, and the term for the construction or reconstruction of the object (Kushchenko 2004).

We have already said that the national civil law does not directly fix the construction of a contract on the real estate development business. Practicing lawyers use the construction of untitled or mixed when drafting such a contract and include investment agreements, a loan, a contract, and a simple partnership in a real estate development contract. In the case of judicial re-qualification of such contracts, the courts and regulatory authorities usually qualify the real estate development contact as an agency (the definition of the Supreme Arbitration Court of the Russian Federation of March 5, 2014 No. VAS-1633/13 in case No. A-40-163120/12 [Supreme Arbitration Court of the Russian Federation 2014]), or an investment agreement built on the model of a simple partnership.

The scientific literature supports the opinion that one of the implementation forms of a real estate development business is the conclusion of an investment contract between the project owner and the developer, in which the parties stipulate the procedure for joint work to achieve the stated goal and participation in structuring the ownership of the built property (Bychkov 2018).

The main regulatory provisions regarding the conclusion of investment contracts and the definition of their content are in the Federal Law of February 25, 1999 No. 39-FL *On Investment Activities in the Russian Federation Carried out in the Form of Capital Investments*, and the above Law of the RSFSR of June 26,1991 No. 1488–1 *On Investment Activities in the RSFSR* (Supreme Soviet of the RSFSR 1991).

The customer is of interest in the investment obligation (State Duma of the Russian Federation 1999), which is the person authorized by the investors to implement the investment project.

It is difficult to classify the developer as a customer in the investment legal relation because the customer does not have a leading role in creating the investment object.

On the contrary, the developer organizes the construction, manages it, and is actively involved in the activities of other project participants, such as contractors. Moreover, the developers are not just involved in solving these issues; they are responsible to the project owner (investor, customer) for these works' timely and quality performance.

However, to consider the developer in the customer's role, it is necessary to determine the legal nature of the obligation, the content of which will be a model for describing the rights and obligations of the customer in such a mixed obligation.

The most unsuitable contractual form, in this case, is a simple partnership agreement because the participants in the real estate development business project have no common goal. Sometimes it is wrong to assign a common goal to the whole project to all participants. Meanwhile, the project owner's goal is to build or reconstruct the property and sell it at a price that exceeds the invested funds' value. The developer's goal is to increase the investment value of the real estate object qualitatively and quantitatively. The final user of the object aims to get the object of interest into ownership.

Furthermore, a number of key provisions of the simple partnership agreement clearly contradict the legal nature of the real estate development business contract. For example, these are the obligation to act jointly, the merger of property and the emergence of common share ownership, and joint liability for common obligations.

Thus, within the framework of the risk-sharing model that we chose within the real estate development business project in construction, the most representative construction for the construction of the real estate development business contract is the agency contract (Article 1005 of the Civil Code of the Russian Federation) (Russian Federation, 1996), according to which one party (agent, developer) is obliged to perform legal and other actions on behalf of the other party (principal, owner) for a fee, but at the expense of the principal or the other party.

The agreement itself determines the scope of activities and terms of reference of the performer and the specifics of economic relations served by the agent. "In view of the ambiguity of the legal meaning of the word "agent," one cannot be satisfied with one name, but it is necessary to consider in each case the legal nature of the relationship to distinguish the true agent from the order or commissioner" (Shershenevich 2003). It is fair to say that an agents are not just a performer under a contract of intermediary services, they are remunerated not just for the quality of the service activities, they "can count on a reward that depends not on the duration of his work, but on its success" (Shershenevich 2003). This seems to be the result of the work of an agent who, in the science of civil law, is considered to lie outside the subject matter of the usual contract for the remunerated provision of services.

Such understanding of the nature of the agent and the subject matter of the agency contract is quite consistent with the performer's competence in the real estate development business contract. The subject of the respective agreement is the actions on the organization of the complex of measures to realize the real estate development business project through management, management, and control of the project at all stages within the framework of achieving the customer's goals. The subject matter includes actions aimed at providing consulting services, legal support, structuring of financial relations, marketing and business planning, investment attraction, formation and administration of contractual relations (conclusion and support of contracts with contractors), representative functions during interaction with representatives of state authorities, local self-government bodies, law enforcement, and regulatory authorities, etc.

By analogy with a classical agency contract, a real estate development business contract is consensual, mutual, reimbursable, has a lasting and fiduciary character. The latter characteristic of a contract acquires when the terms of the agreement contain peremptory norms on the developer's personal fulfillment of the obligations assumed, as well as in solving a number of issues related to the determination of the grounds for termination of the agreement and in assessing the developer's behavior as an agent who failed to fulfill or improperly fulfilled the obligation, that is when the liability arises.

The developer, within the time frame specified in the agreement, carries out the above actions. Moreover, the degree of reliability of these terms may vary in the following ways:

- Due to specific calendar dates;
- Associated with events that will inevitably occur when the parties are more or less involved in the occurrence of the corresponding event;
- Determined at certain stages or stages of the project, respectively, the end of one stage of the project—the beginning of the next.

The developer's remuneration under the contract may consist of monthly payments to the developer's benefit, may be summed up on an accrual basis during the project, and paid immediately upon completion of the project, or be determined as a percentage of the increased value of the real estate development business object during the project, or otherwise.

The main developer's obligation is the efficient and quality activity within the subject matter of agreement with the provision of reports on the work performed in each reporting period, as well as the continuous management and investment audit of the project to minimize possible risks at the relevant stage of the project.

The customer (the project owner) finances the project under permanent management by the developer, or the developer does it within the financing provided by the customer or investors. It is possible to combine the above options within one contractual model. The developer may act as an investor providing the financing for the project implementation.

Concerning the contractual registration of external relations of the developer with the purchaser of the built or reconstructed real estate object (in full or in parts), in practice, there is a bias towards registration of these relations as an investment, which means that the risks of re-qualification of the agreement in case of a dispute are evident.

The legislator focuses on the application of the equity participation agreement in construction, named and put into circulation as a result of the entry into force of the Federal Law of December 30, 2004 No. 214-FL *On Participation in the Share Construction of Apartment Buildings and Other Real Estate Objects and on Amendments to Certain Legislative Acts of the Russian Federation* (Russian Federation 2004b).

The right to raise funds as a part of equity participation in construction is vested in a company which itself or the main company of which or any of the subsidiaries of the main company have at least three years' experience of participation in the construction of apartment buildings with a total area of at least 5,000 square meters in the aggregate, provided that the company has received permits to put the objects into operation as a developer and/or technical customer or general contractor.

It is noteworthy that the provisions of the Federal Law of December 30, 2004 No. 214-FL *On Participation in Equity Construction of Apartment Buildings and Other Real Estate Objects and on Amendments to Some Legislative Acts of the Russian Federation* do not apply to the relations of legal entities or individual entrepreneurs related to investment activities in construction (creation) of real estate objects (including apartment buildings) and not based on the equity participation agreement. The Civil Code of the Russian Federation and legislation of the Russian Federation on investment activity (Russian Federation 2004b) regulate these relations.

It follows from the above provision that the developer, when concluding an agreement with an external participant—the purchaser of the object under construction or a part of it, may use the construction of the equity participation agreement completely legally. Thus, imposing an obligation to comply with all requirements and restrictions stipulated by the Law on equity participation.

Legal entities and individual entrepreneurs have the right to use other contractual structures, formalizing legal ties between themselves when the parties pursue an investment goal and do not want to formalize an obligation in the form of a share participation agreement in construction. A hybrid loan is one of such forms.

Under such a contract, one provides a certain amount of money for the implementation of a specific real estate development business project, and the developer guarantees the return of such provision. The performance of the return obligation may be as money (similar to the original provision) or in another form, such as part of the built real estate.

Such situations are commonly referred to as shimmering cause "the contractual type (exchange of dissimilar benefits in the form of sale or exchange or borrowed type of contract) begins to be uncertain, and is defined and finally fixed, crystallized after the contract. At the time of conclusion of the contract, it is impossible to determine to which contractual type the complex of legal relations of the parties will incline" (Karapetov 2019).

Thus, the parties have agreed on a loan in cash for a specified period of time, with the provision that upon the occurrence of any conditions, such as failure to achieve the volume of construction or the volume of funding for the specified parameters, the same amount of money with interest is returned, and upon the occurrence of other conditions (obtaining a building permit, construction of the foundation, achieving a certain level of investment horizon, etc.) is transferred to the premises under construction.

The lender (as a general rule) can make the choice of execution, although, with the appropriate formulation of clear conditions, this right may also be on the side of the developer (borrower).

The legality of such a contractual construction has repeatedly been the subject of judicial review. Thus, in the Decision of Board for Civil Cases of the Supreme Court of the Russian Federation of November 24, 2015 No. 19-KG15-25, the court concluded that the provisions of the contract on the purposeful use of the funds transferred by the plaintiff, as well as the possibility of the defendant to pay the debt in money or other property (apartment), do not in themselves refute the arguments for the loan (Supreme Court of the Russian Federation 2015).

## 5 Conclusion

Thus, in the course of the study, we draw a number of conclusions, which, in our opinion, should be considered when forming contractual relations between the parties to the real estate development business project.

- The classical understanding of real estate development business comes to the fact that it is a multistage entrepreneurial project aimed at making a profit by creating an added value of the object, which consists of transforming the land or real estate located on it (construction or reconstruction). It is a qualitative and quantitative transformation of the object, the process of progressive real estate development business of the object to increase its value and investment attractiveness.
- The process of real estate development business with legal support of elements of sustainable real estate development business of the urban environment goes, on the one hand, towards the creation of new forms of public–private partnership, and on the other hand, through the use of already known mechanisms and legal forms. The real estate development business as an economic and legal phenomenon possesses the necessary resources to realize of the goals of sustainable real estate development business of the city environment. This should coincide with the possibility of maximum attraction of private investments into construction and improvement of territories with maximum satisfaction of interests and needs of all project participants, which will allow creating favorable conditions for the human activity.

## References

Bychkov AI (2018) Real estate development business in Russia. Industry legal regulation. Infotropic Media, Moscow, Russia

Igonina LL (2002) Investments. Moscow, Russia: Lawyer

Karapetov AG (ed) (2019) Loan, loan, factoring, deposit and account: Article-by-article commentary on Articles 807–860.15 of the Civil Code of the Russian Federation. Moscow, Russia: M-Logos

Kotlyarov MA (2020) Fundamentals of Real estate development business of Real Estate. Publishing House Yurait, Moscow, Russia

Kushchenko VV (2004) Real estate development business. Moscow, Russia: Norma

Livshits VN (2003) Evaluation of efficiency of investment projects in stationary and non-stationary macroeconomic conditions. Economics of Construction 5:2–22

Masur II, Shapiro VD (eds) (2010) Real estate development business of Real estate. Omega-L Publishing House, Moscow, Russia

Maximov SN (2015) Directorate of Real estate development. Moscow, Russia: Prospect

Mayfat AV (2012) Investment agreements: Are there such agreements? (Continuation of discussion). Russian Law Journal 3:166–172
Popondopulo VF, Sheveleva NA (eds) (2015) Public-private partnership in Russia and foreign countries: Legal aspects. Moscow, Russia: "Infotropic Media"
Russian Federation (1996) Civil Code of the Russian Federation (Part Two) (January 26, 1996, No. 14-FZ). Moscow. Russia
Russian Federation (1999) Federal Law "On Investment Activities in the Russian Federation Conducted in the Form of Capital Investments" (February 25, 1999 No. 39-FZ). Moscow, Russia.
Russian Federation (2004a) The urban planning code of the Russian Federation (December 29, 2004a No. 190-FZ). Moscow, Russia
Russian Federation (2004b) Federal Law "On Participation in Equity Construction of Apartment Houses and Other Real Estate Objects and on Amendments to Some Legislative Acts of the Russian Federation" (December 30, 2004b No. 214-FZ). Moscow, Russia
Sharpe W, Alexander G, Bailey J (1999) Investments. Moscow, Russia: Infra-M
Sheremet VV, Pavlyuchenko VM, Shapiro VD (1998–1999) Investment management. Moscow, Russia: Higher School.
Shershenevich GF (2003) Kurstorgovy pravda (Vols. 1–2). Moscow, Russia: Statute.
Soboleva EA (2014) Theoretical justification of Real estate development business as an institute of investment and construction sphere Real estate development business. Economics, Sociology and Law 1:100–104
Supreme Arbitrary Court of the Russian Federation (2014) On refusal to refer the case to the Presidium of the Supreme Arbitration Court of the Russian Federation (March 5, 2014 No. VAS-1633/14). Moscow, Russia.
Supreme Arbitration Court of the Russian Federation (2011) On Certain Issues of Resolving Disputes Arising out of Contracts over Real Estate to be Created or Acquired in the Future (July 11, 2011 No. 54). Moscow, Russia
Supreme Court of the Russian Federation (2015) Determination by the Judicial Collegium for Civil Cases of the Russian Supreme Court (November 24, 2015 No. 19-KG15–25). Moscow, Russia
Supreme Court of the Russian Federation (2019) Determination of the Judicial Collegium for Economic Disputes of the Russian Supreme Court of case number A56–48136/2017 (April 29, 2019 No. 307-KG18–21642). Moscow, Russia.
Supreme Soviet of Russian SFSR (1991) The Law of the RSFSR "On Investment Activities in the RSFSR" (June 26, 1991 No. 14898-1). Moscow, RSFSR
Trutnev EK (2017) Analysis of Law No. 373-FZ on Complex territory Real estate development business. Prop RelatS Russ Fed 7:70–81
Trutnev EK (2019) Urban regulation. Legal support of urban planning activities: Alternative models of legislation and the program for correction of its errors. Institute for Urban Economics, Moscow, Russia
UN Economic Commission for Europe (1998) Convention on access to information, public participation in decision-making and access to justice in environmental matters. Aarhus, Denmark
Vago S, Barkan SE (2018) Law and society. Taylor & Francis, New York, NY

# The Human Development Index as the Key Indicator for Measuring the Quality of Life of the Population in Altai Krai

**Lyudmila G. Glubokova, Polina A. Kosyonkova, Elena V. Silchenko, and Dmitry V. Kokhanenko**

**Abstract** The paper focuses on the basic concepts of human development, which intersects economic growth. For this reason, the most important task of social statistics is to develop an indicator of the quality of life. One such indicator is the Human Development Index (HDI). We outline the role and importance of HDI in assessing the quality of life and present the procedure of calculating the HDI, as well as the main indicators included in the index. The paper compares two procedures of calculating HDI: (1) the arithmetic mean of three indices, used until 2010; and (2) the geometric mean of three indices, used since 2010. We analyzed the UNDP that shows the 2018 ranks of countries by HDI and paid close attention to the top five countries, those that occupy the last places, and Russia. We conducted an analytical review of the HDI and its components (life expectancy index, education index, income index) based on the statistical data concerning the Russian Federation, Siberian Federal District, and Altai Krai. The paper evaluates the indicators of the HDI and its dynamics for 2013–2017.

**Keywords** Human development · Human development index · Life expectancy · Income level

## 1 Introduction

Human development is the main part of the national wealth. It combines all forms of development in the daily life of a person. Its essence is to create a favorable environment for people to achieve their full potential and lead a fruitful, creative life that meets their needs and interests.

L. G. Glubokova (✉)
Altai State University, Barnaul, Russia
e-mail: glg.72@mail.ru

P. A. Kosyonkova · E. V. Silchenko · D. V. Kokhanenko
Altai Branch, Financial University Under the Government of the Russian Federation, Barnaul, Russia

 
S. G. Maximova et al. (eds.), *Advances in Natural, Human-Made, and Coupled Human-Natural Systems Research*, Lecture Notes in Networks and Systems 234,
https://doi.org/10.1007/978-3-030-75483-9_9

Undoubtedly human development intersects the national economic growth. However, the first one is much broader. Economic growth is just a means of improving peoples' lives. The study of living standard is always relevant, regardless of the economic situation in the country, because, in a developed market economy, the effectiveness of economic development is determined by both the material resources and quality of life.

The UN financial organizations, including the World Bank and the Statistical Commission, state believe that people are the real wealth of a nation (UNDP 2019). In 1965, the *United Nations Development Programme* (UNDP) was adopted to improve the quality of life. In 1990, Mahbub ul Haq, a Pakistani economist and the Special Adviser to the UNDP Administrator, led the development of an indicator for economic and social conditions—the Human Development Index (HDI).

The HDI is a complex statistical indicator that includes data on life expectancy, education, and income. This indicator takes on a value in the range between 0 and 1. The closer the value is to 1, the higher the level of human development and the better the living conditions are. The HDI is calculated to rank countries according to the living standard. Experts of the UNDP calculate the HDI every year. This index is published in the *Human Development Report* at least two years after its publication by the national statistical services. The UNDP research aimed to draw attention to policies focused on improving the overall living standard.

Issues of human development are also relevant at the level of the Russian Federation. The Executive order *On National Goals and Strategic Objectives of the Russian Federation through to 2024* highlights the most important goals of national development: (1) ensuring sustainable natural population growth, (2) increasing life expectancy to 78 years, (3) ensuring sustainable growth of real wages, (4) creating a modern and secure digital educational environment to ensure high quality and availability of education (Presidential Executive Office 2018).

## 2 Materials and Methods

The procedure for calculating the HDI has been studied by many scholars, for example, A. P. Buevich, L. G. Kochegarova, O. V. Prokofiev (Buevich 2017; Kochegarova 2017; Prokofiev 2016). Moreover, many papers were devoted to the study of regional HDI (Grekova 2016; Tolcheev 2014).

Until 2010, specialists used the HDI calculation method on the arithmetic mean of three components—life expectancy, education, and income. In 2010, the calculation method changed. The components received the lowest and highest values. The upper limit was set according to the results of previous years. The lower limit was fixed at a level below which human development is impossible. The lower limit for life expectancy index (LEI) is 20 years, the upper limit −83.2 years (UNDP 2019). This is represented in Formula (1).

$$LEI = \frac{x - 20}{83.2 - 20} \tag{1}$$

The calculation method for the education component changed significantly. Specialists now evaluate the educational level based on two values: mean years of schooling (MYSI) and expected years of schooling (EYSI), which have the same weight. In other words, the component takes into account the average number of years of education received by people aged 25 and older and the average number of years of education that can be received by citizens aged 7 to 24. The general education index is calculated according to Formula (2)

$$EI = \sqrt{\frac{MYSI \times EYSI - 0}{0.951 - 0}} \tag{2}$$

In addition, since 2010, the level of income of the population has been estimated on the basis of the gross national income per capita and purchasing power parity. The upper limit of the income index is equal to 108,211 USD, and the lower limit is 163 USD. Many well-known economists use the natural logarithm of income in their calculations. Due to their experience, UNDP experts decided to replace the decimal logarithm with the natural one in the new HDI calculation procedure. At present, the income index is calculated according to Formula (3):

$$II = \frac{\ln(x) - \ln(163)}{\ln(108211) - \ln(163)} \tag{3}$$

The main distinctive feature of the new HDI is that it is based on the geometric mean, which eliminates the compensation of low indicators of one dimension by higher indicators of another dimension. Therefore, the HDI is calculated according to Formula (4):

$$HDI = \sqrt[3]{LET \times EI \times II} \tag{4}$$

At present, the HDI of 189 countries is calculated. Some countries are excluded from calculations due to the lack of statistical data (Somalia, Tuvalu). All countries are divided into four groups, depending on the HDI value: *very high* – (0.8 or more); *high* – (0.7–0.799); *medium* – (0.55–0.699); *low* (less than 0.55) (UNDP 2019).

According to the 2019 Human Development Report *Beyond income, beyond averages, beyond today: Inequalities in human development in the twenty-first century* (UNDP 2019), Russia ranks 49th and is included in the category of countries with a high level of the HDI. The average life expectancy in Russia is 72.9 years (67.8 years for men and 77.8 for women); GNI per capita at purchasing power parity is 24,233 USD; the average and the expected duration of education are 12 and 15.5 years, respectively. The top five countries by HDI are: (1) Norway, (2) Switzerland, (3) Ireland, (4) Germany and China (they share the 4th place), and (5) Australia. In the last few years, Norway had the highest HDI, owing it to the average life expectancy

**Table 1** Countries according to the HDI in 2018

| HDI Rank | Country | Value |
|---|---|---|
| 1 | Norway | 0.954 |
| 2 | Switzerland | 0.946 |
| 3 | Ireland | 0.942 |
| 4 | Germany, China | 0.939 |
| 5 | Australia | 0.938 |
| 49 | Russia | 0.824 |
| 187 | Chad | 0.401 |
| 188 | Central African Republic | 0.381 |
| 189 | Niger | 0.377 |

*Source* UNDP (2019)

of 82.3 years; GNI per capita at purchasing power parity of 68,012 USD; and average and expected school years of 12.6 and 17.9 years, respectively. The countries that rank at the bottom of HDI are Chad, the Central African Republic, and Niger. The average life expectancy of Niger is 60.4 years; the GNI per capita at purchasing power parity is 906 USD; and the average and expected duration of education is 2 and 5.4 years, respectively. Data for 2018 is presented in Table 1.

If one compares the data above with the 2018 UNDP report *Human Development Indicators and Indices: 2018 Statistical Update* (UNDP 2018), one can notice that the HDI of Russia increased by 0.008. The country was still ranked 49th, but the HDI value was 0.816. In 2017, China ranked only 7th. However, Australia had a higher HDI and was ranked 3rd. Niger was still the last country by quality of life, with an HDI of 0.354.

## 3 Results and Discussion

One can also calculate the HDI for regions—the subjects of the Russian Federation in this case. The calculations of the HDI for Altai Krai, Siberian Federal District, and Russia are presented below.

The life expectancy index directly depends on the health and environmental conditions. Life expectancy in Altai Krai is relatively equal to national value but higher than the index for Siberian Federal District (Bobylev and Grigoriev 2017). The index grew slightly, as life expectancy increased by approximately 1 from 2013 to 2018. The data is presented in Table 2. Using Formula 1, we calculated the life expectancy index; Fig. 1 demonstrates the calculation results.

According to Fig. 1, the life expectancy index of Altai Krai tended to grow until 2017. In 2018, the value of this indicator was equal to the previous year. Regional life expectancy values lagged behind the national ones (two-year gap). According to this indicator, Altai Krai is 43rd in the rating of regions of the Russian Federation.

**Table 2** Life expectancy at birth, years

| | 2013 | 2014 | 2015 | 2016 | 2017 | 2018 |
|---|---|---|---|---|---|---|
| Altai Krai | 69.8 | 70.1 | 70.4 | 70.7 | 71.1 | 71.1 |
| Siberian Federal District | 68.6 | 68.6 | 69.3 | 69.8 | 70.5 | 70.7 |
| Russia | 70.8 | 70.9 | 71.4 | 71.9 | 72.7 | 72.9 |

*Source* Federal State Statistic Service (n.d.)

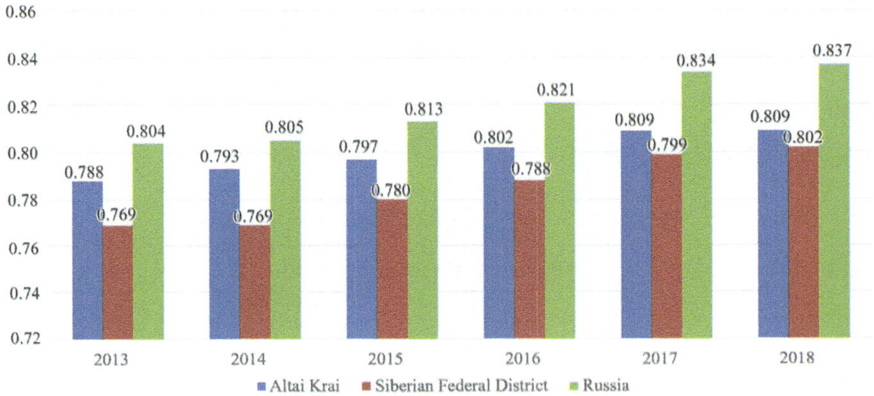


**Fig. 1** Life expectancy index. *Source* Federal State Statistic Service (n.d.)

However, according to the UNDP classification, the value of this index belongs to *very high*.

Life expectancy of women in Altai Krai is much higher than that of men (76.4 and 65.7 years, respectively). The difference is significant—about ten years. According to data of the Federal State Statistic Service of Altai Krai, the death rate of men aged 16 to 59 is 896.6 per 100,000, while the death rate of women is almost four times less –257.7 people per 100,000.

There is no data on the average and expected duration of education for the Altai Krai and Siberian Federal District, which prevents the calculation of the next component. Therefore, the education index values were taken from the reports of the *Analytical Center under the Government of the Russian Federation*, but the latest data were provided for 2016, so the analysis period was changed. The data is presented in Fig. 2. Both the national and regional education indices have a positive trend and high values—they are par with the top ten countries in the world.

In calculating the income index, we substituted the gross national income with the gross regional product—the value of all final goods and services produced in a region during a certain period. The indices for Altai Krai, Siberian Federal District, and the Russian Federation are calculated in Table 3.

Table 3 shows that the income index for Altai Krai and for Siberian Federal District is sustainable. According to the ranking, it corresponds to the *medium* level.

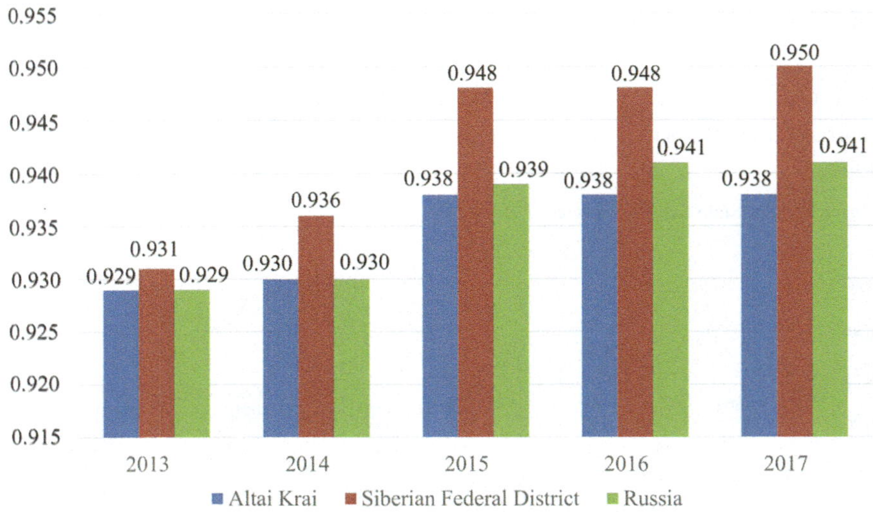


**Fig. 2** Education index. *Source* Bobylev and Grigoriev (2017, 2018)

**Table 3** Income index

| Indicator | 2013 | 2014 | 2015 | 2016 | 2017 |
|---|---|---|---|---|---|
| USD to RUB exchange rate by PPP | 19.42 | 21.01 | 23.59 | 24.36 | 24.34 |
| Altai Krai | | | | | |
| GRP per capita, RUB | 173,763.5 | 186,798.6 | 204,933.1 | 211,658.1 | 215,768.4 |
| GRP per capita, USD | 8,947.7 | 8,890.9 | 8,687.3 | 8,688.8 | 8,864.8 |
| Income index | 0.616 | 0.615 | 0.612 | 0.612 | 0.615 |
| Siberian Federal District | | | | | |
| GRP per capita, RUB | 287,293.8 | 317,784.6 | 353,119.2 | 367,221.8 | 401,808.8 |
| GRP per capita, USD | 14,793.7 | 15,125.4 | 14,969.0 | 15,074.8 | 16,508.2 |
| Income index | 0.694 | 0.697 | 0.696 | 0.697 | 0.711 |
| Russia | | | | | |
| GNI per capita, USD* | 25,330 | 24,900 | 23,400 | 23,410 | 25,080 |
| Income index | 0.777 | 0.774 | 0.764 | 0.764 | 0.775 |

*Source* Federal State Statistic Service (n.d.)

The income index for Russia lies within the range of 0.7–0.799 and corresponds to *a high* value. However, it tended to decrease until 2015 due to lower gross national income per capita. In 2017, the index increased by 0.011.

Table 4 presents the HDI for Altai Krai, Siberian Federal District, and Russia.

Table 4 reveals that the HDI in Altai Krai constantly increased—even in the crisis year of 2015, the HDI rose by 0.003. The HDI value remained in the range of 0.7–0.799 for the entire period, which is considered a *high* value. The components of

**Table 4** The HDI in Altai Krai, Siberian Federal District and Russia

| Indicator | 2013 | 2014 | 2015 | 2016 | 2017 |
|---|---|---|---|---|---|
| Altai Krai | | | | | |
| Life expectancy index | 0.788 | 0.793 | 0.797 | 0.802 | 0.809 |
| Education index | 0.929 | 0.930 | 0.938 | 0.938 | 0.938 |
| Income index | 0.616 | 0.615 | 0.612 | 0.612 | 0.615 |
| HDI | 0.767 | 0.768 | 0.771 | 0.772 | 0.776 |
| Siberian Federal District | | | | | |
| Life expectancy index | 0.769 | 0.769 | 0.780 | 0.788 | 0.799 |
| Education index | 0.931 | 0.936 | 0.948 | 0.948 | 0.950 |
| Income index | 0.694 | 0.697 | 0.696 | 0.697 | 0.711 |
| HDI | 0.793 | 0.793 | 0.798 | 0.804 | 0.814 |
| Russia | | | | | |
| Life expectancy index | 0.804 | 0.805 | 0.813 | 0.821 | 0.834 |
| Education index | 0.929 | 0.930 | 0.939 | 0.941 | 0.941 |
| Income index | 0.777 | 0.774 | 0.764 | 0.764 | 0.775 |
| HDI | 0.834 | 0.834 | 0.836 | 0.839 | 0.847 |

*Source* Compiled by the authors

the HDI are interconnected with its value. In other words, if the income index is the lowest one in the region, it will also be the lowest one in the federal district and the whole country.

## 4 Conclusion

As the analysis shows, the HDI depends on its components—life expectancy, education, and gross national income. The components of the HDI have a direct influence on its total value. Out of the three, the income index was the lowest in Altai Krai, in the Siberian Federal District, and in Russia. Conversely, the education index was the highest on all levels. It is worth noting that the life expectancy index in Altai Krai is *high*, the education index is *very high*, and the income index is *medium*. The main reason for this is the low value of the regional product. For example, the Kemerovo Region (which is also a part of the Siberian Federal District) has approximately the same population as Altai Krai, but its GRP per capita is 316,264 rubles −1.5 times higher than in Altai Krai. This means that the income index hinders the improvement of living standards in Altai Krai. Therefore, it is necessary to develop projects and programs that will improve the financial security of citizens and increase the HDI.

The advantages of the HDI for studying the quality of life are:

- Reflection of the main indicators of the population's standard of living;
- Possibility of application both at the international and regional level;
- Ability to compare observed objects and assess the dynamics of achieved results.

## References


Bobylev SN, Grigoriev LM (eds) (2017) Human development report for the Russian Federation. Environmental priorities for Russia. Moscow, Russia: Analytical Center for the Government of the Russian Federation. Retrieved from http://ac.gov.ru/files/publication/a/15600.pdf

Bobylev SN, Grigoriev LM (eds) (2018) Human development report for the Russian Federation. People and innovation. Moscow, Russia: Analytical Center for the Government of the Russian Federation. Retrieved from http://ac.gov.ru/files/publication/a/19663.pdf

Buevich AP (2017) Human capital as a condition for the formation of the region's innovative economy. Economy and Entrepreneurship 4–1:279–291

Federal State Statistic Service (n.d.) Official website. Retrieved from https://www.gks.ru/

Grekova MV (2016) Modern approaches to assessing the social and economic development of the region. Socio-Economic Research, Humanities, and Law: Theory and Practice 10:16–21

Kochegarova LG (2017) Analysis of relations between economic and social effectiveness in investment development management of Russian regions. Tomsk State University Journal of Economics 5:181–195

Presidential Executive Office (2018) Decree "On national goals and strategic objectives of the Russian Federation through to 2024" (May 7, 2018 No. 204). Moscow, Russia. Retrieved from http://kremlin.ru/events/president/news/57425

Prokofiev OV (2016) Econometric model of the human development index. Symbol of Science 1–1:166–168

Tolcheev VO (2014) Analysis of Russian regions on the base of human development index. Arctic and North 14:70–82

UNDP (2018) Human development indicators and indices: 2018 statistical update. Retrieved from http://hdr.undp.org/sites/default/files/2018_human_development_statistical_update.pdf

UNDP (2019) Human development report 2019. Retrieved from http://hdr.undp.org/sites/default/files/hdr2019.pdf

# Technologies for Organizing Creative Competitions to Support Regional Social Creativity

**Ksenia A. Melekhova, Larisa I. Nekhvyadovich, and Zhanerke N. Shaigozova**

**Abstract** The paper focuses on the features of technology for organizing creative competitions in the regional creative fashion industry. The authors consider the socio-cultural context of the origin and development of creative competitions in the 20th–twenty-first centuries. Moreover, the authors analyze the history of the Altai fashion industry and its current development level. Studying innovative fashion projects as an important condition for developing social creativity in the region determines the research purpose. The authors used such research methods as cultural-historical, comparative, activity-oriented, and systemic, including a socio-cultural approach. The role of competition is determined by the need to introduce new innovative projects and approbate (1) innovative and creative ideas, (2) social initiatives, and (3) creative activities to provide effective and sustainable development of the creative regional economy. Fashion and style contests are studied not only from the standpoint of its organization, but as an innovative form of developing the fashion industry. In the Altai Krai, the fashion industry has its own history, traditions, and art school of modeling. The research novelty lies in consideration of the competition as an open, communicative, and interactive platform for developing professional creativity, providing creative opportunities, ensuring the intellectual development of innovation, which meets the urgent tasks of socio-economic transformation and modernization of the region.

**Keywords** Social creativity · Creative competition · Fashion industry · Fashion product · Innovative project · Organization technology

K. A. Melekhova (✉) · L. I. Nekhvyadovich
Altai State University, Barnaul, Russia
e-mail: kschut@mail.ru

Z. N. Shaigozova
Abai Kazakh National Pedagogical University, Almaty, Kazakhstan



S. G. Maximova et al. (eds.), *Advances in Natural, Human-Made, and Coupled Human-Natural Systems Research*, Lecture Notes in Networks and Systems 234,
https://doi.org/10.1007/978-3-030-75483-9_10

## 1 Introduction

In the Altai Krai, the creative economy needs a creative culture and new types of social and creative communications. In the era of global transformation, cross-cultural interactions, unlimited information fields, and wide communication opportunities, creativity helps society develop effectively. Creativity becomes the basis for sustainable regional development and sets the trends of social modernization. Innovative creative projects will help to (1) ensure the growth of creative mobility in the region, (2) develop social technologies, (3) form a new type of communication based on interactive models of socio-cultural interaction and the corresponding creative ability to respond to the challenges of the changing reality accurately. Creative activity and creative design arise from a socio-cultural context, where the interaction of creative forces, creative ideas, and artistic experience provide the quality promotion of innovation in the Altai Krai. This paper presents the experience of developing regional social creativity by forming a school for the artistic design of costumes and the technology of organizing creative contests in this sphere. The research topic is relevant since the artistic design of the costume is gaining importance in the contemporary socio-cultural context. This fact is evidenced by the active interest in this area among professional Altai fashion designers studying in specialized areas of training at universities, as well as among representatives of related areas in art and design.

Over the past few years, numerous design studios, art laboratories, and specialized creativity centers have emerged in the region. Projects praised not only in Russia but far abroad have been implemented. The variety of fashion theaters in the Altai Krai shows the traditions laid down during the Soviet modeling of the twentieth century. This is also evidenced by professional creative competitions of fashion designers:

- Open interregional competition of young designers "Fashion and Time";
- Professional competition of young designers "Sibirskaya kapel";
- Open city competition of fashion designers "Style. Style? Style!";
- Festival of student creativity "Festa in Altai";
- The regional stage of the international competition of young designers "Russian Silhouette."

## 2 Materials and Methods

The research of the formation of social creativity in fashion and style is connected with the history of the development of this direction and the regional socio-cultural context. Therefore, the consideration of establishing and developing a modeling school in the Altai Krai and analyzing the organization of creative contests in this area are important tasks of this study. The research purpose and objectives determine the research methodology. The study of the development of the regional fashion industry and the significance of the Altai art school of modeling is focused on the cultural-historical approach. The identification of general and special forms and strategies in organizing creative competitions in the region, as well as their specificity compared to others, is of particular importance. This aspect is studied based on the comparative method. The essential model of the technology for organizing the competition as an innovative project aimed at forming social communications is determined by the activity approach. The systemic method, including the socio-cultural approach, formulated the research conclusions.

## 3 Results

To provide a complete picture of the fashion industry in the region, we need to turn to the history of fashion and style. Meeting the needs of people in changing designs of fashionable costume, the solution of the existing problems of modeling clothes for the clothing industry in the twentieth century was entrusted to teams of artists, fashion designers, designers, and technologists of Model Houses. The term "Fashion House" was introduced by Charles Frederick Worth, a French fashion designer of English origin. He is considered the "father" of world fashion. In 1858, together with a partner, combining a creative workshop with a secular salon, Charles Worth opened his model house. In 1871, he became its sole owner. In 1868, he founded the High Fashion Syndicate, an organization uniting the salons that dressed the upper classes of society. The High Fashion Syndicate is still opened. Only members of this organization can be called "couturier" (Selling 2000, p. 15). Charles Worth first started the trend of displaying collections not on mannequins but on real people. Nine royal houses collaborated with couturiers. For example, Charles Worth dressed Maria F. Romanova, the wife of Russian Emperor Alexander III. Before the Russian Revolution of 1917, Russian fashion houses were concentrated in central cities (Moscow and St. Petersburg). The House Brizak, opened in 1855, was the most prestigious one. The clients of House Brizak were the Empress and Grand Duchesses, wives of the courtiers, and maids of honor. The House of Hinduce and the House of Olga Buldenkova were also popular among the court ladies. These fashion houses were closed after the revolution (Kartavaya 2017, pp. 373–375).

The history of Soviet fashion began in the 1930s, and the place of its formation and development was the House of Models, opened in Moscow in 1935. Its activities included the coordination of textile and clothing production, and sewing models for mass consumption. The Central Assortment Cabinet of the Technical Department of the USSR Ministry of Industry developed and introduced clothes of different assortments. In 1958, it was renamed to the All-Union Institute for the Assortment of Light Industry and Clothing Culture of the USSR Ministry of Industry. After the Second World War, there appeared regional Model Houses were subordinated to the central ones. This structure formed the basis of the Soviet hierarchical system for managing the fashion industry in the twentieth century (Bartlett 2011, p. 20). The Soviet fashion, similar to the Soviet economy, was subordinated to the principles of long-term planning, leadership, and centralized supply. Regional Model Houses were supposed to work according to the prevailing principles and attitudes but with a certain degree of creative freedom and individual achievement.

In Siberia, there were two model houses—in Barnaul and Novosibirsk. The Barnaul House of Models was founded in 1966 by the decision of the Executive Committee of the Altai Krai Council of People's Deputies "On the Organization of the Altai House of Fashion of the Directorate of Consumer Services" (October 21, No. 583) (Fund P-834. List 6. File 717. Page 198). The activities of the Altai House of Fashion [AHF] included the development and introduction of modern models of clothing into production, the organization of a systematic display of clothing models to the public, and the order in which orders for the shown clothing models should be accepted by all household enterprises simultaneously with showing them to the population. The AHF was provided with the necessary equipment; appropriate selection and training were carried out. The AHF was entrusted with the work of assisting the regional enterprises in organizing a brigade work, sewing fashionable clothes with one fitting, and introducing uniform methods for designing clothes and taking measurements. The AHF became the coordinating design and technology center for fashion in the region. It also provided practical and methodological advice, carried out educational functions, and provided household services according to the production plan. The grand opening occurred in 1969. Three years after it, the AHF strengthened its reputation as a high-end atelier and experimental-methodical fashion center. In addition to this large and strategically important organization, the following organizations operated in Barnaul:

- "Barnaul Household Plant";
- "Barnaul Service Bureau for Rural Population of the Altai Krai";
- "Indposhiv" sewing factory;
- "Avangard" sewing factory;
- "Dynamo" sewing factory;
- "Progress" sewing factory.

On September 12, 1983, the Regional House of Models [RHM] was created on the basis of the regional experimental laboratory of the "Altaykraishweibyt" production association by the decision of the Executive Committee of the Altai Territorial Council of People's Deputies No. 322 (Fund R-1141. List 4. File 187. Pages 13, 400–401). In 1987, a unique enterprise of an experimental and artistic profile was created on the basis of the Consumer Services Centre. It produced fashionable items, clothing, headdresses, knitwear, and shoes.

In the early 1990s, the national economic conditions changed, and household services, fashion, and style began to restructure their activities and adapt to new realities. At that time, public service enterprises were the first to be privatized. The Regional Model House was restructured into the "RMH Trade and Office Center." Since then, it exists under this brand. As a modeling institute, the RHM ceased its work in 1992. Specialists of the RHM started private activities related to the development of individual creativity of fashion designers, founded private tailor shops, engaged in teaching, and became founders and ideologists of the first creative contests in fashion and style. New areas of activity served as the first steps towards the development of social creativity in the region.

## 4 Discussion

A competitive interest in fashion in the region has developed with the advent of a professional school of costume art. Technologies for organizing creative competitions were created in the 1960s. In May 1966, a regional art council was held in Barnaul to demonstrate sewing and knitwear models for the spring–summer season.

Only professional institutions and specialized enterprises of the region took part in this event, including:

- "Progress" Barnaul factory;
- "Pervomayskaya" Biysk factory;
- "Indposhiv" Barnaul factory;
- "Troitsky" production plant;
- Biysk knitwear factory;
- Novoaltai humpback factory.

In February 1967, in the village of Blagoveshchenka, among 150 models represented by nine enterprises, five models of the Barnaul Indposhiv factory and six models of the Altai Fashion House were announced the best. On April 18–20, ten enterprises of the region took part in the regional Art Council in the city of Rubtsovsk. The models of the Altai Fashion House, "Indposhiv" factory, and "Pervomayaskaya" Biysk factory were recognized as the best and selected for further implementation. These models were selected for the collection of the best samples for subsequent presentation at the Republican seminar held in Krasnoyarsk. Representatives of the "Indposhiv" factory, "Pervomayaskaya" Biysk factory, and Altai Fashion House attended this seminar.

The All-Russian Contest of Fashion Designers was held in November 1989 in Kazan. Its primary purpose was to intensify creative activity, identify artistic individuality, improve the professional skills of fashion designers, and increase the artistic and aesthetic level of products carried out by orders of the population. The competitive relay race in fashion and style was launched in 1991 when a creative competition of young and fashion designers "Smirnoff International Design Award" was organized. Later, it was developed in the design competition "Admiralty Needle" and became the launching platform for new creative projects:

- "Russian Silhouette";
- "Exercise";
- "Velvet Seasons";
- "Podium";
- "Volga Seasons."

In the work “Fashion Competitions in the Russian Fashion Industry,” the member of the Union of Designers of Russia, A. A. Kuleshova, noted, “ < … > these competitions paved the way for the High Fashion Week in Moscow since 1994, which was attended by professionals and first-rate fashion stars” (Kuleshova 2010, p. 125). In the 2000s, competitions of all levels (international, all-Russian, regional, city, district, inter-university, university, etc.) became widespread.

In the Altai Krai, the activity of organizing and conducting competitions of clothing designers began in the 1990s. One of the first event was the open interregional competition of young designers “Fashion and Time” (1994). The competition was organized with the support of the Ministry of Education and Science of the Altai Krai and the Association of Children’s Creative Associations “Golden Needle” (Moscow). Its main goal was to identify and support gifted children, and activate creativity in the art of costume design.

The annual festival of student creativity “Festa in Altai” gained popularity since 1997. Fashion is one of the most popular nominations of the competition.

The founders of the professional contest of young fashion designers “Sibirskaya kapel” (held since 2001) are (1) the Department of Education and Youth Affairs of the Altai Krai, (2) the Youth Affairs Committee of the Barnaul city administration, (3) the Barnaul College of Service and Clothing Design.

The competition aimed to (1) unite the creative potential of young people, (2) increase the prestige of primary and secondary vocational education, (3) increase the level of professional skills of students, and (4) contribute to the formation of a competitive specialist meeting the current requirements and demanded in the labor market. More than 300 authors and groups of authors, as well as about 40 specialized educational institutions, took part in this project.

The jury, as a rule, includes leading experts in the fashion industry. Within the framework of this competition, an interaction was carried out with industrialists and entrepreneurs of the region: (1) leading modeling organizations, (2) large industrial sewing enterprises, and (3) well-known fashion salons in the household sphere.

Open city competition of fashion designers “Style. Style? Style!” (held since 2006) is also focused on supporting creative youth in the region. The main objectives of the project are as follows:

- To create conditions for creative communication;
- To exchange experience and ideas between schools and theaters of fashion;
- To form a common culture and a broad outlook;
- To develop artistic and aesthetic taste in youth;
- To create conditions for spiritual enrichment and self-expression of the young generation;
- To promote social adaptation and vocational guidance of gifted children and youth.

The regional stage of the international competition of young designers “Russian Silhouette” is an innovative project in the region (held since 2019). At the federal level, the Altai Krai was the absolute winner of this competition. The main competition goal is to improve the system of aesthetic education and create conditions for realizing the creative potential of students in clothing design. Students and graduates of Altai universities and colleges aged 18–30 participated in the creative competition, regardless of the form of training and institution profile.

The clothing collection was to consist of five models. It could be designed for women, men, and children. The three stages of the competition revealed the creative potential of the participants gradually. At the first stage ("Sketchy"), the participants demonstrated their sketches, presented the collection concept, and indicated the purpose of its creation and a commercial perspective. The second stage ("On the Hanger") developed the idea of pre-displaying the collection without the participation of fashion models. The third stage ("Demonstration") fully revealed the idea of the collection in five models, which were presented in the form of a classic fashion show. This competition contributed to the creative expression and self-affirmation of the personality of young designers, the development of creative opportunities, and the formation of aesthetic taste. Many participants got a unique opportunity to make a bright debut in the fashion industry.

## 5 Conclusion

Nowadays, the main idea of organizing competitions is developing and strengthening social creativity in the fashion industry. Technologies for organizing creative competitions are aimed at ensuring that each category of participants receives a result. Young fashion designers try to understand their demands and find an employer in the person of an individual customer or a specialized enterprise. Receiving an award can act as a start for their own business or just a PR of a name or brand.

Professional stylists expand their activities, gain investors, and receive honorary titles and new professional competencies. The regional economy receives new opportunities from the organization of such competitions. These events can ensure the expansion of textile production, the introduction of advanced creative ideas, an increase in purchasing power in the fashion sector, and mass media preferences. The competition is an innovative form of activity in the fashion industry and an exchange of professional experience, communication, creative process, and educational activities. Designers need to present their vision of the project and learn the market demand. Creative competitions can be considered an essential tool for the innovation of the socio-cultural environment. The technologies of their organization generate new creative ideas. The regional platform becomes a platform for creative projects, and the competition reveals creative leaders.

## References

Bartlett D (2011) A ghost roaming around Eastern Europe. Moscow, Russia, New Literary Review
Kartavaya M (2017) Russian fashion houses: History and modernity. Young Scientist 6:373–375
Kuleshova A (2010) Fashion competitions in the domestic fashion industry. Innovation 4:124–126

State Archive of Altai Territory. (n.d.). Fund P-834. List 6. File 717. Page 198.
State Archive of Altai Territory. (n.d.). Fund P-834. List 6. File 717. Pages 198–199.
State Archive of Altai Territory. (n.d.). Fund R-1141. List 4. File 187. Pages 13, 400–401.
State Archive of Altai Territory. (n.d.). Fund P-834. List 10. File 3170. Page 66.
Selling, S. (2000). *Fashion. The century of fashion designers 1900–1999* (G. A. Yashina & Yu. A. Bushueva Transl.). Moscow, Russia: Konemann.

# Particularities of Remote Work in Modern Conditions

**Elena V. Kiseleva**

**Abstract** The article discusses topical issues related to the specifics of remote work in the modern period. With the development of information technologies, remote work (outside the employer's location) is becoming increasingly common in the field of labour relations as an atypical form of employment. The dramatic need for remote labour has arisen in recent months around the world, due to the complex epidemiological environment that has engulfed the world in a pandemic. In the article such methods of scientific knowledge as dialectical, logical and technical-legal are used. The study of doctrinal approaches and the practice of applying regulations related to the use of remote labour opens up a number of problems requiring reflection and subsequent changes in the labour legislation of Russia. As a result of the study, advantages and disadvantages of remote work have been identified, both for the employee and for the employer, a new approach to the organization of working time of the remote worker has been proposed, conclusions on the registration of employment agreements with remote workers have been presented, and the need to create safe working conditions for them is justified. The article notes the slight rate of spread of remote labour in Russia due to the imperfections in the legal regulation of remote labour, which leads to the need to change the legal framework for the use of remote labour, to find new approaches to eliminate gaps and contradictions in the labour legislation of Russia.

**Keywords** Remote employment · Remote work · Remote labour · Atypical employment · Distant work · The electronic document circulation

---

E. V. Kiseleva (✉)
Altai State University, Barnaul, Russia
e-mail: elenvik@inbox.ru

 
S. G. Maximova et al. (eds.), *Advances in Natural, Human-Made, and Coupled Human-Natural Systems Research*, Lecture Notes in Networks and Systems 234,
https://doi.org/10.1007/978-3-030-75483-9_11

# 1 Introduction

Global computerization, automation, and informatization lead to significant changes in all spheres of human life and work fields in the 21 century. Remote work is actively used in many foreign countries (USA, Canada, Finland, Sweden, etc.) as a new non-standard form of employment. A stunning breakthrough in the use of remote work occurred in recent months in Russia due to the spread of the COVID-19 pandemic and the self-isolation of citizens. The difficult epidemiological situation led to remote work in many areas of production. There is an increase in such atypical employment forms as remote work (outside the employer's office). Simultaneously, in Russia, remote work as a form of employment is only at the beginning of its formation, unlike many foreign countries.

A separate issue of atypical employment forms was repeatedly discussed in the Russian scientific environment (Azarova 2015; Buyanova 2019; Schur-Trukhanovich 2014). However, the main problem is more related to the imperfections of legal regulation of remote work and the small practice of registration of employment agreements with remote workers. Moreover, in the realities of today's self-isolation, most Russian workers and employers face many technical and legal problems: (1) lack of smooth operation of the Internet, (2) the need to transfer workers to work remotely, (3) recording of work results, (4) limited control over the labor discipline of employees and others. These factors indicate an objective need to study the peculiarities of remote employment in current conditions further.

# 2 Materials and Methods

The authors used the works of scientists in the field of labor law (Azarova 2015; Belitskaya 2017; Buyanova 2019; Korshunova 2015; Lushnikov and Lushnikova 2009; Schur-Trukhanović 2014), provisions of the Labour Code of the Russian Federation on remote work (Russian Federation 2001), and court practice on the stated issue (Golovinsky District Court of Moscow 2018; Kirovskiy District Court of Novosibirsk 2020; Pervorechenskiy District Court of Vladivostok 2018). The methodological base of the research was a set of methods of scientific knowledge: (1) the dialectical method to reveal gaps and contradictions in labor legislation on remote work and its applications; (2) the logical method; and (3) technical and legal methods to formulate proposals to improve labor legislation.

## 3 Results

- Along with advantages, the paper also presents essential shortcomings of remote work: (1) the moral deterioration of workers caused by remoteness from co-workers; (2) due to the lack of control over remote workers, the line between work and household chores is erased; (3) the problems of productivity and quality of work executed by the remote worker have arisen. The employer cannot provide such workers with safe working conditions and also does not bear responsibility for violations of protection of remote workers' rights;
- The research reveals the common and unique features between working conditions of home and remote workers;
- The authors confirm the conclusion on the legal nature of the contract with the remote worker and its classification as atypical forms of employment;
- As a solution to the problem related to working hours, it is proposed to establish in the Russian labor legislation the legal possibility of combining remote and stationary work based on the experience of foreign countries (where most of the time the employee works remotely, and a smaller part works stationary in the employers' office);
- During the study of labor legislation and the practice of registration of employment contracts with remote workers, the authors note that the practice of improper registration of employment contracts with remote workers is frequent, which is confirmed by the materials of judicial practice. If the employment contract is not concluded either in paper or electronic form, it will be impossible to prove the existence of remote work in another way;
- To protect the labor rights of remote workers, the government needs to (1) introduce the obligation of the employer to develop labor safety instructions and rules for remote workers; (2) with the use of electronic document workflow, familiarize them with the instructions via their signature.

## 4 Discussion

More companies worldwide are moving their workers to remote employment, aware of the benefits of remote work. Compared to many foreign countries, remote work is not often used in Russia. The paper explores the problems and reasons that negatively affect the spread of this type of employment in Russia.

In 2015, a survey of 700 SuperJob of HR-services of Russian companies was conducted on the web portal. More than 20% of respondents rated the format of remote work as not effective enough, as "employees have no labor spirit" and often they are "distracted by household chores." Only 12% of Russian employers actively attract remote employees to work, assessing its efficiency and high productivity (Azarova 2015). The shortcomings of remote employment are often the reason for the weak spread of remote forms of employment.

Besides, in Russia, remote work was formed under a limited legal framework and irrationality of its development. For the first time, the peculiarities of regulating the work of remote workers appeared not so long ago, in 2013 with the introduction of Chapter 49.1 into the Labour Code of the Russian Federation.

In the scientific literature, remote labor is classified as atypical employment, which does not have the essential characteristics of a traditional labor legal relationship or its modifications (Motsnaya 2009).

This is an organizational feature of the employment contract, as remote work implies the employees' autonomy in determining the work schedule and labor standards. On the part of the employer, there is no need for constant control; the employee does not have to obey internal labor regulations.

According to A. M. Lushnikov and M. V. Lushnikova, an atypical employment contract is a contract that is distinguished by the novelty of design and content. Its characteristic significantly modifies one to several signs of the employment contract: personal, organizational, or property. The authors consider the following as atypical employment contracts: (1) contracts on agency labor, (2) contracts on remote workers, and (3) contracts on the head of the organization (Lushnikov and Lushnikov 2009).

In Russian legislation, remote work was originally provided only for home workers. At present, working at home workers and remote workers is slightly different. Working at home was introduced mainly for women raising children and people with disabilities. But not all workers working at home could be called home workers under the current legislation. For some categories of employees of intellectual work, there are civil law contracts, which deprive them of the right to protection by labor legislation. Today, remote workers often work as programmers, web designers, lawyers, economists, editors, and proofreaders, who are highly qualified specialists and able to organize their work independently, provide the employer in electronic form with the results of their work: edited texts, computer programs, draft agreements, reports, etc. (Schur-Truhanović 2014). It became possible to conclude an employment agreement for remote work with such workers.

Remote work often has no embodied form, results and may not be related to production as the work of a home worker. Remote workers do not need a stationary workplace. There are some advantages of this job for the employer: (1) low rent, (2) the lack of workplace organization, and (3) its maintenance. Thereby, it increases productivity, the ability to attract the most qualified staff from other regions, and saves wages.

As a rule, the employer includes a condition on a work schedule in an employment contract with a remote employee but cannot control the beginning, end, and work breaks. This means that they can evaluate the work only by the final result, as in civil law relations. This indicates one of the problems related to the absence of an organizational feature in remote work, which allows it to be classified as a labor relationship: compliance with the rules of internal labor regulations.

Meanwhile, after the conclusion of the employment of a contract on remote work, the involved employee is forced to build their own behavior, reporting to the employer, as the latter establishes the rules. This is not a characteristic of civil legal relations but is a clear sign of labor relations and confirms the labor legal nature of the contract with the remote worker.

The following problem arises with the working time of the remote worker. It is the employer's responsibility to account for the working hours of each employee, to monitor the discipline of work, and calculate wages. At the same time, under the current legislation (Russian Federation 2001), a remote worker may independently determine the regime of their working time, unless other conditions are provided for in the employment contract. In fact, employment contracts more often contain only a condition on the length of the working week, without specifying the details of remote work, the provision of daily or weekly reports, or the established time for communication with the employer. As a result, the constant control of the employer, the quality of the employees' work, and labor disputes arise.

One example of court practice shows the negative effects of formalizing a working time condition in an employment agreement with a remote worker. During judicial proceedings in the case No. 2–140/2020 on reinstatement at work, in connection with dismissal for absenteeism, it was established that the text of the employment agreement concluded with plaintiff C. established a 40-h working week, but the working schedule (beginning and end of the working day, lunch break) of the employee was not defined. The defendant of D LLC presented the Acts on the absence of C. at the workplace but did not provide evidence of making calls, sending requests by e-mail, or visiting the employee's location. The court found these acts inadmissible evidence of the employee's absence from the workplace and restored the remote worker to their previous position (Kirovsk District Court of Novosibirsk 2020).

The authors are considering the views of scientists on this issue. T. Y. Korshunova proposes to create the work schedule and accounting of working hours to regulate the terms of the employment agreement to establish the mode of work of a remote worker. For example, a five–day working week with two days off, working hours, the time when the employee must contact the employer, deadlines for the assignments, etc. (Korshunova 2015).

In the case of remote employment, I. J. Belitskaya proposes the solution to this problem. It is necessary to provide the possibility of concluding two different employment contracts: one with a provision on remote labor, and in the other with a provision for the performance of a labor function in the structural division of the employer, but with a special (individual) work schedule (for example, Wednesday, Friday from 10 am to 3 pm) (Belitskaya 2017).

It is more preferably to establish the legal possibility of combining remote and stationary work in Russian labor legislation based on the experience of foreign countries. For example, an employment contract can specify the following schedule: an employee performs their employment duties remotely, except on the 15th and 30th of each month (for personal reporting and handing over materials to the employer). Another example is when an employee can perform their work duties remotely, but in exceptional cases must be physically present in the office (business meeting, a plenary session of management, etc.). Then the employer needs to warn the employee in advance to arrive at the office and fix these provisions in the text of the employment agreement.

It should be noted that there is another problem related to the practice of drawing up an employment agreement with a remote worker, or rather with the registration of personnel documents in electronic form. Labor law allows the maintenance of personnel documentation between the employer and the remote worker completely in a paper format, or even registration without a personal meeting. As a practical matter, such circumstances lead to improper registration of employment relations when the parties do not conclude an employment agreement in a paper or electronic format, limited only to the formulation of tasks and payment of work performed. The situation is complicated if there is no evidence of remote work. In such a case, it will be impossible to confirm the existence of remote work.

One of the typical examples of court practice confirms violations of labor law: an employer does not conclude an employment contract with a remote worker. During judicial proceedings of the Golodinsky District Court of Moscow in the case No. 2–3640/2018 on reinstatement at work and assignment of the obligation to conclude an employment contract on remote work with a software developer, the plaintiff did not present evidence of the performance of remote work: an employment agreement on remote work or other documents of the parties on the mandatory terms of the employment contract. The plaintiff received the tasks through the exchange of information via e-mail, but the court did not consider it as the evidence of an agreement on remote work, as the presence of electronic correspondence itself cannot constitute sufficient grounds for recognition of remote work. The employer submitted to the court the attendance records confirming the existence of an employment relationship. The court granted part of the plaintiffs' claims, except for the employers' obligation to conclude an employment contract with the plaintiff on remote work (Golodinsky District Court of Moscow 2018).

Another legal problem in regulating remote employment is the provision of safe working conditions. The employer performs only three duties to ensure safe working conditions, as set out in Article 212, Paragraphs 2, 17, 20, and 21 of the Labour Code of the Russian Federation. However, some employers do not even comply with them, which is confirmed by the court practice. The employer wrongly believed that it is not their responsibility to conduct the initial workplace induction with the employees of their organization, as the employees were contracted for remote work. In their own way, the employer interpreted the restriction of duties on remote workers based on the requirements of Part 2 of Article 312.3 of the Labour Code. However, the judge clarified the provisions of the current legislation and dismissed the complaint (Pervorechenskiy District Court of Vladivostok 2018).

The authors consider it necessary to add to Article 312.3 the obligation of the employer to develop labor safety instructions and rules for remote workers and using an electronic document workflow, familiarize them with the instructions via their signature.

## 5 Conclusion

One of the features of the development of current labor legislation is its flexibility (not only a simplification of the procedure of admission and dismissal of employees but also flexible payment and working hours, etc.), which is one of the most important points of implementation of employment policy at the European level since the 2000s.

The primary component of this policy is the concept of flexicurity on the principle of the flexibility of the labor market, combined with the social security of workers. Indeed, this principle of flexicurity facilitates the procedure of hiring and dismissing of workers, which ensures the mobility of the workforce, but at the same time increases legislative measures for the assistance of employment and prevention of unemployment (Lushnikova and Kirillova 2018).

Today's global economic crises and the coronavirus pandemic give countries and the scientific community a new attitude towards the issues of flexibility and security in the labor market, the formation and implementation of flexible policies. The search for solutions to these issues is based on national and socio–economic conditions, legal traditions, and world experience. There are also major changes in the legislative sphere in Russia: new types of employment contracts are emerging. At the same time, the functioning of sufficient remote work in current Russia will be possible with a timely change in legislation, eliminating gaps and contradictions existing in the use of remote work.

## References

Azarova EG (2015) Peculiarities of the legal regulation of labour relations of certain categories of employees (Etc.) Institute of Legislation and Comparative Law under the Government of the Russian Federation, Moscow, Russia

Belitskaya IY (2017) Commentary to the labour code of the Russian federation (short, article-by-article) (et al) CONTRACT, Moscow, Russia

Buyanova AV (2019) Remuneration of employees performing work outside the employer's location. Colloquium-J 4–3(28):28–30

Golovinsky District Court of Moscow (2018) Decision (October 22, 2018 No. 02–3640/2018). Moscow, Russia. https://mos-gorsud.ru/rs/golovinskij/services/cases/civil/details/f5bc2929-4759-4bd1-aa9f-88b1adb3acd9

Kirovskiy District Court of Novosibirsk (2020) Decision (March 11, 2020 No. 02–140/2020). Novosibirsk, Russia

Korshunova TY (2015) Peculiarities of the legal regulation of labour relations of certain categories of employees. Institute of Legislation and Comparative Law under the Government of the Russian Federation, Moscow, Russia

Leshukova N (2015) Topical problems of organization of labor of remote workers. Labor Law 8:81–89

Lushnikov AM, Lushnikov MV (2009) Course of labour law, vol 2. Statute, Moscow, Russia

Lushnikova MV, Kirillova LS (2018) The principle of flexicurity in a typical employment agreements. Soc Pension Law 2:46–50

Martynova OV (2016) Remote employment in Russia and prospects for its development. Young Scientist 2(106):542–545

Motsnaya O (2009) Legal regulation of the labor relations at atypical forms of employment. The Personnel Officer the Labor Law for the Personnel Officer 6:20–27

Pervorechensky District Court of Vladivostok (2018) Decision (January 16, 2018 No. 12–52/2018)

Russian Federation (2001) Labour Code of the Russian Federation (December 30, 2001 No. 197; edited by November 9, 2020). Moscow, Russia

Shchur-Trukhanovich LV (2014) The employment contract about remote work: maintenance, features and realization in practice. Personnel of the Works 3:30–35

Zakalyuzhnaya NV (2015) From home work to remote employment: similarities and differences. Legislation 9:49–56

# Characteristic Features of Finance in "Science Cities"

**Elena V. Dyakova, Evgeniy V. Vologdin, and Olga A. Grazhdankina**

**Abstract** This work is dedicated to the research of opportunities and disadvantages of financial self-sufficiency of science cities local government and determination of the ways of local budget growth. By the aid of comparative and financial analyses some external and internal factors have been found. These factors have the influence on income formation of science city local budget on the example of Biysk, Altai Region. The work led to the conclusion that obtaining of science cities financial basis is taking place according to the general Russian standards without regard to their specific focus. Federal transfers to the science cities budgets are insignificant and inefficient. The main directions towards tax and budget self-sufficiency growth have been determined. The Federal Government jointly with regional authorities should establish new development strategies of science cities financing both in research and production economic base of town and also in urban infrastructure. Science cities local government should regularly reveal financial reserves and raise budgetary and tax discipline. Public organizations can play a significant role in science cities financial management.

**Keywords** Science city · Research and production complex · Innovative development · Local government · Subsidy · Finance · Budget · Government co-financing

## 1 Introduction

Current Russian social-economic developments initiated by financial sanctions and high government finance dependence on oil and gas revenues set new government responsibilities. They include innovation development transition, intellectual, scientific, and technical potential stimulation, and high tech production.

E. V. Dyakova (✉) · E. V. Vologdin · O. A. Grazhdankina
Altai State University, Barnaul, Russia
e-mail: elena_dia1@mail.ru



S. G. Maximova et al. (eds.), *Advances in Natural, Human-Made, and Coupled Human-Natural Systems Research*, Lecture Notes in Networks and Systems 234,
https://doi.org/10.1007/978-3-030-75483-9_12

The Decree of the President of the Russian Federation No. 642 of December 1, 2016, "On the strategy of the Russian Federation Science and Technology Development" (Presidential Executive Office 2016), has defined science as one of the preferred directions of national policy.

In this context, a significant role is played by naukograds (science cities) as the powerful centers of science and science-driven industries.

During the prereform period, science cities were under special regulation. Many of them had no names and were not even indicated on the map. Soviet cities with advanced research-and-production infrastructure had excellent government financing. The standard of living of science cities' citizens was significantly higher than of other soviet cities.

However, the transition to market relations has put the existence of city-forming principles of urban economy and science cities themselves at risk. An effective mechanism of science cities government co-financing, the procedure of fund-raising into knowledge-intensive industries is still not developed and basic principles of local budget forming allowing for science cities to be specific, to distribute inter budget transfers, etc. The mentioned circumstances predetermine the importance and necessity of the investigation of modern Russian science cities' finances.

## 2 Materials and Methods

The research basis comprises normative legal acts of the public authorities of the Russian Federation and Altai Krai in the field of social and economic development of science cities and by the studies of national and foreign authors who reveal different aspects of science cities study. The research methods include comparative and financial analyses and induction and deduction methods.

## 3 Results

The local budget providing the achievement of municipality objectives and functions is the financial capability of every local government.

Within the framework of this research, we analyze the science city Biysk. The primary activities of town-forming organizations of a science city are nanosystem industry, biopharmaceutics, rocket, and space industry.

The finances of Biysk form according to the budget and tax legislation of the Russian Federation (Table 1).

According to Table 1, between 2016 and 2018, the revenues of Biysk decreased by 477.1 mln. RUB.

Uncompensated receipts prevail in the revenue structure. They decreased by 586 mln. RUB.

**Table 1** The dynamics of Biysk revenue structure

| Indicators | Sum, RUB mln | | | Percentage, % | | |
|---|---|---|---|---|---|---|
| | 2016 | 2017 | 2018 | 2016 | 2017 | 2018 |
| Tax revenues | 1,004.1 | 1,097.2 | 1,175.1 | 28.0 | 35.8 | 37.7 |
| Including local revenues | 235.5 | 294.7 | 281.7 | 6.6 | 9.6 | 9.0 |
| Non-tax revenues | 252.8 | 214.6 | 190.6 | 7.0 | 7.0 | 6.1 |
| Uncompensated receipts | 2,339.3 | 1,755.2 | 1,753.3 | 65.0 | 57.2 | 56.2 |
| Including science cities subsidies | 86.6 | 119.3 | 83.4 | 2.4 | 3.9 | 2.7 |
| Total revenues | 3,596.2 | 3,067.0 | 3,119.1 | 100.0 | 100.0 | 100.0 |

*Source* Compiled by the authors based on (The Municipality of Biysk, n.d.)

Herewith the subsidies to science city Biysk have an insignificant percentage. They decreased by 3.2 mln. RUB.

Budget tax revenues increased by 171.0 mln. RUB. The local tax percentage does not even reach 10%.

Between 2017 and 2018, the revenues decreased by 13.0 mln. RUB. Subvention increase by 209.4 mln. RUB has led to the percentage reduction of the city budget revenues by 12.3%.

The systematic non-fulfillment of plan indicators both on the revenue side of the budget and in expenditures can be observed in the process of city budget implementation.

The growth of arrears of tax and other compulsory payments is a serious issue (Fig. 1).

Figure 1 shows a sustainable growth rate of local tax arrears in the budget of Biysk between 2015 and 2018 by 62.2 mln. RUB.

During the analyzed period, the town budget has moved from the surplus budget (368.7 mln. RUB in 2016) to the deficit budget (109.8 mln. RUB in 2018).

A short research of a science city Biysk budget has shown that the assigned de jure status does not influence the city finance forming and its level significantly. Biysk local government faces issues that have nationwide character irrespective of municipal uniqueness.

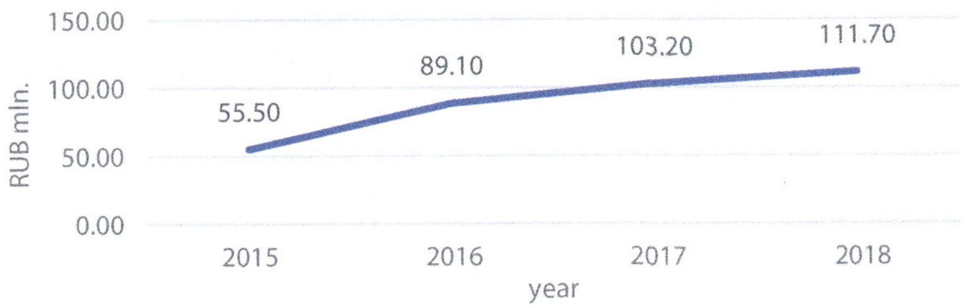


**Fig. 1** The dynamics of local taxes arrears in the budget of Biysk, RUB mln. *Source* Compiled by the authors based on (The Municipality of Biysk, n.d.).

# 4 Discussion

The scientific research in the field of science cities finances shows that state government bodies, local government bodies, research and economic production base, and educational and public organizations of a city are engaged in this sphere (Anttiroiko 2004; Charles 2015; Zyablova 2018).

According to the Federal Law No. 70-FZ of April 4, 1999 (ed. of April 20, 2015), "On the status of Science City of the Russian Federation" (Russian Federation 1999), the federal budget subsidizes the native science cities on research and production complex development and city infrastructure stabilization (Fig. 2).

As it is seen, between 2013 and 2018, inter-budget transfers from the federal budget to science cities decreased by 193.6 mln. RUB with a constant amount of science cities (13).

Part of the funds is distributed based on a contest held by the Ministry of Science and Higher Education. Its procedure and results are not published on the corresponding sites (Proskurnin 2017).

There is no statistical corroboration between the volume of inter-budget transfers to the science cities and their activity results (Maltseva et al. 2016).

In real life, the percentage of science cities budget subsidies makes not more than 10% of the revenues. Nevertheless, the experience of such science cities as Obninsk, Koltsovo, Dubna, and some others shows that even insignificant federal co-financing of a scientific and innovative base with sufficient funds use can attract significant non-budgetary sources (The Non-Profit Partnership "The Union of Russia's Science Cities Development," n.d).

Science cities that have town-forming organizations with security mode have practically no opportunity for investor recruitment. It significantly reduces potential sources of budget income (Abdulmanapov and Medzhidov 2019).

In the process of modern fiscal equalization, the role of inter budget transfers that arrive in science cities on an everyday basis is not definite (Timushev 2020).

Nowadays, the primary tax source of the science cities budgets (50–60%) is the Personal Income Tax. However, the Russian budget and tax legislation attach only

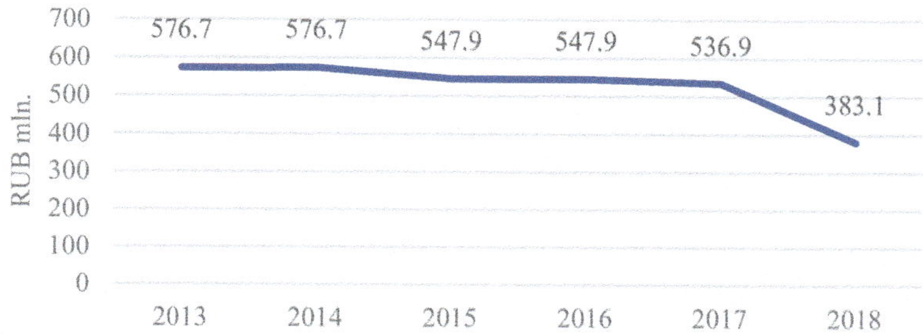


**Fig. 2** The dynamics of inter budget transfers distribution from federal budget to science cities, RUB mln. *Source* Compiled by the authors based on (Ministry of Finance of the Russian Federation, n.d.)

15% of the mentioned tax to local budgets. The rest is transferred into the budgets of Russian constituent entities.

It is necessary to take into consideration a significant regional differentiation of the wages and the Personal Income Tax evasion. These circumstances significantly reduce the local budget income. For a long time, the discussion on the necessity of introducing personal progressive income tax has been held among public authorities and in scientific society (Semyonova 2018).

The most unprofitable tax proceedings into the local budget are local taxes such as individual property tax and land tax (Roshchupkina 2019). The fact is that housing resources that are under the jurisdiction of the Ministry of Defence of the Russian Federation and are rented out to the military and civilians are not liable to taxation. Thus, if there are such objects in science cities, the tax base on the individual property into the local budget will be significantly reduced.

Land taxation causes issues with government cadastral valuation. The shortfall in income is taxes on government-owned organizations, such as scientific research institutes, defense enterprises, and others located on the territory of science cities (Putintsev 2016).

A significant disadvantage in the work of science city local government is the lack of principle of openness. It reveals minimal and superficial information on the financial standing of the city and the budgeting process on the official sites of the administrations.

The lack of full-fledged information on budget planning, results of budget control, current and strategic financial management of municipalities on open access makes the research more difficult (Kotelnikova 2016).

Specific governing bodies are established in some Russian science cities such as Koltsovo, Fryazino, Michurinsk, and others. For example, these bodies are Research and Development Boards, which provide coordination of local government activities, town forming research and production base, and educational establishments (Suhodolov 2016). However, neither on the official site of the non-commercial partnership "Russian Union of Science Cities" nor on the sites of science cities administrations there is information about the results of these activities.

## 5 Conclusion

The research has shown that the financial standing of Russian science cities is determined by a combination of external (direct and indirect) and internal factors.

The external direct factor is the lack of legal boundaries of government co-financing. External indirect factors are taxation, budget, and customs and tariff policy of the government, relating to the finances of all municipal entities of the Russian Federation.

From our point of view, internal factors play a significant role in financing current science cities. Such factors as budget process disadvantages, unskilled municipal workers' level, the formal relationship between administrations and town-forming organizations, and others significantly reduce financial capabilities for city expenditures financing.

It is necessary to implement a set of external and internal measures to enforce science cities' financial capabilities. The government's interest in strengthening Russian science cities' fiscal capacities should play the dominant role.

Resolution of the Government of the Russian Federation No. 538 of April 30, 2019 (ed. of January 28, 2020) "On Measures of State Support for the Establishment and Development of World-Class Scientific Centers" (Government of Russian Federation 2019) has set a relatively new kind of state financial support for world-class scientific centers, grants in the form of subsidies. The procedure of federal grant provision is developed, and the effectiveness criteria of public funds use. It allows the town-forming organizations to enhance their scientific status, get federal financial support, and, thus, strengthen the social and economic situation of science cities.

The primary activity of science cities' administrations is to focus on new growth areas in the local budget. Local government bodies should use all available resources for the strengthening of the financial base.

Resolution of the Government of Altai Krai No. 50 of February 14, 2017 "Strategy of Social and Economic Development of Biysk as a Science City for 2017–2030" (Government of Altai Krai 2017) has set that the foundations of urban economic development are a town-forming research and production complex. Department of Strategic Development and Economics of Administration of Biysk regularly monitors the activities to preserve the science city status.

One should not underestimate the importance of the human factor. It is necessary to attract highly skilled professionals and systematically improve their training level to manage science cities' financial resources effectively.

Public organizations such as science and technology councils can play a crucial role.

Further scientific research of science cities finances should be one of the critical research areas.

## References

Abdulmanapov SG, Medzhidov ZU (2019) Prospects for social and economic development of the special status cities. Vestnik of Chelyabinsk State University 3(425):12–19

Anttiroiko AV (2004) Science cities: their characteristics and future challenges. Int J Technol Manage 28:395–418

Charles D (2015) From technopoles to science cities: characteristics of a new phase of science cities. In: Miao JT (ed) Making 21st century knowledge complexes: technopoles of the world revisited. Routledge, London, UK, pp 82–102

Government of the Altai Region (2017) Resolution of "Strategy of social and economic development of Biysk as a science city for 2017–2030" (February 14, 2017 No. 50). Barnaul, Russia

Government of the Russian Federation (2019) Resolution of "On Measures of State Support for the Establishment and Development of World-class Scientific Centers" (April 30, 2019 No. 538, edited by April 30, 2020). Moscow, Russia

Kotelnikova MV (2016) Planning of local budget tax revenues: tendencies, problems and perspectives. News of Ural State Univ Econ 6:57–72

Maltseva AA, Monachov IA, Narsukova NE, Veselov IN (2016) Science cities of Russia and the United Kingdom: "Cui bono"? Innovations 7(43):42–44. https://cyberleninka.ru/article/n/naukogrady-rossii-i-velikobritanii-cui-bono

Ministry of Finance of the Russian Federation (n.d) Official website. http://www.minfin.ru

Presidential Executive Office (2016) Decree of "On the strategy of the Russian Federation Science and Technology Development" (December 1, 2016 No 642). Russia, Moscow

Proskurnin SD (2017) Prospects for development of science cities as areas of intense scientific and technological development. Region Econ Manage: Electron Sci J 1(49). https://eee-region.ru/article/4909/

Putintsev VP (2016) Problems of local budget of the closed administrative-territorial entities and ways of their approach. Econ Manage Issues 1:67–71

Roshchupkina VV (2019) Specific features of income assessment of entrepreneurship objects for personal customers estate (based on materials of municipal entities of Stavropol Krai). Finance and Credit 5(785):1616–1625

Russian Federation (1999) Federal law of "On the status of Science City of the Russian Federation" (April 7, 1999 No. 70)

Semenova NN (2018) Advance of income assessment of personal customers in Russia in the context of socially oriented taxation policy. Finance and Credit 1:129–142

Suhodolov MI (2016) The interaction between the bodies of local self-government and subjects of innovative activity in science cities. Vestnik of Saint-Petersburg State University 1:53–64. https://cyberleninka.ru/article/n/vzaimodeystvie-organov-mestnogo-samoupravleniya-i-subektov-innovatsionnoy-deyatelnosti-v-naukogradah

The Municipality of Biysk. (n.d). Official website. http://www.biysk22.ru/

The Union of Russia's Science Cities Development (n.d) Official website. http://naukograds.ru/

Timushev EN (2020) Potential restrictions and directions of exploring the local budget decentralization in the regional development of the federal policy. Finance and Credit 1(793):196–212

Zyablova NA (2018) The development of the state policy for the science cities of the Russian Federation. Int Sci Tech J "Theory. Practice. Innovations" 5. http://www.tpinauka.ru/2018/05/Ziablova.pdf

# Sustainable Development and Need Satisfaction: Political and Economic Aspects

**Tamara S. Dobrydneva, Sergey I. Dobrydnev, and Yelena A. Kolos**

**Abstract** The paper is devoted to the research of political-economic aspect of the transition to sustainable development, associated with the social mechanism of need satisfaction. We consider ways to meet the needs of people and their impact on the formation of social relations in historical process. The transition to sustainable development is carried out at the current level of social-economic development, which is characterized by a high level of need satisfaction. Needs are insufficiently studied and underestimated as a factor of socio-economic development in economics. That is why the study of economic mechanisms for meeting needs and their impact on the development of society is an important research area. There are three ways to meet one's needs: (1) own manufacturing, (2) forced redistribution of vital needs, (3) economic reallocation. The needs that increase naturally cause the necessary economic redistribution of the social product, based on the change of forced labor forms. Every way of meeting needs generates a corresponding direction of development of society, including (1) scientific and technological progress, (2) aggressive wars, (3) crime, (4) corruption, (5) the development of social-economic structure. The main directions of sustainable development are closely connected with the methods of need satisfaction. Their sustainability and independence rely on the balance of current methods of need satisfaction.

**Keywords** Sustainable development · Needs · Method of need satisfaction · Forced labor · Social self-coercion

---

T. S. Dobrydneva (✉)
Altai State University, Barnaul, Russia
e-mail: tamara_dobrydneva@mail.ru

S. I. Dobrydnev
Altai State Institute of Culture, Barnaul, Russia

Y. A. Kolos
East Kazakhstan State Technical University Named D. Serikbayev, Ust-Kamenogorsk, Republic of Kazakhstan



S. G. Maximova et al. (eds.), *Advances in Natural, Human-Made, and Coupled Human-Natural Systems Research*, Lecture Notes in Networks and Systems 234,
https://doi.org/10.1007/978-3-030-75483-9_13

# 1 Introduction

In economics, the needs are traditionally considered in the practical applied aspect in the interaction of production, consumption, and effective demand. The issues of the theory of needs and motivation are considered by the following authors: Jean Baudrillard (2006), Galbraith (1973), Maslow and Frager (1987), Alderfer (2011), Weber (1978), Lamben (1996), and many others.

As Jean Baudrillard noted, economists understand need as the utility of available goods (Baudrillard 2006). We can assume that needs should be regarded as an internal characteristic of a person that determines their actions. The impact of needs and ways to meet them on the development direction of society and economic and social forms of behavior of people remains understudied.

Therefore, we consider it important to investigate the interrelation of needs and economic behavior of people in the production and distribution of the product. In addition, it is essential to define how methods of meeting needs affect the formation of (1) public relations, (2) institutions, (3) trends in economic growth, and (4) sustainable development.

The following Russian scientists examined the problems of sustainable development: Atkisson (2015), Bobylev (2011), Krylov (2013), Marfenin (2007), and others. Prerequisites of sustainable human and economic development interaction are considered in the work of Kolos (2006). Still, the factor of needs is not subjected to adequate consideration.

# 2 Materials and Methods

This paper uses general scientific theoretical research methods such as (1) comparative-historical analysis, (2) analysis, (3) synthesis, (4) classification, (5) typologization, (6) analog approach, (7) logical modeling.

# 3 Results

## *3.1 Interrelation of Sustainable Development and Need Satisfaction*

Human behavior focuses on meeting needs. It applies both to the individual and to society in general. Social production serves to meet human needs throughout the entire historical process. The mechanism of growth is determined by the interaction of general laws (the law of growth of labor productivity and the law of growth of needs). Nowadays, humanity realizes that further development should be sustainable, in a way that allows one not to exhaust but to strengthen the potential to meet the human needs of future generations.

Changes related to needs satisfaction are usually considered in the context of contentment with basic needs in developed societies, generating new needs and stimulating consumption on this basis. The concept of sustainable development concept is important since the development itself can be understood not only as economic growth but also as a combination of three directions:

1. Development of production (including exploitation of natural resources, investment, and scientific and technological development);
2. Personal development;
3. Institutional changes.

Each direction is a separate complex area that evolves according to its laws. Currently, these directions are divided into 17 goals. When these areas of social life are considered from the standpoint of need satisfaction, a new research perspective becomes relevant, which leads to new interesting results. As shown below, the main directions of sustainable development are shifting to methods of need satisfaction. Within the historical process, the needs have the same nature. Nevertheless, they have different, specific historical forms.

## 3.2 *Interrelation of Need Satisfaction Methods and the Development of Public Mechanism Directions*

Usually, scientists distinguish the following ways of need satisfaction: (1) self-sufficiency, (2) take away, (3) begging, and (4) exchange. Begging is not a socially significant phenomenon, so it will not be taken into account in the future. Exchange implies equivalent exchange of vital goods by equal owners. However, such goods cannot appear in the equivalent exchange.

The economics considers needs satisfaction as a factor of individual behavior. Nevertheless, they are implemented on the scale of the entire society and give rise to independent directions of its development. Let us consider the historical process briefly from the point of view of the development and ways of satisfying human needs.

Initially, human groups existed as communities. Individual needs were subjected to the demands of community. There were two methods of need satisfaction. The first method is to produce more goods within the community. This is the way to develop production capabilities of people and modern scientific and technological progress. The second method is to take the necessary goods from other people. This is the path of violent redistribution of wealth, which historically existed in the form of external aggressive wars and internal crime. This method underlies the social structures that serve the *war economy*, *criminality economy*, and *corruption economy*, which are also forms of forced redistribution of wealth.

The means of the first method are limited by natural conditions and the weakness of the productive forces. The second method is often preferred and respected since it allows one to acquire things that could not be produced in large quantities in households for a short time. Forced takeoffs primarily occur (1) between communities, (2) inside a community between families, or (3) inside a family. The winner received the primary means of development by force.

With the introduction of exchange, one could get goods that were not produced in the community and to meet the growing demand. Since the exchange mainly occurred between communities and was equivalent, it did not increase the quantity of goods and did not directly affect the satisfaction of individual needs, because the goods belonged to the community, not the family or the person.

The destruction of communities and the transition to family-type society was a revolutionary turning point in the development of need satisfaction mechanisms. The third method appeared: unequal economic exchange.

Some people take possession of goods that were produced by other people. This method is the economic way of distributing wealth in society. Such appropriation can be carried out in non-violent and violent forms of unequal economic exchange. Non-violent forms are (1) debt repayment, (2) payment for the provision of services other than exchange, (3) the performance of social duties, and other forms that are voluntarily conditioned by customs and laws accepted in society.

Violent forms of unequal economic exchange are forced labor in the form of debt-serfdom, family, and common slavery. Such coercion is not a violent way of satisfying needs, since goods belonging to somebody are not taken away.

Wealth is distributed before it is produced and belongs to someone. Later, forced labor was called exploitation and acquired its forms. However, forced labor is a natural, necessary form of human relations since it corresponds to the inner human nature and exists as an objective way to meet needs.

Therefore, the exchange satisfies the needs from the qualitative side as an equivalent method; from the quantitative side—as non-equivalent one.

The third method of satisfying needs through the appropriation of goods produced by other people (the surplus product in classical political economy) is the basis for the development of social relations and society. It is the way of social distribution of wealth that determines economic and social relations.

Figure 1 illustrates the interrelation between need satisfaction methods and social mechanisms for their implementation.

The appropriation of the surplus product requires forced labor. Coercion is inherent in human relationships throughout the history of humanity. Economic and non-economic forms of coercion are covered in many literary sources, especially Marxist ones (Kuzminov et al. 1989). The further development of economic relations is based on a change in the form of forced labor caused by the growth of needs.

There is a natural self-coercion to work in society. The needs of society are met first, after that individual needs are satisfied. The factor of individual needs is suppressed and does not function fully. Such situations occur in current societies with traditional communal lifestyle.

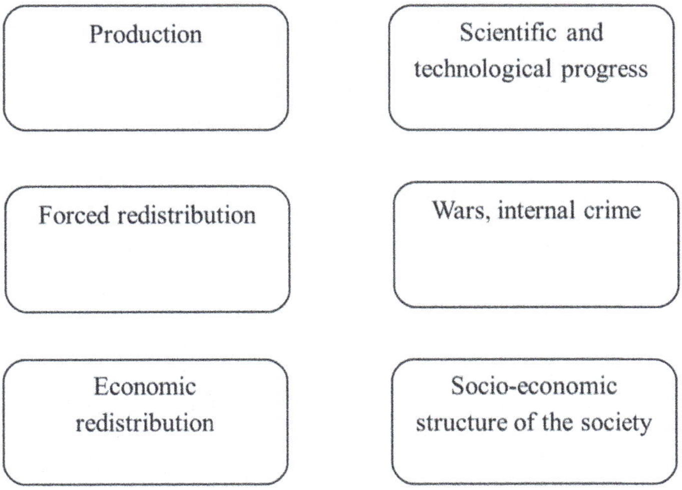


**Fig. 1** The interrelation between need satisfaction methods and social mechanisms for their implementation. *Source* Compiled by the authors

In the transition to a family economy, based on the third method of satisfying needs, an economic system of slavery is formed. This system is based on physical coercion to slavery. The possibilities of satisfying needs are increasing, but so are the desires. To get more products from the slave, one should be interested in it.

Partial coercion to physical labor develops, the slave turns into a dependent peasant along with the establishment of the feudal economic system. The needs of the landowner and the peasant are more fully met, but they grow faster. The efficiency of the peasant labor reaches its limit and then falls. To meet the increased needs, it is necessary to release both the initiative of the productive worker and that of the owner of the production.

Economic forced labor evolves when a person has to be hired not having another alternative. The owner of the production should be able to keep the employee in production. Then, the corresponding economic class and social structure is formed. The needs are met on a much larger scale. There is a new phenomenon related to their satisfaction.

The basic needs of owners and employees appear to be met in developed countries. Moreover, social-economic mechanisms for ensuring the population make work optional. There are different experiments to introduce a universal basic income (Switzerland 2016, Finland 2017–2019, Canada Ontario 2018).

There are new phenomena in the field of labor. More and more people have flexible working hours. The boundaries between work and leisure are becoming less clear. The balance of work and leisure is established by the employees themselves, who tend to be *self-exploited* to obtain higher incomes and a large volume of consumption. Workaholism, *abundant* labor force become a lot of entrepreneurs and managers, not workers. The percentage of creative employees increases, the balance of work and the expectation of inspiration makes practical sense for them.

We assume that social self-coercion to labor makes people work in such a situation. This is an independent economic phenomenon that differs from forced labor, which functions in different areas of social life.

From the point of view of psychology, coercion is an imitation technique that ensures that people act against their will in the interests of a restraining person. We can suggest that coercion includes not only the direct impact on a person but also the creation of conditions under which a person is forced to perform actions required by the restraining person. Modern society has created conditions that force a person to participate in labor activity not for physical survival but for achieving social goals and obtaining social benefits corresponding to a high level of need satisfaction. People are increasingly different not only in how they work but also in how they rest, and this applies to a social form of coercion.

Coercion is defined as self-coercion because incentives to work are not socially required, that is, a person is not subject to the external influence of coercion. People formulate life goals for themselves, choose ways to achieve them, and implements them with certain actions. Therefore, we can say that J. K. Galbraith's thesis about *self-exploitation* of the entrepreneur also covers a wide range of employees (Galbraith 1973).

However, we can see that human goals and values in modern society are being formed, but their behavior is becoming increasingly programmed. Consequently, the base of coercion remains the same, although it represents itself as self-coercion. This phenomenon cannot be described as self-persuasion, but as self-abuse under the influence of the social environment.

The need-oriented approach denies the possibility of non-coercive labor in human society. If labor coercion is an important method of meeting people's needs, it is inevitable and necessary.

In our opinion, such an interrelation of ways to meet the needs and directions of the development of society in the historical process corresponds to the current situation. In general, the development of society is aimed at the close interaction of three methods for need satisfaction: (1) scientific and technological development, (2) forced redistribution, and (3) economic labor coercion. Sustainable development should take this fact into account as a complex development concept.

## 3.3 Interrelation of Need Satisfaction Methods and the Main Vectors of Sustainable Development

The emergence of the concept of sustainable development is necessary in the context of the formation of a new socio-economic reality characterized by the satisfaction of most human needs. Its main goals are not so much economic growth as the social reconstruction of society, the transition to its new forms. That is why two of the three

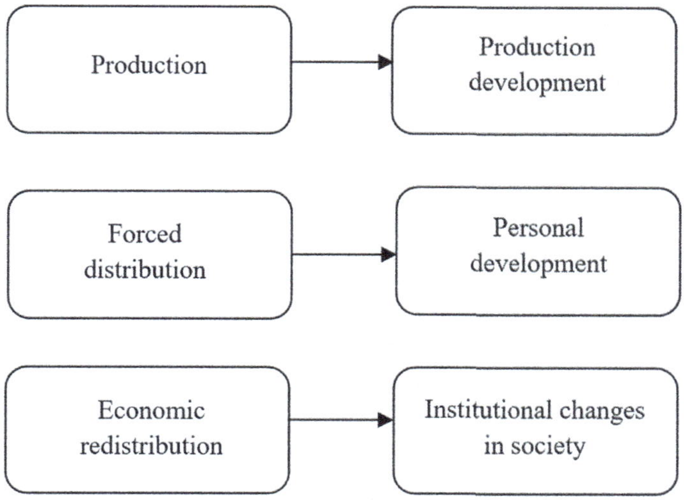


**Fig. 2** Interrelation of need satisfaction methods and the main directions of sustainable development. *Source* Compiled by the authors

directions are not economic. Figure 2 shows the interrelation of need satisfaction methods and the main directions of sustainable development.

Each direction of sustainable development is based on the corresponding method of need satisfaction. On the contrary, each need satisfaction method generates a corresponding development direction.

Whether they are sustainable or not depends on the balance of the current need satisfaction methods.

In our opinion, in order for sustainable development to become real, certain conditions should be met.

First of all, the production should meet the most basic needs, thereby reducing the significance of redistribution for the benefit of groups with social and economic power.

Second, forced redistribution in the form of crime and corruption should be reduced by the moral and psychological qualities of the population. Strict regulation of power, even based on law, is unproductive since it is a form of forced redistribution that generates non-legal forms. The development of the human personality is necessary. Without it, it is difficult to talk about sustainable development at all.

Third, speaking about the forms of forced labor, one should focus on the maintenance of historically promising forms, such as mild forms of (1) economic coercion with reasonable levels of exploitation and wages, (2) social coercion, and, in the long run, self-coercion to labor for societies at different levels of development.

## 4 Discussion

This work comprises a study of the interrelation of the needs and directions of the development of society. We assume that the most important issue related to the insufficiently studied area of needs is whether the search for new methods of need satisfaction (considered as ways to acquire vital goods) is a driving force for the development of economic and social relations between people. We attempted to solve this issue by considering the historical process of changing labor coercion forms from the point of view of needs.

In this regard, the study of the interrelation of needs and economic behavior of people not only in consumption but also in production and distribution of goods is a necessary direction of the development of economic theory. The assumption about motivation to work as a social self-coercion is controversial and requires sound reasoning.

The role of mechanisms for need satisfaction in the transformation of socio-economic relations requires further socio-economic studies.

## 5 Conclusion

Need-oriented approach reveals new prospects for studying the complex problems of sustainable development. The study of the interrelation between people's needs and economic behavior resulted in certain outcomes.

First of all, we identified three main need methods of need satisfaction. They are implemented in the corresponding ways of acquiring vital goods (production, forced and economic appropriation) that are implemented via labor coercion.

Second, we assumed that the methods of needs satisfaction are the basis of the main directions of social development: (1) scientific and technological progress; (2) social and economic forms of forced wealth distribution. Besides, we provided a political and economic reasoning for the main directions of sustainable development in terms of their relationship with need satisfaction methods.

Third, we examined the influence of the genesis of the method of need satisfaction on the development of social relations.

Fourth, we suggested the concept of *social self-coercion to labor* as a form of the current motivation of a person to economic activity.

The theoretical results obtained can be useful for (1) the development of social mechanisms of need satisfaction, (2) effective practical strategies, and (3) actions to ensure sustainable development.

## References

Alderfer CP (2011) The practice of organizational diagnosis: theory and methods. Oxford University Press, New York, NY

Atkisson A (2015) How sustainable development can change the world. Binom, Moscow, Russia

Baudrillard J (2006) Consumer Society: its myths and structures. J. Baudrillard. Translation from French, afterword and notes by E. A. Samarskaya. Republic, Moscow, Russia

Bobylev SN (2011) Modernization of the economy and sustainable development. Ekonomika, Moscow, Russia

Galbraith JK (1973) Economics and the public purpose. Houghton Mifflin Company, Boston, MA

Kolos YA (2006) Social security of the population in Kazakhstan in the context of economic transformation: theory and methodology. Ust-Kamenogorsk, East Kazakhstan State Technical University, Kazakhstan

Krylov AN (2013) Corporate social responsibility: economic models—morality—success—sustainable development. IKAR, Moscow, Russia

Kuzminov YI, Nabiulina ES, Radaev VV, Subbotina TP (1989) Labor expropriation: History and modernity. Ekonomika, Moscow, Russia

Lamben J-J (1996) Strategic marketing. European perspective. Nauka, Saint Petersburg, Russia

Marfenin NN (2007) Sustainable development of humanity. Moscow State University, Moscow, Russia

Maslow A, Frager R (1987) Motivation and personality. Addison-Wesley, New York, NY

Weber M (1978) Economy and society: An outline of interpretive sociology. University of California Press, Berkeley, CA

# Socio-economic Development of Russian Federal Entities in 2019: Multivariate Data Analysis

**Aleksey Yu. Yudintsev and Galina N. Troshkina**

**Abstract** This paper continues the series of our work on the analysis of the socio-economic level differentiation of the subjects of Russia. We study the degree of differences in the socio-economic development of the subjects of Russia using multivariate statistical analysis, based on data published by the Federal State Statistics Service in the statistical bulletin *Information for monitoring the socio-economic situation of the constituent entities of the Russian Federation*, characterizing the economic and social situation of the regions as of 2019. On the basis of factor analysis, we calculate statistically linearly independent factors that determine the level of socio-economic development of the region. Based on the calculated coefficients, we propose a methodology for determining the rating of regional socio-economic development. Through cluster analysis, we analyze the location, composition, and statistical characteristics of clusters in the factor space of the socio-economic development of the subjects of Russia. Moreover, we identified ten special regions which level of socio-economic development significantly differs from the average. The determined size of the calculated clusters indicates that the cluster of special regions is significantly larger than the cluster of other regions. In addition, we carried out a comparative analysis of the level of socio-economic development of the regions of the Siberian Federal District and Altai Krai. Besides, we examined the position of the Siberian Federal District and Altai Krai in the factor space. Furthermore, we calculated regional socio-economic development ratings both for each cluster and for Russia in general.

**Keywords** Subjects · Socio-economic development · Cluster · Cluster analysis

## 1 Introduction

The paper examines the problem of determining the socio-economic differentiation of the subjects of Russia as of 2019 based on data from the Federal State Statistics Service. As a result of factor analysis of the initial data, we formed a two-dimensional

A. Yu. Yudintsev (✉) · G. N. Troshkina
Altai State University, Barnaul, Russia
e-mail: a_yudintsev@mail.ru



S. G. Maximova et al. (eds.), *Advances in Natural, Human-Made, and Coupled Human-Natural Systems Research*, Lecture Notes in Networks and Systems 234,
https://doi.org/10.1007/978-3-030-75483-9_14

factor space. Moreover, we interpretated the content of the factor values and calculated the positions of the subjects in the factor space. After that, we performed clustering using the k-means method. We formed six clusters and determined their composition, size, and position in the factor space. Then, we identified the groups of subjects of Russia with similar parameters of socio-economic development.

## 2 Materials and Methods

To analyze the socio-economic situation of the subjects of Russia for 2019, we applied the methodology used in our previous works (Yudintsev and Troshkina, 2018a, b). We selected the following indicators for 2019:

1. $V_1$—average monthly nominal accrued wages of employees (RUB.);
2. $V_2$—average monthly per capita monetary income of the population (RUB.);
3. $V_3$—average labor forces (thou. ppl.);
4. $V_4$—average number of unemployed (thou. ppl.);
5. $V_5$—volume of investment in fixed assets (mln. RUB.) per annum;
6. $V_6$—retail trade turnover (mln. RUB.) per annum;
7. $V_7$—scope of work performed by a kind of construction (mln. RUB.) per annum;
8. $V_8$—number of employees required for open positions (thou. ppl.)—average per annum;
9. $V_9$—delivered goods of the region's production, works and services performed in the region (mln. RUB.)—average per month (Federal State Statistic Service 2020).

As a result of the standardization procedure, the transition was made to new, standardized variables: $X_1$–$X_9$, which represent the deviations of the original values from the average values of each sample, divided by the standard deviation (1).

$$X_i = \frac{v_i - \overline{V}_i}{S_i} \tag{1}$$

Therefore, all normalized variables will be dimensionless, commensurate, with zero mean values and unit standard deviations. The variables $X_1$–$X_9$, as well as the initial indicators $V_1$–$V_9$, represent a nine-dimensional space of socio-economic indicators, in which each point determines the socio-economic state of the corresponding Russian region. As a measure of regional similarity in the space of socio-economic indicators, we used the Euclidean distance in the space of normalized variables. As the correlation analysis showed, the indicator vectors $X_1$–$X_9$ are not statistically linearly independent (see Table 1).

Therefore, it is necessary to perform factor analysis and make a transition to the space of statistically independent linear combinations of vectors $X_1$–$X_0$—factors. To reduce the dimension of the socio-economic parameters, we applied factor analysis to the array of normalized variables using the principal component method with

**Table 1** The correlation matrix of vectors of the space of socio-economic indicators

| $X_1$ | $X_2$ | $X_3$ | $X_4$ | $X_5$ | $X_6$ | $X_7$ | $X_8$ | $X_9$ |
|---|---|---|---|---|---|---|---|---|
| 1.000 | 0.926 | 0.273 | −0.077 | 0.467 | 0.346 | 0.424 | 0.415 | 0.462 |
| | 1.000 | 0.413 | 0.040 | 0.525 | 0.484 | 0.495 | 0.515 | 0.530 |
| | | 1.000 | 0.663 | 0.841 | 0.976 | 0.830 | 0.951 | 0.887 |
| | | | 1.000 | 0.417 | 0.557 | 0.479 | 0.460 | 0.410 |
| | | | | 1.000 | 0.867 | 0.964 | 0.870 | 0.978 |
| | | | | | 1.000 | 0.823 | 0.974 | 0.913 |
| | | | | | | 1.000 | 0.821 | 0.932 |
| | | | | | | | 1.000 | 0.922 |
| | | | | | | | | 1.000 |

*Source* Compiled by the authors

rotation using the Varimax algorithm. As a result, the factor space was formed from two factors—$F_1$, $F_2$, which are statistically independent variables and can be used as universal indicators of the measure of the socio-economic differentiation of objects. Further, we performed the following cluster analysis procedures in the factor space: tree-type clustering to determine the optimal number of clusters and clustering using the k-means method to calculate the composition and characteristics of the formed clusters.

We used this technique in the preliminary analysis of the array of indicators for all subjects of Russia for a more thorough analysis of the selected groups of regions.

## 3 Results

As a result of a preliminary analysis of the initial array of indicators of the subjects of Russia, we identified two clusters: $C_1$ and $C_2$. Table 2 shows the average values $(\overline{\mathbf{F}}_1, \overline{\mathbf{F}}_2)$ and standard deviations $(\mathbf{S_1}, \mathbf{S_2})$ for the identified clusters.

Cluster $C_1$ includes ten regions which socio-economic development differs significantly from other regions. Figure 1 presents the position of the regions of Russia in the factor space. The subjects included in the $C_1$ cluster are numbered: 1—Moscow, 2—Moscow region, 3—Tyumen region, 4—Saint Petersburg, 5—Chukotka Autonomous Area, 6—Magadan Region, 7—Sakhalin Region, 8—Kamchatka Krai,

**Table 2** Statistical characteristics of clusters $C_1$, $C_2$

| K | $\overline{\mathbf{F}}_1$ | $\mathbf{S}_1$ | $\overline{\mathbf{F}}_2$ | $\mathbf{S}_2$ |
|---|---|---|---|---|
| 1 | 0.4329 | 2.5093 | 2.4098 | 0.9956 |
| 2 | −0.0534 | 0.6317 | −0.2971 | 0.4466 |

*Source* Compiled by the authors

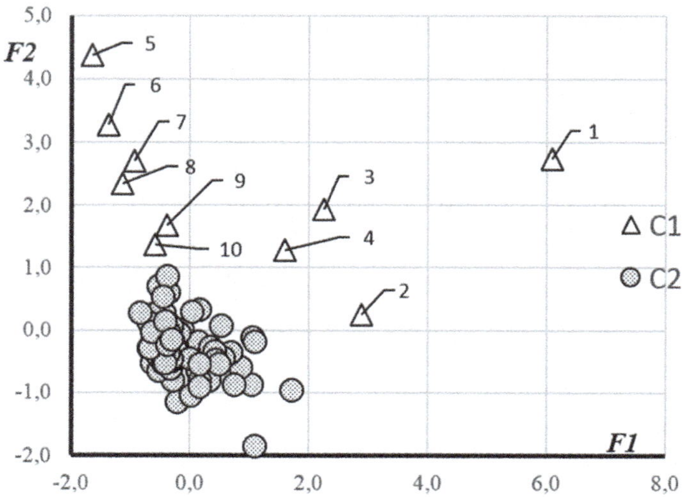


**Fig. 1** The location of the regions of Russia for 2019 in the factor space. *Source* Compiled by the authors

9—Republic of Sakha (Yakutia), 10—Murmansk Region. All regions included in the first cluster are characterized by a high and very high level of income of population—the values of the factor $F_2 > 0$. However, regions 1–4 have a rather high ($F_1 \gg 0$), 5–8—low ($F_1 < 0$), 9–10—close to the average ($F_1 \cong 0$) level of economic development. The remaining 72 regions of Russia constitute a fairly dense cluster $C_2$, and occupy an area in the factor space almost nine times smaller than $C_1$.

Let us consider the structure of the second cluster in more detail. It consists of the remaining 72 regions of Russia. The average values and standard deviations of the initial indicators for the standardization procedure are shown in Table 3.

As a result of using the method of principal components with rotation, according to the Varimax algorithm, we formed two factors: $F_1$ and $F_2$. The composition and factor load factors are provided in Table 4, the significant load values which are large than 0.7 are highlighted in bold. Table 4 shows that the variables $X_3$–$X_9$ have a very strong correlation with the first factor, and the variables $X_1$, $X_2$ correlate with the second factor.

These factors are linear combinations of the source variables. The values of $F_1$, $F_2$ can be calculated from the relations (2, 3).

**Table 3** Mean values and standard deviations of indicators

| i | 1 | 2 | 3 | 4 | 5 | 6 | 7 | 8 | 9 |
|---|---|---|---|---|---|---|---|---|---|
| $\bar{V}_i$ | 35,346 | 27,185 | 795 | 43 | 152,279 | 315,205 | 75,734 | 10 | 75,205 |
| $S_i$ | 6,521.4 | 5,082 | 569.8 | 31.1 | 130,269.6 | 279,109.1 | 69,851.8 | 8.124 | 72,379.3 |

*Source* Compiled by the authors

**Table 4** Factors loadings

| | $X_1$ | $X_2$ | $X_3$ | $X_4$ | $X_5$ | $X_6$ | $X_7$ | $X_8$ | $X_9$ |
|---|---|---|---|---|---|---|---|---|---|
| $F_1$ | 0.053 | 0.393 | **0.981** | **0.851** | **0.844** | **0.950** | **0.864** | **0.835** | **0.842** |
| $F_2$ | **0.918** | **0.778** | 0.100 | −0.158 | 0.420 | 0.186 | 0.307 | 0.310 | 0.410 |

*Source* Compiled by the authors

$$F_1 = -0.171 * X_1 - 0.062 * X_2 + 0.215 * X_3 + 0.235 * X_4 + 0.118 * X_5 + 0.190 * X_6 + 0.145 * X_7 + 0.138 * X_8 + 0.134 * X_9. \quad (2)$$

$$F_2 = 0.591 * X_1 + 0.431 * X_2 - 0.131 * X_3 - 0.274 * X_4 + 0.1057 * X_5 - 0.069 * X_6 + 0.028 * X_7 + 0.036 * X_8 + 0.0995 * X_9 \quad (3)$$

A content interpretation of the calculated factors is a quite difficult task. In this case, we can definitely interpret the second factor—$F_2$. Since it is very strongly correlated with the indicators $X_1$ and $X_2$, it can be called a factor of the level of income in the region. The first factor—$F_1$ correlates with variables $X_3$–$X_9$ and can be called a factor of the level of economic development of the region. Large values of the factors correspond to a higher level of socio-economic development of the region. Consequently, it is possible to introduce a universal rating of the socio-economic development of the region as a sum of the values of factors.

The tree-like cluster analysis shows that the fragmentations into two, three, or six clusters are the most resistant to changing range mergers are. Since the total number of the analyzed regions is quite large (72), it is best to divide them into six clusters. Table 5 shows the position of the subjects of Russia in the factor space ($F_1$, $F_2$) with the indication of the region's belonging to the cluster (column "K") and the intra-cluster rating of the region within the corresponding cluster (column "R"). We calculated the intra-cluster rating as the sum of the deviations of factors from the cluster center or as the difference between the regional rating and the cluster rating, calculated as the sum of the coordinates of the cluster center.

The statistical characteristics of the clusters (positions of cluster centers in the factor space and standard deviations) are shown in Table 6.

Table 6 in column "K" shows the numbers of clusters; columns $\overline{F}_1$, $\overline{F}_2$, show the cluster centers positions; columns $S_1$, $S_2$ provide the values of standard deviations for objects included into the corresponding cluster.

Figure 2 shows the location of the subjects of Russia, obtained by dividing cluster $C_2$ into subclusters $D_1$–$D_6$ in the factor space ($F_1$, $F_2$). The regions included in the Siberian Federal District are numbered as: 1—Altai Republic, 2—Republic of Tuva, 3—Republic of Khakassia, 4—Altai Krai, 5—Krasnoyarsk Krai, 6—Irkutsk region, 7—Kemerovo region, 8—Novosibirsk region, 9—Omsk region, 10—Tomsk region.

**Table 5** Position of the subjects of the Russian Federation in the factor space

| Region | $F_1$ | $F_2$ | K | R | Region | $F_1$ | $F_2$ | K | R |
|---|---|---|---|---|---|---|---|---|---|
| Orenburg region | 0.287 | −0.390 | 1 | 0.13 | Republic of Ingushetia | −0.399 | −1.906 | 3 | −1.03 |
| Saratov region | 0.603 | −1.163 | 1 | −0.33 | Kabardino-Balkaria Republic | −0.400 | −1.356 | 3 | −0.48 |
| Republic of Dagestan | 1.752 | −2.035 | 1 | −0.05 | Karachay-Cherkess Republic | −0.737 | −1.353 | 3 | −0.81 |
| Chechen Republic | 0.027 | −1.439 | 1 | −1.18 | Republic of North Ossetia—Alania | −0.626 | −0.870 | 3 | −0.22 |
| Stavropol Krai | 0.824 | −0.929 | 1 | 0.13 | Altai Republic | −1.091 | −0.551 | 3 | −0.37 |
| Altai Krai | 0.497 | −1.310 | 1 | −0.58 | Tuva Republic | −1.167 | −0.353 | 3 | −0.24 |
| Omsk region | 0.444 | −0.225 | 1 | 0.45 | Kurgan Region | −0.579 | −0.896 | 3 | −0.20 |
| Chelyabinsk reg | 1.479 | −0.441 | 1 | 1.27 | Bryansk Region | −0.421 | −0.355 | 3 | 0.50 |
| Crimea | 0.479 | −0.716 | 1 | −0.01 | Vladimir Region | −0.173 | −0.301 | 3 | 0.80 |
| Volgograd region | 0.668 | −0.723 | 1 | 0.18 | Ivanovo Region | −0.473 | −0.718 | 3 | 0.08 |
| Republic of Buryatia | −0.656 | 0.107 | 2 | −0.43 | Kostroma Region | −0.808 | −0.377 | 3 | 0.09 |
| Transbaikal Krai | −0.613 | 0.513 | 2 | 0.01 | Orel Region | −0.694 | −0.471 | 3 | 0.11 |
| Jewish Region | −1.439 | 0.844 | 2 | −0.48 | Smolensk Region | −0.555 | −0.249 | 3 | 0.47 |
| Karelia | −1.069 | 1.094 | 2 | 0.14 | Tambov Region | −0.468 | −0.408 | 3 | 0.40 |
| Vologda Region | −0.347 | 0.681 | 2 | 0.45 | Republic of Adygeya | −0.894 | −0.148 | 3 | 0.23 |
| Kaliningrad reg | −0.377 | 0.266 | 2 | 0.00 | Republic of Kalmykia | −0.931 | −1.167 | 3 | −0.82 |
| Republic of Khakassia | −1.082 | 0.246 | 2 | −0.72 | Astrakhan Region | −0.514 | −0.143 | 3 | 0.62 |
| Tomsk region | −0.722 | 0.975 | 2 | 0.37 | Primorsky Krai | 0.159 | 1.714 | 4 | 0.09 |
| Belgorod region | 0.080 | 0.520 | 2 | 0.71 | Khabarovsk Terr | −0.624 | 2.738 | 4 | 0.33 |
| Kaluga region | −0.629 | 1.122 | 2 | 0.61 | Amur Region | −0.704 | 1.938 | 4 | −0.55 |
| Kursk region | −0.344 | 0.124 | 2 | −0.11 | Komi Republic | −1.036 | 2.504 | 4 | −0.31 |

(continued)

**Table 5** (continued)

| Region | $F_1$ | $F_2$ | K | R | Region | $F_1$ | $F_2$ | K | R |
|---|---|---|---|---|---|---|---|---|---|
| Lipetsk region | −0.384 | 0.620 | 2 | 0.35 | Arkhangelsk Reg | −0.685 | 2.314 | 4 | −0.15 |
| Ryazan Oblast | −0.518 | 0.006 | 2 | −0.40 | Leningrad Region | 0.575 | 1.793 | 4 | 0.59 |
| Tver region | −0.352 | −0.082 | 2 | −0.32 | Perm Krai | 0.846 | 0.514 | 5 | −0.21 |
| Tula region | −0.101 | 0.458 | 2 | 0.47 | Nizhny Nov. Reg | 1.619 | 0.390 | 5 | 0.44 |
| Yaroslavl region | −0.356 | 0.139 | 2 | −0.10 | Samara Region | 1.360 | 0.108 | 5 | −0.11 |
| Sevastopol city | −1.075 | 0.400 | 2 | −0.56 | Krasnoyarsk Krai | 1.140 | 1.773 | 5 | 1.34 |
| Mari El Republic | −0.759 | −0.837 | 3 | −0.32 | Irkutsk Region | 0.830 | 0.777 | 5 | 0.03 |
| Mordovia | −0.579 | −1.083 | 3 | −0.39 | Kemerovo Region | 0.955 | 0.292 | 5 | −0.33 |
| Republic of Udmurti | −0.226 | −0.238 | 3 | 0.81 | Novosibirsk Region | 0.987 | 0.118 | 5 | −0.47 |
| Chuvash Republic | −0.374 | −1.093 | 3 | −0.19 | Voronezh Region | 0.621 | 0.257 | 5 | −0.70 |
| Kirov region | −0.335 | −0.788 | 3 | 0.15 | Republic of Bashkortostan | 1.886 | 0.025 | 6 | −0.92 |
| Penza region | −0.315 | −0.717 | 3 | 0.24 | Republic of Tatarstan | 2.494 | 1.128 | 6 | 0.79 |
| Ulyanovsk Region | −0.389 | −0.611 | 3 | 0.28 | Sverdlovsk Reg | 2.388 | 1.238 | 6 | 0.80 |
| Novgorod Region | −0.870 | −0.178 | 3 | 0.23 | Krasnodar Krai | 3.442 | −0.119 | 6 | 0.49 |
| Pskov Region | −0.720 | −0.494 | 3 | 0.06 | Rostov Region | 2.170 | −0.505 | 6 | −1.16 |

*Source* Compiled by the authors

**Table 6** Statistical characteristics of clusters $D_1 - D_6$

| К | $\overline{F}_1$ | $S_1$ | $\overline{F}_2$ | $S_2$ | К | $\overline{F}_1$ | $S_1$ | $\overline{F}_2$ | $S_2$ |
|---|---|---|---|---|---|---|---|---|---|
| 1 | 0.7060 | 0.5290 | −0.9369 | 0.5571 | 4 | −0.3859 | 0.6146 | 2.1669 | 0.4144 |
| 2 | −0.5873 | 0.3923 | 0.4724 | 0.3758 | 5 | 1.0447 | 0.3193 | 0.5288 | 0.5483 |
| 3 | −0.5961 | 0.2583 | −0.6792 | 0.4442 | 6 | 2.4761 | 0.5881 | 0.3533 | 0.7830 |

*Source* Compiled by the authors

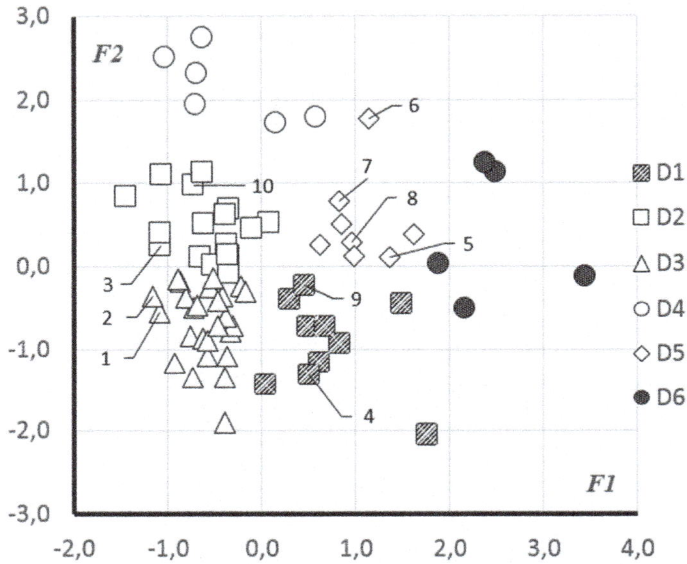


**Fig. 2** The location of cluster $C_2$ regions of the Russian Federation. *Source* Compiled by the authors

## 4 Discussion

The position of the region in the factor space ($F_1$, $F_2$) corresponds to the level of its socio-economic development. Positive values of the factors $F_1$, $F_2$ correspond to the high, and negative—to the low level of economic development of the region or to the level of income of the population, respectively. Clusters $D_1$ and $D_3$ include regions with low $D_2$, $D_5$; $D_6$ average; and $D_4$ high income of the population. At the same time, clusters $D_2$, $D_3$, $D_4$, are characterized by low, $D_1$, $D_5$, average, and $D_6$ high overall levels of economic development. The regions of the Siberian Federal District in 2019 were distributed in different clusters. Therefore, the Altai Republic and the Tuva Republic fell into cluster $D_3$ with low levels of economic development and income and have negative intra-cluster ratings. The Republic of Khakassia is in the middle-income cluster $D_2$, but has a low economic development rating of 0.72. Tomsk Region with the rating of 0.37 fell into the same $D_2$ cluster. Altai Territory and Omsk Region fell into cluster $D_1$—with low income levels, average level of

**Table 7** Regions with the highest and lowest ratings of socio-economic development

| Region | R | Region | R |
|---|---|---|---|
| Sverdlovsk region | 3.63 | Republic of Mordovia | −1.66 |
| Republic of Tatarstan | 3.62 | Kabardino-Balkaria Republic | −1.76 |
| Krasnodar Krai | 3.32 | Karachay-Cherkess Republic | −2.09 |
| Krasnoyarsk Krai | 2.91 | Republic of Kalmykia | −2.10 |
| Leningrad region | 2.37 | Republic of Ingushetia | −2.30 |

*Source* Compiled by the authors

economic development, and with −0.58 and 0.45 ratings, respectively. Krasnoyarsk Territory, Irkutsk Region, Kemerovo Region, and Novosibirsk Region are located in cluster $D_5$ with an average level of economic development and average income level of the population and with ratings of 1.34, 0.03, −0.33, −0.47.

Along with intra-cluster ratings, this approach allows to determine the absolute rating of the region as the sum of both factors. Table 7 shows the five regions with the highest and lowest values of the universal rating of socio-economic development rating. Altai Krai takes the 49th place out of 72.

## 5 Conclusion

The paper provides the analysis of the level of socio-economic development of the subjects of Russia for 2019 using multivariate data analysis. The technique used combines factor and cluster analyses. We revealed that from the entire array of indicators of socio-economic development, two statistically independent factors can be formed: (1) the factor of the overall economic development and (2) the factor of the income level of the region. We proposed a unique method for determining the rating of regional socio-economic development based on the results of factor analysis. The rating is defined as the sum of shifts in the factor space. As a result, the analysis shows that in 2019, from all subjects of Russia, we can identify ten special regions with a level of socio-economic development that significantly differs from that of the main part of Russian regions. The remaining 72 regions are divided into six fairly dense clusters consisting of regions that are close in terms of their socio-economic development. The regions of the Siberian Federal District are highly differentiated by the level of socio-economic development and are included into different clusters with diverse levels of economic development and income. Besides, we calculated the ratings of socio-economic development of the regions and determined that Altai Krai takes the 20th place in the economic development rating, the 67th place in the income rating, and the 49th place out of 72 basic regions of Russia in the overall ranking. The proposed method can be applied to analyze the dynamics of the socio-economic development of regions. To do this, the data arrays for the analyzed periods should have a single structure and be reduced to the prices of the base period. Next, one should apply the standardization procedure to the entire data set and determine the factors. In this generated factor space, one can calculate the change in the ratings of regions. After performing the clustering for certain periods, one can identify the positions, properties, and structure of the clusters depending on the changes in time. This is the task of further research.

## References

Federal State Statistic Service (2020) Information for monitoring the socio-economic status of the constituent entities of the Russian Federation. https://www.gks.ru/storage/mediabank/info-stat-02-2020.rar

Yudintsev AY, Troshkina GN (2018a) The study of differences in socio-economic development of constituent entities of the Russian Federation by methods of multivariate data analysis for 2017. Manage Econ Syst: Sci Electron J 6(112). http://uecs.ru/instrumentalnii-metody-ekonomiki/item/4958--2017

Yudintsev AY, Troshkina GN (2018b) The study of economic security of constituent entities of the Russian Federation by methods of multivariate data analysis for 2016. Manage Econ Syst: Sci Electron J 9(115). http://uecs.ru/instrumentalnii-metody-ekonomiki/item/5096--2016-

# Public Catering Market: Regional Features and Key Growth Drivers

**Iuliia V. Ugarova, Elena N. Bolkhovitina, and Alla Yu. Gorbunova**

**Abstract** Between 2010 and 2018, Russian public catering market has overcome several falls and rises of the market due to certain trends and special features. The paper aims to define the key factors that constrain the growth of public catering regarding regional features and specifying feasible growth drivers for the business sector. The paper compares regional features with national and world trends in catering by employing analytical reviews and statistics from national and regional statistical offices. The formal questionnaire for Altai Krai residents helped to define regional features. The research employed analysis techniques, comparison, statistical techniques, expert evaluation approach, etc. The article presents the findings which reflect regional features and show the most attractive public catering formats for the residents: cafes, bars, and fast food restaurants. The regional consumers show low attendance records for public catering places and have a low average bill. The paper also reveals a conflict between the consumer expectations and real components of quality of service in public catering within the region. In conclusion, the article states major drivers for public catering in the region. Soon one of them will be the format of fast food restaurants. To develop other formats, the region will need to shape some eating choices. The crisis of 2020 has finally shaped another feasible driver in the sector—"*RTE Meals delivery.*"

**Keywords** Public catering factors · Growth driver · Trends · Place format · Consumer preference · Consumer behavior

---

I. V. Ugarova (✉) · E. N. Bolkhovitina
Polzunov Altai State Technical University, Barnaul, Russia
e-mail: julie1207@mail.ru

E. N. Bolkhovitina
e-mail: elenaobraz@mail.ru

A. Yu. Gorbunova
Altai State University, Barnaul, Russia
e-mail: gorbunova-alla@mail.ru

 
S. G. Maximova et al. (eds.), *Advances in Natural, Human-Made, and Coupled Human-Natural Systems Research*, Lecture Notes in Networks and Systems 234,
https://doi.org/10.1007/978-3-030-75483-9_15

# 1 Introduction

Russian public catering market suffered in the last decade. The public catering turnover in current prices shows slightly positive figures (Tables 1 and 2). Compared to 2014–2016, it shows recession both country and region level. We attribute it to the unfavorable foreign policy processes (the food embargo, national currency volatility), contributing substantially to economic results in various sectors, including public catering.

Russian public catering market started to grow slightly since 2017 as it recovered from the recession. However, in 2018, while the all-Russia turnover in public catering had grown, the regional turnover plummeted. Similar trends were in public catering turnover per capita (Table 3). The per capita spending for eating out differs significantly from average Russian rates. Thus, it is essential to define regional features and factors that impact public catering enterprises and determine the key growth drivers.

The public catering market presents the following trends:

- Active transition of the public catering market into more economical formats (so-called market democratization), which has led to a higher demand for the fast food sector, take away, vending and RTE Meals (Ready-To-Eat Meals). The motto of the modern consumer: "*Fast, cheap, tasty*";
- Public catering market is marked by widely spread franchising. This business model has proved highly stable within the crisis. Chain enterprises provide for good performance in non-stable economies owing to their business scale. They may press the suppliers and gain preferences due to bulk deliveries. According to

**Table 1** Dynamics of public catering turnover on the federal level in Russia

| Year | Public catering turnover[a], millions, RUB | | | Inflation rate[b], % | Discount coefficient |
|---|---|---|---|---|---|
| | The Russian Federation (RF) | Siberian Federal District (SFD) | Altai Krai | | |
| 1 | 2 | 3 | 4 | 5 | 6 |
| 2010 | 781,379 | 68,936 | 5,025 | 8.78 | 1.00 |
| 2011 | 903,570 | 80,761 | 6,011 | 6.10 | 1.06 |
| 2012 | 1,019,617 | 94,251 | 6,602 | 6.58 | 1.07 |
| 2013 | 1,131,535 | 104,720 | 7,325 | 6.45 | 1.06 |
| 2014 | 1,234,105 | 113,377 | 8,045 | 11.36 | 1.11 |
| 2015 | 1,308,127 | 118,781 | 8,694 | 12.91 | 1.13 |
| 2016 | 1,348,687 | 128,862 | 9,072 | 5.38 | 1.05 |
| 2017 | 1,434,589 | 139,738 | 9,489 | 2.52 | 1.03 |
| 2018 | 1,527,747 | 134,212 | 9,856 | 4.27 | 1.04 |

[a]*Source* (Federal Service of State Statistics, n.d.)
[b]*Source* (Bureau of Statistics, n.d.)
[c]*Source* Compiled by the authors

**Table 2** Dynamics of public catering turnover on the regional level in Russia

| Year | Public catering turnover in 2010 prices, millions, RUB | | | Chain growth rates for public catering turnover in 2010 prices, % | | |
|---|---|---|---|---|---|---|
| | RF | SFD | Altai Krai | RF | SFD | Altai Krai |
| 1 | 7 | 8 | 9 | 10 | 11 | 12 |
| 2010 | 781,379 | 68,936 | 5,025 | – | – | – |
| 2011 | 851,621 | 76,118 | 5,666 | 8.99 | 10.42 | 12.76 |
| 2012 | 901,666 | 83,347 | 5,839 | 5.88 | 9.50 | 3.05 |
| 2013 | 940,008 | 86,995 | 6,085 | 4.25 | 4.38 | 4.21 |
| 2014 | 920,632 | 84,578 | 6,002 | −2.06 | −2.78 | −1.36 |
| 2015 | 864,274 | 78,478 | 5,744 | −6.12 | −7.21 | −4.30 |
| 2016 | 845,579 | 80,792 | 5,688 | −2.16 | 2.95 | −0.97 |
| 2017 | 877,329 | 85,457 | 5,803 | 3.75 | 5.77 | 2.02 |
| 2018 | 896,039 | 78,717 | 5,780 | 2.13 | −7.89 | −0.40 |

*Source* Compiled by the authors

**Table 3** Dynamics of public catering turnover per capita[a]

| Year | Public catering turnover per capita, RUB | | | Public catering turnover per capita in 2010 prices, RUB | | | Chain growth rates for public catering turnover per capita in 2010 prices, % | | |
|---|---|---|---|---|---|---|---|---|---|
| | RF | SFD | Altai Krai | RF | SFD | Altai Krai | RF | SFD | Altai Krai |
| 2010 | 5,470 | 3,577 | 2,073 | 5,470 | 3,577 | 2,073 | – | – | – |
| 2011 | 6,320 | 4,194 | 2,492 | 5,957 | 3,953 | 2,349 | 8.90 | 10.51 | 13.31 |
| 2012 | 7,120 | 4,891 | 2,748 | 6,296 | 4,325 | 2,430 | 5.69 | 9.41 | 3.45 |
| 2013 | 7,885 | 5,430 | 3,059 | 6,550 | 4,511 | 2,541 | 4.03 | 4.30 | 4.57 |
| 2014 | 8,448 | 5,874 | 3,369 | 6,302 | 4,382 | 2,513 | −3.79 | −2.86 | −1.10 |
| 2015 | 8,935 | 6,149 | 3,652 | 5,903 | 4,063 | 2,413 | −6.33 | −7.28 | −3.98 |
| 2016 | 9,195 | 6,668 | 3,826 | 5,765 | 4,181 | 2,399 | −2.34 | 2.90 | −0.58 |
| 2017 | 9,770 | 7,238 | 4,024 | 5,975 | 4,426 | 2,461 | 3.64 | 5.86 | 2.58 |
| 2018 | 10,405 | 7,802 | 4,209 | 6,103 | 4,576 | 2,469 | 2.14 | 3.39 | 0.33 |

[a]*Source* (Federal Service of State Statistics, n.d.)

experts, the number of closed non-chain enterprises in Russia has exceeded the number of new ones, while chain companies manage to develop and grow;

- Considering the consumer behavior model in the public catering market has become "*economical*" and the claims on service quality are also on the rise, these factors considerably complicate the operation of public catering companies. Thus, the consumer both saves money and wants to spend it efficiently, analyzing his spending;

- Modern consumer is more knowledgeable in nutrition science, and consumer rights protection; they have a great choice and therefore; they are picky and fastidious as far as the range; quality and service go in public catering enterprises. It means the companies have to get more customer-oriented and personalize their services;
- Such trends as "*healthy lifestyle*", vegetarianism and rawism contribute to changes in the public catering menu to the advantage of low-calorie, vegetarian or vegan dishes, using gluten-free, hypo-allergic, eco- and organic products in food production. The consumers are curious about the ingredients, methods of cooking, and caloric value of dishes. Thus, public catering enterprises have to follow this trend: they provide detailed information on the menu, at their websites, mobile apps and train the staff (receptionists, waiters, barmen) to communicate with consumers;
- Public catering enterprises show a great activity increment in the social media. According to the latest research, 39% of Russians browse restaurant sites before visiting. After eating out, about 16% of the respondents share their positive experience in the social media and 12% share the negative one as well (Korchagin 2018). This helps other users to shape their opinion of the place and the restaurant owners get the feedback;
- International research has shown that "eating out" spending in Russia is one of the lowest among developed countries. Between 2010 and 2018, it accounted for 3.1–3.4% of total household spending (Consumer spending of households, 2010–2018);
- New restaurant formats have appeared. They work only for deliveries which supports the trend for Russians to "*eat in*", not "*eat out*".

The aforementioned trends are typical for foreign public catering markets too as foreign experts state in their works (Moller et al. 2019; Pyanikova et al. 2019; Ivkov et al. 2016, 2014).

## 2 Materials and Approaches

We analyzed the statistical data from the Federal Service of State Statistics (Federal Service of State Statistics, n.d.). We also defined boundary conditions to reveal the trends in public catering. The paper identified the development of constraining factors and growth drivers as well. We used analysis, synthesis, comparison, statistical techniques, modeling, expert evaluation approach, description, etc., to address general scientific techniques.

Besides, consumer preferences were evaluated within the public catering market in Altai Krai in July 2018 as a survey based on formal questionnaires. We received 794 questionnaires, and 598 of them (75.31%) were appropriate for analysis. The total sampling size to evaluate consumer preferences within the public catering market in Altai Krai accounted for 598 respondents. The interrogated groups were the following: 49,83% represented the 20–25 age group; 20.74%—aged 26–35; 5.02%—aged 36–45; 24.75% stood for those aged 18–19 and over 46.

## 3 Findings

1. All public catering formats are presented in Altai Krai.
2. Cafes account for 36% of the market structure, bars for 19%, and chain companies for 18% correspondingly. Federal public catering chains prevail among chain companies.
3. Most public catering enterprises do not specialize in a particular cuisine. Also, some enterprises advertise Russian and European cuisines.
4. The average bill claims to be 323 RUB. 54% of enterprises charge between 100 and 180 RUB.
5. Five hundred ninety-eight respondents got involved in evaluating consumer preferences in public catering services. The majority were females aged 20–25.
6. The respondents of various age groups provided a mostly identical opinion, which means the age factor does not directly impact consumer preferences.
7. According to the survey, the average income per capita accounted for 17,521 RUB.
8. The survey defined that "*proximity to the place of work/study/living*" is not regarded as the key choice factor.
9. The respondents in all age groups showed shallow visit frequency for restaurants, bars, clubs, kiosks, and street-food kiosks.
10. The respondents aged between 20 and 25 mostly visit cafes, bars, fast-food restaurants, and cafeterias. With age increment, the preferences shift towards cafeterias and deli stores. The respondents spend between 120 and 350 RUB in popular formats of public catering. It corresponds to the average bill of public catering companies given in the 2GIS system.
11. The respondents in all age groups prefer Russian and European cuisines. With age increment, the preferences shift from Japanese towards Caucasian cuisine.
12. The primary reasons to visit public catering places are staying hungry and meeting friends/colleagues/family. Only an insignificant number of respondents aim to listen to music and dance.
13. The essential criteria to choose a public catering place are the following: cuisine (the quality of dishes), sanitary conditions, quality of service, and the price. The respondents do not care for the entertainment in a public catering place, the variety of wine (cocktail) list, bonus or loyalty programs, how popular the place is, where it is located, and the background music during the visit. It is noteworthy that these criteria are named as negative ones when evaluating the service quality in public catering. There is a clash between consumer expectations and real service quality components in public catering in Altai Krai.
14. The visit frequency for public catering places among men is higher than among women. The latter prefer to eat in. Females prefer cafeteria and deli stores among

public catering places, more rarely cafes, bars, and fast-food restaurants. The following public catering formats are more popular with men: cafeteria, bars, and fast food restaurants. Men prefer Russian and European cuisines, while women opt for Russian and Japanese cuisines. Males spend considerably more than females in clubs, whereas females spend more at food courts, cafes, bars, or fast food restaurants. The respondents of both genders have similar reasons to visit public catering places. However, women would listen to music and dance in public catering places eagerly than men. Women are less satisfied than men with the price-quality ratio in public catering places. Therefore, there is a gender impact on consumer behavior in public catering although it does not look essential.

15. According to the income level, the consumers show the following differences: growing income entails more frequent visits to public catering places (although this growth is insignificant), spending increases. This group with income between 20,600 and 25,500 RUB mostly cares about their spending since they reduce spending on eating out. In all groups ranging between 10,000 and 45,500 RUB, the consumers opt for Russian and Japanese cuisines, followed by European cuisine, whereas the reference group with the income of 45,600 RUB opts for Russian, European, and Italian cuisines. In general, it correlates with age groups, as the two groups according to the income (10,000–15,500 RUB and 15,600–20,500 RUB) consist of mostly young people. The respondents of higher income reside mostly in the regional center, Barnaul. Like the gender factor, the income level only slightly impacts consumer behavior.
16. Due to the research findings, the consumer portrait for public catering services looks as following:

    - A male aged 20–35, with an income is over 17,800 RUB who prefers Russian and European cuisines.
    - A female aged 20–25, with an income is over 16,000 RUB who prefers Russian, European, and Japanese cuisines.
    - Females are less active visitors of public catering places, and after they get married, the frequency even lowers.

17. Some behavioral differences in the groups varied in gender, income level, and age, although their influence is insignificant when defining the consumer behavior pattern for public catering companies. General trends remain unchanged with slight alterations only:

    - Low visit frequency of restaurants, clubs, and kiosks, pavilions, or street-food kiosks.
    - They opt for cafes, bars, fast-food restaurants, cafeterias, and deli stores.
    - Average spending for these services range between 120 and 350 RUB.
    - The main reason to visit public catering places is to stay hungry and meet friends.

- The choice of the place depends on the following parameters: sanitary state, dish quality, price, and quality of service. The respondents are particular about these parameters.

## 4 Discussion

The analysis has revealed the factors constraining the development of public catering to some extent.

Firstly, there are barriers to entering the market. This factor means low investment appeal and many controlling agencies (Ugarova et al. 2019).

Secondly, it is the feedstock factor. Due to inflation and national currency volatility, the prices for the feedstock are rising. The feedstock range is limited, and it may be of low quality or counterfeit. It is hard to find suppliers because of contrasanctions.

The production technologies are the third constraining factor due to costly production and the slow introduction of advanced technologies.

The fourth constraint is the market. The Russians have an inarticulate demand for *"eating out."* Consumer activity decreases due to rising prices for public catering services, the drop in real income level, the decrease in purchasing power, and the reduced number of impulse purchases.

Profitability lowers since the expenditures for the feedstock rise. The state has started regulating alcohol sales, property taxes, and rental rates due to the increased cadastral value of properties that host public catering companies. The structure of public catering shifts towards low-price segments (fast food, take away, and delivery). Vending and RTE Meals are developing, and the demand in these segments is on the rise. Foreign chain companies strengthen their expansion. Consumers are getting more demanding and picky regarding the range, quality, price, and service go. Public catering companies acquire another value (as places to confirm social status, meeting, and short-term distance working).

Relatively low salaries form the fifth constraint and, as a result, high employee turnover. Moreover, the sector depends a lot on the personnel qualification. In the regional market, we lack highly-qualified specialists.

All these factors together indicate the strong impact of negative factors on sector development in the region.

## 5 Conclusion

The research has shown several growth drivers based on the revealed regional features and the findings of comparative analysis for the public catering market in Altai Krai with national and global trends.

Firstly, in the nearest future, the sector driver will appear to be fast food restaurants since they support consumer preferences in the region: takeaways and reasonable prices. This trend will probably last for a long time as the purchasing power decreases.

It is essential for the public catering market that other formats develop as well. However, this is possible only when introducing another growth driver: shaping some eating habits in the region.

The 2020 crisis has completed another driver in the RTE meals delivery. Nowadays, we may state that should the public catering company fail to operate a delivery. It loses the competition, its income, and its stable market position.

## References


Bureau of Statistics (n.d.) The rate of inflation for years. https://www.statbureau.org/ru/russia/inflation-charts-yearly

Federal Service of State Statistics (n.d.) Consumer spending of households. https://www.fedstat.ru/indicator/43228

Federal Service of State Statistics (n.d.) Public catering turnover per capita. https://www.fedstat.ru/indicator/43280

Federal Service of State Statistics (n.d.) Public catering turnover. https://www.fedstat.ru/indicator/31258

Federal Service of State Statistics (n.d.) The turnover of public catering. https://www.fedstat.ru/indicator/31258

Ivkov M, Blesic I, Simat K, Demirović D, Bozic S, Stefanovic V (2016) Innovations in the restaurant industry: an exploratory study. Econ Agric 4(63):1169–1186. https://doi.org/10.5937/ekoPolj1604169I

Ivkov M, Blesic I, Stefanović V, Raljić J (2014) Managing customer satisfaction in the restaurant industry: a word from managers. Econ Themes 52(3):369–378. https://doi.org/10.1515/ethemes-2014-0023

Korchagin YA (2018) Public catering. https://center-yf.ru/data/ip/obshchestvennoe-pitanie-2018.php

Moller B, Voglhuber-Slavinsky A, Dönitz E, Rosa A (2019) 50 trends influencing Europe's food sector by 2035. Fraunhofer Institute for Systems and Innovation Research, Karlsruhe, Germany

Pyanikova EA, Kovaleva AE, Bykovskaya EI (2019) Digital technologies in the public catering industry (regional aspect). Adv Econ, Bus Manage Res 81:482–488. https://doi.org/10.2991/mtde-19.2019.96

Ugarova I, Bolkhovitina EN, Davydenko NI (2019) The development of the regional catering market: The case of the Altai region. Adv Soc Sci, Educ Human Res 364:165–169. https://doi.org/10.2991/icsdcbr-19.2019.35

# Assessment Toolkit for the Economic Integration of Regions

**Tatyana A. Rudakova, Oksana Yu. Rudakova, and Inna N. Sannikova**

**Abstract** This paper discusses the development of tools for assessing the economic integration of regions. The relevance of the study is determined by the complexity of the economic relations of economic entities at various levels of integration processes caused by interpenetration and, in some cases, the merger of economies. These processes are accompanied by the emergence of new non-standard facts of economic activity concerning participants in this process. This causes several difficulties both at the stage of fixing an event, a business transaction, and in the process of generalizing and evaluating the necessary information. According to the authors, the tools for assessing the result of inter-regional interaction should correspond to the purpose of its research and contribute to obtaining a detailed answer to the posed questions. The composition of the basic indicators for assessing the economic development of the region and the result of integration processes should include a limited range of indicators. Absolute indicators should reflect the presence of inter-regional projects and their results in dynamics by stages of implementation, relative—the quality of life of the population. The assessment of inter-regional interaction should be carried out in stages. The assessment tools should correspond to the level at which it becomes necessary to analyze integration processes from the particular to the general: inter-corporate, inter-regional, and inter-country.

**Keywords** Economic integration · Inter-regional cooperation · Region · Assessment toolkit · Levels of integration

## 1 Introduction

Modern economic literature does not give an unambiguous answer to what should be considered "integration": a process, its results, or both.

T. A. Rudakova (✉) · O. Yu. Rudakova · I. N. Sannikova
Altai State University, Barnaul, Russia
e-mail: rta_62@mail.ru



S. G. Maximova et al. (eds.), *Advances in Natural, Human-Made, and Coupled Human-Natural Systems Research*, Lecture Notes in Networks and Systems 234,
https://doi.org/10.1007/978-3-030-75483-9_16

The first mention of the term is contained in the print media of 1620; the semantic content was defined as the combination of separate parts into a single whole. Integration, as a phenomenon, is usually considered in the context of international cooperation. David Greenaway, Robert S. Hain, and Chris Milner argue that the merging of separate economies can be viewed from a position of the state of affairs, while the removal of barriers to international cooperation is considered a process (Greenaway et al. 1995). The opinion of David Henderson boils down to the fact that, along with the process, integration can mean *the result* if all restrictions and obstacles (including political ones) are overcome (Henderson 1992). These statements about the essence of integration can be extrapolated to the inter-regional level since the issue of removing obstacles to the cooperation of separate territorial entities is highly relevant.

This article is devoted to the integration processes of territorial entities of a federal state—the Russian Federation, which, in turn, is a full member of an international integration unit—Eurasian Economic Union. Only coordinated policies of territorial entities in a single federal state in the field of economics and other areas of society can contribute to effective integration processes at the inter-country level.

## 2 Materials and Methods

The use of tools for quantitative and qualitative assessment of inter-regional cooperation is necessary for analysis and research of the results of the process of merging the economies of separate regions of the federal state. This raises the question, "What is considered the result of inter-regional interaction and how to evaluate it?" The answer to the question must be formulated from the perspective of the goal of integration processes for the economy of separate territorial entities, i.e., the desired future progressive changes. As one should know, the concretization of goals through tasks allows for its phased achievement. If the global goal is to designate a reduction in economic development differences, then the priority tasks should include ensuring the regions with equal starting opportunities. Understanding the low likelihood of such a provision, we concretize the task and formulate it "as achieving positive dynamics of indicators that describe the state of the regional economy, social area, and living standards." The complexity of identifying the factors that influence the indicators of interest to the researcher of the problem forms the presence of a significant number of scientific approaches to assessing the results of integration processes. Theoretical and empirical studies of the issues of evaluating the results of integration processes are more focused on the analysis of the results and potential of inter-regional cooperation, the participants of which are separate inter-regional groups and associations that include separate countries or foreign territorial entities within a single state (Azmina 2013; DeLombaerde et al. 2008; DeLombaerde et al. 2011; Lomovtseva et al. 2014; Polikarpova 2011). The extrapolation and use of existing methods for assessing the results of inter-regional cooperation of the regions of the Russian Federation, in the opinion of the authors of this article, is unjustified due to differences in the areas of administration and the goals of these processes, territorial features, resource support and the level of socio-economic development of territorial entities (i.e., participants of inter-regional cooperation).

It should be noted that participants in inter-regional interactions can be both donor regions and recipient regions (i.e., regions that are self-sufficient in their development, whose economy does not require additional financial injections). We can assume that the latter are less interested in inter-regional events. As it seems, the cooperation of the regions in various fields should be in mutually beneficial conditions. Balancing the degrees of economic development of territorial entities is a difficult task since the positive results of inter-regional cooperation can lead to additional growth in the economic performance only of donor regions. In this case, the recipient regions could be disadvantaged.

The authors believe that the main goal of the inter-regional interaction should be socio-economic development. Therefore, one must also solve the problem of its assessment by determining the necessary tools.

## 3 Results

The tools for assessing the result of an interaction should correspond to the purpose of the study and contribute to obtaining a comprehensive answer to the questions posed. What set of indicators will answer questions about the effectiveness of inter-regional interaction? Participants in the economic space with current economic ties with resident organizations or representatives of separate territories already participate in inter-regional cooperation.

Obtaining reliable information on the volumes and results of the interaction can provide a statistical synthesis of data in the context of regional and inter-regional interactions (Fig. 1).

Modern concepts of the regional economy (based on the views of foreign scholars) consider a region as a system of complex nature and structure. It generates new paradigms (the region is a subject of federal education, a corporation, a market, a social community), various scientific theories, and emerging contradictions. Those are based on: (1) the competition at the macro, meso, and micro levels under the influence of ongoing processes of international, inter-regional, and inter-corporate integration, and (2) on mutually beneficial cooperation partners—participants of inter-regional cooperation. As an integral phenomenon of each paradigm, the contradiction is present at all levels of regional cooperation and in the process of inter-country integration, regional integration, and corporate integration.

Rivalry in seeking markets for sales of own products and services is present at all stages of inter-regional interaction, starting with the search for a partner and ending with gaining the trust of buyers of the created values. Note that all this is necessary only to obtain competitive advantages in the future.

If we consider the region is a corporation with the sole aim to motivate the expansion of businesses and involve more entities in the real sector of the economy, then we

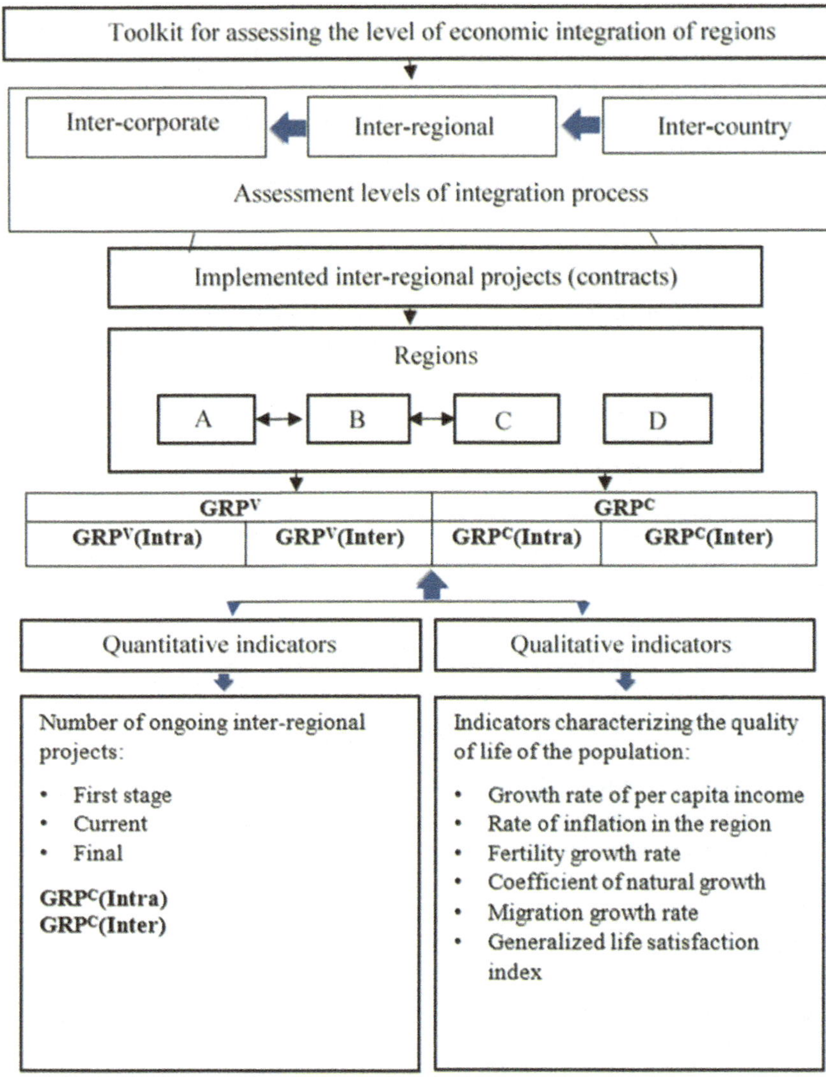


**Fig. 1** Toolkit for assessing the level of economic integration of regions. *Source* Compiled by the authors where: **$GRP^V$(Intra)**—gross regional product in the framework of regional cooperation; **$GRP^V$(Inter)**—gross regional product in the framework of inter-regional cooperation. **$GRP^C$(Intra)**—gross regional product per capita in the framework of regional cooperation. **$GRP^C$(Inter)**—gross regional product per capita in the framework of inter-regional cooperation

face a contradiction regarding the interpretation of the region as a social community that prefers life in a prosperous region from the point of economics and ecology. The presence of a large number of industrial facilities does not contribute to maintaining a balance of interests of investors and the population.

Integration processes, first and foremost, should be considered as a result of the globalization of the economy. Under certain circumstances, globalization can be an obstacle to integration development since the basis of mutually beneficial cooperation of economic entities is the goal of obtaining competitive advantages from the process in the medium and long term. It is worth agreeing with the point of view on the need for a comprehensive study of the economic phenomenon of "inter-regional competition," which changes the functions of the region and makes its role significant in the national production process.

## 4 Discussion

The complexity of the emerging economic relations of economic entities at various levels of integration processes is caused by the interpenetration or merger of economies. It is accompanied by the emergence of new non-standard economic activity facts, difficulties both at the stage of fixation of the event or economic activity, and the process of generalization and evaluation of necessary information. In this situation, one should recall the capabilities of the national accounting system, one of the tasks of which is collecting and summarizing information for various user groups. Automation of the accounting processes in modern conditions allows solving the problem of collecting the database necessary to assess the level of regional economic integration via adjusting the functions, entering additional classification parameters, and summarizing information on the activities of business entities without violating the basic accounting principles.

In our opinion, a limited range of indicators can be included in the basic indicators for assessing the economic development of the region and the result of integration processes (Fig. 1). Absolute indicators point to the presence of inter-regional projects and the result of their implementation that must be analyzed in dynamics. Relative indicators characterize the quality of life of the population: (1) the growth rate of per capita income in comparison with the inflation rate in the region; (2) birth rate; (3) coefficient of natural growth; (4) migration growth rate. Positive or negative dynamics of indicators should be a signal for action for the leadership of the region. The slowdown in the growth of the regional economy harms the quality of life, which is necessarily followed by the outflow of the population to more prosperous regions, a slowdown in the birth rate, and increasing social tension. The favorable consequences of the ongoing positive changes in the economy of the territorial entity will be reflected in the growth of the generalized index of life satisfaction.

## 5 Conclusion

It is much easier to evaluate the results of a project, the planning, and implementation of which was carried out with the participation of several regions. In this case, the assessment toolkit can be a set of indicators that determine the achievement of the stated results in both quantitative and qualitative aspects. Quantitative indicators should reflect certain results in a particular area, the measurement of which is possible in kind, and the growth rates of gross regional product, including per capita.

The composition of qualitative indicators can be expanded depending on the use and the intended recipient of the indicators: (1) assessment of the socio-economic development of the region, (2) assessment of the effectiveness of organizations—business partners, (3) assessment of the quality of life of the regional population, (4) assessment of participation regional leadership in the results of its socio-economic development, (5) assessment of the full involvement of the existing resource potential of the region, (5) the use of infrastructure, etc. Indicators characterizing the involvement of business entities in inter-regional cooperation include:

- Number of inter-regional projects in which the organization participates;
- Number of partner organizations within the region;
- Number of partner organizations outside the region;
- Number of projects at the inter-regional level, in which the organization takes part;
- Number of employees participating in the organization of inter-regional projects;
- Volumes of production capacities involved (number of equipment, $m^2$ of area, land, etc.);
- Volume of products manufactured, work performed, and services rendered for direct consumption, use or application at the market and comparable prices;
- Volume of products manufactured, work performed and provided for direct consumption, use or application as a result of inter-regional cooperation in the market and comparable prices;
- Additional financial resources attracted for the implementation of the project at the inter-regional level.

Statistical data obtained at the level of a separate business entity should be generalized at the regional level. It can be supplemented with information on the infrastructure and administrative resources used:

- Institutional units involved in the implementation of inter-regional projects;
- Institutional units created for the implementation of inter-regional projects;
- Amount of sources of financing inter-regional projects: federal budget funds;
- Financial resource levels
- Amount of sources of financing inter-regional projects: regional budget funds;
- Amount of sources of financing inter-regional projects: funds of organizations.

This means that the assessment of inter-regional interaction should be carried out in stages. Therefore, the assessment tools should correspond to the level at which it becomes necessary to analyze integration processes from particular to general: inter-corporate, inter-regional and inter-country.

**Acknowledgements** The study was carried out with the financial support of the Russian Federal Property Fund in the framework of the scientific project No. 18-010-00135.

## References

Azmina YM (2013) Theoretical and methodological aspects of the study of spatial integration. Bull Volgograd State Univ: Series 3, Econ. Ecol 2(23):80–84

De Lombaerde P, Dorrucci E, Genna GN, Mongelli FP (Eds) (2011) Composite indexes and systems of indicators of regional integration. In: The regional integration manual: quantitative and qualitative methods. Routledge, London, UK; New York, NY, pp. 323–343

De Lombaerde P, Pietrangeli G, Weeratunge C (2008) Systems of indicators for monitoring regional integration processes: where do we stand? The Integr Assess J 8(2):39–67

Greenaway D, Hine RC, Milner C (1995) Vertical and horizontal intra-industry trade: a cross industry analysis for the United Kingdom. Econ J 105(433):1505–1518

Henderson D (1992) International economic integration: progress, prospects and implications. Int Aff 4(68):634

Lomovtseva OA, Shkromada VI, Lebedev AO (2014) Methodological aspects of assessing the integration capabilities of the region. Sci Bull 21(192):12–21

Palmer R (2001) Historical patterns of globalization. The growth of outward linkages of Swedish long-standing transnational corporations 1890s–1990s. Almqvist & Wiksell International, Stockholm, Sweden

Polikarpova MG (2011) Integral assessment of integration activity in the regions of the Russian Federation. Econ, Stat Inform 5:137–141

# The Impact of Sanctions on Stock and Industry Indices of Countries that Initiated Sanctions and Countries Subject to Sanctions

**Mikhail Yu. Sverdlov, Sergey I. Seliverstov, Arkady G. Zinoviev, and Petr I. Kuzmin**

**Abstract** This paper is devoted to the assessment of the impact of sanctions on financial markets initiated by American politicians in the United States, the European Union, Australia, New Zealand, and Canada in March 2014. The stable situation on financial markets during such challenges as sanctions is one of the factors for sustainable economic development. To assess the impact of sanctions, we conducted a correlation analysis of stock and industry indices of countries that introduced sanctions and countries subject to sanctions, including Russia. Correlation analysis of current data allowed us to determine the direction and connections between the data and evaluate its statistical significance in the context of leading stock and industry indices. To conduct a detailed analysis of the impact of sanctions measures, we performed calculations using a market model for evaluating the returns of stock and industry indices, taking into account the leading factor. The results allowed us to draw conclusions about the inefficiency of the sanctions on the Russian financial market. We also found that some stock and industry indices of the countries that initiated the sanctions became less attractive to investors.

**Keywords** Stock index securities · Sanctions · Market model · Stock market volatility · Systematic risk · Non-systematic risk · Risk premium · Average return

M. Yu. Sverdlov · S. I. Seliverstov (✉) · A. G. Zinoviev · P. I. Kuzmin
Altai State University, Barnaul, Russia
e-mail: sergse22@mail.ru

M. Yu. Sverdlov
e-mail: musverdlov@mail.ru

A. G. Zinoviev
e-mail: zinag60@mail.ru

P. I. Kuzmin
e-mail: kuzminpi@mail.ru



S. G. Maximova et al. (eds.), *Advances in Natural, Human-Made, and Coupled Human-Natural Systems Research*, Lecture Notes in Networks and Systems 234,
https://doi.org/10.1007/978-3-030-75483-9_17

# 1 Introduction

Nowadays, the economy of Russia and some other countries, including the Eurasian region (China), is experiencing an unprecedented challenge to its development in the form of sanctions from a group of countries led by the United States. The sanctions were supposed to impact the Russian economy and lead to a rating decrease of industries and enterprises. Many works are devoted to the study of various aspects of sanctions: economic, legal, political, etc.

The research objective is to identify the correlation between the financial and statistical indicators of the stock markets of countries that have applied sanctions and countries subject to them through technical analysis. Moreover, we aim to reveal the degree of impact of such sanctions.

The research is based on theory and methods from the works of foreign and Russian scholars. In his publications, Markovitz G. considers an approach that allows us to study the risks of investment distribution and the correlation and diversification of expected investment returns. *Modern portfolio theory* introduced by H. Markovitz allowed Sharpe W. to build a *Capital Asset Pricing Model* [CAPM] (Vlasov 2016). Based on this model, we can assess financial assets. Such Russian scientists as Aleshina et al. (2016), Komarovskikh (2015), Dmitrenko and Samodurov (2017), Kazantsev (2016), Mirkin (2015) considers the risk factors of investments relevant for the Russian economy in the context of sanctions, the unfavorable geopolitical situation, and the deteriorating conjuncture of global commodity and financial markets. For example, some authors (Aleshina et al. 2016) considered the reasons and directions of the sanctions applied against Russia. They also identified some positive impacts of these measures on the Russian economy: the development of innovation in the long term and import substitution policies in agriculture and food industry. Some papers mentioned the creation of a single financial market by the Eurasian Economic Union against the background of close economic cooperation between Russia and Asian countries. The researchers considered the consequences of the sanctions in the banking sector and the possibility of creating a unified national payment system. In some studies, scholars identified factors accelerating the recovery of the Russian economy and encourage it to a more active investment policy. Until now, no one conducted a comprehensive technical analysis of the financial markets of the world's leading, developing countries and Russia based on up-to-date information.

# 2 Materials and Methods

The research is based on the technical analysis of securities market indicators: returns and risks (Haertfelder et al. 2005; Moshensky 2010). These indicators are calculated using the CAPM market model (Markowitz 1952). The average return, beta coefficients, and growth rates against the background of market indices of countries that initiated sanctions allow us to compare the dynamics of returns of different countries (Berzon 2012; Kogdenko 2010).

We propose a similar approach when comparing risk indicators that characterize the stability and attractiveness of markets for investors (Schwager 2001; Novikov 2000; Buffett 2005). We consider the following risk indicators: (1) standard deviation; (2) proprietary, market and total risks; (3) risk premium, (4) coefficient of variation and determination. A wide range of indicators, calculated on the basis of current data, ensures the research reliability.

## 3 Results

During the initial stage of the research, we conducted a correlation analysis of stock and industry indices of countries that initiated sanctions and countries subject to them, including Russia. We assessed the effectiveness of sanctions against the Russian Federation and their impact on the financial markets of the countries that initiated the sanctions. The purpose of the correlation analysis is to determine the presence, direction, and interdependence between the leading stock and industry indices of countries that initiated sanctions and countries subject to them assessing statistical significance.

The research object is the leading stock and industry indices of the following countries: (1) USA, (2) Russia, (3) China, (4) Germany, (5) France, (6) Japan, (7) Great Britain, (8) Brazil, (9) Canada, (10) Australia, (11) Venezuela, (12) New Zealand. We compare the indicators with each other and with one of the most authoritative indicators of the American securities market, the *S&P 500* stock market index.

In March 2014, U.S. politicians undertook an initiative that the European Union, Australia, New Zealand, and Canada put into effect the first package of sanctions against the Russian Federation. Therefore, we used the research database on stock and industry indices of the listed countries for five years (from March 2014 to December 2019, as the most stable sanctions period) in the form of weekly quotes of their values taken from the internet ("World and industry indices," 2020). In addition to the leading stock indices of the listed countries, in this study, we analyzed the industry indices of the following countries: the United States (5 industry indices), Russia (8), France (6), Venezuela (2), Brazil (7), Germany (7), Canada (7), New Zealand (5), Great Britain (8), Australia (5), China (5), and Japan (7).

Preliminary correlation analysis of the dynamics of quotations showed a weak linear relationship between stock and industry indices of the listed countries. Therefore, to analyze the impact of sanctions on both the Russian securities market and the securities markets of other countries subject to sanctions, we performed further calculations using the market model of returns on stock and industry indices, taking into account the leading factor (Vlasov 2016).

Consequently, the dynamics of the analyzed stock and industry indices of these countries for the period under study were presented in weekly calculated returns for each index. It showed how much [percentage] a particular index for each country increased or decreased compared to the previous week. In other words, we calculated weekly returns on indices using the formula:

$$R_{index} = \frac{Ind_t - Ind_0}{Ind_0} \cdot 100\%$$

where $Ind_0$—index value at the beginning of the week $t$ ($t = 1, 2, \ldots 305$), $Ind_t$—index value at the end of the week.

As a result, we created a database in M.S. Excel for the studied five-year period of weekly calculated returns of stock and industry indices for each country, using the *correlation* function in data analysis. We also built two correlation matrices with automatically defined paired correlation coefficients that characterize the presence, direction, and interdependence between the leading stock indices of the studied countries, including the *S&P 500* (*correlation matrix No. 1*), and between the industry indices of the studied countries, including *S&P 500* (*correlation matrix No. 2*).

The paired correlation coefficients for *correlation matrix No. 1* were positive. This indicates a direct relationship between all stock indices. In other words, as the return on one stock index increases, so does the return on the other stock index. However, they have different correlation indicators. For the sake of objectivity of the close connection between stock indices, we assessed the statistical significance of paired coefficients based on the comparative characteristics of the calculated values of the *Student's t-test* determined for each correlation coefficient with the tabular values. As a result, we found that all paired correlation coefficients, except for the correlation coefficients between the Venezuelan stock index and the stock indices of other countries, are statistically significant. Consequently, the revealed interdependence between the stock indices of all countries, except for Venezuela, is not accidental.

First of all, using the first correlation matrix that characterizes the coefficients of pair correlation between the leading stock indices of the countries under study (both countries that initiated sanctions and countries that are subject to sanctions), we paid particular attention to the presence, direction and tightness of the relationship between the *S&P 500* stock index of the American economy and the stock indices of other countries. The correlation matrix shows that the stock index of the American economy *S&P 500* is closely connected with the stock indices of the countries that initiated the sanctions. The pair correlation coefficient between the *S&P 500* stock index of the American economy and Canadian stock index is 0.76; *S&P 500* and the U.K. stock index: 0.7; *S&P 500* and French stock index: 0.71; *S&P 500* and German stock index: 0.68; *S&P 500* and Japanese stock index: 0.61; *S&P 500* and Australian stock index: 0.57. The same correlation exists between all stock indices of the countries that initiated the sanctions. We found multicollinearity between the stock indices of France, Germany, and the U.K. In other words, there is close interdependence between them, which is confirmed by their entry into the European Union. This indicates a significant influence of the U.S. securities market on the value of the countries that initiated the sanctions and the direct influence of the stock indices of the countries that initiated the sanctions on each other. As a result, we can see the mutual support of the countries that initiated the sanctions in introducing new ones to the countries subject to these sanctions. However, the possible consequences of the imposed sanctions on the financial markets of the countries that initiated the sanctions are unclear. It is necessary to consider the market models and the dependence of the stock indices of countries subject to sanctions on the returns of the stock indices of the countries that initiated the sanctions. In this regard, we have established a correlation between the stock index of the American economy *S&P 500* and the stock indices of countries subject to sanctions. According to the obtained data of the correlation matrix, it is clear that this correlation is rather weak. The pair correlation coefficient between the *S&P 500* stock index of the American economy and the stock indices of Russia and Brazil is 0.4; respectively, with the stock index of China, it is 0.29, and with the stock index of Venezuela, where the pair correlation coefficient is 0.1, there is almost no correlation. These characteristics indicate that the U.S. securities market has an insignificant impact on the stock indices of countries subject to sanctions. However, it is worth paying attention to the correlation between the Russian stock index and those of the countries that initiated sanctions; such as Canada, Germany, France, and the United Kingdom, where the pair correlation coefficients are higher than the pair correlation coefficient between the *S&P 500* stock index of the American economy and the Russian stock index. Therefore, the degree of influence of the stock indices of these countries on the return of the Russian stock index is slightly higher compared to that of the American securities market. Besides, there is a weak correlation between the stock indices of Brazil and China with the stock indices of the countries that initiated the sanctions. This indicates the inefficiency of the sanctions imposed by these countries on the Chinese and Brazilian securities markets since they practically did not affect the stock index returns. We should also pay attention to the correlation between the Venezuelan stock index and stock indices of the countries that initiated the sanctions. Their pair correlation coefficients are almost zero; that is, there is no correlation between them. However, these paired correlation coefficients are not statistically significant since the calculated values of the *Student's t-test* of these paired correlation coefficients do not exceed the tabular values. Therefore, it may be premature to conclude that the Venezuelan stock market is independent of the profitability of the stock indices of the countries that initiated the sanctions.

Based on the correlation matrix under study, we also found a weak correlation between the stock indices of countries subject to sanctions compared to the strong correlation between the stock indices of countries that initiated sanctions. It also allows us to conclude that the mutual independence of stock markets between countries subject to sanctions is still relevant. Moreover, the profitability of stock indices of countries that initiated sanctions in the world securities markets is significant.

In general, as a result of the correlation analysis, we marked little dependence of stock indicators of countries subject to sanctions (including the Russian Federation) on fluctuations in the returns of the stock markets of countries that initiated sanctions (including the United States). This indicates the growing autonomy and stability of securities markets in countries subject to sanctions.

Therefore, based on the correlation analysis of stock indices, we revealed the correlations between the stock indices of the initiator countries and the countries subject to sanctions, taking into account their statistical significance. To assess the overall impact of sanctions on the economies of countries subject to sanctions (including the Russian Federation), we should consider stock and industry indices. We should select market models that characterize the correlation between the returns of stock and industry indices of countries subject to sanctions (including the Russian Federation) from the corresponding market indices of the countries that initiated the sanctions. This approach allows us to assess the impact of sanctions on the economies of countries that are subject to sanctions and analyze the reverse effect for the countries that initiate them.

Now we dwell on the analysis of the *correlation matrix No. 2*, which characterizes the correlation between the industry indices of the studied countries and the stock index of the American economy *S&P 500* with the corresponding industry indices of the studied countries. When analyzing the *correlation matrix No. 2*, we paid special attention to determining the direction and interdependence between the *S&P 500* stock index of the American economy and the industry indices of other countries. Since all paired correlation coefficients are non-zero or positive, we can conclude that the stock index of the American economy *S&P 500* is directly interrelated with the industry indices of the studied countries, that is, with the growth of the returns of the stock index of the American economy *S&P 500*, the returns of the industry indices of the studied countries increases too. When assessing the statistical significance of paired correlation coefficients, which characterizes their objectivity, we also found that almost all paired correlation coefficients between the stock index of the American economy *S&P 500* on the one hand, and the industry indices of the studied countries, on the other hand, are statistically significant. Moreover, their interdependence is logical. There were weak correlation coefficients between the *S&P 500* stock index of the American economy, the financial industry index of Venezuela (*Bursatil Financial*), and the metallurgical industry index of Canada (*S&P TSX Metals & Mining*). There is a strong correlation between the *S&P 500* stock index of the American economy and industry indices of the countries that initiated the sanctions. It is most closely related to such American industry indices as *Dow Jones Financials*, where the correlation coefficient is 0.91, *Dow Jones consumer goods* with a correlation coefficient of 0.83, and *Dow Jones Basic Materials* with a correlation coefficient of 0.81. In addition to the U.S. industry indices, the *S&P 500* stock index of the American economy is closely connected with French industry index *CAC Industrials* (0.72), German industry index *DAX Industrial* (0.69) and with such Canadian industry indices as *S&P/TSX 60* (0,77), *S&P/TSX Canadian Financials* (0.75), *S&P/TSX Canadian Industrials* (0.73). Most of the industry indices of the countries that initiated the sanctions (great Britain, Japan, Canada, Australia, Germany, France, and New Zealand) are close to the average interdependence with the *S&P 500* stock index of the American economy, where the correlation coefficients range from 0.3 to 0.6. In general, we can conclude that there is an average interdependence of the economic sectors of the countries that initiated the sanctions on the American stock market. It indicates a gradual decline in the U.S. leading role in imposing sanctions on individual countries.

The industry indices of the countries that initiated the sanctions are characterized by a medium dependence on the *S&P 500* stock index of the American economy.

At the same time, according to the *correlation matrix No. 2*, the industry indices of countries subject to sanctions, including Russia, show a weak dependence on the stock index of the American economy *S&P 500*. We can see the weakest correlation between the stock index of the American economy *S&P 500* and such Russian industry indices as *MICEX transport*, where the correlation coefficient is 0.16, *MICEX Chemicals* (0.17), *MICEX Metals and Mining* (0.22), *MICEX consumer goods* (0.23). In addition to Russian industry indices, the *S&P 500* stock index of the American economy is weakly correlated with the Venezuelan industry indices *Bursatil Industrial* (0.16) and *Bovespa Electrical Energy* (0.25), *Public Utilities* (0.24), *Real Estate* (0.24); and with some Chinese industry indices: *SSE Industrials* (0.21), *SZSE Agriculture* (0.26), *FTSE China a 600—software & computer services* (0.25), *SSE health care* (0.24). As we can see, almost all industry indices of countries subject to sanctions have a weak correlation with the stock index of the American economy *S&P 500*, where the interdependence between the correlation coefficients ranges from 0.1 to 0.3. In General, there is a weak interdependence of the economic sectors of countries subject to sanctions, including Russia, on the U.S. stock market. We can speak about the relative stability and autonomy of the economic sectors of countries subject to U.S. sanctions.

When assessing the impact of sanctions on the economies of embargoed countries and their effects on the financial and economic condition of the countries-initiators of the sanctions, we found it necessary to justify the interdependence between the industry indices of all the countries under study. According to the *correlation matrix No. 2*, considering the statistical significance of paired correlation coefficients, there is a general trend of the weak correlation between all industry indices of the United States and the Russian Federation. Their correlation coefficient varies from 0.1 to 0.3. Within the same range of paired correlation coefficients, there is a correlation between the U.S. and Brazil industry indices and between the U.S. and China. Consequently, the U.S. industry indices are rather weakly related to the industry indices of countries subject to sanctions. The same weak connection exists between all industry indices of countries subject to sanctions. In this regard, we can confirm the assumption about the inefficiency of U.S. sanctions on the economies of countries subject to sanctions, including Russia. The industry indices of other countries that initiated sanctions, similar to the U.S. indices, are also weakly related to the industry indices of the studied countries subject to sanctions. There is a weak correlation between the industry index of Russia and the industry indices of the countries that initiated the sanctions, where the correlation coefficient varies from 0.1 to 0.3. The same weak relationship is typical for the industry indices of Brazil, Venezuela, and China, with the industry indices of the countries that initiated the sanctions. The sanctions had almost no impact on financial and economic indicators in the form of industry indices. Therefore, the effectiveness of sanctions applied by the United States and other countries was relatively low.

At the same time, there is little interdependence between the industry indices of countries subject to sanctions. This is significantly different from the correlation between the industry indices of the countries that initiated the sanctions, which are much more closely related. A closer interdependence is typical for the industry indices of France and Germany, which paired correlation coefficients range from 0.6 to 0.9. Moreover, we can observe their multicollinearity using five paired correlation coefficients. In particular, such a close relationship occurs between French industry index *CAC Industrials* and German industry index *DAX Industrial*, their correlation coefficient is 0.91; French industry index *CAC Financials* and the German industry index *DAX Banks* correlation coefficient is 0.82; French industry index *CAC Basic Materials* and German industry index *DAX Industrial* correlation coefficient is 0.8. We can also see a close correlation between the U.S. and French industry indices with a variation of the correlation coefficient from 0.4 to 0.7. There is an average correlation between the industry indices of other countries that initiated sanctions. The correlation coefficient between the industry indices of Canada and Australia ranges from 0.3 to 0.6. A similar correlation is found between the U.K. and U.S. industry indices, the United States and Canada, the United States and Germany, the United States and Japan, the United States and Australia, Japan and Germany, Japan and France, and Japan and Canada. However, there is a weak correlation between the industry indices of Japan and the industry indices of the United Kingdom and Australia, where the correlation coefficients vary from 0.2 to 0.4. In addition, there is a very weak correlation between The New Zealand industry index and those of the countries that initiated the sanctions, with a variation in the correlation coefficient from 0.1 to 0.3. This indicates a small role of New Zealand in presenting sanctions measures to countries that are subject to sanctions. In addition, there is a very weak correlation between. Moreover, the main initiators of the sanctions are the United States, Germany, and France. Therefore, when assessing the impact of sanctions on the economy of the countries subject to sanctions, including Russia, in addition to the *S&P 500* stock index of the American economy, we should also take into account other industry indices of the United States (*Dow Jones Financials*), Germany (*DAX Industrial*) and France (*CAC Financials*), identified on the basis of the analysis of the *correlation matrix No. 2*. Moreover, compared to other industry indices, the *S&P 500* stock index of the American economy is more closely related to these industry indices. Furthermore, we can observe a close correlation between the three industry indices we have identified and most of the industry indices of the countries that initiated the sanctions. However, there is a weak correlation between them and the industry indices of countries subject to sanctions, including Russia. Besides, these leading industry indices are not multicollinear and are characterized by a moderate level of correlation. The correlation coefficient between the industry indices of *Dow Jones Financials* and *DAX Industrial* is 0.66; between *Dow Jones Financials* and *CAC Financials*, it is 0.68; between *DAX Industrial* and *CAC Financials*, it is 0.74. Therefore, assessing the effectiveness of sanctions against the Russian Federation and analyzing the impact on the financial markets of the countries that initiated the sanctions, we consider it necessary to use an econometric model of multiple regression. In addition to the leading factor (the stock index of the American economy *S&P 500*), it includes three main industry indices of the USA, Germany, and France: (1) *Dow Jones Financials*, (2) *DAX Industrial*, (3) *CAC Financials*. This will allow us to determine changes in the financial and economic indicators of the economy as a result of the impact of sanctions, as well as to assess the impact of the sanctions on the financial markets of the countries that initiated the sanctions.

## 4 Discussion

Our research approach to assessing the impact of sanctions on the economy of Russia and other countries targeted by them is based on a technical analysis of the securities market indicators, as well as the methods and models described in the works of G. Markovitz (1952), and W. Sharpe (Vlasov 2016). At the same time, the Russian scientific community studied various aspects of the impact of sanctions.

To identify the impact of sanctions on financial indicators, Komarovskikh (2015) used a regression analysis to describe the extent of this impact. Dmitrienko and Samodurov (2017) identified external and internal risk factors for companies' investment activities in Russia.

Kazantsev (2016) assessed the potential negative impact on the Russian economy of such factors as the unfavorable geopolitical situation, deteriorating global commodity markets, sanctions against Russia, and the economic crisis in the country. Under the influence of these factors, the socio-economic development of some Russian regions has slowed down, and the situation has worsened. According to calculations, the economic downturn in Russia, the deteriorating situation on world commodity markets, and financial and economic sanctions applied against the Russian Federation are directly proportional to the dependence of economic entities on foreign capital, external financing, and foreign trade. We have identified regions of the Russian Federation that depend on these factors to varying degrees.

Mandron and Nikonets (2016) presented an approach to identify the impact of several factors on financial market volatility.

Mirkin (2015) assessed the extent of short-term financial risks arising from the sanctions against Russia: (1) credit and market risks, (2) inflation risk, (3) liquidity risk, (4) currency risk. He found that in 2014–2015, sanctions on financial markets had a moderate impact on the risks compared to other negative factors. In the long term, the risks associated with technological sanctions are systemically essential for the financial stability of the Russian economy.

Rozanova and other scholars (2015) assessed the impact of investment on GDP dynamics and opportunities to stimulate investment in a difficult financial market situation under sanctions against Russia. We used the following research methods: (1) method of analysis and synthesis, (2) generalization method, (3) comparison method, (4) grouping method. The empirical basis of the study is statistical data, based on which we analyze structural shifts, trends, and sources of financing for fixed capital investments and determined correlations in the dynamics of investment and GDP for the period from 1990 to 2014. The results of the study are presented through graphical analysis. We made conclusions about the increasing influence of corporate structures on investment in specific economic sectors and outlined the role of the state and the importance of foreign investments. The research novelty lies in the revealed correlation between rapid economic growth and investment in fixed capital if it is properly updated. The multiplier effect is manifested in the innovative modernization of fixed capital.

Koshkina (2015) notes that the refusal of Russian banks to make dollar payments or payments in euros can become a severe economic and financial threat. The sanctions limit the access of American investors to long-term debt instruments placed after July 16, 2014. The integration of Russia and Kazakhstan within the Customs Union and the newly created Eurasian Economic Union is increasing. Therefore, external changes concerning Russia almost automatically affect the economy of Kazakhstan. The banks of Kazakhstan are weakened after the crisis. Consequently, they cannot provide loans on the level that is possible for Russian state banks. The currency market in Kazakhstan is extremely vulnerable due to the hype that has arisen under the influence of rumors about a possible devaluation of the tenge. The paper presents the possible consequences of anti-Russian sanctions on the financial market of Kazakhstan. This research is especially relevant during the difficulties that the financial market of Kazakhstan is currently experiencing. The study of problems and new approaches to their solution will ensure the stability of the financial system of Kazakhstan.

Golovnin (2015) considers the features of the development of national economies at the current stage of financial globalization, which is characterized by the increasing influence of external shocks. His paper analyzes the impact of external shocks of the last decade on the Russian economy: (1) the global economic and financial crisis of 2008–2009, (2) the European debt crisis, (3) quantitative easing tapering in the United States, (4) the political crisis in Ukraine and the annexation of Crimea to Russia, (5) international sanctions against Russia, (6) the fall in oil prices on world markets in 2014–2015. It also evaluates the economic measures applied during these shocks. The conclusion of the paper reveals the delayed response of fiscal policy to external shocks and the shortcomings of monetary policy, including the rejection of currency regulation tools.

In this paper, based on the correlation analysis, we evaluated the interrelation, taking into account their statistical significance, between the leading stock and industry indices of the countries that initiated sanctions and countries subject to sanctions. We also selected the main industry indices of the countries that initiated sanctions affecting the profitability of stock and industry indices of countries subject to sanctions, including Russia. In the future, it will allow us to assess not only the impact of sanctions on the economy of countries that are subject to sanctions but also to analyze the reverse effect for the countries that initiated them.

We believe that the volatility of prices for shares of large national companies, to a certain extent, can serve as an assessment of the socio-economic situation in the state. One of the options for such an assessment is to compare the dynamics of the Moscow exchange index and Russian industry indices with similar indicators of the world's leading countries.

# 5 Conclusion

There was no comprehensive analysis of the securities market indicators in the context of primary and secondary (industry) indices for countries that initiated sanctions and countries subject to them. The study based on correlation analysis and market model of stock and industry index returns, taking into account the leading factor, allowed us to make several conclusions. There is a weak dependence of economic sectors of countries subject to sanctions, including Russia, on fluctuations in the U.S. stock market. The sanctions did not have a significant impact on indirect indicators in the form of industry indices. We can see the relative autonomy and stability of the economy of countries subject to U.S. sanctions.

The goals of the countries that initiated the sanctions measures were not achieved. Furthermore, some of them have worsened the investment attractiveness of their financial instruments and reduced interest in their securities markets.

**Acknowledgements** The paper is based on a study funded by the RFBR in accordance with the research project No. 20-010-00332 *Assessment of the effectiveness of sanctions against the Russian Federation and the impact on the financial markets of the countries that initiated the sanctions*. The team of project performers consists of four people.

# References

Aleshina EV, Martirosyan AE, Tikhomirova AV (2016) Expected positive effects on the development of the Russian economy as a result of the application of sanctions. The Human Socio-Econ Sci 4:151–153

Berzon NI, Arshavsky AY, Buyanova EA (2012) Stock market. Vita Press, Moscow, Russia

Buffett W (2005) Essay on corporate governance, corporate finance, and investing. Alpina Business Books, Moscow, Russia

Dmitrenko VV, Samodurov AA (2017) Risk factors of companies' innovative activities in Russia and methods for it reducing. Upravlencheskoe Konsultirovanie 1(97):119–129

Golovnin MY (2015) The role of external financial shocks in shaping the model of economic development in Russia. VEO of Russia, 195:249–273

Haertfelder M, Lozovskaya E, Hanush E (2005) Fundamental and technical analysis of the securities market. Piter, Moscow, Russia

Kazantsev SV (2016) Assessment of the impact of the economic downturn and anti-Russian sanctions on the regions of the Russian Federation. ECO 5(503):55–70

Kogdenko VG (2010) Analysis of the sustainable growth rate of invested capital in the value chain analysis system. Econ Anal: Theory Prac 13(178):39–46

Komarovskikh YI (2015) The impact of sanctions on the Russian financial market. Ekonomika i Socium 1141–1147

Koshkina OV (2015) Anti-Russian sanctions: possible consequences for the financial market of Kazakhstan. Bull Turan Univ 1(65):162–166

Mandron VV, Nikonets OE (2016) The degree of volatility conditions of the national financial market in crisis conditions. Bull NGIEI 3(58):40–52

Markowitz HM (1952) Portfolio selection. The J Finance 7(1):77–91

Mirkin YM (2015) Influence of sanctions on financial risks of the Russian Economy. Bankovskie Uslugi 8:17–24

Moshensky SZ (2010) Securities market: transformation processes. Ekonomika, Moscow, Russia
Novikov AV (2000) The stock market as a mechanism of investment attraction. NSUEM, Novosibirsk, Russia
Rozanova LI, Moroshkina MV, Tishkov SV (2015) Problems of economic growth in the decline of the investment market conditions. Secur Issues 2:25–42
Schwager J (2001) Technical analysis. Full course. Alpina Publisher, Moscow, Russia
Vlasov DA (2016) Sharpe model as an instrumental basis for asset optimization. Innov Sci 3–1:43–45
World and industry indices (2020) https://ru.investing.com/indices/world-indices

# Protection of Consumers from Unfair Agreement Terms

**Irina V. Kiryushina and Tatyana A. Philippova**

**Abstract** The expansion of international links and cross-border trade has required addressing a number of consumer safety issues at the international level, including ensuring their protection from unfair contract terms. The authors analyzed national and foreign legislation, civil doctrine and jurisprudence that defines approaches to understanding unfair contract terms and the consequences of their inclusion in contracts involving consumers. Foreign experience in this field—legislation of the USA and the European Economic Union—is studied. The concept of unfair contract terms is correlated with the principle of private law on freedom of contract and the design of a public contract and a contract of accession. The paper uses formal legal method, methods of system interpretation and comparative law, legal modeling. Absence in Russian legislation of the concept of unfair contract terms, criteria for their definition and consequences of their inclusion in contracts with consumers is a significant obstacle to the protection of their rights. We concluded we need to improve the legal regulation, to systematize the legislation of international integration of states, to develop the approaches to consolidate common standards for the protection of consumers from unfair terms of contracts and the adoption at the level of the Eurasian Economic Union under special act.

**Keywords** Consumer protection · Unfair conditions of the Treaty · Freedom of the treaty · Accession treaty · Public treaty

## 1 Introduction

In civilism, there is a stable image of the consumer as a weakness of the contract. Entering a relationship with a counterparty (an entrepreneur), they find themselves in a more vulnerable position due to the different kinds of actual inequalities between them: economic, information, intellectual, and others. The challenge is to equalize this inequality. The legislation of most countries provides legal mechanisms to

I. V. Kiryushina (✉) · T. A. Philippova
Altai State University, Barnaul, Russia
e-mail: kirushina_ira@mail.ru



S. G. Maximova et al. (eds.), *Advances in Natural, Human-Made, and Coupled Human-Natural Systems Research*, Lecture Notes in Networks and Systems 234,
https://doi.org/10.1007/978-3-030-75483-9_18

neutralize this inequality, including ones that aim to limit the freedom of contract in relations with consumers. These may include a ban on the inclusion of conditions that are burdensome for the consumer or deprive the consumer of the rights normally granted under a contract of this kind, the construction of a public contract, and a contract of accession, known to Russian domestic law.

In the second half of the last century, foreign legislation adopted special acts regulating the issue of referring the contractual terms to unfair ones. The foreign authors study the following separate aspects of consumer protection from unfair contract terms:

- Unfair contract terms (Willett 2007);
- The protective function of the European Union Directive 93/13/EEC in relations to the participation of consumers (Council of Europe 1993; Reich and Micklitz 2008);
- The impact of the law on unfair contract terms of the EU on the American Internet traders (Dlabay et al. 2009; Winn and Webber 2006).

The absence in the Russian national legislation of the concept of unfair contract terms, clear criteria for their definition, and regulation of the consequences of their inclusion in contracts is a significant obstacle in protecting consumers' rights. As a result, the existing gaps in legal regulation are filled by legal practice. Hence, numerous attempts of the highest courts of Russia to qualify certain conditions of contracts with the participation of consumers as unfair or clearly burdensome are not reflected in the norms of Russian law yet.

The expansion of international relations and cross-border trade further aggravates the problem of including unfair conditions in contracts with the participation of consumers, giving it a supranational character, which is inevitable in the conditions of international legal integration. There is no coincidence that consumer protection is proclaimed in many countries as an element of national security strategy (Filippova et al. 2020). There is no current scientific research devoted to protecting consumers from unfair contractual conditions in the Eurasian space. In this regard, the current analysis of available world experience is relevant.

The specificity of the emerging interregional economic ties in the Eurasian space determines the need to develop common approaches to protect the rights of economic turnover participants, especially the least protected side—consumers. The research concludes that it is necessary to develop and adopt a special regulatory act to control the protection of consumers from unfair terms of contract in Russian national legislation and at the Eurasian Economic Union level, as a form of cooperation in Eurasia. This act should define the criteria categorizing contractual terms as unfair and a mechanism of consumer protection in cross-border transactions.

## 2 Materials and Methods

The formal and legal methods form the methodological basis of current research. They made it possible to carry out a comprehensive analysis of national and international acts, existing law enforcement practices of the Russian Federation, and foreign countries (Council of Europe 1993; Parliament of the United Kingdom 1973; Russian Federation 1992, 2013; United Kingdom Parliament 1977). Relying on the given analysis, the authors generate the general idea of the legal regulation condition in the designated sphere, reveal available lacks, and formulate offers to accept the special normative act regulating consumer protection questions from unfair conditions of contracts, including the Eurasian Economic Union level. The systematic interpretation method allowed to reveal the actual content of the rules governing certain aspects of consumer protection, considering the emerging law enforcement practice.

Since the current Russian legislation does not contain any special acts on the issue of including unfair terms in contracts with the participation of consumers, the authors used the method of comparative jurisprudence. It made it possible to study the existing international experience of the USA and the European Economic Union. These experiences can be adapted to the practice of the Eurasian Economic Union. To study the possible impact of the changes to the current national legislation of Russia and the Eurasian Economic Union, the authors used a legal modeling method. It allowed assuming the positive effect of strengthening consumer protection from unfair contract terms.

## 3 Results

Participants of property turnover, entering into contractual relations, have different means and opportunities to influence the formation and performance of the obligation due to the following reasons: (1) different legal status, (2) financial situation, and (3) professional skills. When these differences acquire the character of non-comparable, and one of the participants gets an opportunity to dictate their will, the task of the legislator is to equalize the parties legally. For example, the weaker party can be provided with special conditions for the conclusion of a contract that initially differ from the usual ones (i.e., changing or terminating the contract, or even increasing the liability of the stronger party for the failure to fulfill an obligation or limiting its liability). This leads to a restriction of the contract freedom but inevitably ensures real equality of the parties.

The Resolution of the Plenum of the Supreme Arbitration Court of the Russian Federation of March 14, 2014 No. 16 "On the freedom of the contract and its limits" (Paragraph 10) specifies the conditions that the court should consider when determining the "weak" side of the contract. The Court determines the actual ratio of negotiating opportunities of the parties, whether joining the proposed terms and conditions was forced, and considers the level of professionalism of the parties in the relevant area (competition in the relevant market, the availability of a real opportunity for the joined party to negotiate or conclude a similar contract with third parties on other terms, etc.), to assess all conditions of the contract, considering the circumstances of the case (Supreme Arbitration Court of Russian Federation 2014). Consumers are also considered as a "weak" part, unable to refuse to include in the contract unfair or burdensome conditions. It is common in financial services when the bank provides the consumer with a pre-prepared draft contract containing points that infringe the consumer rights (Kovalenko and Filippova 2019). Current law and order cannot but pay attention to these affairs, seeking to provide mechanisms for eliminating or reducing the negative consequences of such practices.

As a result, norms limiting contractual freedom concerning the participation of consumers for "equalizing" counterparties, for example, public contract norms, have appeared in Russian legislation. As an exception to the general rule on the inadmissibility of coercion to conclude a contract, enshrined in Paragraph 1 of Article 421 of the Civil Code, the law allows the possibility of coercion of the counterparty (consumer—entrepreneur) to conclude a contract in the case where the contract with the participation of consumers is classified as public (Article 426 of the Civil Code) (Government of the Russian Federation 1994).

In a number of resolutions, the Russian Supreme Court draws the attention of courts to the insignificance of terms of contracts concluded with the consumer does not correspond to acts containing the norms of civil law binding on the parties to the conclusion and execution of public contracts (Article 3; Paragraphs 4 and 5 of Article 426 of the Civil Code of Russia), and the conditions of the transaction in the commission of which was violated by an explicit legislative ban on restricting consumer rights, such as Paragraph 2 of Article 16 of the Law of the Russian Federation on Protection of Consumer Rights, and Article 29 of the Federal Law On Banks and Banking Activities) (Russian Federation 1992, 1994). Later, the Supreme Court stated that the conditions of a public contract that do not meet the requirements set in Paragraph 2 of Article 426 of the Civil Code, as well as the mandatory rules in force at the time of conclusion of the public contract, approved by the Government of the Russian Federation or federal executive bodies authorized by it, are insignificant in the part that worsens the situation of consumers (Supreme Court of Russian Federation 2015, 2018).

Another legislative construction that restricts the contract freedom to protect consumers is the contract of accession. Known in German law as a "dictated" contract, its terms are defined by one party in standard form and could be accepted by the other party only by joining the proposed contract. The acceding party (consumer) has the right to demand the termination or amendment of the contract if it does not contradict the law and other legal acts but deprives this party of rights. Usually, the contracts of this kind exclude or limit the responsibility of the other party for the violation of obligations or contain other evidently burdensome conditions for the acceding party, which they, based on their reasonably understood interests, would not have accepted if they had the opportunity to participate in determining of contract terms (Item 2 of Article 428 of Civil Code of Ukraine). Thus, the possibility to refuse the conclusion of the contract, externally testifying about the recognition of contract freedom, cannot be considered sufficient for its real maintenance to citizens, especially when the right of citizens on protection is not guaranteed properly. The citizen is compelled to agree on the dictated to its conditions (Constitutional Court of Russian Federation 1999).

In this case, the contract conditions do not contradict the law or other legal acts and are accepted by the other party. However, these conditions have the unilateral advantage character, which is a consequence of being formulated unilaterally and can be burdensome for the consumer. The very notion of burdensome contract terms is an evaluation category, which the legislator defines quite broadly: these are terms that a party would not have accepted if it had the opportunity to reject it.

In practice, the protective function of the construction of an accession contract proved to be difficult to use due to the lack of a precise understanding of the "forms and other standard forms." In judicial practice, there were difficulties in qualifying concluded contracts as contracts of accession. Consequently, Article 428 of the Civil Code was supplemented by a rule by which the rules provided in the article are also applicable in cases where the contract (except for an accession contract), the terms of the contract determined by one party, and the other party due to the apparent inequality of negotiating opportunities is put in a position that significantly hinders the harmonization of other content of the individual terms of the contract.

The practice of applying the rules on public contract and the contract of accession shows that the category of "unfair" or "burdensome" terms of the contract needs to be considered more carefully, considering the specificity of participation in such consumer relations.

## 4 Discussion

The foreign legislator is familiar with the institution of unfair contract terms much earlier. The case law of Western European countries on disputes between credit, insurance organizations, and consumers led to the adoption of special laws limiting the use of unfair clauses in contracts on standard terms. For example, the German Law on General Terms of Transactions of 1976, the provisions of which were later incorporated into the German Civil Code (Karimullin 1999). In England, such rules are included in the Fair Trade Act 1973 (Parliament of the United Kingdom 1973) and the Unfair Contract Terms Act 1977 (United Kingdom Parliament 1977).

In the United States, the Magnuson-Moss Act of 1975 was passed, providing a minimum of contractual guarantees for consumers. In particular, under Section 108(c) of the Act, a seller may not make a clause in standard contracts that they use in their practice that excludes warranties of merchantability but may only limit the duration of the contract (United States Congress 1975). Significant efforts were made by the U.S. Federal Trade Commission (FTC) to develop special rules for certain types of services. Based on these rules, states develop their own legislation (Burnham 2006).

In 1993, the Council of Europe approved Directive 93/13/EEC on Unfair Conditions in Contracts with Consumers (Contract Terms). It states that an unfair condition is one that causes a significant imbalance in the rights and obligations of the parties to a contract to the detriment of the consumer. Only the condition that was not subject to a special agreement with the consumer may be considered unfair (Clause 1 of Article 3). The Directive contains an approximate (not exhaustive) list of conditions that may be considered unfair (Council of Europe 1993).

Currently, there are no such acts in the national legislation of Russia. The legal regulation in this area is limited to establishing certain rules, providing exceptions to the principle of freedom of contract, including through the construction of a public contract and a contract of accession, as noted earlier. It fragmentarily determines which of the terms of the contract may be invalidated as infringing on consumers' rights.

## 5 Conclusion

The peculiarities of the legal position of the consumer as a weaker party to the contract led states to adopt several rules aiming to limit the principle of contract freedom concerning their participation and the development of acts limiting the inclusion of "unjust" conditions in the contracts. Russian legal practice lags behind the legislation of other states (the USA and the European Economic Union in this respect), which demonstrates the necessity to develop and adopt a special normative act regulating consumer protection issues from unfair contract terms.

In international legal integration conditions, states should make joint efforts to bring together national laws and practices of consumer protection and develop common approaches to consolidate common standards for protection from unfair terms of contracts. Given the specifics of emerging regional economic ties, it is advisable to adopt supranational legal acts at the Eurasian Economic Union level as well and regulate the issues of ensuring consumer protection from unfair terms of contracts. Adopting such an act will make it possible to create effective mechanisms to protect consumers and ensure their safety by developing legal integration of States within the Eurasian Union.

## References

Burnham W (2006) Introduction to the law and legal system of the United States. Novaya yusticiya, Moscow, Russia

Constitutional Court of the Russian Federation (1999) Ruling of the Constitutional Court of the Russian Federation "On the Case of Verification of the Constitutionality of the Provisions of Part Two of Article 29 of the Federal Law of February 3, 1996 'On Banks and Banking Activities'. In: Veselyashkina OY, Lazarenko NP (eds) Connection with the complaints of citizens (February 23, 1999 No. 4-P). Moscow, Russia

Council of Europe (1993) Directive of European Council "On Unfair Conditions in Contracts with Consumers (Contract Conditions)" (April 5, 1993 No. 93/13/EEC). Luxembourg, Luxembourg
Dlabay LR, Burrow JL, Kleindl B (2009) Intro to business. Cengage learning, Mason, OH
Filippova TA, Zharkenova SB, Kiryushina IV (2020) Consumer protection: international and national aspects Russia-Kazakhstan. Publishing house of the Altai State University, Barnaul, Russia
Government of the Russian Federation (1994) Federal Law "Civil Code of the Russian Federation, Part 1" (July 3, 1994 No. 51-FZ). Moscow, Russia
Government of the Russian Federation (2013) Federal Law "On Consumer Credit (Loan)" (December 21, 2013 No. 353-FZ). Moscow, Russia
Karimullin RI (ed) (1999) German Law: Part 3. Statut, Moscow, Russia
Kovalenko EY, Filippova TA (2019) Legal ensuring of interests of the parties of the mortgage lending agreement. Herald of Omsk University. Series Law, 16(1):96–102
Parliament of the United Kingdom (1973) Fair Trade Act. London, UK
Reich N, Micklitz H-W (2008) Unfair terms in the draft common frame of reference (comments on the occasion of the Tartu conference on recent development in European private law). Juridica Int 14:58–68
Russian Federation (1992) Law of the Russian Federation "On Protection of Consumer Rights" (February 7, 1992 No. 2300–1, edited by December 08, 2020). Moscow, Russia
Supreme Arbitration Court of the Russian Federation (2014) Resolution of the Plenum of Supreme Arbitration Court of the Russian Federation "On Freedom of Contract and its Limits" (March 14, 2014 No. 16). Moscow, Russia
Supreme Court of the Russian Federation (2015) Resolution of the Plenum "On Application by the Courts of Certain Provisions of Section 1, Part 1 of the Civil Code of the Russian Federation" (June 23, 2015 No. 25). Moscow, Russia
Supreme Court of the Russian Federation (2018) Resolution of the Plenum of Supreme Court of the Russian Federation "On Certain Issues of Application of General Provisions of the Civil Code of the Russian Federation on Conclusion and Interpretation of the Contract" (December 25, 2018 No. 49). Moscow, Russia
United Kingdom Parliament (1977) Act of UK Parliament "On unfair contract terms". London, UK
United States Congress (1975) The Magnuson-Moss Warranty Act (January 4, 1975 No. PL 93–637). Washington, D.C.
Willett C (2007) Fairness in consumer contracts: the case of unfair terms. Ashgate Publishing, Leicester, UK
Winn JK, Webber M (2006) The impact of the EU unfair contract terms law on US business-to-consumer internet merchants. Bus Lawyer 62(1):209–228

# Problems and Prospects for the Development of Integration in the Eurasian Economic Union

**Nikolay M. Abramov, Oksana Yu. Rudakova, Tatyana A. Akimochkina, and Larisa A. Bocharova**

**Abstract** The paper focuses on the current problems of interaction between the countries of the Eurasian Economic Union (EAEU) and the prospects for developing integration on the scale of this association. The research relevance is determined by (1) the complexity of the economic relations of various countries, and (2) global trends and crises, significantly impacting the effectiveness of national economies and the results of business activities. The formation of the EAEU is an important decision for all its participants. The EAEU aims to implement a full-fledged integration association focused on the sustainable growth of national economies, contributing to the improvement of the socio-economic climate between the partner countries. Based on the analytical study of the main indicators of foreign trade in EAEU countries, the authors indicate low integration in the indicated union. The authors formulate the main problems that need to be addressed in terms of (1) resolving contradictions between the participating countries; (2) creating conditions for the formation of a single market by unifying the mechanism of removing trade barriers and measures of non-tariff restrictions; (3) changing the structure of exports in the production of goods with higher added value; and (4) ensuring equal access of the EAEU countries to research and innovation projects.

**Keywords** Economic integration · Eurasian Economic Union · Eurasian Economic Commission · Integration problems · Integration prospects

## 1 Introduction

The creation of the Eurasian Economic Union (EAEU) opens up enormous prospects for its participants, which have significant resource potential (4.4% of the world

N. M. Abramov · O. Yu. Rudakova (✉) · T. A. Akimochkina · L. A. Bocharova
Altai State University, Barnaul, Russia
e-mail: rud-oksana@yandex.ru

L. A. Bocharova
e-mail: bocharova@mc.asu.ru



S. G. Maximova et al. (eds.), *Advances in Natural, Human-Made, and Coupled Human-Natural Systems Research*, Lecture Notes in Networks and Systems 234,
https://doi.org/10.1007/978-3-030-75483-9_19

population, 16.4% of the Earth's surface, 25% of the main types of minerals, about 40% of the world natural gas reserves, 25% of coal, 20% of oil, 13% of arable land, and 11% of fresh water) and occupy the 1st place in oil and gas production, 2nd place in the production of mineral fertilizers, 3rd in the production of electricity, and 4th in the production of wheat and coal and steel (Akimov 2014). According to the calculations of the Eurasian Economic Commission (EEC), the cumulative integration effect until 2025 can vary from 17 to 20% of the additional GDP growth for each of the EAEU countries, or about $700 billion in absolute terms (Kuzmina 2017).

In 2015, the union expanded its integration ties with other countries occurred. The EAEU and Vietnam signed an agreement (Eurasian Economic Commission 2015) on the creation of a free trade zone, providing for a gradual reduction of duties, an increase in goods turnover, and the creation of prerequisites for the subsequent integration of the EAEU with the Asia–Pacific countries.

Mongolia, Egypt, Iran, Israel, India, and Serbia, as well as many organizations, show great interest in creating a free trade zone with the EAEU. However, none of the above countries concluded corresponding agreements. These processes should be considered in the context of broader plans to promote a large Eurasian project, where they could participate based on various forms of interaction between the EAEU countries, the Shanghai Cooperation Organization (SCO) and the Association of Southeast Asian Nations (ASEAN).

The publication (January 1, 2018) of a key document, a single Code of the Eurasian Economic Union, became an important event for the participants of the integration association. The entry into force of the Customs Code of the Eurasian Economic Union was delayed due to the fundamental disagreements of the participating countries in their approaches to the principles of customs regulation.

The Customs Code of the Eurasian Economic Union has many positive aspects (the advantage of electronic declaration, the possibility of filing a declaration without supporting documents, etc.). However, it has some drawbacks manifested in excessive bulkiness, detail, and difficult language. The new Customs Code allows reducing the time and cost of processing goods, as well as the number of documents that must be presented in order for goods to cross the border (Member States of the Eurasian Economic Union 2017).

## 2 Materials and Methods

The results of the work performed by the EAEU show the effectiveness of the union. Considering that the EAEU was created as a trade and economic union, the priority indicators should include trade turnover from mutual trade for all member states and each member separately, as well as the trade turnover of foreign trade with third countries for all member states and each member separately. Of particular importance are the value indicators of the commodity structure of exports and imports and the proportion of individual product groups in the volume of trade. This allows partners in the union to judge the product orientation of the economy and make timely adjustments to the development of industrial policy.

The EAEU saw a deterioration in business activity and turnover by 40% in 2019 and by 15% in 2016 (Center for Integration Studies of the Eurasian Development Bank 2019). The reasons for this decline were the fall in oil prices and the sanctions imposed against Russia. These facts affected not only the depreciation of the ruble against the US dollar and the reduction in consumption, but also affected the economies of the EAEU countries.

The situation noticeably improved only by the end of 2017. Trade turnover increased by 26.7% compared to the previous year and amounted to almost $44 billion (Kofner 2019).

According to the statistical report of the EEC (Kofner 2019), for 10 months of 2019, the volume of mutual trade of the EAEU countries decreased by 1.6% compared to the same period in 2018. It amounted to $49,358.7 million. The share of mutual trade in the total foreign trade of the union increased from 13.9 to 14.1%. Compared to the same period in 2018, the structure of mutual trade changed as follows: the contribution of Armenia increased in exports from 1.1 to 1.2%, the Republic of Belarus—from 23 to 24.3%, the contribution of Russia decreased from 64.8 to 63.4%. The contribution of Kazakhstan and Kyrgyzstan did not change and amounted to 10.1 and 1.0%.

The supply of goods from Armenia to the EAEU countries in January–October 2019 increased by 6.1% compared to the same period in 2018. The volume of deliveries amounted to $603.2 million, of which $583.5 million are deliveries to Russia. Over the same period, Belarus increased deliveries to the countries of the trade alliance by 3.8%, the volume exceeded $11.9 billion. Deliveries of goods to Armenia, Kazakhstan, and Russia increased by 38.3, 1.8, and 3.9%. The volume of commodity flows to Kyrgyzstan decreased by 4.7%. The volume of mutual trade of the Republic of Kazakhstan with partners in the union amounted to $4,963.6 million, which is 1.8% less than in the same period last year.

Only Armenia saw an increase in deliveries by 64.5%. Deliveries of goods from Kazakhstan to Russia exceeded $4.4 billion, which corresponds to the last year indicator. Over the same period, deliveries to other EAEU countries significantly decreased: to Belarus—by 15.4%, Kyrgyzstan—by 14.3%.

In the analyzed period, $501.4 million of goods were exported from Kyrgyzstan, which is 2.5% less than in the same period of the previous year. Deliveries of goods to Belarus and Kazakhstan increased by 2.7% and 22%, respectively.

Commodity flows from Russia to the countries of the integration alliance in value terms amounted to $31,313.5 million, which is 3.5% less than in 2018. In the structure of supplies, the largest share falls on the Republic of Belarus—17 billion $581.0 million (−8.4%) and Kazakhstan—$11,319.6 million (+4.4%), the smallest—to the Republic of Armenia—$1,126.2 million (+6.1%) and Kyrgyzstan—1 billion $286.7 million (−6.0%). Deliveries of Russian goods to the market of the EAEU countries decreased due to mineral products by 16.5% (33.1% of Russian total supply in mutual trade). The growth in the value of machinery and equipment of vehicles amounted to 2.8% (18.5%), of chemical products—2.2% (12.3%), and food products and agricultural raw materials—11.8% (9.4%) (Kofner 2019).

Within the analyzed period, there was a tendency to reduce the volume of cooperative supplies within the integration trade association. Cooperation supplies account for almost half of the mutual trade. The volume of the cooperation industry within the framework of the trade association in 2018 amounted to 48% for Russia, Belarus—28%, and Kazakhstan—21% (Eurasian Economic Integration 2019a; b).

## 3 Results

Conclusions on the effectiveness of integration associations can be drawn on the indicators of foreign trade of the EAEU countries with third countries of the world (Table 1).

In the analyzed period, the volume of imports in most EAEU countries exceeds the value of exports forming a negative trade balance. Compared to 2018, there is a decrease in trade in Belarus, Kyrgyzstan, and Russia.

The analysis of the commodity structure of exports and imports of the EAEU countries for the analyzed period showed that the members of the Union mostly export mineral products (41%), and fuel and energy (40%) to the third countries. This is due to the rich natural resource potential of the EAEU countries and the raw material orientation of exports. The largest share in the structure of imported goods is occupied by: machinery, equipment, and vehicles—46%; chemical products—20%; food products and agricultural raw materials—14%. Most of the products of the "machinery, equipment, and vehicles" group are represented by high-tech equipment due to the low level of technical development of the EAEU countries. The same reason is seen in the import of products of chemical industry.

The presence of a high proportion of food products and agricultural raw materials confirms the presence of high demand due to their low cost in the country of origin.

**Table 1** Foreign trade in goods of the Member States of the Eurasian Economic Union with all countries of the world for January–October 2019, million US dollars

| Country | Turnover | Export | Import | Balance | Including January–October, 2018 | | |
|---|---|---|---|---|---|---|---|
| | | | | | Turnover | Export | Import |
| Armenia | 6,418 | 2,145 | 4,273 | −2,128 | 106 | 107 | 106 |
| Belarus | 59,212 | 27,342 | 31,870 | −4,528 | 98 | 96 | 100 |
| Kazakhstan | 79,074 | 47,724 | 31,350 | 16,374 | 101 | 95 | 113 |
| Kyrgyzstan | 5,623 | 1,564 | 4,059 | −2,496 | 98 | 114 | 93 |
| Russia | 549,536 | 348,602 | 200,934 | 147,668 | 97 | 95 | 101 |

*Source* (Eurasian Economic Integration 2019a, b)

From the above analysis, it follows that the EAEU countries mainly export natural resources (mineral, fuel, and energy resources), which account for more than 80% of all exports. Counter deliveries from abroad are represented by machinery, equipment, vehicles, chemical products, food products, and agricultural raw materials, which also account for 80% of imports.

## 4 Discussion

Economic conflicts and restrictions between the EAEU countries hindered the growth of mutual trade and the strengthening of relations between the countries over the past 15 years.

In our opinion, the longest one is the dispute between Russia and Belarus on prices of Russian gas (Expert.ru 2020). Demanding price linking to domestic prices and not receiving preferences, Belarus launched a tariff war with Russia, demanding to unilaterally raise the transit price of Russian energy carriers to Europe. Disagreements of this kind arise over oil supplies and pricing policies.

In 2015, Kazakhstan began discussing restrictions on Russian deliveries of oil products, cars, metal, chicken meat, flour, juices, and confectionery products due to the reduction in prices and the uncompetitiveness of similar products manufactured in Kazakhstan. The situation is developing under the continuous supervision of the EEC and the Ministry of Agriculture of Russia (Strokan and Trutnev 2015).

In February 2017, Kazakhstan imposed a temporary ban on the import of certain products from Kyrgyzstan, including potatoes, apples, persimmons, and pomegranates. The Ministry of Agriculture imposed restrictive measures due to the revealed phytosanitary violations in Kyrgyz products, which served as a significant reason for the temporary ban (Muminov 2019).

At the same time, large batches of agricultural products from Kazakhstan were banned from importing into Russia. Thus, according to the Federal Service for Veterinary and Phytosanitary Surveillance (Rosselkhoznadzor) (Federal Service for Veterinary and Phytosanitary Surveillance 2020), inspectors prevented the import of nine batches of agricultural products of the highest phytosanitary risk with a total weight of 52.8 tons into the Novosibirsk region. Products are delivered from Kyrgyzstan and Kazakhstan for further sale in large Siberian cities. In another case, there was an attempt to bring more than 80 tons of quarantine products (29 tons of dates, 11 tons of raisins, and 41 tons of onions) with invalid phytosanitary certificates from Kazakhstan to the Altai Krai in just one week (from March 4 to 11, 2020). Additionally, Rosselkhoznadzor notes that one-fifth of dairy products imported to Altai in 2019 was counterfeit.

At a meeting on the development of entrepreneurship at the end of 2019, the Governor Viktor Tomenko (Altai Pravda 2020) drew attention to the fact that local entrepreneurs suffer from unfair competition from those who neglect to pay duties on importing products to the territory of the union, and then to the territory of the region.

A serious problem for Russia today is the protection of the domestic market from the import of prohibited goods. For example, the Republic of Kazakhstan established preferences for certain types of goods from the countries of the World Trade Organization (WTO) (Eurasian Economic Commission, 2019a, b) in the form of a reduction in import duties, which have an order of magnitude different from the rates of the Common Customs Tariff of the EAEU for similar product groups. Goods imported from third countries of the world to Kazakhstan are subject to declaration at reduced rates and must remain within its territorial borders. However, this requirement is often violated in practice. Foreign goods are re-exported without any clearance to the remaining territory of the union. To solve this problem, the Federal Customs Service of Russia was forced to form special units to control goods released for domestic consumption. As long as the preference exists, there will be offenses.

The EAEU countries do not participate in the sanction war and could use this situation productively to develop national economies. Thus, Russia found itself face to face with difficulties and problems, and its partners in the union tried to earn dividends instead of showing solidarity, joining forces, and resisting the economic downturn. A trade embargo (Presidential Executive Office 2014) restricting the import into Russia of certain types of food products could well be commented on the deficit not due to food imports from third countries, but due to mutual trade supplies from the EAEU countries.

The current situation is a consequence of the fact that, occupying a dominant position in the EAEU, Russia decided to independently cope with the sanctions and the problems of import substitution without informing its partners in the union. Connecting the EAEU countries to the trade embargo would be a measure of real support for Russia in the sanction war, but would also protect the union's internal market.

The research results confirm that the need for a significant part of the commodity items imported by the member states of the EAEU from third countries of the world can be offset by goods produced in the union. In particular, in 2015–2018, 77% (825 commodity items) of the value of Armenia's imports from third countries may be exported from other EAEU countries—mainly from Russia. Today, the potential share of import substitution of products from the EAEU countries in Belarus is 29% (314 commodity items), in Kazakhstan—29% (322 commodity items), in Kyrgyzstan—58% (676 commodity items), and in Russia—1% (71 commodity items) (Eurasian Economic Commission, 2019a, b).

## 5 Conclusion

Despite all difficulties associated with the foreign economic situation and negotiations ongoing within the framework of the EAEU to resolve labor issues and internal conflicts, the EAEU has a promising future. This fact is evidenced by the tendency towards stabilization or a slight increase in economic indicators in mutual and foreign trade, as well as the creation of conditions for forming a single market.

Currently, a single market for pharmaceutical products has already been created; preparatory work is underway to create a single market for oil, gas, and electricity. An expert group was established to create a single financial market (2020–2025) (Eurasian Economic Commission 2019a, b). Prior to this period, the development of initiatives in the digital economy may become a separate area. For the EAEU countries, the integration of financial markets can provide several advantages in ensuring the economic stability in crisis situations and increasing the efficiency and competitiveness of financial institutions.

Previous economic indicators show a low economic integration within the EAEU. To improve the functioning of integration education, it is proposed to carry out several activities.

Given the unequal speed of economic development in the EAEU countries, the EEC, in its daily work, is recommended to use the international and domestic experience of integration processes, taking into account the features of each country and its economic potential.

To get a greater effect from the single market, it is necessary to unify the mechanism for removing trade barriers and non-tariff restrictions since they act as destabilizing factors in foreign and mutual trade. It is necessary to remove administrative and economic barriers to trade and improve the investment climate in the EAEU countries.

To increase the importance of customs and economic integration and increase the effectiveness of integration education, it is necessary to separate functions. In particular, it is proposed to distinguish two independent areas: (1) customs regulation (customs administration) and (2) economic regulation (creation and regulation of institutional entities of the Eurasian Economic Union). This is especially relevant in the context of forming a full-scale single market for the EAEU on the model of the EU.

The old model of economic development based on commodity export and inflow of cheap investments will not work in modern conditions. Therefore, the EEC should consider measures to (1) change the structure of exports towards production with higher added value, and (2) increase the share of innovative and knowledge-intensive products.

The inclusion of the EAEU countries in the trade embargo would be a measure of real support for Russia in the sanction war. Moreover, it would allow launching the internal market of the union. A trade embargo restricting the import of certain food products into Russia could compensate for the deficit, not by importing food from third countries, but by mutual trade supplies from the EAEU countries. Advertising support of goods and services within the union will contribute to the development of a positive perception of domestic products among consumers.

Ensuring equal access of the Union countries to research and innovation projects will significantly increase the scientific, technical, and intellectual potential of all states with a view to its implementation in the innovative sector of the EAEU economy. Increasing the legal capacity of the EAEU gives confidence in the emergence of a serious player in the international arena, setting itself ambitious tasks to bring the member states of the union to the path of stable and accelerated qualitative growth.

The global financial crisis, caused by the spread of coronavirus infection COVID-19, is a difficult stage in the life of the EAEU countries. Nevertheless, acting professionally and competently, the partner countries can overcome the epidemic with minimal losses.

## References

Akimov AV (2014) Sanctions of the West and responsible persons of Russia: food and tank building. In Khoros VG, Malysheva DB (ed) West-East-Russia. IMEMO RAS, Moscow, Russia, pp 35–43

Altai P (2020) Almost 30 tons of dates were not allowed into the Altai Krai. https://www.ap22.ru/paper/Pochti-30-tonn-finikov-ne-pustili-v-Altayskiy-kray.html

Center for Integration Studies of the Eurasian Development Bank (2019) Eurasian Economic Integration—2019. SRI EDB, Moscow, Russia

Eurasian Economic Commission (2019a) On the results of foreign and mutual trade of the member states of the Eurasian Economic Union. http://www.eurasiancommission.org/ru/act/integr_i_makroec/dep_stat/tradestat/analytics/Documents/express/October2019a.pdf

Eurasian Economic Commission (2015) Free trade agreement between the Eurasian Economic Union and its member states on the one part and the Socialist Republic of Vietnam, on the other part. Republic of Kazakhstan, Burabay

Eurasian Economic Commission (2019b) Eurasian Economic Commission. Facts and figures. http://www.eurasiancommission.org/ru/Documents/3071-%D0%95%D0%AD%D0%9A%20%D0%91%D1%80%D0%BE%D1%88%D1%8E%D1%80%D0%B0%20%D0%A6%D0%98%D0%A4_17.pdf

Expert.ru (2020) Gas conflict with Belarus. https://expert.ru/dossier/story/212578/?n=171

Federal Service for Veterinary and Phytosanitary Surveillance (2020) The Rosselkhoznadzor did not allow more than 52 tons of potentially unsafe products from Kazakhstan and Kyrgyzstan to be imported into the Novosibirsk Region. http://rshn-nso.ru/news/rosselxoznadzor-ne-dopustil-k-vvozu-v-novosibirskuyu-oblast-bolee-52-tonn-potencialno-nebezopasnoj-produkcii-iz-kazaxstana-i-kirgizii

Kofner Y (2019) Mutual and foreign trade of the Eurasian Economic Union. http://eurasian-studies.org/archives/14092

Kuzmina EM (2017) Economic development of the EAEU countries and prospects for economic integration until 2025. Analytical note. https://russiancouncil.ru/papers/EAEU2025-Policybrief-ru.pdf

Member States of the Eurasian Economic Union (2017) Customs Code of the Eurasian Economic Union. Moscow, Russia

Muminov A (2019) Trade wars in the EAEU—Politics or economics? https://kapital.kz/economic/76301/torgovyye-voyny-v-yeaes-politika-ili-ekonomika.html

Presidential Executive Office (2014) Decree "On the Application of Certain Special Economic Measures to Ensure the Security of the Russian Federation" (August 6, 2014 No. 560). Moscow, Russia

Strokan S, Trutnev O (2015) Embargo in Kazakh. Kazakhstan plans to limit imports of Russian goods. Kommersant. https://www.kommersant.ru/doc/2661304

# Auctions with Financial Externalities in Provision of Public Goods

**Dmitriy S. Khvalynskiy**

**Abstract** Auctions are on the rise and have become very useful tool of procurement of goods and services in public sector, because it reduces information rents of the bidders. The competition in the auctions became the basic term of revenue maximization. The author shows that competitive process in public auctions in Russia is losing intensity and there is a threat to the economic security in provision of public goods. The paper overviews research on auction theory and premium auctions, in which a premium is offered to encourage weak bidders to compete aggressively. The author proposes to use premium auctions in Provision of Public Goods in Russia, which may attract more entry. He also attempted to find the better design of the most famous of premium auctions—Amsterdam auction. The author proposes to determine the premium for only the second-highest bidder equal to 50% between the second-highest bid and secret reserve price, or third-highest bid, if it is below the secret reserve price. The proposed auction design may increase the competition and the revenue of the auctions for the auctioneer.

**Keywords** Auctions · Bid · Auction formats · English auction · Premium auction · Amsterdam auction · Procurement

## 1 Introduction

Auctions are on the rise; they have become a useful tool for purchasing goods and services in the public and private sectors. In recent years, their development has been associated with the widespread use of the internet, which has made auctions available to a broader range of bidders. Therefore, the number of bidders is essential for maximizing revenue. The larger the number of bidders, the lower the prices at the purchase auction.

A wide range of products is purchased at various auctions. More than 40% of large firms (with more than $100 million in revenue) surveyed in North America

D. S. Khvalynskiy (✉)
Altai State University, Barnaul, Russia
e-mail: hdms@email.ru

 
S. G. Maximova et al. (eds.), *Advances in Natural, Human-Made, and Coupled Human-Natural Systems Research*, Lecture Notes in Networks and Systems 234,
https://doi.org/10.1007/978-3-030-75483-9_20

use auctions for purchases. In the public sector, auctions are used in the public procurement award procedure. In many countries, a significant proportion of public sector procurement is implemented through public procurement. For example, in OECD countries, public procurement is estimated to account for about 12% of GDP.

Nevertheless, the study of the problem of incentives in public procurement remains urgent, since the basic models are useless. In other words, there is a significant gap between theory and practice. In this paper, we try to find the connection between game-theoretic models of auctions and the real functioning of the public procurement system.

The widespread use of auctions in public procurement requires an increase in their stability and effectiveness; otherwise, there is a threat to economic security in providing public goods (disruption of supplies, overspending, poor-quality execution, etc.).

It seems that auctions with intense competition show the best stability in efficiency. The competition is working just as expected, since on average, bids (standardized) decrease with the increase in the number of actual and potential bidders. For example, according to the European Commission (2017), "Public procurement relies on open competition to ensure the best value for public money. Such a competitive process is either absent or loses its intensity. In 2006–2016, the number of tenders with only one application increased in the OECD countries from 17 to 30%"; in Russia—up to 56%. "The average number of bids per tender in OECD countries has decreased from five to three over the same period." In Russia, the competition for government contracts is lower, and the median bidder count is 1.36.

Bidders tend to be very heterogeneous in size, which is likely to limit competition. That is why attracting more competition through the development of contract rules and auctions is a desirable policy, and auctions with financial externalities demonstrate the most vigorous competition in public procurement.

This paper examines auctions with financial externalities, which occur when the losing bidders benefit from the low price paid to the winning bidder.

## 2 Materials and Methods

We used the methodology of auction theory, which is related to the Nobel Prize winner Wu. Vickrey (1961), who is considered its founder and the first to investigate auction bidding from the point of view of game theory.

The study of auction theory began with the Independent Private Values [IPV] model. In this model, one indivisible object must be purchased from one of the risk-neutral bidders. Each bidder has information about the value of the contract for themselves, but not for other bidders. All values are independently taken from a continuous distribution. We assume that the bidders are competing, so we analyze the auction as a non-cooperative game.

The main result in auction theory is the Revenue Equivalence Theorem, according to which risk-neutral sellers in the independent private values model lead to the same expected revenues for the buyer in Dutch, English, first-price sealed-bid, and Vickrey second-price auctions. Usually, losers do not care about the size of the winner's payout. However, the real procurement auctions usually do not confirm this assumption.

Maasland and Onderstal (2002a, b) studied the case of financial externalities, in which during the procurement auction losing bidders would prefer to reduce the winner's payment. The seller receives more revenue in the second-price auction than at the first-price auction in situations with financial externalities. In second-price procurement auctions, bidders can decrease the winner's payment by lowering their bids. Various literature is devoted to the auctions with financial externalities, or auctions with a "spite motive" (Cooper and Fang 2008; Morgan et al. 2003; Sharma and Sandholm 2010; Lu 2012).

Rothkopf et al. (2003) considered the model of procurement auctions and showed that subsidizing inefficient competitors can reduce the expected cost of the project and increase economic efficiency. Some subsidy is generally better than no subsidy for a wide range of parameters. From a theoretical point of view, the approach of Myerson to the design of the mechanism (Myerson 1981) determined the optimal choice of the seller. The proposed mechanism involves lending to weaker bidders to encourage competition and setting reserve prices for specific bidders. However, in practice, sellers often lack the detailed information needed to select loans and reserve prices optimally; they cannot always discriminate between bidders.

The introduction of the so-called *premium* provokes weak bidders setting an endogenous reserve price for stronger competitors, dramatically affecting the selling price. In a premium auction, the seller offers some "payback," called a premium, to a set of high bidders at the end of the auction. Premium auctions are believed to be superior to standard auctions, such as (1) the first-price sealed-bid auction, (2) the Dutch auction, (3) the English auction, and (4) the Vickrey auction, because bidders compete fiercely for the premium by raising the price paid by the winner. Additional competition may then offset the premium the seller must pay to the runner-up.

## 3 Results

For 3 quarters of 2019, 2.22 million records of procurement worth 6.7 trillion rubles were placed in the unified information system of public procurement of Russia. The cost of tenders with only one bid (or no bids at all) increased to 3.75 trillion rubles (56%) (Ministry of Finance of the Russian Federation 2020a).

In 2018, 3.24 million records of procurement worth 8.3 trillion rubles were placed in the unified information system of public procurement of Russia. The cost of tenders with only one bid (or no bids at all) increased to 3.36 trillion rubles (41%). Of these purchases, 65% were made with only one bid. Simultaneously, since customers made 60% of purchases through an English reverse electronic auction, the largest number of unsuccessful procedures was recognized by the English auction results (Ministry of Finance of the Russian Federation 2019a).

In 2017, 3.16 million records of procurement worth 7.1 trillion rubles were placed in the unified information system of public procurement of Russia. The cost of tenders with only one bid (or no bids at all) increased to 2.49 trillion rubles (35%). In 83.5% of cases, contracts are concluded with one bidder (supplier) (Ministry of Finance of the Russian Federation 2018).

According to the unified information system of public procurement of Russia, the average number of bidders participating in auctions in the first half of 2019 decreased to 1.36 compared to 1.52 in the first half of 2018 (Ministry of Finance of the Russian Federation 2020b).

The average number of bidders participating in auctions, according to the unified information system of public procurement of Russia, in 2018 decreased to 1.64 compared to 1.7 in 2017 (Ministry of Finance of the Russian Federation 2019b).

The competitive process at public auctions in Russia is losing intensity; there is a threat to economic security in providing public goods (disruption of supplies, overspending, poor-quality execution, etc.). In 2017–2019, the number of tenders with only one bid (or no bids at all) increased from 35 to 56%. The average number of bids per auction decreased from 1.7 to 1.36 over the same period, and the largest number of unsuccessful procedures were recognized by the results of the English reverse auction.

Milgrom (2004) theoretically proved that a premium auction could attract more visitors than an English auction, thereby generating more revenue. For example, in a single-lot competitive tendering, it may provide losing bidders a money premium corresponding to the percentage difference between the losing bid and the lower one (Milgrom 2004). This situation means that the best loser (the bidder who submitted the second-lowest price) gets a 30% share of the difference between the third and the second losing price, while the second-highest loser gets a 30% share of the difference between the fourth and the third-highest price offer. However, both premiums and credits give disadvantaged bidders an extra incentive to bid more aggressively. In this case, strong bidders are forced to react more aggressively.

Goeree and Offerman (2004) showed another practical solution to the seller's problem—*the Amsterdam auction*, where a premium encourages weak bidders to compete aggressively. This auction format, which was used to sell houses in Amsterdam, treats all bidders equally and does not rely on detailed information about their distribution of value. The Amsterdam auction is a variant of the premium auction, operating in a certain way. At the first stage, bidders participate in an English auction. The auctioneer raises the price until there are two bidders left. The price at which the last bidder dropped out of the auction is called *the bottom price*. At the second stage, the remaining two bidders submit sealed bids not lower than the bottom (reserve) price. The bidder who submitted the highest bid wins, and the winner and the runner-up receives a premium proportional to the difference between the losing bid in the second stage and the bottom price. K. Goeree and T. Offerman derived strategies for bidders with private values in the symmetric and asymmetric case. They found that a bidding strategy is increasing in value and in the bottom price, which shows that more aggressive bidding of weak bidders provokes strong bidders to bid more aggressively as well. By awarding the premium to the highest losing bidder, the Amsterdam auction favors weak bidders without implementing Myerson's (1981) optimal auction. The researchers analyzed the Amsterdam auction from a theoretical and experimental point of view. The Amsterdam auction works particularly well when bidders are asymmetric. In this case, weak bidders bid more aggressively, hoping to obtain a premium and setting a higher reserve price for strong bidders in the second stage. In a series of experiments, authors compared the standard first-price and English auctions, the optimal auction, and two variants of the Amsterdam auction. There are two types of Amsterdam auctions: first-price and second-price Amsterdam auction. In the first-price Amsterdam auction, the winner pays their bid, while in the second-price Amsterdam auction, the winner pays the runner up's bid in the second stage. K. Goeree and T. Offerman conducted an experiment comparing the results of the Amsterdam auctions with first-price and English auctions. In the case of strong asymmetry, the Amsterdam auction with second-price yields the highest revenue, followed by first-price Amsterdam auction, first-price auction, and English auction. English auction performs particularly badly in the case with strong asymmetry, low revenue, and high variance. On the other hand, first-price and Amsterdam auctions have considerably higher revenue with a small variance. Therefore, in a highly asymmetric environment, Amsterdam auctions perform better than standard ones.

Goeree and Offerman (2004) proved it in theory and laboratory conditions that two variants of the Amsterdam auction bring a higher average revenue than the first-price sealed-bid auction and the English auction in a setting with strong ex-ante bidder asymmetries. With strongly asymmetric bidders, the second-price Amsterdam auction raises substantially more revenues than standard auctions and only slightly less than the optimal auction. Hu et al. (2011) observed in a laboratory experiment that the Amsterdam auction is less favorable for cartel formation and raises more money than the English auction and the first-price sealed-bid auction in the case of strong ex-ante bidder asymmetries. Therefore, the Amsterdam auction is a mechanism that encourages entry into thin markets. Furthermore, there is less chance of collusion at the Amsterdam auction, since cartel members have a strong incentive to cheat to get a premium.

Nevertheless, in a theoretical paper, Hu et al. (2011) discovered that the Amsterdam auction generates a lower revenue than the English auction in the case of risk-averse bidders. Data from a laboratory experiment by Brunner et al. (2014) are consistent with this assumption. The Amsterdam auction can be considered a Pareto improvement over the English auction only if the seller is sufficiently risk-averse (Hu et al. 2017).

We claim that it is impractical to determine the premium for the auction winner, since the savings are the winner's revenue, who is not interested in increasing the auctioneer's revenue. Suppliers are aware of the direct dependence of their revenue on the contract price; they are ready to offer their competitors a premium (usually in the form of a bribe) if it is strictly less than the additional revenue that the supplier will over the standard auction result.

Trying to determine the premium from the third bid, we encourage the first, second, and third bidders to collude. The winner could offer a third party up to their full premium for raising the contract value above the same value.

For example, we have three bidders with contract valuations equal to $V_1 = 3$, $V_2 = 2$, $V_3 = 1$. Consequently, the contract's price will be equal to $E(V_n) = 2$—the second-highest bid. In the Amsterdam auction, one should pay to the highest bidder and the second-highest bidder 50% between the second-highest and third-highest bids—0.5 for each. The expected utility of the third bidder can rise from zero to a strictly positive number due to the minimum bribe he receives from the highest bidder and the second bidder. That is why we can get the following bids in the auction $b_1 = 3$, $b_2 = 2$, $b_3 = 0$, and we will pay the highest bidder and the second-highest bidder 50% between the second-highest and third-highest bids—1.0 for each.

We propose to determine a premium only for the second-highest bidder, equal to 50% between the second-highest bid and the secret reserve price (the price at which the buyer has no obligations, and which is not publicly respected), or the third-highest bid if it is lower than the secret reserve price. This step can be used to deter compliant bidding behavior.

## 4 Discussion

In the seminar work of the Nobel laureate Myerson (1981) on optimal auctions, he stated that the expected income of the weakest bidder is a sufficient statistic for designing an auction. The higher the bidder's utility to conceal the value of the contract, the lower the expected return. Suppose the buyer pays the loser a portion of the auction revenue, the lowest type of expected utility increases from zero to a strictly positive number, because he receives a portion of the bid. Therefore, the losing bidder has no benefit in hiding the bidder's contract value; it lowers the expected contract price and increases the bidder's revenue.

The Amsterdam auction gets the revenue from the third-price auction, since the contract value will be the second-highest bid, and one has to pay the highest bidder and the second bidder up to 100% between the second and third bidders-50% for each bidder. This step is not optimal, because the optimal auction gets the second-highest value.

We argue that determining the premium for the auction winner is impractical, since the expected revenue of the weakest bidder is a sufficient statistic for designing an auction. Suppliers are aware of the direct dependence of their revenue on the contract price and are ready to offer their competitors a premium (usually in the form of a bribe), if it is strictly less than the additional revenue that the supplier will receive in excess of the standard auction result.

The second-highest bidder that we offer has a permanent premium of 50% between the second-highest bid and the secret reserve price, or the third-highest bid if the latter is lower than the secret reserve price. That is why the second-largest bidder has no incentive to bribe the third-largest bidder, since this step will not change the secret reserve price determined by the auctioneer. The highest bidder has no incentive to bribe the second-highest bidder, because they have to give them significantly more than 50% between the second-highest and third-highest value, which is unknown. This behavior can cause the second-highest bid and the third-highest bid to be exchanged and lead to the winner's losses.

For example, we have three bidders with contract valuations $V_1 = 3$, $V_2 = 2$, $V_3 = 1$. The winner $V_1$ pays $V_2$ a bribe equal to 0.6 to get the bid $b_2 = 1$ and get an additional revenue equal to 0.4. However, if the bidder $V_3$ knows about the bribe, he will bid higher than 1 to get the premium, confident that they lose the auction in any case. This case would deprive the winner's additional revenue and could lead to their losses.

## 5 Conclusion

Competition at auctions is the primary condition for maximizing revenue. We identified a significant decrease in competition in public procurement in Russia from 2017 to 2019 and a threat to economic security in providing public goods. The paper suggests using the premium auction in Russian public procurement to attract more participants and increase competition. Additionally, we tried to find the best design of the most famous of the premium auctions—the Amsterdam auction.

We propose determining the premium only for the second-largest bidder, equal to 50% between the second-largest bid and the secret reserve price, or the third-largest bid if it is lower than the secret reserve price. While the losing bidders would prefer the winner to receive as little revenue as possible, the proposed auction design leads to more aggressive bidding and increases competition and auction revenue for the auctioneer. The last result is real due to the ability of direct bidders to influence the level of payment of the winner in the English reverse auction by changing their own bids.

## References

Brunner C, Hu A, Oechssler J (2014) Premium auctions and risk preferences: an experimental study. Games Econom Behav 87:467–484

Cooper D, Fang H (2008) Understanding overbidding in second price auctions, an experimental study. Econ J 118:1572–1595

European Commission (2017) Public procurement strategy. https://ec.europa.eu/growth/single-market/public-procurement/strategy_e0

Goeree K, Offerman T (2004) Notes and comments the Amsterdam auction. Econometrica 72(1):281–294

Hu A, Offerman T, Zou L (2011) Premium auctions and risk preferences. J Econ Theory 146:2420–2439

Hu A, Offerman T, Zou L (2017) How risk sharing may enhance efficiency in English auctions. Econ J 128(610):1235–1256. https://doi.org/10.1111/ecoj.12439

Lu J (2012) Optimal Auctions with asymmetric financial externalities. Games Econom Behav 74:561–575

Maasland E, Onderstal S (2002a) Auctions with financial externalities. CentER Discussion Paper 2002–21:1–32

Maasland E, Onderstal S (2002b) Optimal auctions with financial externalities. CentER Discussion Paper 22:1–17

Milgrom P (2004) Putting auction theory to work. Cambridge University Press, Cambridge, MA

Ministry of finance of the Russian Federation (2018) Analytical report on the results of monitoring purchases, goods and services for state and municipal needs in accordance with Federal law No. 44-FZ of 05.04.2013 "On the contract system for the procurement of goods, works and services for state and municipal needs" for 2017. Moscow, Russia. https://www.minfin.ru/common/upload/library/2018/09/main/svodnyy_analiticheskiy_otchet_-_2017.docx

Ministry of Finance of the Russian Federation (2019a) Analytical report on the results of monitoring purchases, goods and services for state and municipal needs in accordance with Federal law No. 44-FZ of 05.04.2013 "On the contract system for the procurement of goods, works and services for state and municipal needs" for 2018. Moscow, Russia. https://www.minfin.ru/common/upload/library/2019/04/main/Svodnyy_analiticheskiy_otchet_2018_itog.docx

Ministry of finance of the Russian Federation (2019b) Monitoring the application of Federal law No. 223-FZ of 18.07.2011 "On procurement of goods, works, services of separate types of legal entities" in 2018. Moscow, Russia. https://www.minfin.ru/common/upload/library/2019/04/main/223.docx

Ministry of Finance of the Russian Federation (2020a) Analytical report on the results of monitoring purchases, goods and services for state and municipal needs in accordance with Federal law No. 44-FZ of 05.04.2013 "On the contract system for the procurement of goods, works and services for state and municipal needs" for the third quarter of 2019. Moscow, Russia. https://www.minfin.ru/common/upload/library/2019/10/main/3_kvartal.pdf

Ministry of finance of the Russian Federation (2020b) Monitoring the application of Federal law No. 223-FZ of 18.07.2011 "On procurement of goods, works, services of separate types of legal entities" in the first half of 2019. Moscow, Russia. https://www.minfin.ru/common/upload/library/2019/08/main/223-fz_1_polugodie_2019_6.0.docx

Morgan J, Steiglitz K, Reis G (2003) The spite motive and equilibrium bidding in auctions. The B.E. J Econ Anal Policy 2(1):1–27

Myerson R (1981) Optimal auction design. Math Oper Res 6:58–73

Rothkopf MH, Harstad RM, Fu Y (2003) Is subsidizing inefficient bidders actually costly? Manage Sci 49(1):71–84

Sharma A, Sandholm T (2010) Asymmetric spite in auctions. In Fox M, Poole D (eds) Proceedings from the national joint conference on artificial intelligence (AAAI). Atlanta, GA: The AAAI Press

Vickrey W (1961) Counterspeculation, auctions and competitive sealed tenders. J Finance 16:8–37

# Professional Standards as a Tool to Provide for Economic Safety of Businesses and Restraints to Their Implementation

**Elena N. Bolkhovitina, Iuliia V. Ugarova, and Iurii V. Gorbunov**

**Abstract** Between 2019 and 2020 Russia has experienced the introduction of the national qualification system based on professional standards. The National Qualification System (NQS) aims mainly at bringing at compliance supply and demand in the labor market. It must significantly enhance the efficiency of the human resources in the national economy and subsequently, improve economic safety both at micro- and macro levels. The article considers how the NQS and economic safety of the company are interconnected as well as major issues which national businesses face when introducing professional standards into their work. The research is based on the analysis of HR processes in over fifty businesses in Altai Krai and the in-depth interview with specialists. The revealed issues while introducing professional standards have been grouped according to their reason: those which deal with the quality and content of professional standards and enactments of the business sphere; the issues of information and psychological origin caused by the CEO personality and internal issues of the business. The specialists of the Regional center for qualification development possess great experience which allowed them to offer some solutions, implement them and witness positive results.

**Keywords** Professional standard · Economic safety · National qualification system · Human resources · Qualification level

## 1 Introduction

Recently, the issues of national economic security have become quite relevant, especially in the context of permanent economic challenges and crises. Since 2014, Russia has been under constant economic pressure, and in 2020, following the coronavirus pandemic, the global economic crisis broke out.

E. N. Bolkhovitina (✉) · I. V. Ugarova
Polzunov Altai State Technical University, Barnaul, Russia
e-mail: elenaobraz@mail.ru

I. V. Gorbunov
Altai State University, Barnaul, Russia

 
S. G. Maximova et al. (eds.), *Advances in Natural, Human-Made, and Coupled Human-Natural Systems Research*, Lecture Notes in Networks and Systems 234,
https://doi.org/10.1007/978-3-030-75483-9_21

Most approaches consider economic security at three levels: (1) the macro-level (national), (2) the meso level (regional), and (3) the micro-level (entrepreneurial).

This approach can be called hierarchical, since each subsequent level is based on the previous one, and each level has its external and internal factors that protect from threats, which the subject of economic security can manage (Mironova 2016).

The basis of economic security and the successful implementation of socio-economic development of enterprises, regions, and entire Russia is the efficient use of economic resources (Kiselev and Kapitonov 2014). Therefore, a person is the most important intellectual and professional resource (Pecherskaya, Krivtsov & Tarasova, 2019). This is especially important for economic entities, since economic activity shows a person's participation in the labor activity and depends on this person (Savina 2018). It is noteworthy that highly qualified human resources are a means of national competitive advantage in the international market.

One of the main threats to the economic security of the business is the professionalism of the personnel. Numerous studies show a significant imbalance of supply and demand in the labor market of Russia and its regions. Universities and vocational schools regularly graduate specialists. However, according to the employers, they do not meet their requirements. Employers are dissatisfied with the level of qualification of specialists; they are experiencing a shortage of highly qualified personnel.

In this regard, Russia has created a National Qualification System [NQS] and a unified qualification base. The latter is mainly aimed at the compliance of the educational process with the requirements of employers, professional standards, independent qualification assessment, as well as professional and social accreditation of educational programs. The starting point was the Decree "On actions for the realization of the state social policy" (May 7, 2012 No. 597) (Presidential Executive Office, 2012), which defines the priorities of providing jobs with highly qualified personnel.

When developing the Russian qualification system, the experts drew on the experience of the European Union [EU] (Bohlinger 2019; Lester 2015; Young 2008). The EU has a European Qualification System, which allows for the integration of the national qualification systems of the member states; they transfer and accept qualifications and increase the competitiveness of the labor force in the labor market (Pecherskaya et al. 2019).

The Regional Resource Center for Qualification Development [RRCQD] provides advisory support and guidance when introducing the NQS in Altai Krai. Similar centers were established in other Russian regions. The analysis of RRCQD actions and the experience of neighboring regions showed that the NQS introduction in Russia is slowing down, and many issues, challenges, and problems are arising. Some of the issues cannot be solved at the regional level.

Only individual specialists undergo an independent assessment of their qualifications, as a mandatory requirement for them; they cannot work without appropriate documents. Such categories of workers include welders and workers in the elevator industry. Although the independent assessment of qualifications has been in effect since 2016 (Federal Law *"On the Independent qualification assessment"* (July 3, 2016 238-FZ) (Russian Federation, 2016) until May 1, 2020, in Russia, only 60,333 people passed this assessment and received a qualification certificate (data of the

Register of Independent Qualification Assessment [IQA] in the National Qualification Development Agency [NQDA] https://nok-nark.ru/). It accounts for about 0.08% of the labor force in the Russian Federation.

As for professional and social accreditation, it is mainly held by large universities in the central regions of Russia. Regional universities are only getting involved in this process.

From the list of key elements of the NQS, only professional standards are applied at the regional level. Professional standards were introduced in enterprises due to the Decree of the Government of the Russian Federation No. 584 of June 27, 2016 (Russian Federation, 2016). Considering the specific consequences of the decree, mainly municipal, budgetary enterprises and businesses with more than 50% state ownership are implementing professional standards in their work. However, the decree itself does not limit the need to apply professional standards only to these entities.

Commercial organizations are not yet interested in applying professional standards.

All these facts served as the basis for studying the implementation of professional standards in regional companies (as the most active process in the NQS) to identify problems and possible solutions.

## 2 Materials and Methods

We reviewed and analyzed the following materials: (1) regulatory documents on the research topic (Presidential Executive Office 2012; Russian Federation 2016; Russian Federation 2016); and (2) analytical materials of the National Qualifications Development Agency (National Qualifications Development Agency, n.d.).

To solve the research tasks, we applied the following general scientific methods: (1) analysis, (2) synthesis, (3) comparison, (4) method of expert assessment, (5) description, and (6) statistical methods.

Besides, we conducted an in-depth interview with specialists of 100 enterprises of Altai Krai (state ownership of more than 50%) and analyzed the personnel composition of more than 50 enterprises of Altai Krai.

## 3 Results

The research results showed that by the beginning of 2019, about 31% of municipal and state entities in Altai Krai had fully or partially implemented professional standards, and 44% were planning to start (Fig. 1).

Meanwhile, all in-depth interviewers in all the sectors shown in Fig. 1 claimed that specific problems arise during this process. Table 1 summarizes the issues and systematically presents them.

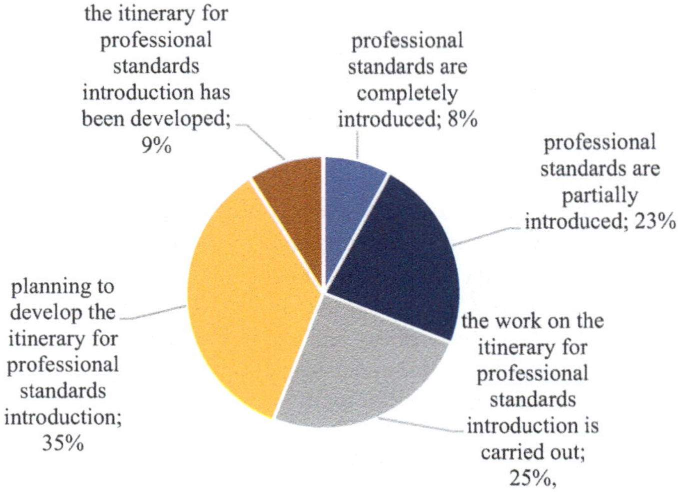


**Fig. 1** The structure of state and municipal entities and businesses with over 50% state ownership in Altai Krai due to the rate of professional standards introduction, %. *Source* Compiled by the authors

Thus, we revealed three groups of problems that can be called external, internal, and those caused by the personality of the CEO or expert. Comparing the results obtained with data from other regions (obtained at international conferences and forums on the development of NQS in Russia), we argue that most of the issues are not regional but national.

## 4 Discussion

Issues related to the quality and content of professional standards and regulations are considered external to companies, since they do not participate in their development. Very often, representatives of the real sector of the economy make claims on the quality of professional standards.

We can define the following issues:

- Lack of essential responsibilities in some fields of professional activities;
- Outdated responsibilities and actions;
- Incorrect wording of responsibilities.

The following issues remain unsolvable both within the company and at the regional level, as they are caused by a legal conflict resulting from the simultaneous application of Unified Qualification References and professional standards. Unfortunately, the explanations of the Ministry of Labor and Social Protection of the Russian Federation and the National Qualification Development Agency do not

**Table 1** The issues arisen when introducing professional standards in the work of Altai Krai companies

| Groups of issues | Revealed issues |
|---|---|
| Quality and content of the professional standards and regulations | Lack of professional standards for some occupations |
| | Some professional standards lack necessary common job descriptions (a vet, vocational staff, etc.) |
| | When changing the job titles due to the professional standard, the employees lose their right for early retirement due to the working conditions (article 57, RF Civil Code) |
| | There is a choice between concurrent Unified Qualification Reference (UQR) and professional standards. Authorities opt for UQR as it states the requirements in a simpler manner and there is no need to sort out levels and sublevels |
| Issues of information and psychological origin caused by the personality of the CEO (specialist) | CEOs and/or specialists are insufficiently ware of professional standards |
| | General misunderstanding of basic processes and regulations |
| | Wait-and-see attitude |
| | Unwillingness to bear responsibility for the decisions to employ professional standards (attempts to find a "fall guy") |
| Internal issues arisen while introducing professional standards | Low quality of job descriptions, when it is impossible to understand what the employee does and, correspondingly, they fail to select the professional standard |
| | Incompliance of duties in professional standards and the current job description for those who work in this professional sphere |
| | Incompliance of employee qualification and the requirements of the professional standard |
| | Need for significant expenses for professional training or advanced training |
| | Having vacancies there is no chance to employ a candidate without the necessary training and qualification level |
| | No understanding what to do if a worker is subjected to several professional standards |
| | Attempts to use the professional standards in order to dismiss an employee or demote them when the authorities are unwilling to train the employee at the company expense |
| | Some CEOs share the following view: "To introduce and implement professional standards means working out an itinerary, nothing else must be done" |
| | When introducing professional standards, a plan was shaped, protocols on standards introduction were made, job titles were changed. However, other personnel documents failed to be changed |

*Source* Compiled by the authors

help solve this problem, since they are not legal. Therefore, these explanations are not convincing to employers and lead to the decision to avoid professional standards due to the issues that arise. Hence, they prefer to work with references.

The issues of psychological and informational origin are related to the next essential management process –resistance to change. Based on our experience and the analysis of implementing professional standards, we must admit that most specialists and managers of companies in the region are "in denial" or "in anger." They have been there for quite a long time and cannot resume on their own. The implementation of changes at such stages shows extremely low efficiency, and the processes are slowed down. This is evidenced by the long period of implementation of the NQS in Russia.

The stage "in denial" is vividly illustrated by the wait-and-see attitude, and it has two options: either wait until the NQS and the introduction of professional standards are abolished, or wait until others are punished and decide to introduce the system in their company. Unresolved problems in this group lead to internal problems.

The problems are caused both by the low quality of personnel processes, including untrained methods of dealing with objections, and the low qualifications of personnel officers, lawyers, and company managers.

External and informational-psychological problems cause inconvenience to specialists and managers, while internal problems can lead to more serious consequences. Among other things, attempts to use professional standards to eliminate unwanted employees may result in legal proceedings. Improper management of personnel and documents when implementing professional standards, changing job descriptions and job titles can lead to fines. One should remember that an incorrect professional standard or qualifications can lead to unnecessary costs for training or professional development of employees. Thus, the company receives higher financial and legal risks for its economic security.

However, implementing professional standards entails some advantages:

- Ability to plan the development of an employee in the company, planning career prospects based on professional standards, as they indicate the transition from level to level and what skills or training are needed for a specialist;
- Absence of a need to build requirements for the company's specialists from scratch, as they are already formed within the standards and can be used and adapted to some specific areas, if necessary;
- Help in determining whether an employee meets the requirements of a job. Moreover, professional standards can be used to evaluate employees, indicating the skills and knowledge an employee should have to perform specific duties;
- Assistance in assessing how well the requirements for employees in the company correspond to the general industrial requirements. Thus, they will learn how competitive the personnel and the company are.

Therefore, even if there are pressing problems with implementing professional standards in the company, there are some benefits that contribute to the professionalism of the personnel, competitiveness in the market, and economic security of companies.

## 5 Conclusion

In our opinion, the issues identified in this study should not be left unresolved, since the implementation of the NQS has been dramatically delayed and, according to experts, with the positive development of the situation, it will take from at least five to eight years. If the identified problems remain, the process may take decades or even stop.

As we said earlier, some issues are unsolvable at the regional level. However, stakeholders regularly make requests and suggestions to change the provisions and guidelines for implementing professional standards.

An important solution to the low quality of professional standards is to involve the business in their development and modification. Currently, this is a relatively rare phenomenon. RRCQD specialists at their seminars talk about the procedure for developing professional standards and show alternatives for involving local employers in this process.

All problems arising from low awareness, lack of understanding of the procedure for implementing professional standards in the company can be solved using various additional educational programs. Their duration and content should correspond to the incoming requests. There are several programs in Altai Krai. The classes involve experts in the independent assessment of qualifications, the implementation of professional standards, and lawyers.

In-depth interviews showed that it is vital to work with company managers separately, as they are often the main *hampering* element in implementing standards. Their main problem is their inability to take responsibility for the decisions they consider imposed. In other words, they are entirely unaware of the essence of professional standards and related processes, and sometimes the qualifications of the CEO are relatively low. This situation requires educational programs for managers, personal coaching, and assistance both in understanding professional standards and in working out possible ways and management techniques to combat objections when implementing standards in the company. These can be guidelines or managerial and psychological coaching, which means engaging highly qualified, well-rounded experts.

## References

Astafeva OV, Mashtakeeva DK, Osipova IA (2019) Council for professional qualifications in the staff training for the digital economy. In Mantulenko V (ed) GCPMED 2018: international scientific conference "global challenges and prospects of the modern economic development". Future Academy, Samara, Russia, pp 852–859

Bohlinger S (2019) Ten years after: the 'success story' of the European qualifications framework. J Educ Work 32(4):393–406. https://doi.org/10.1080/13639080.2019.1646413

European Commission (n.d.) Descriptors defining levels in the European Qualifications Framework (EQF). https://ec.europa.eu/ploteus/en/content/descriptors-page

European Commission (n.d.) Find and compare qualifications frameworks. https://ec.europa.eu/ploteus/en/compare?field_location_selection_tid%5B%5D=471&field_location_selection_tid%5B%5D=472

Government of the Russian Federation (2016a) Resolution "On the peculiarities of applying professional standards in terms of requirements that are mandatory for application by state extra-budgetary funds of the Russian Federation, state or municipal institutions, state or municipal unitary enterprises, as well as state corporations, state companies and business entities, more than fifty percent of the shares whose capital is in state ownership or municipal property (June 27, 2016a, No. 584). Moscow, Russia

Government of the Russian Federation (2016b) Federal Law "On independent qualification assessment" (July 3, 2016b, No. 238-FZ). Moscow, Russia

Kisloshchaev PA, Kapitonova NV (2014) System of criteria, indicators and threats of economic safety. Bull Buryat State Univ 4:12–19

Lester S (2015) The European qualifications framework: a technical critique. Res Post-Compulsory Educ 20(2):159–172. https://doi.org/10.1080/13596748.2015.1030251/

Mironova OA (2016) The concept of economic security and its development. Innovative Dev Econ 6(36):266–269

National Qualifications Development Agency (n.d.) Library. https://nark.ru/lib/?type=2

Office for Official Publications of the European Communities (2008) The European qualifications framework for lifelong learning (EQF). European Commission, Brussels, Belgium. https://doi.org/10.2766/14352

Pecherskaya EP, Krivcov AI, Tarasova TM (2018) Assessment of employees competencies in the system of professional qualifications in the digital economy. Account Anal Audit 6(3):68–75. https://doi.org/10.26794/2408-9303-2019-6-3-68-75

President of Russian Federation (2012) Decree "On measures for the implementation of state social policy" (May 7, 2012, No. 597). Moscow, Russia

Savina TN (2018) Diagnostics of the regional labour market in the format of economic security indicators. Astra Salvensis 6:929–939

Young M (2008) Towards a European qualifications framework: some cautionary observations. J Eur Ind Train 32(2/3):128–137. https://doi.org/10.1108/03090590810861677

# Economic Security of the Russian Federation: The Balance of National Interests and International Cooperation

**Irina I. Ignatovskaya, Nadegda Y. Grinchinko, and Vera S. Zybkova**

**Abstract** The integration of the Russian Federation takes place in various areas, one of which is the sphere of financial relations. The financial system is a category of international, global scale, and it is characterized by deep risks. The increased level of risks in the financial sector is due to both internal and international reasons. Risks depend on objective conditions that affect the economy of States (for example, pandemics), on the actions of States, on their financial policy and its components: banking, monetary, budgetary and tax. In this regard, scientific justification of the consequences, the results of decisions taken, and international cooperation to exclude or minimize the likelihood of negative consequences are important. Many institutions of financial law have been influenced by international standards and rules related to the budget sphere, countering the laundering of proceeds from crime, the system of mandatory insurance of deposits of individuals in banks of the Russian Federation, and ensuring the stability of the banking system and banking activities. It is important that emerging international standards that unify approaches to the legal regulation of financial relations do not jeopardize the economic security and sovereignty of the Russian Federation.

**Keywords** Economic security · Priority of the constitution of the Russian Federation · International agreements · International cooperation in the tax sphere · International cooperation in the budget sphere

## 1 Introduction

Currently, increasing attention is paid to the processes of globalization and integration of economic processes. One cannot ignore them because of the growing contradictions. In this regard, ensuring the interests of Russia and its economic security come to the fore.

I. I. Ignatovskaya (✉) · N. Y. Grinchinko · V. S. Zybkova
Altai State University, Barnaul, Russia
e-mail: Ignatovskaya.asu@mail.ru



S. G. Maximova et al. (eds.), *Advances in Natural, Human-Made, and Coupled Human-Natural Systems Research*, Lecture Notes in Networks and Systems 234,
https://doi.org/10.1007/978-3-030-75483-9_22

Economic security is recognized as the state of protection of the national economy from external and internal threats, which ensures the economic sovereignty of the country, the unity of its economic space, and the conditions for the implementation of strategic national priorities. This is how economic security is defined by Presidential decree No. 208 of May 13, 2017, approved by the economic security Strategy of the Russian Federation for the period up to 2030 (Presidential Executive Office 2017). According to the strategy, objective signs of the destruction of the unipolar world are already clearly visible. The transition to multipolarity is accompanied by the growing geopolitical instability and unsustainable development of the world economy and a sharp aggravation of global competition. There is a desire to redistribute influence in favor of new centers of economic growth and political attraction. Significant changes occur in international law, military-political, and economic fields. All this requires the research for new ways and conditions to protect the economic interests of a sovereign country in the context of international integration.

The authors agree with K. R. Tayupova, who, based on the economic security Strategy, identifies the most urgent external threats to the economic security of the Russian Federation that require improvement of legal regulations:

- Formation of transnational capital that has left the national jurisdiction and is based on offshore territories. This capital is a threat to Russia's economic security since it can undermine the stability of the national monetary unit and the stability of the banking system;
- Increasing the role of transnational corporations (TNCs), which have already transformed into quasi-state institutions and claim to perform the functions of countries;
- Digitalization of economic relations. There are several reasons why there is a lack of positive assessment of technology development. First, the speed of technology development is so rapid that it leads to the problem of technological deflation—the rapid cheapening and spread of innovations that affect GDP (Masood 2016). Second, the globality of technology and often the absence of national legal regulation practically equate with lack of control and, therefore, is at least a potential threat to security (Tayupova 2019).

In addition to these threats, we can add (1) problems with insufficient qualifications and key competencies of domestic specialists of the digital economy; (2) an unbalanced national budget system; (3) insufficient public administration; and (4) a high level of criminalization and corruption in the economic sphere.

These problems must be considered when building the interaction between Russia, foreign countries, and international economic organizations.

## 2 Materials and Methods

The research object is a set of public relations formed in the process of international financial cooperation, as well as the problems of ensuring the economic security of the Russian Federation. The subject is the norms of international agreements regulating financial (primarily tax and budget) relations and their impact on the development of the financial legislation of Russia (Paris convention 1960; Strasbourg Convention 1988; Multilateral Convention 2016).

To conduct the research, the authors use the following set of complementary research methods: system, comparative-legal, formal-legal, and analogy. Their application allows us to consider the object of research holistically and comprehensively.

## 3 Results

The instability of the global financial system makes it necessary to address the issue of priority of international or national law. According to V. D. Zorkin, the provision on the unconditional priority of international agreements concerning national law is not generally accepted. The example of the United States shows the commitment to the priority of their own law of the Anglo-Saxon legal family, where the Supreme Court regularly recognized the supremacy of the Constitution to an international Treaty, and the Congress approved a number of laws designed to prevent tax evasion. However, these laws contradict the requirements of the international tax treaties of the U. S. (Zorkin 2018).

Considering this background, there is no coincidence in the appearance of the proposal of the President of the Russian Federation on the priority of the Constitution over the requirements of international treaties and decisions of international bodies of January 15, 2020, in the message to the Federal Assembly of the Russian Federation (Presidential Executive Office 2020). However, a similar position was expressed in paragraph 2.2. Resolution of the Constitutional Court of the Russian Federation No. 21-P of July 14, 2015 (Constitutional Court of the Russian Federation 2015): as follows from the Constitution, its Articles 4 (part 1), 15 (part 1), and 79, neither the Convention for the Protection of Human Rights and Fundamental Freedoms as an international Treaty of the Russian Federation, nor the legal positions of the European Court of human rights based on it, they do not cancel the priority of the Constitution of the Russian Federation for the Russian legal system. Therefore, they are subject to implementation within this system only if the Constitution is recognized as the highest legal force. Cooperation between the Russian Federation and international actors should be built considering the interests of the country and ensuring the economic security of the Russian Federation.

As a rule, the aim of creating any integration Association is to obtain economic benefits and advantages from mutual cooperation between the participating countries. However, many of the current political, financial, and universal challenges and problems make it urgent to solve other tasks: ensuring stability, peace, and well-being of the world's population. It is important to develop new priorities and principles for meaningful, conscious, and correct interaction within an international association. A convergence of legal regulation of financial relations is fruitful when it corresponds to the general vector of development of the country and its national, socio-cultural, and economic characteristics. The process of harmonization involves the following stages:

- The development of general principles of integration should consider the political, economic, and other characteristics of all participants. The most difficult matter is to achieve a reasonable balance between national–state and international legal values and a clear definition of the principles or cases of necessity and practicability of restrictions of national law;
- Bringing the norms of national law in order according to these principles.

## 4 Discussion

Some processes related to integration can be recognized as positive for Russian national financial and legal systems. The convergence in the public sector contributed to the harmonization of budget accounting, budget transparency for citizens, and financial authorities.

It is important to note that the openness of budget information is one of the priorities of state policy and is regulated by the recommendations of the International Monetary Fund to ensure transparency in the budget and tax sphere. These recommendations are based on the Code of Good practice for Fiscal transparency developed by the International Monetary Fund in April 1998 and updated in 2007 (International Monetary Fund 2007). Starting in 2014, Russia draws up a "Budget for citizens," which is intended to increase the financial literacy of Russians. For example, the budget for citizens in the Federal law on the Federal budget for 2020 and the planning period of 2021 and 2022 is made up of the following items: (1) the main parameters of the Federal budget; (2) a socially oriented budget, where expenditures on socially important national programs are scheduled; (3) a budget for growth (which provides the financing of national programs that stimulate the development of the country's economy) (Ministry of Finance of the Russian Federation 2019). "Budgets for citizens" is also drawn up at the regional level. According to official data from the Ministry of Finance, 8 regions of the Russian Federation, including the Altai region, received the highest score for providing information about the budget and the budget system in an understandable form (82 points), based on the combined assessment of the level. The Republic of Ingushetia has the lowest score on information available in the budget sphere–19 (Ministry of Finance of the Russian Federation 2017). The information about the budget and the budget system of subjects of the Russian Federation is evaluated by 11 criteria (i.e., completeness of the budget process stages, general budget parameters, expenditure of regional budgets within country programs, etc.). Besides, three regions of the Russian Federation (Leningrad, Rostov, and Primorsky Krai) started a pilot project - informational and educational Internet resource "Budget for all." It aims to increase budget literacy and public awareness about the social benefits paid from the regional budget in the region. The project "Interactive budget for citizens" was launched in 67 municipalities of Russia, allowing direct participation in the formation of the draft local budget (Center for Fiscal Policy 2020).

Thus, in practice, the implementation of the principle of openness (transparency) is currently gaining momentum. It facilitates by measures taken by the state. T. A. Vershilo notes that the principle of transparency of budgets will be further developed and institutionalized in the draft Budget code introduced by the Russian Government in 2018 (Government of the Russian Federation 2018). In the draft Budget code, Chap. 18, "Information support of the budget process" is the most interesting. It contains Article 174, "Public information resources," which provides additional requirements for the functioning of the unified portal of the budget system, maintained by the Federal Treasury. Thus, the country in current conditions does everything to ensure that the society can see how the budget is not only formed but also used following the planned and approved financial plan (Vershilo 2018).

The interaction of countries in financial relations has its own peculiarities. For example, it may be necessary to adjust objectively arising financial imbalances in the budget system through the tax system when dealing with tax issues. In Russian national legislation, the term "tax expenses" even appeared and began to be used. Harmonization and unification in international cooperation have certain limits. For various reasons, it is sometimes more appropriate to preserve their national characteristics and priorities. This argument can be supported by China's relations with Hong Kong and Macao. Due to the basic laws of Hong Kong and Macao (special administrative regions), the tax systems and customs territories of mainland China, Hong Kong, and Macao are completely independent (Shepenko 2014). The specifics of tax cooperation are bilateral agreements, although there were multilateral ones in recent years. For example, the Joint Council of Europe-OECD Convention on mutual administrative assistance in tax matters (Strasbourg Convention 1988) applied to 117 countries worldwide and was ratified by Russia in 2014. The signing of this Convention testifies the international integration of Russia in the process of information exchange in the tax sphere. Based on this Convention, the Russian Federal tax service signed a multilateral agreement on the automatic exchange of financial information based on a single reporting standard in 2017. To administrate taxes within the EAEU, its member states signed a Protocol on exchanging information in electronic form between the tax authorities on the amounts of indirect taxes paid.

## 5 Conclusion

Thus, the analysis allows us to conclude that the unification of legislation in the economic sphere has many positive aspects since they allow countries to develop common approaches to the legal regulation of financial relations. International cooperation in the financial sphere is crucial when it is objectively impossible to solve the problems of countering the transfer of non-cash funds to the shadow turnover of cash and the legalization of income obtained by criminal means from economic crimes without combining the efforts of countries. In this case, international agreements aimed to disclose information and organize cooperation between tax and law enforcement agencies are necessary and form the legal basis for international cooperation.

At the same time, when implementing international standards in national legislation, we should not forget about such sovereign interests of the Russian Federation as (1) reducing the critical dependence of the national monetary system on fluctuations in the international financial and commodity markets; (2) reducing the use of foreign currency in economic activities in Russian jurisdiction; (3) developing domestic (national) sources of long-term financial resources, ensuring the accumulation rate, sufficient for the sustainable development of the national economy; (4) improving specialized tools to support lending to strategic sectors of the economy, science–intensive, and high-tech industries; (5) developing the infrastructure of the national financial market, payment system, and the system for transmitting financial messages; (6) ensuring the stability and balance of the budget system of Russia, including state extra-budgetary funds; and (7) improvement of the country financial assets management and debt obligations are tasks that must be addressed to ensure the interests and economic security of the Russian Federation.

**Acknowledgements** Thanks for funding the publication of this article are expressed to the Law Institute of the Altai State University.

## References

Center for fiscal policy (2020) Budget for people and Interactive budget. Moscow, Russian Federation. http://budget-for-people.ru/; http://interactive-budget.ru/

Constitutional Court of the Russian Federation (2015) In the case of check of constitutionality of provisions of article 1 of the Federal law "On ratification of the Convention on the protection of human rights and fundamental freedoms and the Protocols thereto", paragraphs 1 and 2 of article 32 of the Federal law "On international treaties of the Russian Federation" parts of first and fourth of article 11, paragraph 4 of part four of article 392 of the Civil procedure code of the Russian Federation, parts 1 and 4 of article 13, paragraph 4 of part 3 of article 311 of the Arbitration procedural code of the Russian Federation, parts 1 and 4 of article 15, paragraph 4 of part 1 of article 350 of the code of administrative procedure of the Russian Federation and paragraph 2 of part four of article 413 of the Criminal procedure code of the Russian Federation in connection

with the request of a group of deputies of the State Duma (July 14, 2015 No. 21-P). Moscow, Russia. http://www.pravo.gov.ru
Council of Europe-OECD (1988) Convention on mutual administrative assistance in tax matters (January 25, 1988. ETS No. 127, as amended by the Protocol of May 27, 2010) Strasbourg, CE. http://www.pravo.gov.ru
Government of the Russian Federation (2018) Draft Budget code of the Russian Federation (February 7, 2018). Moscow, Russia. https://minfin.gov.ru/ru/perfomance/budget/bud_codex/?id_65=121600-proekt_byudzhetnogo_kodeksa_rossiiskoi_federatsii
Masood E (2016) The great invention: the story of GDP and the making and unmaking of the modern world. Pegasus Books, New York, NY
Ministry of Finance of the Russian Federation (2017) Summary assessment of the level of presentation of information about the budget and the budget system by the subjects of the Russian Federation in an understandable form. Moscow, Russia. https://www.minfin.ru/ru/perfomance/reforms/budget/budget_citizen/?id_65=119658-svodnaya_otsenka_urovnya_predstavleniya_subektami_rossiiskoi_federatsii_informatsii_o_byudzhete_i_byudzhetnoi_sisteme_v_ponyatnoi_forme
Ministry of Finance of the Russian Federation (2019) Budget for citizens to the Federal law on the Federal budget for 2020 and for the planning period of 2021 and 2022 (December 30, 2019). Moscow, Russian Federation. https://www.minfin.ru/ru/perfomance/reforms/budget/budget_citizen/
Multilateral Convention on the implementation of measures related to tax agreements to counteract the erosion of the tax base and the removal of profits from taxation (November 24, 2016). Paris, France. http://www.pravo.gov.ru
Paris convention about the Organization for economic cooperation and development (December 14, 1960) Paris, France. https://base.garant.ru/2564514/
President of the Russian Federation (2017) Decree of the President of the Russian Federation "On the Strategy of economic security of the Russian Federation for the period up to 2030" (May 13, 2017, No. 208). Moscow, Russian Federation. http://www.pravo.gov.ru.
President of the Russian Federation (2020) Message from the President of the Russian Federation to the Federal Assembly of the Russian Federation. Moscow, Russia. http://www.kremlin.ru/events/president/news/62582
Shepenko RA (2014) Sovereignty and its impact on taxation issues. Law Econ 7:48
Tayupova KR (2019) Economic security of Russia in the context of globalization: constitutional and legal aspect. Const Municipal Law 6:34–38
The International Monetary Fund (2007) The code of good practices on transparency in fiscal policy. Washington, D.C. https://www.imf.org/external/np/fad/trans/rus/manualr.pdf
Vershilo TA (2018) Transparency (openness) of budgetary relations in the conditions of digital economy. Financ Law 8:8–11
Zorkin VD (2018) The constitutional court of Russia: doctrine and practice. NORM; INFRA-M, Moscow, Russia

# Xinjiang Production and Construction Corps: Organizational Principles of Urbanization

**Dmitry A. Glazunov, Alexandr B. Kesler, and Zakhar M. Galushkin**

**Abstract** The paper is devoted to the urbanization in the territory of the Xinjiang Production and Construction Corps. The Corps was the largest *tunken* in China [the territory where virgin lands were exploited under state guidance]. The Corps had a special status. It was subordinate to the central authorities. Its structure was borrowed from the army. Council regulated its production. Using official sources, we consider how the Corps was forced to change and switch to new principles of organization under the reforms. The status of settlements was raised to towns and county-level cities (subprefecture-level cities). The introduction of the *shi shi heyi* and *tuan zhen heyi* principles led to the growth of cities and towns in the Corps and Xinjiang. Initially, the *shi shi heyi* principle was successfully tested in the city of Shihezi and then extended to other settlements of the Corps. Urbanization of the Corps had distinctive features. It was characterized by the leading role of agriculture, rigorous management, strict control over the ecological balance, low concentration of population, poor urban environment, and uneven development. The main actors of urbanization were farmers, military personnel, and workers. Such specifics became the engine for the development of the territory of *bingtuan* and the entire Xinjiang Uyghur Autonomous Region [XUAR].

**Keywords** Xinjiang production and construction corps · Tunken · Urbanization · Cities and towns with a unified management system

## 1 Introduction

Xinjiang Production and Construction Corps (XPCC) or *bingtuan* (兵团) originated from the *bingtun* (兵屯) military settlements, which played an essential role in Chinese history starting from the *Han Dynasty* era. In the nineteenth century, *Qing* analogs of such settlements actively developed in the border territories, including Xinjiang. In particular, General Zuo Zongtang initiated a whole range of various

D. A. Glazunov (✉) · A. B. Kesler · Z. M. Galushkin
Altai State University, RussiaBarnaul
e-mail: dm.glazunoff@yandex.ru



S. G. Maximova et al. (eds.), *Advances in Natural, Human-Made, and Coupled Human-Natural Systems Research*, Lecture Notes in Networks and Systems 234,
https://doi.org/10.1007/978-3-030-75483-9_23

types of *tun* (屯; military, civil, commercial, convict, *Manchu*, and *Hui*) during the suppression of the uprising led by Yakub Khan. Such settlements promoted the colonization of the borderlands by immigrants from Inner China.

*Bingtun* continued to develop in the post–imperial period. They proved themselves both in wartime (during the Chinese Civil War, the Anti-Japanese War) and after the formation of the PRC. Military settlements were widespread. Moreover, Mao Zedong had an excellent plan for the development of *bingtun* (Yang Jun, 2004). There were more than 16 *bingtun* in China. The most powerful *bingtun* were located in Inner Mongolia and Heilongjiang (Wei Yingjie, 2013).

We should pay attention to two terms related to the exploitation of virgin and fallow lands: *tunken* (屯垦) and *junken* (军垦). *Tunken* is a specific production type where the country carries out agricultural activities and economic development as the principal land manager. *Tunken* is a country facility, as government agencies interfere in agricultural production and product redistribution. This production type differs significantly from the methods of private management.

The phenomenon known as *junken* is associated with the army, representing the state in the border territories. In wartime, in terms of low economic development, the army relied on strict discipline and subordination. It provided itself with food and created a material base to enhance the efficiency of the service on the border territories. At the same time, the development of *junken* often created conditions for other forms of land exploitation, such as *fantun* (land development by exiles who served their sentences) and *mingtun* (land development by civilians). *Bintun* (production and construction union) as a form of tunken, was developed in peacetime on the border territories of the PRC. Economic development and the exploitation of virgin lands were the main tasks of *bingtuan*.

XPCC was founded in 1952. Many Russian, Chinese, and English scientists studied its history (McMillen Donald H., 1981; Clarke Michael E., 2011; Omelchenko, 2002; Department of Party History Research at XPCC, 2004). The Xinjiang Corps incorporated industrial, administrative, and military aspects.

The structure of the Corps was borrowed from the army. However, grassroots units were defined as production unions or economic entities. According to the country plan, they manufactured products, maintained the financial statements of enterprises, and replaced the supply system of the army used in the past. In other words, with a high concentration of production, they carried out agricultural, industrial, commercial, and other activities to provide the country with a material base.

In 1956, The Ministry of Virgin and Fallow Land Exploitation became the head of the XPCC divisions. The Corps became the largest area for exploiting virgin and fallow lands subordinated to the center. Therefore, it could regulate human rights, real and movable property ownership rights, and rights to the production plan. In Chinese literature, such powers are referred to as the *unified system of four right* [四权统一的制度 *Si quan tongyi de zhidu*] (Wei Yingjie, 2013).

Since the mid—1960s, the internal structure of the Corps began to change under the influence of the border conflicts of the USSR – China and the Cultural Revolution. After these events, the Corps gradually abandoned its original path of development and began to protect the borders. In 1967, it was brought under the authority of the

Xinjiang Military District. The Corps experienced a heavy blow to the economy during the restructuring of the agriculture management model. The balance between production, administrative, and military functions was gradually disrupted, and the Corps became unprofitable, shifting the burden to the government. As a result, in 1975, XPCC was liquidated. Its economic units received the rights of state farms and were transferred to local authorities.

A new era in the history of the Xinjiang Corps came with the beginning of reforms and discoveries. Essentially, the time of *bintun* passed. Many state enterprises and farms involved in the development of border territories became market-driven, independent economic units. They began to play an essential role in the development of Chinese agriculture. Xinjiang was an exception. On December 3, 1981, the Central Committee of the Communist Party of China, the State Council, and the Central Military Commission issued *The Decision to restore the XPCC* (Deng Xiaoping 2004). This decision defined the following tasks of the XPCC: (1) developing agriculture on virgin lands, (2) developing the economy and culture of the peoples of the autonomous region, and (3) defending the country from "hegemonic" aggression and protecting its borders.

## 2 Materials and Methods

The State Council of the PRC clearly defined urbanization principles in its documents. The urbanization process included raising the status of settlements to towns, county—level cities, and prefecture-level cities (State Council, 1984). It was administrative and economic factors that allowed raising the status of settlements.

Large villages (*xian*) became towns (*zhen*) if they had at least 2,000 people employed in the non-agrarian sector of the economy and a population of less than 20,000 people. If the population exceeded 20,000 people, the share of non-agricultural employment should have been at least 10%. The town received the status of a county-level city if it had a non-agricultural population of at least 60 thousand people and a GDP of 200 million yuan. These rules had exceptions: (1) border settlements, (2) minority areas, (3) famous tourist facilities, and (4) transport nodes and ports. County—level cities with a non-agricultural population of 200,000 and a GDP of 2.5 billion yuan (35% of which was in the tertiary sector of the economy) could become prefecture-level cities (Him Chung, 2008).

We wrote about urbanization in Xinjiang in another paper (Glazunov, 2018). Let us dwell on the essential points in the development of urban agglomerations in Xinjiang. The territory of economic development and production consists of the Xinjiang oases, located in a narrow strip between the mountain systems and the southern, western, and northern sides of the basins and deserts. An *oasis economy* requires controlling the development of all cities. There were few large and medium-sized cities in Xinjiang. In Mainland China, large and medium-sized cities with emerging economic zones became the engine of development. Small towns were inefficient for the economy.

In the 2000s, economic units actively received the status of cities. They played a significant role in the urbanization of Xinjiang.

According to statistics and official publications, the Chinese government tried to adhere to its own urbanization principles, granting city status to settlements that met population, economy, and GDP requirements. For example, as of 1988, the population of Shihezi (石河子) was 541,531, and the non—agrarian population was 55.67%. By 2014, these figures had grown to 637,200 and 90.73% (Xinjiang Uyghur Autonomous Region Statistics, 1990; Xinjiang Uyghur Autonomous Region Statistics, 2014).

In modern Xinjiang, there are two forms of government under single state sovereignty: (1) the government of the Xinjiang Uyghur Autonomous Region, (2) the Administration of the Xinjiang Production and Construction Corps (*tunken*).

The research objective is to identify ways to grant city and town statuses to XPCC settlements using official publications (yearbooks, normative acts, etc.). Urbanization, the emergence of settlements, county—level and prefecture-level cities were regulated by law and focused only on the first form of government, namely, on the territory subject to the Xinjiang Uyghur Autonomous Region (XUAR) government. There were no similar laws for XPCC settlements.

In 1981, the restoration of the XPCC had a positive economic effect. It set the development vector for the entire region. Its potential was fulfilled in the early twenty-first century. In 1980, the GDP of the XPCC was 25% of the GDP of XUAR, while in 2013, it was 36%. The Corps was able to reproduce the market development model typical of the regions of Mainland China. It was based on urbanization.

## 3 Results

Nowadays, the XPCC is under the dual leadership of the central government and the autonomous region. The Corps with a subject territory is considered a province, and its economic indicators are specified in the State plan of the State Council. In 1998, the state created the National XPCC Corporation, the largest state enterprise in the PRC, with the function of economic development of the subject territory. The corporation includes agricultural, industrial, transport, and commercial enterprises. Currently, the XPCC is the largest *tunken* in China, which is responsible for protecting state borders and economic tasks. Therefore, its leadership is represented by the party, administrative, military, and economic entities.

Earlier, we noted that the XPCC management structure had the features of a military system that represented a three-level organization:

- *Bingtuan* Corps (兵团);
- 14 *shi*-divisions (师);
- More than 220 *Tuan/nongchang* regiments or economic division (团 (农场)).

The economic units of the Corps were arranged in lines. The *tuan* regiments were initially created outside the traditional economic ecumene of the local population. They were adjacent to the already existing chain of oases, gradually cultivating virgin lands, desert, and semi-desert territories. The *tuan*, located along the lines of natural XUAR oases, formed three arcs.

Historically, the primary responsibility of the Corps was the exploitation of virgin lands, cultivation of lands in the areas of risk farming, and development of animal husbandry. The focus on agriculture prevented the urbanization of settlements until the end of the twentieth century.

The following conditions caused the growth of urban agglomerations on the XPCC territory: (1) market production, (2) relative concentration of the population, and (3) the emergence of non-agrarian economic sectors. However, the XPCC territory was not under the jurisdiction of the autonomous region, so it did not have a seat of government. *Shi shi heyi* system solved this problem.

The *shi*—division is the second-level management structure of the XPCC, which manages production and exercises administrative power in a specific territory. Each division has its headquarters (Table 1).

Table 1 shows us the main way of urbanization of XPCC settlements. Cities with a single (divisional and city) management system or *shi shi heyi* is a specific feature

**Table 1** Basic information about the divisions of XPCC

| Division number | Foundation year | Population (10.000 people) | Number of economic units | Area (10. 000 ha) | Headquarters, 2009 | Headquarters, 2013 |
|---|---|---|---|---|---|---|
| 1st AD | 1953 | 29.12 | 16 | 45.58 | Aksu | Alar* |
| 2nd AD | 1953 | 19.87 | 18 | 93.29 | Korla | Korla** |
| 3rd AD | 1966 | 20.40 | 18 | 80.42 | Kashgar | Tumxuk* |
| 4th AD | 1953 | 22.42 | 21 | 66.41 | Yining | Yining** |
| 5th AD | 1953 | 10.01 | 12 | 39.83 | Bole | Bole** |
| 6th AD | 1953 | 30.21 | 20 | 83.10 | Wujiaqu | Wujiaqu* |
| 7th AD | 1953 | 21.90 | 10 | 59.06 | Kuytun | Kuytun |
| 8th AD | 1953 | 58.44 | 19 | 74.27 | Shihezi* | Shihezi* |
| 9th AD | 1962 | 7.19 | 11 | 42.43 | Emin | Emin |
| 10th AD | 1959 | 7.64 | 11 | 41.39 | Beitun | Beitun* |
| 11th ECD | 1953 | 4.61 | | 11.00 | Urumchi | Urumchi |
| 12th AD | 1982 | 7.08 | 7 | 26.15 | Urumchi | Urumchi |
| 13th AD | 1983 | 7.86 | 11 | 99.85 | Hami | Hami |
| 14th AD | 1982 | 3.48 | 4 | 12.87 | Hotan | Hotan** |

*Note*
* cities with a single (divisional and city) management system
** territories where cities are established with a single (divisional and city) management system, but without the transfer of headquarters
AD—agricultural division, ECD—engineering and construction division
*Source* Statistics Bureau of Xinjiang Production & Construction Corps, 2009

of Xinjiang. This type of city combines the management of the Corps and the city at the party and administrative leadership levels.

Shihezi, the second largest and most economically developed city in XUAR, was the first city to use such a management system. In 1976, the city of Shihezi appeared together with the prefecture of the same name on the basis of the demobilized division of the CPSU. In 1978, Shihezi prefecture was merged with the city. Therefore, along with Urumqi, the second prefecture-level city was created in XUAR. However, in 1985, after the restoration of the XPCC, Shihezi lost its status as a prefecture-level city. The newly formed divisional management system was merged with the city (the head of the eighth division of the XPCC also managed the city). In Chinese literature, such cities are called a subprefecture-level city. That means that they are not included in the established system of territorial and administrative divisions but are subordinate to the management of the autonomous region.

The management performed two sets of duties, working according to the principle *One structure, two tables* (Table 2).

**Table 2** The management structure of Shihezi city 1985

| Post | Person | Note |
|---|---|---|
| Secretary of the City Committee and the 8th agricultural division | Liu Bingzheng (刘炳正) | |
| Deputy Secretaries of the City Committee and the 8th agricultural division | Ren Yuzhi (任友志), Kanji Bieke (坎吉别克), Yang Tingzhu (杨廷柱) | |
| Acting Chairman of the Intermediate People's Court in Shihezi Prefecture | Zeng Jingdong (曾景东) | |
| Acting Senior Prosecutor of the People's Prosecutor's Office in the City of Shihezi | Zhang Zongnan (张宗年) | |
| Commander of the 8th agricultural division | Chen Bing (陈冰) | Combined with management functions in the city of Shihezi |
| Deputy Commanders of the 8th agricultural division | Li Shuxian (李树贤), Zhou Zhongyu (周忠玉), Jiang Yaozhong (姜耀忠), Liao Tianyi (廖天一), He Yunxi (阂永曦) | |
| Commissar | Ren Yuzhi (任友志) | |
| Deputy Commissar | Guo Baoshan (郭宝珊), Lu Zhenou (陆振欧), Wang Zhongjun (王中俊) | |

*Source* Editorial and Codification Committee of the Xinjiang Uyghur Autonomous Region, 1985; Editorial Board of the XPCC Yearbook, 1986

The administrative apparatus of the Party Committee of the 8th agricultural division and the city of Shihezi coincided. In 1983, the 8th agricultural division was subordinated to 18 agricultural divisions (regiments), including 81 rural and livestock farms, 128 mining and industrial enterprises. Shihezi city had four subdistricts (Laoze, Hongshan, Xiangyang, and Xincheng), one neighborhood, and 43 mining and industrial enterprises.

Success in the development of Shihezi had to be adopted by other XPCC settlements. The settlements of 1st AD (Aral in 2002), 3rd AD (Tumshuk in 2004), and 6th AD (Wujiaqu in 2002) received the status of a city with *One structure, two tables system*. In the early 2000s, it looked a bit symbolic, provided that there were two cities in the north and south of Xinjiang.

By 2019, there were already ten similar cities. The following cities were added: Beitun (10th AD) in 2011, Tiemenguan (2nd AD) in 2012, Shuanghe (5th AD) in 2014, Kokdala (4th AD) in 2015, Kunyu (14th AD) in 2016, and Huyanghe (7th AD) in 2019.

The population of the grassroots unit of the *tuan* regiments are part of a particular territory, but due to the unique management system of the *tunken*, they are autonomous compared to the local government. From the Chinese legislation point of view, they have the status of a small community (小社会). In response to the special Corps management system, the Ministry of Civil Affairs and the State Statistic Bureau divided the population of the *tuan* regiments Corps into a *town with administrative bodies* and a *town without administrative bodies*.

The central government had to support and regulate the urbanization of the grassroots unit of the Corps. Therefore, in the second decade of the XUAR development, the following towns appeared: a town with Beiquan administrative bodies and 69 towns without administrative bodies. Beiquan town, a part of Shihezi, precedes a new era, a new policy of *tuan zhen hei* or settlements with a single (regimental and town) system.

## 4 Discussion

As a rule, urbanization is a two-sided phenomenon. However, in the XPCC territories, the state was the main actor in the development of urban settlements. It acted as an investor, providing administrative support, initiating the reconstruction of old agglomerations, and the development of a new type of urban environment. The following factors determined the differences between urbanization and the traditional urban development model: (1) rigorous management, (2) strict control over the ecological balance, (3) low concentration of population, (4) poor urban environment, фтв (5) uneven development. These data are presented in Table 3.

The comprehensive and natural character of the XPCC urbanization is under question. Further research on the urbanization of the XPCC settlements should pay attention to the communities within settlements, the features of urban infrastructure, and the development of urban agglomerations. The urban system of Xinjiang was

**Table 3** Comparison of the main characteristics of different types of urbanization in Xinjiang

| | Normal urbanization | XPCC urbanization |
|---|---|---|
| Development mechanism | Market orientation | Administrative support |
| Objects of changes | Peasants | Military personnel, peasants, and workers |
| Production structure | Agriculture is becoming an additional sector of the economy | Agriculture remains an integral part of the economy |
| Transformation in living conditions | The rural lifestyle is changing towards the urban one | The emergence of a new town, new citizens |
| Management features | Controlled by regional authorities, relatively free management style | Militarized management, strict management style |
| Investments | Private | State |

*Source* (Zhu Lei & Zhang Tan, 2004)

insufficiently studied before. It also concerns the city of Urumqi (Wang Xiaodong & Xie Yan, 2014). Cities and towns of the XPCC have not yet been a separate subject of research.

## 5 Conclusion

XPCC, as a form of *tunken* in the second half of the 20th and early twenty-first centuries, developed unevenly, experiencing periods of creation, flourishing, disbanding, and rebirth. The Corps relied on militarized management, which was strictly limited to the territory under its jurisdiction. The administrative and economic units of the Corps were located on the territory of the XUAR, but they had full autonomy from the regional authorities. As a successful model for the development of regions of Mainland China, urbanization was also implemented in Xinjiang, both in the autonomous region and in the XPCC. The Chinese state introduced the *shi shi heyi* system, which combined city and division types of management. Initially, this system was used in the city of Shihezi in the mid–1980s, and then its experience was adopted by other divisions. In the second decade of the twenty-first century, dual management was also applied to the lower-level units, the tuan regiments. Beiquan town implemented *zhen hei* or settlements with a single (regimental and town) system. From a historical point of view, urbanization has changed the tasks of the XPCC: from protecting the borders of *tunken* to protecting the borders of cities and towns.

## References


Chung H (2008) State regulation and China's administrative system: a spatial perspective. China Rev 8(2):201–230

Clarke ME (2011) Xinjiang and China's rise in Central Asia–a history. Taylor & Francis, New York, N.Y

Deng X (邓小平) (2004) 关于恢复新疆生产建设兵团的两件批示(一九八一年七月、十月) [Two resolutions on the restoration of the XPCC (June, September 1981)]. *党的文献* [Lit Communist Party]6:3–5

Department for party history research at the XPCC. (2004). *新疆生产建设兵团史料选辑* [Collection of materials on the history of SPSK]. 疆人民民版社[Xinjiang People Publishing House], Urumqi, China

Editorial and Codification Committee of the Xinjiang Uyghur Autonomous Region (新疆维吾尔自治区地方志编纂委员会) (1985) 新疆年鉴 (1985) [Xinjiang Statistical Yearbook, 1985]. 新疆人民出版社 [Xinjiang People Publishing House], Urumqi, China

Editorial Board of the XPCC Yearbook (新疆生产建设兵团年鉴编辑部) (1986) 新疆生产建设兵团年鉴. 1986 [*Editorial Board of the XPCC Yearbook, 1986* ɛ.].新疆人民出版社 [Xinjiang People Publishing House], Urumqi, China

Glazunov DA (2018) Features of the urbanization process in Xinjiang (1980s–2000s). Novosibirsk State Pedagogical Univ Bull Ser Hist Philology 18(4): 96–106

McMillen DH (1981) Xinjiang and the production and construction corps: a han organization in a non-han region. Aust J Chin Aff 6:65–96

Omelchenko OA (2002) Socio-economic history of Xinjiang (1949–1978). Altai State University, Barnaul, Russia

State Council (国务院) (1984) *国务院批转民政部关于调整建镇标准的报告的通知.* [Report of the State Council "On approval of the report of the Ministry of Civil Affairs on the regulation of standards for the organization of settlements" (1984, No. 165)]. State Council, Beijing, China

Statistics Bureau of Xinjiang Production& Construction Corps (新疆生产建设兵团统计局). (2009a). *2008新疆生产建设兵团统计年鉴* [*Xinjiang Statistical Yearbook, 2008*]. Beijing, China: 中国统计出版社出版 [China Statistics Press].

Statistics Bureau of Xinjiang Production & Construction Corps (新疆生产建设兵团统计局) (2009b) *2008新疆生产建设兵团统计年鉴* [Xinjiang Statistical Yearbook, 2008]. 中国统计出版社出版 [China Statistics Press], Beijing, China

Wang X [Wang Xiaodong], Xie Y [谢洋] (2014) 历史舆图中的乌鲁木齐——清代至今城市建设发展过程梳理 [Urumqi on historical maps—the Process of urban construction and development from the Qing dynasty to the present days] *建筑学报* [Architecturniy Zhurnal] 3:103–109

Wei Y [魏英杰] (2013) 历史制度主义视角下的新疆生产建设兵团 [XPCC in terms of historical systematization]. *学海* [More Znaniy]6:177–189

Xinjiang Uyghur Autonomous Region Statistics (新疆维吾尔自治区统计局) (1990) *1989新疆统计年鉴* [Xinjiang Statistical Yearbook, 1998].中国统计出版社出版 [China Statistics Press], Beijing, China

Xinjiang Uyghur Autonomous Region Statistics (新疆维吾尔自治区统计局) (2014) *新疆统计年鉴* [Xinjiang Statistical Yearbook, 1988].中国统计出版社出版 [China Statistics Press], Beijing, China

Yang J [杨俊] (2004) 毛泽东屯垦戍边思想与新疆生产建设兵团的建立发展 [Mao Zedong's ideas about the protection of borders and cultivation of lands by residents of military settlements on the borders of China and [their influence] on the formation and development of the Xinjiang Production and Construction Corps]. *军事历史* [Hist Mil Aff]6:5–9

Zhu L [朱磊], Zhang T [张琰] (2004) 新疆兵团城镇化建设问题研究[Studies on the urbanization Problems of XPCC]. *中国农垦经济* [Econ Virgin Fallow Land Develop China]5:24–25

# Conceptual Approaches to Human Capital as the Basis for Social Trust

**Maksim B. Maximov, Daria K. Scheglova, and Olga V. Surtaeva**

**Abstract** The paper presents results of theoretical research about approaches to human capital as a basis for population trust in Russian regions. The main goal was to look through general approaches to form a strategy of empirical analysis of different aspects of human capital in the modern conditions of transforming reality. A basis for theoretical research were works by foreign scientists, describing economic, social and emotional dimensions of human capital and networks between notions of human and social capital, social, human and intellectual capital. Concluded that economic dimension will provide analysis of institutional conditions of formation of human capital, working out of multi-factor model of estimation of potential and security of social trust basing on use of social-economic indexes of development of regions. Sociological dimension will prove new knowledge about qualitative and quantitative characteristics of a human capital and intellectual potential of different social-demographic groups of regions of Russia, comparative analysis of a human capital and intellectual potential of population and territorial (regional) dimensions. Psychological dimension will present subjective characteristics of a human capital, including characteristics of motivation sphere, adaptive strategies, and peculiarities of social and psychological integration of population for transforming social reality.

**Keywords** Human capital · Trust · Social capital · Intellectual capital · Concept · Social trust

## 1 Introduction

The logic of the postindustrial development of humankind determined the "humanistic turn" of the world economy. The goals, strategies, and technologies of the decisions changed, so human is in the focus of the development process, not only as the main subject of consumption and beneficiary of produced goods but as the

M. B. Maximov (✉) · D. K. Scheglova · O. V. Surtaeva
Altai State University, Barnaul, Russia
e-mail: maxbmax@rambler.ru



S. G. Maximova et al. (eds.), *Advances in Natural, Human-Made, and Coupled Human-Natural Systems Research*, Lecture Notes in Networks and Systems 234,
https://doi.org/10.1007/978-3-030-75483-9_24

most important factor and a driver of economic growth. Waves of information revolutions, transboundary information-communication networks, and global information space changed all vital processes, consciousness, and people's behavior and put a knowledge capital in all its manifestations in the forefront.

Human capital is the main advantage of the national economy in Russia. It is not subjected to the devaluation under the influence of internal factors and let to realize technological breakthrough, transform "commodity economy," an extensive model of country economic development, into more intensive, corresponding to the realities of the information society. Migration is one of the important factors, determining general trends in the field of formation and use of human capital in Russia, as the in-country and the international one.

## 2 Materials and Methods

The paper presents theoretical research results on human capital approaches as the basis for population trust in Russia. The main goal is to look through general approaches to form a strategy of empirical analysis of different aspects of human capital. We rely on the works of foreign scientists, describing the economic, social, and emotional dimensions of human capital and interactions between the notions of human and social capital, social, human, and intellectual capital while conducting the theoretical research (Alexander 1992; Bourdieu, Coleman & Fukuyama, 2004; Knack and Keefer 1997; Nahapiet and Ghoshal 1998; Parts 2004; Wright and McMahan 1992).

## 3 Results

Nowadays, people's trust in each other, law and social institutes is considered the weak components of social capital (The third sector analysis..., 2015). According to the data of a survey (2011) by The Civil Society Institute at the Higher School of Economics, 21% of Russians do not trust any forms of public organizations and civic initiatives. Only 37% of the Russians trust certain non-government organizations (Maximova et al. 2018).

During the last decades, trust attracts more attention and analysis in terms of social capital. Bourdieu developed a concept of social capital. By the concept, economic, cultural, political, and social capitals are significant. Thus, society is described as a multidimensional space consisting of multiple independent spheres with their own mechanisms of distribution of influence, connected with different types of capital (Bourdieu 1986).

Coleman considered the phenomenon as a resource, determining the mutual trust possibility, specially created in the form of obligations and expectations, social norms, and channels for transfer of information. Social capital determined the frequency and productivity of social interactions in society (Coleman 1988). Theoretical ideas by Putnam were close to the statements of Coleman. The researcher considered that social capital revealed in social networks, norms, and trust, which form the base for alignment own actions by social subjects to achieve mutual benefit, and is a guarantee for social solidarity. Putnam suggested the form of capital to be the most important for achieving political stability and economic progress than physical and human capital (Putnam 1993). Theories of these two scientists differ about the priority of trust and social capital. Coleman argued that the increasing level of trust leads to the preservation of social capital, but Putnam considers that increasing social capital will increase the level of trust.

Based on the above-mentioned theories, Fukuyama determined social capital as individuals' ability to unite and cooperate activity for a common goal. Formation of social capital is possible whether sociality has a system of values and norms, supported by all its members and under the primacy of group interests above personal ones. Trust is created due to the statement of common values and social confidence in the predictability of people's behavior, consideration of all individuals' interests, and correlation of behavior with social norms (Fukuyama 2004). Thus, the trust reveals in society as an expectation of non-conflict, honest and predictable behavior, and interaction, correlating with coming values, norms, traditions, and customs (Kosyreva and Smirnov 2010).

In social capital theory, a complicated system of stable networks, including those having informal and voluntary character, is a basis for supporting trust in society. Such horizontal networks are characteristic of civil society, creating space for free interaction of people independently from political and social institutes. Mechanisms of solidarity, reproducing the integration of communities with similar interests and harmonizing contradictory positions through common efforts of interacting subjects contribute to the streamlining of these networks. Spontaneously forming social relations presents opportunities for collaboration and mutual help, providing individuals with abilities to realize their own interests and needs with greater productivity. Alexander called these peculiarities of civil society "a place," where the feeling of mutual dependence of social subjects is created (Alexander 1992).

Several examples may demonstrate the importance of trust for social development. Knak and Keefer revealed a statistical correlation between trust and economic success in their research. They concluded interaction between economic growth and the level of trust, based on data on 28 countries. Regression analysis showed that with the increase of the trust, the dynamics of economic growth increases too. Moreover, the scientists revealed a strong dependency between trust level and the part of investments in the gross domestic product (Knack and Keefer 1997). The World Bank research demonstrated correlations between economic growth and income per capita.

The level of trust influences not only economic indexes. La Porta and coauthors found that with the growth of the trust level on one point, the index of effectiveness of juridical decisions grows by 7 points, and the index of corruption decreases by 0.3 points (Nathov 2010). Thus, the connection between the preservation of social capital as a component of common human capital and the formation in society's potential of the trust and dynamics of social development is evident.

During recent years, the use of "human capital" as a term providing a connection between large international organizations (for example, the Organization of the United Nations with its goals of stable development) and the World Bank activities, the World Economic Forum, and other institutes and countries intensified. However, human capital is often considered in the context of individual development of separate countries. Besides, it is considered in the context of a global phenomenon of population growth and life expectancy, requiring correspondent scientific understanding. The acuteness of such an approach is conditioned by new challenges of the twenty-first century:

- Global spread of unemployment;
- Neglect of human abilities;
- Poverty in developing countries;
- Breaks in social conditions of the global community, caused by the capitalist mechanisms functioning despite social-economic ideas requiring for progressive taxes and social entrepreneurship.

In separate countries, economic growth could be estimated through the value of the gross domestic product and, for example, social exclusion, which is not previously considered indexes yet. However, such indexes may be a base for inter-state correlations in the analysis of life quality, progressive development or backwardness, evaluation of the level of achievement of stable development goals, the upward movement in population number and its life expectancy (Omowaye 2019).

A key indicator of inclusive economic growth lies in human development and orients on displacing the focus of the analysis from the microlevel to the macroeconomic understanding of input of qualified specialists into the economy, including capital and labor transfer, access to opportunities, and common *happiness* of the population in the country (i.e., economic well-being). A modern world faces transformations in physical infrastructure on the background of creativity, innovations, and new technologies, but human capital is still the dominant form of capital (Becker 1964). In particular, authors usually consider education as an archetype of human capital. Jaiyeoba (2015) argues that human capital includes education, and health categories; The development of a country, enforcing its human capital is inevitably connected with necessary skills of the population, obtained through education, health, and nutrition. Pakistan economist Mahbubul Haq (1967) described human capital within obtaining freedom, opportunities, and well-being.

We should consider the intellectual development due to education and health that lead to physical "breakthrough" and influence on industrial revolutions and advanced innovations. Thus, such transformations reduce poverty, lead to equality and well-being, and smooth income inequality, create favorable conditions for the development of human capital to a certain extent, and are necessary for progress and economic growth (Jaiyeoba 2015). However, Samans (2018) argues the absence of connections between gross domestic product and inclusion based on data of developing countries or ineffective use of population skills for the economic growth of a nation.

In a broad sense, each form of capital has been attacked by some of the forms of discrimination. While society has a reserve of human capital, it has been intensively or insufficiency used; it becomes demanded in conditions of mass infestations, when the need in it increases the ordinary level and may be supported for a long time because of the development and constant rising of a life quality through educating and supporting the optimal health level (Adelakun 2011). We may find such examples in Japan, Korea, China, and Singapore. There the investments into human capital led to cascade growth of country economics. The abovementioned countries were rapidly developed based on an educated, well-prepared, healthy and hard-working labor force that let them to apply high technologies in mass. The examples rather correlate with the endogenic growth theory, while the way is not always working in some countries (some African countries, for example). There the physical infrastructure is in priority against human capital and not derived from stable development (Address by the chairman…, 2018).

Thus, human capital faces the integral, acquired, and necessary skills of society members, which input into the social development and production of financial results. Despite the difference of such skills, they are based on obtaining, level of development, and ways of application. They are the storage of productive knowledge and skills of a labor force, leading to higher salaries. Human capital can be observed as a non-material asset of skills and force (weight) of the population. It may increase productivity and lead to corresponding growth. It is the ability to create values from the knowledge and skills of an individual. Human capital could be described by the productiveness of the labor force, determination of salaries, production progress, economic opportunities, and transformations. Schultz (1971) considered the population as an option or market asset. However, the scientist described human capital in the context of the necessity of health care services, increased life expectance level and provided endurance and vital force, training, formal educating, educational programs for adults out of the working place, and the mobility of working places.

Human capital is a key driver of a successful economy in the period of neomodernism with the sharp growth in creativity and innovations, such as with salaries and other bonuses, automating, and wide use of robots and artificial intelligence. Herewith, human capital development and the use of scientific data may stimulate the creation of working places and opportunities. Ballot et al. (2001) revealed the influence of human and technological capital on productivity in panel research frameworks among French and Sweden. The evaluation of estimates about expenses for future and present educating became the measurement of joint human capital. Investments in personnel training led to the promotion of a firm in the market in both countries.

We may differently evaluate human capital, including the evaluation of the HPI index, investment coefficient, or use financial indexes of future salaries of the working population. For economic positions, this means that salary usually correlates with the level of qualification—education or work stage—and some other social attributes. It can be interpreted as a secondary product of different types of occupations.

At present, enough scientific works describe the positive connections between human capital development and best productivity. An accent on human capital in organizations reflects how the market price is less dependent on resources than nonmaterial human ones. Empirical works become more complicated and move from evaluating single personnel management measures to comprehensive combinations of HR practices (Fitz-Enz 2002). Such conclusions led some scientists to support "the best practices," arguing the presence of universal identifying factors having a positive effect on labor productivity. Some authors suggest a difficulty in determining a complex set of best practices because of their large unaccounted variations. General trends of these discussions consider existing practices in human capital development as complementary but not the opposite, but best practices are considered as architectural superstructure, having a generalizing effect but in the frameworks of each organization.

A resource-oriented approach is one of the approaches to the consideration of networks between human capital and productivity. It displaces accents from the external positioning in the industry and relative balance for stable efficiency (Wright et al. 2001). Works by Penrose (1959) present a resource-oriented approach in the organization, further continued in the works of Rumelt (1984), Barney (1991, 1995). In particular, he indicated the impotence of forming a set of valuable resources of the organization and their grouping in unique and dynamic ways to avoid its success. A competitive advantage depends on traditional sets, consisting of natural resources, technologies, or economies of scale, which are easy to imitate. The advantage will depend on the value, rare, and hardly reproduced resources. Human capital is an invisible treasure (Itami 1987). Its significance (the totality of human abilities) and how it could be managed in the process of production became the most evident similar to the strategic goals of the organization. From the point of view of a "single" character of human capital: "If the types and levels of skills are unevenly distributing, then one firms may buy talent, and the other—not, such forms of human capital become constant source of competitive advantage, than other things being equal" (Snell, Youndt & Wright, 1996). From the point of view of unique, unrepeatability: there are two reasons why human resources are hardly reproduced: causal uncertainty and partial dependency (Becker and Gerhart 1996). The mechanism of interaction between human resources and policy to create value is rather complicated target mechanism, and the systems of stuff management are rather dependent themselves, second. They base on a constantly changing policy and could not be bought by competitors at market (Becker and Gerhart 1996). Such interdependency between management practices in combination with the context of the activity of certain companies creates considerable barriers for reproduction. On the whole, human resources should be valuable. According to Baxall, they should be "latent in production ability" (Boxall 1996). Furthermore, the advantages of human capital depend on the provision of exclusive talent or "of the best and splendid."

In attention to human capital corresponds consideration in strategic research of "general professional characteristics," in which economic profit belongs to "human skills" (Hamel and Prahalad 1994). However, M. Hummer supposes that "people are the main heritage"—is "the greatest lie of contemporary American business," and the growth of human resources will be "a rhetoric still."

Strategical research obtained development in the 1980s in connection with the phenomenon of "the Japan wonder," which showed success based on a special form of human management and recommendations on how to achieve superiority (Peters and Waterman 1982; Collins and Porras 1994), oriented on development and training employers in the frameworks of a strong supportive culture.

Human capital is "usually understood in the framework of human abilities, knowledge and experience of employers and managers of companies, corresponding to the stated task and have an ability to input into common "reservoir" of knowledge, skills, and experience through individual learning" (Dess and Picken 1999). From this point of view, the notion of social capital is considerably wider than human resources and has connections with "knowledge notion." In the literature on staff management, much attention is paid to the knowledge component, while all the debates concentrated on the level of individual perspective, concerning professional knowledge mainly. However, works in the field displace accents at an individual level and promote the idea about shared group knowledge and its institutionalization in the organization process and everyday life during the last years (Wright et al. 2001).

Concepts and perspectives of human capital emerge from the fact that there is no alternative to knowledge and learning, creativity and innovations, the competitiveness of abilities, and the necessity to seek them in the organizational environment and competitive logics (Rastogi 2000). Such an approach concludes that the accumulation of talented individuals is not enough for the organization; individuals should desire to invest their own experience and skills into the organization and their own position. In other words, individuals should achieve, engage in the organization for effective use of human capital. In addition to human capital, there would be social and organization capital, both in the corresponding context.

Thus, it is important to consider approaches to intellectual capital. OECD determined intellectual capital as "an economic value of two categories of nonmaterial actives of a company"—organization and human capital. Wright and coauthors (Wright et al. 2001) argued that human capital is a factor, including human capital, social, and organization capital. Nahapiet and Ghoshal (1998) considered that intellectual capital is related to "the knowledge and knowledge abilities of the social community such as organizational, intellectual community, professional-practitioners." Gratton and Ghoshal (2003) argued that intellectual capital is considered a part of human capital and social and emotional ones. The majority of authors (Kaplan and Norton 1992; Harvey and Lusch 1999; Stewart 1996; Sveiby 1997) consider intellectual capital has "a wide scales," including human capital as one of the dimensions.

No less important is to consider approaches to the study of social capital. According to Nahapiet and Ghoshal (1998), "a central idea of social capital theories is its connection with value resources to organize social networks, while the most part of this capital in included into mutual contacts." Social capital increases activity efficiency and promotes cooperative behavior. Social relations and social capital significantly influence the development of human capital as intellectual capital. At the individual level, individuals with the best social capital—individuals with strong contacts—have higher salaries than return human capital (Garavan et al. 2001).

At the organization level, social capital is essential. Social capital stimulates intellectual development, influencing the conditions necessary for exchange by what is happening (Garavan et al. 2001).

In the frameworks of social capital theory, authors consider its three general elements:

- structural dimension (network connections, network configuration, and corresponding organization);
- cognitive dimension (shared codes and languages, shared narratives);
- interpersonal dimension (trust, norms, duties, and identifications).

These three dimensions influence intellectual capital, and the approach is connected with resource-oriented with a focus on complex and combinations of resources. Social capital with the accent of connections between individuals creates conditions for contacting individuals unrepeatable, unspoken, rare, and strong. Gratton and Ghoshal considered social capital as based on a dual approach of concepts of socialization ad trust: "the deepness and richness of these ties are potential points of their application, they form a stable pool of knowledge, opportunities or value creativity and judgments" (Gratton and Ghoshal 2003).

A principal role of the theory of organization capital is finding connections between organization resources and creating values for customers, such as stable competitiveness for the firm (Dess and Picken 1999). Such kind of capital includes organization structures and structures of reporting, operation systems, processes, proceedings and statement of purposes, information and communication infrastructure, systems of collection, working out and distribution of resources, processes of taking decisions and information streams, systems of stimulation, control and evaluation of efficiency, organization culture, values, and leadership.

An interdependence between all these components is important for workers to have the motivation to develop and use their own skills and knowledge. Thus, organizational culture has a great influence on recruiting and preserving workers as for the field of obligations. In "War for talents" (1999) was indicated that 58% of workers (the most often chosen alternative) evaluate the strong values and culture of the organization. Supporting a culture with strong corporative goals is a reason for corporative success (Peters and Waterman 1982; Collins and Porras 1994). The second factor, strongly influencing human capital, is a supporting structure, and how to work with personnel is evaluated and managed (Delaney and Huselid 1996).

Relations between human capital, social capital, and organization capital reproduce intellectual capital. In its turn, it influences knowledge inside the organization. Knowledge always was a valuable resource and a subject of attention in the literature in the field of human capital, on aspects of knowledge of generations, inclusion, transfer, and integration, in particular (Wright et al. 2001; Wright and McMahan 1992; Nonaka 1994; Sveiby 1997; Szulanski 1996). Knowledge is conceptualized and characterized in literature in different ways, and the difference between real knowledge (know-how), characterized by non-communicability, and explicit knowledge, subjected to codification in a general way (Nonaka 1994).

## 4 Discussion

Sociological research on human capital in Russian as a basis of trust should base on the interdisciplinary approach stressed using methods and technologies of analysis of human capital and intellectual potential in conditions of transformation of social processes, including the formation of social trust in society, from positions of economic, sociological and psychological knowledge. The economic dimension should analyze the institutional conditions of formation of human capital, working out of a multi-factor model of estimation of the potential and security of social trust based on the use of social-economic indexes of the development of regions. The sociological dimension can prove new knowledge about the qualitative and quantitative characteristics of human capital and intellectual potential of different social-demographic groups in Russia, comparative analysis of a human capital and intellectual potential of the population and territorial (regions) dimensions. The psychological dimension can present subjective characteristics of human capital, including characteristics of motivation sphere, adoptive strategies, and peculiarities of social and psychological integration of population for transforming social reality.

**Acknowledgements** The presented in the article analysis was performed as part of the state task of the Ministry of Science and Higher Education of the Russian Federation FZMW-2020-0001 "Human capital, migration and security: transformation in the new migration environment in Central Asia".

## References

Address by the chairman of Bill and Melinda Gates Foundation (2018) Bill Gates to the Nigerian Government. https://www.gatesfoundation.org/What-We-Do/Global-Health/Discovery-and-Translational-Sciences

Adelakun JO (2011) Human capital development and economic growth in Nigeria. Can Soc Sci J 7(2):207–218

Alexander JC (1992) Democracy and civil society. UCLA, Mimeo, Los Angeles, CA

Ballot G, Fakhfakh F, Taymaz E (2001) Firms' human capital, R&D and performance: a study on French and Swedish firms. Labour Econ 8(4):443–462

Barney J (1991) Firm resources and sustained competitive advantage. J Manag 17:99–120

Barney J (1995) Looking inside for competitive advantage. Acad Manag Exec 9(4):49–61

Becker B, Gerhart B (1996) The impact of human resource management on organisational performance: progress and prospects. Acad Manag J 39:779–801

Becker G (1964) Human capital. University of Chicago Press, Chicago, IL

Bourdieu P (1986) The forms of capital. In: Richardson JG (ed) Handbook of theory and research for the sociology of education. Greenwood Press, New York, NY

Boxall PF (1996) The strategic HRM debate and the resource-based view of the firm. Hum Resour Manag J 6(3):59–75
Coleman JS (1988) Social capital in the creation of human capital. Am J Sociol 94:95–120
Collins J, Porras J (1994) Built to last: successful habits of visionary companies. Harper Business, New York, NY
Delaney JT, Huselid MA (1996) The impact of human resource management practices on perceptions of organizational performance. Acad Manag J 39:949–969
Dess GD, Picken JC (1999) Beyond productivity: how leading companies achieve superior performance by leveraging their human capital. American Management Association, New York, NY
Fitz-Enz J (2002) How to measure human resources management. McGraw-Hill, New York, NY
Fukuyama F (2004) Trust: social goodness and way to the prosperity. AST; Ermak, Moscow, Russia
Garavan TN, Morley M, Gunnigle P, Collins E (2001) Human capital accumulation: the role of human resource development. J Eur Ind Train 25:48–68
Gratton L, Ghoshal S (2003) Managing personal human capital: new ethos for the 'volunteer' employee. Eur Manag J 21:1–10
Hamel G, Prahalad CK (1994) Competing for the future. HBS Press, Boston, MA
Haq M (1967) Tied credits–A quantitative analysis: capital movements and economic development. Palgrave Macmillan, Oxford, UK
Harvey MG, Lusch RF (1999) Balancing the intellectual capital books: Intangible liabilities. Eur Manag J 17:29–41
Itami H (1987) Mobilising invisible assets. HBS Press, Boston, MA
Jaiyeoba S (2015) Human capital investment and economic growth in Nigeria. Int Multi J Organ Leadersh Ethiop 9(1):4–13
Kaplan RS, Norton DP (1992) The balanced scorecard: measures that drive performance. Harv Bus Rev 70(1):71–79
Knack S, Keefer P (1997) Does social capital have an economic payoff? a cross-country investigation. Quart J Econ 112(4):1251–1288
Kosyreva, P. M., & Smirnov, A. I. (2010). Trust and its role in consolidation of the Russian society. In M. K. Gorshkov (Ed.), *Social factors of consolidation of the Russian society: sociological dimension* (pp. 160–199). Moscow, Russia: New Chronograph.
S Maximova O Noyanzina D Omelchenko M Maximova 2018 The trust as a social capital of civil society in contemporary Russia MATEC Web Conferences 212 10004
Nahapiet J, Ghoshal S (1998) Social capital, intellectual capital and the organizational advantage. Acad Manag Rev 23:242–266
Nathov T (2010) Education, social capital and economic development. Questions Econ 8
Nonaka I (1994) A dynamic theory of organizational knowledge creation. Organ Sci 5:14–38
Omowaye G (2019) Further theoretical discourse on human capital development. https://independent.academia.edu/GabrielOmowaye
Parts E (2004) The Role of social capital in economic development of transition countries. Ekonomika 68:4–12
Penrose ET (1959) The theory of the growth of the firm. Wiley, New York, NY
Peters T, Waterman R (1982) In search of excellence. Harper & Row, New York, NY
Putnam R (1993) Making democracy work: civic traditions in modern Italy. Princeton University Press, Princeton, NY
Rastogi PN (2000) Sustaining enterprise competitiveness–is human capital the answer? Hum Syst Manag 19:193–203
Rumelt R (1984) Towards a strategic theory of the firm. In: Lamb R (ed) Competitive strategic management. Prentice-Hall, Englewood Cliffs, CA, pp 556–570
Samans R (2018) A new way to measure economic growth and progress. World Econ Forum. https://www.weforum.org/agenda/2018/01/towards-a-new-measure-of-growth
Schultz TW (1971) Investments in human capital. Macmillan, New York, NY

Snell SA, Youndt MA, Wright PM (1996) Establishing a framework for research in strategic human resource management: Merging resource theory and organizational learning. In: Ferris G (ed) Research in personnel and human resource management. Emerald Publishing Limited, Melbourne, Australia, pp 61–90

Stewart TA (1996) Human resources bites back. Fortune:175

Sveiby KE (1997) The new organisational wealth: Managing and measuring knowledge-based assets. Berret-Koehler, San Francisco, CA

Szulanski G (1996) Exploring internal stickiness: Impediments to the transfer of best practice within the firm. Strateg Manag J 17(Winter Special Issue):27–43

The third sector analysis in Russia: subjects and interactions (2015) "Aventic" company, Moscow, Russia

Wright PM, McMahan GC (1992) Theoretical perspectives for human resource management. J Manag 18:295–320

Wright PM, Dunford BB, Snell SA (2001) Human resources and the resource-based view of the firm. J Manag 27:701–721

# Sustainable Social Systems, Migration Flows, and Social Cohesion

# Demographic Situation in the Region in the Context of Sustainable Development

**Valentina A. Artjukhina, Viktor V. Nagaytsev, and Angelina N. Shrayber**

**Abstract** The paper analyzes the current demographic situation in Altai Krai as a factor of sustainable regional development. We believe that the state of the regional economy and the standard of living are the reasons for the fertility decline and human capital outflow. Birth rate plays an essential role in the stable development of the region and society in general. To assess the demographic situation and birth rates, we used the methodology of the sustainable development concept. To conduct a sociological analysis of the demographic situation in a particular region, we should consider its main indicators: (1) birth rate, (2) death rate, (3) migration flow, (4) marriage and divorce rates in the context of humanity, society and nature unity. Based on the analysis of statistical data, we conclude that soon, the government will not maintain high birth rates without changes in the regional socio-economic situation.

**Keywords** Sustainable development · Demographic situation · Population reproduction · Demographic processes · Regional society

## 1 Introduction

The current global demographic situation is ambiguous. On the one hand, developed countries are characterized by an aging population and a declining birth rate. On the other hand, another problem of developing countries is their overpopulation.

The current demographic situation in Western countries threatens the progressive development of social, economic, and political spheres of life. There is an upward trend in the life expectancy of the population caused by the promotion of a healthy lifestyle and longevity. Nevertheless, declining birth rates and increasing migration flows hinder natural growth and generation renewal.

V. A. Artjukhina · V. V. Nagaytsev · A. N. Shrayber (✉)
Altai State University, Barnaul, Russia
e-mail: 336680@bk.ru

 
S. G. Maximova et al. (eds.), *Advances in Natural, Human-Made, and Coupled Human-Natural Systems Research*, Lecture Notes in Networks and Systems 234,
https://doi.org/10.1007/978-3-030-75483-9_25

In the global context, the threat to sustainable development is related to the over-population of the planet, which destabilizes the ecosystem and leads to the depletion of natural resources, international conflicts, and loss of national identity.

One of the critical problems of Russia is its sparsely populated areas. The decline in the population is due to the high death rate and low birth rate. Since the end of the 1990s, there is no stable dynamics in population growth. Besides, international statistical organizations consider Russia an "outsider" in terms of life expectancy.

Among the various reasons for this situation, we mark (1) high death rates from cardiovascular diseases, respiratory diseases, and neoplasms; (2) low birth rate (which is much lower than the death rate); (3) human capital outflow.

In this situation, the government, solving the problems of preserving the population and state borders, should develop a socio-demographic. Such a policy should provide effective mechanisms to stimulate the birth rate and increase the socio-economic activity of mature and elderly people.

## 2 Materials and Methods

The understanding of complex social processes contributes to the effectiveness of management decisions. Sociological research of the current demographic situation is impossible without the main categories of this problem. The demographic situation is a set of quantitative, and qualitative characteristic of several socio-demographic processes taking place in a territory, determined by (1) population size; (2) population growth; (3) birth rate; (4) death rate; (5) migration flow; marriage and divorce rates, and many other indicators (Rybakovskij, 2005). In our opinion, the key indicators of the demographic situation in a society should include (1) birth rate (the number of births per 1000 inhabitants); (2) death rate (the number of deaths per 1000 inhabitants); (3) migration flow; (4) marriage and divorce rates. These socio-demographic processes determine the sustainable development of society.

We should also consider the demographic situation in the context of sustainable development. Based on the concept proposed by the UN in 1987, sustainable development refers to meeting the basic needs of the current generation without compromising the needs of the future generation (Commission on Environment and Development, 1987). Therefore, the main idea of this concept is to consider development from the point of view of the balance between generations.

Today, many prominent scholars discuss sustainable development. There is a universal methodology for analyzing the economic, social, political, and cultural development of regions and countries in the global socio-ecological system.

A significant contribution to the development of the concept of sustainable development was made by the representatives of various scientific branches: sociology, economics, social philosophy, political science, and many others. Among them, we can find Bateson (1979); Meadows, J. Randers, D. L. Meadows, W. W. Behrens (1972), Todaro and Smith (2003), Taylor (2016), J. Tosun, J. Leininger (2017), P. J.

Cooper, M. Vargas (2004), Brinchuk (2014), Vershilo (2008), Semin (2008), Lukasheva (2009), Begun (2012), A. V. Ivanov, I. V. Fotieva, M. Yu. Shishin (2010), A. G. Ivolga, A. A. Chaplitskaia (2014), and many others.

Methodological basis of sustainable development concept, which is necessary for the sociological analysis of the demographic situation in a particular region, allows us to consider its main indicators: (1) birth rate, (2) death rate, (3) migration flows, (4) marriage and divorce rates in the context of humanity, society and nature unity. On the one hand, the intensity of current socio-demographic processes has a severe impact on society and nature. On the other hand, the quality of human capital, social processes, and the environment largely determine the demographic situation in a certain society. Therefore, such mutual influence requires a comprehensive empirical analysis of the main quantitative and qualitative indicators of the demographic situation in correlation to indicators of social, economic, political, and environmental aspects of society and indicators of quality of life of separate individuals. The results presented below are based on an analysis of official government statistics. In particular, we analyze the demographic indicators of Altai Krai and its municipalities for 2010–2019. The key indicators are (1) death rate, (2) birth rate, and (3) migration flow.

## 3 Results

The current demographic situation in Altai Krai represents the main demographic processes that are typical of Russia. However, it has some unique features concerning socio-economic phenomena and the development of the regional demographic situation, which are reflected in decline in the population of Altai Krai due to high death rates, low birth rates, and migration to other regions. These factors hinder the socio-economic development of the region. Also, the socio-economic status of the residents of Altai Krai determine adverse demographic trends.

Due to the current demographic situation, the decline in the number of people involved in the regional economy may become the main factor limiting the opportunities for sustainable development of society. The demographic situation in Altai Krai demonstrates a correlation between the social and economic factors of sustainable development. As of today, the life, health, and safety of the population is a necessary condition for the sustainable development of society. Consequently, indexes of the demographic situation (birth rate, mortality, and migration flow) are indicators of the sustainable development of society. Based on the management mechanisms of the demographic situation, one can influence various social processes and choose relevant directions.

Recently, we can observe significant changes in the age composition and structure of the regional population. According to the all-Russian population census held in 2010, the "average age" of Altai Krai residents was 39.4 years (in 2002 – 37.7 years).

The "youngest" by "average age" of the population are the German National district and the city of Aleysk (36 years and 36.4 years). The "oldest" by this indicator are Ust-Pristansky (43.7 years), Romanovsky (42.6 years), Yeltsovsky (42.4 years), Kuryinsky, and Ust-Kalmansky (42.3 years) districts of Altai Krai.

The aging population in the region leads to an increase in the number of people outside the working-age group, while the number of people who are younger than the working-age decreases by 15% annually. Therefore, from the beginning of 2020, the generation born in the 1950s is leaving the working-age group, and the generation born in the 1990s, when the birth rate was the lowest for the entire post-war history of Russia, is entering the working-age group. At the same time, the number of children under seven increased by 15%. Consequently, the number of working-age population in this region decreased by 145 thousand people (by 10%). Nowadays, half of the local working-age population is over 35 years. More than that, the increase in the number of elderly people led to an increase in the economic burden on the population (primarily the working-age group).

A sociological analysis of the population reproduction in Altai Krai shows that in 2020 the birth rate decreased by 7% and amounted to 14.5 children born per 1,000 people compared with 2018. However, there should be at least 215 births for every 100 women of fertile age for the renewal of generations. Currently, in the region, the fertility rate or the average number of births per woman is 1.5 against the required 2.15. In the cities, this indicator is slightly lower. The model of the smaller family is inherent in the reproductive behavior of the local population. The birth rate in Altai Krai is relatively low, compared to other Russian regions with similar indicators. Low birth rates in the region can be explained by social, economic, cultural, and medical reasons, such as (1) low standard of living; (2) the distortion of traditional forms of reproductive behavior associated with lifestyle and low fertility rate; (3) reproductive health of people entering the fertile age. The attitude of young people to childbearing leads to a negative trend in the demographic situation of the region. The scholars predict a sharp decline in the birth rate. Postponing the childbirth manifests itself in the increase in the average age of mothers at the birth of their first child. In 2019, this age was 27 years old, whereas, in the 1990s, it was 22 years old. Along with changes in the reproductive behavior, other structural parameters of the regional demographic situation are being corrected, namely, the increase in the birth rate among the elderly population. At present, later motherhood in the region is a normal thing for young people, as it was in Western Europe and the United States. This trend also aggravates the regional demographic situation.

An essential characteristic of the Altai Krai demographic situation is the migration flow. The migration of the population can be explained by the current situation in the labor market. Employment problems in rural areas are especially urgent. Low income and standard of living cause migration flow. Simultaneously, most migrants represent the most active working-age social strata (young people aged 20–30 years and representatives of the social group aged 35–50 years). The outflow of experts from our region leads to a significant decrease in the economic potential of Altai

Krai. Therefore, the number of people who left Altai Krai increased twofold over the past five years (5,612 in 2014 compared to 9,945 in 2018). Simultaneously, there is a downward trend in the influx of Russian-speaking migrants to Altai Krai from neighboring countries (Kazakhstan, Uzbekistan, Kyrgyzstan, Tajikistan).

At present, one can record several migration patterns of residents' behavior: (1) migration to other regions of Siberian Federal District, (2) migration to other Russian regions, (3) international migration (the departure of our citizens abroad). The migration of citizens within the country is marked by the existence of donor and recipient regions. The former include (1) Altai Krai, (2) Omsk region, (3) Kemerovo region, and (4) the Republic of Altai, which are characterized by negative migration trends. The latter are (1) the Novosibirsk region, (2) Tyumen region, and (3) Krasnoyarsk Krai, which involves both labor and educational migration flows. Migration from Altai Krai to other regions of Russia is aimed mainly at (1) Moscow, (2) Saint Petersburg, (3) Krasnodar Krai, (4) Rostov region, and (5) Moscow region. These regions attract most migrants. The majority of migrants are young, educated people, and qualified specialists who are not satisfied with their income and quality of life in Altai Krai. As a result, Altai Krai loses its socio-economic potential and opportunities for further growth. Migration aggravates the demographic situation in the region. Moreover, there is a decrease in the share of the population of reproductive age. Migration from other regions of Russia and other countries may contribute to the stable demographic situation of Altai Krai.

Therefore, the analysis of the general demographic situation in Altai Krai indicates the threat to the sustainable development of the region. The result of such processes can distort the existing distribution by age and ethnicity of the local population.

## 4 Discussion

Improving the demographic situation is becoming an urgent social, medical, economic, and political task of local government. Therefore, sociological studies of the state and development of the demographic situation are of particular importance. We believe that sustainable development requires the reproduction of human capital.

## 5 Conclusion

Today, the government should (1) decrease death rates, (2) increase the life expectancy of the population, and (3) encourage the childbirth. In addition, it is necessary to implement a set of social, political, ideological, educational, and informational measures aimed at meeting the needs of individuals in children. To achieve these goals, one should maintain and improve the reproductive health of women and men. Specific causes that contribute to the high death rates in various social groups require further study. Today, success in the fight against incurable diseases can only be achieved through lifestyle changes, the attitude of people to themselves, and their health. The main threats to the life and health of the population are outside social institutions. They should stick to healthy diet, care for nature, and control their consumer habits. Therefore, to improve the demographic situation in the region, the population should aim to (1) improve health, (2) protect the environment, (3) lead a healthy lifestyle, (4) get rid of harmful habits, creating healthy ones. The demographic situation in the region can change due to profound changes in the attitude of society to their health and life.

Solving the problems of the region, one should focus on population reproduction as the most important indicator of sustainable development of society. In this regard, identifying existing trends will allow us to understand the current demographic situation in society and, consequently, to find possible ways to improve it.

**Acknowledgements** The analysis of the statistics presented in the article is based on Federal State Statistics Service for Altai Krai and the Republic of Altai, published in the public domain (https://akstat.gks.ru/folder/32996).

## References

Bateson G (1979) Mind and nature: a necessary unity (Advances in systems theory, complexity, and the human sciences). Hampton Press, Hampton, VA

Begun TV (2012) Sustainable development: Definition, concept, and factors in the context of single-industry towns. In Shulga OA (ed) Proceedings of the II International Science Conference: Economics, Management, Finance (pp 158–163). Mercury, Perm, Russia

Brinchuk MM (2014) The concept of sustainable development as a methodological basis for civilizational development. State Law 10(SS):15–24

Cooper PJ, Vargas M (2004) Implementing sustainable development: from global policy to local action. Rowman and Littlefield Publishers, Lanham, MD

Ivanov AV, Fotieva IV, Shishin Myu (2010) Spiritual and ecological civilization: Foundations and prospects, 2nd edn. Publishing House of the Altai State Agrarian University, Barnaul, Russia

Ivolga AG, Chaplitskaia AA (2014) Substation of approach to the concept of sustainable development of the regional economy. Biol Res Nat Manag 1–2:145–148

Lukasheva EA (2009) Man, law, and civilizations: Normative-value dimension. NORMA, Moscow, Russia

Meadows DH, Randers J, Meadows DL, Behrens WW (1972) The Limits to growth: a report for the club of Rome's project on the predicament of mankind. Universe Books, New York, NY

Rybakovskij LL (2005) Practical demographics: textbook for universities. Collective monograph. CSP, Moscow, Russia

Semin AN (2008) Features of rural territories sustainable development and improvement of the social and labor sphere of the village in the new regional economy. Izvestiya USUE 2(21):115–119

Taylor SJ (2016) A review of sustainable development principles: centre for environmental studies. University of Pretoria, Pretoria, South Africa

Todaro MP, Smith SC (2003) Economic development, 8th edn. Addison-Wesley, New York, NY
Tosun J, Leininger J (2017) Governing the interlinkages between the sustainable development goals: approaches to attain policy integration. Global Challenges:1700036. https://www.researchgate.net/publication/321037327_Governing_the_Interlinkages_between_the_Sustainable_Development_Goals_Approaches_to_Attain_Policy_Integration
Vershilo ND (2008) Environmental and legal foundations of sustainable development. Formula of Law, Moscow, Russia
World Commission on Environment and Development (1987) Our common future. Oxford University Press, Oxford, UK

# Supporting Foster Families in the Altai Krai as an Innovative Technology in Sustainable Development

**Irina V. Antonovich, Alina O. Shcherbinina, and Angelina R. Ahmedova**

**Abstract** In this study, we conducted empirical social research to identify the directions and components of support for foster families. We employed the help of experts in supporting foster families in the Altai Krai and surveyed the representatives of foster families. We identified the main components of support for foster families: (1) motivation of the target audience, (2) interagency cooperation, (2) legal aspects, (3) personnel structures, (4) methods and forms of work. The novelty of this study lies in the multifaceted analysis of the problem, providing various innovative solutions, and identifying the most effective methods of interaction with foster families.

**Keywords** Sustainable development · Foster families · Sustainable development technologies · Empirical social research · Altai Krai · Social work · Social problem · Support

## 1 Introduction

The scale of industrial, social, and political activities generates global contradictions and new crises. Currently, the government and scientists try to solve two interrelated problems:

1. Influence of technogenic and anthropogenic factors on nature.
2. Growing socio-economic inequality in separate countries.

Social work is the main way of solving social and societal problems. Currently, social work covers an increasingly complex area and level of social problems. Social work is defined as an activity with the purpose of solving the social problems of different population groups and people. Presently, social work with children without parental care focuses on providing in-family upbringing. Family is an indispensable social institution that ensures the full development of the individual.

I. V. Antonovich (✉) · A. O. Shcherbinina · A. R. Ahmedova
Altai State University, Barnaul, Russia
e-mail: irina.antonovich2504@yandex.ru



S. G. Maximova et al. (eds.), *Advances in Natural, Human-Made, and Coupled Human-Natural Systems Research*, Lecture Notes in Networks and Systems 234,
https://doi.org/10.1007/978-3-030-75483-9_26

As a form of upbringing and socialization, foster family care is much closer to biological family upbringing than institutional care.

The socialization potential of a foster family lies in its ability to form a new reliable emotional connection with the child. This connection (or attachment) is a necessary component of psychosocial development and directly affects the success of socialization.

However, foster family care has its own issues and drawbacks. V. V. Barabanova (1995) empirically identified six issues related to functional adaptation to foster family care. Issues are more likely to arise in foster families (Oslon, 2006):

- Where one of the parents is absent;
- Where the age of foster parents is too high;
- Where biological children are absent;
- Where foster parents are ill-informed about the specifics and special needs of a foster child;
- Where foster parents inadequately assess the difficulties of foster parenting.

Analysis of statistical data on the number of children in foster families allows examining the quality of support for foster families [SFF]. In 2013–2016, the number of children in foster families was increasing in Altai Krai. However, in 2017 the number started decreasing. For example, 7,978 children were in foster family care at the end of 2019 (less by 3% compared to 2018: 8,262 children). This trend may be closely related to the increased rate of returns to institutional care. At the same time, 88% of foster care children are currently brought up in foster families. In 2019, 5,515 foster families (more than 400 of which are large families) were in Altai Krai. Work with foster families aims to provide for them, prevent the return of children to institutional care, and preclude the risks of child abuse. Therefore, studying the SFF in the Altai Krai is becoming increasingly important.

The development of foster families depends on the regional economy and regional regulations since the authorities provide remuneration for foster parents and establish the volume and quantity of benefits (Bazarova, 2011).

## 2 Materials and Methods

The empirical social research consisted of two stages. In the first stage, we interviewed eight experts on SFF in the Altai Krai. We used the method of targeted snowball sampling. All interviewees were specialists in the theoretical and practical fields of SFF.

In the second stage, we interviewed several foster families chosen via representative sampling:

- Three foster parents invited to the *Barnaul Industrial District Meeting for Foster Families* on April 24, 2019

- Three foster parents who were clients of the "Altai Regional Center of Psychological, Pedagogical, and Medical-Social Assistance."

The representatives belonged to large families and families with one-two children.

The interview with the experts was *in-depth* and contained four blocks of questions: (1) general information on SFF, (2) regulatory framework of SFF, (3) organizational issues, and (4) expert opinion. The interview with the foster family representatives was *semi-standardized* and contained three blocks: (1) general information, (2) questions on the organization of SFF, (3) personal opinion. The obtained data were processed and analyzed.

## 3 Results and Discussion

Currently, there is no scholarly accepted definition of a foster family. In analytical approaches, the notion is defined in various but interrelated contexts. This study is necessary because foster families often face complex, multifaceted issues, especially during functional adaptation. Therefore, there is a demand for comprehensive and extensive support with the help of specialists and experts (Nekolaeva, 2011).

This study was conducted to reveal the directions and components of SFF. The transcribed texts were analyzed in three main stages: data description, conceptualization, and clustering. At the conceptualization stage, the data were nominated, and the nominations were generalized. We identified five direct components of SFF organization that influence its efficiency. These include: (1) motivation of the target audience, (2) interagency cooperation, (3) legal aspects, (4) personnel, (5) methods and forms of work. It should be noted that the components are interrelated, which emphasizes consistency as one of the main characteristics of SFF.

The first aspect is the motivation of the target audience, which is the foundation for SFF. This component was highlighted as a priority.

When contacting the support service, foster parents need to get qualified assistance from specialists in solving certain issues and problems. Families whose members are motivated for productive work and communication with specialists have great resource potential. Most of the experts in the interview mentioned the importance of foster parent motivation. The direct involvement of foster parents in support influences the correction of child-parent relations and the functional adaptation of the entire family.

The second aspect is the interagency cooperation in SFF organization. Quality SFF requires constructive interaction of specialists from different departments, services, and organizations. In SFF, systematic and comprehensive work with the target group is carried out. Interagency cooperation is also emphasized in the regional legislation on SFF (Main Department of Education and Youth Policy of Altai Krai, 2017). Corresponding legal provisions describe the main powers of institutions at the regional, municipal, and local levels.

Legally speaking, SFF is defined as "...a new type of state aid, which includes comprehensive support and professional activities of social workers". Social support is an interdepartmental and interdisciplinary phenomenon, in which specialists from various departments participate (education, health care, law enforcement, etc.) (Kuzina & Pankova 2019).

Therefore, the inclusion of employees from other spheres of life in SFF organization may serve as an important factor. The inclusion of other specialists should be indirect and consistent, especially in creating a complete picture of a family's social environment and the relations between a family and other social institutions.

The third component, the problems with personnel structures, means a lack of staff support services and other institutions. One expert emphasized the need not only to expand the staff but also to redesign the structure of support services.

Legal aspects are an important component that affects the quality of SFF. This component is currently under development, especially from the point of view of the SFF based on centers for children support. However, in expert opinion, there are gaps in the current regulatory framework. In their opinion, the key point is the lack of standards currently governing such nuances of SFF as the list of services and activities, and time standards.

The last component is the methods and forms of work used in SFF. The use of effective forms and working methods contributes to obtaining quality results and the complete inclusion of a family member in support. The experts noted that consultations with the whole family is the most important form of work, especially if family members can find out the answers to their questions, try to understand each other, and hear the opinion of a competent specialist. Experts also use creative technologies for analyzing and projecting the correct solutions for foster families.

The components mentioned above are related to support directions. Having analyzed the transcripts of interviews with respondents, we identified several directions of SFF.

1. Counseling work, the essence of which is to cover the topics and questions on the difficulties in functional adaptation and life of a foster family. This direction is the most basic one; it serves as the foundation for SFF. Counseling assistance is one of the most popular from the point of view of foster parents and one of the most effective from the point of view of specialists. It should be noted that this direction can be carried out both in the form of separate consultations (if there are no crisis and conflict situations in the family) and as part of a set of measures to address a particular problem. The latter is the second direction of SFF organization.
2. Correction-development work is a separate component aimed at providing psychological and pedagogical assistance, resolving conflict situations in the family, normalizing relationships within the family structure through various forms and methods of work, and development therapy with children. Representatives of foster parents also emphasized the importance of this area.

3. Information and education activities aim to improve the psychological and pedagogical competence of foster parents and popularize psychological and pedagogical knowledge among the target group. Representatives of foster families also noted the effectiveness of these activities.
4. Club activities as a separate direction of SFF organization. It should be noted that foster family clubs are not currently widespread: most experts noted that there are no such structural units in their centers.

## 4 Conclusion

To identify the main directions and components of SFF, we conducted an empirical sociological study in Altai Krai. We partially confirmed the hypothesis that, from the point of view of experts, it would be inexpedient to replace the voluntary SFR with mandatory support. Some experts believe that SFF should be compulsory for some families. Foster family representatives noted a lack of close interdepartmental communication of support services and general educational institutions. According to the results of this study, we identified the main directions of SFF: counseling, correction-development work, information-education work, and club activities. Necessary components of SFF organization include: (1) target audience motivation, (2) interagency cooperation, (3) legal aspects, (4) personnel, and (5) methods and forms of work.

As a result, we made several conclusions:

First, in improving the legislative base of SFF at the regional level, an important aspect is clarifying the support process, and the functions and areas of work of support services.

Second, there is a need for a practice-oriented discipline, which allows forming the SFF competencies in higher education. Therefore, the discipline "Organization of Support for Foster Families" should be included in the mandatory subjects of the 44.03.02 "Psychological and Pedagogical Education" field of study. Presently, SFF services mainly hire specialists with *psychologist*, *teacher-psychologist*, or *social teacher* diplomas.

Third, the additional personnel units must be created among specialists whose job responsibilities only include activities within the framework of the support organization. Moreover, the team of specialists could be expanded, if necessary. One of the possible solutions to this issue can be the development of cooperation with non-profit organizations and the use of grant contests.

Fourth, foster family clubs should be created as additional structural units. Their purpose lies in exchanging experience between foster parents, and creating a comfortable and peaceful atmosphere.

Fifth, information campaigns among the general population and foster families must be improved further.

## References


Barabanova VV (1995) Know the difficulties in the development of children and learn to overcome them. In: Rybinsky EM (ed) Family orphanage: Reality, problems and prospects in modern Russia. Dom, Moscow, Russai

Bazarova EB (2011) A foster family as an institution of socialization of orphans and children without parental care. Bull Buryat State Univ 5:236–243. https://cyberleninka.ru/article/n/zameschayuschaya-semya-kak-institut-sotsializatsii-detey-sirot-i-detey-ostavshihsya-bez-popecheniya-roditeley

Kuzina IG, Pankova NV (2019) Social support as an innovative family-saving technology: from the Primorsky Region experience. Kuban Region Multi-Sector Academy of Vocational Training and Upgrading and Advanced Training of Specialists; Northern Kuban Region Institute for the Humanities and Technologies, Vladivostok, Russia

Nikolaeva EI (2011) Diagnostic tools to help specialists working on the family arrangement of orphaned children. Prospekt Publishing House, Moscow, Russia

Oslon VN (2006) Life of orphaned children: professional foster family. “Genesis”, Moscow, Russia

The Main Department of Education and Youth Policy of the Altai Krai (2017) Order “On the organization of support for foster families in the Altai Krai” (March 7, No. 1224). Barnaul, Russia

# Expert Perception and Risk Evaluation of Migrations in Russia

**Svetlana G. Maximova, Oksana E. Noyanzina, and Daria A. Omelchenko**

**Abstract** In 2018–2019s, authors realized complex sociological research on study about migrations in Russia, including expert questioning in 7 regions of the country. In the article, authors present results of expert evaluation of problem of transit by key persons, making decisions in the sphere of migration. Expert evaluations about regional characteristics and opportunities to consider it as a transit one obtained. Transit migration understood as an intermediate migration in some country (region) with goal of further migrations to other countries or more developed regions of the country, considered as final destination points. Not every region was evaluated as the transit one. Transit movement through Russian borderland of in-country dominantly, with weak redistribution of migration stream into the Western European countries from the European borderlands. Despite common for all regions except Moscow border character of territories of migrants' entry, their different socio-economic and geographic position and other peculiarities, apparently, determine regional societies' capabilities to "capture", attract migrants and become for them the final point of their current migratory route.

**Keywords** Transit · Transit migrants · Transit region · Border region · European and Asian borderlands

## 1 Introduction

Transit migrations are more frequent subjects of interdisciplinary research than others (Pliez, 2013; Düvell, 2014). Therefore, political scientists, sociologists, anthropologists, as well as other representatives of the field, are primarily trying to understand the new migration reality. Besides, the scope of interest of scholars embraces possible ways to *localize* processes of transnationalization of migration policy based on the analysis of their actors, practices, technologies, and discourses, specific to concrete social situations (Fargues, 2011).

S. G. Maximova (✉) · O. E. Noyanzina · D. A. Omelchenko
Altai State University, Barnaul, Russia
e-mail: svet-maximova@yandex.ru



S. G. Maximova et al. (eds.), *Advances in Natural, Human-Made, and Coupled Human-Natural Systems Research*, Lecture Notes in Networks and Systems 234,
https://doi.org/10.1007/978-3-030-75483-9_27

Transit migration is a synthetic concept, uniting notions of transit and migration (Papadopoulou-Kourkoula, 2008). Its content is very dynamic, blurred, and ideologically charged. Very often, transit migration is associated with illegal or return migration, circular migration, and refugees. This notion refers to several forms of supposedly temporary migration and migrants, who are in a state of the move from one country to another, deliberately or forcibly, under the pressure of circumstances, new conditions of living, opportunities, or threats. In a broad sense, transit migration supposes real and imagined resettlements and journeys for long distances and periods, changing directions constantly and being eventually dangerous for a migrating person.

On the one hand, transit migration is an umbrella term, encompassing different categories, such as refugees or labor migrants, which can be applied towards various streams of migration (regular, legal, and well-documented, or irregular, illegal, and hidden, as well as temporary, long-term, or permanent). This expression does not describe singular displacements but those with several stages (from interruption of travels and forced migration to its single trajectory). Some scholars include ambitions and even dreams of *aliens* in receiving countries hoping to venture beyond their limits in its content.

As it was mentioned earlier, the conception of transit migration is relatively new for contemporary science. Its growing popularity may be explained by the intensification of the processes of internalization and externalization of the European Union migration policy, the creation of several projects within the migration control in neighboring countries, and the return migration. Thus, F. Düvell (2006) notes that the concept of transit migration is *very European*. Nevertheless, it also manifested in South-Western Asia, South Africa, and Latin America. In some early publications (e. g., in the global report of the International Organization for Migration for 1994), it was spoken about two million migrants, situating in Central Europe and hoping to move to the West.

Meanwhile, despite significant political and academic attention, there is no single and universally accepted definition of transit migration. A. Papadopoulou (2005) describes transit migration as a stage between emigration and settlement, in contrast to the report of the Assembly of the Inter-Parliamentary Union in Geneva (2005), where transit migrants are defined as foreigners, staying in the country for a certain period and continually trying to move to another country. The UN Economic Commission (1993) identifies transit migration as migration from one country to seek opportunities for emigration to another country as a country of destination. H. Roman (2006) supposes that even after twenty years of living in a country, migrants who have moved to another country can be treated as transit migrants.

Studies show that reducing the research to the social networks of migrants is often insufficient for a comprehensive understanding of migration processes. Most migrants need professional help at different stages of their migratory movement. Moreover, migrants *on the route* create associations (networks) to ensure their safety. Despite complications and high dependency on circumstances, migrants are very mobile and flexible in implementing goals and plans. Their migration trajectories often have spontaneous character and consist of repetitive and temporary movements, intentional and deliberate or involuntary and forced. In reality, migration is much more dynamic than it was initially conceived. From this point of view, the migration process can be considered in terms of mobility, as well as its dynamics and factors, and not only a single and linear process of the movement towards a new and constant place of destination (Skeldon 1997).

With the emergence of the transit migration phenomenon, the interest of theoreticians and practitioners in the migration of people and particularly to the act of the movement itself is growing. Traditionally this movement was almost ignored in migration studies, and the focus of the social research was made on the two discrete steps of migration: (1) decision-making process before migration at the initial stage, and (2) influence of migration on hosting (and receiving) society in the final point of migration (IOM 2015).

Since the development of the holistic approach to migration and its transit phase on the example of several countries (Hamood 2006; Collyer 2010; Temime 1989), scientific research concerned the analysis of migration and life experience of migrants in temporary receiving countries, but it generally considered a reduced number of countries and migrations in specific directions (e. g., from the East to the West) and described only single stages of migratory trajectories and their finish (typically in Europe). That is why such research, even with supposed more in-depth analysis, usually did not examine the organization of the migration process and stimulating or inhibiting factors.

Within the research, we apply a broader approach to transit migration. First, we do not restrict migratory trajectories geographically by a given direction (i. e., the movements of migrants in our focus end not only in Europe but also as a result of past migrations, on the frontier or in other regions of Russia or the countries of Asia). Second, we consider the ways of *actually moving* migrants and those who had previous experience of migration.

## 2 Materials and Methods

In the Russian Federation, as in Europe, transit migration processes had become acute in the 1990s when borders were opened for migrants from the CIS countries, who, using Russia as a transit country, were willing to relocate to European countries. Simultaneously, the activities of various agents, assisting with illegal migration beyond the borders of the former USSR had been increasingly expanded. Russia has become an easy channel for migration in Europe for migrants from the Central Asian countries, China, and Africa. As it was already noted, Russia, as well as Ukraine, has become the leading country for transit to the territory of the European Union.

According to T. Trofimova (2010), choosing Russia as a point for transit is associated primarily with its geographic position, especially to the common land border with the EU, transparent borders with the CIS countries, weak control for the entry of foreign citizens and their stay in the territory of the country, as well as the lack of efficient institutional mechanisms and tools for managing migration processes, high level of corruption, the spread of hidden *shadow* labor practices, and the strong influence of network interactions with diasporic ethnic groups. In particular, in Russia, there are practically unlimited possibilities to participate in the grey (informal) economy (Radaev 2013) and earn enough to pay for the next step of migration.

The Russian Federation is one of the most attractive countries for migration in the world, visited by several million people annually. About 15 million foreign citizens reside on its territory at one time, having different purposes of staying here (education, tourism, private visits, and many others), but people searching for work constitute a separate specific group.

In 2018–2019s, we completed complex sociological research to comprehensively analyze the transit migration phenomenon in Russia, including expert surveys in seven regional territories. It is noticeable that the selected regions take different positions in terms of social and economic development indicators and have different demographic pressure, geographic location, and consequently, different borders. The Murmansk region borders on Norway and Finland, the Pskov region borders on three countries (Belarus, Estonia, and Latvia), the Rostov region borders on Ukraine, the Republic of Dagestan borders on Azerbaijan and Georgia, the Altai Krai and the Orenburg region border on Kazakhstan. Three regions belong to the European borderland zone, and the other three belong to the zone of the Asian borderland. Additionally, the analysis focused on the migration situation in Moscow, the capital city of the country, having maximal migration pressure and being the place of attraction and redistribution of migration flows.

Addressing transit migration issues, 146 experts from seven constituent entities (20 respondents from each region and 26 respondents from the Altai Krai) participated in the survey. The respondents represented the opinion of federal and regional authorities, scientific and academic communities, non-governmental organizations, and representative bodies. The key criterion for including experts in the research group was their participation and involvement in elaboration, support, and implementation of migration policy. The experts took different positions in the organizations they represented. They worked as experts, leading and principal specialists of migration services, senior executives and mid-level managers in regional branches of the Ministry of Internal Affairs, science officers, professors, associate professors, heads of departments, and directors of research centers in leading scientific and academic higher education institutions, deputies and chairpersons of specialized commissions of regional legislative bodies, and leaders of social organizations and ethnic-cultural centers.

The paper aims to analyze the urgency evaluations of transit migration given by the experts in the migration sphere and the key persons making decisions in this field. More than that, the paper aims to answer whether transit exists and whether it is reflected as a significant category in representations of those who try to manage and coordinate migration processes, as well as monitor them from the position of possibilities to ensure control on their dynamics. Another question that arises is how this process influences the social and economic development of the Russian border areas, as well as the national and social security of regional border societies. The research results are the expert assessments of the regions on the totality of indicators and discussion about possibilities of identification of Russian border regions as transit or non-transit (we refer to transit migration as intermediary migration in a country (region) intending to find possibilities for further migration in other countries or more developed regions of the same country considered as places of destination).

## 3 Results

The experts do not think each region is the transit one. In particular, the Rostov region is less correspondent to the notion (40% of the experts consider it as the transit territory), as well as Moscow (45%), while the Orenburg region (in 95% of cases) is a transit region. The same trends are relatively evident in the Murmansk region (80%) and the Pskov region (76.5%). Simultaneously, experts often consider the Altai Krai and the Republic of Dagestan as having transit characteristics. At the same time, the different socio-economic and geographic positions of the observed regions, as well as other peculiarities, will probably determine how regional communities can keep and attract migrants as a final destination point despite the common border character of the territories for the arrival of migrants (except Moscow).

The intensity of transit movements of migrants in the border regions also differs. The evaluation given by the experts allows dividing the regions into three groups according to the transit intensity:

- Intensive, spread transit—the Orenburg region (85%);
- Increasing transit—the Pskov region (47.1% of the evaluations as *the spread* and 38.9% of the evaluations as *weakly spread*);
- Weak, *sluggish* transit—the Murmansk region (80%), the Altai Krai (63.6%), Moscow (60%), and the Rostov region (57.9%).

Different borders predetermine the formation of different senses, identities, risks, and threats in transborder and border areas and possibly different transit directions. Thus, it is considered that transit streams move in the direction to the West, to Western and Eastern Europe. Nevertheless, the presence of the EU borders in some Russian regions allows considering that the transit movement is from the eastern and central regions of Russia to the West. The border accumulates different migrants having experience of several migrations, different countries, and points of origin. However, the same vector of transit (e. g., through the Pskov region to Belarus or further, through Belarus to Western or Eastern Europe, or the Murmansk region to Finland). The regions of the Asian borderland are points of the first stage of transit, and vectors of moving out of them (or through them) are going in different directions.

As for the character (transit/non-transit) of migration processes in the region, the expert evaluations are not so solitary as the composition of migration streams. On the one hand, in most cases, border regions are calm in migration situation (63% is in transit regions and 58.3% is in non-transit regions). On the other hand, migration streams accompany the outflow of the aboriginal population. These concerns transit (28.3%) as non-transit regions (29.2%). The given process occurs together with an

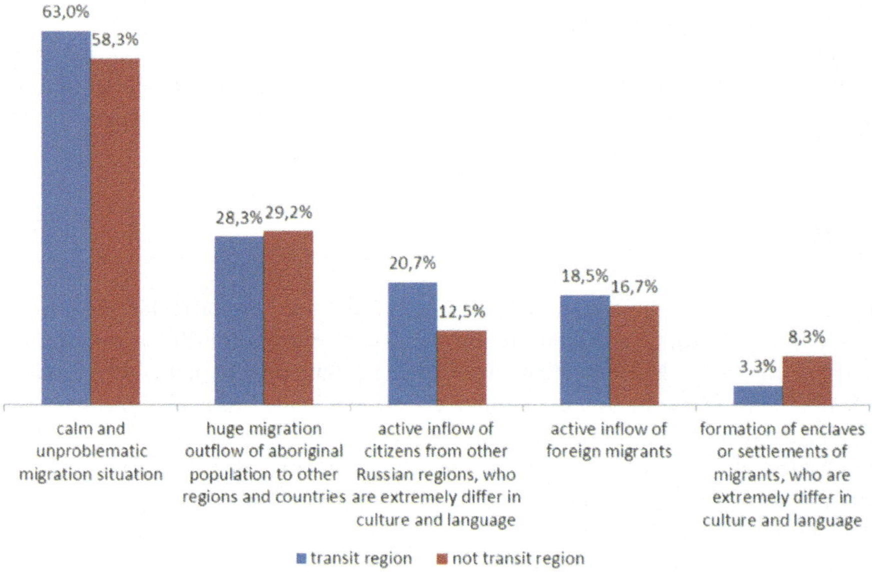


**Fig. 1** Expert characteristics of migration processes composition in borderlands, multiple choices, ($\chi^2$, $p \leq 0.005$), %

active inflow of Russian citizens having alien ethnicity (20.7% in transit regions and 12.5% in non-transit regions) and foreign migrants (18.5% in transit regions and 16.7% in non-transit regions). Simultaneously, the regions rarely have problems with forming ethnic enclaves. Non-transit regions face the problem of enclaves more often (8.3%) than transit regions (3.3%) (Fig. 1).

However, the picture is the following. There is an active process of outflow of the aboriginal population almost in one-third of the border regions. The replacement of the outgoing population (even temporarily, for the further movement) for transit regions occurs with the Russian citizens (even of alien ethnos, delivering untraditional culture to regional societies) and foreigners. However, transit regions dominantly replace the population with foreign labor migrants. For that reason, the phenomenon of forming ethnic settlements became more spread in non-transit regions. Such settlements consist of alien ethnos, poorly adapted to the social life in Russian society, and their identities do not match with the Russian civic identity.

Regions differ from separate countries by the country structure of migration streams and the scale of migration streams. In the regions of intensive, widely spread transit (the Orenburg region), the general stream of migrants consists mostly of citizens of Kazakhstan (100% of the expert choices) and Uzbekistan (90%), as well as Armenia (40%) and Tajikistan (40%) in equal proportions. An insignificant number of citizens of Ukraine and China (5%) come to the Orenburg region.

The Pskov region, which is a region of increasing transit, faces an ethnically diverse and balanced migration stream from Uzbekistan (40%), Tajikistan (35%),

Armenia (30%), Moldova and Ukraine (25%), Kazakhstan (15%), Belarus, Azerbaijan, and China (10%). The weakest migration flow comes from Kyrgyzstan (5%). Another region that belongs to the same group is the Republic of Dagestan. It faces many citizens of Azerbaijan (85%). More than that, it faces a twice weak stream of migrants from Uzbekistan (45%) and Tajikistan (40%), as well as migrants from Kazakhstan (15%) and Ukraine (5%).

The Murmansk region is in the group of weak transit. It faces three ethnically large groups of migrants, dominantly consisting of those from Tajikistan (50%), Uzbekistan (50%), and Ukraine (45%). Citizens of Azerbaijan (30%), Kyrgyzstan (20%), Armenia (15%), Belarus (10%), and Moldova (5%) form the less part of the migration inflow. Altai Krai hosts the flow of migrants dominantly from Uzbekistan (76.9%), Kazakhstan (57.7%), and Tajikistan. The other countries are weakly presented (26.9% from Kyrgyzstan, 15.4% from China, 11.5% from Ukraine, 7.7% from Azerbaijan and Armenia, and 3.8% from Belarus).

Moscow is an absolute center of gravity for a considerable number of migration streams. However, some of them are more evident. The major migration streams are from Azerbaijan and Armenia (47.4%), Ukraine (26.8%), Uzbekistan and Tajikistan (31.6%), Kazakhstan (21.1%), China (15.8%), Kyrgyzstan (10.5%), Belarus and Moldova (5.3%).

In the absolute majority of cases, the Rostov region is an attractive point for the entry of migrants from Ukraine (60%), Armenia (30%), China (15%), Uzbekistan (10%), Tajikistan (10%), Kyrgyzstan (10%), Belarus (5%), and Kazakhstan (5%).

Furthermore, we will try to test our assumptions about different transit migration vectors out of the Russian borderland regions. Most of the transit migrants in Altai Krai (63.6%) go to the European part of Russia or its central regions, more than one-fourth (27.3%) move to the Asian countries, and only a small part (9.3%) seek for the Western European countries. We believe that the main part of the transit flow to the Russian central part consists of citizens of Uzbekistan, Tajikistan, Kazakhstan, and Kyrgyzstan. Most transit migrants in the Orenburg region (90%) continue their movement to the West, to the other central and western regions of the country. They are citizens of Kazakhstan and Uzbekistan. Transit flows out of the Murmansk region are more *European*. Thus, a considerable part of migrants (60%) moves to central Russia, one-third (30%) go to Western Europe, 5% move to Eastern Europe, and another 5% migrate to Asia. Migrants from Ukraine, Tajikistan, and Uzbekistan form most of the transit stream in the Pskov region. General transit directions out of the Pskov region are the Western European countries (55%) and the Russian European part (30%). Another direction is Eastern Europe (5%), as well as Asian countries (5%). In the Republic of Dagestan, transit migrants move to the European part of Russia, and only about 5% go to the Western European countries. In the Rostov region, two further transit directions domain, but the stream of movement to Western Europe is hugely higher (up to 25%). These two transit vectors are actual for Moscow too. However, a part of transit out of the country in the Western direction increases and divides the stream of further movements into two equal parts: 53.3% of transit migrants move to the European part of Russia, and 46.7% go to the countries of Western Europe.

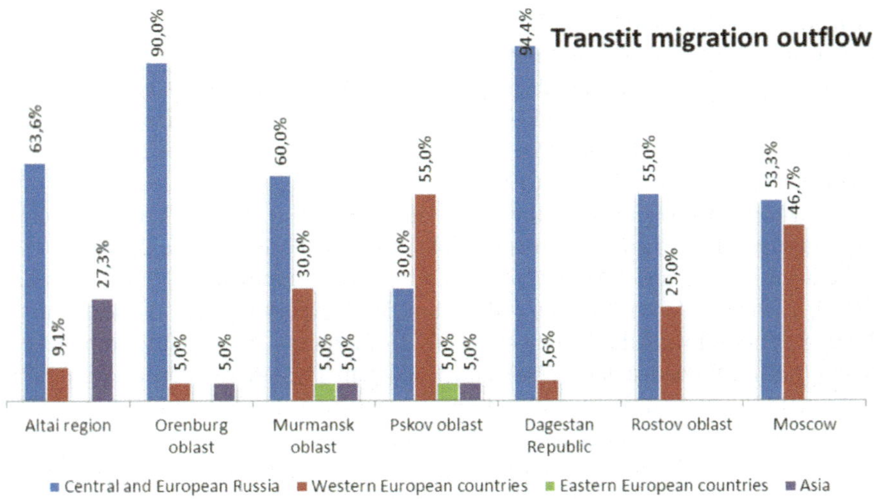


**Fig. 2** Aggregate diagram for the general vector of transit movements out of the borderlands

According to the aggregate figures (Fig. 2), transit movements through the Russian borderland regions are mainly in-country, but with a slight redistribution of migration stream to Western Europe from regions of the European borderland.

Thus, there is an evident redirection in the vector of transit from the Moscow, Rostov, Pskov, and Murmansk regions, which are the European borderlands of Russia, to Western Europe. At the same time, transit from the Orenburg region and Altai Krai (zone of the Asian borders of Russia) occurs to the central and European parts of Russia. The only exception here is the Republic of Dagestan. It borders on the countries of Europe (Georgia and Azerbaijan). Therefore, the ethnic republic redirects almost all migrants to central Russia. These require a separate analysis of peculiarities of living in the republic, relations between migrants and the host community, and the social moods of migrants. It is reasonable to answer what reasons for such coordinated mass transit of migrants are, and what predetermines the channels of movement to the destination considering the migration agents (e.g., diaspora) whether it is due to the unwillingness to stay in the republic for a long time, fast migration with the next stop in another central region of Russia, or *composition* of migrants themselves and its characteristics.

The migration streams significantly form the structure of risks for receiving regions ($\chi^2$, $p \leq 0.005$), which slightly varies in transit and non-transit regions. According to the expert opinions, migration has the following risks:

- Uncontrolled migration inflow from other regions of the world;
- High loading for customs services, inability to protect the state borders and control over foreigners in Russia;
- Uncoordinated migration policy of neighboring countries caused by open mutual borders and economic obligations;

- Blurring between streams of economic migrants and refugees;
- Political manipulation on migration issues (e. g., during the elections, etc.);
- Risks of illegal migration, labor abuse of migrants in the shadow economy;
- Forced prostitution and kidnapping for sexual work;
- Risks of trafficking;
- Risks for the epidemiological situation and spread of infection and virus diseases;
- Formation of closed ethnic groups, having lack of social contacts with the hosting community and creating sources of tension;
- High loading for the system of socials services and social security.

The given risk is equally expressed in the regions of different types of migration and amounts to about 20–25%, with the prevalence of the main threat related to illegal migration and labor abuse of migrants in the shadow economy. However, the transit character of migration streams enforces some specific risks for regional border societies, namely:

- Increase of loading for custom services (29.3% in transit regions and 16.7% in non-transit regions);
- Blurring of migration streams between economic migrants and refugees (23.9% in transit regions and 6.3% in non-transit regions);
- Threats of epidemiology to the population, twice higher spread of infection and virus diseases (38% in transit regions and 18.8% in non-transit regions);
- Increase of loading for social services (20.7% in transit regions and 14.6% in non-transit regions).

In transit regions, local governments are usually weak in control for migration streams: 31.3% of the experts marked uncoordinated migration policy of neighboring countries considering open borders and mutual economic obligations. Additionally, the risks of forming ethnic enclaves and spatial segregation of migrants positively increase in the regions of final migration (39.6% of the expert choices are for non-transit regions, and 26.1% are for transit ones) (Fig. 3).

Furthermore, we tried to reveal factors of transit movements in the observed regions. We have found that some regional peculiarities significantly predetermine transit. Thus, according to the experts, only three features significantly and validly predetermine the types of movement in border regions:

- Peculiarities of the geographical location of a region;
- Socio-economic situation in a region;
- Ethnic composition of the aboriginal population.

Of course, the geography and borders are evident and explainable indexes, influencing the character of migration and determining transit in the first turn. As for socio-economic conditions, the index was not so important in Altai Krai (7.7%), the Pskov region (10%), the Republic of Dagestan (10%), the Murmansk region (15%), and the Rostov region (15%). Simultaneously, it is determinative for the Orenburg region (45%), which is a region of intensive transit, as well as Moscow.

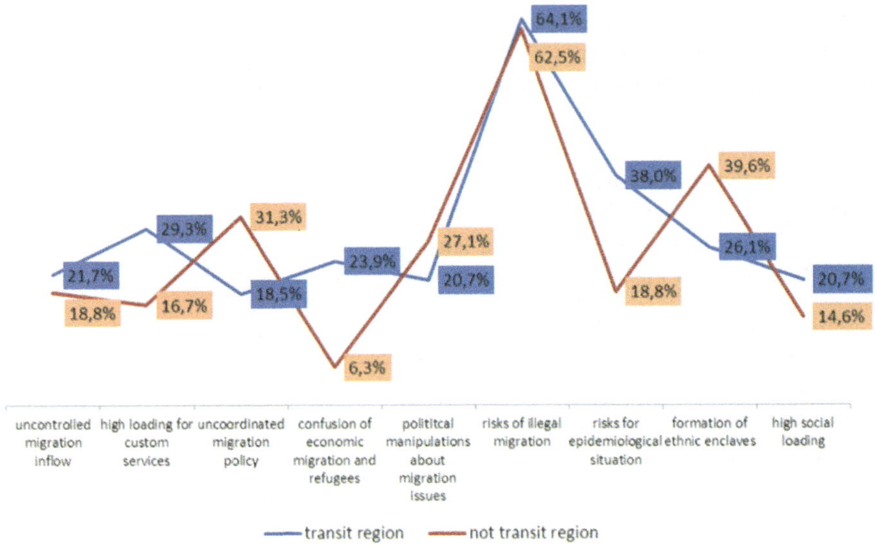


**Fig. 3** Expert evaluations about risks of migration depending on the type of region, ($\chi^2$, $p \leq 0.005$), %

Ethnic composition of the population blurring the border territory of different countries enforces transit movements only in the Rostov and Moscow regions (25% and 42.1%, respectively) even though these characteristics do not act as determinants in the other regions.

Each border region validly faces different risks enforced by transit migrants (Table 1). Thus, the Orenburg region faces the influence of transit on the level of regional security (75%), Altai Krai faces the consumerism of migrants (30.8%), the Murmansk region faces the consumerism (35%), the unwillingness of migrants to assimilate (30%), mobility of ethnic composition of the population (25%), and the influence of migrants on security (25%). As for the Pskov region, the consumerism of migrants (50%) is dominant. In the Republic of Dagestan, it is the influence on the level of security (35%) and consumerism concerning the republic as well (30%). Simultaneously, the Rostov region faces risks that are weak in whole, and there is just slight evidence of the consumerism of migrants (25%). The Moscow region faces the maximal expression of the consumerism of transit migrants (65%).

We should note that the Orenburg region also faces a high threat to security (75%). Moreover, in the Murmansk region, the risks of transforming ethnic composition are incredibly high (25%). As for the risks of blurring Russian culture, they are more evident for the Orenburg region experts (40%). Notably, the unwillingness of migrants to assimilate is expressed as one of the considerable threats almost in all regions (at the level of 15–30%). The risks of low interest of migrants on the well-being of the hosting region are also highly expressed.

Evaluating regional security is the most differentiated in the regions because of transit migration and risks of the transformation of ethnic composition. These two

**Table 1** Expert evaluations about specific risks of transit migrations, regional comparisons, %

| Risk | Altai | Orenburg | Murmansk | Pskov | Dagestan | Rostov | Moscow |
|---|---|---|---|---|---|---|---|
| High influence for regional security | 11.5 | 75.0 | 25.0 | 30.0 | 35.0 | 15.0 | 15.0 |
| Extreme mobility of ethnic and social composition of population | 11.5 | 5.0 | 25.0 | 5.0 | 5.0 | 0.0 | 10.0 |
| The blurring of the Russian culture, distribution of alien ethnic norms and customs | 7.7 | 40.0 | 5.0 | 5.0 | 5.0 | 5.0 | 10.0 |
| Unwillingness of migrants to assimilate, accept norms of the Russian society | 19.2 | 20.0 | 30.0 | 20.0 | 20.0 | 15.0 | 35.0 |
| Consumerism of migrants in relation to the host region as a way station | 30.8 | 10.0 | 35.0 | 50.0 | 30.0 | 25.0 | 65.0 |

specific risks are validly correlated ($\chi^2$, $p \leq 0.005$) with the intensity of migration movements. The more intensive transit is, the higher risk for public security is (78.6% in the high transit regions and 16.7% in the low transit regions). Simultaneously, the less intensity of transit is, the higher risks of changes in the ethnic composition of the hosting region are (53.9% in the low transit regions and 38.5% in the high transit regions).

## 4 Discussion

The notion of transit migration has opportunities, consisting in recognizing the fact that final point of transit migrants can turn at any time into a new point of departure (Agunias 2010), and the places of destination during the displacement can become temporary halts while new vectors of traveling are forming. Hence, we can describe the full trajectory of transit migrants not only a posteriori, when we (and migrants themselves) understand whether this break in migration is temporary or not (Papadopoulou-Kourkoula 2008).

From the perspectives of sociological analysis, such an approach is acceptable and suitable for operationalization. The features of transit are disclosed through the intentions about further migration, migratory status, border crossing, accommodation, and work. This term can be applied to migrants and their practices, in which they are involved in receiving regions (transit migration) and regions where they are found at the time of research (regions/countries of origin and transit). This allows transforming transit from a political category to a scientific one (Düvell and Vogel 2006).

Local experts identify not every border region of Russia as transit. Despite the border character of territories for entering by migrants, which is typical for all regions except Moscow, their different socio-economic and geographic position, as well as other peculiarities, determine the capabilities of regional societies to *capture* and attract migrants, and become the final point of their current migratory route.

The structure of migration flows diverse according to the countries, ethnic groups, and confessions. It differs by the composition and intensity of displacements from definite countries. Of course, neighboring countries usually provide the most significant migration inflow, but it is also evident that inhabitants of several countries, which have no land borders with the Russian regions, prefer to relocate in some particular regions. Perhaps, such *preferences* arise from the ethnic and confessional parameters of regional societies (migrants are attracted by opportunities to find people with shared faith and vast diaspora), particularities of the structure of the regional economy (migrants from certain countries prefer economic domains, which are familiar for them, for example, they move from their agrarian countries to the agricultural regions of Russia and so on).

## 5 Conclusion

The study confirms our hypothesis about the multidirectionality of transit migration vectors in the border regions of Russia. Transit movements through the Russian borderland regions are primarily intra-country with the non-significant distribution of migration flow for the West European countries from the regions having borders with Europe. Thus, in the Moscow, Rostov, Pskov, and Murmansk regions, belonging to the zone of the European frontier, there is a considerable shifting of transit into the countries of Western Europe, whereas in the Orenburg region and Altai Krai (the zone of the Asian frontier), transit shifts into the central and Europeans parts of Russia. One notable exception is for the Republic of Dagestan. Bordering with the countries from the European part (Georgia and Azerbaijan), this ethnic republic reorientates almost all migrants to central Russia. This fact requires further in-depth and thorough analysis from the positions of peculiarities of residency and well-being of migrants in this republic.

**Acknowledgements** Paper prepared in the frameworks of the President's Grant for the State Support of the leading scientific schools of the Russian Federation No. НШ-2632.2020.6 "Return migration and migration policy: adoptive strategies of resettles and host population in the Russian borderlands", 2020–2021.

## References


Agunias RD (2010) Migration's middlemen, regulating recruitment agencies in the Philippines-United Arab Emirates corridor. Migration Policy Institute, Washington, D.C.

Collyer M (2010) Stranded migrants and the fragmented journey. J Refug Stud 23(3):273–293

Düvell F (2006) Irregular immigration in four European Union member states. In: Düvell F (ed) Illegal Immigration in Europe. Palgrave Macmillan, London, UK, pp 43–47

Düvell F (2014) Transit migrations in the European migration spaces politics, determinants and dynamics. In: Düvell F, Molodikova I, Collyer M (eds) Transit Migration in Europe. Amsterdam University Press, Amsterdam, Netherlands, pp 209–236

Düvell F, Vogel D (2006) Polish migrants: Tensions between sociological typologies and state categories. In: Triandafyllidou A (ed) Contemporary Polish Migration in Europe: Complex Patterns of Movement and Settlement. Edwin Mellen Press, Lewiston, NY, pp 267–289

Fargues P (2011) International migration and the demographic transition: A two-way interaction. Int Migr Rev 45(3):588–614

Hamood, S. (2006). *African transit migration through Libya to Europe: The human cost.* Cairo, Egypt: FMRS, AUC.

IOM. (1994). *Transit migration in Hungary.* Geneva, Switzerland: IOM.

IOM. (2015). BRIEF 13: A gender approach to environmental migration. *In IOM Outlook on Migration, Environment and Climate Change* (pp. 103–107). Geneva, Switzerland: IOM.

Papadopoulou A (2005) Exploring the asylum-migration nexus: A case study of transit migrants in Europe. Global Migration Perspectives 23:1–25

Papadopoulou-Kourkoula A (2008) Migration in Europe. Transit migration: The missing link between emigration and settlement. Palgrave Macmillan, London, UK, pp 15–45

Pliez O (2013) Migrants, constructive or disruptive factors for Saharan towns? Sécheresse 24(3):164–170

Radaev V (2013) Does competition eliminate social ties? The case of the Russian retail market. Polish Sociological Review 13(1):63–86

Roman H (2006) Transit migration in Egypt. CARIM research report. European University Institute, Florence, Italy

Skeldon R (1997) Migration and development: A global perspective, 1st edn. Routledge, London, UK

Temime, E. (1989). *Marseille Transit: Les passagers de Belsunce.* Paris, France: Autrement.

Trofimova TI (2010) Phenomenon of transit migration. Izvestiya IGEA 5(73):15–19

UN/ECE. (1993). *International Migration Bulletin, 3.*

Inter-Parliamentary Union. (2005). *Migration and development.* Geneva, Switzerland: IPU.

# Return Migration as a Factor of Sustainable Development in Border Regions of Russia

**Svetlana G. Maximova, Daria A. Omelchenko, and Oksana E. Noyanzina**

**Abstract** The specificity of return migration in Russia rests on historical backgrounds and since 2006 is in large part connected with the intentions of the Russian government to use return migration as a resource of development of regions having increase in population due to low fertility, aging or strong out-migration. The study presented in the article is focused on the analysis of outcomes of the State program of repatriation along with of attitudes of local population toward different political regimes, concerning return migration, studied during social surveys in seven border regions, located in different parts of Russia. It has revealed that general pragmatic orientation of population towards international migration is combined with different political strategies supporting compatriots. Their choice is determined by objective, socio-structural, and subjective, related to culture and education, factors, but their weight and significance differ depending on initial discrepancies in socio-economic, demographic and ethno-cultural characteristics of regional societies.

**Keywords** Return migration · Repatriation · Selective migration policy · Sustainable development · Russian regions · Local perceptions of migration policy strategies

## 1 Introduction

### *1.1 Return Migration as Global Process and Subject of Migration Studies*

Return migration is a multifaceted phenomenon involving different life situations and migration trajectories. Being a middle ground between continuous migration and non-migration, continuous migration and migration, supposing frequent departures and returns (seasonal, contract, and commuting migration), return migration holds a

---

S. G. Maximova · D. A. Omelchenko (✉) · O. E. Noyanzina
Altai State University, Barnaul, Russia
e-mail: daria.omelchenko@mail.ru



S. G. Maximova et al. (eds.), *Advances in Natural, Human-Made, and Coupled Human-Natural Systems Research*, Lecture Notes in Networks and Systems 234,
https://doi.org/10.1007/978-3-030-75483-9_28

special place in migration theory and practice. It is reflected in specific terminology, conceptual approaches, and methods of research.

In general terms, "return migration" describes a move to the country of origin, the place of constant residence, or a previous country of transit after living on the territory of another state (Dustmann and Weiss 2007). In this regard, return migration is placed within a broad context. Since circumstances of migration or departure are not considered, it allows incorporating various migration practices, such as (1) re-emigration, (2) post-educational migration, (3) return of refugees, and (4) deportation of non-documented migrants. The more narrow and operationalized definition underlines the special role of the state to defend national interests and resolve social, economic, demographic, humanitarian, and other issues by managing return migration. It is a voluntary form of international migration, concerning people and their descendants wanting to move back to their country of origin and giving grounds for a simplified procedure of citizenship or residence permit acquisition.

Depending on whether the state (sending or receiving) impacts return migration, it is customary to distinguish between its stimulated and non-stimulated forms. The first occurs when there are special state programs of repatriation, aimed at increasing (1) return migration (such programs exist in Russia, Israel, Greece, Germany, and Kazakhstan) with (2) territorial—when the state gives preferences to channel the migrants' flow in definite regions of the country (Russia, Greece, Israel), or (3) categorical, which is used to maintain the necessary quality of returnees' flow stimulation (Germany). Declarative (non-stimulated) return migration policies exist in many countries, where preferences are only for receiving citizenship (Finland, France, Belarus, Bulgaria, Ukraine, Denmark, Russia—before the adoption of the State Program for assisting voluntary resettlement in the Russian Federation of compatriots living abroad) (Ryazantsev and Grebenyuk 2014). Repatriation programs help emigrants and their descendants to return home or to their historical homeland. The right to move is granted to former citizens who moved abroad for various reasons-war actions, economic crisis, discrimination based on political views, race, gender or religion, etc.

Although the first few and unsystematic attempts to analyze return migration started in the 1960s (Bovenkerk 1974), the intensive elaboration of return migration issues in Western countries occurred in the 1980s, when developed states with significant labor volumes migrants faced their mass exodus to homelands. As a subprocess of international migration, return migration has become a subject of research within different theoretical approaches, often suggesting competing and coming from opposite assumptions and interpretations of the motives and consequences of return migration provisions. The corresponding scientific literature relies on five main paradigms through which the analysis of return migration was guided during recent decades. It has become habitual to claim that all these theories need a critical review in response to the growing diversity of migrants, from economic to refugees and seeking shelter, and changes in the perception of migration, which became an important factor of sustainable development in sending countries (Constant and Massey 2003; Cassarino 2004; Ivakhnuk 2015).

*The neoclassical theory of migration* draws on the analysis of differences in wages in sending and receiving countries and the expectations of migrants to earn more in the country of destination. The return migration is regarded as migration failure, which has not resulted in the expected profit in part because of the non-effective use of human capital. In contrast to this approach, *the new economics of labor migration* [NELM] considers return migration as a "logical outcome of a carefully calculated strategy," determined by the level of migrant's household and resulting from successful achievement of goals, such as higher earnings or financial savings. There is a clear shift from individual-level analysis to a comprehension of the collective interdependence of all participants of the migration process (Stark 1991). The return appears as a natural result of working abroad. A significant part of the migration strategy is related to remittances, allowing diversifying resources of the household and compensating risks related to the absence of sufficient insurance in the countries-recipients.

Within *the structural approach*, the migrant's return is considered not only in the context of individual experience or household but also in social and institutional factors in the countries of origin. Similarly to the new economic approach, it is claimed that remittances influence the decision about return and successful integration. The question about success or failure of migration is analyzed through correlations between expectations of the returnee, economic, and social realities in the country of origin. Situational and contextual factors are the prerequisite for comprehension of how the strength of local traditions and values adjust the expectations and intentions of the migrant after a return. If this adjustment fails, the returnee can undertake a new emigration (Cassarino 2004).

In the late 1980s, the desire to explain the dynamic and sustainable link between sending and receiving countries and bi-directional flows of crossing borders migrants led to the creation of *the transnational paradigm,* borrowing its terminology in international relations (Wessendorf 2007). Transnationalism serves as a conceptual framework for researching strong social and economic ties between donor and recipient countries. Their transnational character stems from regular social contacts sustainable in time (Portes et al. 1999). One important research direction is the analysis of the impact these ties can have on migrants' identities. Unlike structuralists or adepts of the NELM, transnationalism points out that return does not necessarily mean the end of the migration cycle and that migration history continues. Return migration is included in the system of circulating exchanges and relations, facilitating the reintegration of migrants, and promoting the transmission of knowledge and information.

Recent research based on the transnational perspective shows that structural integration into receiving society (namely: (1) behavior in the labor market, (2) education, and (3) maintenance of economic relations with the host country) has a lesser effect on intentions about a return. Investment and social networks in origin are positive associations, whereas socio-cultural integration into receiving country, in contrast, has a negative influence on the return initiative. Thus, there is no uniform return migration process, and the competing theories are rather complementary and mutually reinforcing (De Haas and Fokkema 2011; De Haas et al. 2015). Like the transnational paradigm, *the networking theory* concentrates on the maintenance of social connections between the "former" and the present place of residence, reflecting a migration experience. However, they are considered as supplementary and independent resources used for secure return and reintegration. Scholars working within this approach are mostly interested in network composition and emerging social structures, a configuration of ties, and supporting transborder contacts (Christou 2006; Haug 2008; Reynolds 2010). The actual state of return migration research is characterized by synthesizing and integrating different conceptual approaches, their complementary use for testing empirical hypotheses, and building formalized models based on quantitative-statistical or qualitative-ethnographic data.

## *1.2 Return Migration in Russia: From Spontaneous to Organized Flow*

Unlike Western countries, Russia faced the problem of return migration only in the 1990s, when, after the dissolution of the USSR, the country was flooded by migrants arriving from its former republics. This flow was mostly represented by ethnic Russians and so-called "non-titular" nationalities of the newly independent states. The main reason for migration in that period was to avoid ethnic conflicts, war actions, or harassment by authorities (Ivakhnuk 2011). Sociological studies of this period revealed that Russian regions receiving these migrants were in a crisis and demonstrated their inability to ensure social adaptation and living conditions for settlers, which provoked social tension and conflicts between locals and newcomers. Although many of them were highly educated and skilled professionals favorably disposed towards Russia, and, thus, represented a human capital that could contribute to the regional development, due to the pressure from the local population, unresolved domestic, labor, and economic issues, a large number of compatriots were forced to leave their unhospitable homeland and return to places, where they were chased by reprisal attacks (especially, Uzbekistan and Kyrgyzstan) (Nazarova 2000; Baranova 2005).

By the early 2000s, the intensity of migration caused by unfavorable political processes in the CIS and Baltic countries decreased, whereas temporary labor migration came to the fore. Russia became one of the global centers of migrants' attraction, while active emigration from Russia contributed to an increased number of Russian-speaking communities in foreign countries of the "far abroad." Currently, the main volume of return migration is represented by settlers from the CIS countries. Return migration from other countries is negligible, and the realization of migration potential to a greater extent is determined by factors related to the situation in Russia rather than in countries of current residence. It indicates the need for a more friendly and open policy towards compatriots (Khramova et al. 2017; Maximova et al. 2018). The need to improve migration policy requires consideration of different categories of returnees-labor, political, educational, or displaced people, which have or have not

Russian origins or citizenship but are linked to Russia by life trajectories, perceptions, or experience of living in Russia. The analysis of their adaptive strategies and migration experience along with attitudes of the local population towards return migration will allow making conclusions about determinants, consequences, and opportunities of return migration, not only for the migrants themselves but also for the receiving regions, their sustainable socio-economic and demographic development.

## 2 Materials and Methods

Return migration became the subject of a complex investigation, which included the analysis of statistical data on the results of the state program on rendering assistance to voluntary resettlement to the Russian Federation of compatriots living abroad. The sociological surveys in seven border regions of Russia aimed to study attitudes of the local population towards different migration policy strategies, including return migration.

The state program results are taken from the Federal State Statistics Service (Rosstat) and the Ministry of Internal Affairs website. Sociological surveys were conducted in the Altai Krai, Orenburg Region, Murmansk Region, Pskov Region, Altai Republic, and Republic of Dagestan (quota sampling with a random walk, $n = 500$ in each region, the overall sample $n = 3500$).

Regions with similar socio-economic, demographic, and migration situations were divided into three clusters, and further analysis was fulfilled separately in each cluster. Two main policy strategies concerning return migration - to support returning Russian or Russian speaking people from other countries and support compatriots of different nationalities returning from the CIS countries were compared with other liberal or restrictive migration strategies. The authors revealed factors determining the support of an ethnically and linguistically based strategy or the broader definition of "compatriots" and an internationalist approach. It allowed linking attitudes toward migration with regional developmental characteristics. Statistical methods used in the analysis included frequency analysis, cluster analysis, and binomial logistic regression.

## 3 Results

### *3.1 The Analysis of Outcomes for Russian Regions of the State Program on Rendering Assistance to Voluntary Resettlement of Compatriots Living Abroad*

Return migration in Russia concerns not only ethnic Russians but also those who were born or lived long on the territory of the Russian state and their descendants.

The term "compatriot" first appeared in the Russian law in 1999 and was based on territorial and ethnic grounds that gave rise to considerable criticism related to moral and legal rights of representatives of excluded nationalities, also preserving mental and spiritual ties with Russia, Russian language, and culture (Ryazantsev and Grebenyuk 2014). Difficulties of interpretation and broad public discussion led to changes in legislation concerning the policy of Russia towards its citizens living abroad. In 2010, legislators enacted an extended definition, based on subjective self-identification and acknowledgment of belonging to Russia, justified by objective evidence-professional or public activity, loyalty, and wish to preserve and develop Russian culture and language, to maintain personal and public connections with Russians living abroad. Since 2007, the return migration became more regulated by adopting the state program for assisting compatriots residing abroad in their voluntary resettlement in the Russian Federation.

This multi-purpose program focuses on integrating the economic and social potential of compatriots and the developmental needs of Russian regions with socio-economic perspectives but are threatened by high rates of out-migration. Thus, the program aims to compensate the natural population decrease in Russian regions by attracting people from other countries, raised in the Russian cultural traditions, proficient in the Russian language, and willing to return to Russia. It is claimed that compatriots can best adapt and engage in positive social links with the receiving society. Thus, the program is embedded in the system of targeted interventions aimed at stimulating birth rate and reducing mortality, regulating migration, and, therefore, stabilizing the population size. The return of compatriots itself is not its priority goal.

To attract resettlers, the program provides for significant (at least for a majority of Russians) state guarantees and measures of social support, such as (1) duty-free importation of the household effects, (2) reimbursement for transportation costs, (3) start-up allowances and assistance in getting a job, and (4) educational or medical services. Initially, the program was designed to cover the 2009–2012 period and receive about 300 thousand compatriots, especially ethnic Russians from the near abroad. Unfortunately, the expectations of officials were not justified. Despite an increase in the number of regions-participants, the number of applicants was not significant and consisted of only about 3000 migrants for the first two years. The main issues were related to the reluctance of regions to receive compatriots: there was a lack of affordable housing, vacancies did not correspond to expectations of resettlers—they were relegated to undesirable low-paying jobs; territories for settling were not attractive, compatriots were invited predominately in agricultural regions with undeveloped infrastructure. Among other things, a significant decrease was caused by the influence of the economic crisis and, consequently, an almost fourfold reduction of funding—from 8 billion to 1.8 billion (Ryazantsev et al. 2015).

Meanwhile, the list of regions was extended, and in 2012 it already included 40 regions. Initially, there were three categories of territories-parts of federal subjects with different volumes of state guarantees and social support, where compatriots were attracted according to regional programs. In 2012, the new edition of the state program was adopted, and the program became open-ended. The abovementioned categories of territories were replaced by the "territories of priority resettlement" in the Siberian and Far Eastern federal districts with higher state guarantees and benefits and other territories without a special accent on special parts, accessible for resettlement. New conditions resulted in a dramatic increase in the number of regions-participants.

As of January 1, 2020, the state program was implemented in 76 subjects of the Russian Federation in eight federal districts. In 2019 (the last year with accessible statistical data), 108.5 thousand compatriots and members of their families resettled in Russia. Among them, 59.5 thousand (54.9%) formalized their participation in the program via authorized institutions abroad, and 49.0 thousand (45.1%) applied to authorities in Russia. Overall, over the past 10 years (2010–2019), the number of participants of the program was about 892 thousand people, which is basically the population of a large Russian city; a population with a priori positive attitude towards Russia, with historical and cultural collective memory, identification, and solidarity with their future fellow citizens, motivated to restore the old and create new ties with inhabitants (Toshenko 1997; Vyshnevsky 2000). The largest number of participants was recorded in 2015 –179.6 thousand people (Table 1).

By the number of involved compatriots, the uncontested leader is the Central Federal District, covering over 40% of all program participants, especially in Lipetsk, Kaluga, and Voronezh Regions (more than 40 thousand resettlers in each region). The Siberian Federal District occupies second place, with centers of return migration in the Novosibirsk and Omsk Regions (more than 25 thousand) and the Krasnoyarsk Krai (more than 10 thousand). The Volga Federal District takes the third and fourth places in different years. The most significant return flows are observed in the Kaliningrad Region and the Northwestern Federal District, where the leaders are the Saratov and Nizhny Novgorod Regions. The North Caucasian District has contributed the least, due to the recent participation—from 2014 and until recent times only with one region—the Stavropol Krai, where 7.7 thousand compatriots resettled since 2010. In the Southern Federal District, the situation is similar: there

**Table 1** Number of participants of the Program resettled in Russia in 2010–2019, in thousands of people

| Territory | 2010 | 2012 | 2015 | 2017 | 2018 | 2019 | Total |
|---|---|---|---|---|---|---|---|
| Russian Federation | 11.8 | 56.6 | 179.7 | 118.6 | 107.7 | 108.5 | 894.7 |
| Central federal district | 5.7 | 27.4 | 76.6 | 48.9 | 43.0 | 40.5 | 380.6 |
| Northwestern federal district | 2.8 | 7.8 | 14.5 | 11.6 | 8.7 | 8.6 | 85.7 |
| Southern federal district | 0 | 0.4 | 6.9 | 4.8 | 4,1 | 3.3 | 30.0 |
| North Caucasian federal district | 0 | 0 | 2.6 | 1.3 | 1.0 | 0.9 | 7.7 |
| Volga federal district | 0.4 | 3.7 | 21.2 | 13.7 | 12.1 | 14.0 | 96.1 |
| Ural federal district | 0.2 | 1.1 | 20.3 | 14.1 | 13.5 | 14.6 | 88.9 |
| Siberian federal district | 1.9 | 11.5 | 27.0 | 18.8 | 19.8 | 20.6 | 152.6 |
| Far Eastern federal district | 0.8 | 4.7 | 10.6 | 5.4 | 5.5 | 6.1 | 53.1 |

*Sources* (Federal State Statistics Service (Rosstat), Ministry of Internal Affairs of the Russian Federation)

are four regions, including the Astrakhan Region (since 2015) and the Krasnodar Krai (since 2016). The leader is Volgograd Region (about 16 thousand people). Considering that these regions are not among outsiders and have no need for additional population, they are free to define additional conditions for participation in the program. For instance, in Krasnodar Krai, it is obligatory to have medical or pedagogical education and give consent to live and work in a rural area.

Additionally, it is worth examining the functioning of the program in regions, denominated as the "territories of priority resettlement," situated in the Siberian and Far Eastern Federal Districts. According to statistics, the Irkutsk Region achieved the best results in attracting resettlers—about 20 thousand people moved to this region from 2007 to 2019. The Primorsky Krai is in second place with 16.3 thousand, and the Khabarovsk Krai is in third place (15.1 thousand). The worst positions in this category are taken by the poorest regions with low standards of living—the Zabaykalsky Krai (3 thousand people), the Jewish Autonomous Region (1.7 thousand), and the Republic of Buryatia (1.4 thousand), into which compatriots have no desire to move, despite declared benefits and preferences.

Thus, despite the generally positive effects of the program and its stable project level, it turned out that outcomes vary greatly in different regions. Not all of them can ensure attractive conditions and fulfill their obligations, which casts doubt on the efficiency to achieve the main tasks of the program, including those related to regional sustainable development. Especially it concerns regions with the worst demographic and economic situations, and intense out-migration, difficulties with attracting and retaining compatriots on their territories.

### 3.2 *Attitudes of Population Living in Border Regions Towards Return Migration: Support of Ethnically—Driven and Internationalist Policy Strategies*

Return migration in Russian regions is not only embedded in the general process of international migration. Perceptions of the population about migration and migration regimes towards former citizens and diasporas are interconnected with social moods, evaluations of social security, quality of life, and social relations in the regional society.

Investigating these processes in seven Russian border regions required a preliminary examination of socio-economic conditions and the experience of participation of the region in the state initiatives encouraging loyal expatriates to return to Russia, determining in an evident way the social attitudes of their inhabitants.

The analysis of socio-economic and demographic development (Table 2) allowed to detect three groups of regions, confirmed by statistical clusterization (two-step cluster analysis).

The first cluster was represented by the Murmansk, Rostov, and Orenburg Regions—relatively prosperous and well—developed territories, with high GRP per

**Table 2** Indicators of socio-economic and demographic development in seven regions covered by the research, average values[a]

| Indicator | 1 cluster | | | 2 cluster | | 3 cluster | |
|---|---|---|---|---|---|---|---|
| | MO | RO | OO | PO | AK | RA | RD |
| Number of population (2014–2019), millions of people | 0.8 | 4.2 | 2.0 | 0.6 | 2.4 | 0.2 | 3.0 |
| Rate of natural increase (2014–2018) | −0.4 | −2.6 | −0.7 | −7.5 | −2.4 | 7.4 | 12.2 |
| Total fertility rate (2014–2018) | 1.6 | 1.5 | 1.9 | 1.7 | 1.7 | 2.6 | 2.0 |
| International migration growth (average, 2014–2018) in thousands of people | 1.1 | 5.8 | 2.5 | 0.7 | 2.5 | 0.2 | 0.6 |
| Inter-regional migration growth (2014–2018) in thousands of people | −5.4 | −3.6 | −10.3 | −1.3 | −8.4 | −0.3 | −13.0 |
| Percentage of population or earning less than the minimum living standard (poverty threshold) (2014–2018) | 12.26 | 13.64 | 13.8 | 17.86 | 17.54 | 24.18 | 12.5 |
| GRP per capita 2014–2017, thousand rubles | 527.9 | 285.0 | 387.9 | 215.2 | 204.8 | 197.7 | 190.9 |
| Proportion of economically active people (2014–2018) | 50.48 | 45.74 | 48.76 | 46.38 | 43.76 | 39.94 | 34.94 |
| Unemployment rate (average, 2014–2018) | 7.2 | 5.7 | 4.6 | 6.5 | 7.36 | 11.06 | 11.1 |
| Demographic burden (general) (2014–2019) | 658 | 756 | 785 | 827 | 817 | 837 | 645 |
| Old-age-dependency ratio (2014–2019) | 356 | 466 | 434 | 532 | 478 | 324 | 214 |

[a] Notation of regions: the Murmansk Region—MO, the Pskov Region—PO, the Rostov Region—RO, the Republic of Dagestan—RD, the Orenburg Region—OO, the Republic of Altai—RA, the Altai Krai—AK

*Source* Compiled by the authors based on (Federal State Statistics Service (Rosstat))

capita (285–527 thousand rubles), low unemployment rate, and poverty threshold. These regions possessed a relatively favorable demographic situation with low coefficients of national decrease. Values of demographic pressure were defined by the number of non-working young adults and not by the old-age dependency ratio.

The second cluster included the Pskov Region and the Altai Krai, regions with certain economic issues: dependency on federal budget subventions, low average income level, a high proportion of the population with earnings below the minimum living standard (about 18%), a high demographic burden on the population of working age caused by intensive aging, and considerable population decline (especially in the Pskov Region, where the average rate of natural increase amounted to 7.5).

The third cluster—the national republics of Dagestan and Altai-poor regions with the lowest GRP, high poverty thresholds (in the Republic of Altai—24.2%), and high unemployment level (about 11%). Simultaneously, traditionalist economy and way of life were associated with high fertility and natural population growth, and, consequently, low old-age dependency ratio.

Belonging to a cluster was blended with the time of inclusion in the federal program on repatriation and, at least partially, with program outcomes. Thus, the regions of the first cluster had more experience in elaborating regional programs of assistance (the first programs appeared in 2014), and within the past three years (2017–2019) received more compatriots, especially in comparison with the third cluster: Murmansk Region hosted 1,672 compatriots (2.1 thousand per million of people), Rostov Region—5,290 (1.3 thousand) and the Orenburg Region—4,015 (2 thousand). The Altai Krai and the Pskov Region participated in the federal program since 2016 without any limitations in the choice of place for resettlement, and the number of returned compatriots amounted to 4,760 (2 thousand per million) and 3,450 persons (5.5 thousand per million), respectively. Regions of the third cluster were among the last regions where the federal program was implemented. The Republic of Dagestan adopted its regional program in 2019, and there are already some modest results (141 persons reported in 2019). In the Altai Republic, the program will be implemented in 2020–2021 as a subprogram within the state program on social protection and employment promotion. It aims to attract 30 compatriots by the end of 2021. Thus, in more advanced regions with more competitive economies, return migration was under long governmental control. In the lagging regions, the lack of experience (and real return migration) is combined with their socio-economic issues and development specifics.

The analysis of the results of sociological surveys showed that the local population in all regions, except the Altai Republic, approved a pragmatic migration policy focused on encouraging young and skilled people to immigrate into Russia (especially in the Republic of Dagestan and in the Murmansk Region where this strategy was selected as preferable by more than 40% of respondents). Restrictive policy and a ban on entry were not popular (5–9% of choices), except for two regions from the first cluster—where a significant part of the population was against migrants, regardless of their origin-Rostov (11%) and Orenburg (26.3%) Regions. Liberal strategy, aimed at supporting everybody who wants to live in Russia, was more attractive, and here there was a clear association with socio-economic and demographic indicators. The national republics were among the most hospitable and welcoming to migrants (28–32% of acceptance). In the second cluster regions, the proportion of people supporting the liberal strategy was 22–24%, while in the first cluster—only about 14–16% (Table 2).

The nationalist strategy of return migration, entailing support of ethnic Russians and restricting the entry to other nationalities, received much skepticism. It was evaluated higher in three regions with predominating Russian population—the Altai Krai, Pskov, and Rostov Regions (27–28% of choices). Other regions with the diversified ethnic structure were less supportive of this migration regime: according to the Russian Census of 2010, in the Orenburg oblast where the number of ethnic Russians was less than 75%, 21.2% chose it as preferable, in the national republic with dominant non-Russian population—13–17% only.

The idea to give more support to compatriots from the CIS countries was rather disapproved in the Orenburg and Rostov Regions, where international migration was more intensive (only 11.6% and 15.4%, respectively). In the Pskov Region, it was less preferable than ethnically and linguistically driven strategies (22.6%). In contrast, in the remaining regions, attitudes were similar, and this variant was selected as preferable by 24–29% of participants.

As it was acceptable to choose two preferable strategies, the more differentiated analysis could supplement a simple univariate consideration. Most respondents (65%) chose a single variant, while there were ten possible combinations for the rest. The restrictive policy was the most straightforward. It was chosen as a unique variant in 84%, whereas the liberal strategy was more "friendly" and accompanied by other answers, including repatriation strategies (Table 3).

A comparative analysis allowed to detect four essential patterns of migration policy strategies supported in the regions.

The first pattern was relevant for the Orenburg, Rostov, and Pskov Regions. Participants preferred to support Russian, Russian-speaking migrants, and migrants from the CIS countries (43.4% in the Orenburg Region and 23–25% in two others). In the second place-migrants from the CIS and young, skilled specialists (about 23% in all regions).

In the Murmansk Region, the most preferred combination was to support compatriots from the CIS countries and young migrants (33.6%), while the support of Russians and citizens from ex-USSR was in the second place (22.1%).

**Table 3** Attitudes of population towards different strategies of migration policy, including return migration, %

| Region / Strategy of support | All who wants to stay in Russia | Only Russians and Russian-speaking people | Only compatriots from the CIS countries | Only young and skilled migrants | Restrictive (nobody should be supported) |
|---|---|---|---|---|---|
| The Altai Krai | 21.5 | 28.3 | 28.3 | 28.8 | 6.3 |
| The Pskov Region | 24.2 | 28.5 | 22.6 | 35.9 | 4.9 |
| The Orenburg Region | 15.2 | 21.2 | 11.6 | 36.3 | 26.3 |
| The Murmansk Region | 13.7 | 24.9 | 29.4 | 42.7 | 8.9 |
| The Republic of Altai | 31.6 | 13.2 | 27.3 | 22.7 | 8 |
| The Rostov Region | 16.3 | 27.3 | 15.4 | 33.1 | 11.3 |
| The Republic of Dagestan | 28.0 | 16.6 | 24.0 | 42.2 | 8.5 |

In the Altai Krai, respondents more often combined the support of ethnic Russians and young professionals (25.0%), which corresponded with demographic tendencies related to aging and relative ethnic homogeneity of this region. The second preferred combination was to support Russians and compatriots from the CIS countries (22%).

The fourth pattern was specific for national republics, where the most supported regime was to receive all categories of migrants without any differentiation or look favorably at compatriots from the CIS countries (31–32%). The second place was occupied by the support of migrants from the CIS countries and young professionals (23%). In comparison, a positive attitude towards representatives of the Russian diaspora or Russian-speaking compatriots was demonstrated by 20% of respondents in the Altai Republic and only by 15.7%—in the Republic of Dagestan.

Summing up two return migration strategies, the authors found that they were supported by a total of 43% of respondents. The further analysis was aimed to find significant factors determining population choice, other than already was included in the cluster division. For this purpose, we fulfilled a set of logistic regression analyses in each cluster with political strategy as a dependent variable and possible determinants defined during preliminary bivariate analysis as independent predictors. The list included objective socio-structural factors, such as (1) place of residence, (2) education, (3) gender, (4) belonging to diasporic groups or financial situation in the household, and (5) subjective factors (personal attitudes toward migrants living in the region, evaluation of migration intensity in the place of residence, perception of security and comparison of the socio-economic position of the region with other regions of Russia).

Results of regression modeling in support of Russians and Russian-speaking migrants are presented in Table 4. It shows that there was a single factor significant for regions from all three clusters (in the third cluster, it was marginally significant) associated with the choice of this strategy— the personal attitude towards migrants living in the region. The sign of coefficient indicated that this association was negative ($b_{cluster1} = -0.24^{*}$, $b_{cluster2} = -0.55^{**}$, $b_{cluster3} = -0.23$). It means that friendly relations between migrants and the local population were a deterrent for promoting ethnically-driven return migration. The same effect had the factor of respondent belonging to a diasporic group, also corresponding to non-support of this regime, especially in the two fist clusters ($b_{cluster1} = -0.90^{**}$, $b_{cluster2} = -0.26^{*}$).

Simultaneously, age, education, and financial situation were factors reinforcing the support of Russians to return to their historical homeland, at least in one cluster. Particularly, in favorable regions of the first cluster, there was a tendency to approve such political decisions among wealthier citizens. In the third cluster (national republics), return migration of Russians was welcomed by people with higher education (59.5% of them had a non-Russian ethnic identity; thus, it was not a sign of ethnic favoritism), whereas in the two first clusters with dominant Russian population nationalist decisions were more popular among older generations than young adults. Besides, in the third cluster, there was a tendency to approve return migration of Russians among people with higher evaluations of personal security (reverse scale, hence a negative sign of coefficient).

**Table 4** Logistic regression for support of Russian and Russian-speaking migrants

| Predictor | Cluster 1 | | Cluster 2 | | Cluster 3 | |
|---|---|---|---|---|---|---|
| | B | Exp(B) | B | Exp(B) | B | Exp(B) |
| Place of residence (urban vs rural) | 0.06 | 1.06 | 0.07 | 1.07 | −0.47 | 0.95 |
| Comparison of own region with other regions of Russia | -0.01 | 0.99 | -0.08 | 0.92 | 0.14 | 1.14 |
| Evaluation of migration in the place of residence (predominance of in- or out- migration) | -0.02 | 0.98 | 0.10 | 1.11 | -0.04 | 0,96 |
| Personal attitude towards migrants in the region (5—point scale) | **-0.24*** | **0.79** | **−0.55**** | **0.58** | ***−0.23*** | ***0.79*** |
| Perception of security (5—point scale) | −0.17 | 0.84 | 0.03 | 1.04 | ***−0.27*** | ***0.76*** |
| Household finance | *0.24* | *1.28* | −0.16 | 0.85 | −0.17 | 0.85 |
| Gender | −0.04 | 0.96 | −0.18 | 0.84 | 0.22 | 0.81 |
| Education | −0.11 | 0.89 | −0.15 | 0.86 | **0.56*** | **1.75** |
| Belonging to diaspora group | **−0.90**** | **0.41** | **−0.85**** | **0.43** | −0.11 | 0.90 |
| Age | **0.28*** | **1.32** | **0.26*** | **1.29** | 0.05 | 1.05 |
| Model $\chi^2$ (10) | 41.3 | | 71.6 | | 16.8 | |
| Nagelkerke's $R^2$ | 0.06 | | 0.11 | | 0.04 | |

** – significant at 0.01 level, * – significant at 0.05 level, values in italics are marginally significant

In the second set, concerning the positive endorsement of stimulation of compatriots of different nationalities from the CIS countries, the number and content of factors varied greatly from those detected in previous models.

The most significant results were found in regions of the first cluster. An internationalist approach based on rather civil grounds and common history of returnees than ethnic differentiation was more popular among people living in the rural area ($b_{cluster1} = -0.96$, in comparison with urban inhabitants, for whom the perspective to encourage migration from former soviet republics was less attractive.

At the same time, the support of this strategy was associated with negative results of comparison of own region with other regions of Russia ($b_{cluster1} = 0.26$*), high levels of precepted security ($b_{cluster1} = 0.29$*), higher education ($b_{cluster1} = 0.44$*), and especially—with belonging to diasporic groups most of which were representatives of the so-called "titular" nationalities in the newly independent states.

There were only some statistical tendencies in the other two clusters that should not be neglected, as they were forming some general pattern of perception of return migration in several regions.

In particular, there was a similar effect of evaluation of the socio-economic position of the region with other regions of Russia on the support of internationalist migration policy ($b_{cluster2} = 0.18$, $p < 0,1$), meaning that in the conscience of the population, there was a clear link between economic and demographic issues and their resolution was associated with among other decisions with stimulated return migration.

Instead, the perception of security in the second cluster had the opposite effect on this strategy, as more support was given by those who felt themselves in a less secure position ($b_{cluster2} = 0.19$, $p < 0.1$). In contrast, the first cluster witnessed a differentiated impact of this in different socio-economic and demographic conditions. The factor of education was also reproduced in the third cluster ($b_{cluster3} = 0.38$, $p < 0.1$). Besides, in these regions, two additional tendencies were detected, showing a positive association between favorable attitudes towards compatriots from the CIS countries and migrants living in the region and negative effect of age, meaning that older generations were reluctant to admit more diverse ideas and different ethno-cultural future of the region (Table 5).

**Table 5** Logistic regression for support of compatriots of different nationalities from the CIS countries

| Predictor | Cluster 1 | | Cluster 2 | | Cluster 3 | |
|---|---|---|---|---|---|---|
| | B | Exp(B) | B | Exp(B) | B | Exp(B) |
| Place of residence (urban vs rural) | **−0.96**** | **0.38** | 0.20 | 1.22 | 0.23 | 1.26 |
| Comparison of own region with other regions of Russia | **0.26*** | **1.29** | **0.18** | **1.20** | 0.00 | 1.00 |
| Evaluation of migration in the place of residence (predominance of in- or out-migration) | 0.02 | 1.02 | 0.06 | 1.06 | 0.11 | 1.12 |
| Personal attitude towards migrants in the region (5-point scale) | −0.04 | 0.96 | 0.06 | 1.07 | **0.21** | **1.24** |
| Perception of security (5-point scale) | **−0.29*** | **0.74** | **0.19** | **1.20** | 0.17 | 1.18 |
| Household finance | −0.01 | 0.99 | 0.05 | 1.05 | −0.15 | 0.86 |
| Gender | 0.26 | 1.30 | −0.06 | 0.95 | −0.07 | 0.94 |
| Education | **0.44*** | **1.55** | −0.01 | 0.99 | **0.38** | **1.46** |
| Belonging to diasporas | **0.61**** | **1.84** | 0.20 | 1.23 | 0.40 | 1.49 |
| Age | 0.19 | 1.20 | 0.16 | 1.18 | **−0.23** | **0.79** |
| Model $\chi^2$ (10) | 53.3 | | 13.8 | | 16.4 | |
| Nagelkerke's $R^2$ | 0.08 | | 0.03 | | 0.04 | |

# 4 Discussion

The current analysis results show that there are at least several basic contradictions impeding the more effective realization of migration policy concerning repatriation.

The first consists of the doubtful possibility of increasing population in economically depressed regions by involving compatriots. It runs counter to basic economic theory, postulating that migration occurs when the sum of possible benefits exceeds the sum of risks and costs of resettlement and refusal of migration. In other words, migration should lead to a better, secure, and more comfortable life in economically attractive conditions. In the current case, the choice of the "territory of the priority resettlement" can happen only if the compatriot has worse conditions of life, which narrows the circle of possible participants willing to move into Russia. Besides, regions themselves are still far from dealing adequately with the task of attraction of compatriots, which results in their demands to exclude them from the program or delay in execution. Even if they ensure the initial installation of compatriots, it is almost impossible to stop their further move to more favorable regions.

In successful territories, large migration flows and competition on the labor market allow dictating terms of participation to regulate them more efficiently that is justified. Meanwhile, for compatriots wishing to return to Russia, it is an additional obstacle and a push factor. Consequently, there is a high demand for resettlement in attractive regions, which are excluded from the program or difficult to get into, and low demand for unattractive regions that could not be raised even with additional stimulation. Considering other barriers related to specific requirements needed to prove eligibility for receiving privileges in the program, it is clear that program outcomes are far from ideal and their impact on regional development is relatively invisible.

Attitudes of the local population towards returnees is another crucial factor determining the efficiency of migration policy. Our results show that they differ considerably depending on the socio-economic, demographic, and ethnocultural characteristics of regional society. The more prosperous and attractive a region is for migrants, the more people are against free and unlimited international migration, including return migration of Russian-speaking compatriots of different nationalities. Although the level of manifest migrantophobia and support of repressive policy is rather low, some hotbeds of tension are located in the first cluster regions, especially among the Russian population with a stable financial position, more often supporting ethnically driven strategies. In contrast, the national republics covered in the conducted surveys demonstrate approval of liberal internationalist migration policy regimes. Russian compatriots in these regions are more welcomed by more educated people living in secure conditions.

## 5 Conclusion

It is clear that if Russia wants to increase the flow of return migration, the attraction of compatriots, especially from economically developed countries, should not be directed to close demographic policy gaps. It should be directed towards long-term perspectives of reunification and strengthening positions of the Russian nation, which will require conditions, meeting the expectations of both returnees and the local population. Otherwise, return migration can turn into social tension and re-emigration and create anti-effects of the program.

**Acknowledgements** The reported study was supported by the Russian presidential grant for leading scientific schools "Return migration and migration policy: adaptive strategies of returnees and receiving population in border regions of Russia" No. НШ-2632.2020.6.

## References

Baranova SV (2005) Position of forced migrants in the Sverdlovsk oblast. Sociol Stud 2:139–141
Bovenkerk F (1974) The sociology of return migration: a bibliographic essay. Springer Science & Business Media, Springer Netherlands
Cassarino JP (2004) Theorising return migration: the conceptual approach to return migrants revisited. Int J Multicult Soc (IJMS) 6(2):253–279
Christou A (2006) Crossing boundaries—ethnicizing employment—gendering labor: Gender, ethnicity and social networks in return migration. Soc Cult Geogr 7(1):87–102
Constant A, Massey DS (2003) Self-selection, earnings, and out-migration: A longitudinal study of immigrants to Germany. J Popul Econ 16(4):631–653
De Haas H, Fokkema T (2011) The effects of integration and transnational ties on international return migration intentions. Demogr Res 25:755–782
De Haas H, Fokkema T, Fihri MF (2015) Return migration as failure or success? J Int Migr Integr 16(2):415–429
Dustmann C, Weiss Y (2007) Return migration: Theory and empirical evidence from the UK. Br J Ind Relat 45(2):236–256
Haug S (2008) Migration networks and migration decision-making. J Ethn Migr Stud 34(4):585–605
Ivakhnuk IV (2011) Perspectives of migration policy in Russia: The choice of the right way. Maks Press, Moscow, Russia
Ivakhnuk I (2015) International labour migration management: Global experience and Russian practices. Int J Manag Theory Pract 10:36–44 (In Russ)
Khramova MN, Ryazantsev SV, Pismennaya EE (2017) Return migration of compatriots from the CIS countries to Russia: Probabilities, possibilities, risks. HSE Publishing, Moscow, Russia
Maximova S, Noyanzina O, Omelchenko D, Maximova M (2018) The Russian-speakers in the CIS countries: Migration activity and preservation of the Russian language. MATEC Web Conf 212:10005
Nazarova EA (2000) Peculiarities of migration. Sociol Stud 7:106–111
Portes A, Guarnizo LE, Landolt P (1999) The study of transnationalism: pitfalls and promise of an emergent research field. Ethn Racial Stud 22(2):217–237
Reynolds T (2010) Transnational family relationships, social networks and return migration among British-Caribbean young people. Ethn Racial Stud 33(5):797–815
Ryazantsev SV, Pismennaya EE, Khramova MN (2015) Return migration of compatriots to Russia: Is there any migration potential? Population 2:64–73

Ryazantsev SV, Grebenyuk AA (2014) Our people' abroad. The ethnic Russians, the Russians, the Russian speaking, compatriots: Settlement, integration, and return migration to Russia. ISPI RAN, Moscow, Russia

Stark O (1991) The migration of labor. Basil Blackwell, Cambridge, UK

Toshenko Zh (1997) Post-Soviet space: sovereignty and integration. Ethno-sociological essays. RSHU, Moscow, Russia

Vyshnevsky AG (2000) The dissolution of the USSR: Ethnic migrations and diaspora issue. Soc Sci Contemp World 3:115–130

Wessendorf S (2007) 'Roots migrants': Transnationalism and 'return' among second-generation Italians in Switzerland. J Ethn Migr Stud 33(7):1083–1102

# Demographic Policy of the People's Republic of China: Experience of Legislative Regulation

**Oksana G. Moiseeva, Vitaliy V. Rusanov, and Yuri A. Zelenin**

**Abstract** The paper is dedicated to the legislative regulation of demographic policy in the People's Republic of China [PRC], which started in 1949. Today, the PRC continues to be the most populous country in the world. For over the 70 years of the PRC's existence, Chinese demographic legislation changed a lot. The authors analyze the conditions of directives, decisions of the party and state bodies of the PRC, laws, and constitutions of 1954, 1975, 1978, and 1982, reflecting changes in demographic policy. After a sharp increase in the population in the 1960s and 70s, China proclaimed the "one family–one child" campaign to limit the birth rate. In current conditions, this policy ceased to correspond to the main tasks of the successful development of Chinese society, which led in the beginning to its mitigation and then to its abolition in 2016. The paper analyzes the positive and negative aspects of the Chinese demographic policy, the reasons for its changes, and the prospects for further improvement. Chinese authorities face the difficult task of finding optimal approaches to the legislative regulation of birthrate, considering the experience of other countries. On the one hand, it will help to prevent overpopulation in China, and on the other-it can increase the proportion of the able-bodied population.

**Keywords** People · Republic of China · Population policy · Legislative regulation

## 1 Introduction

China is the most populous country in the world. Since ancient times, the number of Chinese was huge. In 2 AD, the first general population census was carried out. It showed that 60 million people lived in China, which is comparable to the number of Roman Empire inhabitants. During the reign of the Qing Dynasty, a demographic explosion occurred in China. This was facilitated by a change in tax policy, the widespread development of highly productive crops, political stability, and much more. If by the beginning of the seventeenth century, the population was unlikely to

O. G. Moiseeva (✉) · V. V. Rusanov · Y. A. Zelenin
Altai State University, Barnaul, Russia
e-mail: oksanam0iseeva@yandex.ru



S. G. Maximova et al. (eds.), *Advances in Natural, Human-Made, and Coupled Human-Natural Systems Research*, Lecture Notes in Networks and Systems 234,
https://doi.org/10.1007/978-3-030-75483-9_29

exceed 150 million, then by the end of the eighteenth century, it amounted to somewhere around 300 million, and by the middle of the next century, it reached already 400 million (Delnov 2008, p. 500). In 1949, when the formation of the People's Republic of China [PRC] was proclaimed, the number of inhabitants was about 540 million. In 2015, the number of inhabitants was more than 1,367 billion. Over 65 years, the population of China increased by more than 800 million. According to government predictions, by 2020, the population of China will be approximately 1 billion 420 million people ("Russia and China," n.d.). China is characterized by a disproportion between the population and economic potential of the country. Satisfying the vital needs of the large Chinese population leads to the depletion of natural resources, negatively affects the environmental situation in the country, and aggravates social problems, which requires further improvement of the legislative regulation of demographic problems.

The Eurasian countries are characterized by various options for the implementation of demographic policies. There are significant differences between European and Asian states. European countries are characterized by a demographic crisis associated with a low birth rate and population aging. The Asian region countries are characterized by high fertility, the threat of overpopulation, and lack of resources, especially food.

Among the global issues of our time, one of the most difficult is the demographic problem. China, which has experienced population growth at a tremendous pace, and is experiencing serious difficulties with food and natural resources to provide a large number of residents with everything necessary, has already begun to limit birth rates since the middle of the twentieth century. Over the 50 years of implementing the demographic program in the PRC, it accumulated a vast experience that allows us to trace various options for resolving this problem. Gradually, as a result of the active process of legislative regulation of childbearing issues, China managed to create its own effective system of regulating population growth and planned childbearing that meets the needs of the current development of Chinese society. The authors focus on the legislative regulation of demographic processes, analyze various approaches of the Chinese leadership to solve the problems of reducing childbearing. They analyze the current state of the demographic situation in China and the prospects for its further development as well.

Eurasian cooperation is developing very multifaceted. Chinese experience in solving the demographic problem shows the most effective ways to reduce the birth rate, which is an urgent task not only for China but also for other countries of Eurasia.

## 2 Materials and Methods

General information regarding the research issue is presented in works by Veremeychik (2013); Meliksetov (2002); Powell (2012); Van (2018); Delnov (2008); Syupova (2018); and Pochagina (2008). The authors relied on the legal acts, legislation, directives, resolutions of the party and state bodies of the PRC, materials of the

State Statistical Office of China, the Constitution of the PRC of 1954, 1975, 1978, and 1982, and other legislative acts (Kuznetsov 2014). The abovementioned sources allowed us to analyze the main legislative regulation stages of the demographic policy and identify problems and prospects for its further development.

A statistical method allowed us to analyze the causes and results of legislative changes in the demographic policy and quantitatively assess the progress of reforms and their effectiveness. The most important cognitive tool in legal science is the comparative-legal method. It helped to find the criteria for borrowing foreign legal experience (Sorokin 2016, p. 75). Among the special legal research methods, the authors noted the formal legal (dogmatic) method, which allowed to investigate state-legal phenomena in a logical sequence. V. A. Kozlov mentioned that the essence of the formal legal method is to determine legal concepts, identify external signs of legal phenomena, distinguish them from each other, establish classifications, and create logical structures based on legislative concepts and definitions (Kozlov 1989, p. 94). The formal dogmatic method also involves studying the factors and conditions in which legal norms operate and which influence their content. Shershenevich emphasized that jurisprudence, working by a dogmatic method, gives the historical legal development of some people in the context of this moment (Shershenevich 1995, p. 330). A formal legal method allowed us to analyze the Chinese legislation (including constitutional) at various stages of the PRC's development and reveal the reasons for changes in the demographic policy. The research was conducted to develop practical recommendations for further improvement and development of legal acts. Besides, the authors used the method of legal forecasting. Such a method is a system of techniques that make it possible to make a scientifically based forecast about the future state of a legal phenomenon. The authors made a forecast on the development of the demographic situation in China and the prospects for further improvement of the legislation on planned childbearing.

## 3 Results

In the 30–40 s of 20 century, China's population grew at a slow pace. The growth of residents was negatively affected by the Sino-Japanese and civil war. After the PRC's formation, the government set the task to carry out population growth to provide the state with labor. Many Chinese citizens were seen as a source of well-being and prosperity for Chinese society. One of the first legislative acts taken after the formation of the PRC was the Law on Marriage of 1950. The law strengthened the equality of men and women in marriage and established marriage age: for men—20 years, for women—18 years. This law ended the disenfranchised position of a Chinese woman in the family, violating the patriarchal foundations that existed for centuries. The 1950 Law on Marriage laid the democratic (but completely alien to the Chinese mentality) foundations of the institution of marriage and family in China (Pochagina 2008). This law did not address demographic policy issues.

Since the early 1950s, the government limits abortion to increase birth rates. Unlawful abortions were prohibited, but there were formulated conditions under which sterilization was possible.

The first stage of legislative regulation of demographic policy in China began in 1953. According to the population census results of that year, there was a remarkably high increase in PRC citizens. The Constitution of the People's Republic of China of 1954 did not reflect the existence of a demographic problem. Article 96 of the PRC Constitution of 1954 stated that women in the People's Republic of China have equal rights to men; marriage, family, motherhood, and infancy are protected by the state (Kuznetsov 2014, p. 100).

The implementation of demographic policy, development of medicine, reduction of child mortality, and success of industrialization, political and social stability led to an increase in the population by 1957 by 148.6 million people (Skvortsov & Petrov 2015, p. 7). In 1957, the number of people in the PRC was more than 637 million.

Such high rates of population growth caused serious concern among the Chinese leadership. In 1955, the CPC Central Committee proclaimed the party's course to reduce childbearing. Thus, for the first time in history, the PRC launched a campaign to limit fertility. However, at this stage, the campaign was not carried out on a full scale due to the "Great Leap Forward" policy that the government pursued in the same years. The policy envisaged achieving volumes of industrial and agricultural products of the world's leading countries in a very short time. The goal was ambitious: it required a huge number of workers, which led to the liquidation of the childbearing limit program. The Great Leap Forward policy failed. In 1958–1962, the starvation killed more than 10 million Chinese (Meliksetov 2002, p. 667).

The second stage of demographic policy regulation in the PRC is associated with the consequences of the "Great Leap" failure when economic problems worsened. The situation eventually stabilized, and the PRC authorities again returned to the issue of birth control. In 1962, the CPC Central Committee and the PRC State Council adopted the directive on "Concerning the good faith propaganda of childbearing restrictions" ("Directive," 2002). It stated that it is necessary to stimulate birth control in cities and densely populated rural areas and move on to planning births. However, once again, the proclaimed birth control policy was not implemented. The Cultural Revolution began.

The third stage started in 1971 when the PRC's State Council approved the "Report on the work in the field of family planning programs." It indicated that, except for the densely populated areas of national minorities and some other areas, the local authorities should promote voluntary birth planning and late marriage (Veremeychik 2013, p. 475). The government formulated the task of reducing the population growth rate–urban to 10%, rural–to 15% by 1975.

The aggravation of the demographic situation led to creating a special Steering Group for Birth Planning under the State Council. The increase in the number of inhabitants was seen as a threat to the development of the country. The government began a large-scale work to educate the population and promote later marriages, to increase the intervals between the birth of children and the birth of only two children in the family. Article 27 of the 1975 Constitution enshrined provisions on the state's protection of marriage, family, mother and child, similar to the 1954 Constitution.

The ongoing policy has not yielded tangible results. The next stage began in 1978 with an adoption of a new Constitution. Article 53 stated that the state protects and encourages family planning (Kuznetsov 2014, p. 115). For the first time in the country's Constitution, it was stated that in China, there is a demographic problem that needs to be addressed. Chinese researchers see the birth control policy in the country's Constitution as the "watershed" of two fundamentally different periods of the PRC's demographic policy (Dikarev 1989, p. 62).

In 1978, the government proclaimed a course for the "one-child family" considering the new requirements of the demographic policy–the best is to have one child, a maximum of two children (Veremeychik 2013, p. 476). The policy aimed to reduce the population, achieve 1.2 billion people by 2000, and mitigate the negative impact of population growth on the economy, resources, and the environment (Powell 2012, p. 18).

In 1979, the Government Performance Report stated the need to develop measures to encourage couples with one child (Bazhenova 2012, p. 342). Even such an option of moderate demographic reform had a definite result.

In 1980, the Marriage Law was passed (National People's Congress 1980). It mainly contained the provisions of the Marriage Law of 1950, but at the same time, there were innovations. Critical new provisions include an increase in the age of marriage for men and women by two years (22 years for men, 20 years for women), as well as fixing issues of birth planning. Article 16 established the obligation of spouses to plan childbearing. In 2001, the PRC Marriage Law of 1980 was substantially amended, but Article 16 remained unchanged.

A two-year increase in marriage age was supposed to reduce the natural growth of the population. A system of administrative and economic measures was also developed to encourage the birth of one child. Women who decided to have one child were issued a certificate of "excellent birth planning." This gave preferential rights to placement in kindergartens, medical care, and the provision of living space. Families with one child had additional payments of 10% until the child is 14 years old and were exempt from taxes on upbringing and education, etc. (Syupova 2018, p. 277).

In addition to incentive measures, the government also developed sanctions for the birth of a second child. The sanctions varied: a family had to return to the state money received for the first child if the second was born, pay tax for the second child, etc. For violators, there was a fine, which some had to pay for 5 to 15 years in the amount of 10 to 50% of family income (Powell 2012, p. 22).

Article 25 of the 1982 Constitution proclaimed that the state is introducing planned birth rates to bring population growth into line with socio-economic development plans (Kuznetsov 2014, p. 119). Article 49 enshrined spouses' obligation to carry out fertility planning (Kuznetsov 2014, p. 121). Thus, we can note the significant changes in the approaches to the family planning issue comparing with the Constitution of 1978.

The implementation of the "one-child family" course proclaimed by the PRC leadership met the resistance, especially in rural areas. In 1984, the documents adopted by the Central Committee of the CPC spoke of the need to consider local conditions when conducting family planning policies. It was emphasized that family planning should be voluntary, and it was essential to promote family plans to have the desired number of children. Document No 7 of 1984 provided the possibility of developing local, regional legislation of population policy. Regional legislation permitted a second child under certain conditions. There were differences in the implementation of demographic policies between city and country, between different regions and nationalities (Bazhenova 2012, p. 344). This period is characterized by the beginning of the scientific development of demographic problems, laying the foundations for legislative regulation of fertility issues.

Regional legislative acts provided the possibility of having a second child for some categories of citizens. First, it was possible if each spouse is the only child in the family (Kuznetsov 2015). Second, it also was an exception for national minorities. All national autonomous regions were granted the right to conduct their demographic policy. The families who lived in rural areas were allowed to have a second child. Such easing was since most of the rural population is engaged in peasant labor and needs working hands. Rural residents could have a second child, especially if the first is a girl or a disabled person.

In the 1980–1990s, many documents on fertility issues were adopted at the regional level. This made it difficult to control the State Committee for Planned Childbearing (Veremeychik 2013, p. 479).

In the 90 s, there was a further process of regulating family planning issues. The documents stated that by 2000, the population should not exceed 1.3 billion people; by the middle of the next century, there should be no more than 1.6 billion people (Titarenko 2002, p. 112).

Due to the measures taken, the tasks set in 1978 to reduce the birth rate by 2000 were completed. It was possible to reduce the natural population growth by 400 million people. In 2000, the population of China amounted to 1.2 billion people. The standard of living of the population also increased, which had a significant impact on the socio-economic development of the country (Powell 2012, p. 24). The authors analyze the current demographic policy and highlight its negative consequences. First, it is a gender imbalance. According to the results of the sixth census of the PRC in 2010, the ratio of women to men was 100 to 118, respectively (Skvortsov & Petrov 2015, p. 13). Confucius, who regards the state as a large family, believed that relations in the state and family should be built on the principle of sons' respect and ritual. He wrote, "In life, serve them [parents] according to Ritual. In death, bury them according to Ritual. And then, make offerings to them according to Ritual" (Semenenko 1998, p. 59). According to Confucian beliefs, only boys can worship tablets with the names of their ancestors. A son becomes the continuer of the clan and has an obligation to perform the ritual of worshiping the ancestors. Therefore, in the single-child family policy, Chinese families chose one child–a boy. Late abortions and the killing of newborn girls were common. Over decades of childbearing policies, many Chinese have become negative about the very possibility of having children.

The demographic policy of "one family–one child" affected the current Chinese society's moral character (Powell 2012, p. 36). The nature of the relationship between different generations of the family, an only child, a son, a grandson, began to be perceived as a "little emperor," to whom everything is permitted.

Chinese society is aging, which is also a severe problem. Family planning policy led to a sharp reduction in births, a decrease in youth, and an increase in older people. This situation seriously threatens the economic development of China.

Based on the experience of regional regulation of demographic issues, the Law on Population and Birth Planning was adopted in 2001 (National People's Congress 2001). As it was implemented for the first time, the Law can be considered the most important stage of legislative regulation of demographic policy in China. The demographic policy was proclaimed the basic policy of the state: Chinese citizens are responsible for family planning; spouses share responsibility for fertility planning. The law established measures of propaganda and social security to implement population policy goals. Among the other adopted resolutions were: "Measures to manage the family planning of the migrant population," "On the management of technical services for family planning," and "Measures to manage the collection of social tax for raising unplanned children" (Bazhenova 2012, p. 346). Thus, the legislative framework for the population policy implementation in the PRC was created.

In 2006, the CPC Central Committee and the State Council on Strengthening Population and Family Planning Programs adopted the Decision to Comprehensively Solve the Population Problem (Bazhenova 2012, p. 346), with the implementation of which we can speak of a new stage in the legislative regulation of demographic policy. The main objective was to maintain a low birth rate, stimulate families with one child, and provide assistance and support to single-parent families.

The next stage began in 2013. Improved living standards, the situation with gender imbalance, and increased life expectancy led to the gradual softening of the authorities' tough policy of "one family–one child." The plenum of the Central Committee of the CPC decided to allow to have a second child for those families where at least one of the parents was the only child in the family. According to Simonova, such a measure even then was calculated not for a small fraction of the population, but its main part, because during the implementation of the "one family–one child" policy in 2013, there were a majority of such families. (Simonova 2017, p. 117).

Professor at the Institute of Social Development and Public Policy at Shanghai Fudan University, R. Yuan, described the planned birth rate policy as an administrative policy of forced intervention. Modern politics requires adjustment, considering that the contradiction between the rate of economic growth and population growth has softened (Lu 2015, p. 17).

The latest changes in Chinese demographic policy are associated with the plenary session of the CPC Central Committee in 2015. Since 2016, the "one family–one child" was canceled, and payments for the second child were introduced (Vasiliev 2016).

## 4 Discussion

Deputy Chairman of the State Committee for Health and Birth Planning, Wang Pein, noted that a second child's policy is implemented as planned. There are, however, significant difficulties. Women refuse to give birth to a second child due to fear of losing their jobs and material difficulties. China seeks to create a favorable social environment so that families will have a second child ("The new demographic," n.d.).

Experts from the Chinese Academy of Social Sciences noted that a generation born after 1980 has an extremely low desire to have children (Lien 2015). In China, there are a few women of childbearing age due to the context of the "one family–one child" policy when many families preferred to have sons.

Despite the abolition of the single-parent family policy, in 2016, the number of Chinese families who declared their desire to have a second child amounted to only 13% (Lien 2015). In 2016, the birth rate increased by 8% but then began to decline again (Nevelsky 2020). The change in demographic policy is not yet justified. The Chinese leadership is developing a whole system of measures to increase the birth rate. The government carries out an active awareness-raising and propaganda work among young adults on the revival of traditional Confucian values, among which the family occupies a special place. Besides, they are improving the system of healthcare, medical and social assistance to mothers and children.

The researcher, Van, in an article devoted to the analysis of the Chinese demographic policy after the lifting of a second childbirth ban, named the reasons for the rather low population growth rate. Among them is that by the time the policy changed, many already had a second child. In fact, the lifting of restrictions affected only a third of the country's population. Another reason is the tradition over forty years of bringing up and educating an only child. Fears of giving birth to a second child are associated with the high costs of his upbringing and education, housing, as well as the poor environmental situation in China (Van 2018, p. 88). Another problem was considered in work by Kislova and Buyarova. Most Chinese women grew up in a family with one child, which changed their identity. Moving away from patriarchal family planning is one of the new paradigms for Chinese society development (Kislova and Buyarov 2016, p. 348).

## 5 Conclusion

Researchers from the PRC Academy of Sciences proved that the population of 700 million people is the most optimal for the PRC. The maximum permissible number of people in China, for which there are enough resources in the country, is 1.6 billion people (Kireev 2016, p. 19).

Analyzing the history and current state of the PRC, we can conclude that the economic and social development of China depends on an effectively implemented demographic policy. The global nature of world processes indicates the influence of stabilization of the demographic situation in the country on the foreign economic and policy position of the state. The authors conclude that China needs further family planning policies. At the current stage, there is a problem of search for options to stimulate fertility and encourage the birth of a second child. At the same time, it is necessary to prevent excessive population growth. Chinese authorities once again need to find a middle ground. This will require further development of demographic policy and its improvement considering the experience of other countries.

## References


Bazhenova ES (2012) The demographic situation in China. In: Ostrovsky AV, Sham TT (eds) A comparative analysis of the common features and characteristics of the transition period in Russia, China, and Vietnam. Forum, Moscow, Russia, pp 342–353

Central Committee of the Communist Party of China, State Council of China (1962) Directive of the Central Committee of the CPC, State Council of the PRC "Concerning the bona fide propaganda of planned childbearing." Beijing, China. http://news.xinhuanet.com/ziliao/2002-03/04/content_2504428.htm

Delnov AA (2008) China. Eksmo, Algorithm, Moscow, Russia

Dikarev AD (1989) Demographic policy in China. In: Vishnevsky AG (ed) Demographic policy in the modern world. Nauka, Moscow, USSR, pp 47–82

Kireev AA (2016) The demographic dimension of Chinese reform. Russia China Probl Strateg Inter Collect East Center 17:18–21

Kislova EV, Buyarov DV (2016) The influence of politics on solving the demographic problems of the 21st century in China. In: Grigopieva IL (ed) Proceedings of RCHPC'16: Russia and China: History and prospects for cooperation. Blagoveshchensk State Pedagogical University Publishing House, Blagoveshchensk, Russia, pp 347–350

Kozlov VA (1989) The problem of the subject and methodology of the general theory of law. University of Leningrad Publishing House, Leningrad, USSR

Kuznetsov A (2015) Politics China allows all couples to have two children. Politics. https://www.rbc.ru/politics/29/10/2015/5631fdbe9a79476f0eadbf63

Kuznetsov DV (2014) Constitutional acts of China. http://kuznetsov.ucoz.org/books/konstitucionnye_akty_kitaja.pdf

Lien H (2015) China faces a demographic crisis. The Epoch Times. http://inosmi.ru/world/20151005/230642230.html

Lu S (2015) 30 years of planned childbearing policies. China 12:18–20

Meliksetov AV (2002) History of China. Higher School, Moscow, Russia

National People's Congress (1980) Marriage law of China. Beijing, China. http://chinalawinfo.ru/civil_law/law_marriage-2

National People's Congress (2001) The law on population and birth planning of the PRC. Beijing, China. https://chinalaw.center/administrative_law/china_population_family_planning_law_revised_2015_russian/

Nevelsky A (2020) China's declining birth rate threatens its economy. Sheets. https://www.vedomosti.ru/economics/articles/2020/01/22/821283-snizhenie-rozhdaemosti-v-kitae

Pochagina OV (2008) The family: new forms, new values. Otechestvennye Zapiski 3(42):234–246

Powell T (2012) The negative impact of the one child policy on the Chinese society as it relates to the parental support of the aging population. Georgetown University, Washington, D.C.

Russia and China. Asian Illustrated Review (n.d.). https://ruchina.org/naselenie-kitaya.html

Semenenko II (ed) (1998) Confucius: I believe in antiquity. TERRA–Book Club, Republic, Moscow, Russia

Shershenevich GF (1995) General theory of law, vols 1–4. Publishing house Law College of Moscow State University, Moscow, Russia
Simonova AI (2017) Chinese experience in fertility planning policy: results and consequences. In Zakharova LV (ed) Proceedings of IV International conference of young orientalians: Architecture of security and cooperation in East Asia. Federal State Budgetary Institution of Science Institute of the Far East of the Russian Academy of Sciences, Moscow, Russia, pp 114–122
Skvortsov NG, Petrov AV (eds) (2015) Demographic changes and family policy in Russia and China. Asterion, St. Petersburg, Russia
Sorokin VV (ed) (2016) History and methodology of legal science. New format, Barnaul, Russia
Syupova MS (2018) Demographic policy: the experience of foreign countries. Sci Notes TOGU 9(1):272–278
The new demographic policy of China: First results (n.d.) http://www.kremlinrus.ru/news/164/64459/
Titarenko ML (ed) (2002) People's Republic of China: politics, economics, culture. Institute of the Far East, RAS, Moscow, Russia
Van E (2018) Changes in the demographic policy of China: causes, results, prospects. Population 21(1):84–96
Vasiliev A (2016) How China abandoned the "one family-one child rule". Rossiyskaya Gazeta. https://rg.ru/2016/11/24/reg-szfo/kak-kitaj-otkazalsia-ot-pravila-odna-semia-odin-rebenok.html
Veremeychik AS (2013) Legislative regulation of demographic policy in China (on the example of the North-East region). In Kobzev VI (ed) Society and State in China, vol XLIII, Part 2. Federal State Budgetary Institution of Science Institute of Oriental Studies of the Russian Academy of Sciences, Moscow, Russia, pp 474–485

# The Developed Local Self-Government in Strengthening Social Security of Modern Russia

**Olesia L. Kazantseva, Nadezda A. Kandrina, and Lyudmila G. Konovalova**

**Abstract** The purpose of the article is to study the problems of nature definition, the system of public power in modern Russia, as well as the place in this system of local government, the order of its interaction with public authorities, which aims to create conditions for strengthening social security. The analysis of existing constitutional norms and planned changes to the Constitution of the Russian Federation, scientific literature allows to speak about the embedding of local self–government not only in the system of public power, but also in a single vertical of state power. The weakening of the position of local self–government through numerous unsystematic changes in its legal foundations does not contribute to the strengthening of social security in the Russian state, but, on the contrary, undermines the most important form of democracy and reduces the guarantee of constitutional rights and freedoms of citizens, including the right to local self–government. Private and general scientific research methods were used in writing this article. As the main problems, the absence of legally fixed notions of public authority system of public authorities is marked, which generates scientific discussions and different idea about the content of these legal categories, the absence of the general concept of development of local self-government without which its further existence and development as a democratic institution is inconceivable. The unstable legal base directed on the essential restriction of rights of local self-government is narrowing its bases. Accordingly, we proposed to comprehend the current state of local self–government, which is crucial for the development of democratic principles in the current conditions of reform of Russian society, to pursue a clearer and more consistent state policy based on the basic postulates of local self-government, to strengthen its foundations, to provide an appropriate level of guarantees for their implementation, and to combine the joint efforts of the state and local self–government.

**Keywords** Public power · Local government · State power · Public authorities · Social safety and security

O. L. Kazantseva · N. A. Kandrina · L. G. Konovalova (✉)
Altai State University, Barnaul, Russia
e-mail: vaskova82@yandex.ru

 
S. G. Maximova et al. (eds.), *Advances in Natural, Human-Made, and Coupled Human-Natural Systems Research*, Lecture Notes in Networks and Systems 234,
https://doi.org/10.1007/978-3-030-75483-9_30

## 1 Introduction

Local self-government is an essential element of public power and an attribute of modern democratic society. Effective and independent local self-government can serve as a basis for building a civil society, a pillar of statehood, and an essential factor in strengthening social security.

The local self-government system in the Eurasian countries depends on the historical and national traditions, political regimes, national ideas, and government structure. Besides, it is formed due to the presence or absence of the national interest in developed local self-government, which acts as an intermediary between the population and the government. In some countries, a combination of public authorities and local self-government (a combination of local government and local self-government) is used to organize public power at the local level. On the contrary, in other countries, local self-government is autonomous; it is not part of the government apparatus.

The goals of the Eurasian countries are (1) to ensure an adequate level of development of local self-government, (2) to guarantee it, (3) to determine the ratio of public authorities and local self-government in solving the national tasks and tasks of society. These aims are achieved by building an effective interaction between public authorities and local self-government to strengthen social security. Intensifying the importance and effectiveness of local self-government is possible with the active involvement of the population in solving local issues, partnerships with the government, and local self-government protection at the national and international levels.

Russia is a country connecting Europe and Asia, and the current constitutional reform in Russia raises the question of the future of local self-government. It is relevant both for Russia and the international community, since most countries established local self-government at the constitutional level and continued to reform this institution.

There has been enough time since the constitutional establishment of local self-government in Russia to assess its results. However, the accumulated and unresolved problems of the distribution of powers between public power levels, responsibility for their implementation, resource allocation, and corruption cause discontent among scientists and practitioners and the population in general. These factors reduce the positive social activity of citizens and hinder the formation of a full-fledged civil society.

The beginning of 2020 in Russia was marked by another constitutional modernization initiated by the President's Speech to the Federal Assembly on January 15, 2020. The President of the Russian Federation proposed to amend the Constitution (Presidential Executive Office 2020), including issues relating to public power and local self-government. According to these amendments, a unified system of public power is being formed in Russia to include public authorities and local self-government bodies. At the same time, local self-government bodies are guaranteed compensation for additional expenses incurred due to the joint implementation of government functions and powers of national significance. In addition, the amendments establish the possibility of forming local self-government bodies, appointing and dismissing officials of local self-government bodies by public authorities following the federal law. The Constitutional Court positively approved the Law *On Amendments to the Constitution of the Russian Federation* (March 14, 2020) (Russian Federation 2020) of the Russian Federation (Constitutional Court of the Russian Federation 2020), and the amendments themselves were put to a popular vote.

It should be noted that the constitutional reform in Russia was preceded by severe changes in all areas of life (political, economic, social, cultural, etc.). These changes are associated with social risks and threats to the national security of Russia. One of the most critical national security elements is social security, which is not defined in Russian legislation.

The concept of social security is widely presented in scholarship; numerous publications indicate the complex nature of this phenomenon. For example, G. G. Sillaste understood social security as the state of security of an individual, social group, or community against threats that may violate their vital interests, social rights, and freedoms (Sillaste 1998). Social security is understood as "the awareness of hazards, their recognition, development of appropriate measures and practical actions to prevent and minimize threats" (Levashov 2002). The factors affecting the state of integrity and sustainability of social systems were considered by Dürkheim (1991).

To ensure social security in the state, there must be a system of public authorities, which must function based on stable legislation and implement a progressive public social policy (Frolova 2007).

The concepts of *public power* and *public authority* are widely used in Russian legislation; however, they are not defined, which causes scientific discussion. Constitutionalists are characterized by the recognition of two types of public power: government and local self-government. The effectiveness of implementing public policies, including those aimed at strengthening social security, depends on how effectively these types of public authorities interact. The uncertain role of local self-government in the system of public power and a vague attitude of the government to it leads to a significant decrease in the independence of local self-government, undermines its foundations, and as a result, harm social management, the grass-roots level of which is local self-government.

In this regard, it seems relevant to conduct a comprehensive study of local self-government as a factor of strengthening social security at the stage of constitutional reforms in Russia in the field of public power and local self-government, its relationship with the government to anticipate the further development of local self-government in Russia.

## 2 Materials and Methods

We used international standards and studied (1) the existing constitutional provisions on local self-government (Presidential Executive Office 2020); (2) proposed amendments to the Constitution of the Russian Federation regarding the organization of public power and local self-government; (3) the interaction of local government with public authorities (Russian Federation 2020); (4) Federal Law *On General Principles of Local Self-governing Organization in the Russian Federation* (October 6, 2003 No 131-FZ) (Russian Federation 2003); and (5) t their implementation in Russia.

The methodological basis of the research includes general scientific methods of cognition (functional, system, synergetic, analysis, and synthesis) tested by various branches of legal science; and private scientific methods (formal-legal, comparative-legal, modeling, and forecasting).

We used functional and system methods of cognition to identify the essential characteristics of public power, its subtypes (government and local self-government) as a single and complex system, and their interaction in Russia at the current stage. "Cognition possibilities remain extremely limited if they do not rise above the level of unity and spontaneity of the object's existence, unless they explore a specific phenomenon as a system of phenomena…" (Kuzmin 1976).

The use of the synergetic method allowed us to analyze the legal phenomena under study in interrelation, which enabled us to understand the future of local self-government in Russia (Petrov and Zyryanov 2017; Prigozhin and Stengers 1986; Kirdyashova 2003).

The formal-legal method made it possible to identify gaps and conflicts in the legal regulation of the local self-government and its independence. Additionally, we studied the problems of legal regulation of the interaction of public authorities in providing social security.

We applied comparative legal analysis, modeling, and forecasting to examine the current constitutional norms and proposed amendments to the Constitution of the Russian Federation in the field of local self-government.

These methodological tools allowed for a comprehensive approach to studying the state of local self-government in Russia at the current stage of its development. Also, we managed to identify, determine, and systematize the patterns of further development of local self-government in Russia, depending on its legal framework. Besides, we considered the amendments to the Constitution of the Russian Federation and developed a scientific understanding of this problem.

## 3 Results

Studying the constitutional and legal bases of local self-government in Russia over the past decades, it is possible to identify a trend towards a decrease in its independence and increase its dependence on public power. According to many scientists, this situation contradicts the provisions of the Russian Constitution and generally accepted international standards. Based on the analysis of the materials mentioned above, we identified the problems of local self-government in Russia that pose a threat to social security and require priority solutions.

First, public power in modern Russia is efficient enough: there are no legal definitions of the concepts of *public power, the system of public power, bodies of public power*, and forms and principles of interaction between levels of public power. Also, there is no balance between the competence of different public power levels and their material and financial support. Moreover, the distribution of competence between power levels does not contribute to the strengthening of social security and the performance of the functions of the social country. Problems in the legal regulation of public power cause problems in practice, which reduces the effectiveness of public administration and local self-government. Determining the role of local self-government in the system of public power as an independent democratic institution would allow the population to take the initiative in solving vital issues, considering their interests and responsibilities. The constitutional values of statehood and democracy, enshrined in the Basic Law, should be implemented in the interests of individuals and citizens.

Second, de jure local government is significantly different from the de facto government. Initially, the provisions on the independence of local self-government established in the Constitution of the Russian Federation of 1993 were to be developed in the current legislation; therefore, the Federal Law *On General Principles of Organization of Local Self-governing in the Russian Federation* (October 6, 2003 No. 131-FZ) was adopted (hereinafter–Federal Law No. 131) (Russian Federation 2003). Scientific circles felt ambivalent about this law. Some scientists believed that it aimed to centralize power and strengthen authoritarian tendencies (Reddaway and Orttung 2004). Other people positively assessed the Federal Law, noting its timeliness and compliance with reality (Campbell 2006).

Since the adoption of Federal law No. 131, many amendments have been introduced. They intended to (1) significantly restrict the rights of local self-government, (2) prioritize local self-government over the population, (3) complicate the forms of direct implementation of local self-government and the types of participation in it, and (4) expand state intervention in the field of local self-government.

Third, the reforming model of local self-government has many contradictions. The initiation of the reforms carried out by the Center neither considers the regional specifics nor contributes to grass-roots initiatives, resulting in social rejection and protests. This situation restricts the possibility of forming and developing effective relations in local self-government, which is designed to solve local affairs taking into account the interests of the population. However, the population itself, which is legally regarded as the most crucial subject of local self-government, distances itself from local self-government bodies; therefore, it does not consider that the institution of local self-government meets the needs of the population, which significantly reduces the social activity of citizens.

Simultaneously, the effectiveness of local self-government is determined by the degree of involvement of residents in solving vital issues. Furthermore, the interest of citizens in managing resources for solving local affairs forms the basis for the development of local self-government. One of the tasks assigned to local self-government bodies should be involving the population in the implementation of local self-government and increasing the positive social activity of citizens. Having the opportunity to participate in decision-making, realizing their importance, and considering the opinion of the population on the part of the authorities, they will feel responsible for solving local issues. As a result, they can build a constructive dialogue with the authorities and not oppose them. However, the mechanisms of interaction of local self-government bodies with the population are still underdeveloped.

Local self-government in Russia is at the very beginning of its path. Among other things, this situation is due to the low level of public awareness about the possibilities of influencing the activities of local self-government bodies. The existing legislative framework is contradictory. The forms of direct participation in local self-government are complex and often challenging to implement. Consequently, most of them are not put into practice. Local referendums are usually held on self-imposed issues, and there are quite a few of them. Public meetings of citizens are rare. Periodically, they are held in small localities with a population of no more than 100 inhabitants (in some cases, from 100 to 300 inhabitants), who have an active electoral right and exercise the powers of representative bodies of municipalities. This situation is caused by depopulation, which involves the abolition of such rural or urban settlements. The legislative initiative, the vote on the recall of the deputy of the representative body and the elected official of the local government due to the cumbersome procedures are not popular with the population. Rallies and territorial public self-government have become widespread.

Fourth, there is currently a constitutional reform that affects the field of local self-government. The amendments to the Constitution approved by the Constitutional Court of the Russian Federation look relatively harmless; nevertheless, it is not easy to predict the possible consequences of these changes.

According to the amendments to the Constitution of the Russian Federation, it is proposed to include local self-government bodies in the system of public authorities and consolidate the performance of state powers by local self-government bodies in cooperation with public authorities. A single vertical of power has already been built; the proposed amendments merely legitimize the administrative dependence of local self-government bodies.

The desire for de facto regularization can be justified by the desire of the government to strengthen its influence on local self-government. At the legislative level, local self-government bodies may later become an integral part of a single vertical of state power.

At the same time, the Decree of the President of the Russian Federation *On the National Security Strategy of the Russian Federation* (December 31, 2015 No. 683) claims that the long-term national interests are (1) the strengthening of national harmony, (2) political and social stability, (3) the development of democratic institutions, and (4) the improvement of mechanisms for interaction between the government and civil society (Presidential Executive Office 2015). As for the development of local self-government as a democratic institution, the situation with the amendments to the Constitution of the Russian Federation is vague. Much depends on the detailed elaboration of the constitutional provisions in the federal legislation. However, given the current trend of centralization of power and the collapse of local self-government, it is expected that the proposed amendments to the Constitution of the Russian Federation will allow ensuring the direct involvement of local self-government in the implementation of state policy in all areas of public administration.

This situation can lead to greater state control over local government and strengthen the regulatory impact of government on local self-government.

Compensation to local self-government bodies for additional expenses incurred due to decisions of public authorities can also be considered another amendment to the Constitution of the Russian Federation. On the one hand, local self-government bodies will insist on reimbursement of expenses that were not registered but incurred due to the fulfillment of obligations. On the other hand, the state will have the right to hire local self-government to exercise certain powers, thereby subordinating it.

Unfortunately, the practice shows that even the Constitution cannot protect local self-government from the interference of state bodies. For example, the replacement of direct elections of heads of municipalities by the procedure of their competitive selection (when the representative body of the municipality is appointed by the highest official from the candidates proposed by the electoral commission). At the same time, for municipal and urban districts, half of the members of the electoral commission are appointed by the governor of the relevant subject of the Russian Federation; therefore, the elected head will depend not on the residents of the municipality but the regional authorities. Simultaneously, the Constitution of the Russian Federation stipulates that the structure of local government is determined by the population independently, and local governments are not created with state participation; this fact corresponds to the principle of self-organization of local self-government. Nonetheless, in this case, the population cannot directly participate in the election of the head of the municipality. Amendments to the Constitution of the Russian Federation will allow the government to participate in creating local self-government bodies in cases provided for by federal law. As a consequence, the existing practice of electing the heads of municipalities, based on the provisions of Federal Law No. 131 and recognized by many scientists as unconstitutional, will now be established at the constitutional level.

The problem of financial support for municipalities remains unresolved. The amendments to the Constitution of the Russian Federation do not address the issue of strengthening the economic independence of municipalities. At the same time, the government, which is unable to solve social problems (including the lack of necessary material and financial resources), undermines both its authority in the eyes of the population and the entire state. The current trend of consolidation of municipalities can hardly solve the problem of financing municipalities. Moreover, it negatively affects citizens' right to participate in local self-government, creating difficulties in obtaining municipal services for the population.

These examples demonstrate the distancing of the authorities from the population, which creates social tension.

The analysis of the amendments to the Constitution of the Russian Federation allows us to conclude that the future fate of local self-government will depend on the interpretation of the amendments to the Constitution and their practical implementation. We believe that further local self-government reform should (1) improve the quality of life, (2) ensure the rights and freedoms of people, (3) adhere to the principles of social justice, and (4) guarantee an adequate level of social protection and, as a result, social security.

## 4 Discussion

Numerous problems in the field of local self-government prompted discussions on the nature of local government as part of the system of state power, the future of local government, and the definition of the Russian model of local self-government and its *landmark*.

Various authors consider urgent issues of implementing public power, its system, and significance in modern Russia. Avakyan, Chirkin, and Khabrieva made a significant contribution to public power ideas (Avakyan 2009; Chirkin 2013; Khabrieva and Chirkin 2005).

The issue of the model of local self-government remains controversial. For example, Gilchenko noted that the lack of a unified state concept of reforming local self-government on the principles of self-government at the federal level hinders the formation of an effective system of local self-government in Russia. Besides, Emelyanov drew attention to the low level of the political culture of the population. Also, Voronin considered the model with a strong federal-state power, which would overcome regional separatism, to be the most acceptable (Gilchenko 1998; Emelyanov 1997; Voronin 2002).

Avakyan believed that local self-government combines two principles: the state and the public. At the same time, Baglay thought that the nature of local self-government is rather state-based, and, despite its independence, the state should control local self-government (Avakyan 2018; Baglay 1999).

The Russian policy of strengthening the state in the context of ongoing reforms, digitalization, and the formation of the information society is impossible without the grass-roots level of government, the modernization of which will significantly affect the quality of administration in general and strengthen the relationship between the state and civil society.

In the Russian science of constitutional law, local self-government is an independent type of public power. Avakyan defines municipal power as an independent subsystem of public power (Avakysan 2009). At the same time, N. S. Bondar, referring to the legal positions of the Constitutional Court of the Russian Federation, defines public power as a generic concept concerning local self-government (Bondar 2006). The science of constitutional law regards local government as the appropriate level of public authority in the Russian Federation.

Local self-government bodies are not de jure part of the system of public authorities; they are organizationally independent. However, they actively participate in implementing state policy and exercise separate state powers based on the federal and regional laws, which indicates their state nature. President of the Russian Federation Putin noted, “ < … > despite the fact that this level of power is not defined as the state level in the legal sense, in its meaning it is certainly one of the most important, crucial state levels of power, because it works directly with people on a daily, hourly basis” (Presidential Executive Office 2013). Undoubtedly, the problems that need to be solved quickly, the need to finance the implementation of local issues are better seen at the local level. Therefore, there is a real opportunity to consider the peculiarities of municipalities and the interests of the local population. Local self-government is the power exercised by the population and specially formed bodies within the territory of a municipality to resolve issues of local significance, with executive and administrative powers. With the adoption of the Constitution of the Russian Federation and the ratification of the European Charter of Local Self-Government, the nature of local self-government in Russia becomes twofold: public and state. Over the past decades of local self-government reform, there has been an increasing denationalization of local self-government and its interweaving with state power. To determine the future fate of local self-government in Russia, one should define the development vector of its legal bases. This vector should aim to develop local self-government as a democratic institution, increase the level of independence of local self-government, ensure it, and establish partnerships with the public authorities, seeking to strengthen the social security of the population.

## 5 Conclusion

We can conclude that the instability of the legal framework and the continuous reform of local self-governing hinder its development as an independent level of state power. Local self-government plays a critical role in the social security of Russia. The grassroots level of state power is called upon (1) to address the issues of immediate life support for the population of municipalities, (2) to implement measures to ensure security, and (3) to create favorable living conditions for people. The status of local self-government is an indicator of the effectiveness of power in the country and public confidence in the law and the government.

In this regard, it seems necessary (1) to develop the constitutional provisions guaranteeing the organizational and economic independence of local self-government; (2) to design a unified concept of development of local self-government; (3) to conduct public policy in compliance with international law, which enshrines the local self-government as a form of people's power, the most important democratic institution; (4) to enhance the work at all levels of government to involve the population in decision-making.

Above all, the state should (1) ensure social security, (2) protect the population, (3) reduce the level of dissatisfaction of citizens, (4) increase their positive social activity, (5) conduct consistent public policy in the social sector aimed at supporting the population. We believe that civil society institutions should be engaged in this task, since the government and civil society should cooperate to achieve common goals. The proposed measures should reduce social tensions and strengthen social security in Russia.

**Acknowledgements** We express our gratitude to Altai State University for the financial support of our publication.

## References


Avakyan SA (2009) Public authority: constitutional and legal aspects. Bull Tyumen State Univ 2:5–15

Avakyan SA (2018) The structure of public authority in Russia: problems of formation and development. Bull Siberian Law Inst Ministry Intern Aff Russia 4(33):7–13

Baglay MV (1999) Constitutional law of the Russian Federation. NORMA, Moscow, Russia

Bondar NS (2006) The constitutionalization of the socio-economic development of Russian statehood (in the context of decisions of the Constitutional Court of the Russian Federation). Vicor Media, Moscow, Russia

Campbell A (2006) State versus society? Local government and the reconstruction of the Russian state. Local Gov Stud 32(5):659–676

Chirkin VE (2013) Public authority in modern society. Const Municipal Law 4:12–15

Constitutional Court of the Russian Federation (2020) Court opinion "On compliance with the provisions of Chapters 1, 2 and 9 of the Constitution of the Russian Federation that have not entered into force the provisions of the Law of the Russian Federation on amendments to the Constitution of the Russian Federation "On improving regulation of certain issues of the organization and functioning of public authority", as well as on the conformity of the accession procedure with the Constitution by virtue of Article 1 of this Law in connection with the request of the President of the Russian Federation (March 16, 2020 No. 1-Z). Moscow, Russia

Dürkheim E (1991) On the division of social labor: sociology method. Nauka, Moscow, Russia

Emelyanov NA (1997) Local government: problems, searches, solutions. TIGIMUS, Moscow; Tula, Russia

Frolova NA (2007) The ideology of social security and the legal policy of the state: theoretical and historical aspect. AVA-BEST, Moscow, Russia

Gilchenko LV (1998) Reform of local government in Russia: analysis of the state and study of the problems of the formation of local self–governing. In: Alferova EV (ed) State power and local self–governing. Institute of Scientific Information for Social Sciences (INION RAN), Moscow, Russia, pp 8–49

Khabrieva TYa, Chirkin VE (2005) Theory of the modern constitution. NORMA, Moscow, Russia

Kirdyashova EV (2003) On the possibility of using synergetics in the theory of state and law. Rossiyskoe Pravo v Internete 1:3

Kuzmin VP (1976) The principle of systematicity in Marx's theory and methodology. Politizdat, Moscow, Russia

Levashov VK (2002) Globalization and social security. Sociol Stud 3:19–28

Petrov AV, Zyryanov AV (2017) Synergetic approach in the modern legal studies. Bull South Ural State Univ Ser Law 17(4):102–106

Presidential Executive Office (2013) Speech of the President of the Russian Federation at the Meeting of the Council for the Development of Local Self-Government. http://www.kremlin.ru/events/president/news/17397

Presidential Executive Office (2015) Decree of the President of the Russian Federation "On the National Security Strategy of the Russian Federation" (December 31, 2015 No. 683). Moscow, Russia. http://www.kremlin.ru/acts/bank/40391

Presidential Executive Office (2020) Message from the President of the Russian Federation to the Federal Assembly of the Russian Federation (January 15, 2020). Moscow, Russia. http://www.kremlin.ru/events/president/news/62582

Prigozhin I, Stengers I (1986) Order from chaos. Progress, Moscow, Russia

Reddaway P, Orttung RW (ed) (2004) The Dynamics of Russian Politics: Putin's Reform of Federal-Regional Relations, vol 1. Rowman and Littlefield, Lanham, MD

Russian Federation (2003) Federal Law "On the general principles of the organization of local self–governing in the Russian Federation" (October 6, 2003 No. 131-FZ). Moscow, Russia

Russian Federation (2020) Federal Constitutional Law "On improving regulation of certain issues of the organization and functioning of public authority" (March 14, 2020 No. 1-FKZ). Moscow, Russia
Sillaste GG (1998) Woman as an object and subject of social security. Sociol Stud 12:62–72
Voronin A (2002) Back to advice or forward to zemstvos? Municipal Auth 4:100–103

# Social Isolation and Demographic Problems of Border Territories: Theoretical Analysis and Application in Migration Studies

**Syldysmaa A. Saryglar, Ekaterina V. Shakhova, and Olga V. Borisova**

**Abstract** The general goal of the paper is connected with the analysis of aging of the population of modern Russia, identification of the disorganizing factors leading to social exclusion and social inequality of the senior age groups and in turn to conflict situations, social tension in society in general. The problem of removal or mitigation of processes of social exclusion, the analysis of processes of integration of old people in society, adequate to social and economic, political, and cultural realities of the present stage of development, development and realization of the state gerontology policy need in serious scientific providing, and demand both for new approaches and absolutely other scales of development of sociology. Paper bases on theoretical analysis of the existing literature about social exclusion and its correlations with social processes in the modern border societies. Authors empathize three general directions in the development of contemporary science about ageing: study about the ageing as a process in its biological and psychological dimension; institutional approach, focusing on problems of social—economic status and social roles of the elderly people; and, historical—cultural approach, comparing different notions about the ageing and social approaches to estimate it. A question for discussion was stated—how to use knowledge about social exclusion in migration researches in border areas.

**Keywords** Social exclusion · Demography · Migration researches · Border areas · Theoretical understanding · Methodology of research

S. A. Saryglar (✉) · E. V. Shakhova · O. V. Borisova
Altai State University, Barnaul, Russia
e-mail: syldysma93@mail.ru

E. V. Shakhova
e-mail: ews05@yandex.ru

O. V. Borisova
e-mail: olya333es@mail.ru

 
S. G. Maximova et al. (eds.), *Advances in Natural, Human-Made, and Coupled Human-Natural Systems Research*, Lecture Notes in Networks and Systems 234,
https://doi.org/10.1007/978-3-030-75483-9_31

## 1 Introduction

Today, the content and trends in the development of the processes and phenomena that underlie our era are being transformed. One of such processes that significantly affect society, determine its condition and potential is the so-called *aging* of the population.

Russia has a high degree of social differentiation. Therefore, the problem of social inequality is crucial for every citizen and society in general. Currently, a new form of social stratification is emerging. Society is divided into the social mainstream and the socially excluded. The exclusion processes destroy social solidarity and push the excluded to the periphery of the society, making them lose their social connections.

Modernization and urbanization in modern Russia have led to qualitative changes in the image and standard of living of different generations. Moreover, *age segregation* has begun, which is gradually turning the older adults into social outcasts. Current problems of older adults are associated with the following factors:

- Changing nature of work in industrial society compared to agricultural;
- Geographical separation of generations, caused by the growth of large cities employing young people;
- Destruction of the "family production" and the patriarchal multigenerational clan;
- Spread of the nuclear family;
- Drastic acceleration of the pace of technological development which depreciated the experience of older generations;
- Redistribution of *the monopoly of knowledge* in favor of youth;
- Introduction of forced layoffs by age.

In the early 21st century, older adults lost their independence and were driven out of the labor market. They were deprived of the traditional aura of wisdom, which served as a charter of immunity for the previous generation of 70-year-olds and partially compensated them for the inevitable falsehoods of old age.

## 2 Materials and Methods

The main research goal is to analyze the aging of the population of modern Russia. Besides, we seek to identify disorganizing factors leading to social isolation and social inequality in older age groups and, in turn, to conflict situations and social tension. The problem of eliminating or mitigating the processes of social exclusion is crucial. Moreover, it is necessary to analyze the processes of integrating older adults into society that are adequate to the socio-economic, political, and cultural realms of the current development stage. The formation and implementation of the national gerontological policy require significant scientific support, new approaches, and a completely different scale of sociology development.

The paper is based on a theoretical analysis of the existing secondary literature on social isolation and its correlation with social processes in modern society. There are three main directions in the development of the modern science of aging: (1) studying aging as a process in its biological and psychological dimensions; (2) using an institutional approach that focuses on the problems of the socio-economic status and social roles of older adults; and (3) applying a historical-cultural approach that compares different ideas about aging and social approaches to its assessment. These directions have different socio-gerontological knowledge models with the aging process in its socio-historical dynamics as (1) an object and the social position and psychological experience of older adults as (2) a specific age group.

## 3 Results

In recent decades, there has been significant aging of the population. In Russia, in the late 20th–early 21st centuries, the share of people under 60 and older was 17–20%. In 2025, it is expected to reach 27% (Federal State Statistic Service 2020). This demographic process has profound socio-economic consequences, posing new challenges for scientists and practitioners related to the problems of older adults.

The development of the theory of aging and the older adults involves the development of a gerontological paradigm. The formation of a paradigm in gerontology as a social science is associated with a vague understanding of the problems associated with the processes and results of aging and different ways to solve them. Tornston (1982) defines five main paradigms in the modern understanding of aging. The first one is related to the interpretation of social problems in social pathology. From this point of view, the tragedy of aging lies in the individual's non-compliance with the criteria of the *norm* (normality) in physiological, psychological, communicative, labor, and other senses. Despite the way of evaluating criteria of the norm (based on statistical, medical, psychosocial, or moral standards), *pathological* measuring social problems in older adults limit their decisions at the level of the individual without any recommendations on the reconstruction of social relations and structures. The opposite paradigm is morality with the organization of the social *whole* and the relationship of its parts as a starting point. As a result, the collapse of the connection between the social system and its disorganization causes social problems. As for the status of older adults in modern society, there is a disintegration of the traditional lifestyle and the corresponding image of aging as a synonym for life and professional wisdom that stems from industrialization and urbanization. The solution to the social problems of older adults involves specific microstructural changes.

Within the third paradigm, L. Tornston and C. H. Victor believe that social problems are an explicit or implicit conflict of values. The problem of aging lies in the constant conflict between the needs of the industrial market, which requires a new labor force, and the desire of the aging population to actualize the value of their social and personal *ego* through labor (Millar 2007). The fourth paradigm focuses on the emergence of social difficulties through the prism of deviance (Beauvior 1970). The passivity of older adults, their social isolation, exclusion, and similar aging signs can be regarded as variants of deviant behavior. The fifth model is based on the *Theory of Naming*; it interprets the problems of aging as a consequence of its inadequate perception or *labeling* (elderly, weak, sick, poor, etc.).

However, not all existing ontological theories develop within the paradigms mentioned above. The exceptions are attempts to consider aging from the point of view of the life cycle concept.

Scientists pay special attention to the theory of symbolic interactionism, which allows us to empirically record the dynamics of aging, referring to the symbolic aspects of the interaction of older adults with the environment and analyzing the forms of adaptation to changing living conditions.

Some approaches (Beauvior 1970) emphasize the marginality of older adults, their status as social outcasts with low income, and lack of opportunities as their destiny. The followers of this approach pay special attention to the study of the subculture of older adults due to similar living conditions and typical troubles. The weakness lies in the lack of representation of the social heterogeneity of the older adults in society.

Finally, the theory of age stratification by M. W. Riley, R. S. Laufer, and W. L. Bengston understands society as a set of age groups and reflects age differences in opportunities, role functions, rights, and privileges. Chronological age is the basis for stratification. The approach focuses on such problems as (1) the status of older people in society, (2) individual shifts from one age group to another, or (3) mechanisms for the distribution of social and age roles. This position allowed us to characterize older adults in terms of demographic indicators and examine their relationships with other age groups (Baily 2017).

Gerontology reveals socio-psychological approaches to the study of the late period of life: pessimistic approaches, such as *the deficit model of aging, the theory of release from business, and the concept of separation* (Rosen, Neugarten, Cumming and Henry), were replaced by the adoption of limited and modified neo-optimism (Neugarten and Neugarten 1987; Cumming 1970; Rosen 2001; Henry and Cumming 1959).

Pessimism, limited optimism, and other models of aging are not random. They show and deliberately and not theoretically substantiate the unfavorable status of older adults. That is why Russian and foreign gerontology pay particular attention to the problem of social adaptation of older adults as "the way of their adaptation in society due to the new era and the way the society adapts the older adults." (Dichev and Tarasov 1976).

Social adaptation is seen as the result of changes in social, socio-psychological, moral-psychological, economic, and demographic relations between people during their integration into society. In our opinion, specific features of social adaptation include (1) the active participation of consciousness, (2) the impact of human labor on the environment, and (3) the change in the results of social adaptation by a person according to the social conditions of their existence (Romm 2012). In this regard, society itself is an adaptive and adapting system, since human activity is transformative (Maximova et al. 2018).

T. Dihina notes that adaptation can be active (when the individual is inclined to influence the environment to change it) and passive (when the individual is not willing to affect and change anything). The high social status of an individual, their psychological satisfaction with the environment and its elements is an indicator of successful adaptation. In contrast, the shift of an individual to another social environment (marginalization, anomie, deviant behavior) is an indicator of low social adaptation (Dihina 2005).

According to the concept of personality, each individual has a unique lifestyle, the ways of its structuring and organization, on the one hand, and evacuation and conceptualization, on the other. A particular way of action (adaptation) of a person, peculiar to them, is called an adaptive strategy. This strategy characterizes the vector of actions and the involvement of the individual in what is happening and shows different behavior patterns. This strategy is essential for realizing the individual in social life by correlating vital needs with personal activities, values, and ways of self-affirmation (Culliney and Jones 2017). How the members of the society make their living are essential to the society (with their own resources, the resources of society, other people, or the country). Therefore, it is possible to distinguish the adaptive or maladaptive strategies.

The sociological research results indicate various adaptations of different people in similar problematic situations, as well as diverse results of the adaptations. This fact indicates the ability of the individual to adapt and significant differences in the level of !!!near-capacity!!! (Maximova et al. 2018).

Consequently, pluralism in theories and concepts reflects the difficulties in understanding aging as an objective reality and existing opinions as mutually complementary.

Considering the positions of the theoretical directions mentioned above, which are wholly or partially devoted to the problems of aging among older adults, we can conclude that there is no consistency in the approaches to understanding the logical process of human aging. There are different perspectives on the relationship between aging individuals and society. According to Kozlov (1999), all theoretical approaches to the problems of aging and the transition to the group of older adults can be divided into two categories, depending on the answers to the question of whether the continuity of generations is necessary or older adults automatically find themselves over the edge and become a fading marginal stratum. The answer is determined by social policy and social protection directions both in the theoretical and practical sense.

P. Sorokin (1994), the author of the concept of social stratification and social mobility, made a significant contribution to the development of the problems of social isolation. His concept is based on the idea of an uneven distribution of (1) rights and privileges, (2) responsibilities and duties, (3) social values, (4) power or influence among the members of a particular society. Furthermore, there are other approaches to studying the social structure, including those based on objective criteria of social status (position) of the individual and the social group. Such factors as the prestige of the profession or the level of income and education are essential indicators of the generalized social status (the so-called *social position index*). Social inequality is a result of social stratification and the hierarchy of status groups generated by the institutional structure of society (Giddens 2003; Sztompka 1996; Zaslavskaya and Shabanova 2001; Fahmy 2017).

The idea of *social exclusion*, initially described by the deprivation of citizens with disabilities, was discussed in the works of Walsh et al. (2018). They evaluated risk indicators, protective factors, and various political circumstances to overcome them. In a broad sense, social exclusion can be defined as "a process in a result of which individuals or their groups became fully or partly excluded from full participation in the life of society they live [in]" (Hrast et al. 2013).

At the same time, it is possible to distinguish two fundamentally important aspects of the concept of social isolation. First, social isolation is a multidimensional concept. For example, people may be excluded from society because of unemployment, wages, property, minimum consumption, education, quality of life in the country, or citizenship. As a result, the excluded people lack close contact and respect. Nevertheless, the concept of social isolation focuses on the multidimensional nature of deprivation; that is, people can often be deprived of several social factors simultaneously. Therefore, alienation (deprivation) can manifest itself in the economic, social, and political sectors (Myck et al. 2015).

Second, social isolation involves addressing certain relationships of individuals, groups, and processes leading to deprivation. Furthermore, individuals can be excluded from several groups simultaneously (Bonfatti et al. 2015).

Factors of involvement in social exclusion include (1) poverty; (2) subordination in the system of social identity (race, ethnicity, religion, gender); (3) social status (refugees, migrants); (3) demographic characteristics (education, job, age); (3) health status; (4) disability or a stigmatizing disease such as HIV or AIDS.

The social exclusion model developed by the Social Exclusion Knowledge Network (SEKN) presents it as the result of four interdependent factors (social, cultural, economic, and political) at different levels (individual, group, household, local community, country, and the world at large). The multidimensional model underlies the scientific analysis of exclusion. At the same time, the clear connection between exclusion and rights allows us to deal with discrimination based on gender, ethnicity, religion, or health restrictions.

In this regard, it is necessary to discuss the problem of human rights regarding age. The development of modern ideas about human rights evolved from the recognition of personal and political rights, initially proclaimed by the French Revolution, to the emergence of the so-called *third-generation* rights, collective rights, solidarity rights, fundamental right to peace, disarmament, development, the general cultural heritage of the people, humanitarian assistance, and many others (Cardenas 2016).

The problem of analyzing social isolation is one of the most difficult in any study of the status of disadvantaged groups in society. In Western European countries, the problems of social exclusion have a long research history (Abrahamson 1998; Doyle and Gough 1991; Gough 1994; Leonard 1997; Paugam 1996; Touraine 1991) and became one of the general aspects of the analysis of social policy problems. However, in Russia, this problem is only beginning to gain momentum.

Tikhonova (2003) analyzed the problems of social isolation in several projects. Her first work was "Poverty and Social Exclusion in Russia: Regional, Ethno-National and Socio-Cultural Aspects" (INTAS Grant, 98-1439). Project-manager was N. Manning (the University of Nottingham, Great Britain). Besides, a well-known expert on the problems of poverty Prof. D. Veit Wilson (Great Britain) and one of the best experts on the problems of social exclusion, dean of the Sociology Department in Copenhagen (Denmark) Prof. P. Abrahamson, also addressed this issue. N. Tikhonova organized the Russian part of the research and involved Prof. A. Zdravomyslov, Ph.D. N. Davyadova, and Ph.D. A. Tzuriev in the project. Complete research results were presented in the monograph *Poverty and Social Exclusion* (Tihonova 2003). Her second project was "Restructuring of the Welfare State: West and East 1995–1998" (Grant INTAS, 94-3725) (heads N. Manning and O. Shkaratan).

We rely on the paper by P. Abrahamson at the XIV World Sociological Congress in Montreal in 1998 (Abrahamson 1998), which presents a clear and complete description of the problem of social exclusion and its essence in the Western sociological tradition. An abridged but detailed outline version of the paper was published in Russian in the journal *Social Sciences and the Modern World* (Abrahamson 2001). P. Abrahamson defines exclusion as a particular state of *falling out* from the mainstream, position in society and outside it due to (1) the unfavorable position of individuals in terms of education, qualifications, employment, life, and financial resources; (2) the inability of individuals to access the central social institutions that distribute life chances, lower than that of other parts of the population; (3) long-term restrictions.

One can use seven scales to determine the indicators of social exclusion, considering the classification of human rights (formed within the international discourse on social exclusion), which the society should provide, as the most frequent cause of discrimination (Abrahamson 1998; Friedmann 1996; Gough 1994; Beyond the Threshold 1995; Da Costa 1995; De Haan 1998).

The idea of L. Wakquant about marginality, which relies on "individual strategies of self-sufficiency, informal employment, underground trade, and quasi-institutionalized struggling abilities that do not facilitate its precarious state," is quite relevant to the Russian realm (Wacquant 2009).

The next problem is addressed by social exclusion specialists (Room and Britton 2006).

Exclusion is often generated not by unemployment but by a particular type of employment (unstable, differentiated employment), which results in fragmentation and insecurity, rather than a guarantee of security. L. Wakquant characterizes this type of employment as *advanced marginality* (Wacquant 2009).

In this regard, we should mention the features of social isolation given by M. Castels:

> Comparative data show, that in general, in all urban societies the most part of people or their families work to get money, even in poor districts and countries. The question is the following: for which type of job? for which salary? in which conditions? Main mass of the labor power has no constant place of work and circle around different sources of employment (dominantly of random character). Millions of people constantly find and lose paid job, often included into informal activity… Furthermore, losses of stable job, weak positions of workers during making contracts lead to higher level of crisis situations in life of their families: temporary loss of job, personal crisis, illnesses, drug/alcohol addictions, loss of savings, and credits. Many of such crises are interconnected and produce the spiral of social exclusion, going dawn [sic]… (Castels 1998).

Therefore, Castels precisely describes the trigger of the social isolation that is developing in Russia involving millions of people. This is the nature of available work, which does not prevent poverty or alienation and determines the lack of interest among the Russians in getting a proper job to solve problems and successfully integrate into society.

One of the crucial tasks of the study was to determine the ratio of poverty and social exclusion in Russia. Both Russian and international sociologists considered this problem. On the one hand, they indicated that although exclusion involves poverty, it is a much broader concept, which combines disenfranchisement and limited access to the institutions allocating resources (De Haan 2007). On the other hand, poverty does not always mean exclusion, because most of the population may be poor (Paugam 2009).

The concept of P. Tousand was a kind of *bridge* between the concepts of exclusion and poverty. He defined poverty by the standard of living and the nature of the deprivation of the lifestyle itself. Tousand tried to describe a new social phenomenon—alienation; he witnessed its emergence and was the first to understand its meaning through the traditional concept of poverty and gave it a new meaning. According to Tousand, the set of deprivations that act as a measure of poverty included deprivation in social participation, which made it impossible to maintain the traditional standard of living in society. Referring to the concept of *poverty*, Tousand tried to fix and assess the alienation (Taylor 2017).

The international studies addressing the social exclusion of older adults may provide many examples of this phenomenon.

Dmitry Rogozin, Director of the Center for Methodology of Federal Studies of the Russian Presidential Academy of National Economy and Public Administration, spoke about several foreign gerontological studies related to the liberalization of aging.

J. Vincent, in his book *Old Age* (2003), analyzed thousands of studies and analytical reports on aging. He noticed a key dichotomy that defines the status of the older adults as liberalizing from aging and liberalizing of aging. In the first variant, there are practices of avoiding aging; trying to slow down aging; ignoring the changes in the body, perception of the world, and social relations, or simply neglecting them. In the second variant, people found internal resources in aging itself, considered a new opportunity, a specific path, and the final stage of life, without which the existence would lose its full meaning. J. Vincent considers the dominance of medical interpretation a common feature of modern ideas about older age. Medicine predetermined the ideas about aging and displaced other social or religious interpretations. Older age was considered a period of illness and approaching death. The loss of ability to work and the constant deterioration of health, up to complete disability, could only cause a sense of fear and despair. Such a philosophical position has neither a future nor a present. The liberalization of aging is a response to emerging demographic trends. People should not avoid it but look for meaning, exceptional opportunities, and prospects.

It is necessary to free the older adults from the stereotypes accumulated over centuries, reject the technological perspective based on medical diagnoses, and discover a new identity based on years of experience that was not available at a young age. As a paradoxical example of traditional consciousness, Vincent wrote about the liberation from gender stereotypes related to the end of the fertile age, the formation of independence, and self-realization through freedom from the dictates of constant rethinking of the family role and the role of younger generations as an escape from the material problems associated with the upbringing of young adults.

*The Cambridge Handbook of Age and Aging* (Bengtson et al. 2005) is the most comprehensive collection of works representing the current status of gerontology as an interdisciplinary knowledge that combines the social, medical, and biological sciences. This publication covers 80 original works by leading researchers from 60 countries and 5 continents. Unfortunately, the edition does not have any information about Russia. The book is divided into seven parts: (1) introduction and review of theoretical approaches; (2) aging body; (3) aging consciousness; (4) aliening self-understanding (identity); (5) aging and social networks; (6) aging and society; (7) social support program for the older adults. The authors of the book reflected the current state of science, described modern achievements, and identified problematic areas (*points of growth* for science).

First, the biomedical interpretation of older age has maintained the growth of knowledge and formed a base of refutable theories (according to Popper) over the past two decades. Second, the emphasis on medical interpretations has led to the emergence and development of a critical direction that emphasizes social norms, habits, and relations. The social theory of aging seems to be a criticism of the medicalization or physiological stigmatization of the older generation. Third, taking into account the great potential in the social understanding of the third and fourth age, leading experts note the extreme weakness and inconsistency of the theoretical apparatus developed by the humanities. Thus, Malcolm Johnson, editor of the book, pointed out that generalizations of the research, surveys, and social policy estimations, interviews, and ethnographic descriptions of ordinary life dominate in gerontology projects. Theoretical descriptions often are artless and non-heuristic, so many researchers wish to reject them.

Therefore, before the sociologists face a real problem, they try to form a certain communicative field, which can be resolved only by theoretical interpretations. The preservation of the common opinion of the participants on the current situation, the development of joint actions, strategies, and social policies are impossible without a general theoretical basis, as shown in the book.

*The Handbook of Theories of Aging* was edited by highly respected sociologists who study generational issues and aging from a social perspective. Over the past decades, almost every significant international conference has been held with the participation of V. L. Bengston, who often acts as a co-editor or author of papers on the theoretical bases of aging problems. The first edition of the handbook was published in 1998. The second one comprised works that reflect a more radical view of human nature, having fewer connections with the physiological and biological features of the human body.

Despite the logic and development of a *natural* approach to older age, sociology was the central discipline of the publication. Biodemography, immunology of aging, neurophysiology, cognitive plasticity of age, emotional well-being, and other approaches study the integrative role of social gerontology based on the sociological theories about aging. National policy and mature political science cannot be understood without legitimizing multiple perspectives and rejecting the dominant role of the government as the central actor of social transformation and reforms. The future of theories about older age lies in creating explanatory interdisciplinary models of the rejection of aging from accumulated stereotypes, the transition from industrial interpretations to an institutional description of aging, and the scientific anticipation of new ways of organizing social aspects in older age groups.

The founder of behaviorism, the grand-theory of the 1960s, the organizer of the school of experimental psychology, B. Skinner, and his colleague offered practical recommendations for people who should feel free and happy in older age (Skinner and Vaughan 1997). B. Skinner offers a theoretical and analytical understanding of older adults, conceptualizing the position of an open and critical view of the studied phenomenon. B. F. Skinner and M. E. Vaughan consider older adults active agents, responsible for their life and as an object of research analysis, which simultaneously possesses all features and peculiarities of a living creature. Techniques for modifying behavior, changing the social environment, and improving society are possible through the development and implementation of national programs, the mobilization of internal human resources, and changes (possibly through social programming) in their daily practices of dealing with the environment. This direction of humanitarian reinterpretation of human life and its final period is quite significant but stigmatized in Western culture.

The theoretical dichotomy of exclusion and inclusion of old age is central to the conceptualization of aging. The book *From Exclusion to Inclusion in Old Age: A Global Challenge* (Scharf and Keating 2012) consists of ten papers about the general factors limiting the social activity of older adults and the peculiarities of forming an inclusive environment for them. Many factors affect the image and role of the older adults, such as (1) economic and demographic transformations, (2) globalization of social exchanges, (3) migration, (4) changes in the family institution, (5) the organization of a new mobile social field, and (6) actualization of human rights as a new agenda for developed countries. The 1960s was a period of exclusive-inclusive controversy in science. The 1980s was a period of economic instability focusing on exceptional practices. Since the 2000s, according to N. Scharf and N. K. Keating, the emphasis has been placed on creating national programs on aging, the political actualization of the inclusion of older adults in the economic, social, and cultural sectors. Although social exclusion is more likely to be determined by the presence of economic crises and shocks, it cannot be reduced to issues of pension and material security for older adults.

Researcher Ashkara Zaidi examined the exclusion of older adults from the material exchange. Jim Ogg and Silvi Renault analyzed the exclusion of older adults from family relationships, the restriction and minimization of communication with adult children, and the lack of attention from relatives. David Phillips and Cavin Cheng focused on transforming the value system and philosophical positions about older age in modern society. Astrid Stackelberger, Dominique Abrams, and Philippe Chastoney viewed age discrimination as a source of exclusive practices, focusing on policy decisions about protecting human rights or older adults (Older adults... 1991).

The reinterpretation of inclusive and exclusive strategies is conceptualized in multiple solutions, with the older adult as the central figure. Liberation from the dictates of political will, economic determinism, and medical stigmatization is a tirade of the liberal approach, which has become a common trend in developing the social policy of the leading countries for the next decade.

Therefore, despite the similarity with widely used and well-developed deprivation categories, the concept of social exclusion is actively used by Western researchers, is more relevant for studying the aging of the population. First, this field is broader, since it considers both absolute and relative manifestations of deprivation but not covered (or partially covered) by the latest cultural, ethical, economic, and other aspects of life. First, in the process of analysis, social exclusion links the concepts of *poverty* and *employment*. Second, the practical aspect of the concept of social exclusion is of particular importance; it pays the most significant attention to the problems of social justice (the research discusses the specific content and scope of social inequalities between different groups) and the formation of mechanisms for detecting violations of social norms, their restoration, and protection.

The analysis of social exclusion allowed us to identify a particular group of victims of exclusion or, in other words, *exclusants*, who are both in a situation and a state of exclusion. In our view, the latter are people in tension (tense *exclusants*) or apathy (insensitive *exclusants*). To characterize the *exclusants* (poor, unemployed, other socially vulnerable, unprotected, uncompetitive society members, whose wealth is violated or deprived), we can distinguish three groups according to their composition: (1) constant *exclusants* (always in a situation of exclusion during the period under review), (2) new, and (3) former *exclusants*. This classification allowed us to reveal the factors and mechanisms of forming and overcoming social exclusion.

Therefore, considering the presented research trends, we can distinguish certain features. First, international sociology created a significant theoretical, methodological, and empirical basis for the sociological analysis of institutions, processes, and structures of social exclusion or inclusion. Second, there are many conceptual and methodological, practically unresolved issues in the sociological analysis of institutions, processes, and social exclusion structures or inclusion. All these factors require theoretical and methodological rethinking. There are no complete and comprehensive works on the theoretical and methodological problems of social exclusion of older adults, such as the system of categories, technology, typology, drug treatment, or the dynamics of the formation of integrative behavioral models.

## 4 Discussion

Modern researchers note that social exclusion and the maintenance of social security are especially urgent in the border territories located *at the intersection* of geographical, political, economic, ethno-cultural, and confessional borders. The role of the border becomes particularly relevant in the case of the borders of regions that are adjacent or coincide with the borders of other countries (Nazurkina 2011).

From a sociological point of view, the border is considered not a phenomenon of physical or political space but a characteristic that determines the specifics of the area.

Traditionally, the research on the phenomenon of borders is carried out by representatives of geographical science, which studies it to implement the barrier-contact function. The different ratio of these functions forms the following types of border zones:

- Isolated (alienated) border areas—areas between which the national border is impervious to formal or informal contacts;
- Locally interacting border areas—borders with a prevailing barrier function, where cross-border contacts exist, but have mainly social and personal character;
- Actively interacting border areas—areas with a combination of barrier-contact functions and active cross-border cooperation in all sectors (economic, social, cultural, etc.). In such regions, the barrier function filters border flows and protects the internal territory from various negative impacts;
- Interdependent (cooperative) border territories—border territories with a predominance of contact function, where the regions interact in the economic, social, cultural, and other sectors and create close integration and cooperative network structures;
- Open (barrier-free) border areas—border zones with a nominal border without a barrier function, so the border itself is also absent (Osmolovskaya 2016).

Currently, borders are also the research object of political science, economics, sociology, and limology, which studies the functions of borders and border areas.

Limology provides a theoretical understanding of the phenomenon of boundaries. V. A. Kolosova, analyzing the existing theoretical directions in studying political borders, identifies traditional and postmodern approaches. The traditional approaches include (1) historical-cartographic, (2) classificational, (3) functional, and (4) geographical-political. Besides these trends, the phenomenon of borders can be explained by political factors that view them as reflections of the military, economic, and other powers of neighboring countries (Kolosov et al. 2019).

Postmodern theories of the study of borders emphasize their critical role in ensuring the social security of border areas. Therefore, borders (political, administrative, and cultural) represent a special social system affecting all life aspects of society, including the social exclusion of the population.

Certain features of these territories determine the specifics of studying social exclusion and adaptive strategies in border areas.

Cross-border cooperation and economic ties with other countries promote qualitatively new types of socio-economic systems (border areas). According to V. A. Kolosov, the recognition of demographic and social processes occurring in such territories complicates the ethnic structure and identity of residents of the border area, ultimately determining the formation of a particular adaptive behavioral strategy. In addition, these regions become particularly attractive for migration. Simultaneously, migration processes can have a significant impact on the socio-economic security of border areas. According to S. G. Maximova and her co-authors, the residents of border areas associate migration primarily with population growth, the replenishment of the labor market, as well as social tension, a decrease in wages, shortage of experts, and support for low-skilled labor. Hence, a certain opposition of social groups for material goods may give rise to differences in access to economic resources (Maximova et al. 2018). Moreover, empirical studies show that the leading causes of conflicts often lie in the social, economic, and political sectors. Although confessional and cultural factors are important, they play a supporting role.

## 5 Conclusion

During radical economic transformation, the problems of modern Russian society receive much attention from the humanities and social sciences. The steady process of depopulation in Russia is an urgent issue of socio-economic development. The phenomenon of aging of the population in Russia is manifested in (1) a decrease in the birth rate, (2) an increase in the death rate, (3) natural loss of the population, (4) a deterioration in the ratio of marriages and divorces, (5) low life expectancy, (6) migration outflow from rural areas, (7) the *extinction* of rural settlements, and (8) the aging of the population. The low standard of living, low income and wages, poverty, low quality of life, lack of jobs, unemployment, and significant income differentiation of the population exacerbate many social problems and undermine the national security of Russia.

Modern society, which is considered civilized, tries to solve the problems caused by demographic and social changes, taking into account its assessments of the social, economic, and personal potential of social actors. Such problems include the social exclusion of older adults and play a critical role in the aging of humanity. For example, in Russia, on January 1, 2018, the share of the population over the working age was 37.3 million people, or 25.4% of the total population (Federal State Statistic Service 2020). The question of excluding an individual from the active social life according to the age criterion is usually associated with the limitations of their social space and activities. Such restrictions harm both the society, which loses its social and professional experience carriers, and the individuals, who rejected the opportunity to fulfill their potential.

Currently, the social exclusion of older adults represents a vast field for sociological and psychological research. The study of this process in Russian society began only in the 21st century. Before talking about the social exclusion of older adults, we should note several features of this process. As of today, it is necessary to highlight the situation and the state of social exclusion. The situation focuses on objective circumstances resulting in depriving the individual of social life and violating their rights (social, political, and civil). In this case, talking about the causes of social isolation, we focus on external factors.

We should understand what social policy exists in the regions of modern Russia to counter the exclusion or inclusion of older adults in the social sector. Besides, it is necessary to realize what measures are being taken in the Russian regions for the older adults to carry out their work and social activities, preserve their health and feel like full-fledged society members. The exclusion condition is determined by self-identification and individual perception of the situation. In this case, we rely on individual assessments of the population regarding this problem. Considering the causes and characteristics of social exclusion, we should also take into account internal (psychological) factors. Understanding what prevents older adults from recognizing their value and feeling appreciated by society contributes to developing measures to improve the life of this age group. Therefore, the boundaries of this process cannot coincide. Objectively, people may not be in a real social exclusion, while in moral and psychological terms, they do not feel like full-fledged, active citizens. On the contrary, despite the difficulty of obtaining medical and social assistance; low access to cultural, social, and economic benefits; and limited opportunities in the labor market, a person may not feel deprived, like an outcast, showing physical and social activity.

The scientific significance of the research is connected with the development of theoretical, methodological, and technological issues of preventing and overcoming social exclusion of older adults. This issue is becoming increasingly acute for transitive societies (such as the Russian one), since the desire to comply with the principles of developed countries faces a low level of economic development. As a result, the process of *aging* observed in most Russian regions occurs against the background of declining well-being, multidimensional social deprivation, and pronounced social isolation of all social groups of older adults. The main factor that provokes the formation and maintenance of social exclusion of older adults is the problem of employment. In most cases, society does not see the prospects of using the experience of experts accumulated with age. According to the 2017 statistics (Rosstat 2020), about 23% of retirees continue to work in Russia (up to 10 million people). However, during the economic crisis, retirees become the most vulnerable category of employees, and, hence, they are the first to be fired. Therefore, two opposite trends simultaneously collide: the desire to use the experience of older adults, appreciating such experience, and the exclusion of older adults from social life.

First, the study of social exclusion of older adults is of particular importance due to the recurred decline in the standard of living of the majority of the Russian population among older adults. Second, the growing social, financial, and economic differentiation of Russian regions complicates the activities of social service institutions and social protection of the population, especially in agricultural and agro-industrial sectors. Third, it is necessary to take into account the significant changes in the moral and psychological environment in the activities of older adults, who, despite their extensive services to Russian society and the country, find themselves in a difficult financial situation. Fourth, this situation is aggravated by the destruction of the social infrastructure of the peripheral regions of Russia of an agro-industrial or military-industrial nature. Fifth, we should bear in mind the apparent decline in the management of all social processes in Russian society reduced attention to the management of human resources, including institutions of the social sector. Sixth, we should mark the critical changes in the Russian pension system in 2018 associated with an increase in the retirement age.

These aspects are also determined by changes in the functioning and development of the social sector at the federal and regional levels. They fall under the influence of the settlement and municipal system of social protection of older adults. Additionally, these factors depend on the general context of the socio-economic development of Russian regions, including its municipalities.

Therefore, the manifestation of the social isolation of older adults, which occurs in the developed countries, including Russia, requires a comprehensive study and understanding in terms of sociology. Moreover, social policy should take adequate measures to prevent and neutralize possible negative consequences.

**Acknowledgements** Paper prepared in the frameworks of the task of the Ministry of science and higher education of the Russian Federation FZMW-2020-0001 "Human capital, migration and security: transformation in the new migration environment in Central Asia".

## References

Abrahamson P (2001) Social exclusion and poverty. Soc Sci Contemp World 2:158–166

Bailey N (2017) Employment, poverty and social exclusion. In: Bramley G, Bailey N (eds) Poverty and social exclusion in the UK: volume 2 – the dimensions of disadvantage. Policy Press, Bristol, UK, pp 159–178

Beauvior S (1970) The coming of age. André Deutsch Ltd and George Weidenfeld and Nicolson Ltd., London, UK

Bengtson V, Coleman P, Kirkwood T (2005) The Cambridge handbook of age and ageing (Johnson M (ed)). Cambridge handbooks in psychology. Cambridge University Press, Cambridge, UK. https://doi.org/10.1017/CBO9780511610714

Bonfatti A, Celidoni M, Weber G, Borsch-Supan A (2015) Coping with risks during the great recession. In Borsch-Supan A, Kneip T, Litwin H, Myck M, Weber G (eds) Ageing in Europe – supporting policies for an inclusive society. de Gruyter, Boston, MA

Cardenas S (2016) Human rights in comparative politics. In Goodhart M (ed) Human rights politics and practice. Oxford University Press, Oxford, UK

Castells M (1998) End of millennium, the information age: economy, society and culture, vol III. Blackwell, Cambridge, MA; Oxford, UK

Culliney JL, Jones D (2017) The fractal self: science, philosophy, and the evolution of human cooperation. Eurospan, London, UK

De Haan, A. (2007). Reclaiming social policy. Globalization, social exclusion and new poverty reduction strategies. Palgrave Macmillan, London, UK

Dichev TG, Tarasov KE (eds) (1976) Problem of adaptation and human's health. Methodological and social aspects. Medicine, Moscow, USSR

Duhina T (2005) Social adaptation and sociological discourse. Socialno-Gumanitarnye Znaniya 1:297–306

Fahmy E (2017) Poverty, social exclusion and civic engagement. In Bramley G, Bailey N (eds) Poverty and social exclusion in the UK, vol 2, pp 179–199. Policy Press, Bristol, UK

Federal State Statistic Service (2020) Official statistics. Censuses and surveys. Population and census data. https://eng.gks.ru/folder/11654

Henry WE, Cumming E (1959) Personality development in adulthood and old age. J Proj Tech 23(4):383–390. https://doi.org/10.1215/9780822392255

Hrast FM, Kopac Mrak A, Rakar T (2013) Social exclusion of elderly in Central and Eastern Europe. Int J Soc Econ 40(11):971–989

Kozlov AA (1999) Ageing: Social disjunction or totality? (theories and traditions of the western social gerontology). World Psychol 2:80–96

Maximova SG, Noyanzina OE, Omelchenko DA (2018) Model of social exclusion of the elderly people in Siberian Regions. Uspekhi Gerontologii 30(4):579–586

Millar J (2007) Social exclusion and social policy research: defining exclusion. In: Abrams D, Christian D, Gordon D (eds) Multidisciplinary handbook of social exclusion research. Wiley, Chichester, UK, pp 1–15

Myck M, Najsztub M, Oczkowska M (2015) Measuring social deprivation and social exclusion. In: Borsch-Supan A, Kneip T, Litwin H, Myck M, Weber G (eds) Ageing in Europe – supporting policies for an inclusive society. De Gruyter, Berlin, Germany, pp 67–78

Nazurkina MV (2011) Structural levels of regional identity in contemporary Russia. Regionology 4(77):13–19. http://regionsar.ru/ru/node/809

Neugarten BL, Neugarten DA (1987) The changing meanings of age. Psychol Today 21:4–29

Osmolovskaya LG (2016) Functions of a border as a factor of development of border regions and formation of transborder regions. Baltic State Univ Bull 1:45–54

Paugam S (2009) La disqualification sociale. Presses Universitaires de France, Paris, France. https://doi.org/10.3917/puf.paug.2009.01

Room G, Britton N (2006) The dynamics of social exclusion. Int J Soc Welf 15(4):280–289

Rosen J (2001) Rothwell, Walter Henry. Oxford Music Online

Scharf N, Keating NC (eds) (2012) From exclusion to inclusion in old age: a global challenge. Polity Press, Bristol, UK

Skinner BF, Vaughan ME (1997) Enjoy old age: a practical guide. W.W. Norton & Company, New York, NY

Taylor C (2017) Poverty and poverty discourses. Oxford Scholarship Online, Oxford, UK

Tihonova NE (2003) The phenomenon of social exclusion in Russia. Universe of Russia 12(1):36–84. https://mirros.hse.ru/article/view/5293
Tornston L (1982) Gerontology in a dynamic society. In: Hareven TK, Adams KJ (eds) Ageing and life course transitions. Guilfordpress, New York, NY, pp 183–219
Wacquant L (2009) Punishing the poor. Duke University Press, Durham; London, UK
Walsh K, Scharf T, Keating N (2018) Correction to: social exclusion of older persons: a scoping review and conceptual framework. Eur J Ageing 16(1):129–129

# Civic Moods and Civic Engagement of Russians as a Base for Social Trust and Social Capital

**Svetlana G. Maximova, Oksana Es. Noyanizna, and Daria A. Omelchenko**

**Abstract** Paper bases on results of sociological survey in four Russian regions (2019, n = 1598). One of the goals of presented research was the understanding about perception of a true citizen image and its regional variations. During data analysis, the argued the hypothesis about the stable set of characteristics, describing true (ideal) citizen for Russians. These constructs are in the zone of strong positive interdependencies. A true citizen has positive civic senses, is law-abiding and responsible, feels unity and solidarity with the nation, ready to common solution of common public problems, does not hope for the government, and is ready to compromise about own private rights and freedoms in case of emergency. On the junction of this constructs, an image of citizenship have being forming in the consciousness of the Russians. However, in all regions, the image is rather blurring; in the Altai krai and the Kemerovo oblast, it paradoxically combines with high level of civic responsibility while a great lack of civic feelings. All regions had a constant set of constructs, forming the image of a true citizen—far from policy, with weak civic feelings, socially passive, but trying to correspond formal requirement of the state law system.

**Keywords** Social capital · Social trust · Capital · Civil society · Civic moods · Civic engagement

## 1 Introduction

There are many contexts to consider the "civil society," despite it became a widely used term as in political as in scientific discourse and became central in public relations. One of such contexts is the relation between civil society and the state and

S. G. Maximova (✉) · O. Es. Noyanizna · D. A. Omelchenko
Altai State University, Barnaul, Russia
e-mail: svet-maximova@yandex.ru

O. Es. Noyanizna
e-mail: noe@list.ru

 
S. G. Maximova et al. (eds.), *Advances in Natural, Human-Made, and Coupled Human-Natural Systems Research*, Lecture Notes in Networks and Systems 234,
https://doi.org/10.1007/978-3-030-75483-9_32

governmental institutions, the system of local authorities, commercial institutes, and other subjects of public relations (Maximova et al. 2018a, b). Development of the civil society depends on understanding the identity of citizens in the process, their participation in the formation and statement of civic institutes, self-organization, and realization of the potential of social inclusion into associations and civic unions, civic actions, the realization of different scenario of civic engagement, and civic interests (Galkina 2012).

The given discourse refers to multistage aspects in the functioning of civil society, such as formation and content of civic identity, civic and social identifications of population, peculiarities of different type of trust to public institutions, etc. Thus, citizens or their unions must negotiate with state institutes; and the space of the interaction reflects civil society. Here social trust and capital have become a mechanism of construction of group and under-group networks, formal institutions, and organizations. Trust is a base for intergroup interaction, helping elements of civil society to institutionalize social space and social capital (Gezhavina 2015).

An identity, civic and national, especially, is a general element in considering public solidarity, civic unity in social sciences, and an important and effective instrument of measurement. National identity is a characteristic of civic unity in society, the indicator to evaluate ethnic, religious, and race aspects of community functioning and accumulation of its national interests. Moreover, evident national identity provides democratization and independency of the state (Allan 2016). Questions of civic and national identities of contemporary Russians come to the fore of scientific analysis and political management in Russian multicultural and pluralistic in relation to ethnos and religious society.

Researches in the field of civic and national identity face difficulties with the diversity of terms and notions. Thus, such investigations often lay in the field of patriotism as a sense of national loyalty, love to national symbols, specific faith in the superiority of the own nation as a component of the formation of civic relation to "mature nation" (Hurwitz and Peffley 1999; Spinner-Halev and Theiss-Morse 2003; Sullivan et al. 1992). Here we find proximity with the term "nationalism" as a sense of national superiority and need to dominate at the international arena (according to De Figueiredo and Elkins 2003; Karasawa 2002; Kosterman and Feshbach 1989; Mummendey et al. 2001; Sidanius et al. 1997), but it is exactly not a patriotism (Huddy and Khatib 2007).

The discussions about the difference between civic and ethnic, national and ethnic nations are still actual. Probably, the problem is in the peculiarities of social–historical development of countries, some of those (for example, Germany, Western Europe) developed as ethnic nations initially. Unfortunately, there are a few works about theoretical and empirical testing of the dichotomy. Shulman (2002) analyzed data of 15 countries and approved the difference between the construction of national identity in countries of the West and the East. Thus, the western civic nations are more ethnic than recognized, but eastern ethnic nations are rather civic. Furthermore, according to Shulman's evaluations, countries of Central and Eastern Europe are more civic and less ethnic, than countries of Western Europe.

The given conclusions led to the acuteness in considering aspects of the formation of civic and national identity of population, and evaluation of national identity through the social one. In particular, Tajfel based the application of the social identity concept as a theoretical base to evaluate civic and national identity and their influence on intergroup behavior (Tajfel 1982; Tajfel and Turner 1986). Social identity is usually explained as "understanding and psychological sense of objective group membership" (Tajfel 1982). Accordingly, national identity could be determined as a subjective or internalized sense of belonging to the nation. It could be estimated with instruments applied to social identity (Huddy and Khatib 2007).

The identification as a part of the socialization process had been reflected in works by sociologists who try to be distant from psychology, its subject, and conceptual frameworks. The theory of social identity suggests multiple prognoses about national loyalty. Firstly, national identity is not an ideological construct; it is a deep feeling of subjecting belonging to the nation. Secondly, a stable national identity increases the inclusion into the political sphere. According to Turner's self-categorization theory (Turner et al. 1987), one may predict an individual with stable group identity to determine group norms (Turner et al. 1987; Terry et al. 1999). For example, health-preserving behavior depends on whether it is a norm or not in the referent group. Empirically tested, norms of individual behavior change with the change in group norms, such as the behavior of the identifying individuals; the higher the identification the higher the tendency to correspond ideals and prescriptions (Wellen et al. 1998; Terry et al. 1999).

By profiling the self-categorization theories into identity questions, we may conclude about acts of civic participation as central in national identity (for example, in democracies) and forming normative behavior of a "good" citizen (Svensson 2011). Therefore, a stable national identity will enforce political interest and participation, including electoral activity, and civic identity. Thus, Schatz and colleagues revealed a low level of political participation and interest to the policy among uncritical patriots (Schatz et al. 2003).

However, it is important to reveal characteristics and determine the strategy of estimation for the empirical study about problems of civic and national identity. A general problem here is the description of identity by indexes of the ethnicity. Thus, Kymlicka (1999) stressed the necessity to study ethnic and cultural components of the identity. The given analytical scheme includes three variants of consideration of civic/national identity: nation's beliefs, the content of national identity, and factors of formation of national/civic identity. That is, which are the most important criteria to specify one nation from another; and what is the base for the identification of membership? Content of national/civic identity consists of a number of key components: civic (territory, citizenship, will and homogeneity, political ideology, political institutes, and rights), cultural (religion, language, and tradition), and ethnic (origin and race).

Existing literature on problems of civic nations describes five general components of civic identity, but national unity and nation membership originate from the common territory, citizenship, and faith in common political principles or ideology, respect for political institutes, satisfaction about equality in rights, and commitment to be a part of a nation. Cultural identity bases on non-political cultural characteristics: language, religion, and traditions; ethnic identity—on common origin and race, acting as dominant criteria for national identification. These three levels of national identity differ by the level in inclusiveness; "outsiders" find it difficult to correspond to ethnic criteria because they could not choose the own origin. However, it is possible to adopt cultural requirements to enter these or those nations. Some civic requirements (will and homogeneity, for instance) are easy to correspond, but citizenship depends on not only the choice of a potential citizen of the state but on the will of governmental institutions. Exclusive affiliation to a certain territory could be an additional requirement; it is rather hard for some outsiders to correspond it.

Considering problems of civic and national identity requires for the description of different types of citizenship. In this context, it is necessary to study as about peculiarities of a state policy about attitudes of nation members (Brubaker 1992).

## 2 Materials and Methods

Previously, scientists described aspects of content and formation of civic identity (Maximova et al. 2018a, b), and one of the conclusions was the dependency between civic engagement and relation to the state and actual regional specifics, economic condition, and level of life of population at the territory. Furthermore, relation to the state in conditioned by local history, actual regional policy, specifics of coexisting cultures and confessions, and concrete events in regional life. That is why researches about civic identity and transformation of its content in regional correlations have a great potential.

Representations about citizenship have different contexts of interpretation: need in group belonging, political orientations, the process of awareness about citizenship or civic status (Loginova 2010). During the conducted in four Russian regions research,[1] respondents were asked to describe representations about citizenship and civic engagement of "a true citizen" of a country.

By planning the research tasks, we found a semantic approach corresponding to the research goal. Here "a personality is considered as a carrier of a complicated picture of the world, including representations about internal objects and phenomenon" (Petrenko 1983). The task of the semantic approach is to reconstruct the system of references, helping the subject to percept the surrounding world (including themself and the other people). "Psychosemantic models initially have been applied to describe the internal picture of subject's world, but became widely spread to study

[1] Sociological survey (2019) with use of standardized questionnaire in four Russian regions (n = 1,598, people at the age of 18–75 years): the Altai Krai (n = 400), the Transbaikal Krai (n – 400), the Kemerovo oblast (n = 400), and the Orenburg oblast (n = 398). Here and further, we'll use the acronyms: AK—the Altai Krai, ZK—the Transbaikal Krai, KO—the Kemerovo oblast, OO—the Orenburg oblast.

about pictures of the world of "the generalized other" subject (public consciousness); to study social ideas and related social attitudes and behavioral stereotypes, in particular" (Mitina and Petrenko 2001).

Citizenship is usually determined based on of interaction of three aspects: rational, sensitive-emotional, and practical (Russu 1976). Accordingly, indicators of citizenship are civic activity, civic behavior, civic consciousness, and civic senses, describing civic maturity, in combination. Therefore, a set of descriptors for ideal (true) role of a citizen was formed to describe representations about citizenship. The set consisted of 16 bipolar scales, estimated by six-point continuum and reflecting emotional-evaluative, cognitive, and behavioral characteristics of individuals with positive and negative civic identity, social representations, and attitudes about the image of "an ideal citizen":

- D1 "All citizens constitute a unite nation and should feel solidarity to each other"—"Citizens of our country are fragmented; the solidarity is impossible";
- D2 "Citizen must know own rights and duties and struggle against injustice by any means"—"It is impossible to know all laws, a usual citizen unable to prevent illegal acts";
- D3 "A true citizen is always interested in the history of own country and people"—"A good citizen should not necessary know the history of the country well";
- D4 "Citizens must agree to the political course of the country leaders"—"Citizens are not obliged to agree with the government";
- D5 "Participate in elections—is a citizen's duty"—"Elections never resolved anything, there is no point to take part the elections";
- D6 "In case of emergency, all capable citizens must protect own country"—"Only army must protect the country, they are professionals, usual citizens have nothing to do at war";
- D7 "A true citizen is responsible for own acts and behavior"—"Understanding civic responsibility and sense of duty deprive individual of own liberty";
- D8 "All citizens feel proud while rising the state flag and hearing the state anthem"—"State emblem, anthem, and flag are just formal symbols of the state. It is normal not to be enthusiastic while listening to the state anthem or seeing the state flag";
- D9 "Citizens of the country—sounds proudly"—"Citizenship is a formal status, and no emotions here";
- D10 "Individual needs to fell himself as a part of a certain civic community"—"Individual needs to feel a personality, but not a citizen of the country";
- D11 "Citizen must live and work in the country of citizenship"—"There is nothing bad to leave the country in crisis";
- D12 "Citizen must obey the law"—"In some cases, it is possible to get around the law";
- D13 'Each citizen must be active in public-political events, to support any political party, movement'—'Policy is a dirty business, a good citizen may be out of politics';

- D14 "One should take care of own well-being than to think about the well-being of other people and community, above all"—"Each individual should think not only about themself, but about surrounding people and own country";
- D15 "In social problems, there is no sense to change the situation"—"There is no sense in hoping for the government, the most problems could be decided together by joining forces";
- D16 "My interests should be protected from the state"—"Sometimes the state may interfere in private life of own citizens, for the mutual benefit".

"A true citizen" profile was evaluated by analysis of variability of the means by scales of descriptors. Extreme high or low values indicated the significance of the descriptor for the perception of this or those position and marked positive or negative pole of the scale that let to use the corresponding characteristics in data description. The scales by descriptors tend to the negative pole together with the point; the higher the mean points the more evident are negative characteristics of citizenship.

Data possession and analysis were realized in several stages. At the first stage, we constituted a general profile of a role of "a true citizen" basing on the analysis of the means and dispersions above the mean, described regional variations of general profiles. Tentatively, we inverted several scales (D14–D16), converted them to the avoided routine before the survey. At the second stage basing on one-factor dispersion analysis, we revealed differences in the means of scales by regions; a posteriori post hoc tests used to evaluate the statistical significance of variations between pairs of the means to evaluate regional specifics. At the third stage basing on averaged values, we studied latent factors, conditioning perception of the image of "a true citizen" by representatives of different territorial groups. Firstly, we realized the correlation analysis about the interconnection between descriptors that let to reveal semantic proximity and associations in the base of latent factors, basing the perception of categories of ethnicity and citizenship. Then we used the main components method and the Varimax rotation in factor analysis, the factor weights were kept as separate variables by a regression method. At the fourth stage basing on the revealed factors and projection of elements, we built semantic spaces, described specifics of social attitudes and representations of the population of four regions. Additionally, we visualized semantic spaces for each regional sub-sampling basing on factor weights of descriptors. IBM SPSS 25.0 and MS Excel were used to visualize and analyze data.

## 3 Results

Characteristics of citizenship tend to averaged values without any exaggerated features. The most expressed characteristics were responsibility for own behavior and acts (2.17, displacement to the positive pole), interest to the history of own nation (2.46), and obedience to country laws (2.5, displacement to the positive pole). Civic responsibility and obedience to law together with good knowledge of history combine with the highly expressed sense of safety in private interests from the government influence (4.1) and reluctance to compromise for the government necessity, lack of political bias (3.1—"Policy is a dirty business, a good citizen may be out of politics"), individualism (3.55—"One should take care of own well-being than to think about the well-being of other people and community, above all") and disbelief in social solidarity, and social passiveness (3.53—"In social problems, there is no sense to change the situation").

Such characteristics as a sense of compatriotism, unity with the nation, knowledge about civic rights and duties, electoral behavior, readiness to protect the motherland in case of emergency, emotions about citizenship and the state symbols, commitment to the motherland, and will to live there do not depend on social, economic or other circumstances and vary in the average limit (diapason 2.48–3.19 points). Standard deviation testifies about the absence of significant contradictories in evaluations of citizenship.

A combination of the described characteristics makes the image of "a true citizen" emotionally detached, formally related to the attributes of the state and nation, rationalize relation to the civic status as a balance between readiness to correspond government's conditions in exchange for the freedom to protect private interest and social-economic "comfort." Civility is less associated with civic activism and inclusion into political affairs and support for the state policy.

In general, regional profiles of the "a true citizen" position are similar in certain descriptors and direction of their variations to the positive or negative pole. Table 1 presents the mean values by descriptors in regional correlations, but the match in profiles is evident according to the intensity of color in columns. However, there were some regional peculiarities.

In the Altai Krai, the most expressed components of the image were the lack of political interest, the tendency to protect the private interests from the governmental influence, and high level of civic responsibility, while: "Policy is a dirty business, a good citizen may be out of politics" (3.78 points), "My interests should be protected from the state" (3.77 points), and "A true citizen is responsible for own acts and behavior" (1.98 points).

In the Transbaikal Krai, the commitment to protect personal interest, passiveness in the solution of social problems, and care about own well-being became the dominant features of "a true citizen": "My interests should be protected from the state" (3.99 points), "In social problems, there is no sense to change the situation" (3.61), and "One should take care of own well-being than to think about the well-being of other people and community, above all" (3.5). It should be noted, all the dominants were located and the negative pole of the scale, but the other descriptors "balanced" and the averaged level; that is the population of the Transbaikal region is less engaged as in government affairs as in public life and prefer to preserve personal space and good, such people feel convenient to solve the emerging public tasks through personal interest.

The image of "a true citizen" in the Kemerovo oblast was rather similar with the Altai region, but differs with high level of lack of politic interest, exclusion from the policy in whole and from the policy of the state leaders together with high level of

**Table 1** Descriptive statistics of the image of "a true citizen" in regional correlations, means

| Positive pole | . | | ZK | | KO | | OO | | Negative pole |
|---|---|---|---|---|---|---|---|---|---|
| 1 point | M | m | M | m | M | m | M | m | 6 points |
| All citizens constitute a unite nation and should feel solidarity to each other | 2.49 | 1.381 | 2.66 | 1.524 | 2.64 | 1.527 | 3.15 | 1.503 | Citizens of our country are fragmented, the solidarity is impossible |
| Citizen must know own rights and duties and struggle against injustice by any means | 2.73 | 1.608 | 2.77 | 1.562 | 2.89 | 1.735 | 3.24 | 1.522 | It is impossible to know all laws, an usual citizen unable to prevent illegal acts |
| A true citizen is always interested in the history of own country and people | 2.35 | 1.334 | 2.6 | 1.516 | 2.29 | 1.433 | 2.6 | 1.228 | A good citizen should not necessary know the history of the country well |
| Citizens must agree to political course of the country leaders | 3.56 | 1.569 | 3.4 | 1.605 | 3.91 | 1.533 | 3.46 | 1.528 | Citizens are not obliged to the agree with the government |
| Participate in elections – is a citizen's duty | 2.84 | 1.691 | 3.04 | 1.723 | 3.09 | 1.792 | 3.05 | 1.581 | Elections never resolved anything, there is no point to take part the elections |
| In case of emergency, all capable citizens must protect own country | 2.55 | 1.547 | 2.71 | 1.698 | 3.21 | 1.852 | 3.12 | 1.494 | Only army must protect the country, they are professionals, usual citizens have nothing to do at war |
| A true citizen is responsible for own acts and behavior | 1.98 | 1.117 | 2.24 | 1.385 | 1.93 | 1.201 | 2.53 | 1.2 | Understanding civic responsibility and sense of duty deprive individual from his liberty |
| All citizens feel proud while rising the state flag and hearing the state anthem | 2.28 | 1.35 | 2.56 | 1.469 | 2.43 | 1.396 | 2.65 | 1.373 | State emblem, anthem, and flag are just formal symbols of the state. It is normal not to be enthusiastic while listening to the state anthem or seeing the state flag |
| Citizens of the country – sound proudly | 2.36 | 1.375 | 2.75 | 1.564 | 2.48 | 1.423 | 2.96 | 1.373 | Citizenship is a formal status, and no emotions here |
| Individual needs to fell himself as a part of a certain civic community | 2.83 | 1.461 | 2.83 | 1.582 | 3 | 1.595 | 3.26 | 1.301 | Individual needs to feel a personality, but nor a citizen of the country |
| Citizen must live and work in the country of citizenship | 2.92 | 1.543 | 3.34 | 1.702 | 3.22 | 1.588 | 3.29 | 1.391 | There is nothing bad to leave the country in crisis |
| Citizen must obey the law | 2.34 | 1.371 | 2.61 | 1.484 | 2.18 | 1.414 | 2.91 | 1.34 | In some cases, it is possible to get around the la |
| Each citizen must be active in public-political events, to support any political party, movement | 3.78 | 1.461 | 3.25 | 1.501 | 4.1 | 1.52 | 3.67 | 1.504 | Policy is a dirty business, a good citizen may be out of politics |
| One should take care of own well-being than to think about well-being of other people and community, above all | 3.34 | 1.471 | 3.5 | 1.64 | 3.5 | 1.362 | 3.84 | 1.18 | Each individual should think not only about himself, but about surrounding people and own country |
| In social problems, there are no any sense to change situation | 3.24 | 1.436 | 3.61 | 1.51 | 3.55 | 1.269 | 3.71 | 1.26 | There is no sense in hoping for the government, the most problems could be decided together by joining forces |
| My interests should be protected from the state | 3.77 | 1.414 | 3.99 | 1.561 | 3.71 | 1.39 | 4.57 | 1.267 | Sometimes the state may interfere in private life of own citizens, for the mutual benefit |

*Source* Compiled by the authors

responsibility for own civic behavior: "Citizens must agree to political course of the country leaders" (3.91 points), "Policy is a dirty business, a good citizen may be out of politics" (4.1), "My interests should be protected from the state" (3.7), and "A true citizen is responsible for own acts and behavior" (1.93).

Representations of respondents in the Orenburg Oblast, in their turn, were similar with evaluations, obtained in the Transbaikal Krai; they balance in the negative space of the scales: evidently express willing to protect private life from the government together with individualization of interests and exclusion from active social actions: "One should take care of own well-being than to think about the well-being of other people and community, above all" (3.84), "In social problems, there is no sense to change the situation" (3.71), and "My interests should be protected from the state" (4.57) (Table 1).

Taking into consideration the displacement of evaluation of the center of the scale, the image of "a true citizen" in all regions was rather vague, and in the Altai Krai and the Kemerovo Oblast, while excluded from civic feelings, it paradoxically combines with high level of civic responsibility. Meanwhile, all regions have a constant set of constructs, forming the image of "a true citizen"—far from the policy, with weak civic feelings, socially passive, but trying to follow formal requirements of the state law system.

At the next stage, we realized a one-factor dispersion analysis; all descriptors demonstrated good discriminative sensitivity that let to follow specifics of the estimated position of "a true citizen" (all descriptors had statistically significance by the Fischer F-criteria). The results of correlation analysis demonstrate significant and the strongest positive and negative interconnections between indexes of citizenship, solidarity with nation, and estimations about knowledge of civic rights and duties ($r = 0.967$, $p \leq 0.05$), care about surroundings and the country ($r = 0.995$, $p \leq 0.01$), and tolerance to the governmental influence on private life ($r = 0.951$, $p \leq 0.05$).

The manifestation of citizenship through good knowledge and respect for civil rights and duties has a strong positive correlation with the necessity of unite and solidary nation ($r = 0.967$, $p \leq 0.05$), willing to be and have a sense of belonging to a unite civic community ($r = 0.989$, $p \leq 0.05$), and confidence in the necessity to think about society but not only about self ($r = 0.963$, $p \leq 0.01$).

Civic responsibility for behavior and other acts has a straight correlation with the evaluation of pride about citizenship ($r = 0.96$, $p \leq 0.05$), law-abiding ($r = 0.989$, $p \leq 0.05$), and readiness to agree with suppressing of private rights for the state interests ($r = 0.979$, $p \leq 0.05$).

Evaluations of the pride about the state symbols, their emotional perception during the estimation of the image of "a true citizen" strongly correlated with evaluations of the importance of civil status ($r = 0.977$, $p \leq 0.05$) and confidence in the opportunity of solidary the solution of social problems ($r = 0.958$, $p \leq 0.05$).

Evaluations of the pride about the citizenship of the country have strong interdependence with civic responsibility ($r = 0.96$, $p \leq 0.05$) and emotional perception of state symbols and civic feelings ($r = 0.977$, $p \leq 0.01$).

Commitment to the motherland, willing to live on the territory, and contribute to its development has a true correlation with evaluations about the opportunity of the solidary decision of social problems (r = 0.962, p ≤ 0.05), but the law-abiding—with civic responsibility (r = 0.989, p ≤ 0.05) and tolerance to the governmental influence of private life for common well-being (r = 0.960, p ≤ 0.05).

Thus, we found one more proof for the existence of a stable set of characteristics, describing true (ideal) citizen and true citizenship for the Russian. Furthermore, these constructs are in the pole of strong positive interdependencies. A true citizen experience positive civic feeling, a law-abiding and responsible, feels unity and solidarity with the nation, is ready to engage in common solution of public problems; they do not hope for a joint decision of common social problems and do not hope for government; they are ready to compromise by rights and private interests for necessity. Just the combination of these constructs forms an image of citizenship for contemporary Russians.

Results of analysis demonstrated the significant interdependencies that mean the existence of latent variables, determining the perception of the image of the ideal (true) citizen of Russia and acting as semantic determinants of citizenship. After factor analysis at a general sampling of aggregate data, we specified three significant factors, describing 66.6, 22.6, and 10.8% of dispersion of included into the model variables (100%, in total) (Table 2).

**Table 2** Results of factor analysis

Rotation component matrix[a]

| | **Component** | | |
|---|---|---|---|
| | **1** | **2** | **3** |
| **D1 Unite nation** | **0.913** | 0.214 | 0.346 |
| **D2 Knowledge about rights** | **0.962** | -0.025 | 0.271 |
| **D3 Interest to history** | 0.382 | **0.859** | 0.341 |
| **D4 Solidarity with policy** | -0.181 | **-0.973** | 0.145 |
| **D5 Participation in elections** | 0.212 | -0.126 | **0.969** |
| **D6 Protection of the country** | 0.527 | -0.561 | **0.638** |
| **D7 Civic responsibility** | **0.743** | 0.615 | 0.264 |
| **D8 Proud by national symbols** | 0.566 | 0.448 | **0.692** |
| **D9 Proud by citizenship** | **0.663** | 0.537 | 0.522 |
| **D10 Civic community** | **0.947** | -0.171 | 0.271 |
| **D11 Residence** | 0.227 | 0.277 | **0.934** |
| **D12 Law-abiding** | 0.698 | **0.699** | 0.154 |
| **D13 Civic activism** | 0.148 | **-0.977** | -0.156 |
| **D14 Collectivism** | **0.884** | 0.179 | 0.432 |
| **D15 Solidarity** | 0.482 | 0.218 | **0.849** |
| **D16 Violation of private interests** | **0.863** | 0.472 | 0.178 |

*Source* Compiled by the authors

The first factor ($\lambda = 10.6$) includes variables with high positive loadings in the range of 0.66–0.96, describing civic feelings, civil rights and duties: unity and solidarity with nation, knowledge and abiding of laws, and emotional relation to civic status, sense of social solidarity, willing to protect collective interests, and relation to compromise by rights and private interests for state necessity.

The sense of the second factor ($\lambda = 3.6$) formed by two groups of variables: with high positive loadings (0.7–0.86), describing knowledge and interest to the history of the nation and law-abiding, and with high negative loadings (−0.97 to −0.98), describing civic activism and solidarity with the governmental policy.

The third factor ($\lambda = 1.7$) with loadings in the range of 0.6–0.97 included variables mostly of activity, the behavioral manifestation of citizenship: electoral activity, readiness to protect the motherland, pride about national attributes, country residence, and social solidarity.

Figure 1 presents a semantic space, built with coordinates on factor values for four regions by two first factors. Geometry configuration (location) of regions in the space let to evaluate the similarity in the perception of the image of a true citizen in sense of notions and meanings of the main factors—the factor of "civic feelings and cognitive component of citizenship" and factor of "civic behavior."

Thus, the Orenburg Oblast is in the zone of positive values (scales by descriptors with the increasing of the point strived to the negative pole of characteristics) by both factors (civic feelings, rights, and civic behavior). People are maximally detached from the government and tend to individual values and private aspects of behavior in society.

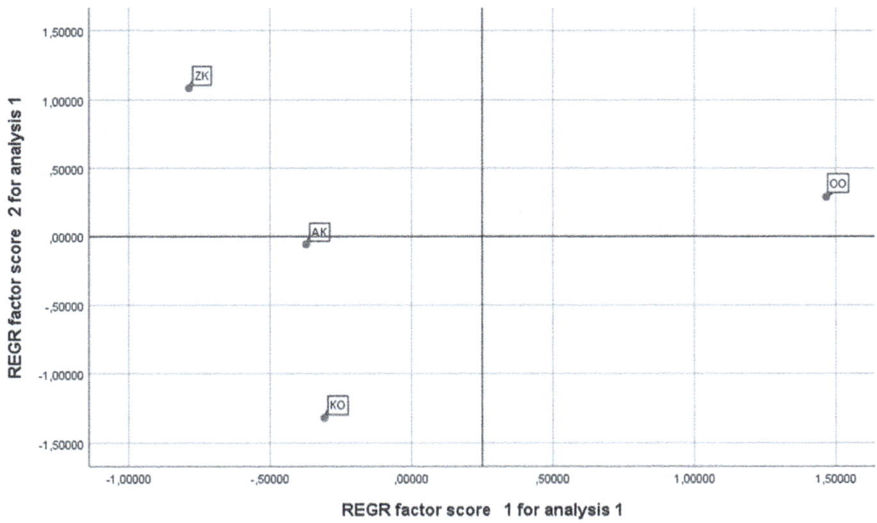


**Fig. 1** Semantic space for four regions, based on revealed factors of citizenship. *Source* Compiled by the authors

The Kemerovo Oblast and the Altai Krai are situating in one segment of semantic space, but the Altai Krai balance on the axis of the factor of civic feelings, whereas respondents in the Kemerovo Oblast reveal as high civic feelings as civic activism, but in the Altai Krai, respondents having expressed civic feelings, are marked by lack civic activity. In contrast, in the Transbaikal Krai population is considerably passive.

Further, let us consider the geometric configuration of descriptors in the semantic space for the population of each region of the survey. Four factors, forming the sense of perception of the civic images, were revealed for each region basing on results of factor analysis on aggregated regional sub-samplings.

For the Altai Krai, the four factors described 26.7, 11.0, 8.0, and 7.8% of dispersion, in total 55.3% of dispersion. The first factor ($\lambda = 4.3$) entered (with positive loadings at the range from 0.43 to 0.77) civic feelings and behavior (protection of the country by citizens, pride about national symbols and citizenship, sense of civic community, residence, and law-abiding). The second factor ($\lambda = 1.8$) (0.5–0.7) included cognitive and behavioral characteristics of citizenship, reflecting loyalty to the state system (unity with the nation, knowledge about rights and duties, solidarity with politics, electoral activity, and civic activism). The sense of the third factor ($\lambda = 1.5$) was determined by two variables—interest to the history and citizenship and civic responsibility; the fourth ($\lambda = 1.3$) —three variables, describing relation to the participation of citizens in the state and public affairs, relation to collective means of problem solution, solidarity and readiness to compromise by private interests.

For the Transbaikal Krai, the four factors described 29.7%, 11.3%, 9.4%, and 7.8% of dispersion, 56.6% of dispersion in total. The first factor ($\lambda = 4.7$), look like the Altai Krai, included civic feelings and behavior (protection of the country, pride about national symbols, residence, and law-abiding) with positive loadings (range from 0.45 to 0.8), and variable of civic responsibility added here in this region. The second factor ($\lambda = 1.8$) (0.5–0.75) was formed by cognitive characteristics of citizenship and electoral activity (unity with the nation, knowledge of rights and duties, interest to the history, and participation in elections). The sense of the third factor ($\lambda = 1.3$) is determined by two variables—solidarity with the government's policy and civic activism, the fourth ($\lambda = 1.2$)—three variables, describing relation to the citizen's participation in the state and public affairs, relation to collective means of solution general problems, solidarity and readiness to compromise with private interests.

For the Kemerovo Oblast, the four factors described 33.0%, 10.7%, 8.1%, and 7.0% of dispersion, a total of 58.9% of dispersion. The sense of the first factor ($\lambda = 5.3$) was also formed by variables, reflecting civic behavior and feelings: protection of the country in case of an emergency, civic responsibility, pride about national symbols, citizenship, sense of belonging to the nation, residence, and law-abiding. The second factor ($\lambda = 1.7$) included cognitive characteristics of citizenship with loadings in the range from 0.5 to 0.7, similar to the Transbaikal Krai (unity with the nation, knowledge about rights and duties, interest to the history of the nation). It should be noted, the sense of the given factor is not formed by the electoral activity, unlike the second region. The third factor ($\lambda = 1.3$) (0.5–0.8)—loyalty to the government and civic activism. The fourth factor also had a similar sense loading (λ = 1.1, 0.65–0.73)—three variables, describing relation to the participation of the citizen in the state and community affairs, relation to collective means of problem solution, solidarity, and ready to compromise by private rights.

Quite another sense lays in the base of representations about the citizenship of respondents of the Orenburg Oblast. The four factors described a considerable part of dispersion—66.4%: 38.6, 12.2, 8.9, and 6.37% correspondingly. Thus, the first factor (λ = 6.2) included variables, dominantly describing civic emotions of a true citizen (perception of unity and solidarity of a nation, knowledge about rights and duties, interest to the history of own nation, protection of the country, and pride about national symbols and citizenship). The second factor (λ = 1.9) was formed by variables, related to behavioral characteristics of citizenship and commitment to the territory and policy of the state with loadings in the range from 0.5 to 0.7. The third factor (λ = 1.4) was determined by three descriptors—electoral activity (0.7), civic responsibility (0.7), and law-abiding (0.8), and the fourth one (λ = 1.1)—by the sense of collectivism (0.8) and social solidarity (0.7).

In the Altai Krai, civic feelings and cognitive-behavioral characteristics of citizenship are the determining factors; in the Transbaikal Krai—civic feelings, cognitive component, and electoral behavior; in the Kemerovo Oblast—civic behavior and feelings, first, cognitive components, then; in the Orenburg Oblast—civic feelings, civic behavior, and commitment to the state and its territory. Further Figs. 2, 3, 4 and 5 demonstrate the disposition of descriptors of citizenship in the semantic spaces by axes of two main factors for the each of regions.

For the Altai Krai, almost all characteristics of citizenship were located in the zone of positive values of factors; characteristics of the solidarity in solution of the

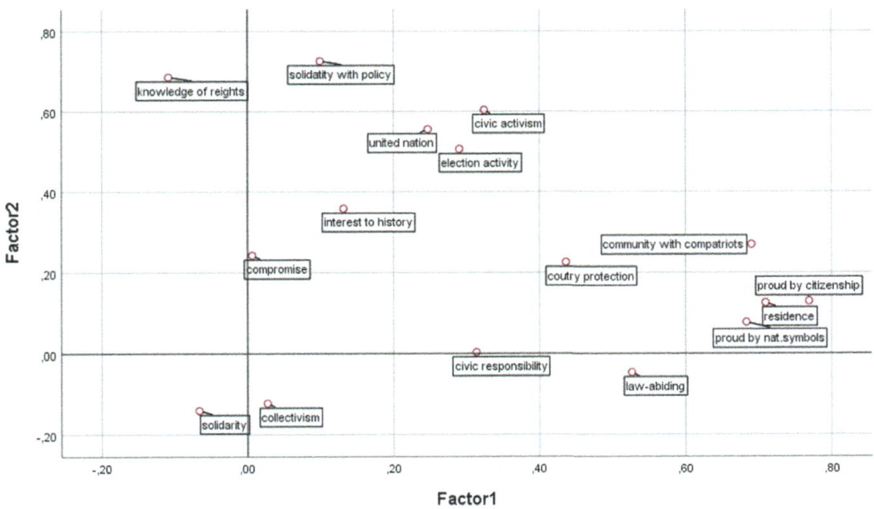


**Fig. 2** Semantic space of descriptors for factors of citizenship in the Altai Krai. *Source* Compiled by the authors

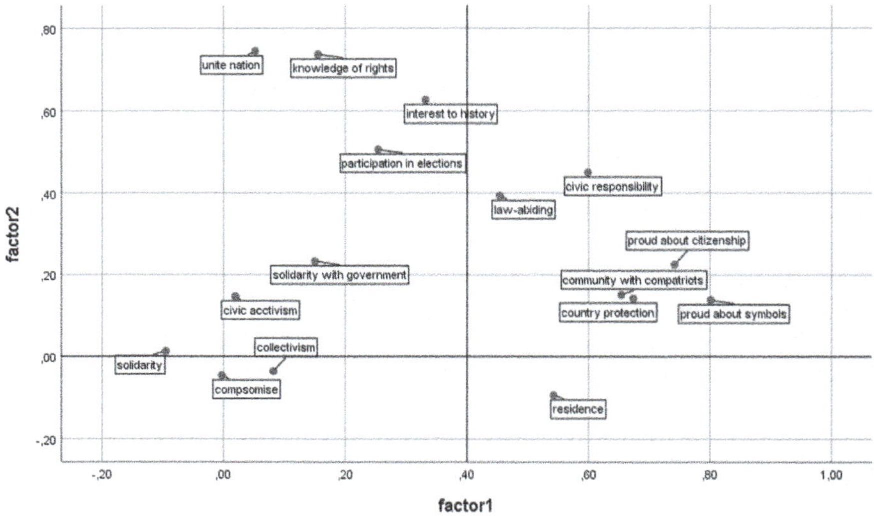


**Fig. 3** Semantic space of descriptors for factors of citizenship in the Transbaikal Krai. *Source* Compiled by the authors

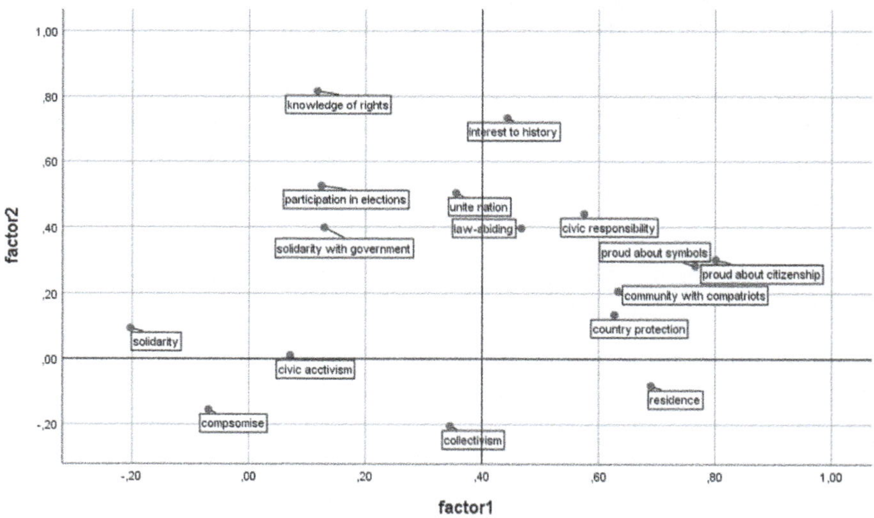


**Fig. 4** Semantic space of descriptors for factors of citizenship in the Kemerovo oblast. *Source* Compiled by the authors

social problems (D15 "In social problems, there is no sense to change the situation"—"There is no sense in hoping for the government, the most problems could be decided together by joining forces") placed in the zone of negative values by both factors; variable, describing knowledge about civil rights and duties, was located in the zone

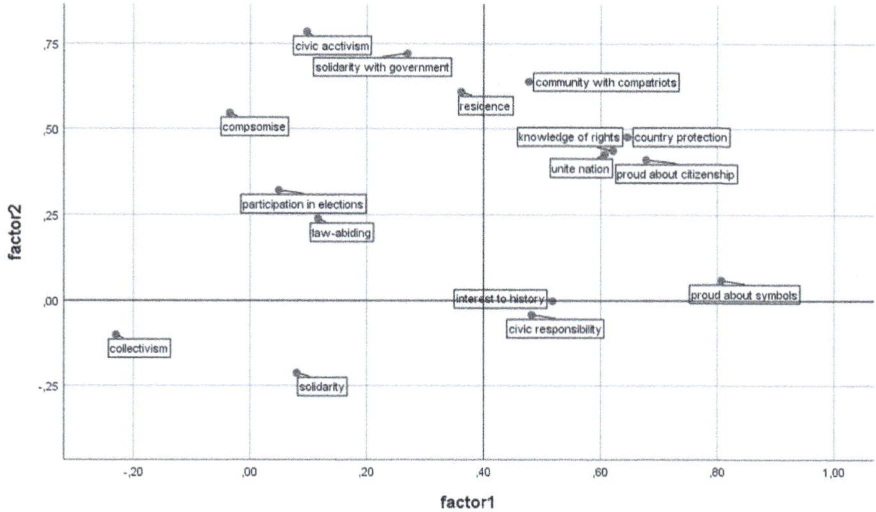


**Fig. 5** Semantic space of descriptors for factors of citizenship in the Orenburg oblast. *Source* Compiled by the authors

of negative values by the first factor, collectivism, care about social well-being—by the second factor.

In the Transbaikal Krai, there was a similar configuration, none of the factors located in the zone of negative values by both factors; however, characteristics of solidarity in the solution of public problems set up in the negative zone by the first factor, but the commitment to the motherland, care about social problems and readiness to compromise in case of emergency for social well-being—by the second factor.

Semantic space for the Kemerovo Oblast had a configuration, similar to the Transbaikal Krai, in whole. However, characteristics, related to the readiness to compromise for state necessity shifted to the zone of negative values by both factors (D16 “My interests should be protected from the state”—“Sometimes the state may interfere in the private life of the citizens, for the mutual benefit”).

Population of the Orenburg Oblast, as it was mentioned before, quite differs by social moods and evaluations in contrast to the population of other regions. In this case, characteristics of readiness to agree with the priority or low prioritization of private interest shifted to the zone of high positive values by the second factor, while collectivism, readiness for collective actions—to the zone of negative values by both factors. We shall notice other position in the space of descriptor, explaining the sense of civic responsibility—it is shifted to the zone of negative values by the second factor (commitment to the institutes of the state and governmental policy). Also, such characteristics as civic responsibility and law-abiding, which were semantically close for the first three regions, separated in the space of coordinates for the Orenburg

Oblast. These may testify about social moods than the population considers abiding by the law because of formal the unwillingness to have a conflict with state institutes but not because of the agreement with the existing rules.

## 4 Discussion

By trying to interpret the obtained data, we should notice that we just revealed different tendencies on the civic moods in four Russian regions. We only fixed the tendencies, requiring for the deeper analysis of reasons of polarization of public opinion in different regions with taking into consideration specifics of border regions, social-economic features, specifics of functioning of civil society, and its state support. It is another factor for correlation analysis of differences in policy of regional governments on how to promote the development of civil society, form patriotic senses, etc.

## 5 Conclusion

The image of a true citizen in representations of contemporary Russian is rather blurred; it has not any expressed characteristics to dominate and determine citizenship. The obtained evaluations varied about averaged values. Responsibility for own behavior and acts, interest to the history of the nation and law-abiding, obligation to follow all governmental rules are the most expressed features of a true citizen. In the image of a true citizen, civic responsibility and law-abiding together with good knowledge of national history combine with a rather evident tendency to the sense of safety of private interests, lack of confidence in social solidarity, and social passiveness. Such characteristics as a sense of compatriotism, unity with the nation, knowledge about civic rights and duties, electoral behavior, readiness to protect the country in case of emergency, emotional perception of citizenship and national symbols, commitment to the motherland, and willing to live in its territory despite social, economic or other circumstances vary in the range of averaged values. We should notice the absence of significant contradictories in the evaluation of characteristics of citizenship.

The totality of the described characteristics attaches an emotional sense of detachment, formal relation to the attributes of the state and nation to the image of a true Russian citizen, rationalize relation to the civic status as a balance between readiness to follow governmental requirements in exchange for private life safety and social-economic comfort. Citizenship was less associated with civic activism and inclusion into political affairs and support to the political course of the state.

During the data analysis, they argued the hypothesis about the stable set of characteristics, describing true (ideal) citizen for Russians. Furthermore, these constructs are in the zone of strong positive interdependencies. A true citizen has positive civic senses, is law-abiding and responsible, feels unity and solidarity with the nation, ready to the common solution of common public problems, does not hop for the government, and is ready to compromise about own private rights and freedoms in case of emergency. On the junction of these constructs, an image of citizenship has formed in the consciousness of the Russians.

However, in all regions, the image is rather blurring; in the Altai Krai and the Kemerovo Oblast, it paradoxically combines with a high level of civic responsibility while a great lack of civic feelings. All regions had a constant set of constructs, forming the image of a true citizen—far from policy, with weak civic feelings, socially passive, but trying to correspond formal requirement of the state law system.

Social opinion about the necessity of activism in social and political spheres considerably differs on the observed regions as an indicator of citizenship: population in the Kemerovo Oblast is more radical and detached from the social-political sphere; the second group was formed by similar evaluations respondents of the Orenburg Oblast and the Altai Krai; the population of the Transbaikal Krai was in the third group.

Descriptor D16 "My interests should be protected from the state"—"Sometimes the state may interfere in the private life of own citizens, for the mutual benefit" describes the most expressed feature of citizenship and divides regions into three groups: respondents in the Kemerovo Oblast and the Altai Krai were in the first group as the most tolerant to the compromise about private rights for the state interests; evaluations of people of the Altai Krai and the Transbaikal Krai cross in the second group; the third group include evaluations of people from the Orenburg Oblast—people here are very sensitive to the necessity to protect own private rights to be safe from the governmental influence.

The Kemerovo Oblast and the Altai Krai are in the same segment of semantic space of the factor values, but the Altai Krai balance on the axis of factor of civic feelings, while people of the Kemerovo Oblast determine as high civic feelings as civic activism in the image of a true citizen. So, in the Altai Krai, the expressed civic feelings combined with the lack of political activity. Despite the Transbaikal Krai—the population is considerably more passive here.

Thus, regional variations in perception of civil society are the following: in the Altai Krai civic feelings and cognitive-behavioral characteristics are the determining ones; in the Transbaikal Krai—civic feelings, cognitive component and electoral behavior; in the Kemerovo Oblast—civic behavior, first, and cognitive components; in the Orenburg Oblast—civic feelings, civic behavior, and commitment to the state institutes and the national territory.

**Acknowledgements** Paper prepared in the frameworks of the task of the Ministry of science and higher education of the Russian Federation FZMW-2020-0001 "Human capital, migration and security: transformation in the new migration environment in Central Asia".

## References

Allan BB (2016) Recovering discourses of national identity. In: Hopf T, Allan B (eds) Making identity count. Oxford University Press, Oxford, UK, pp 20–45

Brubaker R (1992) Citizenship and nationhood in France and Germany. Harvard University Press, Cambridge, MA

De Figueiredo RJP, Elkins J (2003) Are patriots bigots? An inquiry into the vices of in-group pride. Am J Polit Sci 47(1):171–188

Galkina EV (2012) Theoretical concepts of the civil society and Russian political practice. Contemp Res Soc Probl 6:14

Gezhavina TA (2015) Regional civil society: conditions and factors of existing. Probl Territ' Dev 3(77)

Huddy L, Khatib N (2007) American patriotism, national identity, and political involvement. Am J Polit Sci 51(1):63–77

Hurwitz J, Peffley M (1999) International attitudes. In: Robinson JP (ed) Measures of political attitudes. Academic, Wrightsman, San Diego, CA, pp 533–590

Karasawa M (2002) Patriotism, nationalism, and internationalism among Japanese citizens: an etic-emic approach. Polit Psychol 23(4):645–666

Kosterman R, Feshbach S (1989) Toward a measure of patriotic and nationalistic attitudes. Polit Psychol 10(2):257–274

Kymlicka W (1999) Misunderstanding nationalism. In: Beiner R (ed) Theorizing nationalism. State University of New York Press, Albany, NY, pp 131–140

Loginova AA (2010) Formation of civic identity of pupils by means of internet-projects. Dissertation of Candidate of Pedagogical Sciences. Povolzhskaya gosudarstvennaya socialno-gumanitarnaya academia, Samara, Russia

Maximova S, Noyanzina O, Omelchenko D (2018a) Civil society development in Russian borderlands: indexes and predispositions. SHS Web Conf 55:02030

Maximova S, Omelchenko D, Noyanzina O (2018b) The analysis of national identification of Russians through images of meta-ethnic groups: the case of four borderland regions. Pertanika J Soc Sci & Humanit 26(4):2747–2768

Mitina OV, Peternko VF (2001) Russians and Americans: stereotypes of behavior (psychosemantic analysis). Sociol Res 8:22–38

Mummendey A, Klink A, Brown R (2001) Nationalism and patriotism: national identification and out-group rejection. Br J Soc Psychol 40(1):159–172

Petrenko VF (1983) Bases of psychosemantics. Piter, Moscow, Leningrad, USSR

Russu IN (1976) Educating citizenship at pupils at secondary school. Shtinitsa, Kishinev, Moldova

Schatz R, Staub E, Lavine H (2003) On the varieties of national attachment: blind versus constructive patriotism. Polit Psychol 20:151–174

Shulman S (2002) Challenging the civic/ethnic and West/East dichotomies in the study of nationalism. Comp Polit Stud 35(5):554–585

Sidanius J, Feshbach S, Levin S, Pratto F (1997) The interface between ethnic and national attachment. Public Opin Q 61(1):103–133

Spinner-Halev J, Theiss-Morse E (2003) National identity and self-esteem. Perspect Polit 1(3):515–632

Sullivan JL, Fried A, Dietz MG (1992) Patriotism, politics, and the presidential election of 1988. Am J Polit Sci 36(2):200–234

Svensson J (2011) Theorizing citizenships in late modern ICT societies. TripleC 9(2):644–656

Tajfel H (1982) Social identity of intergroup relations. Annu Rev Psychol 33:1–39

Tajfel H, Turner J (1986) The social identity theory of intergroup behavior. In: Worchel S, Austin WG (eds) Psychology of intergroup relations. Nelson-Hall, Chicago, IL, pp 7–24

Terry DJ, Hogg MA, Duck JM (1999) Group membership, social identity and attitudes. In: Abrams D, Hogg MA (eds) Social identity and cognition. Blackwell Publishers, Oxford, pp 280–314

Turner JC, Hogg MA, Oakes PJ, Reicher SD, Wetherell M (1987) Rediscovering the social group. Blackwell, Oxford, UK

Wellen JM, Hogg MA, Terry DJ (1998) Group norms and attitude-behavior consistency: the role of group salience and mood. Group Dyn 2(1):48–56

# Mixed Migration Streams and Different Realities—How Do Migrants Perceive Security of Hosting Communities?

**Svetlana G. Maximova, Oksana E. Noyanizna, and Daria A. Omelchenko**

**Abstract** The purpose of the paper is to consider problems of self-perception of security and safety by transit migrants in Russian border regions. In the research (2018–2019) participated representatives of migrant's community in seven Russian regions: the Altai region (n = 319), the Orenburg oblast (n = 100), the Murmansk oblast (n = 100), the Pskov oblast (n = 100), the Republic of Altai (n = 20), the Republic of Dagestan (n = 100), and the Rostov oblast (n = 94), 843 respondents, total. Concluded, that socially economic conditions of life in hosting community in border society will be the determining factor of general satisfaction of stay in Russia. That is why migrants constantly move, while having been foreigners in new homelands. If the migration had an evident advantage, increased migrant's and his/her family life level, migrant will come for a work again and again, he/she will move across Russia and abroad to search for best job, domestic and housing conditions. In such case, social-psychological difficulties related to movement, breaking of family, relative and friendship networks down, and unfavorable relation with aboriginals relegate to the background, and transit migrant became ready to accept unsafety, social inequality and injustice, and to risk by personal security. It should be noted, that the residence at the territories of ethnic republics of Russia as objectively as subjectively less satisfying for migrants but the residence in socially and economically favorable regions objectively forms higher level of satisfaction about life in whole and its separate aspects.

**Keywords** Migration · Transit migrations · Transit regions · Security · Social safety · Perception

---

S. G. Maximova (✉) · O. E. Noyanizna · D. A. Omelchenko
Altai State University, Barnaul, Russia
e-mail: svet-maximova@yandex.ru

O. E. Noyanizna
e-mail: noe@list.ru

 
S. G. Maximova et al. (eds.), *Advances in Natural, Human-Made, and Coupled Human-Natural Systems Research*, Lecture Notes in Networks and Systems 234,
https://doi.org/10.1007/978-3-030-75483-9_33

## 1 Introduction

Migration contributes to the formation of the human capital of the migrants' countries of origin. For example, there is ample evidence of the contribution of remittances from host countries to education, health, and everyday consumption in the countries of origin (Adams 2005; Nagarajan 2009; Valero-Gil 2008). Migrant children can get a better education and life prospects. In general, migration can change existing social norms, as it can form the intention to continue education (Yang and Choi 2007).

Studies in rural areas of Pakistan suggest a connection between temporary migration and school attendance, especially among girls (Mansuri 2006). In addition to the direct impact of well-being, migration increases the level of health knowledge, decreases the infant mortality rate, and increases the number of healthy infants in general (e.g., in Mexico) (Papademetriou et al. 2009). When migrants return from temporary migration, they bring more effective health care methods, such as clean drinking water and compliance with sanitary regulations (D'Emilio et al., 2007). On the other hand, migration is a challenge and a threat for migrants, because some labor activities are associated with occupational health risks (e.g., occupational injuries) (Kahn et al. 2003). Increasing mobility facilitates the spread of such diseases as HIV (Brummer 2002; Decosas et al. 1995; Lurie 2000; Lurie et al. 2003). For example, Kane and co-authors claimed that 27% of migrant men were HIV-positive, while the proportion of non-migrant men who were HIV-positive was about 1%. Sexually transmitted diseases are also widespread among migrants and their permanent partners in the country of origin (Kahn et al. 2003).

At its best, migration can be a valuable experience, driven by an interest in improving the well-being of the family. However, moving to another country often involves high emotional costs (D'Emilio et al. 2007). Temporary cyclical migration often leads to family breakdown, fragmentation of social networks, and psychological stress (Kahn et al. 2003). The emotional impact concerns both migrants and their families. In this regard, low-income families are especially vulnerable, because they cannot allow joint emigration, and the departure of one person often disrupts family structures and relationships. According to D'Emilio et al. (2007), the more significant the gap between parents and children, the lower the authority and governing role of parents in the family as common *providers* of love and material care. Other family members begin to displace the parents, or the children themselves take on parental functions. Feelings of ostracism, isolation, and loss haunt children without parental care and cannot be compensated for by monetary donations from abroad. Of course, modern technologies, such as e-mail and phones, allow multinational families to exist and, to a certain extent, maintain family networks (UNDP 2009).

While migration is usually a voluntary and planned individual choice, the reality often falls short of the initial expectations. Often, migrants' expectations are not met, and migrants are exploited (especially young women) (Kebede 2001). The abuse of migrants by intermediaries or recruitment agencies is an urgent problem, increasing in proportion to the scale of migration. Migration is becoming more and more commercialized. Moreover, many migrants work in conditions close to slavery, and the international labor market turns into a trade in labor contracts and abuse. This situation leads to the need for further regulation of the activities of recruitment agencies during the entire migration process (Agunias 2010).

When migrants enter the country, they need to be integrated into the labor market, which increases competition for existing positions. Such attitudes are reinforced during the economic crisis. There is a growing awareness among the population and politicians of the potential negative impact of immigration on the opportunities of the natives; emigrants are becoming *scapegoats*, blamed for the rising unemployment (Papademetriou et al. 2009).

In the international community, migration issues have become crucial for ensuring the socio-political stability, economic, and demographic development of the migrants' countries of origin and destination. In recent decades, vast flows of migrants to Russia have become an integral part of the socio-political and economic environment and social discourse. Almost every CIS member state has slightly increased the donation of its citizens to Russia over the past ten years. Only in the first 9 months of 2019, 11.6 million migrants entered Russia (in January–November 2018—11.1 million). Discussions on migration address various issues, including trends in labor migration in Russia and how the national economic and demographic development economically determines them.

Migration is one of the issues that Russian society and its labor market are facing. At the same time, international migration is partly determined by the self-perception and perceptions of migrants. These ideas go beyond the widespread cliche; therefore, there is no clarity about the attitude toward migration, the influence of myths about migration on its dynamics (Maximova et al. 2018, 2019).

Transit migrants often follow traditional and sometimes "ancient" roots within the established migration patterns, for example, in the Russian space. Transit migration can duplicate emigration from a transit country, as we see in Ukraine, where migrants from non-EU countries, together with natives, move as labor migrants through Poland to Spain or the UK. Sometimes the migration flow may intersect with cyclical migrations from Portugal or pass through specific (ethnic) networks. Some researchers (Duvel, for example) claimed the existence of interconnected transnational ethnic communities in different countries that provide safe migration corridors. Agents are involved in networks; they provide short complex root sections and determine the future root and the duration of stops in countries at the root for migrants. After all, migrants are often "stuck" in non-European countries due to lack of money, inability to cross the border, or dependence on agents. The various aspects of such movements (departure from the country of origin, entry, and temporary residence in the country of transit, entry into the country of destination) can be either permanent or discrete. Usually, if there are problems in the destination countries (unbearable political or economic conditions) that drive them out of the country, migrants try to succeed in other countries.

## 2 Materials and Methods

As part of our research, we surveyed migrants, that is, foreign citizens temporarily residing in Russia for work, education, or other reasons. The study (2018–2019) involved representatives of migrant communities in seven Russian regions: Altai Krai (n = 319), the Orenburg Region (n = 100), the Murmansk Region (n = 100), the Pskov Region (n = 100), the Altai Republic (n = 20), the Republic of Dagestan (n = 100) and the Rostov Region (n = 94), the total number was 843 respondents. In some regions (Rostov, Pskov, Murmansk, Orenburg regions and the Republic of Dagestan), all migrants lived in cities; in Altai Krai, 13.2% of migrants lived in rural areas, in the Altai Republic—45%.

It was not easy to determine whether a particular migrant was in transit. Sometimes the intentions of further migration remain unknown not only for the researcher but also for the migrant; in some cases, the intentions are situational, or there is simply a desire to move on.

Based on the responses, we formed a sub-sample of migrants in a transit situation for three reasons. First, the migrants indicated their place of origin (another country or another Russian region). The first reason for including a migrant in the sub-sample was the entry from another region, not from the country of origin; at the time of the survey, the migrant was, at least, at the second stage of transit (hereafter—*actual transit migrants*). Second, we asked the migrants to indicate the duration of stay in the given region. Third, we asked the migrants who do not want to permanently live in the region of residence to describe the desired migration trajectories. Of course, most migrants would like to return to their country of origin, but the other part (minority) prefers (no later than in three years) to move to another Russian region or another country in the future.

Based on the variables describing the implemented and desired movements, we calculated a variable and a sub-sample for migrants in the next stage of transit or those who intend to take the next step of migration no later than three years from the current migration stage. Hence, we have a description of migrants in a situation (who passed at least two stages of migration) or a state (those who want to move to the second or next stage of migration). We designated both groups of migrants as transit migrants.

Most transit migrants in the regional sub-sample were registered in Altai Krai (28.5%), the Republic of Dagestan (23.4%), and the Pskov Region (16.5%), the Altai Republic (12%), and the Orenburg Region (10.1%). The least transit migrants were in the Murmansk Region (5.7%) and the Rostov Region (3.8%). Consequently, we considered this distribution only at the trend level, taking into account that the transit criteria were not included in the number of quota indices when planning the overall research sample. Further analysis is based only on data on a sub-sample of transit migrants.

The paper aims to consider the problems of self-perception of safety and security of transit migrants in Russian border regions.

## 3 Results

Most (60.8%) of the actual transit migrants arrived in the Russian border region from abroad (not from the countries of origin), the transit points were as follows: Armenia, Azerbaijan, Belarus, Iran, Kazakhstan, China, Kyrgyzstan, Moldova, Romania, the United States, Tajikistan, Turkmenistan, Turkey, Uzbekistan, and Ukraine. A third (39.2%) moved to the Russian border regions from other Russian regions; that is, they were at the next stage of transit (possibly the last one).

The transit regions for this group of migrants were (1) the Sverdlovsk Region, (2) the Komi Republic, (3) Krasnoyarsk Krai, (3) Moscow, (4) the Moscow Region, (4) the Novosibirsk Region, (5) the Omsk Region, (6) the Republic of Altai, (6) the Samara Region, (6) Saint Petersburg, (7) Tatarstan, (7) the Tyumen Region, (8) Khabarovsk Krai.

As for transit migrants who wanted to leave the border region not to the country of origin within the next three years, 86.4% wanted to move to the Russian regions as attractive enough to work there. Most of the transit migrants from the Russian border region sought to move to the central and western regions of the country or the south of Russia. Most of the respondents wanted to move to the capitals, megacities: Moscow and Saint Petersburg. The capital of Tatarstan, Kazan; Krasnodar Krai and the cities of Krasnodar and Sochi, the Republic of Crimea; the Novosibirsk, Tomsk, and Sverdlovsk regions, and the city of Surgut are extremely attractive for migrants in Russia. Some of the migrants said that they want to move to the *big city*. Here we found confirmation of our hypothesis—in their dreams, as the destination, migrants see economically favorable regions of Russia, which are regions of intensive transit that receive most of the flow of labor migration to Russia.

Just over 13% of migrants wanted to move from Russia to other countries (Belarus, Korea, Romania, Turkey); nevertheless, they often had no idea about a particular country; they want to live in Russia.

Some of the actual transit migrants did not want to stay in the Russian border regions, seeing them as a stopover, and wanted to go further. We should note the following trends: (1) migrants who moved around Russia wanted to continue moving within the country (75%), only a quarter of them wanted to go abroad; (2) migrants who came to Russia from another country (not the country of origin) and moved through the countries several times with intentions of both international and internal transit accounted for 45.5 and 54.5%, respectively.

Not every transit migrant wanted to obtain Russian citizenship (only 36.2%). However, migration to Russia and work there have a positive impact on the well-being of migrant families. In total, 66.5% of them noted that the financial situation of their families improved after moving to Russia; 27% could not find the desired increase in well-being; and 6.6% slightly worsened the financial situation of their families.

The attitude of transit migrants who wanted to move on to the next migration stage is noteworthy. Analyzing the literature, we clarified the research data on Russian citizenship, which increases the chances of migrants to move on successfully. Sometimes

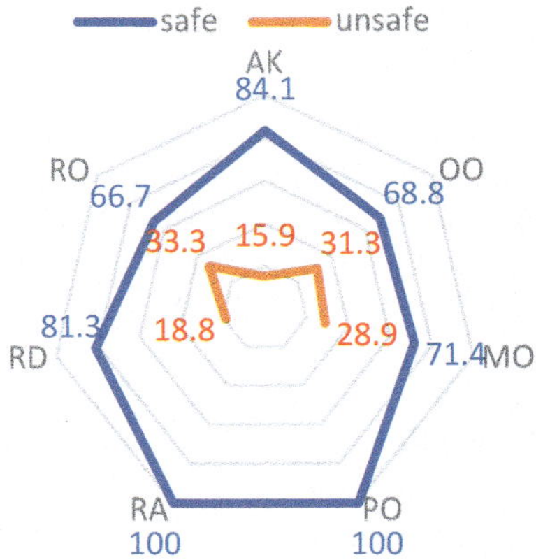


**Fig. 1** Self-evaluations about safety of transit migrants in regional correlations, $\chi^2$, $p \leq 0.005$, %. *Source* Compiled by the authors

the main reason for coming to Russia was the opportunity to obtain a Russian passport. Thus, most migrants who wanted to move through the territory of Russia want to obtain Russian citizenship (72.7%), while 45.5% of them wanted to go abroad.

In addition to geographical and socio-economic attractiveness, some factors affect the overall satisfaction and comfort of staying in Russia, such as social security, public security, the attitudes of natives, conflicts, and the general perception of personal security. Based on the generalization of subjective assessment of the personal security of migrants, we have identified groups of migrants who are secure or insecure in each region. The most secure regions were the Pskov Region and the Republic of Altai: almost all migrants feel secure there. The second place was occupied by Altai Krai, where most migrants felt secure (84.1%). In the third place was the Republic of Dagestan (81.3%). At the same time, a third (33.3%) of migrants in the Rostov Region felt in danger; in the Orenburg Region—31.3%; in the Murmansk Region—28.6%. Therefore, different regions of both the Asian and European Russian border regions generate various threats for transit migrants. Thus, we can assume that security threats can act as driving factors in the region (Fig. 1).

Figure 1 shows that self-assessment of a secure life in the host community correlates with assessments of the relationship between the natives and migrants. Summing up the opinions of migrants on the relations with the natives, we identified groups of migrants who mark these relations as mostly negative and vice versa.

Such assessments probably contribute to the overall security of staying in Russian regions (Fig. 2). In the Altai Republic, we recorded the most neutral attitude toward migrants: all respondents reacted positively to contacts with the local population. Altai Krai occupied the second place (93% of positive responses), followed by the Pskov Region with 76% of positive responses, and the Republic of Dagestan (77.1%). The Rostov Region was included in the group of regions with an extremely negative attitude towards migrants (56.3%) and the Orenburg Region, where assessments equally divided migrants into two groups. The Murmansk Region had a third of the migrants who have a negative attitude to the local population.

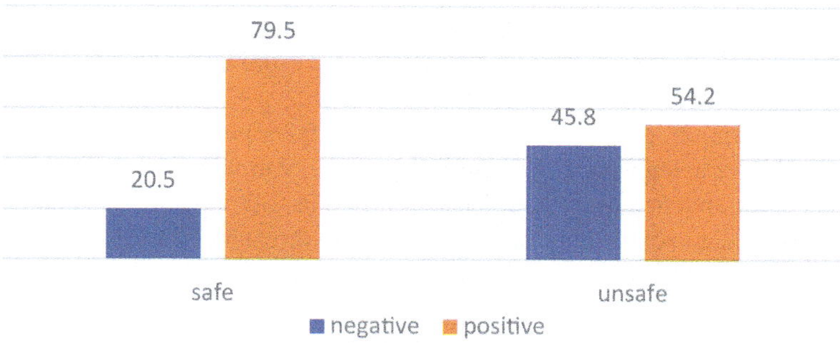


**Fig. 2** Evaluation of relations between aboriginal population and migrants in regional correlations, $\chi^2$, $p \leq 0.005$, %. *Source* Compiled by the authors

We tried to predict the possible negative or positive attitude towards migrants and the formation of a sense of security or insecurity based on the classification tree method. We used the following variables as dependents: (1) personal attitude towards migrants, (2) personal security assessments (for convenience, we summarized positive and relatively positive, negative and rather negative assessments to shift the forecasts to one or another pole). The predictors were (1) age group, (2) children, (3) marital status, (4) education, (5) material well-being, (6) gender status, (7) security ratings (in one case), (8) ratings about the attitude of residents (in the other case), (9) attitude to religion, (10) region of residence, (11) Russian language proficiency, (12) denomination, and (13) friends and relatives in the region of residence. None of the predictors were included in the model in both cases (Fig. 3).

Next, we considered how the risks of personal security of transit migrants in Russian border regions are objectively manifested. Study participants spoke about cases of violation of human rights in Russia: (1) if the employer did not give the documents; (2) if the employer did not pay for the work done or not paid the entire salary, kept at work against one's will, and forced to work without a contract; (3) if the

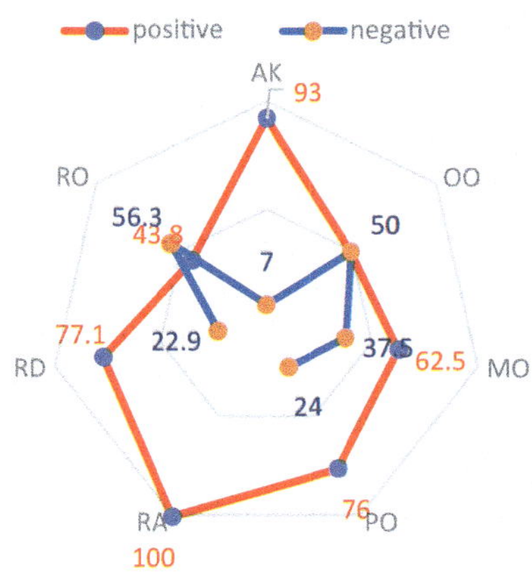


**Fig. 3** Evaluations about personal security of transit migrants against evaluations with aboriginal population, %. *Source* Compiled by the authors

employer used physical violence, intimidated by the police of someone or something else, and sexually harassed. All these violations are criminally and administratively punishable in the Russian Federation.

During migration, a fifth of transit migrants (20.3%) experienced the violation mentioned above at least once. The most common violation was partial salary, faced by 8.1% of respondents. Moreover, 3.8% of migrants did not get their salary in full; 3.2% suffered from sexual harassment; 3.2% were threatened; 2.5% were forced to work without a contract; 0.6% suffered from physical violence; and 1.9% faced a situation where the employer took out their documents as collateral.

The most dangerous territory for transit migrants was the Republic of Dagestan, where more than half of the respondents experienced violations of rights (51.4%). A similar situation was observed in the Orenburg Region, where 43.8% of migrants had at least once been severely violated; in the Murmansk and Rostov regions, more than a fifth of migrants faced the same problems—22.2 and 21.1%, respectively; however, in Altai Krai, Pskov Region, and the Altai Republic, none of the respondents faced problems in the field of work.

We understood the experience of migrants, forming their ideas about the local population based on the consensus scores on the following statements with a four-point scale: (1) "I consider the natives unfair and unfriendly to people of my ethnicity"; (2)"I feel that the natives do not accept me"; (3) "I was teased or insulted because of my ethnicity"; (4) "I was threatened and attacked because of my ethnicity."

We summed up the proportions of migrants who agree and are likely to agree with at least one of the above statements. As a result, we found that a fifth of transit migrants (20.9%) had a negative experience in the host community.

The spread of the facts of the negative attitude of the natives to migrants differs significantly ($\chi^2$, $p \leq 0.005$) by region. As a consequence, almost half of migrants in the Murmansk Region (44.4%), more than a third of migrants in the Orenburg Region (37.5%), and the Altai Republic (33.3%) experienced negative attitudes from the local population. In Altai Krai, the Pskov Region, and the Rostov Region, about a fifth of the migrants (20, 23.1, and 21.1%, respectively) suffered from the negative attitude of the local population. The population of the Republic of Dagestan happened to be the most tolerant: only 5.4% of respondents noted a negative trend in relations.

Therefore, the fear of migrants is expressed in this attitude towards migrants, which makes them feel rejected by Russian people (the average value in the sample was 2 points), injustice, and unfriendliness (the average value in the sample was 1.94 points). In Russian border regions, insults against migrants were rare (1.82 points), and cases of attacks on migrants due to their ethnicity also occurred quite rare (1.65 points). It is necessary to clarify the most common forms of negative attitudes of the indigenous population of the border region towards transit migrants, which are subjective and depend on the characteristics of the perception of the natives by the migrants themselves. Less common facts of negative attitudes were objective and reflected the real conflicts between migrants and the host community (Fig. 4).

The negative attitude of the local population and regional communities towards migrants was not always expressed in the willingness to violate their labor rights or personal security. Comparing the cases of negative attitudes and the facts of violations

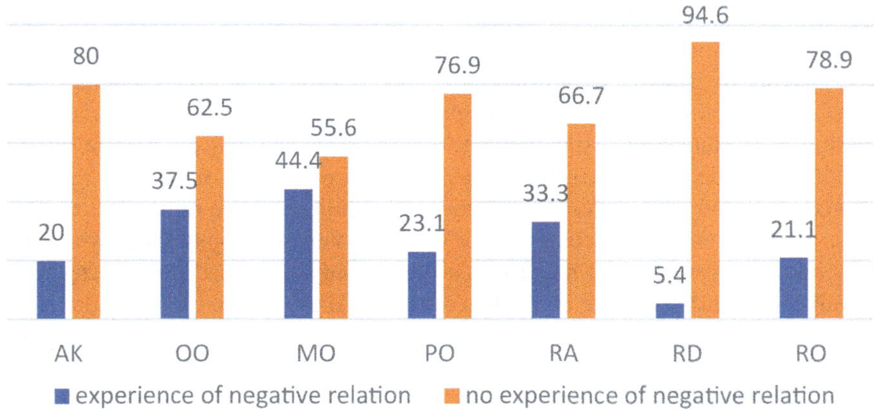


**Fig. 4** Relation of aboriginal population to migrants in regional correlations, $\chi^2$, $p \leq 0.005$, %. *Source* Compiled by the authors

of the rights of migrants, we found that in the Republic of Dagestan, migrants are the *weakest and vulnerable* in their rights; however, they rarely noted the negative attitude of the host community. On the contrary, in the Murmansk Region, the negative context of relations with the host community had little impact on the spread of various criminal cases of harassment of rights both in work and personal security. For example, in the Rostov region, some migrants who faced security threats were equal to those who noted negative relations with the host community. In Altai Krai, the Pskov Region, and the Altai Republic, about a third to a fifth of negative assessments were recorded. Nevertheless, there were no cases of infringement of rights.

Based on the classification tree method, we tried to predict the characteristics of migrants associated with exposure to risks in work and personal relations with the employer. The predictors used were variables that characterize (1) the location of the respondent, (2) the experience of labor migration (number of migrations to work), (3) the presence of relatives and friends in the host community, (4) the place of work, (5) education, (6) marital status, (7) religion and attitude to religion, (8) gender status, (9) the cases of negative interaction with the natives, (10) the assessment of attitudes towards migrants. The only variable included in the model was the region of temporary residence of migrants (the dependent variable is the facts of human rights violations).

Thus, living in the Republic of Dagestan, Rostov, Murmansk, and Orenburg regions increases the risk of infringement of rights (60.5% of cases of infringement of rights).

Considering widespread negative background in the relations with the local population in the subjective assessment of migrants, not real cases of conflicts with the natives, we tested several hypotheses about significant dependencies in the perception of public opinion about migrants on (1) the region where they live, (2) Russian language proficiency, (3) health status, (4) country of origin, (5) stage and the experience of migration, (6) work, (7) obtaining a patent for a career in the region, (8) engaging in crime (selling drugs, sex work), (9) the debt (loans, credits), (10) education, (11) marital status, (12) gender status, (12) age, (13) assessment of the possibility of mutual understanding between migrants and natives. As a result, we identified significant ($\chi 2$, $p \leq 0.005$) dependences of the risk of negative relations between migrants and the local population. The more negative ratings we receive, the more often migrants face negativity from the natives. Among those who rated their relationships with the natives as negative, only 4.2% had no experience of negative relationships, while 16.1% had conflicts. Among those who considered the relationship most likely to be negative, 14.4% had no negative experience, but 32.3% experienced conflict. Among those who consider the relationship between migrants and natives most likely to be positive, 48.3% had no conflicts with the natives, while 25.8% had them. Among those who consider the host community friendly, 33.1% had no conflicts with the natives, but 25.8% had them.

The migrants, who felt insecure in Russia, more often experienced negative relations with the natives (significant dependence). Migrants who felt secure had no conflicts with the natives (32.5%), but 23.3% had them. Among those migrants who were most likely to be secure, 53.8% never had conflicts with the natives, but 43.3% had them. Among those who felt insecure in the host community, only 9.4% had no conflicts, but one in three (30%) experienced negative attitudes from the natives.

Table 1 shows the distribution of satisfaction with various aspects of life associated with accommodation in the host community in different groups of transit migrants.

First, the satisfaction ratings had noticeable regional differences. In general, life in Russian regions is quite satisfactory for migrants living in the Pskov region (4.27) and the Rostov Region (4.05), slightly less in the Republic of Dagestan.

In two border regions, migrants were satisfied with their work opportunities (Rostov region—4.11 points, Pskov Region—4.1 points); lower job satisfaction was in the Altai Republic (2.83 points).

The level of wages in the Russian border regions was less satisfactory for transit migrants, estimates of this factor were low, but the lowest one was observed in the Republic of Altai (2.17) and the Republic of Dagestan (2.79), with the highest average in the Rostov Region (3.95) and the Pskov Region (3.88).

Border regions provide migrants with good opportunities for internal conditions: in all regions, the satisfaction level ranged from 4.11 (Rostov Region, maximum value) to 3.0 (the Altai Republic, minimum value).

As for relations with colleagues and employers, the ratings on the constituted practices were relatively high: from 4.65 points in the Pskov Region to 3.33 points in the Republic of Altai.

Despite migration without a family, most migrants often had satisfactory relations with relatives and ranged from 3.55 points (The Republic of Dagestan) to 4.65 points (Pskov Region).

We found differences in the assessments of satisfaction with the support of the religious community. In the Altai Republic, there was a low level of satisfaction (2.0 points); similar situations were recorded in the Republic of Dagestan (2.77 points)

**Table 1** Distribution of evaluations about satisfaction by different life aspects in various groups of transit migrants, means

| | Life in whole | Work in whole | Salary | Living conditions | Relations with colleagues | Relations with family and relatives | Support of religious community | Support of ethnic community | Support of religious community |
|---|---|---|---|---|---|---|---|---|---|
| Region | | | | | | | | | |
| AK | 3.51 | 3.63 | 3.24 | 3.38 | 3.75 | 4.40 | 3.61 | 3.58 | 3.73 |
| OO | 3.75 | 3.29 | 3.07 | 3.63 | 3.47 | 4.31 | 2.87 | 3.56 | 3.75 |
| MO | 3.89 | 3.89 | 3.56 | 3.67 | 3.89 | 4.11 | 3.67 | 3.44 | 3.89 |
| PO | 4.27 | 4.10 | 3.88 | 4.04 | 4.23 | 4.65 | 4.64 | 4.38 | 3.96 |
| RA | 3.50 | 2.83 | 2.17 | 3.00 | 3.33 | 4.50 | 2.00 | 1.67 | 3.83 |
| RD | 3.18 | 3.15 | 2.79 | 3.09 | 3.41 | 3.55 | 2.77 | 2.90 | 3.79 |
| RO | 4.05 | 4.11 | 3.95 | 4.11 | 3.89 | 4.00 | 3.44 | 3.06 | 3.94 |
| Actual transit | | | | | | | | | |
| International transit | 3.70 | 3.65 | 3.16 | 3.50 | 3.70 | 4.05 | 3.26 | 3.28 | 3.93 |
| Domestic transit | 3.48 | 3.52 | 3.10 | 3.31 | 3.75 | 3.96 | 3.60 | 3.50 | 3.75 |
| Intents of transit | | | | | | | | | |
| Domestic transit | 3.72 | 3.57 | 3.45 | 3.65 | 3.68 | 4.27 | 3.44 | 3.49 | 3.93 |
| International transit | 3.43 | 3.38 | 2.92 | 3.00 | 3.08 | 3.54 | 3.08 | 2.62 | 2.46 |
| Security | | | | | | | | | |
| Safe | 3.85 | 3.77 | 3.45 | 3.66 | 3.82 | 4.25 | 3.52 | 3.50 | 3.93 |
| Unsafe | 3.08 | 2.84 | 2.60 | 3.04 | 3.29 | 4.00 | 2.88 | 3.00 | 3.28 |

(continued)

**Table 1** (continued)

| | Life in whole | Work in whole | Salary | Living conditions | Relations with colleagues | Relations with family and relatives | Support of religious community | Support of ethnic community | Support of religious community |
|---|---|---|---|---|---|---|---|---|---|
| *Relation to migrants* | | | | | | | | | |
| Negative | 3.51 | 3.41 | 3.03 | 3.47 | 3.53 | 4.17 | 3.23 | 2.86 | 3.53 |
| Positive | 3.70 | 3.64 | 3.36 | 3.55 | 3.78 | 4.18 | 3.44 | 3.59 | 3.93 |

*Source* Compiled by the authors

and the Orenburg Region (2.87 points), but the highest satisfaction with the support of migrants of the religious community was observed in the Pskov Region (4.64 points).

In the ethnic republics covered by the study, there was a low satisfaction with the support of the diaspora: 1.67 points in the Republic of Altai, 2.9 in the Republic of Dagestan. In other multi-ethnic regions, migrants were satisfied with the support of the diaspora.

The health status of migrants was poor, dependent on any other factors, and satisfied migrants in a similar way in all survey regions.

Thus, migration to the territory of the Russian ethnic regions was objectively and subjectively less satisfactory for transit migrants. Living in more socially and economically favorable regions of Russia objectively generates a higher level of satisfaction with life in general and its aspects.

Migrants with international transit experience were more satisfied with the support of local communities, such as the diaspora, hometown associations, and religious communities in Russia. Nevertheless, the aspects of life conditioned by socio-economic circumstances (work, salary, housing) were far less satisfactory. Migrants moving across Russia from their country of origin were more positive in their general assessments of life. We found a similar distribution of opinions among transit migrants with different intentions. In general, those migrants who wanted to move to other Russian regions were more satisfied with all aspects of life, and those migrants who wanted to leave Russia at the next migration stage were less satisfied. Perhaps such dissatisfaction with the general economic and socio-psychological needs was the primary motive for the further movement of transit migrants.

The feeling of insecurity in the Russian border regions has also significantly reduced satisfaction with all aspects of life in the country. Simultaneously, the perception of the host community as unfriendly to a lesser extent affects the assessment of life satisfaction. Probably, if the attitude of the host community was understood as insufficiently friendly, based on personal perception or subjective feelings, and not on the cases of conflict, then economic factors came to the fore, as well as the low and absorbing importance of the opinions of the local population.

## 4 Discussion

Border regions differ significantly in attracting migrants and have unique characteristics stimulating transit migration and migration flows. Each region has its national borders and is a *point* of internal entry of migrants into the country. Nevertheless, each region has its specifics in the context of the study. The presence of family or friendly ties in the region was the most critical factor when choosing a region for migration to Russia. Migrants go there because of *agents* that make the migration transparent and secure. Such ties are crucial for those who go to the ethnic regions, such as the Republic of Dagestan and the Republic of Altai.

In the three border regions (the Republic of Dagestan, Pskov Region, and Altai Krai), the opportunities to get a long-distance job are wider, up to the moment of migration. In other words, immediately after arrival, the migrant has not yet had a job. However, after arriving in the Rostov Region, it was easier to get a job, so the region was more attractive for a quarter of transit migrants. The worst conditions for employment were in the Pskov Region. The favorable climate and nature are a common attraction factor of the Altai Republic, while this factor was relatively insignificant and weak in other regions. According to transit migrants, the Republic provided comfortable living conditions compared to other border regions (except for the Rostov Region). The Altai Republic was more favorable for good relations between representatives of different ethnic groups, while in other regions, this factor was weakly expressed, and in Altai Krai, it was not observed at all.

According to the migrants, only three border regions can attract incoming migrants with various opportunities: the Orenburg Region, the Murmansk Region, and the Rostov Region. For example, it was easy to prepare the necessary documents in the migration service of Altai Krai. The Orenburg Region was geographically convenient for moving through the central and southern regions of Russia and Altai Krai—to the most economically profitable Novosibirsk Region and Krasnoyarsk Krai. The attractiveness of the Orenburg Region is enhanced by the opportunities for transit and further movement abroad; none of the regions had such an advantage.

In addition to geographical and socio-economic attractiveness, there are factors affecting the overall satisfaction and comfort of staying in the Russian host community, such as (1) public security, (2) conflict, and (3) a general sense of personal security. The most prosperous regions, according to some responses, were the Pskov Region and the Altai Republic. In these regions, all migrants felt secure. Altai Krai took the second place, with the majority of migrants being secure. The Republic of Dagestan occupied the third place. However, a third of transit migrants consider themselves in danger during their stay in the Rostov, Orenburg, and Murmansk regions (28.6%). Thus, different regions of both Asian and European border areas pose various threats for transit migrants. We can assume that these risks to personal security can serve as the factor of *pushing out* of the region.

Migrants' self-assessments of the security of their stay in the host community correlate with assessments of the attitude of the local population towards migrants. The Rostov Region was included in the group of regions with the most negative attitude of the natives (56.3%). In the Orenburg Region, estimates were divided equally (by 50%). In the Murmansk Region, more than a third of the respondents consider their relations with the natives negative.

During migration, a fifth of transit migrants (20.3%) faced at least one type of violation in work. The most common violation was non-payment of all earned money, faced by 8.2% of migrants; while 3.8% of migrants did not receive any salary at all; 3.2% were sexually harassed; 3.2% were threatened; 2.5% were forced to get a job without a contract; 0.6% suffered from physical violence; 0.6% were held at work against their will; and the employer took the documents of 1.9% of migrants as collateral.

The most dangerous territory was the Republic of Dagestan, where half of the migrants were violated in their labor rights (51.4%). A similar situation was recorded in the Orenburg Region: 43.8% of transit migrants noted at least one violation of their rights during their stay in Russia. About a fifth of migrants faced such problems in the Murmansk and Rostov regions; however, in Altai Krai, Pskov Region, and the Altai Republic, none of the respondents were violated in their rights while working in Russia.

Fear of migrants and feelings that the local population does not like migrants and is unfriendly to them are common in Russian border regions. Sometimes migrants face insults. Moreover, there were cases of attacks on migrants, although rarely. The most frequent forms of negative attitudes of the natives are somewhat subjective and depend on the characteristics of the perception of the population by the migrant himself. Nevertheless, the less common cases of negative attitudes are objective and reflect real conflicts between the local population and migrants.

The negative attitude of the local population towards migrants was not always expressed in the desire to infringe on their rights or personal security. Comparing the cases of negative attitudes with the cases of violation of rights, we found that in the Republic of Dagestan, migrants were the *weakest and vulnerable* in their rights, but they did not feel the negative attitude of the natives. On the contrary, in the Murmansk Region, the negative relations with the host community did not affect the spread of various cases of criminal harassment of migrants, both in work and in the personal security of migrants. In the Rostov Region, the share of migrants who faced security threats was equal to those who noted negative relations with the natives. In Altai Krai, Pskov Region, and the Republic of Altai, according to the predominance of negative assessments of social relations from the fifth to the third part of migrants, there were no violations.

Therefore, the risk of violation of labor rights was caused by living in the Republic of Dagestan, the Rostov Region, the Murmansk Region, and the Orenburg Region (60.5% of violations of rights).

## 5 Conclusion

Despite the facts of violation of labor rights of migrants, threats to personal security, not always tolerant and friendly relations with the natives, the insignificant contribution of transit migration to the well-being and quality of life of the migrant household, migrants continue to move both in the Russian regions and return to Russia from their countries of origin to get a job.

The socio-economic conditions of the host community in the border region can be a determining factor in the overall satisfaction with staying in Russia. Therefore, migrants are moving, being foreigners in their new homeland. If migration had a clear advantage, increasing the standard of living of migrants and their families, the migrant would come to work, again and again, moving around Russia and abroad to search for better work, living, and housing conditions. In this case, the socio-psychological difficulties associated with displacement, the destruction of the family, kinship, friendship ties, and negative relations with the natives fade away. Therefore, transit migrants become ready to accept insecurity, social inequality, injustice, and risk personal security. Living in the ethnic regions of Russia is both objectively and subjectively less satisfying for migrants. Nevertheless, living in socially and economically favorable regions generates a higher level of satisfaction with life in general and its aspects.

**Acknowledgements** Paper prepared in the frameworks of the task of the Ministry of science and higher education of the Russian Federation FZMW-2020-0001 "Human capital, migration and security: transformation in the new migration environment in Central Asia".

## References

Adams RH (2005) Do international migration and remittances reduce poverty in developing countries? World Dev 33(10):1645–1669

Agunias RD (2010) Migration's middlemen, regulating recruitment agencies in the Philippines-United Arab Emirates corridor. Migration Policy Institute, Washington, D.C.

Brummer D (2002) Labor migration and HIV/AIDS in Southern Africa. IOM Regional Office for Southern Africa, Pretoria, Southern Africa

D'Emilio AL, Cordero B, Bainvel B, Skoog C, Comini D, Gough J, Dias M, Saab R, Kilbane T (2007) The impact of international migration: Children left behind in selected countries of Latin America and the Caribbean. Division of Policy and Planning, United Nations Children's Fund (UNICEF), New York, NY

Decosas J, Kane F, Anarfi J, Sodji KD, Wagner HU (1995) Migration and AIDS. Lancet 1(346):8978, 826–828

Düvell F (2006) Crossing the fringes of Europe: transit migration in the EU's neighbourhood. Centre on Migration, Policy and Society (COMPAS), University of Oxford, Oxford, UK

Düvell F, Vogel D (2006) Polish migrants: tensions between sociological typologies and state categories. In: Triandafyllidou A (ed) Contemporary polish migration in Europe: complex patterns of movement and settlement. Edwin Mellen Press, Lewiston, NY, pp 267–289

Kahn K, Collison M, Tollman S, Wolff B, Garenne M, Clark S (2003) Health consequences of migration: evidence from South Africa's rural northeast (Agincourt). In: Clark S (ed) Paper prepared for conference on African migration in comparative perspective. Institute of Behavioural Sciences, University of Colorado at Boulder United States of America & Agincourt Health and Population Unit, University of the Witwatersrand South Africa, Johannesburg, Sounth Africa, pp 4–7

Kebede E (2001) Ethiopia: an assessment of the international labour migration situation: the case of female migrants. In: GENPRO working paper, vol 3. Gender Promotion Program, International Labour Office, Geneva, Switzerland

Lurie M, Williams BG, Zuma K, Mkaya-Mwanburi D, Garnett GP, Sturm AW, Sweat MD, Gittelsohn J, Abdool-Karrim SS (2003) The impact of migration on HIV-1 transmission in South Africa: a study of migrant and non-migrant men and their partners. Sex Transm Dis 30(2):149–156

Lurie M (2000) Migration and AIDS in Southern Africa: a review. S Afr J Sci 96:343–347

Mansuri G (2006) Migration, sex bias, and child growth in rural Pakistan. In: Policy research working paper, vol 3946. World Bank, Washington, D.C.

Maximova SG, Morkovkina AG, Omelchenko DA, Noyanzina OE, Maximova MM (2018) Gender peculiarities of the attitude of Russians to migration. In: Procedia – 4th EER international conference on social sciences and interdisciplinary studies. Atlantis Press, Palermo, Sicily, Italy, pp 72–80

Maximova S, Omelchenko DA, Noyanzina OE (2019) Social risks and vulnerability of labor migrants in the Russian borderland. In: Bataev DKS, Gapurov ShA, Osmaev AD, Akaev VKh, Idigova LM, Ovkhadov MR, Salgiriev AR, Betilmerzaeva MM (eds) The European proceedings of social & behavioural sciences EpSBS. Future Academy, Tomsk, Russia, pp 73–80

Migration data portal (2020) Migration data portal the bigger picture. https://migrationdataportal.org/?i=stock_abs_&t=2019

Nagarajan S (2009) Migration, remittances, and household health: evidence from South Africa. Dissertation Candidate of Technical Sciences. The George Washington University, Washington DC

Papademetriou D, Sumption M, Somerville W (2009) Migration and the economic downturn: what to expect in the European Union. Transatlantic Council on Migration, Migration Policy Institute, Washington, DC

United Nations Development Program (UNDP) (2009) Overcoming barriers: human mobility and development. United Nations, New York, NY

Valero-Gil J (2008) Remittances and the household's expenditure on health. In: MPRA paper, vol 9572. University Library of Munich, Munich, Germany

Yang D, Choi H (2007) Are remittances insurance? evidence from rainfall shocks in the Philippines. World Bank Econ Rev 21(2):219–248

# Specifics and Factors of Social Adaptation and Integration of Migrants in Russian Border Regions

**Syldysmaa A. Saryglar and Svetlana G. Maximova**

**Abstract** The paper presents a theoretical and sociological analysis of social adaptation and integration of migrants. We turn to the classical model of acculturation, consider the types of adaptation, and focus on analyzing the essence of social adaptation to determine the processes of adaptation and integration of migrants. Recently, Russia has seen an increase in migration. The majority of migrants arriving in Russia are CIS citizens. We conducted several sociological studies and interviewed 449 foreign migrants in 2018–2019. The research results emphasize the integration tendencies of migrants. They strive to integrate into Russian society, obey laws, observe behavior rules, and honor values. Howey, they also continue to honor the traditions and values of their culture. We also emphasize the importance of migrants' social capital. The presence of social connections and migrant communities determines migration and contributes to adaptation and integration. The factor model has shown that a considerable role in the social integration of migrants in Russia is played by (1) language proficiency, (2) the activities of public organizations to protect the rights of migrants, (3) the health of the migrants, and (4) relationships with the local population, in particular, colleagues and employers.

**Keywords** Migration · Adaptation of migrants · Integration of migrants · Social adaptation · Social integration · Russia · CIS

## 1 Introduction

The migration increase and the constant influx of nonethnic migrants are becoming a global challenge and require constant measures to adapt and integrate migrants. Poorly timed adoption of such measures can further aggravate social, economic, and ethnic-confessional problems in society.

The problem of adaptation and integration of migrants occupies a pivotal position among researchers. Adaptive behavior of migrants—aimed at adapting to the

S. A. Saryglar (✉) · S. G. Maximova
Altai State University, Barnaul, Russia
e-mail: syldysma93@mail.ru

 
S. G. Maximova et al. (eds.), *Advances in Natural, Human-Made, and Coupled Human-Natural Systems Research*, Lecture Notes in Networks and Systems 234,
https://doi.org/10.1007/978-3-030-75483-9_34

climatic, socio-economic, political, and other conditions of the host country—and their further integration in the new society are essential components of migration behavior (Saryglar and Maksimova 2020). Migration behavior includes the migration act itself, activities to prepare for migration, and adaptation and integration in settlement places.

Adaptation and integration of migrants into the host society occupy a central position in J. W. Berry's acculturation model (Berry 1980). Based on two indicators—(1) preserving the original culture of migrants and (2) cross-cultural contacts between migrants and the local population—J. W. Berry identifies four types of migrant acculturation: (1) assimilation, (2) separation, (3) marginalization, and (4) integration (Saryglar and Maksimova 2020).

Assimilation is distancing oneself from the original culture, traditions, and values and accepting the culture of another society. One can differentiate two directions of assimilation (Endryushko 2017). They are as follows:

- The acceptance of the culture of the majority (or dominant group) by the minority;
- The inclusion of all society members with different ethnic, social, and economic characteristics into the new society.

Assimilation is focused on erasing the ethno-cultural dissimilarity of the host society, which is based on the so-called "ethnic color blindness," which does not distinguish between skin colors and officially considers different races and ethnicities equal (Tsapenko 2018).

Separation implies that migrants support their own culture and avoid any contact with the other, particularly the culture of the host society. Separation may be (1) forced (by nonacceptance and displacement by the host society) and (2) spontaneous (desirable for the minority, allowing them to exist outside the dominant group and lead a relatively separate life even in a new environment). The separation of nonethnic migrants can lead to the formation of migrant and territorial enclaves in the host environment. This can threaten the political and socio-economic stability of the country (Mukomel 2011).

Marginalization is the state of an individual on the border of two different, sometimes contradictory, cultures. This can lead to a loss of identity, a state of confusion, a search for the place and role, and identity definition or formation. When migrants are marginalized, they lose touch with their culture and do not identify with the culture of the host society (perhaps for the time being).

The study of marginalization began in the 1920s and was directly related to migration processes. This term was first used by R. E. Park in his work "Human Migration and the Marginal Man" (Park 1928). A marginal person is a person (migrant) who is in a crisis due to relocation and transition to a new society. The marginal person faces the moral upheaval caused by the clash of different cultures and subsequent changes. R. E. Park noted that we might best study civilization and progress by analyzing the consciousness of the marginalized individual, in which the fusion of different cultures shapes new changes and formations.

Interaction between migrants and the local population is considered successful when there is a cultural exchange with the preservation of elements of their respective cultures. Integration is the most profitable and suitable form of acculturation. It involves adapting newcomers and locals to each other, which ensures the harmonious functioning of the social system and the stability of relationships between its members (Jiménez 2011; Le Thi 2017).

An obligatory condition and initial stage of social integration is adaptation. Adaptation is the process of adapting a person to the conditions of a new social environment. There are several types of adaptation: (1) economic, (2) political-legal, (3) social (or socio-cultural), (4) psychological, (5) geographical, etc.

Economic adaptation involves adaptation to the local labor market, new payment conditions for their labor, various payment types, and adaptation to housing and material surroundings.

Political-legal adaptation is aimed at forming ideas about (1) state authorities, (2) the political life of the country, and (3) state policy (in particular, concerning migrants).

Psychological adaptation is characterized by changes in both positive and negative directions for all subjects of the adaptation process, measured through value-semantic, emotional, motivational, and communicative characteristics (Konstantinov 2018).

Geographical adaptation refers to the adaptation of a migrant to new geographical, natural, and climatic conditions of residence.

Social adaptation involves adjusting to a new cultural environment, fully or partially overcoming the language barrier, and direct interaction with a foreign culture. Social adaptation is an active interaction between an adaptor (individual or social group) and the social environment (Ledeneva 2014). It is the process of mutual harmonization of the individual with the social, political, moral, and material conditions of the external environment. The success of social adaptation is determined by the level of consistency of the personal and social system of expectations, assessments, opportunities, goals, and values (Petrova 2013).

Data from the Federal State Statistics Service of Russia show that the migration increase was 285,103 people in 2019 (Federal State Statistics Service 2020). The majority of migrants came from CIS countries. This tendency has been observed for several years. For example, last year, the total number of newcomers was 701,234 people, out of which 617,997 came from the CIS countries. Since its foundation, the CIS has become one of the transnational social spaces that later became an active migration system with Russia as its core.

Dissatisfaction with vital needs in the territory of residence and the availability of opportunities to meet them outside this territory, as well as knowledge of the situation inside the living space and outside it, are the main conditions for migration processes (Korobov 2003). Any migration (internal or external) is caused by specific changes in the adaptor's life, both external circumstances and personal characteristics of the migrant. Therefore, there is a question of finding out the factors that influence the formation and development of migration behavior, including the processes of adaptation and integration of migrants in the places of settlement.

## 2 Materials and Methods

The research goal is to study the features and factors of social adaptation and integration of migrants based on sociological research.

In 2018–2019, the Altai State University conducted many sociological studies in six Russian border regions. A total of 449 migrants from Uzbekistan (52.2%), Tajikistan (23.5%), Azerbaijan (4.3%), Kyrgyzstan (3.9%), Kazakhstan (3.6%), Ukraine (3.6%), and other countries were interviewed. Among the migrants who took part in the study, 88.2% were men, and 11.8% were women. The average age of migrants is 35 years.

To analyze the specifics of social adaptation and integration of migrants, we used the method of frequency and two-dimensional data analysis. To identify the factors of social integration of migrants, we performed a regression analysis, for which an 11-point integration scale was calculated.

## 3 Results and Discussion

To identify the features of social adaptation and integration of migrants in the new cross-cultural environment, we performed a frequency analysis of the data ($\chi^2$, $p \leq 0.05$).

The research results indicate the integration tendencies of cross-cultural communication in Russian regions. The majority of migrants (84.3%) understand the need to integrate into Russian society, support the norms and rules of behavior adopted in it, and respect the values and cultural traditions of peoples. However, this group also indicates that it is necessary to continue to honor the traditions and values of their culture. Integration with the local population is actively carried out at informal events, when, for example, an adaptor devotes his free time to communicating with new people and getting acquainted with their culture. The majority of migrants in Russian regions prefer to spend their free time with their fellow countrymen and representatives of the local community (71.4%) and establish friendly relations with members of the new society (71.2%).

A significant contribution to the formation of migration behavior and the further adaptation and integration of migrants is made by their social capital—the presence of social connections that ensure a "soft" entry into new territory and inclusion in an unfamiliar culture. In most cases, social connections or migrant communities play a significant role in choosing a destination (Sue et al. 2019). The research results confirm this statement. A third of the migrants came to the Russian regions because they had relatives, acquaintances, and friends there. Besides, half of the migrants found work through relatives, friends, and acquaintances. Representatives of ethnic diasporas living in Russian regions play an essential role in the life of migrants, helping them migrate, adapt, and integrate. Thus, half of the migrants noted that they frequently interact with local diasporas and turn to them to solve their problems.

Based on the linear regression analysis, we constructed a factor model of the social integration of migrants ($R^2 = 0.554$). The influence of independent variables explains 55.4% of the dependent variable (integration) variance. According to the model, the significant factors of social integration of migrants included such indicators as proficiency in the Russian language ($b = 0.374$; $p \leq 0.003$), trust in public organizations that protect the rights of migrants ($b = 0.612$; $p \leq 0.020$), satisfaction with health ($b = 0.424$; $p \leq 0.030$,) and trust in the employer ($b = 0.392$; $p \leq 0.033$).

Most of the migrants coming to Russia are labor migrants. The primary purpose of staying in Russia is work (88.6%). The main reason for the arrival in certain Russian regions is "the availability of a workplace." For migrants, relationships with the employer and colleagues become essential. Sometimes this is the only type of communication during their stay in the country. Trusting relationships with colleagues and employers creates a favorable basis for the inclusion of migrants in society and their further integration. Therefore, the higher the trust in colleagues, the more actively migrants integrate into society.

The integration of migrants in a new society depends on the knowledge of the language of this society. The ability to communicate in the official language of the host country is an essential factor not only in the professional development of migrants but also in their adaptation and integration in the new society (Madziva, McGrath and Thondhlana 2016). Our research also confirms these factors—the proficiency in Russian increases the chances for successful adaptation and integration. Migrants with an advanced Russian language were more integrated than those with intermediate or starter levels, and vice versa.

As noted above, knowledge of the language of the host society is the key to successful adaptation and integration of migrants. Ignorance of the language creates additional obstacles, starting with paperwork and ending with establishing contacts with residents. Insufficient knowledge of the language, in most cases, is typical for migrant workers who come and find work with the help of acquaintances and friends living in the region or having experience working in the region. Thus, a third of migrants with a low level of Russian prefer to spend their free time only with their fellow countrymen (34.6%). Among migrants with high language proficiency, only 8.1% prefer to communicate only with their fellow countrymen. Despite this, migrants note the need to take additional measures to help them learn the language and teach their children (26.2%).

The next factor of social adaptation and integration of migrants is the satisfaction with their health. The deterioration of health reduces the chances of integration in the new society and the act of migration itself (Kovalzhina 2019). The World Health Organization defines human health as a state of complete physical and psychosocial well-being.

As our model shows, trust in public organizations protecting the rights of migrants is also one of the main factors of the social integration of migrants. It indicates the positive impact of civil society institutions on the adaptation and integration of migrants. Public organizations representing the interests of migrants are, in most cases, national and cultural associations. Having a significant level of trust from the migrants and the local population, such organizations ensure the interaction of local communities and accumulate financial, organizational, and technological resources.

## 4 Conclusion

The study of the adaptation and integration of migrants is a topical issue and will remain so for many years to come. The number of migrants increases every year; in some regions, migration flows are difficult to control. State programs and measures to control and regulate migration flows ensure the safety of territory and population. However, it is even more important to preserve cross-cultural harmony by guiding migrants and the local population towards integration.

The results of sociological studies conducted in border regions in 2018–2019 confirm the integration potential of migrants in Russia. Migrants accept the laws and values of Russian society and honor their own culture because, for the majority, migration became possible due to social ties, particularly ties with ethnic diasporas in Russia.

The factor model obtained as a result of regression analysis allowed us to identify the most significant factors of adaptation and integration of migrants. It includes knowledge of the Russian language, trust in public organizations, satisfaction with health, and trust in an employer. The obtained data do not fully reflect the state of migration processes. The problem of adaptation and integration of migrants is not limited to one study and requires a comprehensive approach. The data indicate the presence and development of integration processes in Russian regions and suggest new comprehensive research using in-depth interviews and the psycho-semantic method.

**Acknowledgements** The research was supported by the grant RFBR No. 19-311-90039 "Post-graduate students. Adaptive strategies of migrants from the CIS countries from the perspective of social exclusion/acceptance of the local community of the Russian border region" (2019–2021).

## References

Berry JW (1980) Acculturation as varieties of adaptation. In: Padilla A (ed) Acculturation: theory, models, and findings. Westview Press, Boulder, CO, pp 9–25

Endryushko AA (2017) Theoretical approaches towards examining the adaptation of migrants to the host society: foreign practices. Bull Inst Sociol 4(8):46–70

Federal State Statistics Service (2020) Official statistics. Population. Demography. https://www.gks.ru/folder/12781.

Jiménez TR (2011) Immigrants in the United States: how well are they integrating into society? https://www.wilsoncenter.org/sites/default/files/media/documents/event/integration-Jimenez.pdf

Konstantinov, VV (2018) Socio-psychological adaptation of migrants in the host multicultural society. Dissertation of Doctor of Psychological Sciences. Saratov State University, Saratov, Russia

Korobov AA (2003) Structural-genetic approach to the analysis of population migration (theoretical aspect). Bull Saratov State Socio-Econ Univ 6:97–101

Kovalzhina LS (2019) Health inequalities: a sociological discourse. Sociodynamics 6:1–6. https://doi.org/10.25136/2409-7144.2019.6.29868

Ledeneva VYu (2014) Social adaptation and integration of migrants in modern Russian society. Dissertation of Doctor of Sociological Sciences. Russian Presidential Academy of National Economy and Public Administration, Moscow, Russia

Le Thi M (2017) Social capital, migration, and social integration. GATR J Int Inst Knowl Manag 3:353–363. https://doi.org/10.17501/icedu.2017.3135

Madziva R, McGrath S, Thondhlana J (2016) Communicating employability: The role of communicative competence for Zimbabwean highly skilled migrants in the UK. J Int Migr Integr 17(1):235–252

Mukomel VI (2011) Integration of migrants: challenges, policies, and social practices. Univ Russia. Sociol Ethnol 1:34–50. https://cyberleninka.ru/article/n/integratsiya-migrantov-vyzovy-politika-sotsialnye-praktiki

Park R (1928) Human migration and the marginal man. Am J Sociol 33(6):881–893

Petrova OS (2013) Control of social adaptation of labor migrants in Saint Petersburg: a sociological analysis. Dissertation of Candidate of Social Sciences. Herzen State Pedagogical University of Russia, St. Petersburg, Russia

Saryglar SA, Maksimova SG (2020) Migration behavior: Strategies and practices of social integration, specifics of adaptive behavior of migrants in the Asian borderlands of Russia. Soc Secur Insights 3(1):37–48. https://doi.org/10.14258/ssi(2020)1-02

Sue AC, Riosmena F, Pree JL (2019) The influence of social networks, social capital, and the ethnic community on the US destination choices of Mexican migrant men. J Ethn Migr Stud 45(13):2468–2488. https://doi.org/10.1080/1369183X.2018.1447364

Tsapenko TV (2018) Search for new approaches to the socio-cultural integration of migrants. Demogr Rev 5(4):125–149. https://cyberleninka.ru/article/n/poiski-novyh-podhodov-k-sotsiokulturnoy-integratsii-migrantov

# Human Capital of International Migrants in Border Regions of Russia: What Component Really Matters?

**Svetlana G. Maximova, Daria A. Omelchenko, Oksana E. Noyanzina, and Evelina O. Ebeling**

**Abstract** The orientation of the Russian economy and social policy to the innovative transformation and the development of the national human capital is challenged by the acute demographic issues and peculiarities of migration processes. Border regions are particularly involved and suffer from the disbalance of export and import of the human capital, changing the potential of development of the border territories. The sociological research, conducted among international migrants in six border regions of Russia with different socio-economic development and ethnic composition, allowed to reveal main characteristics of imported human capital and its impact on individual well-being. It was shown that there are some general trends and regional peculiarities, related to institutional conditions.

**Keywords** Human capital · Educational capital · Intellectual capital · Language capital · Cultural competencies · International migrants · Border regions of Russia · Modeling of the human capital efficiency

## 1 Introduction

### *1.1 HC Theory: Principal Provisions and Current State*

The general economic shift to the *innovative path*, the reorientation to high technology, the growing role of migration, internationalization of education, healthcare, and labor market aroused considerable interest among scholars in the formation, reproduction, and efficient use of human capital [HC]. More than 400,000 scientific publications on HC have appeared on the Google Scholar platform in the last three years. *CrossRef* reports on 3,000 books, 237,000 journal articles, 795 special issues

S. G. Maximova · D. A. Omelchenko (✉) · O. E. Noyanzina · E. O. Ebeling
Altai State University, Barnaul, Russia
e-mail: daria.omelchenko@mail.ru

O. E. Noyanzina
e-mail: noe@list.ru



S. G. Maximova et al. (eds.), *Advances in Natural, Human-Made, and Coupled Human-Natural Systems Research*, Lecture Notes in Networks and Systems 234,
https://doi.org/10.1007/978-3-030-75483-9_35

of journals, more than 25,000 books of conference proceedings, and 2,700 databases, where HC appears as an essential indicator.

Meanwhile, the idea that knowledge and experience can be transformed into financial wealth is not new. The theory of HC originates in the classical works of political economy by A. Smith, W. Petty, A. Marshall, K. Marx, L. Walras, J. M. Clark, J. R. McCulloch, D. Ricardo, and many others. The formation of a separate and independent concept occurred in the late 1950s and early 1960s. The interpretation of human actions and decisions not only in the field of wealth creation and material goods based on economic categories has become one of the most popular approaches in scientific research developed by representatives of the Chicago school and other American and British economists—Schultz (1961), Mincer (1958), Denison (1962), Kuznets (1966), Thurow (1970), Solow (1988), Lucas (1990), and others. The theory was crystallized based on neoclassical theory and institutionalism, and later it incorporated the theories of endogenous growth and other particularistic theories associated with the categories *human potential*, *human resources*, *sustainable development*, *knowledge economy*. Its popularity and relevance are due to its timely appearance as a response to the transformation of economic life and productive forces. HC rapidly became a strategic reserve, increasing labor productivity and contributing to the development of society.

Currently, HC is a crucial criterion for explaining the national wealth of countries, regions, and organizations. Versatility and heuristic concepts allow for its detailed analysis at different levels: (1) an individual (micro-level-individual HC); (2) an individual enterprise or group of companies (meso-level-organizational HC); (3) a region or territory, the country as a whole (macro—level-national HC); and (4) the entire humanity (global HC).

The main postulate of the HC theory is based on comparing the economic efficiency of different individuals, determined by a wide range of factors—from physical health and temperament features to broadly defined cultural factors. The first studies of HC used narrow definitions, including in its structure only the knowledge, skills, and competencies acquired due to receiving principal education and used as a resource for generating income (Becker 2003; Graham et al. 2003). At the same time, the structure usually includes two main components: (1) material capital (innate properties and abilities, such as health, physical strength, and endurance) and (2) intangible HC (acquired intangible capabilities, such as education, know-how, and qualifications) (Kuznetsov and Mikhasova 2010). Many system reviews (Kapelushnikov 2008; Tsyrenova 2006; Bourdieu 1986) cite the well-known phrase of G. Becker, where he argues that investment contributing to the formation of HC includes the costs of education, on-the-job training, health expenditures, migration, and information costs.

Many studies focus on the ability of HC to reproduce and the need for energy and purposeful activity to create it. Unlike unskilled labor, which does not require intellectual effort, skilled labor is more productive due to investment that increases its output. These investments are usually called HC. Meanwhile, definitions of HC in the social sciences usually do not refer to its value-relevant nature. Thus, when comparing the physical and human forms of capital, J. Coleman emphasizes that it is created due to the transformation of human nature, when people acquire skills through which they can master new tools and activities (Coleman 2001, p. 304).

Today, scholars are increasingly using a broad approach that describes HC through various personal traits and properties, value and worldview orientations, and peculiarities of self-perception affecting labor productivity. It is not only seen as an economic factor but the factor that affects society. HC is understood as the most educated and intellectually developed part of the labor force, the achievements of science, technology, and management, the living and working conditions that ensure more sustainable, efficient, and rational functioning of all social institutions (Korchagin 2011). Ployhart and Moliterno (2011) emphasize the importance of cognitive and emotional-communicative properties and skills, such as self-control, sociality, responsibility, creativity, openness to new experiences, and self-confidence. Handel (2003, p. 154) underlines the transformation of the attitude of modern employers to the recruitment of personnel. Such personal qualities as a positive attitude, willingness to cooperate, and communication skills, which in many cases are assessed higher than formal educational and professional requirements, came to the fore. Some authors define HC not only through education and health but also through a *common culture*, including the concept of personal freedom (Iliynsky 1996; Rimashevskaya 2004). Nevertheless, despite similarities, there is a clear difference between human and cultural capital. V. Radaev (2002) highlights the reflexive nature of HC, its logical character, distinguishing it from cultural capital, which is poorly reflected and embodied mainly in the characteristics of habitus (dispositions, physical, and practical abilities). Besides, HC actively participates in social stratification and the division of professional groups by the level of education and qualifications of their bearers (Radaev 2002).

The ambiguity of the definition is largely due to the interdisciplinary nature of the concept and the simultaneous use of traditional (narrower and more specific) and modern (extensive) approaches. Moreover, there is no consensus on assessing HC, identifying the components and factors that promote or hinder its formation and accumulation. This ambivalence gives rise to numerous attempts to capture the essence of HC by mixing the methods of analyzing the effectiveness of investment in education, health care, vocational training, migration, childbearing, and care (Koritsky 2010).

There are several basic approaches. Indirect approaches assess HC residually, assuming that the value of the benefits that the capital will bring over its life cycle will be equal to the current monetary value of the underlying asset. Direct approaches assess HC based on its various components. The three most critical direct methods include (1) an indicator-based approach that assesses HC based on educational performance indicators (adult literacy, school enrolment, average years of study, etc.); (2) a cost-based approach that focuses on past efforts, expenditures, and investments in HC; and (3) an income-based approach that concentrates on analyzing future gains and returns from HC (United Nations 2016). These approaches are not free from criticism, which emphasizes the practical impossibility of detecting direct relations between HC, its indicators, and economic growth at the individual and macroeconomic levels. Many studies questioned the effectiveness of traditional HC assessment and demonstrated that its importance was overestimated. Relying on these works, many researchers adopted a revisionist approach, where HC was treated as a common variable generating economic achievements (Cohen and Soto 2007). In Russian scholarship, the criticism of the HC concept is based on recognition of the incompleteness of the notion, limitations of the principle of rational expectations, and utility maximization. It is proposed to replace the HC theory with broader conceptions, where equality and increased opportunities for self-realization and creative work are recognized as key goals for public welfare and sustainable development. Despite methodological issues, the HC perspective remains very sought-after, especially in investigations of innovative economics and development of territories, where the HC is considered one of the vital factors (Grachev et al. 2016). The analysis of HC migration in cross-border activities constitutes an important research direction, questioning the influence of migration on its development and transformation, both positive and negative, both for individuals and society.

### *1.2 The Concept of HC in Migration Studies*

The HC model occupies a leading position in migration theories, especially those related to labor migration. The migration process is seen as an investment in HC, during which the individual bears the corresponding costs, which are paid off and converted into benefits and opportunities after moving. Hence, migration is mentioned in the HC theory of G. Becker as a basic activity affecting "future monetary and psychic income by increasing resources in people" (Becker 1994, p. 11). Therefore, migration was developed as the core element in the HC migration models by L. A. Sjaastad and B. Chiswick. In the most straightforward formulation, the logic of these models is as follows: an individual compares the costs and potential income from moving to another country in the decision-making process, and, if benefits outweigh, makes a positive decision about migration, and if not, remains in the country of origin, which allows using the concept of HC for modeling migration processes (Moiseenko and Chudinovskikh 2000).

Despite the merits of HC theory in explaining the empirical facts of labor migration and the adaptation of workers to new working conditions, the practical use of these models encountered difficulties due to the inconsistency of the modeling results with real facts. The main complaints were related to the individual's ability to predict migration results and their uncertainty. The situation is much more difficult. The models just revealed the failure of the theory of migration behavior, especially its provisions on individual decision-making. Modern migration theories, including the rapidly developing paradigm of transnationalism, showed that migration could become part of a lifestyle and give rise to transnational social spaces. Within the development of HC theory, several concepts were further developed, which consider migration in household activities and migration as a consolidated investment decision of the entire family (Kolosnytsyna and Suvorova 2005). The relationship between migration and HC relates to the economic well-being of the nation, including poverty reduction, income equalization, and redistribution, or the negative consequences and opportunities for brain drain and gain effects (Mahroum 2000; Dustmann et al. 2011; Cervantes and Guellec 2002).

Migration, both internal and international, is one of the most critical factors determining the main trends in the formation, accumulation, and use of HC in Russia. The outflow of the working-age population (especially young people with higher education) from villages to large cities, and from cities to attraction centers and foreign countries, and the influx of migrants with a specific educational and cultural background is the central problem of the redistribution of the national economy, reducing its quantity and quality.

The problems of depopulation and migration have become more acute in the border regions of Russia, especially in its Far Eastern part. Unfortunately, the tools to support regional development are not enough to preserve and develop HC of Russian territories (Soboleva et al. 2014; Khaliy 2014). Due to their greater vulnerability, migrants are often trapped in areas of illegal or semi-legal and, most often, low-skilled work. Experts note high risks of exploitation, discrimination, and social exclusion among international migrants (Maximova et al. 2019a, b). The specifics of the profession and work practice do not contribute to mobility and investment in the national economy. The standard procedures of professional identification and self-determination are violated, and the mechanism that allows professional mobility support is blocked. Language barriers, cultural differences, and the peculiarities of the education, also lead to the devaluation of the migrant status, especially among non-working members of their families. There is a need to obtain up-to-date scientific information, assess, and analyze the scale, factors, and conditions of migration processes in the border areas of Russia. Besides, one should take into account their connection with the human and intellectual capital of migrants. This paper contributes to analyzing HC of international migrants living in border regions with different socio-economic development and geographical location. Also, we clarify which elements become more noticeable and influence personal well-being in a particular region.

## 2 Materials and Methods

The empirical base included formalized sociological surveys with foreign migrants conducted in 2019 in 6 border regions of Russia: (1) Altai Krai, (2) the Orenburg Region, (3) the Murmansk Region, (4) the Pskov Region, (5) the Republic of Dagestan, and (6) the Rostov Region ($n = 100$ in each region). The main selection criterion was residence time in Russia (no more than five years, including a short-term return to the country of origin). The distribution of migrants by country of citizenship corresponded to the official statistics on the main labor force donors: 19.3% of respondents were citizens of Uzbekistan; 16.5%—Kazakhstan; 12.5%—Tajikistan; 10.2%—Kyrgyzstan; 5.0–7.0% of migrants had the citizenship of Armenia, Azerbaijan, and Ukraine; 1.0–3.0%—Belarus, Moldova, and Georgia; other countries were represented by about 3% of respondents. The respondents' age ranged from 18 to 73 years (with an average of 31.2 years and the median—28 years); the gender distribution was 67.1% of men and 32.9% of women. A considerable part of our informants (40.1%) were temporary labor migrants wishing to return to their homelands after contract expiration or seasonal work; 1.7% were transit migrants considering Russia as a stopover between the countries of origin and destination. The rest of the migrants immigrated with the family of a migrant worker, who came to Russia for educational purposes, private visits, trips, trade, and shopping as a purpose of stay. However, up to 60% of this group of migrants also worked, but illegally. We should note that 61.8% of the total of respondents, 62.6%—among labor migrants, and 82.6%—among accompanying relatives would like to obtain Russian citizenship, and 30%—to move their families to Russia, which, of course, affected the nature of the responses and characterized the integration intentions, which are also crucial for HC.

The choice of regions was due to the need to study the border regions with different geographical locations, administrative structure, ethno-social characteristics, and socio-economic and demographic development. All six regions were located in different parts of Russia and belonged to separate federal districts—North-Western (Murmansk and Pskov regions), Southern (Rostov Region), Volga (Orenburg Region), Siberian (Altai Krai), and North-Caucasian (Republic of Dagestan). The Republic of Dagestan is a national republic with a large proportion of the non-Russian population, specific ethnic and religious situation. The Murmansk, Rostov, and Orenburg regions were among the relatively prosperous and developed regions with high GRP, low unemployment rate and poverty threshold, and a favorable demographic situation. The Pskov Region and Altai Krai had several economic problems, such as (1) dependence on federal budget subventions, (2) low level of income, (3) a high proportion of the population with income below the subsistence minimum (about 18%), (4) a high demographic burden on the working-age population due to intensive aging, and (5) a significant population decline. Dagestan had the lowest GRP, high poverty thresholds, and high unemployment rate (about 11%). Simultaneously, the traditional economy and way of life in this region are associated with a high birth rate and natural population growth, and therefore, a low old-age dependency ratio.

We carried out statistical analysis using univariate and multivariate methods, including (1) regression analysis, (2) modeling of the relationship between migrants' satisfaction with their economic situation and various components of their HC, and (3) monitoring of socio-demographic parameters.

## 3 Results

### *3.1 Characteristics of the HC of International Migrants: Educational, Professional, Linguistic, and Cultural Competencies*

The most common way of assessing HC volume and quality is to focus on the benefits accruing to individuals through education. Hence, any approach, direct or indirect, necessarily includes indicators characterizing the highest level of education achieved.

Official statistics and demographic studies show that in the 1990s and 2000s, the educational level of immigrants and the Russian population was quite similar. However, the situation has changed significantly. Migrants arriving in Russia in recent years were less educated than their compatriots (middle-aged and older adults). Such a considerable outflow of relatively low-skilled people may hinder their integration into society, although this may be useful from an economic point of view and allow them to meet market needs, which could not be satisfied by the Russian population (Mkrtchyan and Florinskaya 2018). According to the results obtained, the main part of migrants (about 40–45% in all regions, in the Pskov Region—19%) had primary or secondary vocational education; about 30% (49% in the Pskov Region and only 22% in the Republic of Dagestan) had higher education; and about a fourth part (32% in the Pskov Region and 35% in the Republic of Dagestan) had only primary, basic, or general secondary education (Table 1). Thus, foreign HC imported to the border regions of Russia was primarily characterized by a predominance of manual labor jobs and low-skilled, unprivileged workers, which correspond to general trends.

The relationship between education and the age of migrants noted above was only partially confirmed. There were much more people with higher education among migrants over 61 (40.0%) than among 40–60-year-olds (27.1%) and 26–39-year-olds (29.6%). However, the largest share of people with higher education was among young migrants under 25 (41.2%), which may have represented some new trends due to the development of international cooperation in education, including the functioning of network universities (CIS, SCO, BRICS), which appeared in the late 2000s–early 2010s, and educational programs, aimed at attracting young people from Central Asia. At the same time, the education of migrants did not depend on the social, urban, or rural origin of the respondents, which indicated relatively equal access to educational services in the donor countries of the labor force, without considering the quality of education.

Despite the widespread demand for cheap and low-skilled migrant labor, people with higher education had more financial opportunities: 37.4% of them reported a very good financial situation, allowing them to make expensive purchases and increase savings; while in the categories of secondary or primary and secondary vocational education, this financial situation was typical for 16.3–22.2%. Similarly, education was a significant factor in reducing poverty and material deprivation: only

**Table 1** Characteristics of international migrants' human capital in six border regions of Russia, %

| Characteristics of the human capital | | Regions* | | | | | | |
|---|---|---|---|---|---|---|---|---|
| | | AK | OO | MO | PO | RD | RO | Total |
| Level of education | Primary, basic or general secondary education | 24.0 | 24.0 | 26.0 | 32.0 | 35.0 | 22.0 | 25.0 |
| | Initial or intermediate vocational education | 44.0 | 44.0 | 42.0 | 19.0 | 43.0 | 46.0 | 41.0 |
| | Partially completed or completed higher education | 33.0 | 31.0 | 32.0 | 49.0 | 22.0 | 31.0 | 34.0 |
| Experience of work in other countries | Yes | | 15.0 | 10.6 | | 35.6 | 21.1 | 18,4 |
| | No | | 85.0 | 89.4 | | 64.4 | 78.9 | 81.6 |
| Work status | Employed | 52.2 | 70.7 | 79.0 | 55.4 | 60.0 | 58.5 | 60.1 |
| | Unemployed | 47.8 | 29.3 | 21.0 | 44.6 | 40.0 | 41.5 | 39.9 |
| Type of organization | Private person | 60.7 | 58.6 | 42.9 | 5.4 | 61.1 | 58.7 | 50.6 |
| | Budgetary (state) organization | 12.3 | 21.4 | 22.1 | 7.1 | 11.1 | 9.5 | 14.6 |
| | Commercial (private) organization | 27.0 | 8.6 | 13.0 | 87.5 | 8.3 | 14.3 | 25.7 |
| | Self-employment | | 11.4 | 22.1 | | 19.4 | 17.5 | 9.0 |
| Health satisfaction | Mean | 3.8 | 3.8 | 3.8 | 4.1 | 3.6 | 3.9 | 3.8 |
| | Proportion of well satisfied | 65.4 | 64.0 | 67.7 | 79.1 | 61.9 | 68.5 | 67.8 |
| Status of the Russian language | First and unique native language | 18.3 | 4.0 | 10.0 | 24.0 | 17.8 | 44.7 | 19.0 |
| | Second native language | 55.3 | 65.0 | 46.0 | 33.0 | 37.8 | 30.9 | 49.2 |
| | Foreign language | 26.0 | 31.0 | 44.0 | 40.0 | 42.2 | 24.5 | 31.2 |
| Language competency level | Basic (only simple phrases and slow speech) | 15.9 | 40.6 | 6.8 | 4.7 | 23.1 | 19.6 | 15.6 |
| | Intermediate (good understanding and fluency with some mistakes) | 41.9 | 56.3 | 67.8 | 24.7 | 46.2 | 45.7 | 44.3 |
| | Advanced (ability to read original literature, understand quick speech and carry on a conversation) | 42.2 | 3.1 | 25.4 | 70.6 | 30.8 | 34.8 | 40.1 |

(continued)

**Table 1** (continued)

| Characteristics of the human capital | | Regions* | | | | | | |
|---|---|---|---|---|---|---|---|---|
| | | AK | OO | MO | PO | RD | RO | Total |
| Knowledge of the Russian culture | Zero | 9.7 | 30.0 | 12.5 | 6.3 | 29.5 | 10.8 | 13.3 |
| | Bad | 28.0 | 41.0 | 32.3 | 22.1 | 36.4 | 22,6 | 29.0 |
| | Mediocre | 38.4 | 22.0 | 34.4 | 44.2 | 27.3 | 48.4 | 37.3 |
| | Good | 23.9 | 7.0 | 20.8 | 27.4 | 6.8 | 18.3 | 20.4 |

* *Notation of regions:* AK—the Altai Krai, OO—the Orenburg Region, MO –the Murmansk Region, PO—the Pskov Region, RD—the Republic of Dagestan, RO—the Rostov Region
*Source* Compiled by the authors

**Table 2** Standardized coefficients of regression models, determining relationship between income satisfaction and different components of the human capital of international migrants in border regions of Russia*

| Predictors/Regions | AK | OO | MO | PO | RD | RO |
|---|---|---|---|---|---|---|
| Education attainment | 0.21** | 0.28** | −0.09 | 0.36* | 0.22 | 0.15 |
| State of health (satisfaction) | 0.47** | 0.25* | 0.29* | 0.43** | −0.08 | 0.05 |
| Knowledge of the Russian culture | −0.12 | 0.16 | 0.27* | 0.37* | −0.01 | −0.12 |
| Previous experience of living in Russia | −0.02 | 0.02 | 0.30** | 0.01 | 0.14 | −0.35 |
| Experience of work in other countries | | −0.13 | −0.03 | | −0.47 | −0.19 |
| Length of stay in Russia | 0,13 | 0.06 | −0.08 | −0.38** | 0.05 | −0.12 |
| Age | 0.10 | *−0.21* | *−0.24* | −0.04 | 0.18 | 0.18 |
| Gender | −0.03 | −0.31** | 0.15 | −0.04 | 0.16 | −0.05 |
| Adjusted R squared | 0.23 | 0.29 | 0.20 | 0.39 | −0.14 | 0.12 |

* —$p < 0.05$, ** $p < 0.01$, coefficients in italics—only marginal significance ($p < 0.1$)
*Source* Compiled by the authors

4.7% of migrants with higher education found themselves in a difficult life situation, while in other educational groups, this figure was significantly higher (18.1–19.6%). Simultaneously, educational resources and their benefits were distributed unevenly and had spatial (national), gender, and age characteristics. In addition to the already mentioned disproportions of highly educated migrants in different age groups, there was a significant discrepancy in educational achievements among migrants who arrived from different CIS countries. Therefore, migrants from Belarus, Ukraine, Moldova, and Kazakhstan were more educated (37.7 and 43.9% of them had higher education) than migrants from Kyrgyzstan and Armenia (30–31.6%), and especially from Tajikistan (27.3%) and Azerbaijan (20.8%). Among the migrants from Armenia and Kyrgyzstan, the highest share had only primary, basic, or general secondary education (33–38%, the lowest level was among migrants from Kazakhstan—17.9%). The comparative analysis showed that women were more educated (40.8%—with higher education, 37.9%—with primary and secondary vocational education, and 21.3%—only with secondary education) than men (30.2%, 42.3%, and 27.4% respectively, $\chi^2$, z-criterion, $p < 0.05$).

The level of education is not only valuable in itself; it is often a signal to employers of migrant learnability (Di Stasio 2014), while the recognition of formal qualifications is highly dependent on spatial, social, and cultural inequalities (Lulle et al. 2019). Professional skills and experience are less formalized and more challenging to grasp; nevertheless, they are tangible and equally important HC component. The experience of migration helps to learn about new conditions and jobs. In our sample at the time of the study, not all migrants worked—only about 60% according to aggregate data from all six regions (52.2%—in Altai Krai, 70.7%—in the Orenburg Region, 79.0%—in the Murmansk Region, 55.4%—in the Pskov Region, 60.0%—in the Republic of Dagestan, 58.5%—in the Rostov Region). The unemployed respondents were students or non-working members of migrant families. Further analysis was carried out only in the group of employed migrants.

The fact of the division of professional niches between migrants and the local population and their intersection is well-known and does not need proof. Previous studies have repeatedly identified the main sectors of the economy and areas where migrant labor in Russia was in high demand, such as (1) wholesale and retail trade; (2) repair of transport and household appliances, (3) construction; (4) social, municipal, and household services; (5) hotel business, (6) public catering; (7) transport, and (8) communication services (Mukomel 2017). According to our data, migrants in the border regions were most often employed in construction (34%), retail trade (18.3%), public catering (10.4%), and industry (10.4%). The remaining areas were represented by less than 10% of respondents in each area. The variation of migrant jobs by area in individual regions was similar with several deviations. Thus, in Altai Krai, the main economic sectors occupied by migrants were construction (46.1%), retail trade (13.8%), and public catering (12.0%). In the Orenburg Region, more than a third of migrants worked in retail trade, in construction—only 15.7%, in transport infrastructure –11.4%, in agriculture—11.4%, and in non-industrial household services—10.0%. In the Murmansk Region, the situation was similar, except for the differences in the order and fourth position: retail trade (32.5%), construction (24.7%), public catering (16.9%), and transport (9.1%). In the Rostov Region (the second agricultural region of Russia), about 40% of the respondents worked in construction, 17.5% in retailing, while only about 20%—in the agriculture and transport domains. There were two main fields in the Pskov Region where migrants were represented most: industry (59.6%) and construction (21.1%). In the Republic of Dagestan, 42.4% of the respondents, worked in construction; 12.1%—in the educational system; and about 9% in each sphere—in industry, agriculture, and transport.

The migration provoked significant changes in the professional orientations of employees, which led to the depreciation of their former values—about half of the respondents in the United States were forced to change their field of activity after moving to Russia. Maintaining specialization was typical for the Orenburg and

Rostov regions, where about 60% of migrants worked in the same field as before the migration; while in the Republic of Dagestan, only 36.7% were able to maintain their specialization. The builders and retailers worked in similar places in more than 70% of cases. Only 25% of migrants who worked in industrial enterprises could get a job in the industrial sector. At best, 18.8% of such workers found work in the construction sector, and 43.8% had to become retailers.

The distribution of the fields of activity of migrants has a lot in common. Nevertheless, their industry affiliation in different regions is rather specific. Like V. Mukomel, in 2017, we noted the predominance of employment by individuals, 42.9–61.1% of jobs in all regions, except for the Pskov Region, where only 5.4% of migrants worked for a private person. The level of migrants working in budgetary (state) organizations ranged from 7.7% in the Pskov Region to 21–22% in the Orenburg and Murmansk regions. Commercial organizations recruited international migrants much more often in the Pskov Region (87.5%) than in Altai krai (27%); while in the Murmansk, Rostov, and Orenburg regions and the Republic of Dagestan, the commercial area was relatively closed for international workers (8.3–13.4%) or, more likely, the employment of migrants in this field took shadow and informal manifestations.

We also found significant regional differences in the work experience of migrants abroad (in countries other than Russia), contributing to better chances of positioning migrants in the labor market. This survey was conducted only in four regions, the largest share of migrants with foreign experience was in the Republic of Dagestan (35.6%), where the majority of these migrants worked in construction (33.3%), education (16.7%), and retail trade (16.7%). In the Rostov Region, every fifth migrant had work experience abroad, with 50% of them in retail trade; in the Orenburg Region, this category was represented by only 15% of respondents (33.3% of them worked in transport infrastructure, the rest—in other areas with a smaller share); in the Murmansk Region—only 10.6% (60%—construction, 20%—agriculture, and 20%—retailing).

Being internally embodied in a person, HC is associated with health behavior; health demand models have shown their interdependence (Bolin et al. 2017). We assessed the level of health satisfaction as an indicator of the material (physical) HC component, given that the majority of international migrants were employed in low-skilled positions that require good physical health. The average values of the 5-point scale of health satisfaction (from *not completely satisfied* to *completely satisfied*) ranged from 3.55 in the Republic of Dagestan to 3.92 in the Rostov Region and 4.08 in the Pskov Region (there are significant differences only between these regions due to significant age differences, while other values occupied an intermediate position, as shown by unilateral ANOVA and post hoc tests); the average value for all six regions was 3.8, which is significantly higher than the average value of the scale itself (t-test, $p < 0.001$). Moreover, the analysis of generalized frequencies also confirmed that 67.5% of respondents were satisfied, and 11.4% were dissatisfied with their physical condition. Naturally, the level of health satisfaction depended on age, and younger migrants had higher scores ($m = 4.02$ in the group under 25 years, $m = 3.77$ in the group 26–39 years) than older ones ($m = 3.57$ in the group over 40 years, all differences are statistically significant, $p < 0.05$, one-way ANOVA with post-hoc tests). Other factors that determined the variation in satisfaction levels were gender and place of residence. Men were generally more satisfied with their health (m = 3.9) than women (m = 3.71), and urban migrants were more satisfied (m = 3.85) than those from rural areas (m = 3.42, t-test, $p < 0.05$). These factors were significant even after controlling for age in the regression analysis.

In contrast to the explicit competencies related to the level of education, physical condition, and professional skills, cultural competencies as HC indicators are *softer* and more *silent*; they are usually not formally evaluated but certainly play an essential role in reducing post-migration distress and acculturation (Bhugra et al. 2021). Recent studies show that migration leads to new ways of producing and reproducing cultural capital, including actions to mobilize, adopt and affirm it, and that cultural capital is linked to the power relations of both the country of origin and the country of migration. Moreover, migrants have their own migrant-specific cultural capital that reproduces gender, ethnic, and class inequalities (Erel 2010).

When studying the cultural components of HC, one should consider the results of HC together with language problems. Dustmann (1999) offered a synthesized view of language skills and communication as a particular form of *language capital* (Dustmann 1999). Many migrants from the CIS countries in Russia retained a good knowledge of Russian as a legacy of common Russian history. Nonetheless, many studies, including our own, show a decline in Russian language proficiency among the new generations of migrants arriving from the newly independent states, which hinders the use of educational capital and provokes problems with their adaptation and integration into the host society (Maximova et al. 2019a, b).

Indeed, we found significant differences in the knowledge of Russian among migrants from Russian border regions: from 4% of migrants in the Orenburg Region to 44.7% in the Rostov Region spoke Russian as their unique native language (in Altai Krai—18.3%, in the Murmansk Region—10%, in the Pskov Region—24%, in the Republic of Dagestan—17.8%). This variation was primarily due to differences in the ethnic identification and citizenship of international migrants in various regions. While 46.2% of citizens of Belarus, Ukraine, or Moldova (about 40% of whom worked in the Rostov Region) spoke Russian as their native language, the corresponding share among migrants from Kazakhstan was 34.4% (80% of these migrants were interviewed in Altai Krai, which has the largest border with this country). Only 10% of migrants from Kyrgyzstan (37.2% of them were interviewed in the Pskov Region and 30.8%—in Altai Krai); 11.5%—from Azerbaijan (mainly from Altai Krai, the Murmansk Region, and the Republic of Dagestan); 6.6%—from Armenia (respondents mainly in the Orenburg, Murmansk, and Rostov regions) spoke Russian at the native level. Meanwhile, for about 50% of international migrants, regardless of their native country, Russian was the second language, which also revealed a good knowledge. From 24.5% (Rostov Region) to 44.0% (Murmansk Region). Migrants from Tajikistan (49.5%), Kyrgyzstan (43.8%), Armenia (37.7%), and Azerbaijan (32,7%) considered Russian a foreign language. In the latter category, 15.6% of the entire sample (from 4.7% in the Pskov Region to 40.6% in the Orenburg Region) had only an elementary language proficiency in Russian; and 44.6% (from 24.7% in the Pskov Region to 67.8% in the Murmansk Region)—an intermediate, linguistic factor had an important and differentiated impact on the management of migrants' HC in the border regions. The generation gap in using the Russian language was statistically significant among migrants from Kyrgyzstan and Tajikistan. In the former country, Russian was already a foreign language for 61.1% of respondents aged 18–25, and only for 8.3% of the same age—the only native language. The latter country has seen a similar trend, demonstrating the considerable changes in the language and cultural policies of many post-communist countries, resulting in difficulties in ensuring good knowledge of the Russian language for those wishing to work in Russia. Only 27.3% of migrants over 40 identified Russian as a foreign language, while in younger groups this figure was much higher (52.4–60.6%).

Another equally important characteristic of the cultural capital of migrants is the knowledge of Russian culture, history, traditions, and the peculiarities of everyday life that make up the informal background that helps the migrant to navigate in Russian society, and the formal requirement to obtain Russian citizenship. When asked about their knowledge of Russian culture, history, politics, show business, or media, only 20.4% of international migrants in the 6 border regions reported a good general knowledge of Russian culture, while 42% admitted that they knew it poorly or did not know it at all. International migrants were more familiar with Russian popular culture—TV shows and films (28.4% of high ratings), actors and actresses, show business stars (27.5%). At the same time, good knowledge of Russian history, political leaders, and national heroes indicated a more in-depth immersion, and the higher intellectual level was characteristic of only 14–19% of respondents, especially over 40-year-olds with higher education (one-way ANOVA with post hoc tests).

### *3.2 Modeling the Efficiency of International Migrant Management in Russian Border Regions: Typical and Specific Regional Determinants*

Given that migration itself is considered an important activity for investing in the HC and improving well-being, we need to know whether migration to Russia helped improve the financial situation of migrants.

The research showed that 20.5% of participants in all regions reported a significant increase in their well-being after migration; 41.9% noted that the economic position improved slightly; 31.9% said it did not change. Only 5.7% of the respondents were not satisfied with the new economic conditions of life and said that their financial situation deteriorated, of which 1.8% reported a significant deterioration. Hence, more than 60% of foreign migrants managed to improve their living conditions due to migration, which corresponds to the general provisions of the HC theory and human migration models. Meanwhile, for a significant part of the respondents, the visit to Russia did not bring tangible economic results. This situation can indicate (1) failed and unrealistic expectations and plans, (2) the lack and limited information about opportunities to improve their financial situation, or (3) the forced nature of migration. These factors question the rational nature of the behavior of migrants and problematize the possibility of transferring HC, professional competencies, and skills across national borders, as well as consider the dynamics of HC only in economic terms, which is consistent with the latest research on HC and mobility (Lulle et al. 2019).

In the further analysis, we used the income satisfaction scale (5-point Likert scale) as an indicator of the migrant's work performance and migration success. Of course, this indicator was subjective and had many connotations that were not taken into account. At the same time, in our sample, the satisfaction with income significantly correlated with the satisfaction with the life of migrants ($r = 0.6$, $p < 0.01$); therefore, it could be used as a subjective assessment of the objective financial situation, and as a broader concept that characterizes the quality of life of migrants.

The preliminary bivariate analysis confirmed the relationship between the level of education and income satisfaction. People with higher education were more satisfied with their earnings ($m = 3.86$) than migrants with other educational achievements ($m = 3.35$–$3.41$, one-way ANOVA, $p < 0.05$). The correlation with health status was quite moderate, but statistically significant ($r = 0.41$, $p < 0.01$) and higher in the group with general secondary education or lower level of education ($r = 0.48$, $p < 0.01$), as well as among women ($r = 0.52$, $p < 0.01$). Work experience in other countries had a negative correlation with income satisfaction. Migrants who previously worked in other countries were less satisfied with their earnings in Russia ($m = 3.2$) than those who did not have such experience ($m = 3.7$, $p < 0.01$, t-test).

We found no empirical evidence that income satisfaction is related to status and Russian language proficiency; therefore, we excluded these predictors from the further regression analysis. However, we did not mean that language was not significant. We believe that there were too many migrants in our sample with good language proficiency who could answer our questions. This fact allowed us to calculate the impact of this factor. Knowledge of Russian culture was very modest but significantly correlated with income satisfaction ($r = 0.14$, $p < 0.01$), while the relationship with specific cultural indicators was detected.

Apart from the predictors listed above, we tested hypotheses about the role of previous experience of living in Russia and the duration of stay as characteristics and possible determinants of HC effectiveness.

We calculated the models for each region and identified unique determinants relevant for most, if not all, border regions, that characterize regional peculiarities.

We found that the main factor, the significance of which was reproduced in four of the six regions (except for the Republic of Dagestan and the Rostov Region), was health satisfaction, especially significant for migrants living in Altai Krai and the Pskov Region—border regions with socio-economic and demographic problems and a low standard of living. The effect of education was positive and significant in three regions—Altai krai ($\beta = 0.21$, $p < 0.01$), ($\beta = 0.21$, $p < 0.01$), the Orenburg Region ($\beta = 0.28$, $p < 0.01$), and the Pskov Region ($\beta = 0.36$, $p < 0.05$).

Other predictors were only relevant for one or two regions. Hence, the general knowledge of Russian culture determined the earnings of migrants in the Murmansk region ($\beta = 0.27$, $p < 0.05$) and in the Pskov region ($\beta = 0.37$, $p < 0.05$). Previous experience of living in Russia was a significant predictor of the economic well-being of migrants in the Murmansk region ($\beta = 0.30$, $p < 0.01$) and the duration of stay—in the Pskov Region. We should note that the latter relationship was negative ($\beta = -0.38$, $p < 0.01$), which meant that income expectations grew as more experience was gained; however, this process did not change the financial situation. The effect of the experience in other countries was negative, but very small and insignificant when other factors are considered.

Only in the Orenburg Region did we find a significant relationship between income satisfaction and gender. Taking into account the variable encoding, it was indicative of discriminatory practices against migrant women in this region ($\beta = -0.31$, $p < 0.01$). The age factor had an indirect, significant, and negative impact in two regions—the Orenburg Region ($\beta = 0.21$, $p < 0.1$) and the Murmansk Region ($\beta = 0.24$, $p < 0.1$). Its negative orientation meant that better jobs and earnings became more accessible to young and strong migrants.

The quality of our models was quite modest, and they explained from 20 to 40% of the variation of the dependent variable in four regions but did not explain anything in two regions—the Republic of Dagestan and the Rostov Region.

Additional correlation analysis showed that in these regions, income satisfaction was more determined by the development of particularistic components of social capital, rather than by the individual HC characteristics. Thus, in the Republic of Dagestan, the amount of earnings depended on favorable relations with colleagues and superiors ($r = 0.61$, $p < 0.01$), support for ethnic associations, communication with compatriots ($r = 0.56$, $p < 0.01$), and trust in ethnic diasporas ($r = 0.46$, $p < 0.05$). In the Rostov Region, the list of significant correlates included the level of trust in the employer ($r = 0.52$, $p < 0.01$), public organizations that defend the interests of international migrants ($r = 0.52$, $p < 0.01$), the diplomatic mission of the country ($r = 0.46$, $p < 0.01$), religious organizations ($r = 0.44$, $p < 0.01$) and the ethnic diaspora ($r = 0.43$, $p < 0.01$), relations with colleagues and superiors ($r = 0.42$, $p < 0.01$), and trust in police and security agencies ($r = 0.36$, $p < 0.01$). $< 0.01$).

## 4 Discussion

In current conditions of society development, HC is an essential factor of socio-economic growth and a priority of economic policy. At the same time, its relevance and impact depend on the requirements of society for education and skilled labor; they are also determined by the institutional environment that forms the conditions for individual and organizational economic activity. As for HC in Russia, many scientists share pessimistic views and doubts that HC is in demand in the modern Russian economy due to the low quality of social institutions, reduced labor demand, and the deformation of its structure (Gimpelson 2016). The institutional conditions in the health, education, and labor relations systems indicate severe problems in the effective use of HC and the lack of opportunities for its transformation into the real income of the population. When we focus on the link between the HC and international migration, things get even more complicated.

The number of international migrants imported into Russia is not uniform, with a predominance of physical labor and a few highly qualified and educated migrants. It is devalued in low-skilled temporary, provisional, or informal professions, harming the employee and the country. Moreover, this situation can lead to the underutilization of knowledge and competencies and block the development of the HC potential (Orlova 2017).

The research results illustrate some well-known trends in the impact of international migration on the quality of the criminal code in Russian regions and give rise to reflections on its role in determining the well-being of international migrants, social security, and the effectiveness of migration policy in the border areas of Russia.

## 5 Conclusion

The peculiarities of migrants' residence in the border regions are determined by their educational, professional, ethnic, linguistic, and cultural characteristics, which largely depend on the implementation of migration, language, and cultural policy in international relations between Russia and the country of origin. These features form individual migration trajectories and contribute to the development or degradation of the individual HC, its ability to improve living conditions.

Our research has shown that the situation in the border regions is not the same, although there are several common determinants. In addition to the evident and formal elements related to education and professional skills, the effectiveness of the migrant's activities is characterized by the importance of cultural components that contribute to acculturation in Russian society (especially in ethnically homogeneous Russian regions). In economically prosperous areas, HC formation and use face social inequality, age, and gender discrimination. In the southern regions, the mechanism for the formation and development of individual HC is blocked and compensated by particularistic social capital and trust relations with private and public institutions.

In current conditions, the opportunities for effective use of labor resources by migrants are limited to a few areas that allow meeting acute financial needs but lead to *patch dependence* supported by informal networks of migrants, and only systemic shifts affecting economic and innovative development and regulation of labor relations can help to overcome it.

**Acknowledgements** Paper prepared in the frameworks of the task of the Ministry of science and higher education of the Russian Federation FZMW-2020-0001 "Human capital, migration and security: transformation in the new migration environment in Central Asia".

## References

Becker GS (1994) Introduction to the first edition. In: Becker GS (ed) Human capital: a theoretical and empirical analysis with special reference to education, 3rd edn. The University of Chicago Press, Chicago, IL, pp 11–14
Becker GS (2003) Human behavior: Economic approach. selected papers on economic theory. HSE, Moscow, Russia
Bhugra D, Watson C, Ventriglio A (2021) Migration, cultural capital and acculturation. Int Rev Psychiatry 33(1–2):26–131. https://doi.org/10.1080/09540261.2020.1733786
Bolin K, Lindgren B, Grossman M, Gyrd-Hansen D, Iversen T, Kaestner R, Sindelar J (eds) (2017) Human capital and health behavior Emerald Group Publishing, Bingley, UK
Bourdieu P (1986) The forms of capital. In Richardson J (ed) Handbook of theory and research for the sociology of education. Greenwood, New York, NY, pp 241–258
Cervantes M, Guellec D (2002) The brain drain: old myths, new realities. OECD Obs 230:40–45
Cohen D, Soto M (2007) Growth and human capital: good data, good results. J Econ Growth 12(1):51–76. https://doi.org/10.1007/s10887-007-9011-5
Coleman J (2001) Social and human capital. Contemp Sci Mod 3:121–139
Denison EF (1962) Sources of economic growth in the United States and the alternatives before us. Supplementary Paper No. 13. Committee for Economic Development, New York, NY
Di Stasio V (2014) Education as a signal of trainability: results from a vignette study with Italian employers. Eur Sociol Rev 30(6):796–809
Dustmann C (1999) Temporary migration, human capital and language fluency of migrants. Scand J Econ 101(2):297–314
Dustmann C, Fadlon I, Weiss Y (2011) Return migration, human capital accumulation and the brain drain. J Dev Econ 95(1):58–67
Erel U (2010) Migrating cultural capital: Bourdieu in migration studies. Sociology 44(4):642–660
Gimpelson VE (2016) Does the Russian economy need human capital? Ten doubts. Russ Educ & Soc 58(11):696–717
Grachev S, Donichev O, Malkova T (2016) Human capital as the resource of innovation development of the region. Econ Anal Theory Pract 5(452):64–77
Graham B, Baxter RE, Evan D (2003) The penguin dictionary of economics. Penguin Books, London, UK
Handel MJ (2003) Skills mismatch in the labor market. Ann Rev Sociol 1(29):135–165
Iliynsky IV (1996) Investments in the future: education in investment reproduction. SPbUEF, St. Petersburg, Russia
Kapelushnikov RI (2008) Notice about national human capital. Preprint WP3/2008/01. HSE, Moscow, Russia
Khaliy IA (2014) Modern Russian borderland: General characteristics. Vestn Inst Sotziologii 5(4):23–33
Kolosnytsyna MG, Suvorova IK (2005) International labor migration: theoretical basis and regulative policy. HSE Econ J 9(4):543–565
Korchagin YuA (2011) Human capital – intensive socio-economic factor of the development of person, economy, society and state. HSE, Moscow, Russia
Koritsky AV (2010) Human capital as a factor of economic growth in the Russian regions. Siberian University of Consumer Cooperation, Novosibirsk, Russia
Kuznets SS (1966) Modern economic growth. Yale University Press, New Haven, CT
Kuznetsov YuA, Michasova OV (2010) Human capital: formation, measurement, impact on economic growth. Econ Anal: Theory Pract 26(191):21–33
Lucas RE (1990) Why doesn't capital flow from rich to poor countries? Am Econ Rev 2(80):92–96
Lulle A, Janta H, Emilsson H (2019) Introduction to the special issue: European youth migration: human capital outcomes, skills and competences. J Ethnic Migr Stud (Special Issue: Human Capital). https://doi.org/10.1080/1369183x.2019.1679407

Mahroum S (2000) Highly skilled globetrotters: mapping the international migration of human capital. R&D Manag 30(1):23–32
Maximova SG, Omelchenko DA, Noyanzina OE (2019a) Social risks and vulnerability of labor migrants in the Russian borderland. In: The European proceedings of social & behavioural sciences EpSBS, pp 72–80. https://doi.org/10.15405/epsbs.2019.04.02.10
Maximova SG, Shermatova SN, Omelchenko DA, Noyanzina OE (2019b) Institutionalization of network mechanisms of the Russian world functioning: aspects of preservation and advancement of the Russian language and culture abroad. Soc & Secur Insights 1(4):16–37. https://doi.org/10.14258/ssi(2018)4-01
Mincer J (1958) Investment in human capital and personal income distribution. J Polit Econ 4(66):281–302
Mkrtchyan N, Florinskaya Yu (2018) Skilled migration into Russia: balance of losses and benefits. Econ Dev Russia **25**(2):60–63
Moiseenko VM, Chudinoskikh OS (2000) Theory of human capital and migration studies in Russia. Probl Forecast 4:124–137
Mukomel VI (2017) Migrants at the Russian labor market: occupations, mobility, intensity of labor and wages. Stat Econ 14(6):69–79. https://doi.org/10.21686/2500-3925-2017-6-69-79
Orlova NA (2017) Non-standard forms of employment and changes in human capital: analysis of the unskilled labor migration to Russia. Monit Public Opin: Econ Soc Chang 1(137):156–170
Ployhart RE, Moliterno TP (2011) Emergence of the human capital resource: a multilevel model. Acad Manag Rev 1(36):127–150
Radaev VV (2002) The notion of capital, forms of capitals and their conversion. Econ Sociol 3(4):96–111
Rimashevskaya NM (2004) Human capital of Russia and issues of "preservation of population." Russ Econ Rev 9–10:22–40
Schultz TW (1961) Investment in human capital. Am Econ Rev 51:1–17
Soboleva SV, Grigoriev YuA, Smirnova NE, Chudaeva OV (2014) Features of population formation of Siberian border territories. ECO J 44(11):20–35. https://doi.org/10.30680/ECO0131-7652-2014-11-20-35
Solow RM (1988) Growth theory and after. Am Econ Rev 78(3):307–317
Thurow L (1970) Investment in human capital. Wadsworth Publishing Company, Belmont, CA
Tsyrenova AA (2006) Development of human capital in the conditions of transformation of institutional environment. VSTGU, Ulan-Ude, Russia
United Nations (2016) Guide on measuring human capital. United Nations Publishing, Geneva, Switzerland

# Diaspora Policies in Russian Border Areas: Segregation or Integration?

**Maria I. Cherepanova, Svetlana G. Maximova, and Syldysmaa A. Sariglar**

**Abstract** In this study, the authors examine national diasporas' social policies regarding immigrants from Asian countries. The authors conducted surveys with migrants and interviewed the diasporas' representatives. The authors concluded that in the studied regions, the migrants are most likely to integrate than segregate. Moreover, the authors described the role of diasporas in migrant acculturation.

**Keywords** Segregation · Integration · Diasporas · Migrants · Migration policy

## 1 Introduction

In Russia's border areas, like in the rest of the world, the diaspora overtakes the traditional, somewhat segregated ways of migrant settlement. It forms new ways of acculturation for both long-term residents and newly arrived people.

The authors define the term "diaspora acculturation strategy" as a system of social means for forming a new migrant identity. This identity harmonically combines the immigrants' internal cultural values with the external cultural values of host regions.

In this study, the authors analyzed the following criteria:

1. The perceived diaspora credibility;
2. Ethnic self-identification assessment;
3. Subjective evaluation of immigrants' quality of life in Russia;
4. The intensity of interaction between immigrants and diasporas;
5. The evaluation of acculturation demand.

Diaspora credibility is one of the main criteria that form its acculturation strategy. On average, only a fifth of all immigrants trusts their diasporas completely. Every third immigrant answered "average" in the question about diaspora credibility. This

M. I. Cherepanova (✉) · S. G. Maximova · S. A. Sariglar
Altai State University, Barnaul, Russia
e-mail: cher_67@mail.ru



S. G. Maximova et al. (eds.), *Advances in Natural, Human-Made, and Coupled Human-Natural Systems Research*, Lecture Notes in Networks and Systems 234,
https://doi.org/10.1007/978-3-030-75483-9_36

means that their trust in diasporas is contextual. Every fifth migrant mistrusts their diaspora, stating that this social institute is not enough for the immigrant acculturation in Russia.

## 2 Materials and Methods

This study is based on the "Transit Migration, Transit Region, and Migration Policy of Russia: Security and Eurasian Integration" sociological survey, conducted in 2017–2019. In this survey, 3,600 migrants from the national diasporas of six Russian regions were sampled. The leaders of national diasporas and associations were interviewed ($n = 600$). The authors employed mathematical-statistical data and used the methods of modeling and regressive analysis.

## 3 Results

The highest index of diaspora credibility was registered in the Pskov Region, where most immigrants acculturated and integrated into local communities. Moreover, most immigrants in Pskov Region speak highly of their life in Russia. However, there are no statistically significant differences in diaspora credibility indexes between the regions. Migrants in all regions display an average, situational level of trust in diasporas. The influence of diaspora on the immigrant acculturation depends mostly on their ethnic fluidity and their desire to study the host regions' culture.

In all the studied regions, the migrants were most likely to integrate than segregate, combining their own cultural values with the host regions' culture. However, the integration indexes differ between the regions. In Dagestan, every fifth immigrant aims to preserve their own identity and reject Russian cultural traditions. In Orenburg and Murmansk regions, only a fifth of immigrants tends to do so. The immigrants in Altai Krai, Pskov Region, and Rostov Region are more likely to assimilate. Only 6% of immigrants to the Rostov Region reject Russian cultural values. Most likely, the immigrants that refuse to integrate view Russia as a lucrative economic and technological resource. This behavior is somewhat similar to exploitation colonialism (Voronov and Dmitriev 2016; Cherepanova and Sariglar 2019). This fuels social unrest and extremism.

Most immigrants tend to value individual social activities, the social capital of their ethnicity, and familial relations more than the diaspora's assistance. Therefore, diaspora acculturation strategies must pay attention to the individual's psychology.

One of the main functions of diasporas is assisting in immigrant acculturation. More than half of the diaspora leaders agree that there is a need for a universal system for immigrant acculturation measures. Every third leader believes that just selective measures must be undertaken to help only the struggling immigrants. A small number of experts (5.3%) believe that no measures are required. More than half of the diaspora leaders in the Rostov Region, Orenburg Region, Dagestan, and Altai Krai were in favor of integration measures. In Murmansk and Pskov Regions, a bit less than half of the leaders favored these measures. The diaspora leaders favored selective measures in Pskov Region, Murmansk Region, Altai Krai (38.5%), Dagestan (30.1%), and Orenburg Region (25.1%).

Some diaspora leaders disfavored the integration measures altogether in Pskov Region (~10%), Murmansk Region (~10%), Dagestan (5.1%), Rostov Region (5.1%), and Altai Krai (3.8%). The increased migration in these regions partially causes these opinions. For example, in Altai Krai in 2017–2018, around 7,000 working-age citizens left the region, while almost 6,458 immigrants arrived. Per one person with higher or vocational education that left the region, almost five immigrants without education and with a low level of Russian language have arrived. Most immigrants (65%) arrive from CIS countries and neighboring countries, mainly from Tajikistan, Uzbekistan, Armenia, and Kyrgyzstan.

In 2018, 20,573 people arrived in the Rostov Region; 3,238 in Dagestan; 2,452 in Pskov Region; 3,295 in Murmansk Region; and around 1,600 in Orenburg Region. Most of these people arrived from Kazakhstan, Tajikistan, Armenia, and Ukraine. In the Murmansk Region, the immigrant flow from China increased by 40%. Moreover, the flow of Ukrainian refugees has significantly increased immigration to these regions. However, the Ukrainian population is ethnically and culturally similar to the Russian one, which eased their integration processes.

According to the diaspora leaders, some of the most popular acculturation measures are Russian language education; legal education; cross-cultural communication; popularization of ethnic and religious tolerance; job search assistance. The least popular measures included house search assistance; domestic problem solving; informing on the activities of human rights non-profit organizations; education assistance; competence development assistance; psychological consulting; stress relief counseling; acculturation behavior development.

In Altai Krai, the diaspora leaders believe that Russian language education is the most socially important measure since it forms the basis for all other measures. This measure is popular with the immigrants in Altai Krai, Dagestan, Murmansk Region, Rostov Region, and Pskov Region. In the Orenburg Region, the immigrants are more likely to need help with legal education, employment, and labor law consultations. A third of all diaspora leaders state the importance of Russian language education for cross-cultural communication. Every fifth leader believes that immigrants need house search assistance.

In Murmansk Regions, the diaspora leaders stress the importance of Russian language education, legal education, and information on non-profit human rights organization (35.0%). A fifth of the region's diaspora leaders believes that it is important for immigrants to find a job, get an education or additional qualifications. The head of the Murmansk Region Migration Service stated that, due to an extensive system of measures, the crime among immigrants is under control (Cherepanova and Sariglar 2019).

The diaspora leaders in Pskov Region mainly value Russian language education, legal counseling, and cross-cultural education. A third of all leaders note the importance of job search assistance and labor law consultation.

In Dagestan, the most popular integration means were Russian language education, judicial education, job search assistance, and house search assistance. Such means as psychological counseling, education assistance, according to the diaspora leaders, were not in-demand. This may be mainly caused by the national and ethnic characteristics of the region. In the Rostov Region, most immigrants valued Russian language education, cross-cultural communication events, and the events to popularize ethnic and religious tolerance.

Generally speaking, the data suggests that diaspora communities aim to integrate and acculturate new immigrants, although these processes' specifics may differ between regions.

## 4 Discussion

Different types of acculturation correspond to the different levels of ethnic identity building. A high level of cultural integration facilitates cross-cultural communication and helps immigrants adapt to the new environment.

Most regional diaspora leaders (48.9%) believe that the type of "partial ethnic integration" is predominant in their region. This type presupposes a minor syncretism between two ethnic identities, where the original ethnic identity is still predominant. A third of all leaders state that acculturation follows the "full ethnic integration" pattern in their region. According to the diaspora leaders, only a tenth of all immigrants may fully assimilate into the host culture, fully or partially rejecting their original ethnic identity. Only a minor fraction of immigrants tends to marginalize (1.1%) and segregate (2.3%), meaning that there is a demand for acculturation in these regions.

The current migration policies must be based on objective communication mechanisms and a conscious, purposeful system for managing relations and technologies, information, and communication processes (Voronov and Dmitriev 2016; Kuznetsov 2017).

## 5 Conclusion

Modern diasporas are a powerful tool in immigrant acculturation.

The most important acculturation indexes are the perceived diaspora credibility and the positive impact of acculturation in interacting with the host population.

The processes of enclavization and self-segregation hinder acculturation and integration.

Regional diasporas play a significant part in the immigrant acculturation and help in forming the multifaceted ethnic identities.

The diaspora acculturation processes differ between the regions, but overall patterns and trends are universal for all studied regions.

Some migrants tend to create ethnic and economic niches, refusing to integrate into society. They view their host territory and culture as a source of personal economic gain.

These tendencies create ethnic barriers, hinder the adaptation and acculturation of other migrants, and raise social tensions.

**Acknowledgements** This study was conducted as a part of the project НШ-2632.2020.6 "Return Migration and Migration Policy: Acculturation Strategies for Migrants and Host Populations in Russian Border Regions".

## References

Cherepanova M, Sariglar S (2019) Social sustainability and migration in the transit region of Russia. Adv Soc Sci Educ Humanit Res 364:737–741

Kuznetsov IM (2017) Balance of interethnic attitudes as an indicator of the state of interethnic relations. Universe Russ 26(1):58–80

Voronov VV, Dmitriev AV (2016) Regional problems of diaspora communities: an expert evaluation. Novyy Khronograf, Moscow, Russia

# Ethnic Stratification: Formation of Enclaves of Migrant Adaptation in the Transboundary Territories of Russia

**Maria I. Cherepanova, Svetlana G. Maximova, and Syldysmaa A. Sariglar**

**Abstract** The paper focuses on the process of migrant enclavization during their adaptation in the host communities of six Russian regions. The paper describes the attitude of the expert community of regions to enclavization. The duality in the assessment of this phenomenon is revealed. The conditions and factors of ethnic stratification are determined. Regional models of adaptation of migrants in the context of their interaction with diasporas with a strong ethnic stratification in the territory of residence are presented.

**Keywords** Adaptation · Assimilation · Diasporas · Enclave · Ethnic stratification · Migrants · Migration policy

## 1 Introduction

The problem of ethnic stratification and enclave formation is one of the most acute in migration. This process reduces or makes it impossible to adapt and integrate those arriving in the local community (Akhiezer 2007; Horst 2008, 2013). The main barriers are closeness and self-sufficiency, which minimizes the communicative potential in interaction with the local population. Arriving migrant groups are marked with quasi-primordialist self-organization strategies leading to the formation of outwardly closed communities marked by ethnicity (Akayev 2013; Bedrik 2007). This separation in the arrival regions is caused by the unfriendly attitude of the local population, especially the young one. Additionally, the high cohesion of the non-ethnic population against the background of the disunity of the Russian population acts as an additional risk factor.

M. I. Cherepanova (✉) · S. G. Maximova · S. A. Sariglar
Altai State University, Barnaul, Russia
e-mail: cher_67@mail.ru



S. G. Maximova et al. (eds.), *Advances in Natural, Human-Made, and Coupled Human-Natural Systems Research*, Lecture Notes in Networks and Systems 234,
https://doi.org/10.1007/978-3-030-75483-9_37

## 2 Materials and Methods

The empirical basis of this study is formed by the results of sociological study ?Transit migration, transit regions, and migration policy of Russia: security and Eurasian integration (2017?2019).? Migrants and representatives of national diasporas from six Russian regions were interviewed. The sample consisted of 3,600 people. Expert interviews with leaders of national associations (n?=?600) were analyzed. The relevance of the described results and trends is achieved by using mathematical statistics, modeling methods, and regression analysis.

## 3 Results

More than a third of all respondents indicated that they were more likely to view national enclaves negatively. One-third of the representatives of diasporas are negative. Every tenth member of the diaspora is more likely to have a positive attitude towards this process. A comparative analysis of estimates of the population and diasporas recorded the expected greater loyalty of the latter to the problem of national enclaves (Brubaker 2012; Cherepanova and Sariglar 2019).

The essence of the process of enclave formation is the formation of a special adaptation scenario, which forms special attitudes of migrants aimed at preserving cultural patterns of behavior, values, and norms. As a result, there is no deep process of mastering the specifics of life in the arrival region (Clifford 2014; Fedorova 2014).

The analysis of regional results revealed that the most remarkable intolerance to Diaspora enclaves is shown in the Pskov, Murmansk, Rostov regions, and the Altai Krai. The obtained results are consistent with other identified trends. They allow us to predict the presence of hidden inter-ethnic problems in these regions (Glick Schiller 2009; Ho 2011; Horst 2006). The areas with a high concentration of immigrants are most tolerated in the Republic of Dagestan.

Adaptive strategies of diasporas are implemented through specific practices in cooperation with regional and municipal authorities. In this regard, the role of diaspora and communities significantly increases in regional societies (Jeffery and Murison 2011; Khenkin 2008; Sirazetdinova 2016).

We will present significant social practices that optimize the process of adaptation of migrants.

Representatives of diasporas indicated the demand for the following assistance to migrants:

- Assistance in learning Russian by foreigners and their family members (46.5%);
- Fight against crime (26.1%);
- Mutual acquaintance with the culture of ethnic groups (34.3%);
- Development of official relations with donor countries (29.6%); etc.

The minimum number of respondents (3.8%) believes there is no need to do anything about migrants. This fact emphasizes the understanding of the social significance of integration and adaptation measures. The results emphasize a much more pronounced need for all measures than the opinion of the regional population. However, the respondents consider it two times less necessary to take measures aimed at combating violations in migration. Thus, we recorded a discrepancy between the estimates of population and immigrants in the context of hazards and risks associated with migration that may reflect stereotyping of the population or underestimating the real dangers from representatives of diasporas, their faith in the law-abiding behavior of their representatives in the host regions.

A certain regional specificity of responses about the demand for assistance to migrants is revealed. For example, in the Altai Krai, the most popular areas of assistance are teaching Russian and familiarizing the population and migrants with the culture (Fedorova 2015; Grillo and Riccio 2004; Horst et al. 2010).

In the Murmansk region, the most critical areas are teaching the Russian language, fighting crime, and teaching newly arrived migrants the culture, customs, and values. For the Pskov region, familiarity with the culture of the population and migrants is essential and the development of multilateral good-neighborly relations between the donor and recipient regions. Dagestan assists in teaching Russian (34.9%) and protecting migrant interests and rights (28.7%), etc. The migrants of the Rostov region believe that almost all areas of assistance to migrants are equally important.

The diaspora representatives clearly understand the role of diaspora in the integration of migrants in the region. Every second member of the diasporas believes that diasporas are an important social institution that significantly optimizes the adaptation of newcomers. Simultaneously, every fourth migrant believes that diasporas do not play a role. Approximately the same number of respondents found it difficult to determine the role of diasporas. Every tenth member of the diaspora does not see the socializing role of the diaspora and notes the negative disintegrating value of the diaspora for migrants. It is noteworthy that, compared with the population response, diaspora members more often indicate the positive role of this public institution based on a better understanding of the diaspora role in any modern society.

Representatives of diasporas are optimistic about the role of this social institution in the Altai Krai (40.7%) and the Rostov region (49.4%). In the Pskov and Orenburg region, the respondents believe that, under certain conditions, diasporas can significantly hinder integration. However, compared to the population responses, members of diasporas appreciate the role and importance of national and ethnic communities much more positively. This appreciation is usually based on practical experience in adapting and integrating migrants in the Russian regions.

# 4 Discussion

To determine the social determinants that have the greatest and decisive influence on the formation of conditions for effective adaptation of migrants, we carried out a regression analysis of statistical data characterizing the features of naturalization of migrants? socio-economic life and the specifics of interaction and social attitudes towards ethnic diasporas.

The indicator of a dependent variable is the level of trust in diasporas, including the positive pole (high level of trust in local diasporas) and the negative pole (low level/lack of trust in diasporas).

The following factors are established as independent variables:

- Satisfaction with the support of the national association and communication with people coming from the same country;
- Manifestations of everyday social practices in interaction with people coming from the same country and diasporas, their intensity during the adaptation;
- Assessment of motivation to accept Russian citizenship;
- Assessment of the security of residence in Russia;
- Frequency of interaction with diasporas during integration, etc.

Regression analysis included the analysis of statistical indicators such as R?(R-square of Cox and Snell)?the coefficient of difference in prevalence; B?the coefficient of the interaction of variables; p?the significance.

One of the main conditions for effective adaptation of migrants was a high level of quality and intensity of interaction with ethnic diasporas during naturalization in everyday life in Russia.

# 5 Conclusion

In conclusion, we must note two basic processes. The first process is more pronounced. It is aimed at increasing the priority of cultural and national values of the host society among migrants. Therefore, this process ensures effective adaptation. The second process is expressed to varying degrees in the regions. It relates to the non-adaptive process of using the environment as a temporary economic and technological resource and the lack of explicit recognition of its cultural and value norms. Simultaneously, there is a reproduction of their traditional norms and rejection of attempts at acculturation. The analysis highlighted the sustainable factors inherent in the adaptation models of each of the analyzed cross-border territories (Cherepanova and Sariglar 2019).

## 5.1 *The Model of Adaptation of the West Siberian Diaspora (on the Example of the Altai Krai)*

The specificity of adaptation strategies is that positive role of diasporas in migrant adaptation is expressed in the region. However, during the interaction of migrants with the local population, the institute of diasporas has some unrealized social potential. In particular, the organization of adaptive courses is supported by only one in ten diaspora members. The issue of protecting the rights and interests of migrants is almost not given due attention. The level of trust in the diaspora institution is relatively high for every second migrant. In general, an indicator of the importance of diasporas is the prevalence of a constructivist identification strategy in the region, which is manifested in the positive dynamics of ethnicity towards acculturation and the predominance in the region of optimal patterns of residence of migrants (adaptation, integration), and the absence of pronounced processes of marginalization and maladaptation of the non-ethnic population. Initially, the ethnically diverse population of the Altai Krai (about 120 ethnic groups) is psychologically stated as ?we? complementary to the arriving close ethnic groups (Russians, Ukrainians, Belarusians, Tatars). People from Central Asia and the Caucasus are often referred to as ?strangers.? The relative confrontation with these migrants is associated more with psychological discomfort and does not have a religious or national connotation. Two groups of migrants are forming in the region?old-age groups that have lived in the region for five or more years and newly arriving groups of migrants who need support and adaptation. There is a low risk of ethnic conflicts in the region. Nevertheless, there are also ?quasi-mardialist strategies for self-organization? (Akhiezer 2007) of migrants and the relatively loyal attitude of diasporas to this problem. Outwardly closed national groups identify ethnicity as social capital in the competition between newcomers and local populations.

In general, optimistic scenarios for the integration of migrants are being implemented in the region, primarily due to the institutional support for the diasporas.

## 5.2 *The Model of Adaptation of the North Caucasian Diaspora (on the Example of the Republic of Dagestan)*

Conflicting trends were identified in the region. On the one hand, Dagestan has one of the highest levels of trust and a positive role of diasporas in migrant adaptation. A positive result is that two-thirds of migrants show a sufficient level of adaptation and integration into the local community.

On the other hand, in the region, one in three migrants experiences significant problems in adapting or forming negative living patterns, such as marginalization or isolation. In this regard, the region has the highest demand for measures to protect migrants? rights and freedoms. The potential of the regional diasporas is not fully realized in efforts to optimize the interaction between migrants and the local community. The region has one of the highest levels of aversion to the process of enclavization, which may have predictive estimates of the risk of ethnic conflict.

## 5.3 The Model of Adaptation of the North-Western Diaspora (on the Example of the Murmansk and Pskov Regions)

Constructivist models of migrant adaptation are common in the Murmansk region. A reasonably high integration potential is being realized in the region. However, as in other territories, the social capital of diasporas needs to be implemented in practices that support the interests of migrants and help them interact with the local population of the region.

The Pskov region was equally dominated by constructivist quasi-free models of migrant adaptation. Only one in two migrants trust diasporas and assess their positive role in adaptation. The region has the most apparent rejection of the process enclavization.

## 5.4 The Model of Adaptation of the Volga Diaspora (on the Example of the Orenburg Region)

Constructivist models of migrant adaptation prevail in the region, which is a natural result of pronounced integration processes. The region is dominated by a high level of confidence in diasporas and an optimistic assessment of their positive potential.

## 5.5 The Model of Adaptation of the South Russian Diaspora (on the Example of the Rostov Region)

Two mutual processes were identified in the region. They are expressed multidirectional and equally. The first is marked with the spread of constructivist models of adaptation and the dynamics of ethnicity towards acculturation. The second process is associated with the pursuit of national norms and traditions. A high level of loyalty of local diasporas to the process of enclavization was revealed, which can be predicted as a particular risk of national tension.

**Acknowledgements** The paper was prepared as part of the project NSH-2632.2020.6 ?Return migration and migration policy: Adaptive strategies of migrants and the receiving population in the border regions of Russia.?

## References

Akayev BH (2013) Islam in Europe: state, adaptation, prospects. Islam Stud 4:4?8

Akhiezer AS (2007) Territorial migration?the realization of the need for the fullness of being. Soc Sci Mod 3:141?149

Bedrik A (2007) Labor adaptation of ethnic migrants in conditions of social risks: the example of the Rostov (Dissertation of candidate of social sciences). Southern Federal University, Rostov-on-Don, Russia

Brubaker R (2012) Ethnicity without groups. https://id.hse.ru/data/2013/05/15/1245582947/Brubaker_Press%20ca??.pdf

Cherepanova M, Sariglar S (2019) Social sustainability and migration in the transit region of Russia. Adv Soc Sci, Educ Humlties Res 364:737?741

Clifford J (2014) Diasporas. Cult Anthropol 9:302?338

Fedorova YE (2014) Islam in the modern European community: stereotypes and reality. Philos Thought 7:99?125

Fedorova YuE (2015) Women?s issues in Islam: discourse on the status of women in modern Islamic society. Philos Thought 10:102?128

Grillo R, Riccio B (2004) Translocal development: Italy?Senegal. Popul Space Place 10:99?111

Ho E (2011) Claiming the diaspora: elite mobility, sending state strategies. Prog Hum Geogr 40:1?16

Horst C (2006) Transnational nomads: how somalis cope with refugee life in the dadaab camps of Kenya. Berghahn Books, Oxford, New York, NY, UK

Horst C (2008) A monopoly on international tourism and refugee camps. Afr Spectr 43:121?131

Horst C (2013) The depoliticization of diasporas from the Horn of Africa: from refugees to transnational aid workers. Afr Stud 72:228?245

Horst C, Ezzati R, Guglielmo M, Saggiomo V, Warnecke A, Mezzetti P, Laakso L (2010) Participation of diasporas in peacebuilding and development: a handbook for practitioners and policymakers. Peace Research Institute Oslo, Oslo, Norway

Jeffery L, Murison J (2011) The temporal, social, spatial, and legal dimensions of return and onward migration. Popul Space Place 17(2):131?139

Khenkin S (2008) Muslims in Spain: problems of adaptation. World Econ Int Relat 3:48?58

Schiller NG (2009) A global perspective on migration and development. Soc Anal 53:14?37

Sirazetdinova MF (2016) The image of a Muslim woman and the manipulation of consciousness. In: Frolova IV, Gazizova LI (eds) The status of a Muslim woman in a polyconfessional society: history and modernity. Mir Pechati, Ufa, Russia, pp 195?206

Voronov VV, Dmitriev AV (2016) Regional problems of the diaspora of the new-earth communities: expert dimension. New Chronograph, Moscow, Russia

# Confessional Factors in the Process of Migrants' Social and Cultural Adaptation in Altai Krai

**Mirra V. Kashaeva**

**Abstract** The author considers the role of religion in the social and cultural adaptation process of Altai Krai ethnic communities. The study shows that Central Asian migrants see religion as one of the factors of ethnic and cultural identity. At the same time, there is a clear tendency of migrants to create non-denominational structures within the ethnic community. The leaders of ethnic communities, by creating various confessional catechizes, inviting imams from their permanent residence to give lectures, contribute to forming radical trends within the communities. One of the alarming factors of the migrants' ethnic and cultural adaptation is youths' isolation. Identifying themselves with ethnos' traditional religion, they create their own communication groups (mainly in social networks) and do not strive to go back to the established religious practice.

**Keywords** Evolution of Islam in Siberia · Altai Krai ethnic communities · Confessional adaptation of migrants · Social and cultural adaptation of migrants

## 1 Introduction

The migration processes that began with the USSR collapse, changing their target bases, retain their intensity at the present stage. During the first five years after the Soviet Union collapsed, the migration flow structure was dominated by the returning Russian population to the places of permanent residence of their relatives and friends. Since the beginning of the 2000s, due to political and economic instability in the Former Soviet Union countries, citizens of Uzbekistan, Tajikistan, Kyrgyzstan, and Kazakhstan began to predominate in the migration environment. As a result of the difficult military and political situation in the North Caucasus republics, the internal flow of migration to the regions with operating centers of ethnic culture that helped the social and cultural adaptation process increased. Altai Krai, on the one hand, is labor-surplus in terms of the population employment structure. On the other—initial

M. V. Kashaeva (✉)
Altai State University, Barnaul, Russia
e-mail: mirra2407@mail.ru

 
S. G. Maximova et al. (eds.), *Advances in Natural, Human-Made, and Coupled Human-Natural Systems Research*, Lecture Notes in Networks and Systems 234,
https://doi.org/10.1007/978-3-030-75483-9_38

adaptation in Altai Krai is most attractive for migrants due to the relatively low housing and food costs and the functioning of a wide range of ethnic and cultural communities. One of the factors of the migrants' social and cultural adaptation is the confessional factor since religion in the mindset of the believer organizes a way of relating to the world as a whole and to each person separately, forms life-purpose and value-oriented dominants of behavior, regulate intrafamilial and methods of social communication.

There is a consideration in the Russian scientific community about the practical merging of confessional and ethnicity and the allocation of territorial communities in the Russian Muslim Ummah structure (e.g., Volga Muslims, Central Asian Muslims, Caucasian Muslims) (Singizov 2016). Since the main way to increase the size of ethnic communities is the external migration, it is worth focusing on the fact that the arriving migrants incorporating into the local community are looking mostly for a geographically and ethnically close circle of adherents than for a mosque (Fatkhullin 2016). In this regard, the actual study contributes to the solution of regional and country security issues, identifying methods and mechanisms of control and interaction with the Islamized part of the migration flow.

## 2 Materials and Methods

The author studies the role of the confessional factor in the process of social adaptation based on interviews with the leaders of the ethnic communities of Altai Krai: center of national culture "Vainakh," Altai regional social organization "Tajik Diaspora," Altai regional social organization "Center of Uzbek culture - Batyr," local ethnocultural autonomy of Kazakhs "Asyl Mura"; Altai regional ethnocultural organization "Azerbaijan"; Altai regional social organization "Union of Armenians of Altai Krai"; as well as local Jewish ethnocultural autonomy. During interviews with the leaders of ethnic communities, the author posed the following questions:

1. What are the specific characteristics (age, degree, and occupation) of your national community members attending the mosque?
2. Have members of your community experienced the restrictions of freedom of religion?
3. Do you consider modern society as tolerant?
4. Have you heard any complaints from community representatives about manifestations of xenophobia?
5. What prevention methods of extremist activity does your community use?
6. What inquiries have you heard from ethnic community members regarding ways of spiritual development?
7. Have you ever heard complaints from community representatives about the problems of internal confessional communication?

During the research, the author used the following methods:

1. The observation method helped to identify the specific confessional characteristics inherent to separate ethnic communities;
2. A comparative analysis method allowed us to specify the structural features of the Muslim ummah in Altai Krai.

## 3 Results and Discussion

The dynamics of ethno-social processes in modern Russian society are characterized by a high external and internal migration level. Migrants are not only bringing a cheap labor resource to various regions of the Russian Federation, but also introduce the traditions, norms, and religion of their motherland. These days, the Russian scientific community is trying to identify the radicalization factors of the religious life of ethnic communities caused by the migration influx. Researchers note the changing ethnic composition as one of the factors that sharply change the mood among the Siberian Muslims. The internal migration in Russia increases the number of North Caucasus Muslims. In the Muslim ummah of the Khanty-Mansiysk Autonomous Okrug, as noted by T. R. Singizov, there has been an increase in Caucasian and Central Asian Muslims by 45%, and the growth of the total number of ethnic Muslims in the region continues (Singizov 2016). The author sees the dangerous tendency in the fact that Caucasian Muslims adhere to a more politicized form of Islam, do not accept Tatar imams' service, and call for imams from their midst who perform services and ritual practices in Arabic. The consequence of this situation is a surge in extremist organizations' activity. Another dangerous trend associated with the extremist sentiment manifestation in ethno-confessional communities is the increasing number of unofficial religious communities.

The researcher A. P. Yarkov claims two contradictory tendencies in the Siberian Muslim ummah: radical and conservative. In his opinion, the Adat form of Islam in Siberia exists due to the natural and social specifics of the environment. Focused on the Islamic and customary law norms as well as on regional social specifics, Islam of the Siberian ummah is not inclined to manifest ideological and cult radicalization. Due to the number and specificity of community activities in the Siberian region, A. P. Yarkov says, according to official statistics, there are six officially registered and 73 more unregistered communities in the Novosibirsk region that canonically differ from the traditional ones (Yarkov 2014). A. V. Martynenko and T. D. Nadkin note the influence of organizational processes on the Islamic doctrine and ideology peculiarities. They consider the possibility of Caucasian and Central Asian preachers becoming radical elements in the republics of the Volga region. The dangerous tendencies of splitting the Muslim communities are enhanced by the influence of various centralized muftiats competing for leadership in religious organizations (Martynenko and Nadkin 2015). Thus, ethnic migration within Russia and other countries is the factor of radicalization and criminalization of religious life. The active spread of Islam in Altai Krai occurred at the beginning of the twentieth century. At that time, as the leading confession, Orthodoxy was actively supported by the government to involve the foreign population. Further events related to the implementation of religious provisions of the Soviet government's policy do not allow Islam to establish or develop any significant institutional forms. The development of Islam in Altai Krai followed up the migration processes of the early 1990s. Simultaneously, the migration brings two forms of Islam—first, the traditional form, expressed in the mosque building, forming of a community, and establishing of a catechism school; second, a politicized Islam ideology, which manifests itself in active missionary work and literature distribution which format and content are far from the classical versions. As noted earlier, because of the merging tendency of the ethnic and confessional factors, ethnic communities' structures become a filter for the entering radical, politicized forms of Islam.

Currently, there is a mosque in Altai Krai, as well as the Armenian Apostolic Church parish. In this regard, Kazakhs professing Islam, as well as Armenians who belong to the Apostolic, Orthodox, and Catholic Churches, can freely meet their religious needs. The attempt to revive Orthodox Judaism in 2006–2012 did not form a stable confessional structure. Therefore, religious tradition preservation occurs within the Local Jewish National and Cultural Autonomy framework. The leaders of the ethnic communities highlight the fact that there are no restrictions on freedom of religion. However, Russian society's perception of the national and confessional characteristics of the studied ethnic groups cannot be characterized optimistically. The leaders of local ethnocultural autonomy of Kazakhs "Asyl Mura" and the Jewish autonomy noted the xenophobic and antisemitic manifestations. In this situation, ethnic communities see the way out in educational programs and events organized through the interaction with cultural centers, museums, and archives that aimed to overcome xenophobic and antisemitic manifestations and prevent the manifestations of extremism.

An example of the positive influence of the confessional factor on social and cultural adaptation processes can be the Armenian Apostolic Church's activities and the regional social organization "Union of Armenians of Altai Krai." The establishment of the Armenian Apostolic Church structure began in 2001–2002. At first, without their own building yet, the Armenian parishioners practiced in the Catholic Church's premises, provided by a priest. It was when the Armenian ethnic and confessional consolidation took place in Altai Krai. By 2004, the Armenian diaspora managed to build their own church complex. Within its framework were held classes related to catechesis and the transmission of cultural traditions.

As the researchers note, when the Armenian diaspora found itself outside its territory, it was forced to create effective mechanisms to organize its representatives, since, both at the beginning and the end of the twentieth century, the Armenian migrations were caused by violent reasons (Haroyan and Klyushnikov 2015). Ethnic Armenians who have become a part of Russian society remain committed to their native language and cultural traditions. Still, this process follows studying and using the Russian language and communication within the Russian continuum (Aghajanyan 2017). The Armenian bilingualism has specific features depending on gender and age. Usually, the younger generation and women speak fluent Russian, sometimes with no accent. It contributes to better adaptation in Russian society, avoiding negative accentuations associated with ethnicity. Thus, we cannot speak of assimilation among ethnic Armenians but can point out effective cultural adaptation in Russian society. The consequence of this phenomenon is the long-term presence of ethnic Armenians in Altai Krai, the successful creation of their ethnic and confessional structures that contribute to both its preservation and adaptation in the Russian environment.

The ethnic situation in Altai Krai is characterized by the presence of ethnic communities that have been partially assimilated (typically language assimilation) but kept its cultural and religious adherence. In Altai Krai, such communities include the center of national culture "Vainakh," Altai regional ethnocultural organization—"Azerbaijan." As the vice president of the "Vainakh" noted, modern changing community size processes are closely related to the Chechen Republic's socio-economic recovery processes. An influx from the Chechen Republic in the early 2000s resulted in 200 to 500 people remaining in Altai every year. At the turn of the first decade of the twenty-first century, the influx of Chechens to Altai stopped due to the improvement in the Chechen Republic's social situation. Chechen diaspora is currently moving back to the Chechen Republic and leaving for the Russian Federation's central region, which is primarily due to economic factors. The Chechens, who are well integrated into the local society, have their own business and know the Russian language well, remain in Altai Krai and Barnaul. The chairman of the Altai regional ethnocultural organization "Azerbaijan" characterized the ethno-social situation as favorable and contributing to Azerbaijani integration into the local society. During the conversation, he emphasized that Altai Krai economic situation currently does not contribute to Azerbaijani citizens' interest in this region, while it does in the Novosibirsk and Tyumen regions. The Azerbaijani diaspora in Altai Krai, and Barnaul has not significantly changed in numbers. Azerbaijanis who are living in Altai since the mid-1990s have formed niches for socio-economic and cultural adaptation. Although most of the Azerbaijani diaspora representatives know the Azerbaijani language well and do not use it daily, the processes of losing their native language as a constant communication tool are taking place among young adults studying in the Russian educational establishment. Both leaders of "Vainakh" and "Azerbaijan" marked that the integral and effective reproduction of ethnic culture in the diasporas is preserved due to the cultural centers' activities, as well as the involvement of significant numbers of studied ethnic representatives in Islam structures operating in Altai Krai.

The condition and development of the confessional situation in most of the studied diasporas are due to the fact that the only unifying religion for them is Islam and that a significant part of the believers is parishioners of the Barnaul Cathedral Mosque. The leaders also noted the absence of visible obstacles in the realization of the freedom of religion right. In particular: missionary activity restriction, infringement of religious practices due to doctrine specificity, restraining the rituals prescribed by the religious norms.

In general, a favorable situation is characterized by certain concealed tendencies. The Uzbek and Tajik diasporas' leaders declared the need to build another mosque, both in Barnaul and Slavgorod. Speaking about why the mentioned cities need new churches, the Uzbek diaspora's head declared that the existing complexes could not accommodate everyone, especially during religious holidays (day and night). The head of the Tajik diaspora noted another reason. Focusing on the fact that the mosque as a ritual complex promotes religious education, he pointed out the impossibility of visiting the Barnaul Cathedral Mosque by imams from Tajikistan. The topic of ethno-confessional separation within one mosque in scientific literature is of particular relevance. Singizov T. R. says that the increase of North Caucasian imams in the Khanty-Mansiysk Autonomous Okrug generates a counterweight to Tatarstan's and Bashkortostan's imams. North Caucasian imams concentrate around representatives of their ethnic environment, creating more politicized forms of Islam (Singizov 2016). The leaders of the Chechen and Azerbaijani diasporas are particularly concerned about the confessional indifference among young adults. A holistic and effective reproduction of the confessional tradition is possible with constant communication within a religious community. In this regard, the leaders stressed that young adults have little interest in active participation in religious life events. Attending both the Chechen and Azerbaijani community events, the younger generation demonstrates their passion for communication projects on social networks, preferring virtual communication to real. This fact limits the influence on young people's views within ethnic and confessional ideology and complicates tracking the possible spread of radical ideologies.

The ethnocultural diversity preservation within the Russian civil identity framework is not possible without a tolerant attitude to interethnic interaction. Due to the historical conditions, Russian civic identity is polyethnic; therefore, maintaining mutual understanding and tolerance towards society's non-dominant ethnic groups is the direct task of government authorities and law enforcement. The government authorities are responsible for protecting the cultures of ethnic communities and minorities (Kusmidinova 2016).

The specific perception of other ethnic groups by the dominant ethnos in Altai Krai has two factors. The first can be characterized as the factor of historical presence. It determines the tolerant attitude of the dominant ethnic group towards national minorities based on the limitation period of social contact, mutually beneficial economic and industrial ties, and the historically cultural continuum. The second factor is the dynamics of ethno-social processes, which testifies continuing changes in the ethnic communities' structure and explains the reasons for ongoing changes. It declares that the success of the ethno-social processes dynamics depends on the stability of the socio-economic adaptation of migratory communities and representatives of the dominant ethnic group. In case of unfavorable development of the socio-economic situation, the rise of unemployment, low profitability, ineffectiveness of regional economic policy, and external migration are perceived as aggravating factors of the socio-economic situation and resisting the dominant group development. The crisis manifestation tendencies are the growing rejection of recently arrived representatives of a different ethnic environment.

## 4 Conclusion

Thus, the confessional factor appears to be consolidating for all the studied national communities. Close attention to the politicization of the Islamic religious tradition stimulates the ethnic community leaders to actively interact with government bodies. The leaders of all three national communities, Uzbek, Tajik, and Chechen specify that there is intolerance towards these communities' representatives in the region despite the more secure social environment. The Uzbek and Tajik diaspora leaders noted the manifestation of intolerant attitude in connection with the migration circumstances. The researchers analyzed the causes of xenophobia and distinguished such a phenomenon as migrantophobia and called it a reason for migrants' desire to lead a "parasitic" lifestyle (Tabasaransky and Melikov 2016). Even though the national communities of Altai Krai independently solve the migrants' employment problems, and, in turn, migrants do not consider Altai Krai as a place of permanent residence, the leaders note various manifestations of intolerant attitude weekly. Revealing the problematic aspects of mutual communication, the Chechen diaspora vice president stressed that the aggravating factor of xenophobia and intolerance is the conflict-generating behavior of the dominant ethnic group's youth. Within the framework of various community meetings, the community leader and its most authoritative members conduct conversations to prevent a further escalation of interethnic rejection. In view of such circumstances, national-cultural autonomies are not only organizations whose activities are associated with minimizing the conflict-generating behavior of their representatives, but also with opposition to the demonstrative manifestation of the representatives' of their ethnic group. The national community leaders associate the prospects for improving interethnic interaction with further intensive cooperation with government authorities and law enforcement and joint work with the museum and archive funds and educational institutions. Joint efforts of various social institutions should aim to demonstrate the ethnic communities' contribution to the development of the region and country and to create a holistic view of the ethnic communities' role in maintaining a safe social environment.

## References

Aghajanyan LA (2017) The "Union of Armenian youth" in the structure of the Armenian diaspora of the Samara. Samara Sci Bull 4:193–197

Fatkhullin YuA (2016) The influence of national and cultural associations on the process of social adaptation to the local community. Modern Educ Pract Soc Spirit Values 3:209–212

Haroyan AS, Klyushnikov SS (2015) Modern forms of interaction between national communities and local government (On the example of the Armenian community of Rostov-on-Don). State Munic Adm Sci Notes SKAGS 3:392–396

Kusmidinova MKh (2016) The role of national societies in the formation of regional culture on the example of the Astrakhan region. Caspian Region Polit Econ Cult 2(2):243–250

Martynenko AV, Nadkin TD (2015) Islam in the republics of the Middle Volga region of the post-Soviet period (On the example of Mordovia, Chuvashia, Mari El). Clio 5:117–119

Singizov TR (2016) The dynamics of the composition of the muslim ummah in the Khanty-Mansiysk Autonomus Region-Yugra. Vestnik Tomskogo Gosudarstvennogo Universiteta. Filosofiya-Sotsiologia-Politologiya, 1:113–123

Tabasaransky RS, Melikov IM (2016) Migrantophobia and Xenophobia as phenomena of mass psychology. Soc Policy Sociol 1:178–185

Yarkov AP (2014) Islamic extremism in Western Siberia: the basis, history and forms of manifestation, methods of neutralization. In Kuryshev IV (ed) History of everyday life in Western Siberia at the end of the 19th and beginning of the 21st century. Ishim, Tyumen, Russia, pp 164–190

# Taking Action for Greater Security and Effective Partnerships for Sustainable Development

# Social Well-Being as an Integral Characteristic of the Secure Functioning and Sustainable Development in Regions

**Maria I. Cherepanova, Svetlana G. Maximova, and Tatiana V. Sirotina**

**Abstract** The paper focuses on the analysis of living conditions, the interdependence of social well-being, and the social security of the Russian population. We describe the social, economic determinants of social security of the population, which are essential for regional sustainable development. In addition, we study life satisfaction in the context of social security using a structural–functional approach. As a result, we made conclusions about the correlation between subjective assessments of security in the region and a low evaluation of the potential of Russian social institutions. Low social confidence in the central institutions leads to the decline of social security in the region. Social well-being is the intersection of the moods of social groups and individuals, identifying current interests, intentions, and goals. High levels of social well-being anticipate behavioral practices. Moreover, the social well-being of the population could promote integration or initiate dysfunctional behavioral practices. Therefore, this phenomenon regulates social and political processes. We can see that social attitudes and well-being underlie the current mass movements determine their nature and direction due to historical experience.

**Keywords** Security · Social well-being · Sustainable development · Public confidence · Social policy · Functioning · Region

## 1 Introduction

Today, Russian society is undergoing a transition period. The mass consciousness of Russian citizens plays a crucial role in assessing the effectiveness of current political and social events.

The large population of Russia in conditions of continuous social transformation causes socio-psychological stresses that affect individuals and entire social groups and society. According to sociological and psychological surveys, more than half of Russian citizens live in a constant state of stress (Benson et al. 2019). In this regard,

M. I. Cherepanova (✉) · S. G. Maximova · T. V. Sirotina
Altai State University, Barnaul, Russia
e-mail: cher_67@mail.ru



S. G. Maximova et al. (eds.), *Advances in Natural, Human-Made, and Coupled Human-Natural Systems Research*, Lecture Notes in Networks and Systems 234,
https://doi.org/10.1007/978-3-030-75483-9_39

the problem of regulating social processes is relevant. The main indicator of this regulation is the social well-being of citizens. In the context of social security, well-being acts as (1) a systemic dominant of mass consciousness, (2) the embodiment of certain values and attitudes, and (3) an indicator of readiness for their implementation.

Social well-being is the intersection of the moods of social groups and individuals, identifying current interests, intentions, and goals. A high level of social well-being anticipates behavioral practices.

Besides, scholarship on social security focuses on social well-being in the context of its creative or destructive nature. In this regard, it is necessary to monitor this phenomenon to ensure prompt forecasting and regulate its manifestations (Chugunenko and Bobkova 2013).

Social well-being significantly affects the economic behavior of citizens due to their social status, level of life satisfaction, prestige, and many other factors. Therefore, the dynamics and direction of social well-being indicate changes in the structure of society (Diener and Suh 2000; Gallagher et al. 2009). Analysis of the social well-being of citizens should be focused on the implementation of national policies aimed at stabilizing society. At the same time, social policy should be based on the interrelation between social well-being, relationships, institutional factors, and the consequences of social reforms in the main areas of people's lives. In this regard, we present the analysis of indicators of social well-being and security of the population of Altai Krai.

## 2 Materials and Methods

We conducted a study of factors and mechanisms of trust formation in the system of maintaining social security in the Russian border regions in Altai Krai (2018–2019) on a multistage stratified sample, n = 1,200, among respondents aged 18–70. Also, we measured social well-being and trust based on classical scales and a questionnaire. After that, we performed an initial analysis of descriptive statistics.

## 3 Results

Almost half of the regional population is pessimistic about their standard of living, which indicates low socio-economic development. The high migration outflow from the region, when people "vote with their feet" on life in Altai Krai, confirms this trend. Annually, about seven thousand people leave the region, migrating to the European part of Russia (Maximova et al. 2019).

Only one in eight residents of the region is fully satisfied with their career. Statistics show that the market of Altai Krai is characterized by social tension. The region has one of the highest levels of hidden unemployment in Russia (Nemirovsky and Nemirovskaya 2012). Labor market problems are aggravated by the lack of demand for labor and its deficit due to (1) the imbalance of supply and demand, (2) lack of qualified personnel, and (3) low labor mobility of the population. The agricultural sector of the region is experiencing a labor market crisis. There is little investment in high-tech manufacturing. Therefore, most people find low-skilled and low-paid jobs unattractive. More than that, small business development in the region is minimal. Therefore, we can observe an imbalance in the labor and education markets.

Wage satisfaction reflects the living conditions of the population. Every fourth resident of the region is not satisfied at all or is not satisfied enough with their salary. Only a tenth of the residents are fully satisfied with their payment, and every fourth representative of the region is relatively satisfied with it. According to statistics, the average salary in Altai Krai in 2019 was 25,000 rubles. Altai Krai is one of the six poorest Russian regions in terms of wages.

The average salary in the region is almost half the average salary in Russia.

According to studies, the economic concerns of Russian citizens about low income are quite reasonable. Last year, the real income of the population decreased by 0.2%. The study of quality of life in Russia demonstrated that the rating of benefits, indicating a decent level of life includes: (1) having a house or an apartment, (2) access to quality healthcare, (3) financial ability to get prestigious education, (4) balanced nutrition (including eating out), (5) variety of leisure activities, (6) modern household appliances, (7) high paying job, (8) car and ability to travel (Naci and Ioannidis 2015).

Satisfaction with professional growth opportunities in the region is identical to job satisfaction and reflects the low level of economic development. Most residents of the region are dissatisfied with this aspect of their lives. Only 8.9% of residents are completely satisfied with their career, while 34.5% are relatively satisfied with their career. According to experts, the personnel problem in the region is caused by the inefficient system of remuneration and motivation of personnel.

Assessing their daily routine, almost half of the residents reported low (9.7%) or insufficient (39.2%) satisfaction with life in general, while 39.6% were rather satisfied.

According to a previous assessment of their socio-economic and financial situation, about 10% of residents had a high standard of living and, consequently, high satisfaction with different life aspects.

Residents of the region are pessimistic about the quality of products on the consumer market. This trend is paradoxical because the region focuses on agriculture and can provide the population with eco-friendly products. Only one in five residents of the region were relatively satisfied with this aspect (20.6%). The rest of the population is wholly dissatisfied with the quality of products. Food quality is traditionally considered one of the most critical components of public health. However, due to the low standard of living, the population cannot afford quality food. According to official statistics, the morbidity and mortality rate in the region is high. Mortality from all causes accounts for 14.9% per 1,000 people, according to Rosstat data for the four months of 2019. Simultaneously, the planned value for 2019 is 13.5% per 1,000 people (Maximova et al. 2019).

Therefore, the sociological survey results indicate a marked social, financial, economic, and status stratification of thy residents. According to national statistics, a significant number of residents live in poverty. There is also widespread poverty among the working population.

The analysis of the presented research results indicates a relatively low satisfaction of residents with the main aspects of their life. Every second resident of the region is entirely or not completely satisfied with the level of security in the region. Only 13.9% of the residents are fully satisfied with this factor. The data obtained correlate with the results on the satisfaction of the residents with their salary, career, and life in general, and indicate low ontological security and uncertainty about the future.

The low satisfaction of the residents of the region with the situation in the country reflects the unfavorable situation of their life in the region. Most respondents (80.9%) are not entirely satisfied or not satisfied at all with the situation in the country. Only 19.6% of the residents of Altai Krai are quite satisfied with the situation in the country. Every second respondent positively assesses their standard of living in the region, including their career and salary. Thus, there is a positive correlation between subjective satisfaction with the main aspects of life and institutional and social satisfaction with the country's situation.

Analysis of the results shows that a significant group of the Altai Krai population is experiencing internal conflict. The desire to succeed in life, improve social status is faced with the impossibility and the resulting dissatisfaction with this area of life. This phenomenon indicates social frustration and increasing pessimism among the population.

## 4 Discussion

Comparing the results of the assessment of personal security with social security in the region, we can assume that the subjective-emotional component of the sense of security is identical to the objective-cognitive component of social security. Therefore, the level of social well-being of the population contributes significantly to the overall level of security (Galinha and Pais-Ribeiro 2011).

The presented results reflect identical processes in assessing the cognitive and emotional level of security. Most of the Altai Krai residents do not assess ontological security in their region. Almost half of the population (49.9%) is not entirely satisfied with their personal security, 36.5% are rather satisfied, and 19% of residents are fully satisfied with the level of security in the region.

Every second resident of the region has not faced any violations of their rights and freedoms.

However, 17.1% of the respondents experienced discrimination based on age, and 13.5% of residents faced it based on financial status. One in ten people experienced gender-based abuse. Violations of citizens' rights on other grounds are rather uncommon. Thus, young people or retirees and low-income citizens of the region most often experience alienation from society in various aspects, in violation of their rights and freedoms.

The results obtained reflect a certain ambivalence of the civil position of the population and a certain level of social passivity in protecting their rights. Although about 40% of the Altai Krai residents are not satisfied with their security, half of them (21.6%) believe they are not in danger. First of all, in case of danger, people turn to their friends and relatives (17.2%). The same number of citizens applies to the court and the police in such cases (16.7%). About 16.1% of the respondents do nothing, demonstrating social passivity. Almost the same number of residents found it difficult to answer, perhaps showing a high level of passivity in solving their personal security problems. On the other hand, there is a low level of trust in social services, which rarely helps solve personal problems.

The results show that a low level of institutional trust prevails in the region, confirmed by the rare appeal to relevant organizations to solve urgent problems. Besides, the level of interpersonal trust is also low.

Therefore, a low level of trust in the region indicates an unrealized sense of personal and ontological security.

Residents of the region are pessimistic about the dynamics of deterioration of life in the region. Almost every second resident of the region (45.7%) believes that there are many problems. One in three does not see any changes (35.2%). Only 6.5% of the residents are optimistic about life in the region. One in eight residents of Altai Krai (12.9%) found it difficult to assess the regional situation. Thus, most residents of the region negatively assess the quality of their life in the region. Uncertainty is one of the main risk factors in Altai Krai. The uncertainty of life, as a rule, is caused by various fears. According to the research, the main fears of Russian citizens are related to the illness or death of their loved ones, threats of military conflict, and arbitrary power. Simultaneously, compared with 2017, the number of citizens' fears almost doubled in 2018.

According to the results of a study conducted by the Public Opinion Fund, the rating of fear was led by fears for children and grandchildren, losses, severe health problems, reduced income, terrorism, and uncertainty in life.

The lack of institutional confidence in law enforcement and security agencies has been identified in the region.

The level of institutional trust of citizens in essential services, such as (1) banks, (2) mass media, (3) property rights protection agencies, is low and contributes to a specific nature of trust.

Residents rated the work of law enforcement agencies by five points (18.8%) and four points (18.2%). About 16.5% of respondents rated the effectiveness of these services by at least one point. Only 2.6% of residents showed maximum satisfaction (ten points) with the effectiveness of law enforcement agencies. The results obtained indicate a lack of institutional trust in the law enforcement and security agencies of Altai Krai.

As for the system of protecting economic rights and freedoms in the region, the citizens of Altai Krai showed some loyalty and at the same time ambivalence. Thus, 26.8% of residents rated the effectiveness of this system as average. However, about

50% of residents rated this service at least from one to four points. The population noted that current conditions do not stimulate small and medium-sized businesses in the region. Altai Krai Department for Entrepreneurship and Market Infrastructure notes a downward trend in the number of people employed in small and medium-sized businesses.

Every third resident of the region believes that the banking system is reliable to an average degree (five points). Every tenth potential customer of banking services evaluates this system by at least one point. Slightly more residents rate the system by four and six points. Overall institutional trust in the banking system is low and contributes to a low culture of general trust.

Citizens are extremely pessimistic about the effectiveness of ordinary citizens' participation in the discussion and control of government decisions. Every fifth resident of the region gave one point to this system. Almost the same number of citizens showed some loyalty to this issue and gave four points. Thus, more than half of the residents demonstrate a low level of trust, indicating the absence of dialogue between the population and the authorities. In addition, according to the residents of the region, civil society activities are relatively low.

Assessment of interpersonal trust in society varies greatly. Most residents believe that this phenomenon is expressed in an average degree (five points). On the other hand, every second person rated trust, cooperation, and mutual assistance from one to four points. Only one in five citizens considers trust more expressed (from six to ten points). Therefore, according to the Altai Krai residents, interpersonal trust is low. In our opinion, this situation reflects the low satisfaction of the population with their standard of living, social well-being, a high level of uncertainty, and pessimistic forecasts. The results obtained are consistent with repeatedly proven theses that the level of trust is inversely proportional to the degree of social well-being in the region (Tov and Diener 2013).

Prevention of interethnic conflicts is assessed quite positively by residents. Every fifth person considers it to be expressed in an average degree (five points). Almost the same number of respondents rate it at six points. On the other hand, about 37% of citizens rated this direction from one to four points. Every fourth resident rated the effectiveness of prevention of interethnic conflicts by seven to ten points. Currently, the region is home to about 120 different ethnic groups. The nine largest ethnic groups are:

1. Russians (92.0%);
2. Germans (3.0%);
3. Ukrainians (2.0%);
4. Kazakhs (0.4%);
5. Belarusians (0.3%);
6. Armenians (0.3%);
7. Tatars (0.3%);
8. Mordvins (0.2%);
9. Azerbaijani (0.2%) (Source: Official website of Altai Krai).

Therefore, experts note that the region has high interethnic potential (Connor 1994).

Citizens of the region showed low satisfaction with the legislative power and laws to protect their interests. Every fifth resident rated the effectiveness of this direction by at least one point. More than half of the population rated the effectiveness of this aspect of social life from two to five points. However, every fifth resident rated it more positively (from six to ten points). The diversity of opinions on this issue indicates that society does not have a single opinion on the authorities. The revealed facts may reflect a particular conflict potential in the region and indicate low social confidence in local authorities.

The analysis of this issue allowed us to identify two opposite groups. One of them (18.5%) represents residents who believe that public solidarity in the region is minimal (one point). Another group of citizens (22.8%) believes that the population has an average integration potential (five points). According to this group, the population can unite to solve socially significant problems. Thus, more than half of the residents of Altai Krai (53.7%) are optimistic about the possibility of uniting people in socially significant situations.

## 5 Conclusion

To sum up, we can say that half of the population of Altai Krai is pessimistic about their standard of living, which indicates low socio-economic development of the region and a low level of social well-being.

Objective indicators of the economic and political situation in the region correlate with low self-assessment of citizens' well-being. Many population groups are not sufficiently satisfied with their salary, careers, opportunities for personal growth, and improvement of their social status.

The region has a marked social, financial, economic, and status stratification of the residents.

Low satisfaction of residents with their social well-being and life, in general, indicate low ontological security and uncertainty about the future.

There is a positive correlation between the attitude to the main aspects of life and institutional and social indicators that determine the situation in Russia.

Analysis of assessments of cognitive and emotional levels shows that most Altai Krai residents do not appreciate the social institutions responsible for security in the region.

Uncertainty is one of the main risk factors in Altai Krai and a source of various fears and anxieties.

Changing the dynamics of the well-being of citizens should solve the critical problems of Altai Krai.

The relevance of socially significant decisions of regional authorities is a necessary condition for optimizing the social well-being of citizens, as one of the main components of the stability of the region.

**Acknowledgements** This work was supported by RFBR Grant No. 19-011-00417 "Factors and mechanisms of trust formation in the system of social security preservation in the border regions of Russia."

## References

Benson T, Sladen J, Liles A, Potts HWW (2019) Personal wellbeing score (PWS)—a short version of ONS4: development and validation in social prescribing. BMJ Open Qual 8:e000394. https://doi.org/10.1136/bmjoq-2018-000394

Chugunenko VM, Bobkova EM (2013) New trends in the study of social well- being of the population. Sociol Stud 1:16–24

Connor W (1994) Ethnonationalism: the quest for understanding. Princeton University Press, Prinston, NJ

Diener E, Suh EM (eds) (2000) Culture and subjective well-being. A Bradford Book; The MIT Press, Cambridge, MA

Galinha I, Pais-Ribeiro JL (2011) Cognitive, affective and contextual predictors of subjective wellbeing. Int J Wellbeing 2(1):34–53

Gallagher MW, Lopez ShJ, Preacher KJ (2009) The hierarchical structure of Well-being. J Pers 77(4):1025–1050

Maximova S, Surtaeva O, Cherepanova M (2019) Migration policy as a factor in ensuring social security: expert opinion in cross-border regions of Russia. Adv Soc Sci, Educ Human Res 364:737–741

Naci H, Ioannidis JPA (2015) Evaluation of wellness determinants and interventions by citizen scientists. The J Am Med Assoc 314(2):121–122

Nemirovsky VG, Nemirovskaya AV (2012) Feeling of insecurity from the social dangers as a basis of typology of regions (based on sociological research in Western and Eastern Siberia). Monit Public Opinion 1(107):113–127

Tov and Diener, , 2013 Tov W, Diener E (2013) Subjective well-being. Research collection school of social sciences, 1395. https://doi.org/10.1002/9781118339893.wbeccp518

# Criminal Procedural Resources in National Security Against Internal Threats in Eurasian States

**Sergei I. Davidov, Maria A. Neymark, and Evgeni N. Petukhov**

**Abstract** The aim of the research is to analyze national legislations in the states united by the Eurasian legal space that can regulate legal activities to ensure national security from internal threats and decrease the level of security of persons involved in criminal proceedings. The aim of the study was also to establish the problems arising in the course of application of security measures to ensure the rights and interests of the said persons and search for ways to resolve them. The special attention is given to an estimation of norms of criminal procedure legislations of the states which are a part of the Eurasian Union, protecting the legitimate rights and interests of participants of criminal proceedings, and also the provisions containing in the Agreement "About protection of participants of criminal proceedings" of the states-participants of the Commonwealth of Independent States, in national laws "About the state protection of participants of criminal proceedings" and the State programs accepted on their basis "Safety of participants of criminal proceedings." The novelty of the research is determined by the consideration and assessment of the degree of mutual influence of the legislations of the Eurasian Union countries with the purpose of their improvement to increase the effectiveness of ensuring the security of persons participating in criminal proceedings. We grounded the expediency of the further convergence in the norms of Russian and other Eurasian states legislation in the protection of rights and interests of persons, participating in the criminal proceedings. The main directions of improvement of criminal procedure regulation are singled out and recommendations on optimal use of criminal procedure resources are formulated in order to ensure effective security of participants of criminal proceedings from threats to their life and health inside the state.

**Keywords** Internal threats · National security · Criminal process · Resources · Personal security

S. I. Davidov (✉) · M. A. Neymark · E. N. Petukhov
Altai State University, Barnaul, Russia
e-mail: davidov_ord@mail.ru



S. G. Maximova et al. (eds.), *Advances in Natural, Human-Made, and Coupled Human-Natural Systems Research*, Lecture Notes in Networks and Systems 234,
https://doi.org/10.1007/978-3-030-75483-9_40

## 1 Introduction

The relevance of the given work is due the necessity of researching a problem of legal maintenance of safety of the person from the threats arising in the sphere of criminal proceedings on the territory of the states of the Eurasian space in connection:

Firstly, the lack of full protection of participants in criminal proceedings both in the Russian Federation and in other countries of the Eurasian space, due to the fact that in every case of a crime committed against a person, this person, especially once the verdict has passed, ends up unprotected, as security measures provided by the state are being withdrawn;

Secondly, the legislators underestimate the accumulated knowledge developed both by scientists of criminology in Russia and by scientists from other Eurasian countries in the development of concepts that ensure the satisfaction of the Eurasian public in the needs of the criminal process in order to ensure the national security in these countries;

Thirdly, legal non-adjustment of several relations connected with questions of ensuring of the safety of participants of criminal proceedings both in national legislations of the countries of Eurasian space and in joint international documents accepted by them; the Law "On state protection of the people participating in criminal proceedings" of the Republic of Kazakhstan contains even provisions with corruption factors (Parliament of the Republic of Kazakhstan 2000).

Fourthly, the adoption of international legal documents, which are a benchmark for national legislations of the Eurasian countries, regulating the sphere of ensuring the safety of people participating in criminal proceedings.

The laws regulating the grounds and procedure for utilizing security measures against threats to participants in criminal proceedings in the Eurasian Union countries should be fully balanced and internally consistent. At the present, the issue of developing a unified security concept in the sphere of criminal procedural legal relations is more urgent than ever. Therefore, in our opinion, it is important to consolidate the conceptual provisions of national security in the legal regulation of the imposition (purpose, objectives) of criminal proceedings, so that the protection of participants in criminal proceedings serves as a fundamental start of criminal proceedings in all countries that are part of the Eurasian space.

To achieve the established task, it is necessary to analyze national legislation and legislation of other countries of the Eurasian space in the field of protection of participants of criminal proceedings from threats to their life, health, and property, and to try and find ways of its improvement in order to increase efficiency of this protection. The task can only be solved by using the comparative research of international and Russian experience in the legal regulation of the mentioned sphere of activity.

The novelty of the article is defined by the fact that so far, no scientific analysis has covered the provisions of the relevant legislation of the countries that are part of the Eurasian space from the standpoint of ensuring the safety of participants in criminal proceedings with the resources of the criminal process.

## 2 Material and Methods

On the methodological level in the science of criminal process when comprehending the essence of legislation regulating the issues of legal support of national security from internal and external threats, the concept of security is defined and interpreted almost unambiguously. It includes an absence of danger, condition and measure of protection of the subject from threats and damage and is used concerning to various processes—both natural and social. Security as a condition and strategy of protection aims ultimately at the survival of the social system, the individual, society, and the State.

Hence, the main objects of security are the individuals, their rights, and freedoms, including in criminal proceedings.

Research of the legislation of the Eurasian countries in the sphere of ensuring national security in general and security of participants of criminal proceedings in particular in this region with the resources of criminal proceedings is conditioned by the ability of its results in the form of generalized analysis and experience to contribute to scientific activity on explanation, description, and prognosis of legal phenomena peculiar to a successful resolution of the problem of ensuring the security of the individual in criminal cases with the resources of criminal proceedings.

The normative base of research consist of:

- The interstate agreement on protection of participants of criminal proceedings accepted by the participating members of the Commonwealth of Independent States and signed on November 28, 2006 (Russian Federation 2008);
- National state and international legal documents containing provisions on protecting participants of criminal proceedings (Russian Federation 2010); Laws on the state protection of participants of criminal proceedings (Russian Federation 2004); Criminal Procedural Codes (Russian Federation 2001) and other documents (Presidential Executive Office 2015).

The empirical basis of the research includes the materials of criminal cases, the results of interrogations of the preliminary investigation officers, statistical data on the application of security measures to the participants of criminal proceedings on the territory of the countries of the Eurasian space (Government of the Russian Federation 2018).

We are using the data from methodological works (Langbroek et al. 2017), the importance of which is increasing in the current conditions characterized by numerous studies of legal problems around the world. We also believe that the legal research is of both normative and empirical nature and considered the usefulness and relevance of dichotomy between such normative and empirical research (Christiani 2016). The selected methods made it possible to study and compare the legislation of federal and unitary states of the Eurasian space, to reveal the formal, abstract identity of separate fragments of this legislation, to identify gaps in it, similarities and peculiarities in the regulation of homogeneous social relations related to the safety of participants in criminal proceedings and, as a result, to propose the optimal variant of their regulation in the relevant national legislation.

## 3 Results

In Russian Federation and other Eurasian states, a legal basis for national security is being actively formed in the area of ensuring the protection of participants in criminal proceedings from the threats to their (their relatives') wellbeing. At the present, it includes a set of normative legal acts at various levels. At the legislative level, this sphere is regulated by the provisions of the interstate Agreement "On the protection of participants in criminal proceedings," norms of national legislation in the form of the Criminal Procedure Codes, laws "On the state protection of participants in criminal proceedings," as well as State programs "Insurance of the safety of victims, witnesses and other participants in criminal proceedings." To implement the application of the specified legal provisions contained in the listed acts, for the safety of protected persons from federal budgets of the countries, substantial financial resources are allocated annually.

Proceeding from this, the legal provision of national security of Russia and other states that are part of the Eurasian space is regulated by a rather extensive system of legal norms and principles of implementation of the protection policy of the citizens, the society and the state from external and internal threats. At the same time, criminal procedure law, with its resources, occupies a special place in the system of legal support for national security against the threats to the rights and interests of participants in criminal proceedings within these countries.

The main areas of ensuring national security through criminal procedure should be defined as the following:

1. Taking into account the fundamental importance of ensuring the safety of the individual, the state, and society as a whole, it is necessary for a proper understanding and implementation of this important governmental task to fix this concept of security in the criminal procedural legislation of the countries of the Eurasian space, including this concept of security in the purpose (goals, objectives) of criminal proceedings (Article 6 of the Criminal Procedural Code of the Russian Federation) along with the protection of the rights and legitimate interests of individuals and organizations, the interests of society and the state, and the committed(?) crimes;
2. Taking into account the experience of law-making in the sphere of criminal proceedings of the Republic of Kazakhstan, the legislators of Russia, Uzbekistan and Tajikistan should adopt the experience of the legislators of Kazakhstan in the issue of legal support of the activities of authorities conducting criminal proceedings when taking security measures against the offenders in criminal proceedings. Namely, to develop and consolidate the provisions regulating the protection of participants of national criminal proceedings in one chapter of the

Criminal Procedural Legislation. At the same time, to stipulate in this chapter a norm fixing criminal procedure obligation of the body conducting criminal proceedings—to provide safety of all participants of criminal proceedings in cases of real threats to their (their close ones) health, life, and property. As well as the obligation to decide on further application of security measures or their total or partial cancellation simultaneously with the suspension of the preliminary investigation, termination of the criminal case or criminal prosecution.

We believe that only with such regulation of the purpose (purpose, tasks) of criminal proceedings and security measures will the modern concept of national security, the concept of ensuring the security of individuals, society, and the state in criminal proceedings be properly implemented.

Besides, the experience of the Russian, Uzbek, and Tajik legislators in enshrining in the Federal Laws "On state protection of victims, witnesses and other participants in criminal proceedings" such participants in the stage of initiation of criminal proceedings as an applicant, an eyewitness or victim of a crime, which are subject to the right to ensure their safety, should, in our opinion, be extended to the similar national legislation of the Republic of Kazakhstan.

In the legal activities of the states of the Eurasian Union, we also pay attention to the need to exclude corruption factors that take place in the analyzed acts of national legislation. For example, in the Law "On State Protection of Persons Participating in Criminal Proceedings" of the Republic of Kazakhstan, where the concept of "sufficient data" used in article 22 in the absence of criteria for establishing their admissibility to be the basis for applying security measures exceeds the limits of discretionary powers of an investigator (inquirer), carrying out proceedings on a criminal case (Parliament of the Republic of Kazakhstan 2000).

Provided, that the decision to apply and cancel security measures affects both the rights, freedoms, and legitimate interests of the various actors in criminal proceedings, as well as the interests of the various State bodies, we propose that the right to make such decisions would be reserved for the prosecutorial authorities during pre-trial criminal proceedings. We believe that the special status of the prosecution services in the State system will make it possible to ensure and protect the rights and freedoms of any participant in criminal proceedings and to guarantee an appropriate level of coordination among law enforcement agencies in combating crime.

The following recommendations should be considered by law enforcement in efforts to apply security measures to participants in criminal proceedings in the countries of the Eurasian Union:

1. If there is a need to forcibly bring in a protected person, the action should be entrusted to the official who initiated the procedure of security measures.
2. It makes sense to apply criminal procedure security measures, provided that the person against whom they are to be applied has not been previously known (unknown) to the suspect or the accused. In these situations, the application of special security measures (non-procedural) should be considered.

3. The disclosure of a "pseudonym" (personal data) of the classified participant of criminal proceedings is only possible if the classified person provides to the court a written consent after consulting with a qualified legal adviser.

## 4 Discussion

The legislation of all countries of the world confirms that national security reflects the state of vital activity of the state in which the development of the individual, society, and the state is secured and national interests are protected from internal and external threats. Undoubtedly, the resources of the criminal process play an important role in the system of ensuring the national security of any country. They are aimed at protecting the rights, freedoms, and interests of participants in criminal proceedings from threats within the State that arise in the course of the State activities related to criminal proceedings.

In the Russian language, the meaning of the word "resource" is established as something that can be used, spent, and be in stock (Zholkovskiy 1964). In French, the word "resource" means an auxiliary measure, so, using these meanings of the word "resource" in our study, we associate it with the criminal process as a secondary means of ensuring national security.

In general, however, criminal proceedings, as the means of ensuring national security and as a systemic phenomenon, are implemented in this area through the application of their structure of criminal procedural means, referred to in legislative and theoretical sources as security measures.

In this form, as scientists note correctly, criminal proceedings while ensuring national security serves as a way of communication between the subject of government (state) and the object of management influence (Agutin and Panshin 2016).

As a result, the role of criminal proceedings in the system of ensuring national security of the countries of the Eurasian space is to implement tasks related to the protection of constitutional rights and freedoms, as well as to guarantee the personal security of each person involved in the orbit of criminal proceedings through the proper performance of the said activity (duty) of the state representative in this area.

Numerous legal acts of the Eurasian Union demonstrate the importance and necessity of state activities in the field of ensuring national and personal security, including the resources of the criminal process.

At the interstate level, an agreement on the protection of participants in criminal proceedings was adopted by the members of the Commonwealth of Independent States and signed on 28 November 2006. This document is the legal basis for the protection of participants in judicial proceedings conducted in one State party and of other people liable for protection on the territory of another State party to the treaty. The Agreement sets out the forms of cooperation between the parties in ensuring the safety of protected persons. A list of protective measures has been defined. Grounds for refusing to apply protective measures have also been established. For example, if the protected person has committed an offense on the territory of the requested party, or if the implementation of the protection measures may prejudice the interests of that party.

Each country has adopted and maintains laws on security, on State protection of participants in criminal proceedings, and criminal procedure codes. Based on these laws, decrees of the Presidents of these countries approving national security strategies, programs defining national interests, and domestic and foreign policy measures aimed at strengthening the national security of the country have been developed and come into force.

As we see, the mentioned domestic normative legal acts and documents of other states which are a part of the Eurasian Union, contain a considerable number of provisions focused on the implementation of measures of the legal protection of participants of criminal proceedings from infringements on their (their relatives') well-being. It follows from law enforcement experience that, primarily, such security measures are necessary when serious crimes are investigated: intentional homicide, terrorism, extremism, and other crimes committed by organized criminal groups, gangs, and communities.

In Russian Federation, the legal basis for ensuring the safety of participants in criminal proceedings is provided by the provisions of the Code of Criminal Procedure (Russian Federation 2001), the Federal Security Act (Russian Federation 2010), the Federal Strategic Planning Act (Russian Federation 2014), the Federal Act on State protection of judges, law enforcement and regulatory officials (Russian Federation 1995), the Federal Act on State protection of victims, witnesses and other participants in criminal proceedings (Russian Federation 2004) and the Presidential Decree on the Strategy of national security in the Russian Federation (Presidential Executive Office 2015). The period of validity of this Strategy is from 2019 to 2023.

Per the Strategy until 2023, it is established that the national security must achieve the level of protection for individuals, society, and the State against internal and external threats that makes it possible to ensure constitutional rights, freedoms, decent quality, and standard of living for citizens and their security.

The insurance of the national security includes implementation of legal and other measures and instruments at the disposal of the State authorities, including in the form of criminal process resources aimed at countering threats.

Franciforov (2018) noted that the national interests of Russia relate to the interests of the individual, which are based on the conditions of the legal State, which is the Russian Federation.

Hence, the main directions of ensuring the State, public and personal security are strengthening the role of the state as a guarantee of individual security and property rights, as well as improving legal regulation of crime prevention, including those related to participants in criminal proceedings.

Data from the Main Information and Analytical Centre of the Ministry of Internal Affairs of the Russian Federation (the "Goszaschita-FL" form) demonstrate that the State protects participants in criminal proceedings in an appropriate manner. For example, it was only in 2016 that investigators of the investigation units of the territorial bodies of the Ministry of Internal Affairs of Russia for the Udmurt Republic

(28), Chelyabinsk (20), Rostov (19), Kemerovo (19), and Kostroma (14) regions issued rulings on the implementation of the State protection to victims, witnesses, and other participants of criminal proceedings.

Under the Government Decision on the approval of the State program to ensure the safety of victims, witnesses and other participants in criminal proceedings for 2019–2023, the following measures have been taken (2018) The financial support for the Program implementation for these years from the federal budget amounts to 1,059,256.1 RUB.

In Kazakhstan, the protection of participants in criminal proceedings is governed by the rules of special chapter 12 of the Code of Criminal Procedure (President of the Republic of Kazakhstan 2017) and the Act on the State protection of people participating in criminal proceedings (Parliament of the Republic of Kazakhstan 2000). These acts establish a system of measures to protect the life, health, property, legal rights, and interests of people participating in criminal proceedings, members of their families and close relatives, and to ensure their safety to prevent unlawful interference in criminal proceedings.

A special unit was established within the Ministry of Internal Affairs to ensure this activity. Following Articles 95 and 96 of the Code of Criminal Procedure of the Republic of Kazakhstan (President of the Republic of Kazakhstan 2017), the body conducting the criminal proceedings is obliged to establish what are the conditions for receiving state protection. It also must take security measures in respect of such participants in criminal proceedings. However, participants of pre-trial proceedings such as an applicant and an eyewitness fall out of this list. Statistics show that in 2016, the internal affairs agencies of the Republic directly applied security measures to participants in criminal proceedings in 477 cases, in 2017 in 363 cases, and in the first 7 months of 2018 in 211 cases.

In Uzbekistan, issues related to ensuring the safety of participants in criminal proceedings are regulated by article 270 of the Code of Criminal Procedure (Supreme Council of the Republic of Uzbekistan 1994) and the provisions of the Act on the protection of victims, witnesses, and other participants in criminal proceedings signed by the President (President of the Republic of Uzbekistan 2019). The main purpose of these legal and regulatory instruments is to establish a system of measures for the protection of victims, witnesses, and other parties involved with the criminal proceedings, including security and social protection measures used in the event of a threat to the life, health or property of persons. By the law, if necessary, persons covered by the witness protection program may change their appearance and passport data and block access to information about the person being protected. These protection measures are applied only in cases of serious or particularly serious crimes, and changes in appearance are made only upon application or with the consent of the protected person. Measures for State protection of participants in criminal proceedings also include protection of the home, provision of personal protection, issuance of technical means of protection, and change of place of residence and work. The Act supports the establishment of a special state program, which is to be financed exclusively from the budget.

In Tajikistan, the legal basis for ensuring the safety of participants in criminal proceedings is the provisions of the Code of Criminal Procedure (article 12) (President of Tajikistan 2009) and the provisions of the Law on State Protection of Participants in Criminal Proceedings (President of Tajikistan 2010). This legislation defines the safety and social support measures and the procedure for their implementation.

Under the Act, multiple security measures could be applied simultaneously to the protected person: personal protection, protection of the home and property, special-means of personal protection, systematic warnings, temporary relocation to another area, change of identity, change of educational institution, non-disclosure of information regarding the person protected and application of additional security measures to the protected person.

Protection is granted by the ruling of the court, the procurator, or investigating authorities. The internal affairs, national security, and anti-corruption agencies are responsible for ensuring security measures. The bodies established by the Government of the Republic are providing social support. State protection measures could be withdrawn through a written request by the person concerned or by the decision of the respective authorities.

To ensure the state protection for the participants in criminal proceedings, the Government of Tajikistan green-lit a government program. This program is financed from budgetary funds and other sources of funding provided for by the legislation of the Republic. For example, from 2013 to 2016, TJS 600,4190.8 was allocated from the country's budget for the implementation of this program.

The protective measures of participants in criminal proceedings in the legal environment and on the pages of special legal literature are divided into procedural (criminal procedure) and other (non-procedural). The latter are usually regulated by the Federal Laws "On State Protection of Participants in Criminal Proceedings." They include personal protection, protection of housing and property; special means of individual protection, warnings; insurance of the confidentiality of information regarding the protected person; relocation; change of identity; temporary placement at a secure location, and others.

Traditionally, criminal procedural security measures include those listed in article 11 of the Code of Criminal Procedure of the Russian Federation in the form of special rules of conduct of investigative and judicial actions and the recording of their course and results (Russian Federation 2001).

It should be noted that the legal regulation of the use of security measures with consideration to participants in criminal proceedings, as well as the problems of their implementation in the practical activities of the bodies conducting criminal proceedings, have so far been a subject of scientific discussions.

Among the most significant works dedicated to the research of the aforementioned problem one should mention the works of processional scientists who proposed the concept of personal security in criminal proceedings (Epikhin 2004), developed theoretical provisions concerning the provision of security of the subjects of criminal proceedings, formulated the correct concept of the internal threat to national security in the sphere of criminal proceedings: "Intentional actions of the accused or other persons inducing the execution of the national security measures in criminal proceedings" (Brusnitsyn 2002, 2018; Zaytsev 2001).

It is also worth mentioning foreign authors (Haroon 2014; Kuanaliyeva 2013), their recent research has been devoted to the problems of the national criminal procedure legislation of Kazakhstan and Uzbekistan and its reform in the area of enhancing the observance of international legal norms in the field of protection of individual rights in criminal proceedings.

Based on our research subject, we will consider the criminal procedural security measures aimed at ensuring the rights, freedoms, and interests of participants in criminal proceedings and their relatives in further detail (mostly using the example of Russian legal practice).

As the scientists note, investigation of a criminal case itself should be carried out in conditions safe for the participants of criminal proceedings.

At the same time, they explain that the official, who investigates a criminal case, had there been a real threat to the life, health, and property of the applicant is obliged not only to accept the relevant application, register it and check it but also to take all the appropriate measures provided by the legislator (Kolosovich et al. 2014).

Thus part 9 of Article 66 of the Code of Criminal Procedure of the Russian Federation establishes the right of an investigator, when necessary to ensure the safety of the victim, his representative, a witness, or their relatives, not to reveal their identity on the record of the investigative action in which they participate (Russian Federation 2001). To that end, the investigator, with the consent of the head of the investigative body, rules out the reasons for the decision to keep the data confidential, provides an alias to the participant in the investigation, and a sample of their signature, which they will use in the record of the investigation actions carried out with their participation. The implementation of such security measures under the Code of Criminal Procedure may begin at the stage when criminal proceedings are initiated.

To obtain the evidence in the event of a threat of violence, extortion, and other criminal acts against the victim, witness, or their close relatives or close associates, the Code of Criminal Procedure provides for the conduct of investigations such as monitoring and recording of negotiations (article 186, paragraph 2) (Russian Federation 2001). Also, per with part 8 of the article 193 of the Code of Criminal Procedure of the Russian Federation, to ensure the safety of the person testifying, by the ruling of the investigating body their identification process may be carried out under conditions of excluding visual representation of the person in question.

To ensure the safety of a suspect or the accused person with whom a pre-trial agreement on cooperation with his or her close relatives and close associates has been concluded had there been a threat to their life and health, the investigating body may decide to file a petition for the conclusion of a pre-trial agreement on cooperation with the suspect or accused in accordance with the decision of the investigating body. Also, in the case of the said defendant, the materials of the criminal case identifying him or her could be retrieved from the initiated case and attached to the criminal case against the defended participant, which has been placed in separate proceedings (paragraph 4, part 1, article 154 of the Code of Criminal Procedure).

In special cases, security measures may be taken at the request of the Attorney General or his deputy. The venue of a criminal case under articles 205, 205.1, 205.2, 206, 209, 211, 277–279, and 360 of the Criminal Code may be changed (Russian Federation 1996). In addition, paragraph 4 of part 4 of article 205.1, 205.2, 206, 209, 211, 277–279, and 360 of the Criminal Code Russian Federation (Russian Federation 1996). Article 35, paragraph 2, of the Code of Criminal Procedure of the Russian Federation, provides for the possibility in case of a real threat to the personal safety of the participants in the proceedings or their close relatives, for the Supreme Court of the Russian Federation to take a decision to refer a criminal case for consideration to a district (naval) military court local to the crime committed.

When there is a need to ensure the safety of participants in court proceedings and their close relatives, the court, guided by paragraph 4 of part 4 of the article. 2, Art. 241 of the CCrimP of Russia may decide to have a court hearing conducted in closed session.

When there is a need to ensure the safety of a witness, their close relatives, and close persons, the court can, without disclosing the true identity of the witness, interrogate them under conditions excluding the visual observation of a witness by other participants in the proceedings, which is the subject of a court ruling (part 5 of article 278) (Russian Federation 2001).

The use of videoconference is also utilized by Russian criminal procedure law when security measures are taken to ensure that the disclosure of personal data of a participant in a court hearing is excluded.

It appears that the security measures considered in the criminal procedure are highly important to secure the exercise of rights, freedoms, legitimate interests, and personal safety of participants in the criminal proceedings and their relatives. However, there are problems of legal, organizational, and tactical nature of law enforcement of these measures in the countries of the Eurasian Union.

We interviewed a total of 80% of 120 investigators and interrogators, who believe that the security measures of Russian criminal procedure legislation are not effective. They also explained that people involved in their criminal cases also do not believe in the effectiveness of these measures and attempt to evade participation in a criminal case.

Indeed, a question arises regarding the effectiveness of security measures when the participants in a criminal case knew each other beforehand or when the parties have made a reasonable request for disclosure of true information about the person giving evidence due to the need to protect the defendant or establish any circumstances essential to the case, the court, guided by Part 6 of Article 278 of the CPC of the RF, may decide on declassification of such data. In that situation, as scientists note (Davydov and Petukhov 2018), there is simply a "bluff" on the part of an official who persuaded a participant in pre-trial proceedings to apply security measures to him.

A significant problem affecting the effectiveness of the application of criminal procedural security measures, related to the lack of the necessary resources in the investigation and interrogation agencies to conduct identification in the absence of visual contact between the identifier and the one being identified. Moreover, from the results of the survey of participants in court hearings in criminal cases, it has been established that, at the stages of court proceedings, the application of criminal procedural security measures has actually stopped, as the judges, referring to their procedural independence, refused to keep on creating conditions for the conduct of court proceedings under the regime of minimal contact between participants in criminal proceedings.

Moving on with the problem of subjects competent to apply criminal procedural security measures, it should be noted that nowadays the problem of increasing the efficiency of law enforcement activity in general and criminal proceedings is rather urgent, as the unjustified application of security measures leads to additional resource costs.

At the same time, the question of the validity of their application is also at the discretion of the investigator. The paradoxical situation is seen as the fact that the actual basis for the decision on implementation of security measures is the data received and verified by the body that carries out these measures, and the cancellation of these measures at the discretion of the subject that carries out the criminal proceedings.

As we can see, questions about the figures for making relevant decisions within the framework of criminal proceedings, their competence, and powers are rather complicated. Given that the decision on security measures affects the rights, freedoms, and legitimate interests of citizens participating in criminal proceedings, as well as the interests of various government agencies, we support the proposal of researchers to assign such powers only to the prosecutor's office. We believe that this will lead to the more effective protection of the rights, freedoms, interests and personal safety of persons involved in criminal proceedings (Krasilnikov 2018).

The problems identified, even in such numbers, are sufficient to show how negatively such problems affect the established practice of applying security measures in the countries of the Eurasian Union.

Considering that before the perestroika (renovation) period of Russia's development, there was no institution of security measures in criminal proceedings, nowadays there is a critical attitude to the existence of criminal and procedural security measures in Russia in the legal environment of processionalist scholars. They consider the institute of personal security measures as an innovation in modern criminal proceedings to be a populist decision of the legislator, a tribute to fashion borrowed from foreign colleagues, as they believe that the successful implementation of the norms of this institution in the conditions of modern Russian reality is impossible.

In this sense, a rather radical position is taken by several Russian legal academics, who conclude that the criminal procedure doctrine conducted in the CPC of the Russian Federation, in its world view, is one of the manifestations of liberalism, which is generally based on the Western model of thinking, distinguished by its perception and understanding of the world. Having analyzed the National Security Strategy of the Russian Federation approved by Presidential Decree No. 683 of 31 December 2015 (Presidential Executive Office 2015), they conclude that it is based on a liberal ideology that undermines the traditional spiritual and moral values of the Russian people in the area of criminal procedure (Agutin et al. 2017).

Other countries of the Eurasian Union, also have issues with national legislative acts, as to their imperfection in the sphere of ensuring national security from threats inside these states to participants of criminal proceedings. For example, Kazakhstan has long been in the process of developing and enshrining legal safeguards that satisfy the notion of a fair trial in the relevant acts. These include the effective protection of participants in criminal proceedings who are subjected to violence, threats, and other undue influence. However, as noted by Kazakh researchers, although that procedural mechanisms for the protection of participants in criminal proceedings were established in the current Code of Criminal Procedure of the Republic of Kazakhstan (President of the Republic of Kazakhstan 2017), some of serious problems of a non-procedural nature have remained, which should have been addressed by the Law "On State protection of persons participating in criminal proceedings" adopted on 5 July 2000 (Parliament of the Republic of Kazakhstan 2000). The intention was to consider both Kazakh realities and international legal practice and legislation. However, during the analysis of the Act regarding corruption, the experts found provisions in article 22, "Grounds for the application of security measures," indicating possible manifestations of corruption in the form of broad discretionary powers. Since the concept of "sufficient data" used in the Law in the absence of criteria for establishing its admissibility, acting as a basis for applying security measures, exceeds the limits of discretionary powers of the authorities conducting the criminal proceedings (Shaimurunova 2007).

This problem could be resolved by limiting the powers of the body conducting the criminal proceedings, allowing it to determine "sufficient data" at its discretion to decide on the appointment of security measures.

The issue is also called regarding the content of article 3 of the Law of the Republic of Kazakhstan "On State Protection of Persons Participating in Criminal Proceedings" (Parliament of the Republic of Kazakhstan 2000) concerning the possibility of applying security measures to the applicant, as such subject is not mentioned in this law. However, their place among the subjects of criminal proceedings was found in article 181 of the Code of Criminal Procedure of the Republic of Kazakhstan (President of the Republic of Kazakhstan 2017). We believe that they, as well as other participants of pre-trial proceedings, could be subjected to threats and pressure from criminals.

We believe that the experience of Tajikistan (part 2 of Article 2 of the Law "On State Protection of Participants in Criminal Proceedings" (President of Tajikistan 2010)), Uzbekistan (Article 4 of the Law "On Protection of Victims, Witnesses and Other Participants in Criminal Proceedings") and Russia (Article 2 of the Federal Law "On State Protection of Victims, Witnesses and Other Participants in Criminal Proceedings" (President of the Republic of Uzbekistan 2019)), and part 1.1. of Article 1.1. of the Law "On State Protection of Victims, Witnesses and Other Participants in Criminal Proceedings" (Russian Federation 2004) should be used here 144 of the Code of Criminal Procedure of the Russian Federation (Russian Federation 2001), where it is permitted by law, if necessary, to ensure the personal safety of persons participating in the proceedings during verification of a crime report.

In general, the problem of the absence of a separate chapter in most of the criminal procedure codes of the European Union countries, which would systematically contain all provisions on measures to ensure the safety of participants in criminal proceedings, deserves attention. We believe this position is correct, and it is confirmed in the Code of Criminal Procedure.

In that regard, scientists point out that international law should also be more wide-spread and that there should be a mechanism for implementing international law into national law (Zaytsev and Yepihin 2016). For example, by implementing provisions in national legislation on the judge's questioning of a protected person outside the courtroom with the subsequent announcement of their testimony or on the demonstration in a courtroom of video recordings of the testimony given by the protected person at the stage of preliminary investigation without them being summoned to a court session (Zaytsev and Yepihin 2016), as well as on enshrining in criminal procedure legislation the obligation of the investigative officials and the court to decide on the extension of security measures or their full or partial abolition.

## 5 Conclusion

In general, both domestic and national legislations of the other countries of the Eurasian space are sufficiently filled with effective criminal procedural resources, which at an appropriate level productively provide the legal regulation in the protection of participants in criminal proceedings from threats to their rights, life, health, and property.

We have suggested improving the regulatory framework of Eurasian Union countries in the protection of human and civil rights and freedoms, ensuring protection from threats to the individual in criminal proceedings. We have also proposed recommendations for the law enforcement activities within the framework of legal guarantees to ensure the safety of participants in criminal proceedings.

## References

Agutin AV, Panshin VI (2016) Understanding the resources of the criminal proceedings as a component of institutional mechanisms ensuring national security. Vestn Mosc Univ Minist Intern Aff Russ 4:10–15

Agutin AV, Goshulyak VV, Sintsov GV (2017) The national security strategy of the Russian Federation and the ideology of Russian criminal legal proceedings. Russ J Criminol 11(4):817–823

Brusnitsyn LV (2002) Theoretical and legal bases and world experience of ensuring safety of persons assisting criminal justice. Dissertation of Doctor of Juridical Sciences. University of prosecutor's office of the Russian Federation, Moscow, Russia

Brusnitsyn LV (2018) Post-criminal impact and some aspects of participation of victims and witnesses in criminal proceedings under a pseudonym. Judic Auth Crim Process 1:71–86

Christiani TA (2016) Normative and empirical research methods: their usefulness and relevance in the study of law as an object. Procedia Soc Behav Sci 219:201–207

Davydov SI, Petukhov EN (2018) Problems of protection of witnesses from disclosing party confidants: criminal procedure aspect. Izv Altay State Univ 3(101):39–42

Epikhin AY (2004) The concept of personal security in criminal proceedings. Syktyvkar State University, Syktyvkar, Russia

Franciforov YuV (2018) Inviolability of the person in criminal proceedings as a guarantee of the national security strategy. Legal Concept 17(2):10–16

Government of the Russian Federation (2018) Resolution of the Government of the Russian Federation "On approval of the state programme for the safety of victims, witnesses and other participants in criminal proceedings for 2019–2023", 25 October 2018, No. 1272. Moscow, Russia

Haroon A (2014) Ensuring protection human rights in criminal procedure legal legislation of Uzbekistan. https://dnd.com.pk/ensuring-protection-human-rights-criminal-procedure-legal-legislation-uzbekistan/77942

Kolosovich MS, Kolosovich OS, Smolyakov PP (2014) State protection of participants of a criminal court procedure as means of ensuring their rights and lawful interests. Actual Probl Russ Law 10(47):2303–2309

Krasilnikov AV (2018) The criminal proceedings participants' safety: law regulation and the subjects of support. Proc Manag Acad Minist Inter Russ 1(45):68–72

Kuanaliyeva GA (2013) Some issues of international legal standards compliance in the area of individual rights in criminal proceedings of the Republic of Kazakhstan. Procedia – Soc Behav Sci 81:552–557

Langbroek P, van den Bos K, Simon Thomas M, Milo M, van Rossum W (2017) Editorial methodology of legal research: challenges and opportunities. Utrecht Law Rev 13(3):1–8. https://doi.org/10.18352/ulr.411

Parliament of the Republic of Kazakhstan (2000) Law of the Republic of Kazakhstan "On state protection of persons participating in criminal proceedings", 5 July 2000, No. 72-II. Astana, Kazakhstan

President of the Republic of Kazakhstan (2017) Criminal procedure code of the Republic of Kazakhstan, 4 July 2017, No. 231-V. Astana, Kazakhstan

President of Tajikistan (2009) Law of the Republic of Tajikistan "On the adoption and implementation of the criminal procedure code of the Republic of Tajikistan", 3 December 2009, No. 564. Dushanbe, Tajikistan

President of Tajikistan (2010) Law of the Republic of Tajikistan "On state protection of participants in criminal proceedings", 29 December 2010, No. 644. Dushanbe, Tajikistan

President of the Republic of Uzbekistan (2019) Act of the Republic of Uzbekistan "On the protection of victims, witnesses and other participants in criminal proceedings", 14 January 2019, No. ZRU-515. Tashkent, Uzbekistan

Presidential Executive Office (2015) Decree of the President of the Russian Federation "On the national security strategy of the Russian Federation", 31 December 2015, No. 683. Moscow, Russia

Russian Federation (1995) Federal Act "On state protection of judges, law enforcement and regulatory officials", 20 April 1995, No. 45-FZ. Moscow, Russia

Russian Federation (1996) Criminal code of the Russian Federation, 13 June 1996, No. 63-FZ. Moscow, Russia

Russian Federation (2001) Criminal procedure code of the Russian Federation, 18 December 2001, No. 174-FZ. Moscow, Russia

Russian Federation (2004) Federal Law "On state protection of victims, witnesses and other participants in criminal proceedings", 20 August 2004, No. 119-FZ. Moscow, Russia

Russian Federation (2008) Federal Law "On ratification of the agreement on protection of participants in criminal proceedings", 25 December 2008, No. 275-FZ. Moscow, Russia

Russian Federation (2010) Federal Law "On security", 28 December 2010, No. 390-FZ. Moscow, Russia

Russian Federation (2014) Federal Law "On strategic planning in the Russian Federation", 28 June 2014, No. 172-FZ. Moscow, Russia

Shaimurunova AK (2007) Conclusion on the presence of corruption norms in the law of the Republic of Kazakhstan dated July 5 2000 "On state protection of persons participating in criminal proceedings." Bull Inst Legis Repub Kazakhstan 3(7):140–142

Supreme Council of the Republic of Uzbekistan (1994) Act of the Republic of Uzbekistan "On the code of criminal procedure of the Republic of Uzbekistan", 22 September 1994, No. 2013-XII. Tashkent, Uzbekistan

Zaytsev OA (2001) State protection of participants of criminal proceedings. Exam, Moscow, Russia

Zaytsev O, Yepihin A (2016) State protection and safety of participants in Russian criminal proceedings. Intern Secur 8(1):283–291

Zholkovskiy AK (1964) Lexicon of expedient activity. Mach Transl Appl Linguist 8:67–103

# Social Responsibility, Trust and Security in the Russian Border Region

**Maria I. Cherepanova, Svetlana G. Maximova, and Sildisma A. Sariglar**

**Abstract** The purpose of the study is to analyze the interdependence of various components of civil society with the emerging level of institutional trust and security of the population in the Altai Krai. The article describes expert assessments of social responsibility, activism, civic solidarity, the level of implementation of public initiatives, etc., which together reproduce a certain level of institutional trust in the region. The polyparadigmatic nature of the study of the phenomenon of social responsibility in the context of politics, psychology, sociology, and behavioral practices allowed us to give specific and fragmentary descriptions of the mechanism of this phenomenon in the regional community. Conclusions are made that the most important factor in overcoming distrust for Russian society and its territories is to optimize the development of democracy in society, as well as to increase the economic and social well-being of the population in the context of the formation and strengthening of civil society, the growth of social responsibility of the government and each social actor. According to experts, trust in the government plays a crucial role in the formation of ontological security. On the other hand, the low level of institutional trust in basic social structures indicates the spread of a low culture of trust and reproduces a reduced sense of security in the region. The identified trends are an indicator of the insufficient level of responsibility of the government itself in building the trust of citizens. These results confirm the scholars' ideas about the close relationship between the economy, which is the basis of material well-being, and generalized trust and security of citizens.

**Keywords** Security · Power · Civil society · Democratic institutions · Trust · Social responsibility · Social well-being · Public initiatives

M. I. Cherepanova (✉) · S. G. Maximova · S. A. Sariglar
Altai State University, Barnaul, Russia
e-mail: cher_67@mail.ru



S. G. Maximova et al. (eds.), *Advances in Natural, Human-Made, and Coupled Human-Natural Systems Research*, Lecture Notes in Networks and Systems 234,
https://doi.org/10.1007/978-3-030-75483-9_41

# 1 Introduction

The analysis of modern sociological and political science research at the present stage of social development of the Russian society is characterized by the pluralism of opinions regarding democratic freedoms. Scientists note the lack of unity in the dialogue between social actors and the government (Fukuyama 2008). This situation is caused by unequal opportunities for the implementation of state power. Low integration of the public and government minimizes civil activity or reduces its effectiveness. Also, the specificity of modern political processes is the lack of effective mechanisms for integration between parties and civil society actors. Political scientists argue that the historical features of our state and the characteristic features of the political process in modern Russia make it significantly difficult to increase the level of civil activity, and hence the development of civil society institutions (Barinov 2017). The civil activity of the population is a systemic phenomenon that includes such important components as social responsibility, readiness for social solidarity, the degree of participation in NKOs, etc.

At the same time, the applied nature of the study of social responsibility is intended to identify general and special manifestations of this phenomenon, the analysis of the interdependence of objective conditions and subjective components of its implementation in modern Russian society. The essence of social responsibility in scientific discourse is overcoming individualism and implementing prosocial behavior that includes compassion, selflessness, and altruism. Social responsibility should be based on the needs of citizens for public welfare. The spread of successful social practices characterized by the high level of responsibility contributes to increasing the readiness of citizens to care for and help others in society. The essence of social responsibility lies in the actor's awareness of the results of own activities for other people and society as a whole. Focus on the implementation of social expectations and concern for others. The peculiarity of social responsibility is the initiative of citizens, prompted by internal motives of participation in socially significant affairs. Social responsibility has the character of a subjective prescription (Boytsova 2012). The social mechanism of responsible behavior includes the subject, personal and behavioral levels. The subject level includes goals and motives, assessment of the consequences of decisions made.

The personal level activates the meaning of social initiative, the manifestation of volunteerism in implementing of socially significant events. Responsible behavior is the highest manifestation of orientation to the public good in various spheres of human activity.

The problem of studying the social capital of a society is related to the phenomenon of trust.

Trust, in a broad sense, is considered one of the basic indicators of social stability. The study of social trust is included in the scientific legacy of almost any well-known sociologist. For example, Weber's (1978) contribution to the study of this problem is the idea of "formal rationality" as the basis of a legal society in overcoming significant social challenges.

It is relevant to analyze modern concepts of trust in the context of risk society theories and consider structural–functional and neoevolution approaches. The essence of the latter approach is to analyze social trust as a construct of the social order of functioning of a social institution, as well as everyday interactions in conditions of high conflictogenicity (Magomedov 2009).

Within the neoinstitutional approach, social trust is considered as the result of the interaction of social reciprocity and a sense of control (Green and Hansen 2016). Other representatives of neoinstitutionalism actualize social reputation and its role in social trust.

According to Fukuyama, trust is the expectation of predictability of the others' behavior by accepted norms in society (Fukuyama 2003). According to Seligman, lack of trust is not able to provide a long-term perspective of power and social order in any country in the world (Seligman 2002).

There are many current theories of trust. For example, Russian researchers consider generalized trust to be a function of interpersonal trust at the subject level and an indicator of social capital and the optimal social climate in society. This approach actualizes the macro-level analysis of this phenomenon. In the context of social psychology, generalized trust is the result of an individual's attitudes about the willingness to trust in interpersonal communication. Russian social psychologists find an analogy between trust and belief or consider it a manifestation of human emotion. In another context, its vague essence is emphasized, but it has considerable power (Bakhtina 2011).

According to Bayer's socio-psychological concept, trust is originally a person's inherent belief in the goodwill of others. The essence of current socio-psychological approaches to social trust analysis is to reduce this phenomenon to a personalized emotional and psychological state.

In the second half of the twentieth century, the political science direction of the study of social trust intensified. The particular contribution was made by comparative transitology, as well as global studies in the context of the postmodern development of the world. At the same time, the idea of constructing social trust as a macrohistorical phenomenon of modernity, developed.

Trust in the context of a risk society was studied by Giddens, Luman, Seligman, Fukuyama, Shtompka, and others. In this context, this is not only a reaction to uncertainty and risk but also the condition for minimization and neutralization of danger. According to Giddens, security constructs a balance of trust and acceptable risk (Giddens 2005). The scientist singled out personified trust in people and trust in abstract systems, which include symbolic signs (for example, money as an exchange tool, means of political legitimation) and expert systems that organize our material and social environment (Giddens 2005).

According to researchers, the social nature of trust is changing significantly in a global risk society. According to Giddens, trust is formed as a constructive response to risk (Giddens 2011).

Trust constructs peculiar configurations, horizontal and vertical social ties form a sense of group identity, ownership of society determines the nature of solidarity and cooperation.

Modern scholars agree that trust is a system-forming social capital factor (Fukuyama 2008). If human capital includes a combination of abilities, knowledge, skills, communicative potential, and is used for personal purposes, then social capital exists, firstly, to build optimal social relations. According to the ideas of R. Putnam, social capital can increase the optimality of society by increasing the coordination of joint actions.

Despite the abundance of approaches to the study of social trust, all of them are characterized by the presence of methodological contradictions in the analysis of the essence of this phenomenon and the features of its current transformation. There are two concepts in the study of trust.

The first considers this phenomenon as an institution of a traditional society, which regulates the totality of interpersonal interactions. The second approach considers trust as the outcome of a post-industrial society, as well as a condition for the optimal socio-economic development of a safe society.

Trust is associated with the problems of self-presentation and self-representation of an individual in the conditions of activity in a social group. This is the intersection of the likelihood of the individual to behave by the expectations and social norms.

In the context of increasing pluralism of modern society, trust organizes, regulates social and cultural diversity. This explains in current Russia the growth and prevalence of mutual trust, the growth of group solidarity, and relations in the context of ethnicity, religion, gender, and common views on different aspects of everyday life.

In today's world, the realization of trust is complicated by circumstances such as the constant threats of war, terrorist acts, and environmental disasters. However, structural and role differentiation, an increase in the complexity of public relations, pluralism of social roles, etc., are deepening. According to research, a decrease in social trust in society is an indicator of increasing social tension, mass emigration, brain drain, flight of capital, etc. (Ivanov and Danilov 2014; Maximova et al. 2019). Besides, the low culture of trust in the vertical vector context is manifested in the growth of absenteeism, minimization of support for social institutions, and low satisfaction with economic and social policies. These social conditions in society lead to the expansion of the so-called "gray zone" (Zabolotnaya 2003).

Formed in society, social capital induces generalized trust. At the same time, institutional trust contributes to the legitimation of legal and political institutions. At the same time, horizontal trust is the basic for the involvement of the individual and individual social groups. Institutional trust determines the activities of complex social systems in the process of globalization and modernization. In particular, the mechanism of formation of institutional trust reflects the specifics of the organization of life, perception, and regulation of the relationship between the individual and society. In this context, the heterogeneity of institutional systems is an indicator of low coherence of the social order and forms a low level of trust. A similar situation is typical for current Russia, as the historical result of administrative violence and forced adaptation to the mobilization system of institutions.

Most modern scholars testify to the fundamental relationship between trust and the level of development of democratic institutions (Boytsova 2012). Generalized trust is the basis of the democratization of society, correlates with a high level of education, and inversely proportional to social stratification and rising poverty. In addition, generalized trust is often recorded against the background of increased life expectancy and the implementation of humanistic values in society.

There are various factors that shape trust. Therefore, individual trust is often based on optimism, self-control, and willingness to cooperate with others in the process of interpersonal interactions (Allport 1960; Cattell 1950). Confidence can grow amid increased incomes, increased social status, and satisfaction with basic aspects of life. Increased happiness and decreased anxiety can also be attributed to increased confidence.

At the meso-social level, the trust will be strengthened through participation in volunteer organizations, strengthening family and friendship ties, increased participation in public actions, etc. (Putnnam and Fukuyama 2008).

Social theories of trust testify to its relationship with the growth of social conflicts and are indicators of satisfaction with the functioning of political institutions, democratic freedoms, and the safety of society (Maximova et al. 2019).

For example, with a high level of generalized trust in society, legitimate state support is formed that reproduces the corresponding level of trust.

With a high level of interpersonal trust and a low level of trust in social institutions, informal ties play a stabilizing role in the generally ineffective public administration system. A decrease in horizontal trust with an increase in vertical trust is usually an indicator of the low efficiency of civil society. Finally, the low level of all types of trust indicates a deep systemic crisis of power, the lack of necessary coordination of all social actors.

According to Shtompka, the culture of trust is a function of a democratic system of government (Shtompka 2001 and 2002). As the author points out, democracy uses the institutionalization of distrust. Autocracy strives for the direct institutionalization of trust and turns it into a demand "weighed down by serious formal sanctions" (Shtompka 2001 and 2002).

Such a society is characterized by delegitimization, which demoralizes the members of society and stimulates individualism and the growth of deviations in society. This phenomenon, in particular, is characteristic of transforming post-Soviet societies.

As scientists testify to the transitivity of society, the "defectiveness of democracies," the decline in horizontal and vertical trust correlates with significant social polarization and the high conflict potential of society.

In modern Russian society, as well as throughout the world, "erosion" of ontological security is taking place (Giddens 2011). The growth of the dangers and uncertainties of the current society is formed by the crisis and transitivity of public life, irrationality, and blurring of the boundaries of scientific and ordinary knowledge.

As one of the mechanisms to reduce the uncertainty of the modern world, routinization is associated with the continuity of the world, integrity, and personal self-identity autonomy.

The unpredictable current world violates the routine basis of being and is the basis of psychological tension, fear. It leads to the reconstruction of personality. According to E. Giddens, the routine of everyday life is being tested. Institutional order underlies confidence in social institutions. Modern researchers have identified the relationship between institutional trust and the level of fear and anxiety in the conditions of life in society. On the other hand, trust is always realized in conditions of risk and uncertainty, and even the existence of an institutional order does not guarantee the complete absence of dangers and threats (Shtompka 2001 and 2002).

Consequently, anxiety states, negative social moods of significant social groups of the population indicate objective dangers, which are the basis of low trust and ontological security.

Thus, the permanent crisis of social institutions and the transitivity of social life distort the routine of everyday life, reducing generalized trust. The destabilization of social relations violates the reproduction of safe practices, contributes to the growth of suspicion, mistrust, anxious expectations, etc. The decrease in the protective function of social institutions activates the growth of uncertainty in everyday life. However, the relative stability of the functioning of the institutions of society will underlie confidence in predictability in a certain security of life.

Nevertheless, as Luman showed in his research, some factors destroy the reversibility and continuity of everyday life (Luman 2007). This is one of the objective mechanisms for reducing generalized trust, which prevents the minimization of informal channels of social interaction. Also, with the spread of a conservative bureaucratic hierarchical system, fears, discontent, and aggression in society will grow.

Despite the relative autonomy of humans' everyday life, institutional dysfunctionality shakes the boundaries of ontological security and reconstructs the individual's basic trust in the world around them. Researchers distinguish differentiated conditions and factors of decreasing institutional trust. One of the basic conditions is constant social change. The dynamic imbalance of social development creates uncertainty in assessing life safety limits in such a society. As E. Giddens points out in his books, innovations always form fear of novelty, undermining familiar foundations, and a sense of threat to familiar values, status, and self-identity.

Concluding the review of the factors that form the basic level of trust, it is necessary to summarize the objective global conditions for its formation, such as the spontaneity of the processes, characteristic of any complex social system.

Nevertheless, optimal institutional mechanisms for regulating social life can minimize a sense of danger and risk.

## 2 Materials and Methods

The study of factors and mechanisms of building confidence in the system of maintaining social security in the border regions of Russia was carried out in the Altai Krai (2018–2019), a multi-stage stratified sample, n = 1.200, and the age of respondents was from 18 to 70 years. Social security assessment and generalized trust were measured based on classical scales and questionnaire questions. An initial analysis of descriptive statistics was conducted.

## 3 Results

The expert community rather ambivalently assesses the social and civic activity of the inhabitants of the region. Every third expert (32.7%) notes the severity of this quality is at a low level. Although the opinions of others are differentiated, approximately every tenth expert assesses these manifestations more highly. Together, about half of the experts rated the severity of civic and social activity at an average level. It should be noted that the opinion of experts regarding civic engagement is much more optimistic than the self-esteem of the population on the effectiveness of this area, which indicates some inconsistency in the opinions of the authorities and the public themselves. Experts assessed the severity of the social responsibility of residents in the region relatively positively. Every fourth specialist noted the implementation of this quality to a low level, that is about the same on average and above average points. Together, more than half of the experts rated social responsibility quite positively.

Civic solidarity is represented in the region among its degree. In practice, every second expert emphasizes the expressiveness of civic solidarity. Other experts' opinions are fairly evenly distributed over all estimates, which indicates a lack of unity in the assessment of this important aspect.

The reproductive function of social solidarity is integrating of diverse social systems with the differentiated level of the hierarchy. The degree of social solidarity is an indicator of favorable forecasts for the functioning of any social system, its coherence, and integrity. Consequently, solidarity dysfunctions threaten integration and indicate a crisis of generalized trust. Analysis of the problems of social solidarity actualizes the strength of the country's ruling group in the reproduction of society, stimulating the functioning of the relevant social institutions. Thus, the unrealized level of social solidarity in society indicates a crisis in social relations and a violation of social reproduction effectiveness.

The severity of public initiatives in the region is not expressed enough. According to the expert community in the region, the average level of realization of social activity of citizens prevails.

The social activity of citizens is an indicator of the prevalence of social subjectivity, its forms, level, and nature of implementation in society. Social activity strategies include a wide range of behavioral practices, from active participation in social life to social exclusion and exclusion from socially significant problems of society. According to sociologists, in the face of growing social inequality and territorial heterogeneity, a widespread culture of distrust of basic social institutions, social activity can evolve from creative and constructive forms to civilized protest and social rebellion. The ambiguity of assessments of social activity in Russian society reflects a situation of ambivalence. We can say that there is an arrhythmia of social activity in society, which is characterized by periods of "a relative lull, then a wave roll" (Guha-Khasnobis et al. 2006).

According to experts, among the population of the region, moods of indifference and apathy prevail. Every fourth expert emphasized this fact. Every sixth expert was more optimistic. About the same number of experts were pessimistic and appreciated the severity of social passivity of citizens almost as much as possible. Comparing the data of self-esteem by the population with their social mood, it should be noted the objectivity of the views of experts in connection with the comparability of the opinions of citizens themselves.

Most experts are pessimistic about the severity of the diverse negative sentiments prevailing in the region. More than half of the experts believe that pessimism, apathy, lack of faith in the future, developed in the region are almost as pronounced in the region. These estimates are consistent with the opinion of the population and testify to the objectivity and systematic nature of deprivation processes in the province.

The severity of anxiety and fear is identical to previous results. The prevalence of these emotions indicates low ontological security in the region, which is consistent with the opinion of the population itself. Consistency of these opinions also indicates a systemic crisis in the region, the presence of a variety of social risks, and challenges. Our regional data are consistent with the results of the WCIOM study. According to the results of an all-Russian study, the fear associated with a decrease and loss of income excites more than half of citizens (63%) as many citizens are concerned about the growth of social injustice. These fears are more characteristic of people of mature and old age. One of the common fears was the fear of losing free medical care. It is noteworthy that the youth was characterized by the fear of disorder in the country, holding protests (Zhbankov 2010).

A qualitative analysis of experts' opinions regarding the understanding and significance of various factors that determine the level of solidarity and civic engagement has shown a systematic view of experts. Therefore, in their opinion, many social factors are important for this phenomenon. However, two times more experts recognized the priority and importance of welfare in the formation of solidarity in society than experts acknowledging the importance of all other indicators. The contribution of the level of security, culture, ecology, economic development, faith injustice, and a sense of collectivism identified the majority of experts as significant.

The overwhelming majority of experts in the region believe that the population is poorly involved in the activities of non-commercial public organizations. Every fifth specialist was more optimistic and considered the population more active in terms of the activities of non-commercial public organizations, but not to a sufficient degree. Thus, the development level in the region of civil society has significant unrealized potential, which is consistent with the opinion of the population itself.

A qualitative analysis of the answers of Altai Krai experts to the question of what measures should be taken to increase the participation of the region's population in the activities of public organizations showed that the activity of the population in social activities can stimulate a range of activities, such as campaigning, widespread informational coverage of best and effective social practices, cooperation training, monetary rewards and other systemic activities of various kinds. It is noteworthy that every tenth expert could not name the measures necessary to increase the effectiveness of the social activity, showing a certain social passivity.

Experts and the population are not satisfied with the objectivity and independence of the media. Most experts evaluate the implementation of these components minimally. However, every third expert in the aggregate evaluates the media's effectiveness and freedom more highly, from 6 to 10 points inclusive. The opinions of experts regarding the legality and independence of the courts revealed polar assessments. So, the level below the average was emphasized in the aggregate by about (70%) representatives of the expert community. Every third expert assesses the rule of law in the region more optimistic. In general, assessments of the legality of courts by experts are more optimistic than the population. The results obtained indicate a significant lack of trust in the institution of legal proceedings in the region, which enhances the feeling of unpredictability and arbitrariness, reducing the ontological safety of the population.

The expert community believes that the power of the region's governing authorities is fully realized in the region. Every fifth expert evaluated this direction almost as much as possible. About the same number of expert representatives gave less optimistic estimates. Thus, according to experts, the regional authorities effectively exercise their power.

Every second expert believes that the region does not express the exclusion of citizens in obtaining services and goods on a national and religious basis. About 120 different ethnic groups peacefully coexist in the region. The monitoring results to assess interethnic relations indicate a minimal severity of conflicts between representatives of different ethnic groups and religions in the region, particularly in the field of obtaining rights and freedoms.

More than half of the experts noted the region of serious material stratification and exclusion of citizens in obtaining public services and benefits, depending on their income level. The high level of social stratification in the region prevents the formation of a sense of justice and, accordingly, complicates social trust development.

Every tenth expert evaluated the effectiveness of protection of property rights in the region as minimal. More than optimistic in their assessments were about half of the representatives of the expert community. Thus, the potential for protecting property rights is not fully realized and requires increased efficiency. In this context, the sense of defenselessness contributes to the growth of dangers and accompanying anxieties and fears, which is inversely proportional to the feeling of social institutional trust.

Most experts believe that the potential of anti-corruption measures in the region is insufficient, and the activists rated the effectiveness of this area below average. Approximately one in three experts believes that anti-corruption activities are useful in the region.

One-fifth of the experts noted the presence of equality in the field of educational services in the region. However, half of the experts evaluated the implementation of this area quite pessimistic and set a minimum level.

The availability and quality of medical services in the region, according to experts, is not sufficiently implemented. So, a quarter of experts rated the effectiveness of this area by 3 points. Every third expert evaluated this aspect even more pessimistic. About half of the experts evaluated the effectiveness of implementing medical services to the population in the region on average. The data obtained are consistent with the Altai Krai population's opinion about the insufficient availability of high-tech medical care, which confirms the objectivity of the identified trends.

According to the experts, the effectiveness of public control over the activities of the authorities is insufficient. About a fifth of the experts were optimistic that public control over the decisions of the authorities in the region is well developed.

Every second expert believes that the rights of voters in the region are provided with a minimum. A third of experts consider the severity of this aspect to a moderate degree. Only every sixth representative of the expert community is optimistic about the effectiveness of voter protection.

Regarding the assessment of control over compliance with laws in the region, the experts were divided into two equal groups, which is an indicator of the ambivalence of these positions. Almost every second expert considers the severity of this control to be minimal. Every sixth expert evaluates the effectiveness of the control at an average level.

Experts assess the safety of citizens in the region more optimistic than the population itself. Every fourth expert believes that life in the region is quite safe. The opinion of other experts was evenly distributed according to all differentiated estimates. Comparing the satisfaction of the residents of the region with their safety, we can distinguish the contradiction between the opinions of experts and the feelings of the population of the region. This contradiction indicates the absence of social unity in a given context.

According to the latest WCIOM polls, indicators of declining life safety in Russia are fears of rising social stratification, declining income levels, and fear of losing free medicine.

A fifth of experts assess the severity of the powers of local authorities in the region at a minimum level. However, every fourth expert assesses the powers of local authorities optimistically.

Most experts evaluate the receipt of social services to the population at a low level. A fifth of specialists believes that such services are implemented in the region quite successfully.

According to experts, the region does not fully consider public opinion when adopted by authorities. Regarding the publicity of the authorities, the opinions of experts are not agreed upon, as are the estimates of the population itself. This mismatch is an indicator of insufficient social solidarity in the region.

A fifth of experts assess the severity of the possibility of conflict potential in the region to a moderate degree. The same number of experts consider the severity of potential protests in the province to be minimal. However, a third of experts note a high level of probability of conflicts and protests in the region. The protest potential in the region is an indicator of the lack of confidence in the main life support institutions. The results of all-Russian studies show that fear of unrest and protest is characteristic of young people under 25 years old. The anxiety of the older population is slightly expressed in this regard (Barinov 2017).

The opinions of experts regarding the integration potential of citizens and their ability to participate in managing society in the region are quite contradictory.

Almost equal groups of experts evaluate the severity of this direction differentially.

## 4 Discussion

For the current Russian society, the problem of regenerating generalized trust as a social background for optimizing post-industrial transformation is extremely significant. Among the factors of restoring social trust, Russian scientists note the stabilization of the active role of social institutions, the dissemination of spiritual values, such as the common good, interethnic tolerance, social justice, etc. (Magomedov 2009).

The internalization of the described values into the institutional system of Russian society can become a social condition for the growth of trust between society and the state, people, and authorities (Magomedov 2009, p. 79). However, the decline in social trust in Russian society and its regions contributed to a compensatory increase in loyalty to ethnic, family, and friendly groups, thus identifying the crisis of all-Russian identity. There is, as it were, the closure of a circle of trust at a personalized level. A mistrust climate is being formed between representatives of different ethnic groups, religions, adherents of different values, and stereotypes.

Social trust is woven into the economic discourse. The crisis of competitiveness of the Russian economy and Russian society in the context of other countries requires a systematic study of social trust as a socio-humanitarian and normative phenomenon (Stolyar 2007).

One of the objective conditions for reducing social trust in Russian society was transforming the political system from authoritarian to democratic system change. Russian scientists define the specifics of social trust in Russian society as a set of normative reciprocal expectations aimed at increasing the stability, predictability, and transparency of social life in the context of nationwide transformational transformations (Magomedov 2009, p. 89).

The prevailing culture of mistrust that has developed in Russian society, according to expert opinion, hinders the effective socio-economic development of the country, minimizes the implementation of trans-regional and trans-territorial social networks (Magomedov 2009, p. 189). The transformation of social trust in modern conditions is that the crisis of political and ideological institutions characteristic of the Soviet era contributed to the gradual disintegration and disunity. In turn, social segmentation contributed to the spread of compensation systems, such as corruption and quasi-market relations (Magomedov 2017). These conditions had a significant impact on the style of everyday life, the implementation of rational and irrational actions of people of different age, national and regional groups in modern Russian society. For example, the decline in social trust stimulated the xenophobization of Russian society. A peculiar externalization of generalized trust occurred. Functional substitutes were spreading, which transferred trusting relationships outside of Russian society. The atomization and individualization of Russian society is the basic social background for the formation of a credit of trust, which would help to overcome depersonalization and impersonality in the implementation of social practices and everyday life.

The reinstitutionalization of trust in Russia reflects the following processes in a society:

- Institutional deformation and reformation;
- Crisis phenomena in the spiritual and moral sphere;
- Decrease in the consolidation potential of the Russian people.

Besides, this process is determined by the specifics of the historical development of Russian society and the deformation of the type of post-Soviet man. At the same time, experts note the mismatch of low institutional trust with an unjustifiably high public trust level.

The institutional mechanism for building trust includes institutions of socialization, government agencies, and the media. According to the Russian scientists, these trust-building agents act in the conditions of a de-ideologized and value-deformed social space and produce an increase in institutional risks (Merry 2011). The basic institutions for the restructuring of trust are the family, education, and the state. The culture and ethics of trust in society is, above all, a function of Russia.

Violation of stability and habitual way of life reinforces the uncertainty of various aspects of significant social groups in Russia. The decline in living standards, constant readiness, and expectation of a crisis in the financial and banking systems contribute to the erosion of values. Pessimistic social moods are forming, which is an indicator of reduced social trust.

In our opinion, social trust has a regionally differentiated character. Thus, the intensification of social risks and the low level of modernization of the economy in the border regions of Russia, and the relatively high standard of living and social progress in the central regions of the country form fundamentally different levels of trust in the authorities and social institutions. Current Russia is characterized by significant economic, social, ethnocultural territorial heterogeneity. Consequently, the level of social risks formed, the degree of perception of a decrease in ontological security and public confidence in the broad sense, is different. The significance of trust as an economic phenomenon is that "it reduces transaction costs in exchange relations and affects economic efficiency (Veselov et al. 2004).

However, the basic social conditions that form trust, but having regional specifics and the context of its formation, are highlighted. Domestic sociologists highlight the growth vector of well-being from north to south and from east to west of the country. In this context, regional profiles of social well-being are formed (Yanitsky 2003).

The phenomenon of trust reflects the optimality of the model for constructing social and economic relations (Shtompka 2008).

According to the results of a sociological study conducted in the regions of the country, more than half of Russians believe that one should be careful in dealing with people. Only every fifth person believes that people can be trusted. Consequently, the Russians consider a significant mass of their fellow citizens to be strangers. Only every fourth resident of Russia believed that the crisis did not affect the level of confidence. At the same time, half of the respondents felt a decrease in confidence in the context of another crisis in society (CIVICUS 2012). In such a society, there is no coherence in anticipation of the future, which is being replaced by a variety of political and everyday practices of survival (Dankin 2006). On the other hand, the unifying discourse is the safety of survival and stabilization of the existing situation. The results obtained indicate the prevalence of sensations of the potential danger of the world, the low possibility of trust in it.

Under these conditions, existential distrust is an indicator of insufficient ontological security. Social integration in such conditions is achieved through protection from a common enemy. Douglas (1986) formed it as "a condemning hazard model."

The social resources of such a society include economic or coercive leverage. The emerging counter-resource blocks the self-organization of social forces. It is characteristic that a public organization is characterized by an increase in apathy and cynicism, the absence of guarantees of the rights and freedoms of citizens. The everyday practices of the population are starting to include consumerism and a crisis of work ethics. Connections, acquaintances, luck, finally, coercion, and violence, but not everyday work, bring benefits (Viktorov 2008). Creation as a fundamental form of social action and, therefore, as a sociological category, loses its meaning (Zhbankov 2010). According to the expert community, the specificity of the riskogenicity of Russian society lies in the delayed nature of the response to the changed situation and the insufficient level of sociocultural reflection (Kuznetsov 2009).

The historical nature of the decline in the ontological security of post-Soviet people has stabilized universal mistrust and fears in everyday life in response to its insecurity and unpredictability. According to the scientist, diffuse fears and anxieties are an indicator of the constant conservative blockage of institutional changes." In such conditions, society and its economic environment reproduce the maximum uncertainty and distrust of social institutions, and the prospects for their future changes.

A characteristic feature of the described social systems is deinstitutionalization, which reflects the constant transitivity in a crisis. In this case, the formal regulators seem to turn off and lose their basic functions.

Thus, in a Russian risk society, guarantees of security and social protection of citizens before the law are reduced. Extended riskogenicity impedes optimal reform of Russian society. Besides, the modern space of Russia strengthens its mosaic and heterogeneity, producing the differential nature of trust of various social groups. For example, ethnocultural diversity growth can stimulate distrust and xenophobia in other social groups, etc. In this case, trust in otherness acts as a necessary tolerance. Recognizing the otherness of other people expects similar confidence in themselves.

According to international studies, only every third-social actor experience generalized trust in Russia. According to the World Values Survey (WVS 2010–2014), only 27.8% of Russians believe that people can be trusted. Similar data were obtained for post-Soviet countries. Similar data were presented in the report "Trust Barometer" (Edelman 2001–2020), published by an international research company, based on the neoinstitutional approach and the theory of the structure of Giddens.

The analysis of social trust and institutional order in society in the context of comparing the sociocultural characteristics of Western and Eastern cultures revealed that the level of general trust is higher in developed countries and lower in countries of catching up development. On the other hand, countries with a high level of autocracy, such as China, show a relatively high trust level. Whereas "partially democratic countries" are characterized by lower levels of horizontal and vertical trust. This trend indicates the presence of significant social stratification and the high conflict potential of such societies.

The paradoxical phenomenon of the very high generalized and institutional trust of modern China is explained by the unique combination of economic successes acting against the backdrop of the implementation of specific sociocultural values and a mentality based on the priority of the public good. In this context, China is a paradoxical autocratic exception to the universal axiom of democratization.

## 5 Conclusion

The study revealed the ambivalence of assessments of social and civic activity of the Russian region residents, which emphasizes the presence of unrealized potential in the formation of civil society.

The expert community of the region positively assesses the civic activity of the population, which contradicts the public's self-assessments. This phenomenon emphasizes the isolation of the opinions of representatives of the authorities and people.

Most experts rated civic solidarity as moderate, which gives hope for the positive dynamics of this social process in a typical Russian region.

The expert community notes the presence in the region of an average level of realization of citizens' social activity, reflecting positive processes and the prevalence of social subjectivity.

The pessimism of assessments of the severity of negative moods prevailing in the region indicates a reduced socio-economic reserve of life in the region and provokes pessimism, apathy, and lack of faith in the future. These estimates are consistent with the opinion of the population and testify to the objectivity and systematic nature of deprivation processes in the province.

The prevalence of negative emotions indicates low ontological security in the region, complicating the manifestations of social subjectivity and activity.

The population of a typical Russian region is weakly involved in the activities of NKOs. Thus, the level of development in the region of civil society has significant unrealized potential.

The expert community believes that the region fully implements the authority of the region's governing authorities, which is a positive trend that inspires hope for the growth of social solidarity and civil society in the region.

The regional profiles of the reproduction of the level of trust and models of irrational forms of behavior of the population are revealed. Thus, according to monitoring data on assessing social risks in six border regions of Russia, it was revealed that the reduction in ontological security depends on a combination of economic, social, psychological well-being, and is also determined by the characteristics of social practices of adaptive behavior of the population of the regions. All this is realized in a special ethnocultural context. In particular, in the regions of the Siberian Federal District, the predominant factor in the reproduction of irrational behavioral practices is the economic factor, characterized by poverty, low level of material well-being, lack or low level of hope for a change in material prospects. Thus, in our opinion, it is advisable to consider social well-being as a predictor in the system of reproduction of the level of social trust, the growth of dangers and risks in modern Russian society.

An analysis of subjective assessments, various aspects of social well-being, and social responsibility of the population allows us to develop models for the reproduction of various types of social risks, which are indicators of social trust in the region. A low level of social trust stimulates deviant and asocial processes in the context of everyday interactions. Cynicism as a norm of communication, prevalent in post-Soviet society, fatigue from government intervention, and general social disappointment gradually destroy the prevailing confidence patterns and generate increased individual fears and concerns.

**Acknowledgements** This work was supported by RFFI Grant No. 19-011-00417: "Factors and mechanisms of trust formation in the system of social security preservation in the border regions of Russia."

## References

Allport GW (1960) Personality and social encounter. Beacon Press, Boston, MA

Bakhtina MA (2011) Trust in the state as a factor in increasing its effectiveness. J Instal Stud 3(3):57–65

Barinov DN (2017) Ontological safety and its borders in modern society. Publishing house ACT, Moscow, Russia

Boytsova OY (2012) Confidence in the state as a theoretical problem. Power 4:103–106

Cattell RB (1950) Personality: a systematic theoretical and factual study. McGraw Hill, New York, NY

CIVICUS (2012) State of civil society 2011. Retrieved from https://www.civicus.org/downloads/2011StateOfCivilSocietyReport/State_of_civil_society_2011-web.pdf

Dankin DM (2006) A supporting element of the architecture of the world. On the role of trust in the model of the future world order. Secur Eurasia 4(26):91–107

Douglas M (1986) Risk acceptability according to the social sciences. Russell Sage Foundation, New York, NY

Edelman (2001–2020). 20 years of trust. Retrieved from https://www.edelman.com/20yearsoftrust/

Fukuyama F (2003) The great break. LLC Publishing house ACT, Moscow, Russia
Fukuyama F (2008) Trust: social virtues and the path to prosperity. LLC Publishing house ACT, Moscow, Russia
Giddens E (2005) Organization of society: essay on the theory of structure. Academic Prospect, Moscow, Russia
Giddens E (2011) Consequences of Art Nouveau. Praxis, Moscow, Russia
Green S, Hansen FS (2016) Russia and the search for ontological security. Academic Prospect, Moscow, Russia
Guha-Khasnobis B, Kanbur R, Ostrom E (2006) Linking the formal and informal economy: concepts and policies. Oxford University Press, Oxford, UK
Ivanov AV, Danilov SA (2014) Social trust and the institutional order of society in the sociocultural context of Western and Eastern cultures: a comparative analysis. Fundam Res 11–8:1852–1859
Kuznetsov VN (2009) Sociology of security. KDU, Moscow, Russia
Luman N (2007) Social systems. Essay on the general theory. Nauka, St. Petersburg
Magomedov MG (2009) Social trust in Russian society: a sociological analysis (Dissertation of Doctor of Sociological Sciences). Southern Federal University, Rostov, Russia
Maximova S, Surtaeva O, Cherepanova M (2019) Migration policy as a factor in ensuring social security: expert opinion in cross-border regions of Russia. Adv Soc Sci Educ Humlties Res 364:737–741
Merry YV (2011) Trust and justice: moral foundations of a modern economic society. Aspect Press, Moscow, Russia
Seligman A (2002) The problem of trust. Idea Press, Moscow, Russia
Shtompka P (2001) Cultural trauma in a post-communist society (second article). Sociol Stud 2:3–12
Shtompka P (2002) Trust: a sociological theory. Sociol Rev 2(3):30–41
Shtompka P (2008) Sociology. Analysis of modern society. Logos, Moscow, Russia
Stolyar VY (2007) Trust as a tool for managing global risk. Bull Russ State Pedagog Univ 82–1:330–336
Veselov YuV, Kapustkina EV, Minina VN, Sinyutin MV, Stoyanov K, Helman K-U, Schrader H (2004) Economics and sociology of trust. Russian Sociological Society named after M.M. Kovalevsky, St. Petersburg, Russia
Viktorov A (2008) Introduction to safety sociology. Moscow, Russia: “Canon +” ROOI “Rehabilitation”
Weber M (1978) In: Roth G, Wittich C (eds) Economy and society, vols 1–2. University of California Press, Berkeley, CA
World Values Survey [WVS] (2010–2014) Source of open empirical data on the worldview, values of people. Retrieved from http://www.worldvaluessurvey.org
Yanitsky ON (2003) Sociology of risk: key ideas. World Russ 1:3–35
Zabolotnaya GM (2003) The confidence phenomenon and its social functions. Vestn RUDN Ser Sociol 4–5:79–85
Zhbankov A (2010) Trust as a security factor. Power 8:19–21

# Amendments to the Constitution of the Russian Federation Through the Prism of Social Security

**Lyudmila G. Konovalova, Nadezda A. Kandrina, and Olesia L. Kazantseva**

**Abstract** We studied the changes in the Constitution of Russia, as threats to the social security in the Russian Federation. We carried out the analysis of normative acts, law enforcement practice and relevant scientific literature. The main problems with amendments to the Constitution of the Russian Federation were the frequency with which they were adopted, the lack of substantiation, the similarity of their wording, the excessive simplification of the adoption procedures, the focus on authoritarianism and the creation of obstacles to the succession of powers and the principle of separation of powers. The link between the processes of constitutional change and social security issues based on appealing to foreign experience is demonstrated. Negative social cataclysms are proposed to be solved by world jurisprudence using the theory of constitutionalism and constitutional engineering. There is a variety of approaches in the definition of constitutionalism, but in the work constitutionalism is understood as a process of significant influence of the constitution on social relations through the principles: recognition and guarantee of human rights, separation of powers, people's sovereignty, election and responsibility of officials, the highest legal force and stability of the constitution. Reference to the theory of constitutional engineering is made to motivate the protection of the constitution from unjustified changes in its text. The article also referred to the concept of "unconstitutional" constitutional amendment, which is actively applied in foreign countries. Special attention we paid to the analysis of different views on the prospects of changing the Constitution of the Russian Federation. Three main approaches were considered: calls to constitutionalize Russia's national specifics, a proposal to strengthen European values in the Russian Constitution and justification of the need for its stability. The authors suggested to conserve the stability for the text of Russian basic law.

**Keywords** Constitution of Russia · Constitutional engineering · Constitutional justice · Constitutional amendments

---

L. G. Konovalova · N. A. Kandrina · O. L. Kazantseva (✉)
Altai State University, Barnaul, Russia
e-mail: verwaltung@mail.ru

 
S. G. Maximova et al. (eds.), *Advances in Natural, Human-Made, and Coupled Human-Natural Systems Research*, Lecture Notes in Networks and Systems 234,
https://doi.org/10.1007/978-3-030-75483-9_42

# 1 Introduction

In the scientific literature of various world countries, social security is presented as a capacious social and legal category. For example, Sillaste understands social security as the state of security of an individual, social group, or community against threats that may violate their vital interests, social rights, and freedoms (Sillaste 1998). The integrative nature of social security is proved, which makes it possible to understand it as a necessary reality, which in its qualitative aspect presupposes the functioning of mechanisms and structures for the development of the foundations of the life structure of an individual and society, the achievement of a sustainable social order, and the establishment of the legal, spiritual, and moral values (Frolova 2007). Public safety in scientific works is defined as a system of social relations and legal norms regulating these relations to ensure public peace, inviolability of life and health of the population, normal work and leisure of citizens, normal activity of state and public organizations, institutions, and enterprises (Sukharev 1984). Scientists substantiate the idea that it is necessary to form a social security system in any country of Eurasia, which should function based on a stable legal mechanism, a well-thought out, and socially responsible policy of the state (Frolova 2007).

In Russia, as the largest country on the Eurasian continent, social (public) security is not legally defined. In the modern Russian legislation, public security acts as a legal term for defining the sphere of activity of executive authorities and is singled out as a generic object of the infringement of specific types of crimes and administrative torts. It is mentioned in the "National Security Strategy of the Russian Federation" as one of the strategic national priorities (Presidential Executive Office 2015).

At the same time, considering the approaches available in the world to the perception of social security, its direct correlation with the country's basic law becomes evident. The fact is that the constitution of any country currently has the function of defining the basic values of the structure of society and the state, as well as the function of maintaining the system and stability of the entire legal field of the country. In this connection, even a new term has appeared in the bowels of the world science of constitutional law, which characterizes the high significance of the constitution for the rule of law, justice and stability of law and order, maintenance of social security—constitutionalism. Constitutionalism means the fact of the existence of a constitution and its active influence on the political life of a country, the supremacy and defining the role of a constitution (written or unwritten) as the basic law in the system of the current legislation, the mediocrity of political relations by constitutional and legal norms, the constitutional recognition of rights and freedoms of an individual, the legal nature of relations between a citizen and the state (Nersesyants 1983). The principles or attributes of constitutionalism are multifaceted in legal science but generally overlap with the attributes of the rule of law that are the following:

- Recognition and guarantee of human rights;
- Separation of powers;
- Popular sovereignty;
- Election and responsibility of officials;

- Supreme legal force and stability of the constitution (Bellamy 2007; Dick Howard 1992).

The emergence of this term is linked to the world history of social conflict resolution and statehood development. Only by the eighteenth century, along with democratic revolutions and the struggle for citizens' rights, the idea of establishing a clear hierarchy of normative acts headed by the basic law, which officials were strictly forbidden to violate and revise unilaterally. This idea served as a basis for the formation of constitutionalism (Dick Howard 1992).

Correspondingly, the question of the relationship between the requirement of stability of the constitution and the need to change or supplement it under the influence of changing social relations is important for constitutionalism. By virtue of the accumulated world experience on this issue, constitutional law offers the theory of constitutional engineering (constitutional design), a new direction that investigates the processes of accepting and changing constitutions. Constitutional engineering is based on the rationalistic belief that "good" constitutions can be developed and can construct an effective state (Von Beyme 2001). Nevertheless, in general, constitutional engineering specialists compare the adoption of new constitutions or introduce significant changes to existing basic laws to a revolution. At these moments, the entire legal system of the state is unbalanced to reorient it. Therefore, the world science of constitutional law recommends not to abuse constitutional revision procedures (Shustov 2017).

In Russia, the theories of constitutionalism and constitutional engineering become particularly relevant due to repeated amendments to the Constitution of the Russian Federation of 1993, which followed after 2009, contrary to the original guidelines for its stability, and in the light of the preparation for the adoption in 2020, at the initiative of the President of the Russian Federation, extensive amendments to the Constitution. The scale and frequency of amendments to the Basic Law of the Russian Federation have already made it possible to think about their impact on social security in the country. Therefore, it seems timely to analyze the Russian constitutional amendments through their social security prism considering the international experience. Russia is a country that connects Europe and Asia, and the constitutional reform currently being carried out in Russia actualizes the future fate of constitutionalism. It is of interest not only to Russia but also to the international community, as it demonstrates the practice of some deviation from the values of constitutionalism recognized in the world.

## 2 Materials and Methods

We studied the texts of Constitutions of different countries, regulating rules of making changes in basic laws of states (Federal Republic of Brazil 2001; The Republic of Hungary 1949; Queen of the United Kingdom, Prime Minister of Canada, Minister of Justice and Attorney General of Canada and Registrar General of Canada 1982; The

Portuguese Republic 1976; The United States of America 2001). Particular attention was paid to the analysis of the practice of constitutional courts in various countries regarding constitutional amendments, including those found in scientific sources (Carothers 2009; Conference of European Constitutional Courts 2008; Constitutional Court of the Russian Federation 2015, 2020; European Court of Human Rights 2006; Nussberger 2016; Siehr 2008). A significant amount of scientific literature on constitutionalism, constitutional engineering, and the role of constitutions in the current state has been used (Aranovsky and Knyazev 2013; Bourdon et al. 1980; Colón-Ríos 2013; Dixon and Landau 2015; Elster 2000; Ely 1980; Fallon 1997; Garlicki and Garlicka 2014; Holmes 2012; Kammen 1987; Landau 2013; McBain 1927; Sajo and Uitz 2017; Toniatti 2016; Waldron 1998). Separate consideration was given to the constitutional legislation of Russia, the practice of its application, and the opinions of Russian specialists and citizens on changes to the Constitution of the Russian Federation (Alebastrova 2020; Yeskin 2020; Avakyan 2007; Denisov 2012; Dmitriev 2013; Galanina and Starikova 2020; Gosuslugi Online 2020; Gritsenko 2011; Kenenova 2013; Kondrashev 2013; Medushevsky 2013; Morshchakova 2020; Presidential Executive Office 2020; Russian Federation 1995, 2008a, b, 2014a, b, 2020; Starikova 2020; Tasalov 2011; The New Times 2020; Velichko 1999; Vinogradov 2008; Zaikin 2014, 2018).

Current methods of cognition form the methodological basis of the research, in particular, general scientific (functional, systemic, synergetic, analysis, and synthesis), general logical (method of ascent from abstract to concrete, from general to private), private-scientific (for example, formal legal, comparative law, method of legal forecasting, etc.).

Functional, systemic, and general logical methods of cognition were applied to reveal the essential characteristics of social security, constitutionalism when considering a complex of proposed amendments to the Constitution of the Russian Federation through the prism of basic constitutional and legal theories. "The possibilities of cognition remain extremely limited if they do not rise above the level of unity and spontaneity of the object's existence, unless they investigate a specific phenomenon as a system of phenomena…" (Kuzmin 1976).

The application of the synergetic method made it possible to analyze the studied legal phenomena in interaction with each other, which helped to correlate constitutional reform in Russia with social security in Eurasia, as well as with the world experience (Petrov and Zyryanov 2017).

Formal, prognostic, and legal comparison methods were used in analyzing the impact of specific constitutional norms in Russia and foreign countries on social relations, revealed inconsistency between the goals of constitutional reform in Russia and its expected consequences, and demonstrated "scholasticism" of proposed constitutional amendments.

The methodological tools allow a comprehensive approach to the study of the impact of constitutional amendments in Russia on social security in the country and compare this experience with the world theory of constitutionalism.

## 3 Results

Contrary to the original emphasis on the stabilizing role of the Basic Law of the country, the Constitution of 1993 of the Russian Federation could not avoid changes affecting the term of the Administration of the President of the Russian Federation and deputies of the State Duma, control powers of the Parliament with regard to the Government of the Russian Federation, the judicial system and the Prosecutor's Office, as well as the procedure for forming the Council of Federation (Russian Federation 2008a, b, 2014a, b). Although politically, these changes have gone quite "easily," without significant opposition resistance, the scientific community has received them ambiguously and often critically, seeing in the novelties the desire for authoritarian statehood and the unreasonable encroachment on the "sacred text" of the Constitution (Dmitriev 2013; Zaikin 2018). The amendments are criticized for the fact that in addition to increasing the term of the Administration of the President of the Russian Federation and the State Duma of the Federal Assembly of the Russian Federation, as well as the liquidation of the voluntary and highly professional Supreme Arbitrazh Court of the Russian Federation do not carry a substantial component of the annual reports of the Government of the Russian Federation to the State Duma, the role of the Prosecutor's Office could have been quite regulated at the level of the law, the amendments on the members of the Federation Council, in general, are a mystery to jurisprudence since no scientist has understood the meaning of the appearance of this n". Excessive conciseness of the explanatory notes accompanying the initiative of the constitutional amendments, ignoring the negative opinions of the subjects of the Russian Federation on the relevant bills, and the speed with which decisions are made are bewildering (Tasalov 2011; Zaikin 2014). Many researchers see the reason for the simplicity of such controversial amendments as the fact that the current state mechanism of Russia does not imply either open comprehensive discussion, criticism, or prompt correction of decisions developed by the Presidential Administration (Denisov 2012; Kondrashev 2013).

The next amendments to the Constitution of the Russian Federation (Presidential Executive Office 2020) on the priority of constitutional norms over international law, on additional requirements to the Head of State and other officials and employees, on the procedure of formation of the Government of the Russian Federation and some expansion of the functions of the Parliament, on the provision of the Constitutional Court of the Russian Federation and constitutional consolidation of the status of the State Council of the Russian Federation were quite unexpected for the legal community during the Address to the Federal Assembly of the Russian Federation on January 15, 2020. A working group was set up to prepare proposals for amendments to the Constitution of the Russian Federation with the involvement of the public, and as a result, the initiatives have grown to about 200 amendments to the Constitution. To date, the State Duma has already adopted a law on the amendment to the Constitution of March 14, 2020, approved by the Federation Council and signed by the President of the Russian Federation. (Russian Federation 2020), which received a positive opinion from the Constitutional Court of the Russian Federation (Constitutional Court of the

Russian Federation 2020). The final list of amendments awaits its submission for national voting, originally scheduled for April 22, 2020, but was postponed due to the coronavirus epidemic (Gosuslugi Online 2020).

Once again, we see an unprecedented rush in adopting of the amendments and general approval of all authorities. This time, the initiators seek to create an impression of the scale of discussion of the proposed initiatives: discussions about the amendments are held in public councils, on television, radio, and official Internet sites. At the same time, from the point of view of professional jurisprudence, the process of adopting regular amendments to the Constitution of the Russian Federation raises both procedural and substantive issues.

Procedurally, there is an outrage at the conceptual weakness of the initiatives. The explanatory note to the bill on making such large-scale amendments to the Basic Law seems superficial, weakly revealing their motivation and validity, with only three text pages. It criticizes the unprofessional composition of the working group on the proposal of amendments to the Constitution, the excessive speed and ill-conceived procedure for the adoption of amendments, the doubtful legitimacy and legal logic of the new form of public approval of the amendments—the popular vote, which was not known before the current constitutional and legal practice in Russia. There is a significant influence of official propaganda, which uses not quite correct methods of information technologies, as well as the political incompetence of all subjects involved in the adoption process, including the Constitutional Court of Russia. The very logic of placing amendments on the chapters of the Constitution that do not correspond to either the meaning or the title of the respective chapters is criticized (Alebastrova 2020; Galanina and Starikova 2020; Yeskin 2020). The scope of the amendments introduced and how they were formulated are impressive. As it is known, the basic laws of states are characterized by thesis formulations, which allow the formulation of the main goals of the state in a capacious and principled way. Amendments to the Constitution of the Russian Federation in 2020 are formulated in a very lengthy manner, and repeated repetition of the same material is allowed. For example, a ban on persons occupying a public office and civil servants to have foreign citizenship and foreign accounts could be formulated in one norm. However, it is separately set forth for the President of the Russian Federation, deputies of the State Duma, the Commissioner for Human Rights, judges, and other officials. Russia has already had a negative experience of transforming the Basic Law into a "loose" and haphazard set of articles by introducing amendments to the Constitution of the RSFSR of 1978. Therefore, difficulties in the direct application of constitutional provisions are expected, which will be due to the excessively large volume of amendments introduced, the similarity of the introduced formulations, and their contradiction to the existing constitutional norms.

From the substantive point of view, the initiatives under consideration are inappropriate because they propose to consolidate either what already exists or what is more appropriate to regulate at the level of other acts or what should not be contained in the Constitution as a matter of principle. Thus, it is thought that amendments in the social sphere are already covered by the provisions on the social nature of the state existing in the Constitution; the State Council is already legally functioning, and the amendments being introduced do not fix the specifics of its presumed constitutional status; the priority of the Russian Constitution over international acts is already practically universally recognized in Russian jurisprudence and supported by the provisions of the Federal Law *On International Treaties of the Russian Federation,* and the obligation assumed by the Constitutional Court of the Russian Federation to resolve the issue of non-compliance with the provisions of this Law. Patriotic provisions restricting the legal relations of certain officials with foreign states could well be spelled out in ordinary laws. Amendments to the requirements to the President of the Russian Federation raise the question of the admissibility of a new countdown of the presidential powers of the incumbent head of state in connection with the legal "zeroing" of the procedural rules for filling the post, as it has happened in relation to senior officials of the constituent entities of the Russian Federation. There are serious concerns about initiatives to reduce the number of judges of the Constitutional Court of the Russian Federation and the possibility for the President of the Russian Federation to submit to the Council of Federation representations on the termination of the judges' competences in connection with the commission of "unworthy" misconduct, as well as the establishment of the inviolability of the President of the Russian Federation who terminated own powers.

What is presented in the official mass media as "expansion of parliamentary control" and "strengthening of the role of the Constitutional Court of the Russian Federation" in the legislative process is assessed by experts as a further expansion of the power prerogatives of the President of the Russian Federation and a strengthening of authoritarian tendencies of the statehood development (The New Times 2020). At the same time, in the relations between the President, executive, and legislative power, despite the introduction of mechanisms for the approval of certain members of the Government by the State Duma and its participation in the appointment of "power" ministers, little has changed. The President will still be able to dissolve the State Duma in case of any dissent. The President of the Russian Federation retains unbalanced large-scale powers to create a structure of executive bodies, as well as restrictions contained in the legislation on parliamentary control over the activities of the President of the Russian Federation, remain unchanged. On the contrary, significant changes are expected in the President's interaction with the judicial branch of power, which encroach on the independence of this branch of power through the tools of retiring judges.

This fact is particularly depressing against the background of the generally recognized role of courts in ensuring the rule of law and constitutionalism through norm control and constitutional justice. The arguments of foreign scholars have prevailed in world jurisprudence, emphasizing the special importance of constitutional control authorities in protecting the will of the people against possible encroachments by the Parliament and the executive branch of power in case of their disrespect for the Basic Law (Bourdon et al. 1980). Moreover, since the publication in 1927 of "The Living Constitution: An Analysis of the Realities and Legends of our Fundamental Law" by the American Professor McBain (Gritsenko 2011), the theory of the so-called "living constitution" has gained popularity, whose representatives prove changes in the meaning of the constitution overtime under the direct influence of interpretations coming from constitutional justice (Kammen 1987; McBain 1927; Toniatti 2016). This approach is in line with contemporary Pan-European views on the supremacy of the constitution. Thus, the General Report of the XIV Congress of the Conference of European Constitutional Courts, held in Vilnius from 3 to 6 June 2008, states that only an active position of a constitutional court provides a real, not supposed, implementation of the principle of supremacy of the constitution (Conferences of European Constitutional Courts 2008).

## 4 Discussion

The next amendments to the Constitution of the Russian Federation are essentially conducted in the absence of a real scientific discussion. Highly professional constitutionalists and recognized legal scholars prefer to emphasize ignoring the ongoing changes, as they consider them to be contrary to the foundations of constitutionalism and believe that talking about them now means supporting the initiated unjustified movement towards the revision of the Russian Constitution (Morshchakova 2020; Starikova 2020).

Before the 2020 initiatives were implemented, there was a debate in the Russian jurisprudence about the fate of Russian constitutionalism. The variety of proposed options for the constitutional development of Russia can be reduced to three basic approaches. The first is characterized by the denial of the liberal-democratic vector of development, which is acceptable to Russia, and calls for the return of specific historical (national) forms of the state structure. Proponents of such a vector appeal to Russian sobornost, autocracy, and ethnicity, to Orthodoxy, to the belief in truth rather than the law attributed to the nation, searching for a certain national idea, to the need for duties to dominate human rights (Velichko 1999). The second approach to constitutional development unites supporters of modernization through strengthening the European vector of constitutionalism. Representatives of this position are inclined to the effectiveness of a rapid and decisive transformation of statehood along with the European model. In particular, it is proposed the following:

- Increase the importance of the concept of legal statehood;
- Ensure full economic competition;
- Increase the importance of the institutions of direct democracy;
- Make the parliament more representative;
- Switch to real multiparty democracy;
- Achieve the full implementation of the principle of separation of powers;
- Limit the prerogatives of the president;
- Make federalism more efficient;
- Activate local self-government institutions;
- Reform and legitimize the judicial system (Avakyan 2007).

Our third and shared approach to constitutional development is based on defending the unchanged text of the Russian Constitution. Its proponents recognize the problems of modern Russian constitutionalism related to the imbalance of the separation of powers, weakness of civil society, difficulty in political participation, social and economic difficulties, and formalism of federalism and local self-government, but consider it fundamental to stabilize constitutional regulation to overcome such negative phenomena. Regarding to the numerous claims to the wording of the Russian Constitution, representatives of this approach respond that there are no ideal constitutions and those successful in the practical constitutionalism of the country are not the main suppliers of brilliant constitutional formulas and rather admire the attachment to their sometimes-outdated rules. They assess the content of the current Constitution of the Russian Federation as fully compliant with the ideals of constitutionalism and, accordingly, as worthy of respect (Aranovsky and Knyazev 2013; Medushevsky 2013; Vinogradov 2008).

In this regard, foreign experts in the field of constitutional law often accuse Russia, along with other Eastern European countries, of adopting "unconstitutional amendments to the constitution" that are the amendments that change the essence of constitutionalism, aimed at retaining the power of the political elite with authoritarian tendencies (Dixon and Landau 2015; Garlicki and Garlicka 2014). The doctrine of "unconstitutional constitutional change" has been adopted by the constitutional courts of individual countries and has already had positive results. In Colombia, for example, the doctrine prevented constitutional amendments that could have allowed the Head of State to remain in power indefinitely. In India, the doctrine helped to understand the anti-democratic effects of the state of emergency (Colón-Ríos 2013).

## 5 Conclusion

We concluded that the process of numerous and ill-considered amendments to the Constitution of the Russian Federation is beginning to touch upon social security issues. In the expansion of the authoritarian foundations of the Russian statehood, in the unprecedented extension of the term of the Administration of the Head of State and the authority, in the consolidation of numerous similar formulations, in the simplified procedures for the adoption of amendments, there is a gross violation of the value of constitutionalism and the very principle of the rule of law. The world experience shows tragic examples of ignoring constitutionalism postulates, including bloody revolutions, totalitarian systems, etc.

There are discussions about the fate of the Russian Constitution echo international practice. The fact is that for all its positivity, the theory of constitutionalism today does not look as optimistic as it was at the time of its inception, since the idea of improving social reality through the introduction of "good" constitutional principles comes up against the difficulties of building so-called "new" democracies in Eastern Europe, Africa, Latin America, as well as the problems of globalization. The view has emerged that constitutionalism is undergoing a crisis. For example,

A. Sajo argues that against the background of manipulation of citizens' constitutional emotions, unfair use of constitutional formalism, nationalism, false patriotism, populism, rationalized parliamentarism, and other problems, constitutionalism loses its appeal in the eyes of society (Sajo and Uitz 2017).

However, we agree with those researchers who believe that, despite the threats, constitutionalism is an absolute value, needs to be preserved and protected, is not in violation of the rule of law, and assumes reception outside Europe (Kenenova 2013). Of course, it is impossible to solve a state's problems and create an ideal legal order by adopting a "good" constitution. However, among other components, such as economy, traditions, culture, etc., adherence to constitutional values can improve the efficiency of the state, so we support the stability of the text of the Russian Constitution.

**Acknowledgements** We express our gratitude to Altai State University for the financial support of our research.

## References

Alebastrova I (2020) Political conclusion: why the constitutional court agreed to amend the constitution. Accessed from https://www.rbc.ru/opinions/politics/27/03/2020/5e731e869a794796b32e4f3b

Aranovsky KV, Knyazev SD (2013) The role of the constitution in political-legal formation of Russia: original circumstances and modern expectations. Comp Const Rev 3(94):50–59

Avakyan SA (2007) Country needs a new Constitution. EJ Lawyer 50:12

Bellamy R (2007) Political constitutionalism: a republican defense of the constitutionality of democracy. Cambridge University Press, Cambridge, UK

Bourdon J, Pontier JM, Ricci JC (1980) Droit constitutionnel et institutions politiques, vol 1. Éditions scientifiques et juridiques, Paris, France

Carothers T (2009) Rule of law temptations. Pro Et Contra 3–4(46):139–151

Colón-Ríos JI (2013) Beyond parliamentary sovereignty and judicial supremacy: the doctrine of implicit limits to constitutional reform in Latin America. Vic Univ Wellingt Law Rev 44(3):521–534

Conference of European Constitutional Courts (2008) General Report of the XIV Congress of the Conference of European Constitutional Courts. Constitutional justice. Bulletin of the Conference of the Constitutional Control Organs of the countries of new democracy, vol 2–3. Center for Constitutional Law of the Republic of Armenia, Yerevan, Armenia

Constitutional Court of the Russian Federation (2015) The decision of the constitutional Court of the Russian Federation In the case of verifying the constitutionality of the provisions of Article 1 of the Federal Law 'On Ratification of the Convention for the Protection of Human Rights and Fundamental Freedoms and its Protocols', paragraphs 1 and 2 of Article 32 of the Federal Law 'On International Treaties of the Russian Federation', parts one and the fourth article 11, paragraph 4 of the fourth part of article 392 of the Civil Procedure Code of the Russian Federation, parts 1 and 4 of article 13, paragraph 4 of part 3 of article 311 of the Arbitration Procedure Code of the Russian Federation, parts 1 and 4 of article 15, paragraph 4 of part 1 of article 350 of the Code of Administrative Court Russian Federation and clause 2 of part four of Article 413 of the Code of Criminal Procedure in connection with the request of a group of deputies of the State Duma, July 14, 2015 No. 21-P. Moscow, Russia

Constitutional Court of the Russian Federation (2020) Conclusion of the constitutional court of the Russian Federation "On compliance with the provisions of Chapters 1, 2 and 9 of the Constitution of the Russian Federation that have not entered into force the provisions of the Law of the Russian Federation on amendments to the Constitution of the Russian Federation 'On improving regulation of certain issues of the organization and functioning of public authority', as well as on the conformity of the accession procedure with the Constitution by virtue of Article 1 of this Law in connection with the request of the President of the Russian Federation, March 16, 2020 No. 1-Z. Moscow, Russia

Denisov SA (2012) Imitation of the constitutional order in Russia. Const Munic Law 10:2–8

Dick Howard AE (1992) Constitutionalism. The rule of law. Moscow, Russia, Progress, pp 53–79

Dixon R, Landau D (2015) Transnational constitutionalism and a limited doctrine of unconstitutional constitutional amendment. Comp Const Rev 2(111):32–63

Dmitriev YuA (2013) Can you increase the democratic potential of the current constitution of the Russian Federation? State Law 12:45–53

Elster J (2000) Ulysses unbound: Studies in rationality, precommitment, and constraints. Cambridge University Press, Cambridge, UK, New York, NY

Ely JH (1980) Democracy and distrust: a theory of judicial review. Harvard University Press, Cambridge, MA

European Court of Human Rights (2006) Resolution of the European Court of Human Rights "The Case of the Moscow Branch of the Salvation Army against Russia (05 Oct 2006 No. 72881/01). Strasbourg, France

Fallon RN Jr (1997) "The rule of law" as a concept in constitutional discourse. Columbia Law Rev 97(1):1–56

Federal Republic of Brazil (2001) The constitution of the federative Republic of Brazil. In Dubrovin VN (ed) Constitution of foreign countries. Yurlitinform, Moscow, Russia, pp 288–380

Frolova NA (2007) The ideology of social security and the legal policy of the state: theoretical and historical aspect. AVA-BEST, Russia, Moscow

Galanina A, Starikova M (2020) Amendments came under friendly fire. Kommersant 19:3

Garlicki L, Garlicka ZA (2014) Unconstitutional constitutional amendments: does the problem exist and if so, is there a solution to be found? Comp Const Rev 1(98):86–99

Gosuslugi Online (2020) Amendments to the Constitution of the Russian Federation 2020: List of amendments. Retrieved from https://gosuslugi-online.ru/popravki-v-konstituczіyu-rf-2020-spisok-izmenenij/

Gritsenko EV (2011) The boundaries of the constitutional court's rulemaking. Comp Const Rev 5(84):107–130

Holmes S (2012) Constitution and constitutionalism. Comp Const Rev 3(88):56–84

Kammen M (1987) "A vehicle of life": the founders' intentions and American perceptions of their living constitution. Proc Am Philos Soc 131(3):325–340

Kenenova IP (2013) The crisis of the western legal tradition and the Russian tradition of power from the perspective of comparative constitutional law. Comp Const Rev 1(92):13–29

Kondrashev AA (2013) The current political regime of Russia: restrictions on basic civil rights and freedoms through legislative innovations and law enforcement practice. Const Munic Law 2:10–17

Kuzmin VP (1976) The principle of systematicity in Marx's theory and methodology. Politizdat, Moscow, USSR

Landau D (2013) Abusive constitutionalism. Univ Calif Davis Law Rev 47(1):189–260

McBain HL (1927) The living constitution: a consideration of the realities and legends of our fundamental law. The Workers Education Bureau Press, New York, NY

Medushevsky AN (2013) Constitutional principles of 1993: formation, results and perspectives of realization. Comp Const Rev 1(92):30–44

Morshchakova T (2020) Text and reality. Novaya gazeta, 24. Accessed from https://novayagazeta.ru/articles/2020/03/06/84197-tekst-i-realnost

Nersesyants VS (ed) (1983) The history of bourgeois constitutionalism of the XVII-XVIII centuries. Nauka, Moscow, Russia

Nussberger A (2016) Independence of the judiciary and the rule of law in the jurisprudence of the European Court of Human Rights. Comp Const Rev 2(111):142–151

Petrov AV, Zyryanov AV (2017) Synergetic approach in the modern legal studies. Bull South Ural State Univ Ser Law 17(4):102–106

Presidential Executive Office (2015) Decree On the National Security Strategy of the Russian Federation. (31 Dec 2015 No. 683). Moscow, Russia

Presidential Executive Office (2020) Presidential address to the Federal Assembly (Jan 15, 2020). Russia, Moscow

Queen of the United Kingdom, Prime Minister of Canada, Minister of Justice and Attorney General of Canada and Registrar General of Canada (1982) Constitutional acts of Canada. The proclamation of the Constitution Act 1982, 17 April 1982. Accessed from http://www.concourt.am/armenian/legal_resources/world_constitutions/constit/canada/canada-r.htm

Republic of Hungary (1949) The Constitution of the Republic of Hungary, 20 Aug 1949. Budapest, Hungary. Accessed from https://worldconstitutions.ru/?p=789

Russian Federation (1995) Federal law On international treaties of the Russian Federation. (15 July, 1995 No. 101-FZ). Moscow, Russia

Russian Federation. (2008a). Federal law on amendments to the Constitution of the Russian Federation "On changing the term of office of the President of the Russian Federation and the State Duma, 30 Dec, 2008a No. 6-FKZ). Moscow, Russia

Russian Federation (2008b) Law of the Russian Federation on the amendment to the Constitution of the Russian Federation On the control powers of the State Duma in relation to the Government of the Russian Federation. (30 Dec 30, 2008b No. 7-FKZ). Moscow, Russia

Russian Federation (2014a) Law of the Russian Federation on the amendment to the Constitution of the Russian Federation On the Supreme Court of the Russian Federation and the prosecutor's office of the Russian Federation. (5 Feb 5, 2014a No. 2-FKZ). Moscow, Russia

Russian Federation (2014b) Law of the Russian Federation on the amendment to the Constitution of the Russian Federation On the Council of the Federation of the Federal Assembly of the Russian Federation. (21 July, 2014b No. 11-FKZ). Moscow, Russia

Russian Federation (2020) Law of the Russian Federation on the amendment to the Constitution of the Russian Federation On improving regulation of certain issues of the organization and functioning of public authority. (14 March 2020 No. 1-FKZ). Moscow, Russia

Sajo A, Uitz R (2017) The constitution of freedom: an introduction to legal constitutionalism. Oxford University Press, Oxford, UK

Shustov DG (2017) Essentia constitutionis: constitution of the Russian federation—a look through the centuries of constitutional theory. Comp Const Rev 3(118):71–90

Siehr A (2008) Symbolic legislation and the need for legislative jurisprudence: the example of the Federal Republic of Germany. Legisprudence 2(3):271–305

Sillaste GG (1998) Woman as an object and subject of social security. Sotsiologicheskie Issledovaniya 12:62–72

Starikova M (2020) In the minority, but not in insult. Henry Reznik and Tamara Morshchakova refused to cooperate with the HRC working group on amending the Constitution. Kommersant 15:3. Accessed from https://www.kommersant.ru/doc/4234591

Sukharev AYa (ed) (1984) Encyclopedic law dictionary. Soviet Encyclopedia, Moscow, USSR

Tasalov FS (2011) The invisible house in the XXI century: is it possible to reform the federation council? Comp Const Rev 1(80):33–46

The New Times (2020) Constitutionalist Tamara Morshchakova on the decision of the Constitutional Court's to reset Putin's terms. Accessed from https://newtimes.ru/articles/detail/192183?fcc

The Portuguese Republic (1976) The Constitution of the Portuguese Republic, April 2, 1976. Lisbon, Portugal. Retrieved from http://www.concourt.am/armenian/legal_resources/world_constitutions/constit/portugal/portug-r.htm

Toniatti R (2016) A living judiciary for a living law: constitutional transitions and governmental checks and balances. Comp Const Rev 5(114):97–104

United States of America (2001) The Constitution of the United States of America. In: Dubrovin VN (ed) Constitution of foreign countries. Yurlitinform, Moscow, Russia, pp 240–258

Velichko AM (1999) State ideals of Russia and the West. Parallels of legal cultures. Publishing House of Institute of Law, St. Petersburg, Russia

Vinogradov V (2008) The Russian constitution: 15 years later. Comp Const Rev 3(64):59–68

Von Beyme K (2001) Institutional engineering and transition to democracy. In: Zielonka J (ed) Democratic consolidation in Eastern Europe, vol 1. Oxford University Press, Oxford, UK, pp 1–3

Waldron J (1998) Judicial review and the conditions of democracy. J Polit Philos 6(4):335–355

Yeskin AA (2020) Fatal amendments to the Constitution of the Russian Federation. Accessed from https://www.9111.ru/questions/777777777819556/

Zaikin SS (2014) The Supreme Arbitration Court: Data deleted. Part 1. Silence and consent. Comp Const Rev 3(100):54–71

Zaikin SS (2018) Landscapes of democracy an inconspicuous constitutional amendment or what the representatives of federation in the council of the federation of Russia are for. In: Narutto SV (ed) Questions of federalism: a collection of scientific papers, vol 4. Prospect, Moscow, Russia, pp 205–237

# The Birth of Russian Extremism: The Beginning of National Security in Russian Society

**Maksim G. Kolokoltsev and Natalya Yu. Kolokoltseva**

**Abstract** The paper is devoted to the study of extremist organizations in Russian society. The research objective is to identify the first extremist communities and determine extremism formation factors in Russia. Based on the analysis of works of public figures and archives, we studied the emergence and spread of terrorism in Russia. Historians of the Soviet period claimed that terror was part of the Narodniks' program. This conclusion did not contribute to an objective study of the Narodnik movement. Extremism did not stand out as a specific phenomenon among the Russian social movements. According to the Narodnik ideology, the people had to determine the way of Russian society transformation. The Narodniks condemned terror. The centralization of power and political struggle led to an increase in extremism in the country. In this paper, we draw conclusions about the causes of terror and possible ways to solve the problem. The rigid centralization of power provoked violence. National identification, marginalization, alienation, and split society reinforced the protests in Russian society. In the 1860s, the first extremist groups were formed, and terrorist attacks took place.

**Keywords** Social movement · Narodnichestvo · Terror · Extremism · National identity · Monarchism · Anarchism · Nihilism · Raznochintsy · Karakozovtsy · Marxism

## 1 Introduction

Terrorism and extremism are essential global problems of our time. Scholars study the criminal activities of extremist organizations and the possibilities of preventing terror consequences (Belsky 2017, p. 479). The paper focuses on the economic

M. G. Kolokoltsev
Altai State Agrarian University, Barnaul, Russia

N. Yu. Kolokoltseva (✉)
Altai State University, Barnaul, Russia
e-mail: n_kolokolceva@mail.ru



S. G. Maximova et al. (eds.), *Advances in Natural, Human-Made, and Coupled Human-Natural Systems Research*, Lecture Notes in Networks and Systems 234,
https://doi.org/10.1007/978-3-030-75483-9_43

and legal aspects of extremism. The researchers most often deal with international terrorism (Chernyadeva 2017, p. 384). However, it is equally important to study the history of the formation and development of extremism, as well as socio-political and ideological causes of terror in Russia. The definitions of concepts and factors of the development of these phenomena are debatable. Therefore, identifying and studying the origins of extremism in Russia, we can reveal its nature. Modern English journals contain such statements as, "The ideas of the Narodnik and anarchist Bakunin should have appeared a century later" (Peter 2008, pp. 115–128).

In the second half of the nineteenth century, the social movements in Russia gradually split into liberal, conservative, and radical parties. Ideological principles of liberalism, conservatism, radicalism, and extremism overlap; their differences are conditional. Narodnichestvo took shape. Extremist organizations took the ideas of this movement as a call to action.

In the mid-nineteenth century, political thought focused on the ideas of anarchism. In the Oxford University journals, we can find the statement that the anarchists were not only Bakunin and Kropotkin, but also Marx, and even Lenin (Pancho 2005). The American researcher A. Kelly considered Herzen a greater anarchist than Bakunin (Kelly 1989, p. 128). However, European historians believe that anarchism does not determine the ideology of the Russian followers of the anarchist Bakunin (Walicki 1989, p. 91). The researchers relate anarchism to liberalism (Naarden 1995, p. 52).

At the same time, the Narodnik ideology of building a chain of command from the bottom up is the basis of social security and cooperation of all society members. Our task was to study extremism, determine its ideological principles, place in national politics, and factors of its formation.

## 2 Materials and Methods

In the second half of the nineteenth century, during modernization, the national identity of Russian society became an urgent problem. Methodological theory of modernization of Maine, Tönnies, and Weber regarded the transition problems from traditional to industrial society. The Russian intelligentsia was looking for a rational and effective way to transform the imperial political system. During the period of rigid centralization of power, freedom, individualization, and self-organization (typical for modernization) manifested themselves in the spread of terrorist views.

However, modernization theory is limited by the Eurocentrism principle. The national social movements regard themselves as the opposition to the violent introduction of Western European culture. Danilevsky, Spengler, and Toynbee contrasted the theory of modernization to the theory of cultural-historical types. According to this theory, every civilization develops according to the unique historical path. Sociologist S. Huntington presented terrorism as a response of traditional societies to the challenges of liberal civilization (Huntington 2003). Russian radical social movements were looking for an authentic way of national modernization.

The study relies on the analysis of publications (letters and memoirs) of participants of the Russian social movements of the late nineteenth century. Historical-genetic and biographical methods of studying letters (those of Bakunin, Kavelin; Memoirs of Herzen, Panteleev, Eliseev, Khudyakov, Kovalik, and Lopatin) allowed us to identify the development stages of the social movements, and to determine the criteria for dividing them into groups (Bakunin 1906; Kavelin 1892; Panteleev 1958; Yeliseyev 1933; Debogoriy-Mokrievich 1906; Vinogradov 1959; Lopatin 1922).

The comparative-historical method in the study of proclamations issued in Russia and abroad in the 1860s enabled us to determine the ideological priorities of public figures and understand the specifics of the political views of its members (Rudnitskaya 1997). Using historical-system method, we structured the obtained data based on the interactions of public figures of the late nineteenth century.

Besides, we used the materials of the State Archive of the Russian Federation (Fund 95, List 1, File 38, Page 189) and the Central Archive of the Republic of Kazakhstan (Fund 64, List 1, Book 3, File 5326, Page 131). We confirmed our assumptions using content analysis and cross-validation methods.

## 3 Results

National identity of Russian society in the nineteenth century revealed and aggravated social conflicts that the government did not solve, trying to preserve its privileged position. The existing illegitimate political system led to hidden unmanageable consequences. The terror resulted from the dysfunction of the existing social system. There was no such an experience or conflict resolution mechanism. This caused its suppression or transition to hidden forms. As a result, the terror aroused with redoubled force.

Narodnichestvo emerged in the process of national self-identity, seeking to unite all social strata. The followers of this movement believed that only national unity and common interests could radically change the Russian political system. Collegiality and the search for possible ways to unite all social groups characterize the Narodnik ideology.

Russian mentality relied on the idea of fraternity, the messianic role of Russia in uniting all peoples. However, the defeat in the Crimean War in 1956 revealed the uneven development of different countries. Russian progressive society became more aware of the threat to its national security. This situation generated and strengthened extremist views in society.

In the middle of the nineteenth century, the self-organization of the outskirts of the vast empire began; however, Saint Petersburg suppressed it. The emerging elite could not express their interests and needs. As a result, the politics was radicalized. The Narodniks planned to engage in political activities, despite their negative attitude toward the constitutional monarchy. "Russia has its own elements of representability; its internal needs to make them real is growing" (Ogarev 1863, p. 1366).

Conservatism, liberalism, and radicalism began to take shape in Russia. They differed in their attitude to power. For example, Bakunin accused Herzen and Ogarev of inventing the utopian doctrine, which claimed that it was possible to solve the social issues in Russia without eliminating the existing regime. Bakunin emphasized, "In the name of practicality, you have created an impossible theory about a social revolution without a political revolution." In his opinion, such a theory was not reasonable, since "<…> both coups are inseparable. Moreover, they are a single unit" (Bakunin 1906, p. 283).

The political engagement of Russian society increased. For example, the Narodnik N. I. Zhukovsky, who was in exile since 1862, sent publications of the *Volnaya Russkaya Printing House* to Russia and propagandized the peasant revolution. The success of this event depended on the fact that people would be introduced to science, learning various political factors" (Zhukovsky 1955, p. 133). Zhukovsky, following the leader of Narodnichestvo Bakunin, realized the significance of political struggle. Later, he lived in Switzerland and took part in the activities of the *young emigration*.

Repressions against the democratic intelligentsia, as well as Chernyshevsky, Mikhailov, Serno-Solovyevich, Shchapov destroyed the faith in the policy of Alexander II. The authors of the proclamations did not trust the peasant reform of 1861. They began to persuade the peasants to demand land and freedom from the emperor. Terror was supposed to have a socio-psychological impact on society.

Resocialization and social identity during the modernization and bourgeois-democratic reforms of the 1860s gave rise to a new type of deviation called terror. There was no established system of norms and previously unknown ideological patterns appeared. Then, Chernyshevsky created a professional revolutionary image embodied in Karakozov, and later in Nechaev (Rudnitskaya 1997, p. 173).

The terrorist act of 1866 against the tsar increased alienation and split Narodnichestvo. In his memoirs, Russian writer and publicist Boborykin stated, "By 1865, radical youth was embittered and the underground movement intensified, which caused the following attempt of Karakozov 1866" (Boborykin 1965, p. 324). In 1866, in the investigative testimony, Ishutin reported his connections in 1864 with the liberals Boborykin, Saltykov-Shchedrin (Klevensky and Kotelnikov 1928, p. 317). Later, their relations with Herzen collapsed. Extremism in Russia was born.

The political form of the Narodnik ideology was rationally changed according to the interests of all social strata. Terror methods of struggle were excluded. In the obituary of P. G Zaichnevsky, published in Geneva in 1896, in the *Materials for the History of the Russian Revolutionary Socialist Movements*, written by his contemporaries, his views were characterized as follows, "Pyotr Grigoryevich was, as they said at the time, a 'centralist,' and as for the people, his activity should have been purely agitational. Of course, all demonstrations and terror were condemned by them as a direct hindrance to the 'Organization" (Kozmin 1931).

In the 1860s, the Narodniks did not want to abolish the monarchy. Moreover, Lopatin denied the idea of a conspiracy against the emperor. Assessing the circumstances of the terror of 1866, he established the *One Ruble Society*. This organization continued to propagate the ideas of Narodnichestvo. The prematurity of such actions could lead to greater government terror (Lopatin 1922). Lopatin called on rural teachers to collect evidence of anti-government propaganda among the people.

In the mid-1860s, an extremist organization began to develop and started the political activity. In Russian historiography, there was an opinion that Karakozov's extremists did not contact with the Narodnik members of *Zemlya i Volya* (*Land and Liberty*) Researcher Vilenskaya (Vilenskaya 1969, p. 112) suggested that Karakozov was still related to them.

Initially, Karakozov took part in Narodnichestvo. In 1862, at the requests of Kelsiev and London emigrants, he sent the *Veche* newspaper to the provinces of Russia (State Archive of the Russian Federation, n.d.). Kelsiev published those materials in the *Kolokol* newspaper. His views on the role of old believers in the social movement coincided with the beliefs of Bakunin. The old believers considered that the existing government had abandoned the true faith and the holy throne. From a religious point of view, the murder of a tyrant was sacred.

By the mid-1860s, the Narodniks, who regarded conspiracy as the primary political engagement method, formed a group. The Narodnik idea of combining social and political revolution contributed to this event. However, the Narodniks believed that a political coup would not change the country, since the political power would belong to another ruling elite. Bakunin wrote, " <…> such a bloody revolution, based on the creation of a highly centralized revolutionary state, would inevitably lead to the military dictatorship of the new ruler." (Bakunin 1906, p. 164). Therefore, conspiratorial views were not popular. The followers of the extremist Ishutin denied the existence of extremist movements.

Then, the conspirators refused to use extremist methods to achieve their goals. Zaitsevskii served as Secretary of the *Zemstvo Board*, a department for peasant affairs. This post allowed him to cooperate with local people and influence their work. We found the information about the activities of Sichnevogo in the memoirs of Rusanov, "Revolutionary, socialist, and political conspirator. However, he was not so negative about the constitution and liberalism, unlike most of the revolutionaries of that time" (Kozmin 1923).

Dostoevsky condemned the revolutionary propaganda of the movement *Young Russia*. He called the followers of Zaichnevsky "liberals who preached the disintegration of Russia into the united states". In his opinion, "The whole world will soon be renewed due to the Russian thought. However, for this great event to occur, the political right and primacy of the great Russian tribe over the Slavic one must be finally and irrevocably accomplished" (Dostoevsky 2008). Then Dostoevsky went on a Russian mission to Constantinople, which was to become the center of the future Slavic Orthodox world.

The alternative nature of the Narodnik program caused much controversy. In this regard, Kulchitsky noted the common views of Bakunin and the organization *Young Russia* in terms of combining social and political coups, which differed in tactics (Kulchitsky 1908). Bakunin rejected and condemned the extremist ideas of *Young Russia* and prepared the young emigration for an uprising. Competent in this matter, Lenin wrote about the liberation movement of the early twentieth century, "Terror was the result, sign, and companion of disbelief in the uprising, the lack of conditions for it" (Lenin 1960). The Narodniks called for a social transformation of the Russian imperial system. They believed that terror would not lead to significant changes.

In 1848, Bakunin was accused of terrorism (Lemke 1906). However, Steklov and Dragomanov, the first researchers who studied the life of leader of the Russian social movement, concluded that Bakunin did not support terror (Steklov 1927, p. 317). Bakunin propagated and agitated people. Violence and terror could alienate them even more. Bakunin wrote, "Therefore, the victory of the Jacobins or Blanquists would mean the death of the revolution" (Bakunin 1906, p. 164). The Narodniks were against terror. On the contrary, they promoted the non-violent evolution of the political system and increased the literacy of the people.

The prevailing ideas about the need for cultural and educational work excluded terrorist methods. In this regard, Kovalskaya, a member of the Narodnichstvo, wrote that the shot of Karakozov caused a heated debate in their group. The attempt to assassinate the emperor was considered a moral and theoretical problem (Kovalskaya 1931). For example, Shelgunov wrote about various means of achieving political goals. However, he noted that the choice was always left to the people (Shelgunov 1881). The leader of the Narodnik movement propagated, "The emancipation of the workers is the business of the workers themselves. This is the basis of our union" (Bakunin 1920, p. 14).

In 1862, Kelsiev, in a letter to Salias declared the need to "call to the people" as soon as possible. He participated in the student uprising of 1861. Afterwords, he was arrested and exiled first to Verkhoturye (from where he escaped), and later imprisoned in the Peter and Paul fortress. Khudyakov, in his stories, referred not to regicide but a violent coup. Utin wrote about the tactics of the *Zemlya i Volya* group as follows, "Of course, a single shot, could not be part of the program of those people…" (Smirnov 1959, p. 341). Kozmin concluded that not all *the Sixtiers* could be called the Narodniks.

The leaders of the *One Ruble Society* Lopatin and Volkhovsky took an active part in the work of political groups in the mid-1860s. The work of this organization marked the end of the first decade of Narodnichestvo. Vilenskaya investigated the ways of foreign publications delivery, studying the memoirs of Lopatin. For example, the *Malorossiyskaya obshchina* (Little Russian Community) of his friend Volkhovsky existed in the mid-1860s.

The historian Bazanov found that the predecessor of Volkhovsky and Lopatin was Khudyakov (Bazanov 1974, p. 65). Moreover, Lopatin, promoting the ideas of Narodnik and sociologist N. D. Nozhin and condemning the views of Karakozov wrote, "The plan of the so-called *Karakozov conspiracy* at first assigned a very small role to the people in forcibly changing their fate" (Lopatin 1922, p. 130). Michael Hughes, an English historian, correctly understood the ideology of Felix Volkhovsky. In his opinion, Felix was convinced of the vast potential of Russian society, thinking that it could be realized only "if the Russian bureaucracy would not kill freedom" (Hughes 2009).

Tolerance, typical for the Narodnik ideology, allowed for a dialogue between various political groups. Furthermore, Nechkina proved the fact that the *Young Russia* joined *Zemlya i Volya* (Nechkina 1980, pp. 378–390). These organizations constructively resolved tactical differences. The proclamation *Freedom* printed by the *Zemlya i Volya* was published in the newspaper *Kolokol* on June 1, 1863. It explained that "the society has managed to unite the members of those separate groups that have repeatedly demonstrated what they are capable of" (Bakunin 1863).

Extremism in Russia developed independently. Ballod, a member of the social movement of the early 1860s, described a group of the extremists during the student uprisings of 1861 (Cheshikhin-Vetrinsky 1923, p. 215). Educated youth picked up the ideas of the progressive nobility of the first quarter of the nineteenth century.

Social groups and communities were not structured in Russian society. That is why marginalization has led to an increase in the number of Raznochintsy. The education reforms caused the mass training of various ranks of the intelligentsia expressing social protests. University professors taught the Raznochintsy. They were aware of the need for radical change in the context of a political crisis. However, the students were expelled for the protests.

Expelled students challenged their fate and created groups with their own ideologies. They added tens of thousands to the army of professional revolutionaries. However, their nihilism led to the mass terror. Nihilism, as an ideology of denial, was beneficial for terrorists. The old system of norms was collapsing. The old ideological guidelines in the context of overdue reforms were challenged.

In 1862–1863, in Heidelberg, where Russian students studied, Karakozov communicated with Cherkesov and Serno-Solovyevich. A fragment from the letter of Herzen to Cherkesov of February 8, 1863, evidenced this fact, "If you know Vladimirov (if not, do not get acquainted with him), thank him for the letter and tell him that we do not need books from Heidelberg yet" (Smirnov 1959, p. 742). Vladimirov was an alias of Karakozov (Vilenskaya 1969, p. 93). Herzen tried to prevent the influence of Karakozov on the fighting mood of Russian youth.

In 1861, Tkachev, the leader of the conspiratorial section in the Narodnik movement of the 1870s, entered the law faculty of Saint Petersburg University. In the fall, he took part in student uprisings. Tkachev was an active member of the Ishutin organization focused on the world political revolution. He approached *Smorgon*, the union from which the future Nechayevtsy [followers of Nechaev] came out" (Rudnitskaya 1997, pp. 180–200). An extremist movement separate from Narodnichestvo was developing.

Extremist groups started operating in Russia. The former student of Kazan University, member of Orlov mentioned a community in Russia, the main goal of which was to form revolutionary sects provoking the people's anti-government protests, by whatever means (Dyakov 1963, p. 372).

An extremist group of the nihilist Nodkin corresponded with foreign agitators (Bogucharsky 1906, p. 347). Later, he was arrested and exiled to the steppe Governor-General's settlement, where in July 1888, the police have increased their surveillance of him (Central Archive of the Republic of Kazakhstan, n.d). The extremist movement emerged during the development of Narodnichestvo, which increased its influence in the 1870s and resumed by the beginning of the twentieth century.

The substitution of the names of the extremist group (Ishutintsy and Karakozovtsy) in the testimony given by the Narodniks for the case of Karakozov calls into question the existence of the leader of extremism. Liberal journalist Kozlinina reported on the lack of any leadership qualities in the personality of Ishutin (Kozlinina 1913, p. 312). Moreover, even the name of his organization did not exist. In his autobiography, Khudyakov recalls that Ishutin was not the head of the organization (Khudyakov 1930, p. 214). The *Organization* and *Hell* participating in the Karakozov case were the Narodnik organizations characterized by terrorist views.

Terrorism became more active in 1869. The extremist movement took shape in the second half of the 1870s. Conservative politician Meshchersky, recalling the split of the *Zemlya i Volya* in 1879, stated that its branch *Chernyi Peredel* was Narodnik, and *Narodnaya Volya* was a terrorist organization (Meshchersky 1898, p. 514). In Russian historiography, the organization *Chernyi Peredel* is called Bakunist (Tvardovskaya 1965, p. 63). Members of the *Narodnaya Volya* recalled the ideas of Chernyshevsky and the actions of Karakozov. Members of this organization unsuccessfully propagandized the peasantry and included terror in their program. During the 1870s, the Narodnik movement changed its ideology.

Extremist ideology became part of the organization, and the actions began in the late nineteenth century. From March 8, 1883, to January 16, 1885, a group of people was sent to Kokchetav under police oversight for participating in the *Narodnaya Volya*: Zinaida Zatsepina, Nikolay Yakimov, Nadezhda Andrushchenko, and other extremists. They passed through the city of Omsk to Akmola regional administration. A case was brought against them. In the mid-1880s, the Minister of Internal Affairs, Lieutenant General Orzhevsky, according to the Minister of Justice, in his letter to the Governor of Akmola in this case, calls these people members of the revolutionary social movement, referring to their belonging to its terrorist faction (Central Archive of the Republic of Kazakhstan, n.d).

The terror in the country increased along with the spread of Marxism in Russia. The Marxist idea of class struggle was well established among the people already familiar with the ideas of anarchism and utopian socialism.

The desperate desire to change something in society enhanced terror, caused by the emotional breakdown of people. It was a normal reaction to the humiliation and the result of increasing contradictions in society. In the eyes of society, terror became morally justified and heroic.

According to *the Seventiers*, the terror and conspiracy could be carried out without the participation of the people. In the early twentieth century, the slogan of the political revolution and the idea of subordination of the social revolution to the political one was declared. The Bolsheviks adopted this ideological postulate and, having come to power, began terrorizing their opponents. The legitimate use of violence by the state increased the terror used as a means of protest in society. The difference between the white and the red terror was the fact that the latter became a state terror.

## 4 Discussion

Marxism allowed researchers to study the economic basis of socio-political relations and class struggle. The Russian historiography of the Soviet period exaggerated the revolutionary nature of the Narodnik movement. Troitsky, noting the similarities between the Narodnik movement and Marxism, stated that terror prevailed in the Narodnik movement as a result of the empire crisis. Rudnitskaya, adhering to the Lenin concept, proved the existence of terrorist methods of struggle due to the utopic nature of the Narodnik ideas. However, Tvardovskaya noted that terror was not the primary method of achieving the goal.

The Narodnik methodology specifies the economic and socio-psychological aspects of social relations. It allows us to take a comprehensive look at the social movements. Bakunist and publicist of the first period of social movement research Debogoriy-Mokrievich (1906, p. 600), studied Narodnichestvo in the South of Russia; he noted that it was developing towards liberal constitutionalism and terrorism. This movement, using the ideas of anarchism, became the antipode of the state.

The criteria for classifying the development stages of the Narodniks are their goals and the involvement of the people in the national political processes. Historians usually associate the beginning of the extremist movement with the creation of the *Narodnaya Volya* (Kozmin 1923). However, there are also opposite points of view. Teodorovich (1933), a former member of the Narodnik movement, identified the following leaders of this movement: Karakozov, Nechaev, Tkacheva. In his opinion, this direction was finally formed as a result of the activities of the *Narodnaya Volya* organization. Analysis of the contemporaries' memoirs confirms the absence of numerous terrorists and extremists among them.

Liberalism in modern historiography is considered a democratic movement of Russian society. Therefore, there is no strict differentiation between radicalism, liberalism, and extremism. Revolutionary and terrorist methods of struggle were equated. However, the historian Mokshin did not share this opinion. He wrote that after the defeat of *Narodnaya Volya*, thousands of young people, encouraged by the idea of an unpaid debt to the people, went to villages for *cultural work* (Mokshin 2015). Nevertheless, according to Mokshin, who studied Narodnichestvo, the Narodnik organizations are compared with radical terrorist organizations operating in Spain, Ireland, and Palestine (Mokshin 2015).

## 5 Conclusion

The centralization of state power gave rise to violence. There were two types of terror: government and anti-government. The political struggle for power and its increasing centralization led to extremism. The state wanted to destroy the will of the people disloyal to the political regime. Violence begat violence. The Narodniks tried to avoid violent struggle methods. Later, the Narodnik movement became the basis of *Tolstoyism* with its fundamental idea of non-resistance to evil by violence.

The peaceful union of the upper social strata with the peasantry became the goal of Narodnichestvo. Its representatives tried to reconcile the warring social groups based on a single national culture. Therefore, terror was excluded from the ideology of classical Narodnichestvo. The Narodniks realized that a political coup and a change of power of political leaders and parties would not change the existing socio-political relations.

Narodnichestvo could unite liberals, radicals, and conservatives in the political course and had a theoretical basis for this. In the understanding of the Narodniks, the historical process is variable. Evolutionary social changes should be made by the entire society based on the political culture of the Russian people.

Narodnichestvo introduced a consolidating element into *Obshchee Delo* movement. The attempt of Karakozov on the emperor changed the plans of the movement's leaders in the early 1860s. The ruling elite's reluctance to fully implement reforms and resolve the crisis and civilizational split led to the spread of terrorism. The marginalization of Russian society contributed to the expansion of the intelligentsia stratum and the spread of nihilism. Extremist groups appeared.

## References

Bakunin MA (1863) Svoboda. Kolokol 164:1350

Bakunin MA (1906) Letters to A. I. Herzen and N. P. Ogarev. V. Wroblewski Publisher, Saint-Petersburg, Russia

Bakunin MA (1920) Selected works, vol. 4. RSFSR, Golos Truda

Bazanov VG (1974) The Russian revolutionary Democrats and narodoznanie. Sovetskij pisatel. Leningrad department, Leningrad, USSR

Belsky VYu (2017) Terrorism in historical retrospect and current conditions. UNITY–DANA, Moscow, Russia

Boborykin PD (1965) Memories. For half a century, vol. 1. Hudozhestvennaya literature, Moscow, USSR

Bogucharsky V (ed) (1906). State crimes in Russia in the 19th century. Russkaya Skoropechatnya, Saint Petersburg, Russian Empire

Central State Archive of the Republic of Kazakhstan. (n.d.). (Fund 369, List 1, Book 1, File 398, p 46)

Central State Archive of the Republic of Kazakhstan. (n.d.). (Fund 64 List. 1 Book 3 File 5326. p 131)

Chernyadeva NA (2017) International terrorism. The origin, evolution, and current issues of the legal counteraction. Prospekt, Moscow, Russia

Cheshikhin-Vetrinsky VE (1923) N. G. Chernyshevsky, 1828–1889. Petrograd, RSFSR: State printing house named after comrade Zinoviev.

Debogoriy-Mokrievich V (1906) Memories. Svobodnyj trud, Saint-Petersburg, Russian Empire

Dostoevsky FM (2008) Essays, letters, documents: dictionary-reference. Pushkin House, Saint Petersburg, Russia

Dyakov V (ed) (1963) The Russian-polish revolutionary relations (Vol. 2). Publishing house of the USSR Academy of Sciences, Moscow, USSR

Hughes M (2009) Searching for the Soul of Russia: British Perceptions of Russia during the First World War. Twentieth-19British History 4, p. 78.
Huntington S (2003) Clash of civilizations. AST, Moscow, Russia
Kavelin KD (1892) Letters of K. D. Kavelin and I. S. Turgenev to A. I. Herzen. Ukrainian Printing House, Moscow, Russia
Kelly A (1989) Mikhail Bakunin: a study of the psychology and politics of utopianism. Yale Univ. Press, New Haven, London, UK
Khudyakov IA (1930) Zapiski karakozovtsa. Molodaya gvardiya, Moscow, Leningrad, USSR
Klevensky MM, Kotelnikov KG (eds) (1928) Attempt of Karakozov. Verbatim report on the case of D. Karakozov, I. Khudyakov, N. Ishutin, and others (Vol. 1). Centralarchive Publishing house, Moscow, USSR
Kovalik SF (1928) The revolutionary movement of the 1870s and the process of the 193s. Publishing house of the Society of Former Political Prisoners and Exiled Settlers, Moscow, USSR
Kovalskaya E (1931) About the letter to V. Malyutina (to the paper by Ya. Malyutina D. Bauman, The history of the Kharkiv revolutionary groups). Katorga and Ssylka, 4, p 141
Kozlinina EI (1913) For half a century. Notes of the oldest Russian journalist 1862–1912 Memoirs, essays, and characteristics. Publishing house of the N. Berdonosov, F. Prigorin, and Co. Trading house, Tambov, Russia
Kozmin BP (1923) Political processes of the 1860s: materials on the history of the revolutionary movement in Russia. State publishing house, Moscow, Leningrad, USSR
Kozmin BP (1931) P. G. Zaitsevskii in prison, settlement in the exile (1863–1872). Katorga and Ssylka, Book 8–9, p 113
Kulchitsky L (1908) History of the Russian revolutionary movement. Electropech. Ya. Levenshtein, Saint Petersburg, Russia
Lemke M (1906) Bakunin's regicides. False report of the alleged bribery of two Poles, Vigovsky brothers, by Bakunin for the murder of Nicholas the 1st. 1848. Biloe, 5, p 34
Lenin VI (1960) Complete works, vol 12. Gospolitizdat, Moscow, USSR
Lopatin GA (1922) Autobiography. Letters. Articles and poems. State publishing house, Petrograd, RSFSR
Meshchersky VP (1898) My memoirs (1865–1881), vol 1–2. Printing House of Knyaz V. P. Meshchersky, Saint Petersburg, Russia
Mokshin GN (2015) Narodnichestvo as an ideology of the authentic modernization of Russia. Bull Voronezh State Univ Ser: Hist Polit Sci Sociol 4:77–80
Naarden B (1995) Socialist Europe and revolutionary Russia perception and prejudice 1848–1923. Cambridge University Press, Cambridge, UK
Nechkina MV (1980) Meeting of two generations. From the history of the Russian revolutionary movement of the late '50s–early '60s. 19th century. Nauka, Moscow, USSR
Ogarev NP (1863) The constitution and the Zemsky Sobor. Kolokol 166:1366
Pancho CM (2005) Third world revolution and the history of Hollywood cinema. Am Lit Hist 17(3):78
Panteleev LF (1958) Memories. Hudozhestvennaya literature, Moscow, Russia
Peter C (2008) Ludwig Feuerbach and German Radicalism. Ger Hist 26(1):67. (Oxford University)
Rudnitskaya EL (1997) Propaganda campaign of Nechaev and Bakunin 1869–1870. In: Rudnitskaya EL (ed) Revolutionary radicalism in Russia: nineteenth century. Archeographic center, Moscow, Russia, pp 173–306
Shelgunov N (1881) Int Rev Delo 3:145
Smirnov AF (1959) Two letters from Herzen to A. A. Cherkesov and A. A. Serno-Solovyevich. In: Vinogradov VV (ed) Literary legacy, vol 67. Revolutionary democrats: new material. Publishing house of the USSR Academy of Sciences, Moscow, USSR, p 742
State Archive of the Russian Federation. (n.d.) (Fund 95, List 1, File 38, p 189)
Steklov YuM (1927) M.A. Bakunin His life and work, vol 2. State Publishing House, Moscow, Leningrad, USSR

Teodorovich IA (1933) From bakuninism to babuvizm. In: Teodorovich IA (ed) Popov M. R. Notes of zemlevolets. Publishing house of the Society of Former Political Prisoners and Exiled Settlers, Moscow, USSR, p 7
Tvardovskaya VA (1965) The second revolutionary situation in Russia and the struggle of the Narodnaya Volya. In: Ivanov LM (ed) Social movement in post-reform Russia. Nauka, Moscow, Russia, pp 61–93
Vilenskaya ES (1969) I. A. Khudyakov. Molodaya gvardiya, Moscow, USSR
Vinogradov VV (ed) (1959) Literary legacy (Vol. 67): revolutionary democrats: new materials. Publishing house of the USSR Academy of Sciences, Moscow, USSR
Walicki A (1989) The controversy over capitalism studies in the social philosophy of the Russian populists. Univ. of Notre Dame Press, Notre Dame, IN
Yeliseyev GZ (1933) The sixties. Memories. Academia, Moscow, Leningrad, USSR
Zhukovsky NI (1955) Letters. In: Rudnitskaya EL (ed) Literary legacy, vol 62. Publishing house of the USSR Academy of Sciences, Moscow, USSR

# Unresolved Disagreements as a Pre-conflict Stage of Individual Labor Dispute in the Context of National Security

**Julia A. Mikhailenko, Inna A. Prasolova, and Daria A. Statsenko**

**Abstract** This paper focuses on unresolved disagreements as a pre-conflict stage of an individual labor dispute. We used dialectical, logical, formal-legal, and comparative-legal methods of research. The analysis of scholarly opinions and regulatory legal acts revealed the shortcomings in the legal regulation of the pre-jurisdictional procedure for resolving disagreements that could later become individual labor disputes. We prove that it is necessary to include a norm establishing the procedure for settling disagreements before the occurrence of an individual labor dispute in the Labor Code of the Russian Federation. We analyzed the law provisions (statutory regulations) that allow the parties to attract a mediator (intermediary) at this stage. Moreover, we identified the gaps in the legal regulation of this field. In particular, we proved the need for providing mediator services for employees for free and the demand for the State control of the legality of pre-conflict stage mediation agreements. We propose providing a shortened-term mediation procedure for the settlement of pre-dispute differences. Considering the identified shortcomings, we formulated proposals for amending the current legislation, which will contribute to the effective resolution non-conflict resolution of disagreements between an employee and an employer.

**Keywords** Unresolved disagreements · Individual labor dispute · Conflict · Negotiations · Mediator

## 1 Introduction

In labor relations, the employee and the employer inevitably face various social conflicts, most of which are individual labor disputes. The pre-conflict stage of individual labor disputes consists of unresolved disagreements. In the theory of labor law, there are discussions about the concept of the terms "individual labor dispute," "unresolved disagreements," and the means for resolving conflicts. Some authors tend to

J. A. Mikhailenko · I. A. Prasolova · D. A. Statsenko (✉)
Altai State University, Barnaul, Russia
e-mail: dariastatsenko01@gmail.com

 
S. G. Maximova et al. (eds.), *Advances in Natural, Human-Made, and Coupled Human-Natural Systems Research*, Lecture Notes in Networks and Systems 234,
https://doi.org/10.1007/978-3-030-75483-9_44

believe that the terms "individual labor dispute" and "unresolved disagreements" are equivalent (Skobelkin et al. 2002). On the contrary, others consider unresolved disagreements to be a type of labor dispute (Motsnaya 2017). The definition of the concept of labor dispute through the category "unresolved disagreements" is also debatable (Petrov 2014; Bizyukov 2011).

The resolution of individual labor disputes via conventional ways (in jurisdictional order) requires considerable time and financial expenses and does not always lead to the intended effect. This has created a need for an effective and equitable resolution of unresolved differences in the pre-conflict phase of the individual labor dispute through negotiation. Currently, there is a tendency to involve an intermediary at this stage. A mediator may act as such an intermediary. Mediation allows not only to resolve the conflict but also to achieve a balance of interests of the parties and come to a mutually beneficial solution (Lektorsky 2009; Dyakovych 2019). However, the scientific literature also indicates the shortcomings of the legal regulation of this procedure in relation to labor conflicts (Rashidova and Darchinyan 2016; Demidova and Tatarnikova 2018).

Even though the Federal Law "On the Alternative Dispute Resolution Procedure with the Participation of a Mediator (Mediation Procedure)" (Law No. 193) was adopted in 2010, the practice of using mediation to resolve individual labor disputes in Russia has not yet been established, unlike other countries of Eurasia, where conciliation and arbitration are effectively used to resolve disagreements (Russian Federation 2010). The legislative changes proposed in this article will improve the procedure for resolving unresolved disagreements at a pre-conflict stage of an individual labor dispute. Since Russian legislation on mediation is not devoid of imperfections and gaps, the study of this topic seems relevant.

## 2 Materials and Methods

Scholarship of Russian and foreign scientists in the field of labor law and conflict resolution, the provisions of the Labor Code of the Russian Federation on individual labor disputes, and the Law No. 193 served as the materials for the study (Russian Federation 2001). The regulation of individual labor disputes and the application of mediation was studied by Alexander, Cruz, Dyakovych, Dolinskaya, Filipova, Lektorsky and others (Alexander 2008; Zaitseva et al. 2019; Dyakovych 2019; Dolinskaya 2011; Filipova 2018; Lektorsky 2009).

In this study, we employed the dialectical, logical, comparative-legal, and formal-legal methods. The dialectical method allowed us to incorporate the theories and methods of conflict resolution in jurisprudence, reveal the relevant gaps and contradictions in the legislation. The logical method was used in generating proposals and conclusions about the regulation of dispute resolution at the pre-conflict stage of an individual labor dispute. The comparative-legal method allowed us to compare the legal regulation of labor dispute resolution in different countries. The formal-legal method was used to identify the definitions of a labor dispute, the pre-conflict stage of a labor dispute, etc. Also, it helped in formulating the proposals for improving labor legislation and the rules of mediation.

## 3 Results

As a result, we formulated several proposals for amending the current Russian legislation to improve the resolution of unresolved disagreements at the pre-conflict stage of an individual labor dispute. These proposals are the following:

(1) Include the following provision in Chap. 60 of the Labor Code of the Russian Federation: "In the case of disagreements between the employee and the employer concerning the application of the laws and other regulations containing standards of the labor law, the collective agreement, the agreement, the local statutory act, the employment contract, the worker has the right to apply to the employer with a written statement about negotiation concerning their settlement. The employer is obliged to enter into negotiations within seven calendar days from the date of receipt of the said application. The parties can attract the intermediary (mediator) for negotiations. Negotiations, including mediation, shall not exceed thirty days";
(2) Establish rules for mediation agreements that were signed at the pre-dispute stage, providing control over the legality of such agreements to identify the conditions that could harm employees or unlawfully worsen their status. The control could be provided by sending agreements for registration with the relevant labor government body;
(3) Add an exception to Part 5 of Article 11 of the Law No. 193, which currently prevents the mediator from making proposals on the settlement of the dispute, unless otherwise agreed by the parties;
(4) Exempt employees from paying for mediation procedure;
(5) Provide for the shortened terms for the mediation procedure at the pre-conflict stage of an individual labor dispute in Article 13 of the Law No. 193. The mediation procedure at this stage should not exceed thirty days.

## 4 Discussion

Item 1 of Article 381 of the Labor Code of the Russian Federation defines the term *individual labor dispute* as "unresolved disagreement between an employer and worker on issues of the application of the labor legislation and other legal regulatory acts containing labor law norms, collective negotiations and other agreements, local normative acts, or labor contracts (including those relating to establishing or changing specific conditions of labor), on which a petition is made to the individual labor dispute review body" (Russian Federation 2001).

There are several views about the difference between the terms "individual labor dispute" and "conflict" in jurisprudence. Thus, some authors tend to believe that the terms "individual labor dispute" and "unresolved disagreements" are equivalent concepts (Skobelkin et al. 2002). Other researchers state that "conflict" is one of the forms of dispute, and the terms "conflict" and "dispute" relate to each other as a part and a whole (Motsnaya 2017). It appears that the individual labor dispute is a social conflict and goes through the same stages of development: (1) pre-conflict stage; (2) the conflict, which includes three stages—beginning, development, completion; (3) post-conflict stage.

It is very important to highlight the pre-conflict stage of an individual labor dispute. A pre-conflict situation is a latent stage of the conflict; it is easier to resolve it in a confrontation-free way until it has developed into the conflict itself. According to Item 1 of Article 381 of the Labor Code of the Russian Federation, an individual labor dispute begins from the moment an employee or an employer sends a petition to the review body of individual labor disputes (Russian Federation 2001). Therefore, unresolved disagreements are a pre-conflict stage of an individual labor dispute.

One of the problems in the jurisprudence of labor law is the whether it is appropriate to define an individual labor dispute as *unresolved disagreement*. Petrov notes that if the disagreement is settled, it does not exist. Therefore, it makes no sense to define a labor dispute as *unresolved disagreement*. The existence of disagreement between the parties to the employment relationship automatically leads to a labor dispute (Petrov 2014).

According to Bizyukov, the terms "disagreement" and "labor dispute" should be differentiated. The existence of disagreement does not always lead to a labor dispute. In his opinion, disagreement is a pre-conflict stage, the outcome of which depends on the will of the parties (Bizyukov 2011). We support this view because an individual labor dispute begins from the moment of sending a petition to the relevant jurisdictional bodies. Accordingly, the current contradictions between the employee and the employer up to that point are unresolved differences.

At the same time, the Labor Code of the Russian Federation does not regulate the procedure for settling disagreements until an employee or an employer petitions to the review body. Item 2 of Article 385 of the Labor Code of the Russian Federation establishes that "an individual labor dispute shall be heard by the labor dispute commission unless the worker has, independently or through a representative, resolved their disagreement in direct negotiations with the employer" (Russian Federation 2001). It seems that this rule has a declarative nature because it does not oblige an employee or an employer to attempt a preliminary settlement of differences and negotiations.

At the time, Perederin proposed to amend the rule by establishing the order and timing of an employee's appeal to the employer with a statement on the preliminary settlement of disagreements. Moreover, he proposed to oblige an employer to consider the written application of the employee within the prescribed period and to give a written response to it (Perederin 2000). We must note that direct negotiations are a more appropriate way to resolve disagreements. These negotiations can be conducted both exclusively between the employee and the employer or with the involvement of an independent intermediary. According to Item 1 of *R092—Voluntary Conciliation and Arbitration Recommendation* of the International Labor Organization (ILO), voluntary conciliation bodies, appropriate to national conditions, must be created to help prevent and resolve labor conflicts between employers and workers (International Labor Organization 1951).

Today, an intermediary can act as a mediator. Over the previous decade, mediation as a way of alternative dispute settlement has expanded significantly in Russia (Filipova 2018). Until 2008, only out-of-court mediation could be used. In 2008, judicial mediation began as a sort of legal experiment, which was later enshrined in the Law No. 193 (Russian Federation 2010).

"Mediation procedure is a means of dispute resolution with the assistance of a mediator on the basis of the voluntary consent of the parties with the aim of reaching a mutually acceptable decision" (Article 2 of the Law No. 193). As noted in contemporary secondary literature, the main advantage of mediation is the possibility to resolve the conflict themselves via a mutually beneficial agreement. (Alexander 2008).

Mediation began to develop in the second half of the twentieth century. It occurred in the countries of Common law (the United States, Australia, Great Britain), then spread to the rest of Europe. Mediation was initially applied in family dispute resolution. As time passed, the scope of mediation expanded and began to affect commercial and public areas (Shamlikashvili 2008).

Mediation for settlement of labor disputes is extremely popular outside of Russia. A significant amount of attention is paid to the pre-dispute settlement of labor conflicts. For example, mediation is used to resolve all labor disputes in Canada and Argentina. In South Korea, mediation for the settlement of labor disputes is one of the possible alternative dispute resolution procedures. Labor courts or administrative bodies for labor dispute resolution exist in several European countries. The conciliation and arbitration proceedings occupy a special place in the law of these countries. These processes can be independent actions or serve as a part of the judicial process. For example, German legislation establishes a mandatory stage of reconciliation. For this purpose, the *Bundesverband Mediation in Wirtschaft und Arbeitswelt* (*Federal Association for Mediation in Business and the World of Work*) was established there. In the UK, conciliation and arbitration are also used in the initial stage of dispute settlement. *The Advisory, Conciliation, and Arbitration Service* of the UK assists in settlement of labor disputes, aiming to improve the legal position of employees and employers and find a balance between their interests (Zaitseva et al. 2019).

As Filipova notes, mediation as a procedure of alternative dispute regulation in Russian practice is only at the stage of its formation. Today, the use of mediation in Russia is not the norm, but it is gradually gaining prominence in law practice (Filipova 2015).

According to Item 2, Article 1 of the Law No. 193, the mediation procedure applies to disputes arising from the legal relations of employment. Collective labor disputes and disputes that may affect the rights and interests of third parties are exceptions to this rule (Paragraph 5 of Article 2 of the Law No. 193). Item 2, Article 7 of the Law No. 193 states, "Mediation procedure maybe applied in case of occurrence of dispute both prior to recourse to state court or arbitration and after the beginning of litigation or arbitration proceeding, as well as on the proposal of the judge or the arbitration judge."

Individual labor disputes can be divided, according to the subject, into *law disputes* and *interest disputes.*

In the first case, the labor disputes arise from the application of regulations, collective agreements, agreements, local regulations, and the employment contract—they are disputes caused by a real or alleged violation of labor law. *Law disputes* are resolved by jurisdictional bodies, including the courts, and therefore are also called *claim disputes. Interest disputes* arise from new or changing existing working conditions—they are a conflict of interests between the parties, and none of them has the right to insist on the fulfillment of their requirements or the obligation to comply with them. Such disputes are settled in conciliation (without the jurisdictional bodies), and therefore they are called *non-claim disputes.*

According to the Law No. 193, mediation can be applied to *law disputes* and *interest disputes.* However, there is reason to assume that the application of this procedure is preferable to the settlement of disputes on working conditions not mediated by the law—*interest disputes.* The current labor legislation does not establish a procedure for preventing and resolving such disputes. Part 1, Article 381 of the Labor Code of the Russian Federation provides the basis for individual labor disputes only concerning the legal provisions, including the establishment and modification of working conditions regulated by the law. The only way of resolving *interest disputes* is via conciliation—negotiations between an employee and an employer. With the adoption of the Law No. 193, the parties of the labor relationship can effectively resolve differences arising from changes in working conditions with the help of a mediator.

Mediation under the Law No. 193 may also be applied to *law disputes*, which some scholars see as problematic. If the parties can make concessions to each other in the civil law, in the labor law, an employee sometimes cannot make concessions to the employer. Consequently, mediation cannot be applied to some *law disputes* cannot be applied, as it can violate rights and guarantees of employee rights (Dolinskaya 2011).

For example, the terms of the mediation agreement will not be valid if an employee agreed to only receive a part of the salary, or to be restored in the position without being paid for the lost salary. Item 2, Article 9 of the Labor Code of the Russian Federation states, "Collective agreements, agreements, labor contracts shall not contain terms that limit the rights of employees or reduce the level of guarantees in comparison with those established by the labor legislation and other normative legal acts containing labor law norms." The legality of a mediated agreement can be controlled via confirmation by a court. This agreement is then treated as an amicable settlement agreement (Part 3 of Article 12 of the Law No. 193). However, there also should be rules for establishing the legality of disputes resolved at the pre-jurisdictional stage. Special attention should be paid to identifying the conditions that could unlawfully harm an employee or worsen their status. Such rules could, for example, mandate to send the agreements for registration to the relevant labor government body, as in the case with collective agreements and agreements.

However, labor disputes often arise from the ignorance of labor legislation (by both employees and employers) or a deliberate violation of the law (usually on the part of employers). Apparently, the mediators should be the wrongfulness of the committed acts (if an offense did occur) and instruct to restore the rights so that the case would not be brought to trial. It will allow the employer to save time and money. If one were to consider labor disputes as a social conflict, then a mediation agreement can mitigate the consequences of the conflict and normalize the relationship between the employee and the employer.

However, a mediator cannot perform such a role, according to Item 5 of Article 11 of the Law No. 193. It establishes that the mediator is not entitled to make proposals for the settlement of a dispute unless agreed otherwise by the parties. We believe that Item 11, Article 11 of the Law No. 193 about *law disputes* should be amended and contain an exception.

The availability of a mediation procedure for an employee is a serious issue. According to Part 1 of Article 10 of the Law No. 193, mediation procedures are carried out by paid or free mediators. However, the parties pay for the mediation equally, unless agreed otherwise (Part 2 of Article 10 of the Law No. 193). On the contrary, Item 3 of *R092—Voluntary Conciliation and Arbitration Recommendation* states that conciliation procedure should be free and expeditious—all ways of protection should be provided to employees free of charge (International Labor Organization 1951). The parties could be allowed to seek the assistance of a mediator, for example, from the Federal Service for Labor and Employment (Rostrud), which forms lists of labor arbitrators and intermediaries.

Article 13 of the Law No. 193 stipulates that an agreement on the mediation procedure should determine the time frame of mediation. At the same time, the mediator and the parties must take all possible measures to ensure that the procedure is concluded within 60 days. In all cases, the mediation should not exceed 180 days; otherwise, the case goes to the court or arbitration.

In the case of labor disputes, the period of mediation for the settlement of disputes should be shortened since the limitation period in labor disputes is only three months (unlike civil law disputes, where the limitation period is three years). Therefore, if mediation results were unsatisfactory, an employee should be able to apply to the jurisdictional bodies for the protection of their rights. In our opinion, the time limit for mediation procedures for labor disputes should be only thirty days.

Therefore, we propose to include the following norm in Chap. 60 of the Labor Code of the Russian Federation—"In the case of disagreements between the employee and the employer concerning the application of the laws and other regulations containing standards of the labor law, the collective agreement, the agreement, the local statutory act, the employment contract, the worker has the right to apply to the employer with a written statement about negotiation concerning their settlement. The employer is obliged to enter into negotiations within seven calendar days from the date of receipt of the said application. The parties can attract the intermediary (mediator) for negotiations. Negotiations, including mediation, shall not exceed thirty days".

## 5 Conclusion

There is an obvious need to expand and improve the pre-jurisdictional ways of solving disagreements that could become individual labor disputes. Mediation plays an important role in solving these disagreements since a professional mediator can help the parties to come to an agreement in a pre-confrontation stage.

We proposed several changes to the legal regulation of dispute resolution at the pre-confrontation stage of individual labor disputes. These changes would contribute to the widespread use of dispute settlement with the help of a mediator. This, in turn, will reduce the burden on jurisdictional bodies and save time and money for both employees and employers.

## References


Alexander N (2008) Mediation and the art of regulation. QUT Law Justice J 8(1). https://doi.org/10.5204/qutlr.v8i1.60

Bizyukov PV (2011) Individual labor conflicts: Can the workers protect their rights alone. Center for Social and Labor Rights, Moscow, Russia

Demidova SE, Tatarnikova SN (2018) Regarding the application of mediation in labor relations. Labor Soc Relats 6:53–62

Dolinskaya LM (2011) On the concept of the applicability of mediation to labor relations. Laws Russ Exp, Anal Pract 7:48–56

Dyakovych M (2019) Mediation as means of rights and interests protection by notaries. Astra Salvensis S1:435–441

Filipova IA (2015) Mediation in labor disputes: the experience of the United States, Germany and France and its possible use in reforming Russian legislation. Attorney 7:34–39

Filipova IA (2018) Mediation in collective labor disputes: Limitations and opportunities (taking into account foreign experience). J Russ Law 5:113–124

International Labor Organization (1951) Recommendation No. 52 On voluntary conciliation and arbitration (29 June 1951 No. 51). Geneva, Switzerland.

Lektorsky VA (2009) Mediation and discourse: mediation as a means of collective activity. In: Sannino A, Daniels H, Gutiérrez KD (eds) Learning and expanding with activity theory. Cambridge University Press, Los Angeles, CA, pp 75–87

Motsnaya OV (2017) Conflict of interest as a prerequisite for a labor dispute. J Russ Law 11:87–95

Perederin SV (2000) Procedural mechanism for ensuring the labor rights of employees. Publishing House of Voronezh University, Voronezh, Russia

Petrov AY (2014) Individual labor disputes: fundamental aspects of labor law. Right. J High Sch Econ 2:36–47

Rashidova AI, Darchinyan ZM (2016) Effectiveness of mediation proceedings in employment relations. Actual Probl Russ Law 8(69):114–120

Russian Federation (2001) The labor code of the Russian Federation (21 Dec 2001 No. 197-FL). Moscow, Russia

Russian Federation (2010) Federal Law On the alternative dispute resolution procedure with the participation of a mediator (mediation procedure) (27 July 2010 No. 193-FL). Moscow, Russia
Shamlikashvili TA (2008) Mediation secrets. Law Econ 2:92–94
Skobelkin VN, Perederin SV, Chucha SYu, Semenyuta NN (2002) Labor procedural law. Publishing House of Voronezh University, Voronezh, Russia
Zaitseva L, Gomes E, Racheva S, Cruz V (2019) Intermediary in a collective labor dispute resolution. BRICS Law J 6(2):33–59. https://doi.org/10.21684/2412-2343-2019-6-2-33-59

# Promising Development of Scientific Cooperation in Eurasia: Comparative Legal Research of a Jury Trial in CIS Countries

**Nina A. Dudko**

**Abstract** The paper focuses on the promising directions and forms of scientific cooperation in criminal proceedings in the Eurasian legal space. One of the promising areas of international comparative legal research is the jury trial, which successfully operates in the CIS countries. The need for further improvement and development of the jury trial in Russia and the Republic of Kazakhstan, as well as the planned integration of the jury trial in the Kyrgyz Republic, determine the relevance and scientific novelty of the study. The research used general and specific methods of scientific knowledge. The main research method is the specific scientific method of comparative legal study. The place of jury trial in the judicial system of the CIS is specified. The author revealed the general trend in the development of the jury trial in Russia, Kazakhstan, and the Kyrgyzstan—the establishment of a jury trial on the district level of the legal system. Based on the analysis of normative, scientific, and theoretical terms and judicial practice in criminal cases involving jurors, the author formulated and justified the following conclusions. Comparative legal research of the CIS jury trials is relevant and necessary for their further improvement and development. The comparable features of a jury trial allowed conducting a scientifically based comparative study and defining the general trend in the development of a jury trial in the CIS countries. The results of comparative legal research can be used to improve legislation and judicial practice. The expansion of scientific cooperation and its development will increase the effectiveness of comparative legal research of the CIS jury trials.

**Keywords** Eurasia commonwealth of independent states · Scientific cooperation · Comparative legal research · Jury trial

N. A. Dudko (✉)
Law Institute, Altai State University, Barnaul, Russia
e-mail: dudkona@yandex.ru



S. G. Maximova et al. (eds.), *Advances in Natural, Human-Made, and Coupled Human-Natural Systems Research*, Lecture Notes in Networks and Systems 234,
https://doi.org/10.1007/978-3-030-75483-9_45

# 1 Introduction

International cooperation is carried out through various directions and forms of interaction. It is necessary to analyze their normative and theoretical basis when studying the directions and forms of cooperation between them.

As a rule, the basic terms on interaction and cooperation between countries are legally assigned at the interstate level in interstate documents.

For example, when creating the Commonwealth of Independent States, the post-Soviet countries designated equal and mutually beneficial cooperation as the main goal of the CIS formation. "The High Contracting Parties will develop equal and mutually beneficial cooperation of their people and countries in politics, economy, culture, education, public healthcare, environment, science, and trade, as well as in humanitarian and other fields, to promote broad information exchange and conscientiously and strictly observe mutual obligations. The parties consider it necessary to conclude agreements on cooperation in the mentioned areas (Article 4 of the Agreement *On the Establishing of the Commonwealth of Independent States* (December 8, 1991) (Commonwealth of Independent States 1991).

The CIS member states identified the protection of human rights and freedoms as the most crucial area of cooperation. The heads of the CIS countries considered the urgency of the problems related to the human rights observance. They declared the fulfillment of obligations on human rights and basic freedoms arising from international treaties and agreements to which the USSR was a party (Paragraph 1 of the CIS Adgreement 1993) (Commonwealth of Independent States 1993a).

At the same time, it was declared that member states would take the necessary measures to fulfill their obligations to protect human rights and basic freedoms. Firstly, the member states developed and concluded the Convention of the Commonwealth of Independent States on the basic human rights and liberties (Paragraph 5 of the 1993 CIS declaration). Secondly, states adjusted their national legislation according to obligations under international human rights treaties (Paragraph 6 of the 1993 CIS declaration) (Commonwealth of Independent States 1993a).

In the following years, these measures were successfully implemented.

The CIS Convention on Human Rights and Fundamental Freedoms held in 1995 proclaimed that the CIS members would effectively implement obligations to protect human rights and fundamental freedoms. "In the spirit of collective efforts," the Contracting Parties uphold the ideals of freedom and the supremacy of the law, prevent the violation of human rights and fundamental freedoms, develop traditions of tolerance and friendship among people, and strengthen the civil peace and harmony" (Commonwealth of Independent States 1995). The CIS Convention enshrines the need for "concentrating collective efforts," i.e., only the cooperation of countries can ensure the protection of human rights.

In accordance with the basic principles, the Convention enshrined fundamental human rights and freedoms (Articles 1–29). Considering the subject of this study, the following conventional rights are the most significant:

- Universal right to life (Article 2);
- Right to due process of law (Article 6);
- Right of every citizen to participate in the management and conduct of public affairs directly and through democratically elected representatives (Article 29).

These rights are in the first place when the courts consider criminal cases with the participation of jurors.

The corresponding changes were made to the national legislation as well. The most significant changes were related to the cancellation of the institution of people's assessors and the establishment of the jury trial.

To ensure the implementation of the right to a fair trial, it was necessary to reform of the judicial systems of the CIS countries and resolve the issue of the expediency of establishing a jury trial to consider and resolve criminal cases. The jury trial is considered one of the most democratic forms of legal proceedings, which provides the participation of civil representatives in the judicature and protects the legal rights and interests of those accused of committing crimes.

The establishment of a jury trial as one of the directions of judicial reform in the post-Soviet states was envisaged by the Russian Federation, Georgia, the Republic of Belarus, Ukraine, the Republic of Azerbaijan, the Kyrgyz Republic, and the Republic of Kazakhstan.

Nowadays, in the CIS countries, the jury operates only in the following countries:

- Russian Federation since November 1, 1993 (Supreme Soviet of Russian Federation 1993);
- Kazakhstan since January 1, 2007 (Republic of Kazakhstan 1997a);
- Ukraine since November 19, 2012 (Republic of Ukraine 2012).

The Kyrgyz Republic postponed the establishment of jury trial until January 1, 2025 (Kyrgyz Republic 2019). In 2015, the Republic of Azerbaijan legally refused the jury trial (Republic of Azerbaijan 2015).

One can conclude that the jury trial has received considerable recognition and use in the CIS countries—Russia, Kazakhstan, and Kyrgyzstan. It can be explained by the commonality of historically formed legal, scientific, moral, and ethical traditions. Accordingly, the development of bilateral and independent cooperation with the CIS members, primarily the Republic of Kazakhstan and the Kyrgyz Republic, was declared a priority of Russian foreign relations (Presidential Executive Office 2016).

The listed international and national documents confirm that the interaction and cooperation of the CIS members is based on a sufficiently developed regulatory and legal framework. However, it is necessary to have a scientific and theoretical base to ensure effective international interaction and cooperation.

International and national normative legal acts consider scientific cooperation independent are of international cooperation. For example, scientific cooperation is included in the list of possible areas of engagement in (1) the Agreement *On the Establishing of the Commonwealth of Independent States* of 1991 (Commonwealth of Independent State, 1991), (2) the agreement of the CIS members *On the Cooperation in the Protection of Copyright and Related Rights* (Commonwealth of Independent

States 1993b), (3) the Foreign Policy Concept of the Russian Federation (Presidential Executive Office 2016).

The jury trial in the CIS countries is successfully functioning and is constantly being improved. The results of international theoretical comparative legal research and generalized judicial practice can further develop national forms of jury trials. The improvement and development of the jury trial can be effective only based on the analysis of the organizational experience and activities of various models and national forms of jury trials.

Therefore, the topic and results of this study are of theoretical and practical importance for the development of equal and mutually beneficial cooperation between the CIS countries, which determines the relevance and scientific novelty of the study.

Comparative legal research is one of the most important of scientific activity areas and scientific cooperation of modern countries, including Russia and other CIS members.

"Comparative jurisprudence in Russia, as well as comparative studies throughout the post-Soviet space, is experiencing difficulties associated with overcoming the consequences of the secluded and isolated nature of legal systems and negative trends caused by the interrupted national legal traditions. The processes of borrowing and adapting the institutional legal experience of foreign countries become one of the main tasks of jurisprudence in the post-Soviet space" (Nemytina 2013).

The legal theory distinguishes comparative jurisprudence as an independent area of scientific knowledge and comparative legal research as a type of scientific activity (Nemytina 2013). Comparative legal studies are one of the parts of comparative jurisprudence.

Comparative legal research aims to improve legislation and law enforcement practice. Such studies "represent an irreplaceable tool for improving national law"; their importance and significance are increasing every year (Levchenko 2017). Comparative studies are actively carried out in all law branches based on unified methodological recommendations on directions, forms, and legal techniques of comparative legal analysis. Hence, comparative legal research is important for the development of the criminal procedure, including the jury trial institution.

The results of comparative legal studies on the jury trial of foreign countries were widely used during its introduction and subsequent monitoring in Russia and Kazakhstan, the following steps were held:

- Arranging the international conferences and round tables (Yurist, n.d.; Grib 2005; Russian State University of Justice 2017; Sinyukov et al. 2018);
- Participating foreign researchers in the preparation of methodological and scientific publications (Thaiman 1994; Kirillova 1998; Burnham (n.d.); Burnham 1996);
- Publishing the results of joint research in collaboration with foreign scientists (Kovalev and Smirnov 2014; Kovalev and Suleymenova 2010);
- Studying the conclusions and proposals of foreign authors on the jury trial (Thaiman 1995, 2004; Batyrbaev 2017).

It is advisable to continue the study and use it to develop proposals for the further growth and improvement of the existing jury trial in Russia and Kazakhstan, and prepare the criminal cases processing with the participation of jurors in Kyrgyzstan. The use of various forms of scientific cooperation will increase the efficiency and quality of comparative legal research of jury trials in the CIS.

The study aims to identify directions and forms of scientific cooperation development in criminal proceedings in the Eurasian legal space. Thus, it is necessary to formulate and substantiate the answers to the following questions:

- What is the importance of scientific cooperation for the development of interaction and cooperation between the CIS members?
- Is the comparative legal study of the institution of jury trial in Russia, Kazakhstan, and Kyrgyzstan relevant?
- What is the common comparable feature of the jury trial under the legislation of Russia, Kazakhstan, and Kyrgyzstan that allows conducting a scientifically grounded comparative legal study?
- Can the results of this comparative legal study be used to enhance the efficiency of the functioning and further improvement of the jury trial?
- What are advisable forms of scientific cooperation are advisable in carrying out a comparative legal study of the organization and activities of the jury trial in the CIS?

## 2 Materials and Methods

The research materials include:

1. International documents and normative legal acts of the CIS countries on human rights and fundamental freedoms:
   (a) Agreement establishing the Commonwealth of Independent States (Commonwealth of Independent States 1991);
   (b) Agreement of the heads of Commonwealth of Independent States on the international obligations of human rights and fundamental freedoms (Commonwealth of Independent States 1993a);
   (c) Agreement of the CIS countries *On the Cooperation in the Protection of Copyright and Related Rights* (Commonwealth of Independent States 1993b);
   (d) Commonwealth of Independent States Convention on human rights and fundamental freedoms (Commonwealth of Independent States 1995).
2. The legislation of the CIS members on the judicial system (Supreme Soviet of the Russian Federation 1993; Kyrgyz Republic 2019; Russian Federation 2011);
3. Criminal procedure codes of the CIS members, regulating the organization and functioning of the jury trial (The Republic of Ukraine 2012; the Kyrgyz Republic 1999a; The Kyrgyz Republic 2017; Milli Majlis of the Azerbaijan Republic 2015;

The Republic of Kazakhstan 1997a; The Republic of Kazakhstan 2014; Supreme Soviet of the Russian Federation 1993; Russian Federation 2001);
4. Statistical data of the Judicial Department at the Supreme Court of the Russian Federation for the consideration of criminal cases by courts with the participation of jurors (Judicial department at the Supreme Court of the Russian Federation 2017, 2018, 2019);
5. General theoretical (Varlamova 2014; Levchenko 2017; Nemytina 2013) and criminal procedural scientific literature on the research topic (Batyrbaev 2017; Burnham 1996; Bobotov 1995; Kirillova 1998; Kovalev and Suleymenova 2010; Kovalev and Smirnov 2014; Nasonov 2016; Thaiman 1995; Thaiman 2004);
6. Publications of the author on the organization and activities of the jury (Dudko 2018; Dudko and Neymark 2018; Dudko et al. 2019).

The methodological basis of the study is made up of general and specific methods of scientific knowledge. Using the statistical analysis, we studied the judicial practice data of the processing of criminal cases with the participation of jurors. The formal legal method was used to analyze national normative legal acts of the CIS member states and the provisions of international instruments. The main research method is the specific scientific method of comparative legal study, which allows to identify general and specific features of the jury trial functioning in the post-Soviet space of the Central Asian region.

## 3 Results

The jury trial is a legal institution with a centuries-long history (Bobotov 1995). However, it is still relevant to the theory and practice of criminal justice in Russia. The relevance of the study of the organization and activities of the jury trial is due to the fact that the judicial practice of considering criminal cases with the participation of jurors constantly reveals the problems that require legislative resolution.

The issue of a jury trial was repeatedly discussed at the highest national level of Russia:

- At the meeting of the President of the Russian Federation with members of the Council for the Development of Civil Society and Human Rights and ombudsmen for human rights, the rights of the child, and the protection of the rights of entrepreneurs on December 5, 2014 (Presidential Executive Office 2014);
- In the Presidential Address to the Federal Assembly of December 3, 2015 (Presidential Executive Office 2015);
- At the meeting of the Council for the Development of Civil Society and Human Rights on December 10, 2019, it was re-discussed (Presidential Executive Office 2018).

Following these discussions, the President of the Russian Federation signed Instructions on December 8, 2015 (Presidential Executive Office 2015) and January

29, 2020 (Presidential Executive Office 2020), providing the need to expand the use of the institution of jurors and increase the jurisdiction of the jury trial. In the pursuance of the Orders of the President of the Russian Federation of 2015, the jury trial was introduced on June 1, 2018, in the district courts (Russian Federation 2016). The Instructions of the President of the Russian Federation of January 29, 2020, recommended the Supreme Court of the Russian Federation to consider the issue of expanding the elements of a crime, the cases of which are within the jurisdiction of the jury trial and also make appropriate proposals due June 1, 2020 (Presidential Executive Office 2020).

Historically, various models of jury trials were formed. They have common "essential features" and specific production features depending on the country and the legal system of application (Nasonov 2016).

The research subject is one of the general provisions that characterizes the jury trial—the place of the jury trial in the structure of judicial systems in the CIS countries. The subject of comparative legal research should be compatible objects, which is the main methodological principle of comparative legal research (Varlamova 2014). According to Varlamova (2014), research objects (legal institutions) should be compared in terms of the function performed and the development level.

In Russia and Kazakhstan, the jury trial meets the requirement of comparability of the research objects. Russia and Kazakhstan were the first CIS members to establish a jury trial. In Russia and Kazakhstan, jury trials are successfully operated and are consistently improved (Dudko et al. 2019). Taking this fact and the prospects of a jury trial in Kyrgyzstan into account, we consider it appropriate to carry out a comparative study of a jury trial under the laws of the Russian Federation, the Republic of Kazakhstan, and the Kyrgyz Republic.

The issue of the place of jury trial in the judicial system has theoretical and practical significance. The procedure for passing the case through the courts depends on the level of the judicial system in which criminal cases are considered with the participation of a jury trial.

The Russian, Kazakh, and Kyrgyz legislation analysis allow us to conclude that its place was significantly adjusted in the judicial system after establishing the jury trial. This conclusion is confirmed by the following documents:

- Laws of the Criminal Procedural Code of the RSFSR of 1960 as amended in 1993;
- Criminal Procedural Code of the Russian Federation as amended in 2001;
- Criminal Procedural Code of the Russian Federation as amended in 2016 (Supreme Soviet of Russian Federation 1993; Russian Federation 2001);
- Criminal Procedural Code of the Republic of Kazakhstan as amended in 2006;
- Criminal Procedural Code of the Republic of Kazakhstan as amended in 2009;
- Criminal Procedural Code of the Republic of Kazakhstan (Republic of Kazakhstan 2014);
- Criminal Procedural Code of the Kyrgyz Republic as amended in 2007 (Kyrgyz Republic 1999a);
- Criminal Procedural Code of the Kyrgyz Republic as amended in 2012 (Kyrgyz Republic 1999b);

- Criminal Procedural Code of the Kyrgyz Republic as amended in 2017 (Kyrgyz Republic 2017).

In Russia, from November 1, 1993, to June 1, 2018, criminal cases with the participation of jurors were considered only by regional and territorial courts, courts of cities of federal importance, courts of autonomous regions, courts of autonomous districts, and Supreme courts of republics (Supreme Soviet of the Russian Federation 1993; Russian Federation 2001).

Since June 1, 2018, district courts have also been empowered to consider criminal cases with the participation of jurors (Russian Federation 2016). By the explanatory note to the federal law draft, these changes were aimed to expand the use of the institution of jurors for further development and growth of the democratic foundations of criminal justice and increase the transparency of justice and public credit in the court. The participation of jurors in district courts will strengthen the status of such courts as the main link in the Russian judicial system closest to the population (State Duma of Russian Federation 2016).

In the Republic of Kazakhstan, from January 1, 2007 to January 1, 2010, criminal cases with the participation of jurors were considered by regional and equivalent courts—the Capital court of the Republic of Kazakhstan and city courts of cities of republican status (Republic of Kazakhstan 1997b). Since January 1, 2010, these powers were transferred to the specialized inter-district criminal courts (Republic of Kazakhstan 2014).

In the Kyrgyz Republic, the jury trial was established in 2007. It was planned to process criminal cases with the participation of jurors in regional courts and the Bishkek city court starting from 2012 (Kyrgyz Republic 2019). Since 2012, the introduction of the jury trial has been sequentially postponed until 2015, 2019, and 2025, with the transfer of powers of the jury trial to the inter-district courts (Kyrgyz Republic 1999b, 2017, 2019).

Thus, there is a general tendency to transfer the jury trial from the territorial (regional) to the district level of the judicial system. This reform is aimed at expanding the population's access to this form of criminal justice. Compared with Kazakhstan, Russia has not yet accumulated sufficient experience in processing cases in district courts with the participation of jurors. For the Russian Federation, we consider it efficient to use the legislative, practical, and scientific-theoretical experience of the Republic of Kazakhstan. This determines the relevance and scientific novelty of this study and substantiates the need to continue the comparative legal analysis of the jury trial of the CIS members.

## 4 Discussion

In Russia, the question of the advisability of a jury trial in district courts was actively discussed before its introduction. Discussions continue even after June 1, 2018. The subject of discussion also includes the efficiency of the processing of criminal cases with the participation of jurors in district courts regarding the emerging judicial practice (Russian State University of Justice 2017; Sinyukov et al. 2018); Dudko 2018; Dudko and Neimark 2018).

The introduction of jury trials in district courts was aimed at expanding the use of the institution of jurors (Russian Federation 2001). Statistical data on the judicial practice of considering criminal cases with the participation of jurors are posted on the website of the Judicial Department under the Supreme Court of the Russian Federation. Russian courts with the participation of jurors processed 373 cases in 2017, 280 cases in 2018 (including 86 cases in district courts), and 717 cases in 2019 (including 474 cases in district courts) (Judicial Department at the Supreme Court of the Russian Federation 2017, 2018, 2019). According to the statistical data, the number of criminal cases considered by the jury in 2019 almost doubled compared to 2017.

At the same time, the judicial practice of processing criminal cases with the participation of jurors in district courts revealed some problems that require resolution and elimination. First, these are organizational, personnel, material, and technical problems (lack of specifically equipped courtrooms and deliberations, rooms for jurors, problems with the attendance of candidates for jurors, etc.). Additional material costs and organizational measures are required to eliminate these problems. We consider it expedient to use the judicial experience of the Republic of Kazakhstan and the Kyrgyz Republic in the consideration of criminal cases with the participation of jurors in inter-district courts. We propose to envisage the consideration of such criminal cases not in every district court, but in the most prepared ones. Such district courts can be given the powers of an inter-district court (inter-district specialized criminal court) to consider criminal cases with the participation of jurors. Some district courts can receive the status of inter-district courts without the reorganization of the Russian judicial system. The Law *On Courts of General Jurisdiction in the Russian Federation* includes district, inter-district, and specialized courts (Russian Federation 2011).

The changes proposed in the legislation to create the inter-district specialized criminal courts for considering criminal cases with the participation of jurors could help to eliminate organizational problems and prevent possible issues with the formation of lists of candidates for the jury and the formation of a jury court.

## 5 Conclusion

The comparative legal study of international and national normative legal acts of the CIS members and the study of the judicial practice of the jury and scientific literature allow us to conclude that the research goal was achieved.

Scientific cooperation is an independent area of international cooperation. It is regulated by the normative legal acts of the CIS member states.

Jury trials are successfully operating in Russia and Kazakhstan. However, it requires further development and improvement. The jury trial was established in

Kyrgyzstan, but its introduction was postponed. Therefore, comparative legal studies of the jury trials in the CIS are relevant and should be continued.

This study analyzed the place of a jury trial in the judicial systems of the CIS countries. The analysis proved the validity of the establishing of a jury trial at the district level of the judicial system.

The organizational and functioning questions of the jury trial should be a subject of constant comparative legal research by scholars and practitioners in criminal justice. It is necessary to identify trends in the further development and improvement of the jury trials in the CIS countries. The research results can be used in legislative activity and judicial practice.

We consider international comparative legal studies of various legal institutions, primarily the jury trial, reasonable and promising for developing and expanding of scientific cooperation between the CIS countries. Scientific cooperation can be carried out in the form of scientific and practical conferences, round tables, seminars, mutual scientific internships, and joint publications.

## References

Batyrbaev BS (2017) What model does the jury trial of the Kyrgyz Republic refer to in the system of the world's courts? Alatoo Acad Stud 4:189–195

Bobotov SV (1995) Where did the jury come from? Russian Law Academy of the Ministry of Justice of the Russian Federation, Moscow, Russia

Burnham W (1996) Russian and American jury trial. Comparative legal analysis. In: Burnham W, Reshetnikova IV, Proshlyakov AD (eds) Judicial advocacy. Saint Petersburg State University, Saint Petersburg, Russia, pp 114–140

Burnham W (n.d.). The jury trial. Moscow Independent Institute of International Law, Moscow, Russia

Commonwealth of Independent States (1991) *Treaty On the establishing of the commonwealth of independent states* (8 Dec 1991). Republic of Belarus, Minsk. Accessed from https://www.consultant.ru/document/cons_doc_LAW_30726/

Commonwealth of Independent States (1993a) The declaration of the heads of commonwealth of independent states on international obligations in the field of human rights and fundamental freedoms (24 Sept 1993a). Moscow, Russia. Accessed from https://www.consultant.ru/cons/cgi/online.cgi?req=doc&base=INT&n=216#016089988039679803

Commonwealth of Independent States (1993b) Treaty On the Cooperation in the Protection of Copyright and Related Rights (24 Sept 1993b). Moscow, Russia. Accessed from https://www.consultant.ru/document/cons_doc_LAW_5865/

Commonwealth of Independent States (1995) The convention of the heads of commonwealth of independent states on international obligations in the field of human rights and fundamental freedoms (26 May 1995). Belarus, Minsk. Accessed from https://www.consultant.ru/cons/cgi/online.cgi?req=doc&base=LAW&n=6966&fld=134&dst=1000000001,0&rnd=0.025448196 24800665#048486002816422935

Dudko NA (2018) On the jury trial in district courts. In: Andreeva OI, Trubnikova TV, Chadnova IV (eds) Legal issues of strengthening Russian statehood. Tomsk State University Publishing House, Tomsk, Russia, pp 153–160

Dudko NA, Neymark MA (2018) Constitutional court of the Russian Federation on jury trial in district courts. Izvestiya of Altai State University. Leg Sci 6(104):70–73

Dudko N, Neymark M, Petuchov E (2019) Trial jury in Russian Federation and the Republic of Kazakhstan: comparative legal analysis. Religación. Revista De Ciencias Sociales Y Humanidades 4(18):90–97
Grib VV (ed) (2005) Discussion of theoretical and practical issues on the introduction of the jury trial in the Republic of Kazakhstan. Accessed from https://journal.zakon.kz/203686-kruglyjj-stol-po-obsuzhdeniju.html
Judicial Department at the Supreme Court of the Russian Federation (2017) Judicial statistics data. Form No. 1 Report on the work of courts of general jurisdiction on the consideration of criminal cases in the first instance. Accessed from https://www.cdep.ru/index.php?id=79&item=4476
Judicial Department at the Supreme Court of the Russian Federation (2018) Judicial statistics data. Form No. 1 Report on the work of courts of general jurisdiction on the consideration of criminal cases in the first instance. Accessed from https://www.cdep.ru/index.php?id=79&item=4891
Judicial Department at the Supreme Court of the Russian Federation (2019) Judicial statistics data. Form No. 1 Report on the work of courts of general jurisdiction on the consideration of criminal cases in the first instance. Accessed from https://www.cdep.ru/index.php?id=79&item=5258
Kirillova NP (1998) The jury trial in Russia and world experience: Published under the Russian-Dutch project Modern standards in criminal law and criminal procedure. Saint Petersburg Law Institute of the Prosecutor General's Office of the Russian Federation, Saint Petersburg, Russia
Kovalev N, Smirnov A (2014) The nature of the Russian trial by jury: Jurata patriae or raison d'etat. Eur J Crime, Crim Law Crim Justice 22:115–133
Kovalev N, Suleymenova G (2010) New kazakhstani quasi-jury system: challenges, trens and reform. Int J Law Crime Justice 38:261–278
Kyrgyz Republic (1999a) Criminal Procedure Code of the Kyrgyz Republic (30 June 1999a No. 62, edited by 25 June 2007 No. 91). Bishkek, Kyrgyz Republic
Kyrgyz Republic (1999b) Criminal Procedure Code of the Kyrgyz Republic (30 June 1999b No. 62, edited by 6 Aug 2012 No. 150). Bishkek, Kyrgyz Republic
Kyrgyz Republic (2017) Code of criminal procedure of the Kyrgyz Republic (2 Feb 2017 No.20). Bishkek, Kyrgyz Republic
Kyrgyz Republic (2019) Law On jurors in the courts of the Kyrgyz Republic (15 July 2009 No. 215, edited by 26 Dec 2019 No. 146). Bishkek, Kyrgyz Republic
Levchenko IV (2017) The role of comparative legal research in determining of the vector of state development. Baikal Res J 8(1):25. https://doi.org/10.17150/2411-6262.2017.8(1).25
Nasonov SA (2016) Historical models of the jury trials. R. Valent, Moscow, Russia
Nemytina MV (2013) Comparative legal research and typology of the legal culture. RUDN J Law 4:166–173
Presidential Executive Office (2014) Meeting with members of the Council for Civil Society and Human Rights and federal and regional human rights commissioners. Accessed from http://en.kremlin.ru/events/president/transcripts/47179
Presidential Executive Office (2015) Presidential address to the federal assembly (3 Dec 2015). Accessed from http://kremlin.ru/events/president/transcripts/50864
Presidential Executive Office (2016) Decree On approval of the Foreign Policy Concept of the Russian Federation (30 Nov 2016 No. 640). Moscow, Russia
Presidential Executive Office (2018) Meeting of the Council for Civil Society and Human Rights (11 Dec 2018). Accessed from http://www.kremlin.ru/events/president/news/59374
Presidential Executive Office (2020) The list of instructions following the meeting of the Council for the Development of Civil Society and Human Rights and the meeting with ombudsmen for human rights. Accessed from http://www.kremlin.ru/acts/assignments/orders/62700
Republic of Azerbaijan (2015) Law On Amendments to the Criminal Procedure Code of the Azerbaijan Republic (13 Feb 2015 No. 1194-IVQD). Baku, Republic of Azerbaijan
Republic of Kazakhstan (1997a) Criminal procedure Code of the Republic of Kazakhstan (13 Dec 1997a № 206-I, edited by 16 Jan 2006). Astana, Republic of Kazakhstan
Republic of Kazakhstan. (1997b). Criminal Procedure Code of the Republic of Kazakhstan (13 Dec 1997b № 206-I, edited by 10 Dec 2009). Astana, Republic of Kazakhstan

Republic of Kazakhstan (2014) Criminal Procedure Code of the Republic of Kazakhstan (4 July 2014 No. 231-V ZRK, edited by November 16, 2020). Republic of Kazakhstan, Astana
Republic of Ukraine (2012) The Criminal Procedural Code of Ukraine (13 April 2012 No. 4651-VI, edited by 21 July 2020). Kiev, Ukraine
Russian Federation (2001) The Criminal Procedure Code of the Russian Federation (12 Dec 2001 No. 174-FZ, edited by 8 Dec 2020). Moscow, Russia
Russian Federation (2011) Federal Constitutional Law On Courts of General Jurisdiction in the Russian Federation (7 Feb 2011 No. 1-FKZ, edited by 06 March 2019). Moscow, Russia
Russian Federation (2016) Federal Law On Amendments to the Criminal Procedure Code of the Russian Federation in Connection with the Expansion of the Application of the Institute of Jurors (23 June 2016 No. 190-FZ, edited by 29 Dec 2017) Moscow, Russia
Russian State University of Justice (2017) Materials of the international scientific and practical conference Reform of the jury in the Russian Federation at the present stage: Problems and prospects for implementation. Fatherland, Kazan, Russia
Sinyukov VN, Rossinskaya ER, Gracheva YuV, Matskevich IM, Voskobitova LA, Ishchenko EP, Komissarova YaV (eds) (2018) Materials of the XIV international scientific and practical conference (Kutafin Readings): round table evolution of the jury trial in Russia. RG-Press, Moscow, Russia
State Duma of Russian Federation (2016) Explanatory note to the draft federal law On amendments to the criminal procedure code of the Russian Federation in connection with the expansion of the application of the institute of jurors. Moscow, Russia. Accessed from https://old.duma.gov.ru/systems/law/?number=258213-7&sort=date
Supreme Court of the Russian Federation (2019) Report of the Chairman of the Supreme Court of the Russian Federation at a meeting of judges of courts of general jurisdiction and commercial courts of the Russian Federation [Video]. Supreme Court of the Russian Federation. https://vsrf.ru/press_center/video_archive/27607/
Supreme Soviet of Russian Federation (1993) Law On Amendments and Additions to the Law of the RSFSR 'On the Judicial System', the Criminal Procedure Code of the RSFSR, the Criminal Code of the RSFSR and the RSFSR Code on administrative offenses (16 July 1993 No. 5451/1–1). Moscow, Russia
Thaiman S (1995) Jury trial in modern Russia through the eyes of an American lawyer. State Law 2:67–76
Thaiman S (2004) Introduction of public participation in the administration of justice in the Republic of Kazakhstan. Russ Justice 6:61–69
Thaiman S (ed) (1994) The jury trial: a guide for judges. Russian Law Academy of the Ministry of Justice of the Russian Federation, Moscow, Russia
Varlamova NV (2014) The object, ends, and principles of the comparative legal studies. RUDN J Law 1:120–128
Yurist. (n.d.) International round table "Introduction of the jury in the Republic of Kazakhstan: Problems of the optimal model of public participation in the administration of justice. Accessed from https://online.zakon.kz/Document/?doc_id=30027353

# Interethnic Conflicts in Local Society as a Threat to Social Security

**Victor V. Nagaytsev, Anatoliy N. Domashev, and Daria A. Mitskikh**

**Abstract** The paper focuses on the inter-ethnic conflict in one of the regions of Siberia—Altai Krai, as a real threat to the social security of the regional society. In 2018–2020, we conducted a sociological study of this conflict, and we present it in this research. Besides, we analyzed the following components of inter-ethnic tension and strife in the region: (1) the frequency of various inter-ethnic conflicts and clashes in the life of citizens; (2) the areas of human activity in which such collisions occur most often; (3) the most common ways of resolving inter-ethnic conflicts used by the local population of the region; (4) the existence of real threats to social security of the region from the escalation of inter-ethnic conflicts; (5) the assessment of the well-being of inter-ethnic relations that have developed in the regional society in recent year, provided by the population. We found that the inter-ethnic conflict in the region is ad hoc and situational. Moreover, it is entirely preventable and regulated. As a result, we concluded that the lack of effective mechanisms for managing inter-ethnic conflicts in society could lead to the most adverse social consequences, a real threat to its social security. The conducted study has shown that the inter-ethnic conflict in local society is quite favorable, calm, with no outbursts of nationalism, inter-ethnic tension, or conflicts, ensuring social security for society.

**Keywords** Social security · Inter-ethnic conflicts · Local society · Local tension · Region

## 1 Introduction

Inter-ethnic conflicts have become an acute and intractable problem for humanity in the twenty-first century. Any modern society includes a large number of people of various nationalities. With such a diverse ethnic composition of society, inter-ethnic conflicts are inevitable in almost all interaction areas.

V. V. Nagaytsev · A. N. Domashev · D. A. Mitskikh (✉)
Altai State University, Barnaul, Russia
e-mail: mickih@mail.ru



S. G. Maximova et al. (eds.), *Advances in Natural, Human-Made, and Coupled Human-Natural Systems Research*, Lecture Notes in Networks and Systems 234,
https://doi.org/10.1007/978-3-030-75483-9_46

Ethnic conflict is exceptional as a local social system in which the interaction of subjects of different nationalities is characterized by open or disguised resistance, confrontation, and opposition, which is caused by contradictions in the interests of the subjects (Esteban et al. 2012). Such a conflict can be managed by preserving the generally accepted forms of its manifestation and development (Cordell and Wolff 2009).

Modern researchers understand the conflict as a natural phenomenon in any social system, a source of its progressive development and proper functioning (Ackerman 2003). We pay special attention to interpersonal, family, labor, national, and inter-ethnic conflicts. Researchers and regional communities study such conflicts. In this regard, the management of inter-ethnic conflicts at the level of local society is crucial. The lack of effective mechanisms for managing inter-ethnic conflicts in society can lead to the most unfavorable consequences, threatening the social security of the region. Inter-ethnic tensions and conflicts are intensified by (1) the wide flow of educational and labor migration in the regions of the country, (2) the emergence of ethnic enclaves in the regions, (3) the activity of socio-political, religious, and nationalist groups.

Numerous studies show that there is always a certain background level of inter-ethnic tension and conflict in the local society. Real threats to social security appear when this level becomes critical. In such a situation, it is necessary to create unique mechanisms to minimize threats in this area.

## 2 Materials and Methods

We assessed inter-ethnic strife in the regional society and the degree of its impact on social security according to a large-scale sociological study within the framework of the project *Development and promotion of social measures aimed at reducing the level of social tension and conflict among the population of Altai Krai in 2018–2020*.

The study used several data collection methods: (1) traditional document analysis; (2) secondary data analysis; (3) mass survey of the population of Altai Krai in 2018 (1220 respondents) and 2019 (600 respondents) by the method of standardized interviewing; (4) four focus groups with representatives of different ethnic groups; (5) expert survey.

## 3 Results

The study shows that inter-ethnic conflicts are an essential part of the daily life of regional society. First of all, we identified public attitudes and the general socio-psychological environment in society.

Table 1 shows that most respondents (43%) assessed the existing socio-psychological environment in the region as negative rather than positive. Another

**Table 1** Distribution of answers to the question "In your opinion, what socio-psychological environment currently exists in Altai Krai?" (The share of the total number of respondents)

| Alternative responses | % |
|---|---|
| Definitely positive | 4 |
| Positive rather than negative | 31 |
| Negative rather than positive | 43 |
| Definitely negative | 13 |
| Not sure | 9 |
| Total | 100 |

*Source* Compiled by the authors

13% of the respondents consider the existing environment to be definitely negative or not sure about the answer at all (9%). Only 4% rated the environment as definitely positive or rather positive (31%). Such a distribution of answers indicates that the respondents recognize the socio-psychological environment in the region as negative. Only a third of the respondents believe that the environment in regional society is positive.

According to Table 2, most respondents (74%) say that sometimes they still face inter-ethnic conflicts in their lives. Quite often, such conflicts occur among 15% of the regional population. Whereas 2% of the respondents continuously find themselves in conflict situations with representatives of other ethnicities. Another 6% of the respondents say that such conflicts never happen in their lives (rural residents and older adults claim this more often than other social groups).

Most often, the residents of Altai Krai have inter-ethnic conflicts in such areas of life as (1) labor and industrial relations (29%); (2) trade and services (26%); (3) business and business relations (18%); (4) friendship (12%); children's education (9%); (5) family relations (6%); (7) the neighborhood and daily life (5%).

In an inter-ethnic conflict, respondents prefer to (1) find a compromise solution (35%); (2) try to prove their point to their opponents at any cost (22%); (3) do not contradict their opponent (19%); (4) ignore the actions of their opponent (13%); (5) resort to the help of another person (3%).

Table 3 indicates that the vast majority of respondents (56%) quite rarely feel any threat to themselves or their families from people of other nationalities. However,

**Table 2** Distribution of answers to the question: "How often do ethnic clashes and conflicts occur in your life?" (the share of the total number of respondents)

| Alternative responses | % |
|---|---|
| Constantly | 2 |
| Often | 15 |
| Sometimes | 74 |
| Never | 6 |
| Not sure | 3 |
| Total | 100 |

*Source* Compiled by the authors

**Table 3** Distribution of answers to the question: "Please, tell us if you feel any threat to yourself or your family from people of other nationalities?" (The share of the total number of respondents)

| Alternative responses | % |
|---|---|
| Often | 8 |
| Quite often | 29 |
| Quite rarely | 56 |
| Hardly ever | 4 |
| Not sure | 3 |
| Total | 100 |

*Source* Compiled by the authors

**Table 4** Distribution of answers to the question: "Do you think that the interethnic relations that have developed in the regional society in the current period are generally quite favorable?" (The share of the total number of respondents)

| Alternative responses | % |
|---|---|
| Yes | 62 |
| No | 33 |
| Not sure | 5 |
| Total | 100 |

*Source* Compiled by the authors

almost a third of the respondents surveyed (29%) quite often encounter threats coming from individuals or groups representing different nationalities. At the same time, 8% of respondents very often face various threats from citizens of other ethnicities. Only 4% of respondents hardly ever experienced such a situation.

As shown in Table 4, most respondents (62%) claim that the inter-ethnic relations in the regional society are generally quite favorable. A third of the respondents consider these relations unfavorable. Another 5% of the respondents found it difficult to answer this question.

## 4 Discussion

Modern sociology offers two scientific directions, depending on the attitude of their representatives to the functions of conflict in the social system (Yang 2000). Representatives of the first direction, which can be described as *consensus theory*, distinguish such categories as *stability*, *sustainability*, and *progressive development*, and refer to conflicts arising in society as negative phenomena and processes (Nakao 2009; Wolff 2006; King 2001; Gilley 2004; Esman 2004). Representatives of the second direction, which is called the *theory of positive conflict*, propose to consider conflict as a social phenomenon that, in addition to some negative consequences, has benefits for the social system (Coser 1956; Galtung 1969; Horowitz 2001; Pearson 2001; Lobell and Mauceri 2004; Brubaker et al. 2004). For example, Frederic Pearson, a well-known American sociologist, presents the social conflict as "a dynamic type of interaction between subjects that cannot be completely eliminated" (Pearson 2001).

The concept of inter-ethnic conflict in society is not yet universally recognized as a scientific category of sociology. In our opinion, nowadays, it is essential to record not only specific cases of inter-ethnic clashes between citizens but also (1) their participants, (2) causes, (3) history of occurrence, (4) typology. There is a correlation between various conflicts, which impacts society and its social security. As a result, numerous social conflicts form one independent phenomenon. Indeed, we will not be able to avoid analyzing individual inter-ethnic conflicts.

In Russia, ethnic conflictology as an independent branch of knowledge is still at its formation stage. However, the point of view of ethnic conflictology is most consistent with the realities of modern Russian society, when conflicts help different subjects reach an agreement, lead to positive changes in society, and solve many problems. As of today, there is no single methodological approach or expert opinion to understanding the phenomenon of inter-ethnic conflict. Sociologists agree on the content of the main inter-ethnic conflict elements, while they differ significantly in determining its causes and consequences.

Modern researchers understand inter-ethnic conflict as an inevitable phenomenon in any local society (Brown 2001). It is necessary to develop as many constructive functions of inter-ethnic conflict in society as possible.

## 5 Conclusion

This study considers inter-ethnic conflicts arising in Altai Krai as a real threat to the social security of the region. We analyzed the following components of inter-ethnic conflict in the region: (1) the population's assessment of the general socio-psychological environment in society; (2) the frequency of inter-ethnic conflicts and clashes in the life of citizens; (3) the areas of life in which such collisions occur most often; (4) the most common methods of conflict resolution; and (5) threats from inter-ethnic conflicts.

Most of the population of the region periodically faces various inter-ethnic conflicts. It is noteworthy that Altai Krai residents rather negatively assess the current general socio-psychological environment in society, which indicates the predominance of pessimistic moods. Most often, inter-ethnic conflicts arise among the residents of the region in such areas of life as (1) labor and industrial relations; (2) trade and services; (3) business and business relations; (4) education of children; (5) neighborhood and daily life. According to the opinion of the regional population, the most acceptable way to resolve the conflict is finding a mutually beneficial solution or compromise.

Inter-ethnic conflict as a social phenomenon manifests itself through specific forms of confrontation and opposition of representatives of different nationalities due to differences in ethnic cultures, styles of behavior, existing stereotypes, or other misunderstandings. The main reasons for inter-ethnic conflicts in society are

(1) interpersonal incompatibility of people of different national origin; (2) lack of understanding of the traditions of other cultures; (3) insufficient communication between representatives of some ethnic groups; everyday issues of neighborly life; inequality of different ethnic groups; (4) personal characteristics of representatives of ethnic communities. Nevertheless, the inter-ethnic conflict has a local, ad hoc, and situational nature and can be fully managed and regulated. In the case of an escalation of the conflict, the regulatory mechanisms of society may not cope with maintaining harmony in the relations between the key social actors of the region. Such a situation will inevitably lead to the transformation of the entire social system.

The research showed that inter-ethnic relations in the regional society of Altai Krai at the current development stage are quite favorable. There are no bursts of nationalism, inter-ethnic tension, and conflicts, ensuring social security in the region.

## References

Ackerman A (2003) The idea and practice of conflict prevention. J Peace Res 40(3):339–347
Brown ME (2001) Ethnic and internal conflicts: causes and implications. Turbulent peace: the challenges of managing international conflict. Institute of Peace Press, Washington, DC, pp 209–226
Brubaker R, Loveman M, Stamatov P (2004) Ethnicity as cognition. Theory Soc 33(1):31–64
Cordell K, Wolff S (2009) Ethnic conflict: causes, consequences, and responses. Polity Press, Cambridge, UK
Coser LA (1956) The functions of social conflict. Free Press, New York, NY
Esman MJ (2004) An introduction to ethnic conflict. Polity Press, Cambridge, UK
Esteban J, Mayoral L, Ray D (2012) Ethnicity and conflict: theory and facts. Science 336(6083):858–865
Galtung J (1969) Violence, peace, and peace research. J Peace Res 6(3):167–191
Gilley B (2004) Against the concept of ethnic conflict. Third World Q 25(6):1155–1166
Horowitz DL (2001) Ethnic groups in conflict. University of California Press, Berkeley, CA
King C (2001) The benefits of ethnic war: understanding Eurasia's unrecognized states. World Polit 53(4):524–552
Lobell SE, Mauceri P (2004) Ethnic conflict and international politics: explaining diffusion and escalation. Palgrave Macmillan, Basingstoke, UK
Nakao K (2009) Creation of social order in ethnic conflict. J Theor Polit 21(3):365–394
Pearson FS (2001) Dimensions of conflict resolution in ethnopolitical disputes. J Peace Res 38(3):275–287
Wolff S (2006) Ethnic conflict: a global perspective. Oxford University Press, New York, NY
Yang PQ (2000) Ethnic studies: issues and approaches. State University of New York Press, New York, NY

# Personal Security of the Population of the Khabarovsk Region

**Anastasya S. Spirina, Syldysmaa A. Saryglar, and Olga V. Borisova**

**Abstract** The purpose of this article is to determine the theoretical significance of personal and social security of the population, as well as to identify factors and features of the formation of a subjective feeling of personal security among the population of the border region. The empirical basis of the analysis was the data of a sociological study of the population of the border regions of Russia. The main analysis methods are frequency data analysis and regression analysis. The relationship between some socio-demographic indicators and a sense of satisfaction with the personal security of the population was revealed—the higher the level of education of the population and the higher their assessment of their financial situation, the higher the population assesses satisfaction with their personal security. The regression analysis made it possible to identify factors that can influence the formation of a subjective feeling of satisfaction with the personal security of the population of the border region. The identified factors were characterized as "general" and "specific". The first group included such factors as institutional and interpersonal trust, as well as satisfaction of their job and their marital status. The second group included such a factor as the level of satisfaction with the ability of the population to observe their national customs/traditions.

**Keywords** Social security · Personal security · Regression analysis · Border regions of Russia · Trust

## 1 Introduction

Human security is assessed as protection against numerous, diverse, and contemporary threats and problems people face. There are various types of threats: economic, food, environmental, social, political, national, health threats, personal, etc. All these aspects are an integral part of the social life of the population. The security of these segments is interconnected with social and personal security: on the one hand, these

A. S. Spirina · S. A. Saryglar · O. V. Borisova (✉)
Altai State University, Barnaul, Russia
e-mail: olya333es@mail.ru



S. G. Maximova et al. (eds.), *Advances in Natural, Human-Made, and Coupled Human-Natural Systems Research*, Lecture Notes in Networks and Systems 234,
https://doi.org/10.1007/978-3-030-75483-9_47

segments are influencing human security. On the other – they are influenced by human security (Popović-Mančević and Ruzic 2017).

Social security is an integral concept, where various security indicators (national, economic, environmental, etc.) are closely intertwined. The range of security objects is becoming wider due to the inclusion in the composition of all vital institutions, relations, and processes of the social sphere of society. Besides, in the framework of social security, attention is paid to the importance and need to ensure the security of many entities, including the country, civil society, and individuals. In this regard, social security is considered as a multi-level phenomenon in which three levels can be distinguished: institutional, group, and personal. At each level, the relevance and severity of the problem require constant attention and analysis of the state, including sociological monitoring (Shahmatova 2016).

The semantic content of social security is made up of such indicators as the quality of life and standard of living. The quality of life includes the protection of health, nutrition, clothing, education, culture, communications, free movement and access to information, recreation, etc. The standard of living includes indicators of social development, a living wage, incomes and expenses of the population, an increase in consumer prices, etc. These components made it possible to consider social security as a protection against real and potential threats and as a specific toolbox for assessing the implementation of social policy goals (Podolskaya and Nazarkina 2016).

Social security research aims to study the state and ability of the country and society to operate the social sphere effectively, prevent destructive phenomena and processes, preserve and develop the conditions, means, and methods of socializing people, observing their lifestyle, welfare, inalienable rights, and freedoms in society, spiritual and moral values state. Social security is currently a comprehensive description of the state of such elements as a country, economic and social institutions, social communities, and individuals (Nikulina and Kolodiev 2017).

A critical indicator of social security is the individual's satisfaction with personal security. This indicator underlies the actions of individuals to identify, prevent, mitigate, eliminate and repel dangers and threats that can destroy them, deprive fundamental material and spiritual values, cause unacceptable damage, and close the path to survival and development (Maksimova et al. 2012).

The concept of personal security was initially more developed in the legal and political science aspects, where it is disclosed primarily through civil rights and individual freedoms. Despite the differences in subjects and problems, the desire for interdisciplinary interaction contributes to the fact that various security concepts consider the fact that a person, along with legal ones, also has private, civil, economic, and other rights and freedoms, the protection of whose contributes to human security (Shlykova 2018). Personal security is a state of protection not only of life, health, freedoms, and human rights, but also is the security of their spiritual, social, economic, political, environmental, and other spheres of life (Rumachik 2011). The concept of personal security primarily reflects the subjective side of the individual, characterizing the state of protection of an individual from threats, which are formed at the level of their personal perceptions and needs (Zubok and Chuprov 2017).

The feeling of personal security depends on individual and social circumstances. The subjective and objective elements of personal security are interconnected with physical, material, and financial security, and security related to social relations, cultural and institutional security (Forbes-Mewett 2019). A subjective assessment of personal security is one of the factors determining social well-being—a high assessment of personal security is an important adaptation resource, and a low one acts as a barrier to adaptation (Shlykova 2018).

## 2 Materials and Methods

The empirical base is the data of a sociological study of social risks and security in the context of the transformation of migration processes in the Asian border of Russia. The research subject is the features of perception of satisfaction with personal security by the Khabarovsk region population. The research aims to identify factors in the formation of a subjective feeling of personal security among the Khabarovsk region population. The analysis of the specifics of satisfaction with personal security is carried out using the following indicators:

- Presence or absence of problems in the region (decrease in the standard of living of people, unemployment, lack of prospects for youth, the security of citizens, healthcare, corruption, ecology, deterioration of relationships between people of different nationalities, increasing hostility between different religious movements, growing hostility between indigenous peoples and migrants, etc.);
- Satisfaction with some aspects of life (life, job, marital status, material status, housing conditions, health, freedom of expression religious views and values, the ability to comply with one's national customs and traditions, the ability to preserve, study, and develop national culture and language, etc.);
- Socio—demographic indicators (gender, age, financial situation, level of education, etc.);
- Interpersonal trust (the head of the organization, colleagues, relatives, friends/acquaintances, neighbors, people of their nationality, people of their religion, etc.);
- Institutional trust (in the President of the Russian Federation, Government, State Duma, Governor, Municipal authorities, army, court, police, etc.).

Quantitative data is processed using the statistical processing software SPSS Statistics 23.0. Statistical conclusions were based on the results of frequency data analysis and regression analysis. These methods allow us to identify factors that influence the formation of a subjective feeling of satisfaction with the personal security of the population and determine indicators that can predict the response values of the variable of interest to us.

## 3 Results

An assessment of the general situation in the region shows that for most of the population (60.3%) of the Khabarovsk Krai, the region is considered "prosperous and developed." 68.5% of the population notes a sense of security, and almost a third of the population does not feel safe.

Satisfaction with personal security is also at a reasonably high level—77% of the population indicate that they are satisfied with their personal security. A third is completely satisfied with their personal security (33.1%). 23% of the population is not satisfied with their personal security.

Assessment of the satisfaction with their personal security among men and women is almost identical—80.2% of men and 74.6% of women are satisfied with personal security. Statistically, significant differences are found in the assessment of the population's satisfaction with their personal security, depending on education ($\chi^2 = 0.287$, $p \geq 0.05$) and financial situation ($\chi^2 = 0.306$, $p \geq 0.05$). Populations with incomplete higher education are more satisfied with their personal security—81.5–83.7%. A smaller proportion of the population with additional secondary and secondary general education feels their personal security—68.6–63.9%. People with basic additional education note that they are not satisfied with their personal security.

Satisfaction among the population who noted their financial situation as "above average" is 85.5% (53% of them are fully satisfied). Among the population with an "average" level of financial situation, the satisfaction level with their personal security is 74.4% (the percentage of those completely satisfied here is much lower—24.4%). The lowest level of satisfaction with their personal security is in the group with a financial situation "below average"—63.9%.

The obtained data allow us to conclude that the higher the level of education of the population and the higher their assessment of their financial situation, the higher the population evaluates satisfaction with their personal security.

When conducting a binary logistic regression, an indicator of the subjective feeling of satisfaction with personal security was used as a dependent variable in the regression model. The constructed model allows us to classify 73.5% of respondents. The obtained solution identifies factors in the formation of a subjective feeling of satisfaction with the personal security of the population. The model was brought based on the inverse step—by—step method of eliminating variables; the best quality of the model was achieved at the 23rd step. Table 1 shows the statistical significance of the independent variables.

The most significant factor in the subjective feeling of satisfaction with personal security for the Khabarovsk Krai population is satisfied with the ability to observe their national customs and traditions. The probability of getting into a subgroup that is satisfied with their personal security increases by 42.8 times due to a high level of satisfaction with the ability to observe their national customs and traditions.

The second most important factor is the trust in friends and acquaintances, whose high values increase the probability of getting into a subgroup satisfied with their personal security by 19 times. In the third place is the factor of police trust, high values

**Table 1** Variables in the regression equation

| Predictors | B | Wald. | Relevance | Exp (B) |
|---|---|---|---|---|
| Lack of the problem "unemployment" | −1.465 | 4.243 | 0.039 | 0.231 |
| Lack of the problem "lack of prospects for youth" | −1.581 | 4.595 | 0.032 | 0.206 |
| Lack of the problem "poor quality healthcare" | −1.749 | 5.198 | 0.023 | 0.174 |
| Satisfaction of job | 1.445 | 5.056 | 0.025 | 4.242 |
| Satisfaction of marital status | 1.367 | 3.156 | 0.076 | 3.922 |
| Satisfaction of the opportunity to comply with their national customs, traditions | 3.756 | 14.379 | 0.000 | 42,773 |
| Trust to colleagues | 1.510 | 4.421 | 0.035 | 4.525 |
| Trust to relatives | −5.112 | 5.911 | 0.015 | 0.006 |
| Trust to friends | 2.946 | 4.208 | 0.040 | 19.030 |
| Trust to the President of the Russian Federation | −1.431 | 4.060 | 0.044 | 0.239 |
| Trust to police | 1.769 | 5.863 | 0.015 | 5.864 |

*Source* Compiled by the authors

of which increase the population's satisfaction with personal security by 5.9 times. At approximately the same level are such factors as (1) confidence in colleagues (4.5 times), (2) satisfaction of job (4.2 times), and (3) satisfaction of marital status (4 times) contributing to increased satisfaction with personal security.

Also, relying on the results of the obtained model, the authors identified factors that could reduce personal security satisfaction among the Khabarovsk Krai population. First, it is such problems as unemployment (0.2), lack of prospects for young adults (0.2), and low quality of health care (0.2), as well as a low level of trust in relatives (0.01), and the President (0.2).

## 4 Discussion

The general characteristic of the prosperity and security of the Khabarovsk Krai, from the point of population, can be described as "positive"—most of the population feels safe and evaluates their region as prosperous and developed. The population's satisfaction with their personal security includes many components and is closely interconnected with various indicators.

The most significant factor is the level of the population's satisfaction with the ability to comply with their national customs and traditions. It can be explained by the specifics of the territorial location of the region. The border regions of Russia have their own characteristics, which in particular are manifested in a state of interethnic relations. In the Khabarovsk Krai, interethnic relations are generally characterized as "positive." However, there are some manifestations of the negative attitude of residents towards other nationalities, mainly due to the inability of people of other nationalities to consider the customs and norms of behavior adopted in Russia (Shakhova 2018). In this regard, the factor of population satisfaction with the opportunity to observe their national customs and traditions can be assigned to the group of "specific" factors.

Another significant factor influencing satisfaction with personal security is trust. A number of studies support the interconnection of trust and security at various levels. Institutional trust is seen as an important parameter of integration, cohesion, and social security of a society (Maksimova et al. 2017). In the current research, among the included social institutions, the trust in the police turned out to be the most significant. On the one hand, it can be explained by the main function of the police, which consists precisely in the protection and security of the population. On the other hand, among all the institutions that ensure public security, this institution is the "closest" one if a person needs help or under threat.

Less significant factors are satisfaction with the job, marital status, the relationship between satisfaction with personal security and material status, and level of education. These factors can be assigned to the group of "general" factors.

The authors evidence the importance of the economic factor influence on the feeling of security. The level of economic security of an individual affects the quality of human and social capital, determining the ability of a particular person to use the economic environment to ensure a decent standard of living. The individual economic component is understood as the stability of the individual's income sufficient to maintain worthy living standards now and in the near future (Tamošiūnienė and Munteanu 2015). The implementation of the individual economic component, in turn, occurs precisely at the expense of the individual's labor activity. Also, some studies confirm the existence of a significant differentiation in assessments of the feeling of security depending on the level of education of citizens[1] (Yudina et al. 2017).

## 5 Conclusion

Personal security can be considered as one of the components of social security, which is ensured by the formation of a current social environment around a person working to improve their health, education, housing, and working conditions, to increase income and personal competitiveness. In addition to external conditions, an individual's sense of security at the individual—individual level is important. An analysis of the population's satisfaction with the personal security of the Khabarovsk Krai was a part of the research. The authors revealed the interconnection between satisfaction with personal security and such indicators as the level of education and assessment of one's financial situation. The higher the level of education of the population and the higher their assessment of their financial situation, the higher the

[1] A quantitative survey of the population in federal subjects of Russia (85 regions), including the Republic of Crimea and Sevastopol. The total sample size is 48,800 respondents.

population evaluates satisfaction with their personal security. A regression analysis was carried out using the binary logistic regression method to comparatively assess the degree of influence of various factors on forming a subjective feeling of personal security of the population. This method allowed us to build a model of the dependence of satisfaction with personal security and determine the "general" and "specific" factors.

**Acknowledgements** The publication was prepared as part of the implementation of the RFBR grant No. 19-011-00417 "Factors and mechanisms for building confidence in the system of maintaining social security in the border regions of Russia" (2019–2021).

## References

Forbes-Mewett H (2019) Personal security. The new security: Individual, community and cultural experiences. Palgrave Macmillan, London, UK, pp 217–236

Maksimova SG, Goncharova NP, Noyanzina OE (2012) Features of risk perception in the structure of assessing personal and social security. Izvestiya Altajskogo Gosudarstvennogo Universiteta 2-1:211–215

Maksimova SG, Noyanzina OE, Omelchenko DA, Maksimova MM, Surtaeva OV (2017) Assessment of the features of interpersonal and institutional trust of the population of the Altai region. Vestnik Altajskogo Gosudarstvennogo Agrarnogo Universiteta 11(157):188–192

Nikulina MA, Kolodiev MY (2017) Social security in the Russian Federation: criteria, threats, areas of support. Istoricheskie, Filosofskie, Politicheskie i Yuridicheskie Nauki, Kul'turologiya i Iskusstvovedenie. Voprosy Teorii i Praktiki 4:153–157

Podolskaya EA, Nazarkina VN (2016) Social security: the nature, threats and ways to ensure. Sociologicheskie Issledovaniya 11:133–139

Popović-Mančević M, Ruzic M (2017) Individual security in academic discourse. Nauka Bezbednost Policija 22(1):127–140

Rumachik IA (2011) Personal—Social—National security: an informational aspect. Socialno-Gumanitarnye Znaniya 12:257–262

Shahmatova NV (2016) Social security in Russian society: gender cut. In: Bronnikova YuO (ed) Materials of the V international scientific conference: conflicts in the modern world: international, state and interpersonal dimension. Pero, Moscow, Russia

Shakhova EV (2018) Features of interethnic relations in the Khabarovsk Territory as a border region of the Russian Federation. Soc Secur Insights 1(4):198–212

Shlykova EV (2018) Subjective assessment of personal safety as an indicator of adaptability to a risky environment. Sociologicheskij Zhurnal 24(3):56–75

Yudina TN, Frolova EV, Tanatova DK, Dolgorukova IV, Rodimushkina OV (2017) Personal security and victim concerns. Zhurnal Sociologii i Social'noj Antropologii 20(1):114–127

Zubok YA, Chuprov VI (2017) Threats in a transforming living environment as a factor of social risks: forecasting and regulation. Sociologicheskie Issledovaniya 5:57–67

# Female Criminality as a Threat to National Security: An Interdisciplinary Reconceptualization

**Lyubov Yu Kiryushina**

**Abstract** The current research is of undoubted relevance since females also commit selfish and violent crimes. The fact that women have started committing crimes that threaten national security is of particular concern. The most socially dangerous crimes include illicit trafficking in drugs and psychotropic substances and crimes of extremism and terrorism. The paper aims to justify (1) the need for a comprehensive approach for studying female crime as a distinct crime type and (2) the use of interdisciplinary research methods (historical method, method of comparative law, method of interview and conversation, psychological analysis of criminal cases, and analysis of criminal statistics). The scientific novelty of the research consists in the interdisciplinary approach to the problem of female crime. I have conducted research to improve the effectiveness of the investigation, establish the essential role of women in the committed crime, identify all accomplices, and develop measures aimed at crime prevention. Using interdisciplinary methods helps identify aspects that deserve further attention and research in criminology, sociology, and psychology. Detecting and using the identity of a female offender will contribute to the development of crime investigation. I have substantiated the need to use international experience, including theoretical developments aimed to prevent and combat female crime.

**Keywords** Female crime · Gender stereotype · Full circle stereotype · Crime investigation · Crime prevention

## 1 Introduction

Practice shows that women committing crimes use specific mechanisms and ways to commit them compared to men. I have noted within my dissertation research on *Woman Personality in a Crime Mechanism and its Significance for the Forensic Methods of Investigating Separate Types of Crimes* that investigating crimes

L. Y. Kiryushina (✉)
Altai State University, Barnaul, Russia
e-mail: love-kiryushina@yandex.ru

 
S. G. Maximova et al. (eds.), *Advances in Natural, Human-Made, and Coupled Human-Natural Systems Research*, Lecture Notes in Networks and Systems 234,
https://doi.org/10.1007/978-3-030-75483-9_48

committed by women has its peculiarities. In particular, I have noted in the abstract of the dissertation research that, as a rule, women think over and plan the mechanism of a crime they commit compared to male criminals (Kiryushina 2007). In addition, particular psychological, psychophysiological, and psycho-emotional states, processes, and properties of female criminals that affect the formation of gender stereotypes existing in society explain some problems in investigating and solving such crimes (Kiryushina 2007).

Criminologists consider female crimes as a particular type of crime. They study it along with such types of crimes as violent, self-serving, and juvenile delinquency. This is reflected in studying the relevant discipline in law schools. In turn, criminology as an academic discipline does not consider female crimes and methods for investigating crimes committed by women as a separate independent scientific board despite the existing research and the great theoretical and practical significance of this problem.

A special biological and social function of women in society (namely, their role as mothers) is, to some extent, a prerequisite for the normal development and functioning of society. Therefore, committing crimes by women and prescribing punishments associated with their isolation from society is already a factor that creates preconditions for violating social security in a particular country and the Eurasian space. The fact that women are involved in crimes of extremism and illegal drug trafficking and even act as organizers is particularly alarming. Thus, 736 women are serving their sentences only in Altai Krai. Among them, 310 women have committed crimes against public security and order, five women who committed crimes against state authorities, and 175 of them have committed complicity crimes (Federal Penitentiary Service in Altai Krai 2019).

I suppose that researching problems in investigating and preventing crimes committed by women should be interdisciplinary and with the use of interdisciplinary research methods of criminalistics, criminology, sociology, psychology, and its branch—legal psychology.

The scientific novelty of the research is in the attempt to use interdisciplinary research methods aimed to understand the causes and motives of crimes committed by women, new methods and mechanisms of committing crimes, and the principles and patterns of forming and organizing criminal groups that commit multi-episodic crimes.

## 2 Materials and Methods

Within the study, I used socio-demographic and legal information concerning women convicted and serving sentences in penal institutions in Altai Krai, statistical data provided by the Office of the Prosecutor General of the Russian Federation on the legal statistics portal, periodicals (including those of Altai Krai), studies of Russian and foreign authors, and regulatory legal acts aimed to regulate the issues in this area (Federal Penitentiary Service in Altai Krai 2019).

I used various methods in my research. The historical method made it possible to do the following: (1) investigate and identify the main types of female offenders; (2) determine the range of crimes, committing which was most typical for women; (3) reveal the most typical methods for preparing, committing, and concealing crime traces typical for women, which allowed one to draw a conclusion and put forward the version about the personality and gender of a criminal; and (4) trace the dynamics of using crime instruments and objects used as tools by female offenders.

I used the method of analyzing crime statistics to study the following parameters: (1) socio-demographic characteristics of women serving a sentence; (2) structure and nature of crimes committed; (3) methods of preparing for committing and concealing the traces of crimes; and (4) gender characteristics of a group crime and interrelation in a group; (5) degree of influence of male accomplices on involving women in committing crimes; and (6) connection of domestic crimes with the involvement of men into committing crimes.

I also used the method of psychological analysis of criminal cases to study the psychological characteristics of women who committed crimes (including those serving sentences associated with isolation from society). Applying this research method allowed me to establish psychological prerequisites for forming criminal groups and mechanisms for managing them in cases where a woman is an organizer and a leader.

Analyzing the periodical press made it possible to identify and analyze the main tendencies that manifest themselves in female crimes covered by the periodical press (including publications in Altai Krai). According to the predesigned scheme, I investigated the following: (1) age of criminals; (2) level of their education; (3) marital status; (4) nature of a crime committed; (5) method of committing a crime; (6) motives and goals; and (7) forms and ways to counteract law enforcement and correctional authorities.

Interviews and conversations are confidential communication with the subject under research (in this case, with women serving a sentence associated with their isolation from society). This method allowed me to collect initial indicative research and clarify the results obtained by other methods. The method of interrogation was a variant of the conversation method I used. It represented a dialogue with a person convicted according to a predesigned questionnaire. Applying this method was possible due to my work as a member of the public oversight commission of Altai Krai on public control of ensuring human rights in places of detention and on assistance to persons in places of detention since 2016. From 2016 to 2020, as a member of this commission, I visited various detention facilities of Altai Krai, including Corrective colony No. 6 and Colony settlement No. 7 in Shipunovo (Altai Krai, Russia), and Corrective colony No. 11 in Novoaltaisk (Altai Krai, Russia), where women are serving sentences associated with their isolation from society. According to the Federal Law No. 76-FZ of June 10, 2008, *On Public Control of Ensuring Human Rights in Places of Detention and on Assistance to Persons in Places of Detention* (with amendments and additions) (Russian Federation 2008), I participated in a personal reception of persons serving a sentence. Moreover, I responded to the written applications of women serving a sentence associated with their isolation from society.

The method of comparative law allowed me to generalize and use the experience aimed to develop the following recommendations: (1) recommendations on tactics of operational search and investigative actions for investigating crimes of extremism; and (2) recommendations on the prevention of this type of crimes in cross-border regions of foreign countries. Some Russian researchers, such as Komov (1992), Khlopkov (2006), and Shiian (2005) studied these issues.

## 3 Results

Lombroso, an Italian prison doctor, psychiatrist, and criminologist, was one of the first researchers who drew attention to female crimes as a particular type of crime worthy of independent research. He is known to be the author of the theory of a *natural* criminal. Despite the criticism, this theory has many supporters. There are its followers that prove some provisions of this scientific theory.

Thus, the theory of a criminal was formulated. Moreover, the generalization of practical material led to the book *Criminal Woman, the Prostitute, and the Normal Woman* published by Lombroso at the turn of the 19th–20th centuries. The book reflected the relationship between the moral and psychological qualities of women and their crimes (murder, robbery, theft, etc.). In the book, the author noted the impact of these qualities on the motive, the choice of a crime subject, the way women prepared and committed crimes, and most importantly, the postcrime behavior of women, including repentance or its complete absence. In this work, Lombroso managed to do the following: (1) identify various types of female criminals (considering the moral and psychological qualities based on the theory of the biosocial roots of a crime); and (2) distinguish criminal types (congenital and casual criminals, criminals of passion, and suicides) (Lombroso 1995).

Other researchers also conducted interesting studies. In the last quarter of the nineteenth century, Tarde drew attention to the physiological influence of gender on a crime (Tarde 2004). Within the research, he found that gender had a strong physiological effect on a crime. He also concluded that the rates of female crimes were significantly lower than the rates of male ones. In 1880, there were five times more male than female convicts and six times more male defendants than female ones. At that time, 54 out of 100 boys and only 17 out of 100 girls were punished for fighting.

According to the Russian researchers, judicial investigators had 33,298 criminal cases of murder in progress in 1908. Among people sentenced to hard labor, 11.5% were women, and 57% of these women were killers (Kolokolov 2002). The analysis of statistical data and psychological analysis of criminal cases allowed me to conclude that at this stage, women committed a wide range of crimes and that their victims included not only close relatives (husbands, cohabitants, children, and elderly parents) and acquaintances, but also strangers. The choice of victims was determined by the place, time, and manner of a crime. In committing crimes that threaten national security (such as extremism, terrorism, and illicit trafficking of narcotic drugs and psychotropic substances), an indefinite number of persons become victims. More than that, such crimes damage the interests of society and the state.

The results obtained by using the research methods such as psychological analysis of criminal cases, interviews and conversations, analysis of crime statistics and periodicals, and comparative law allowed me to note a change in the distribution of roles of accomplices in criminal groups involving women. Every year, female masculinity (Kozyreva 2002) studied in the works of psychologists and sociologists is manifested in crimes committed by women who copy and unknowingly use male patterns of behavior and thinking.

The methods I used showed that male patterns of behavior and psychological qualities reflected in the studies of psychologists replaced traditional psychological female qualities (such as empathy, femininity, romanticism, conformity, shyness, fearfulness, anxiety, lack of self-confidence, exposure to frustrations, kindness, soft-heartedness, etc.). Compared with male organizers and leaders of criminal groups, female criminals who actively participate in creating and leading criminal groups use different mechanisms of managing accomplices. The research on leadership made by psychologists showed that women were less active than men in the presence of the opposite gender, and they claimed the role of a spontaneous leader.

For example, Jorstad (1996), a Norwegian psychotherapist, concluded that leadership was not a priority and a part of the value system for most women. Johnson (1994) found that female leaders lost to male leaders because they had less power, influence, and resources.

Using historical and comparative methods showed that until the late 1980s and early 1990s, women had the role of the so-called *baits*, or accomplices or co-actors, while men played an active role as perpetrators, organizers, or leaders of a criminal group. Nevertheless, by the mid-1990s and early 2000s, women actively participated in crime planning or organizing and leading criminal groups. Within the research conducted in 2002, Kozyreva compared male and female individuals in terms of aggressiveness regarding masculinity and femininity. The research results were different from the ideas in psychology about the greater aggressiveness of men and less aggressiveness of women. Kozyreva observed a high level of hetero-aggression among women with pronounced masculinity (Kozyreva 2002).

Interviews and conversations, as well as the psychological analysis of criminal cases and analysis of criminal statistics, showed that in committing violent crimes, women chose ways inherent to men (such as striking multiple victims, strangling victims, etc.).

The research in differential psychology also showed that masculine females were not inferior to men in their spatial abilities and were close to them in their mathematical thinking (Ilyin 2006), which one should consider when conducting investigations involving women.

## 4 Discussion

The problems of women participating in crimes with a high level of public danger and threatening national security are an essential topic for discussion. There are three interdisciplinary areas that Russian and foreign scholars, practitioners, and members of human rights organizations should consider in further research and scientific discussions on female crime:

1. Criminalistics direction;
2. Criminological and sociological direction;
3. Psychological direction.

Criminalistics direction implies the following: (1) developing recommendations aimed to investigate crimes and reveal all episodes of crimes and accomplices; (2) preparing recommendations aimed to overcome countermeasures in the investigation of crimes related to illicit trafficking in narcotic drugs and psychotropic substances; and (3) developing a criminal profile of criminals involved in extremist crimes.

Criminological and sociological direction. Since criminology originated as a doctrine about crime as one of the social phenomena, it studies criminal, social, and economic reasons for forming the personality, committing crimes, and social prerequisites for the growth of certain types of crimes. As a result, it seems expedient to consider the issues of preventing crimes committed by women on the basis of legal and sociological studies. Certain types of crimes committed by women in transborder regions represent the most significant social danger. Illegal trafficking in narcotic drugs and psychotropic substances, as well as crimes of extremism, occupies a special place. Using recommendations developed in sociology on the characteristics of a small group (including criminal ones) (such as functions of a leader, procedure of distributing roles within a group, level of cohesion, causes and procedure of conflict resolution, etc.) will make it possible to develop tactical methods for conducting investigative and operational search actions aimed to identify accomplices. I believe that establishing the essential role of each accomplice will contribute to a lawful, reasonable, and fair verdict and fair punishment.

Psychological direction. Identifying the true motives and goals of crimes (especially the so-called *unconscious motives* not fixed in the criminal law doctrine and not considered in imposing punishment) will allow one to develop recommendations aimed to do the following: (1) establish psychological contact with a suspected person during interrogation; (2) identify psychological prerequisites of forming criminal groups; (3) identify psychological prerequisites of manipulative forms of communication during the preparation for committing a crime and distribution of roles within a group. The legal assessment aims to establish the essential role of a female criminal in a criminal group and overcome gender stereotypes in preliminary investigation and court proceedings. This is not only of the act committed by a female but also of the public danger of a personality threatening national security. Moreover, this is of the punishment assignment associated with these circumstances and related or not related (in case of alternative punishment) to isolation from society. I should note that it is necessary to conduct further research on the psychological prerequisites for committing crimes by women in complicity with male opinion. An increased sense of conformity combined with emotional instability and impulsiveness becomes a prerequisite for committing a single criminal act. It also becomes a decisive factor influencing female participation in criminal groups with a mixed-gender identity. The social danger of such criminals lies in the fact that they commit multiple crimes, including those related to the illegal possession and distribution of narcotic drugs and psychotropic substances, and crimes of extremism and terrorism.

## 5 Conclusion

Without belittling the public danger of any crime, I believe that crimes related to the illegal circulation of narcotic drugs and psychotropic substances and crimes of extremism and terrorism are of the most significant social danger. It seems that these crimes represent a particular social danger, especially those committed by women or with their participation since the social institution of family and marriage is destroyed. The role of the mother-woman is completely leveled. Consequently, there is a threat of deviant behavior of children deprived of attention and love of their mother and prerequisites for socially unacceptable forms of behavior of children and adolescents (such as homelessness, begging, and juvenile delinquency).

I believe that one should develop measures aimed to prevent such crimes involving the resources of law enforcement agencies and services, which ensure the security of the country, and public and human rights organizations.

## References

Federal Penitentiary Service in Altai Krai (2019) Annual statistics for 2019 [Internal document]. Federal Penitentiary Service in Altai Krai, Barnaul, Russia

Ilyin E (2006) Differential psychology of man and woman. Piter, Saint Petersburg, Russia

Johnson C (1994) Gender, legitimate authority, and leader-subordinate conversations. Am Sociol Rev 59(1):122–135

Jorstad J (1996) Narcissism and leadership: some difference in male and female leaders. Leadersh Org Dev J 17(6):17–23

Khlopkov AS (2006) Features in investigating murders committed by women (Dissertation of Candidate of Law). Saint Petersburg Institute of Law, Saint Petersburg, Russia

Kiryushina LYu (2007) Woman Personality in a crime mechanism and its significance for the forensic methods of investigating separate types of crimes (Dissertation of Candidate of Law Science). Altai State University, Barnaul, Russia

Kolokolov V (2002) Judicial system of the Kursk region 100 years ago. Russ Judge 7:45–47

Komov AYu (1992) Features in investigating violent crimes committed by women (Dissertation of Candidate of Law Science). Voronezh State University, Voronezh, Russia

Kozyreva E (2002) The effect of masculinity-femininity on the manifestation of auto- and hetero-aggression. In: Chesnokov VB (ed) Proceedings from international scientific conference. Saint Petersburg State University, Saint Petersburg, Russia, pp 208–210

Lombroso C (1995) Genius and insanity: composition. Republica, Moscow, Russia

Russian Federation (2008) Federal Law On Public Control of Ensuring Human Rights in Places of Detention and on Assistance to Persons in Places of Detention (10 June 2008, No. 76-FZ). Moscow, Russia

Shiian VI (2005) Prevention of self-serving crimes committed by women (Dissertation of Candidate of Law). All-Russian Research Institute in the Ministry of Internal Affairs of the Russian Federation, Moscow, Russia

Tarde G (2004) Criminal and crime: comparative crime. Infra-M, Moscow, Russia

# Transformation of Eurasian Scientific and Technical Cooperation Within the SCO Framework

**Anton A. Vasiliev, Yulia V. Pechatnova, and Meruert B. Muratkhanova**

**Abstract** The purpose of our research was to study the transformation of international Eurasian scientific and technical interaction within the framework of the Shanghai Cooperation Organization (hereinafter—SCO). The methodological basis of the study is formed by comparative legal and formal-legal methods. The method of comparative legal analysis is aimed at comparing the legal regulation of scientific and technical cooperation in the SCO member states. The method of formal and legal analysis was used to interpret legal acts of the SCO member-states aimed at developing legal and institutional basis for mutually beneficial cooperation in science and technology. The relevance of the study is predetermined by the growing influence of the development of science and technology within the framework of interstate organizations. The novelty of the work is due to the small degree of study of the problem of international Eurasian scientific and technical cooperation. The study analyzed ways to intensify international scientific and technical cooperation within the SCO, including the establishment of commissions or councils on scientific and technical cooperation, development of model legal acts aimed at creating a legal framework for the development of international relations in the field of science, the formation of special funds to finance research, as well as the adoption of a new agreement on scientific and technical cooperation between the SCO member-states. As a result of the research, the authors conclude that in order to develop international Eurasian scientific and technical cooperation, a number of legal, institutional and organizational and financial measures are required.

**Keywords** Eurasian space · Scientific and technical cooperation · SCO

A. A. Vasiliev · Y. V. Pechatnova (✉)
Altai State University, Barnaul, Russia
e-mail: jp_0707@mail.ru

M. B. Muratkhanova
L. N. Gumilyov, Eurasian National University, Astana, Kazakhstan



S. G. Maximova et al. (eds.), *Advances in Natural, Human-Made, and Coupled Human-Natural Systems Research*, Lecture Notes in Networks and Systems 234,
https://doi.org/10.1007/978-3-030-75483-9_49

## 1 Introduction

The Shanghai Cooperation Organization (SCO) was initially an international organization with a distinct interstate alliance aimed at addressing security issues in Asia. Gradually, other aspects of international cooperation became the subject of mutually beneficial cooperation among the SCO member states. Hence, the Agreement on Scientific and Technical Cooperation was signed in 2013 between the SCO member states. The Agreement laid down the principles and forms of scientific and technical cooperation within the SCO framework (Shanghai Cooperation Organization 2013). In institutional terms, the scientific and technical cooperation of the SCO member states was entrusted to a particular interstate working group.

Despite the expansion of spheres and topics for interstate cooperation within the SCO framework in seven years since the adoption of the SCO Agreement on Scientific and Technical Cooperation, relations in the field of science and technology continued to be exclusively bilateral between the SCO member states. In its essence, scientific and technical ties were being established between Russia and each individual SCO country. Multilateral agreements and scientific and technical ties had not been established over the years.

The SCO Agreement on Scientific and Technical Cooperation is of general and declarative nature. Unfortunately, after the conclusion of this agreement, it is impossible to confirm integration and cooperation in science between all SCO members. Undoubtedly, many obstacles and factors have a negative impact on the intensification of international scientific and technical cooperation within the SCO framework: (1) geopolitical tension, (2) sanctions war; (3) border, religious, ethnic conflicts; (4) uneven socio-economic development, (5) the condition of science in the SCO member states, and much more.

To a great extent, the development of international scientific and technical cooperation between Russia and foreign partners is constrained by (1) the obsolete Federal Law No. 127-FZ of 23.08.1996 "On Science and State Scientific and Technical Policy" (Russian Federation 1996), (2) the lack of unified legal approaches and the concept of international scientific and technical cooperation, and (3) bureaucratic and restrictive rules regarding scientific and technical cooperation. In this regard, Russia still lacks a clear legal basis for developing scientific and technical, and other links between its constituent entities and border regions of foreign countries.

Simultaneously, it is clear that mutual understanding between different nations, development of economic, cultural, and political contacts is possible based on sustainable and effective scientific and technical relations between states. Science is a serious tool of soft power in the modern world that ensures peace and understanding between nations by neutralizing international tensions and conflicts. As one of the modern types of diplomacy, scientific diplomacy can have a positive impact on the integration processes of the SCO member states (Bukalova 2018).

In this regard, further development of the SCO in the field of science and technology implies intensification of international scientific and technical cooperation, creation of a necessary legal and institutional framework for intensification of international scientific and technical ties.

## 2 Materials and Methods

The research uses comparative-legal and formal-legal methods to study ways of scientific and technical cooperation development. The comparative-legal analysis method aims to compare legal regulation of scientific and technical cooperation in the SCO member states. The formal-legal analysis method interprets legal acts that regulate the development of the legal and institutional framework for building mutually beneficial international cooperation in science and technology.

The goal to intensify international scientific and technical cooperation within regional international organizations, including the SCO, is defined by the Concept of International Scientific and Technical Cooperation of the Russian Federation (Government of Russian Federation 2019). The mentioned document defines the objectives of cooperation in the framework of regional international organizations as follows:

- Harmonization of norms, rules, and coordination of scientific and technical activities to ensure the goals, objectives, and priorities of international scientific and technical cooperation of Russia and implementation of the global scientific and technical agenda related to overcoming the "Big Challenges;"
- Initiating and implementing international scientific and technical programs and projects with the participation of Russia.

One of the areas of intensification of international scientific and technical cooperation within the SCO framework can be the establishment of a special commission or a separate council for international scientific and technical cooperation.

The existing working group is not a permanent institution and cannot provide a systematic and continuous line of scientific and technical contacts. The working group meets once a year. "The Action Plan for Implementation of the Agreement between the Governments of SCO Member States on Scientific and Technical Cooperation for the Period from 2016 to 2020" has been approved to implement the SCO Agreement on Scientific and Technical Cooperation. Nevertheless, we have to point out that the SCO working group has not yet brought tangible results.

The competence of such a commission or a council of the SCO can include such issues as:

- Coordination of international scientific and technical cooperation of the SCO states;

- Development of the SCO documents and model legal acts in the sphere of science and technology (e.g., export of scientific and technical products, intellectual property protection issues, creation of scientific collaborations, special scientific and technical and innovation zones, etc.);
- The SCO Education and Science Ministers' meeting organization;
- Identification of relevant scientific topics of basic and applied research for the SCO member states based on consultations with the scientific community;
- Initiation of competitions for research and project funding;
- Conduct of analytical studies for decision-making by the SCO governing bodies.

Similar councils and commissions have been established under other international organizations, such as the Eurasian Economic Community (EurAsEC), the European Union, the CIS, the BRICS, and others. For example, a high-tech center was established as part of the EurAsEC in 2009. The BRICS National Research Committee has been in operation since 2012 (Kniazev 2020).

From an institutional standpoint, such an SCO commission can work closely with an expert council of representatives of the academic community of the SCO states. The Expert Council can become an expert and analytical institute that will generally accompany the work of the SCO bodies in all areas of international cooperation.

Another significant way to intensify scientific and technical ties in the SCO can be a model law on international scientific and technical cooperation for the SCO member states, which can provide a legal framework (principles, forms, directions of cooperation, issues of export of scientific and technical products, protection of the results of intellectual activity, the regime of select areas of scientific and innovation character, etc.).

There is no doubt that the essential condition for successful scientific and technical relations is the existence of common research topics that affect each of the SCO member states and are of universal importance. The identification and support of research on common scientific issues for the SCO member states will contribute to the scientific dialogue.

Thus, it is of universal importance for the SCO member states to study the causes and forms of battling terrorism and extremism as one of the most severe threats to national and international security. The definition of joint subjects is possible based on permanent scientific contacts at the level of the academic community of various countries in the form of scientific conferences, exhibitions, official visits, and competitions to order common subjects of scientific research.

Creating a special fund to support research and innovation or joint international scientific contests by national foundations of SCO member states can play a significant role in enhancing international scientific and technical cooperation among the SCO member states. Regardless of the organizational form, support for scientific and technical cooperation of the SCO member states can take shape in the following formats:

- Research grants to international collaborators;
- Funding mega-science projects;
- Funding scientific conferences for the exchange of scientific and technical success;

- Private and public sector investment in research and innovation;
- Supporting academic mobility of scientists and students to exchange experience and develop scientific contacts within the SCO framework.

Finally, intensification of international scientific and technical cooperation in the SCO is possible by concluding a new, more specific agreement on scientific and technical cooperation. Such an agreement can envisage the mentioned above legal and institutional forms and instruments of scientific and technical cooperation:

- The establishment of a permanent commission on scientific and technical cooperation;
- The establishment of the SCO Center of Expertise and Analysis;
- The establishment of a special fund to support science and innovation for joint SCO scientific and innovation projects;
- Addressing issues of intellectual property rights protection;
- Legal regulation of exporting scientific and technical products and technologies;
- Simplified procedures for academic mobility and employment of foreign scholars.

## 3 Results

Consequently, in order to improve the efficiency of international Eurasian scientific and technical cooperation within the SCO, we propose the following measures.

First of all, it is crucial to establish commissions and councils for international scientific and technological cooperation to address organizational issues.

Second, the development of a model law to provide a uniform legal framework for regulating science and technology in the SCO member states is necessary.

Third, it is advisable to establish special funds to finance research and support innovation development.

Fourth, it is needed to shift the topic of research to universal problems specific to each of the SCO member states in order to build a multilateral and interested dialogue.

Finally, all the above mentioned organizational, legal, financial, and scientific nature measures should be put in writing by adopting a new, more specific agreement on scientific and technical cooperation within the SCO framework.

## 4 Discussion

The issue of international scientific and technological cooperation is often covered in the scientific literature. As a rule, the scientists unanimously recognize the importance and effectiveness of the development of science and education in the framework of interstate organizations.

For example, Kiselev notes the priority of the principle of interested participation in international cooperation—the interest of the state and scientists in conducting international scientific and technical research (Kiselev 2014).

Scientific works on the participation of Russian scientific organizations in international scientific and technical activities emphasize the importance of financial support for joint scientific research and illustrate the widespread use of this measure in other integrated associations (Pikalova et al. 2017). For example, the BRICS is implementing international competitions for research grants under the auspices of the EurAsEC (Shugurov 2019).

## 5 Conclusion

In conclusion, we declare that the development of international scientific and technical cooperation of SCO is only possible through the implementation of many legal (a new agreement on international scientific and technical cooperation, a model law on international scientific and technical cooperation, a new law on science), institutional (establishment of a permanent commission on scientific and technical cooperation), and organizational and financial measures (establishment of a fund for the support of science and innovation, organization of international competitions for granting).

**Acknowledgements** The work was carried out with the financial support of the Russian Foundation for Basic Research (scientific project No. 18-29-15011 "Principles, sources and peculiarities of legal regulation of international scientific and technical cooperation and international integration in the field of research and technological development in Russia and foreign countries of the Shanghai Cooperation Organization").

## References

Bukalova S (2018) Scientific diplomacy: the essence and role in the system of international relations. World Policy 2:95–103

Government of the Russian Federation (2019) The Concept of International Scientific and Technical Cooperation of the Russian Federation was approved by the Government of the Russian Federation (8 Feb 2019 No TG-P8-952). The Government of the Russian Federation, Moscow, Russia

Kiselev VN (2014) International scientific and technical cooperation of the Russian Federation: Brief review and development issues. SpetcKniga, Moscow, Russia

Kniazev P (2020) Partnership for the future. Accessed http://www.nkibrics.ru/posts/show/5e7a128062726974ca490000

Pikalova EG, Nasybulina AV, Sokolov AG (2017) Methodological guide for participation of Russian scientific organizations and Universities in international scientific and technical activities. National Research University Higher School of Economics, Moscow, Russia

Russian Federation (1996) Federal law on science and state science and technology policy (23 Aug 1996 No. 127-FZ). Collection of Legislation of the Russian Federation, Moscow, Russia

Shanghai Cooperation Organization (2013) Agreement between the governments of the Shanghai Cooperation Organization member states on scientific and technical cooperation. Bishkek, Kyrgyz Republic, Collection of Russian legislation

Shugurov MV (2019) International dimension of scientific and technological integration of the Eurasian economic community in the context of sustainable strategy: political and legal framework. Lex Russica 8(153):148–163

# Human Capital in the National Security System of the Russian Society

**Maria I. Cherepanova, Svetlana G. Maximova, and Daria K. Scheglova**

**Abstract** The purpose of the research is to analyze the socio-cultural component of the human capital of the population of the border territories of modern Russian society in the context of the national security system of the country. The article describes the value attitude to health depending on the gender, social status, educational status, financial status, and socio-cultural experience of the individual. The basic components of depopulation processes in the regions are analyzed, including the level of alcoholism, tobacco Smoking, drug addiction, etc. The components of depopulation processes, the specificity of their causes and existence are presented. The complex nature of the demographic crisis in the Russian border region is described. Conclusions are made about the reduced reserves of life-saving behavior in the regions. The socio-demographic groups of the population that could become carriers of progressive trends related to the growth of awareness and the benefits of self-preservation behavior and a healthy lifestyle are identified. It is proved that the significance of health correlates with the significance of life as the highest ontological value. Maximum health value is the main protective factor. Anti-vital behavior patterns are based on low values of the vital value of health and life. One of the key factors in the formation of a new life-saving behavior is the high level of responsibility of the population for their life, preservation, and health promotion. It is the high level of responsibility of the person for their life that stimulates the growth of the importance of taking care of their health. A significant part of the population of the studied border regions demonstrate an instrumental, consumer-oriented attitude to their health. At the same time, every second resident differs in the presence of chronic diseases and an insufficient level of General health. Thus, the background for the formation of ineffective patterns of self-preserving behavior and autoaggressive tendencies has been identified in the region. Human capital, its demographic component, and the improvement of its reproductive mechanisms will contribute to the formation and development of a modern society in the Russian border area.

---

M. I. Cherepanova (✉) · S. G. Maximova · D. K. Scheglova
Altai State University, Barnaul, Russia
e-mail: cher_67@mail.ru

 
S. G. Maximova et al. (eds.), *Advances in Natural, Human-Made, and Coupled Human-Natural Systems Research*, Lecture Notes in Networks and Systems 234,
https://doi.org/10.1007/978-3-030-75483-9_50

**Keywords** Social security · Demographic security · Human capital · Health · Value of life · Self-preserving behavior · Demographic crisis · Population reproduction

## 1 Introduction

The multi-valued scientific discourse of the concept of "national security" continues to be supplemented with new social content.

Differential types of social security include objective indicators, such as the need of social institutions, groups, and individual social actors in a set of systemic measures to ensure safe life. The ultimate goal is to preserve unexpendable resources and fundamental human values.

Protection of life, health, and social well-being become the primary goals of the social security system.

The essence of such a system consists in creating effective mechanisms for regulating the life and socialization of the population.

The central component of social security is the demographic situation in the country, the system of which determines the specifics of demographic security.

In modern Russian society, systemic and specific mechanisms for the deterioration of the demographic situation are relevant.

The constant prevalence of mortality rate, the high mortality rate of the young working-age population, and the high percentage of unnatural exogenous deaths in the overall mortality structure are complex and distinguish Russia from other countries (Vishnevsky 2015).

Many researchers call the current situation a demographic catastrophe that provokes new civilizational challenges to the population of Russia and the future development of the entire country. Every year since 2010, the natural population decline in Russia has exceeded one million people. According to researchers' forecasts, this level may double by 2025 (Shklyaruk 2015).

Therefore, the fundamental priority of government policy should be to ensure the systemic social security of the functioning and reproduction of human capital.

Social security implies being a comprehensive system of measures aimed at social actors' life stability and efficiency.

We support Russian researchers' view that human capital is determined by the totality of knowledge, educational and professional potential of the entire population of the country, and the individual spiritual and psycho-physical components (Iontsev 2014).

Human capital, the quality of life of the population, the effectiveness of education, health, and culture systems form the basis for ensuring the demographic security of the country.

Indicators of destabilization of the foundations of human capital are such factors as (1) the tendency to deteriorate demographic indicators and indicators of national health, (2) the decrease in the quality of education and science, and (3) the growth of social and economic differentiation of the population. The mentioned indicators reduce the systemic quality of life of the Russian population.

The demographic crisis manifests itself in the population aging (significant structural changes in the age composition of the population of the country). In particular, citizens over 65 years of age will exceed more than half of the population of the country by 2030 (Iontsev and Magomedova 2015).

Population aging is a global issue. It is noteworthy that this process is functioning at the fastest pace in Russia in particular. For instance, the percentage of older persons is about 25% in Germany and almost 30% in Japan (Iontsev 2015).

Furthermore, Russia is characterized by increased mortality and reduced life expectancy in older age groups.

Another negative tendency of the human capital deterioration of the Russian population is associated with the rapid aging and death of the economically active population. Researchers predict a noticeable decrease in the percentage of young and non-disabled citizens by 2030. The issues of young people's attitude to health problems, fundamental values, healthy lifestyle, and motivation for a long and productive life become relevant.

The presented tendencies cause increased tension in the regional and national labor market, reducing the speed of economic development of society as a whole. The labor shortage is rapidly increasing, creating new challenges and risks for Russian society.

In general, the Russian scientific community notes the progress of depopulation processes that are steadily changing the socio-demographic characteristics of the population for the worse and ensuring a decrease in the quality of the human capital.

According to the statistics provided by the World Health Organization [WHO] in 2018, Russia ranked 97th globally in terms of population health (Coleman 2014).

Apart from that, the ratings of alcohol, tobacco, and drug consumption in the country are increasing. Simultaneously, the analysis of the index of environmental efficiency and rational use of natural resources provided Russia with only 106th place in the world (Demeny 2014).

## 2 Materials and Methods

The research information base included the results and materials of several sociological studies conducted in Russian border regions from 2009 to 2020 (Maximova et al. 2019).

We conducted the study in Altai Krai, the Novosibirsk region, Krasnoyarsk Krai, the Republic of Buryatia, the Kaliningrad region, Stavropol Krai, etc.

The research focused on analyzing various social risks that determined the tendencies of change and the quality of human capital in Russian border regions. The total sample of the study included approximately 4,200 people.

The paper presented a fragment of the results of sociological studies of the interdependence of health values with the social and behavioral characteristics of the life of the population of Russian border regions (Maximova et al. 2019).

## 3 Results

Depopulation processes are especially characteristic of certain Russian regions, which traditionally are territories with a low level of socio-economic development and infrastructure, and a large percentage of the rural population, high mortality rate, etc. Unfortunately, Russian border regions can be classified as such. These tendencies are multidimensional and complex, including, precisely, the economic aspect of the development of modern society. However, despite the importance of financial investments in increasing the effectiveness of Russian health care, economy, and social welfare of the population, modern researchers point out "specific social archaisms," which are socio-cultural characteristics of the behavior of Russian citizens of lacking "life-saving behavior" (Vishnevsky 2015).

The assessment analysis of their health significance shows that the highest level of health is attributed to non-rural citizens, as opposed to the rural population. Sex groups also differentiate the value of one's health. Thus, among the population of both sexes, the same number of men and women have the lowest and the highest rates of their health self-evaluation. However, the proportion of women linearly prevails over the proportion of men in the medium–high level of health self-evaluation. The average woman values health more, and, therefore, life than the average man. At the same time, there are "perfectionists" and "outsiders" in both sex groups. We can predict a certain tendency for a decrease in the life-saving behavior of women. Perhaps this is due to problematic gender trends in the modern world associated with the masculinization of women and the feminization of society.

The educational level of the population stimulates the growth of the value of health. Its maximum value prevails among people with higher professional education, which is eight times more than among people with primary general education and one and a half times more than among graduates of vocational schools and colleges. A similar trend is associated with the status and nature of the professional activity. Health is most significant for professional workers with higher professional education. Students of vocational schools, colleges, and unemployed people value health at a low level. At the same time, a certain number of highly skilled professional workers appear to value their health highly. Perhaps, a new group of people emerges in the Russian society which aims at a positive lifestyle and fundamental values. People who grew up and socialized in a complete family with both parents are more focused on taking care of their health than divorced, those in a civil marriage, etc. The economic context of life-preserving behavior in Russian border regions is that the highest assessments of the importance of health, and, hence, the focus on a healthy lifestyle, prevail among the behavior of those who have an average level of material well-being. The significance of health is notably at a lower level for people with severe financial difficulties. The social background of life-preserving behavior in a region is determined due to the prevalence of different moods and the well-being of the population. Health is most important for people who feel confident and optimistic about the future or feel relatively normal. The group with minimal health ratings often includes people who experience apathy, indifference, intense emotional stress, and irritation. Only every tenth resident can evaluate their current health as "very good" in the studied regions.

At the same time, we reveal the reverse pattern—the value of health is increasing due to its general deterioration. People begin to take care of their health and value it only when the losses are significant enough, which confirms the low focus of the population on prophylactic measures for health improvement. The revealed relationship between the health value of the population of the regions and the likelihood of addictive behaviors, including alcohol and drug abuse, is apparent. Thus, in the group of people who use drugs and alcohol, 25 times fewer people evaluate their health as highly as possible. The psychological, personal, and behavioral levels can hardly be overestimated in forming a new life-preserving behavior because the importance of health correlates with the significance of life as the highest ontological value. There is no suicide risk among people who rated the value of health as highly as possible.

On the contrary, a suicidal risk is present in a group with low values and a group of those who have difficulties evaluating their health. One of the critical aspects of the new life-preserving behavior is the high level of responsibility of the population for their lives and the preservation and promotion of health. It is the high level of responsibility of a person themselves for their life that stimulates the growth of the importance of taking care for the health. In the group of people who believe society and the state should be held responsible for the citizens' health and life, there are statistically fewer maximum evaluations of the significance of health. The presented results expose low reserves of life-preserving behavior in the regions and also identify groups of people who can become carriers of progressive trends characterized by increased awareness and benefits of self-preserving behavior and a healthy lifestyle.

## 4 Discussion

Assessment of the quality and potential of human capital requires an analysis of the reproductive health of the population of the border regions.

According to our research, more than half of the population (about 70%) shows significant loyalty to abortion and sex outside of marriage. The mentioned trends are even more pronounced in the younger groups of the Russian population. Moreover, a tenth of the population of Russian border regions admits the presence of homosexual relationships. Adding to that, a quarter of the population is quite loyal to the issue of prostitution, which is a cause for sexually transmitted diseases, including AIDS. The factors mentioned above confirm the extremely negative tendencies of the reproductive potential of the country.

Every tenth resident of the border regions can implement self-oriented aggressive behavior patterns, such as self-harming, making demonstrative and genuine attempts at suicidal behavior.

Modern society is called the risk society. The overwhelming majority of the population is tolerant to risk. People choose extreme sports and hazardous professions, significantly risking their health and life.

The prevalence of various types of addictive behavior is a determinant of the low quality of human capital in Russian border regions. More than half of the population is tolerant to the consumption of alcoholic beverages. Low-alcohol beverages are consumed in significant quantities by approximately 80% of the population, increasing the risk of spreading beer alcoholism. It mostly relates to the younger groups.

For instance, a third of young people tend to consume energy cocktails. A significant indicator is the frequency of alcohol consumption, ranging from three times per month to one time per week and daily for every twentieth young person in the regions.

Simultaneously, the officially recorded level of alcoholism in the country does not always represent the full picture. The results of surveys are characterized by a high level of stigmatization of the mentioned phenomenon.

The motivation of young people to consume alcohol reflects socially significant spiritual and cultural components of human capital. For example, a quarter of the young population of the border regions uses alcohol to improve mood, improve communication quality and facilitate relationships with loved ones. Every third resident of the border regions uses alcohol to relax and relieve tension, ten percent—to overcome fears and anxieties. However, most respondents consume alcohol to preserve traditions, which emphasizes the national characteristics of alcoholism in the country.

The emotional background of human capital is reflected in indicators of the prevailing social mood. The results show that half of the population is characterized by a reduced level of mood, the predominance of uncertainty about the future, anxiety, and pessimism. To compensate for these conditions, the population often resorts to alcohol and other drugs.

The potential risk of drug addiction is typical for 5% of the population of Russian border regions. Every fifth resident of the border regions abuses sleeping pills, and every second uses tobacco products.

The analysis of their self-evaluation of the psycho-physical state and health level shows that only a tenth of the population of the border regions has perfect health and high psychophysiological potential. Every third resident demonstrates good health. More than half of the population has average, not very good health and reduced psychophysiological potential. More than that, every tenth resident of the border regions complains of an extremely low health level. Cardiovascular diseases, spine diseases, neurological disorders, and mental pathologies predominate among the population of the studied regions.

The revealed low level of health of the population is primarily due to the spread of consumer attitudes of the population to their psychophysiological potential and low readiness for a healthy lifestyle. Thus, there is a decrease in the regions in the ontological value of health, ineffective behavioral strategies concerning health and life, and a tendency for increasing self-oriented aggression.

Simultaneously, the formation of a robust human capital of the country should be based on favorable demographic trends, reproduction of a healthy and active population, the promotion of optimistic reproductive attitudes, etc.

The population number is becoming one of the most important indicators of the effectiveness of the government in a broad sense. Current examples are China and the United States, in contrast to the countries of Europe and Russia.

According to the demographic statistics, Russia's share of the world population decreased from 4.1% in 1950 to 2.0% in 2014 and is predicted to be approximately 1.2% of the world population in 2050 (Lokosov and Rybakovskiy 2014).

The crisis of human capital becomes a fundamental condition for developing an effective society in Russia and its regions. These trends increase in the border regions, namely the Far Eastern, Siberian, Ural, and Northwestern Federal districts.

It is crucial to prevent unnecessary depopulation as a result of natural processes and the process of building an effective self-preservation potential of the country.

## 5 Conclusion

The formation of a robust human capital requires a systematic improvement of all social and economic foundations of the well-being of the population. It also requires the development and promotion of positive life-saving strategies among the Russian population.

It is important to overcome the crisis of socio-cultural attitudes many representatives of the population that reduce the value of life and health in favor of the immediate needs of the individual and society. It is necessary to create motivation for long and socially active life, especially in the younger generation.

Despite the inertia of adverse demographic trends of the Russian population, the task of optimizing human capital is quite feasible.

It is important to combine the efforts of the government, society, primary social institutions, and individual social actors in order to solve the mentioned issue.

It is essential to understand the urgency of systemic actions to increase the birth rate (at the level of 1.9–2.0 healthy children per woman). It is also crucial to reduce the mortality rate (primarily by avoiding preventable losses).

An urgent task is to form citizens' motivation for a spiritually and physically healthy and prosperous human life.

Human capital should become the most important value, which will be the key to the national security of Russia and its regions.

**Acknowledgements** The publication was performed as part of the state task of the Ministry of science and higher education of the Russian Federation FZMW-2020-0001 "Human capital, migration and security: transformation in the new migration environment in Central Asia."

## References

Coleman D (2014) Europe at the cross-roads. must europe's population and workforce depend on new theories and concepts. international migration: ten years after Cairo. Scientific series International Migration of Population: Russia and the Contemporary World. MAX Press, Moscow, Russia

Iontsev VA (2014) Demographic aspects of the stagnation of the Russian economy. In: Magomedova AG (ed) Interdisciplinary research society's economy. MAX Press, Moscow, Russia, pp 18–28

Iontsev VA, Magomedova AG (2015) Demographic aspects of human capital development in Russia and its regions. Regional Economy 3:89–102

Lokosov VV, Rybakovskiy LL (2014) Migratory processes in Russia. Ekon-inform Publ, Moscow, Russia

Maximova S, Surtaeva O, Cherepanova M (2019) Migration policy as a factor in ensuring social security: expert opinion in cross-border regions of Russia. Advances in social science. Education and Humanities Research 364:737–741

Shklyaruk VY (2015) Formation of self-preserving behavior in the conditions of Russia's transition to expanded population reproduction. MAX Press, Moscow, Russia

Vishnevsky A (2015) Time of demographic changes. Publishing house of the Higher school of Economics, Moscow, Russia

# Interdependence Between Social Security and Social Exclusion of Population in the Altai Krai

**Olga V. Surtaeva, Svetlana G. Maximova, and Daria A. Omelchenko**

**Abstract** The relevance of the research topic lies in the transformation processes in the sphere of social safety which occurred in recent decades in the Russian regions. Scientific implication is based on the lack of knowledge about the relationship between social security and social exclusion of the population living in Russian regions. The paper aims to reveal and analyze the peculiarities of social exclusion of the population and their dependence on subjective perceptions and objective conditions of security in one of the Russian border regions—the Altai Krai. Based on the data from the sociological survey conducted in 2019 in the Altai region (n = 621) and two-step cluster analysis and Mann–Whitney U-tests, the authors reveal that the social well-being of population of the Altai Krai is not homogeneous and could be divided into three main vectors. The first vector is formed by coherent evaluations of objective conditions and subjective feelings of social security. The second vector is represented by subjective perceptions of social security. Finally, the third vector is represented by objective evaluations of the conditions responsible for social security in the region. These three vectors determine the peculiarities of social exclusion of population of the Altai Krai and give rise to their unique configuration.

**Keywords** Social security · Social well-being · Border region · Altai region

## 1 Introduction

Nowadays, many scholars investigate different aspects, kinds, and levels of security. Defining the term "security," scholars often refer to the opposite notions—"risk" (Luhmann 2007; Yanitsky 2003), "threat" (Alekhnovitch 2001), or "danger" (Serebryannikov and Khlopiev 1996), the relationship of which is determined by the authors' position and chosen research methodology. So far, the range of approaches dealing with security and its integral component, social security, is very extensive, which impedes efforts to outline their frameworks and contents. At the same time,

O. V. Surtaeva (✉) · S. G. Maximova · D. A. Omelchenko
Altai State University, Barnaul, Russia
e-mail: bubuka_s@mail.ru



S. G. Maximova et al. (eds.), *Advances in Natural, Human-Made, and Coupled Human-Natural Systems Research*, Lecture Notes in Networks and Systems 234,
https://doi.org/10.1007/978-3-030-75483-9_51

social security includes positive and value-related features of social relations and involves ensuring the sustainable development and social well-being of the citizenry.

In the international scientific community, the meaning of "social security" as it is defined and understood in the Russian scientific works is developed within the theoretical framework of the "human security" (Hentz 2010; King and Murray 2001–2002; Krause 2006; McCormack 2011). Scholars attempt to overcome narrow views and limited understanding of security issues by adapting this term for analytical purposes and inserting into its content not only technical issues related to the provision of a secure country, but also the identification of potentially secure social development, moral and ethical aspects of security, and the role of the government in ensuring humanitarian impact, creating secure conditions, guaranteeing the stability of the country, and protecting citizens.

The state of social security depends on objective factors—real threats and imperfections in the social system, as well as on subjective perceptions of personal security by the members of society. Social security is deeply affected by the level of wealth and prosperity in various areas of public life. It is strictly related to the level of social exclusion of citizens: the level of social inclusion will be satisfactory in satisfactory conditions of indicators of social security.

Hence, social security is one, but not the only significant factor of social well-being that can be analyzed through the peculiarities of the social exclusion of the population. However, more accurate estimates require complex research, relying on multiple indicators measuring a greater number of areas with manifested social exclusion.

The research aims to reveal the features of social isolation of the population in the Altai Krai, depending on the subjective perception of security and objective conditions of security in the region. Empirically, the research is based on the data obtained during the sociological survey on social security and well-being in the border regions of Russia.

## 2 Materials and Methods

The collection of data was based on the multi-stage sampling with quotas controlling for the parameters of gender, age, and settlement type. The authors calculated the sampling based on the data of the Federal State Statistics Service (Rosstat) (2020) on the number of people aged 15–75 living in the Russian regions. The sample size consisted of 621 respondents from the Altai Krai. The data was gathered with standardized face-to-face interviews at the place of residence.

The quantitative data analysis was performed using *IBM SPSS Statistics 22.0* statistical software. Comparative evaluation of parameters of social exclusion in subgroups of respondents was fulfilled by means of the Mann–Whitney U-test.

The independent variables included the indicator of the subjective feeling of security (coded as "I don't feel secure (0)" and "I feel secure (1)") and the indicator of the objective conditions for security in the region (coded as "conditions are not formed, they are insufficient (0)" and "conditions are well formed, satisfactory (1)").

The values of the indicators describing the objective conditions were calculated based on the results of two-step cluster analysis, which revealed two clusters of respondents: (1) people believing that conditions for security in the Altai Krai are rather satisfactory (63.4%) and (2) people considering these conditions insufficient (36.6%). Predictors of cluster analysis represented sixteen 10-point scales characterizing the efficiency of security conditions in the Altai Krai. These scales are as follows:

1. Responses to natural disasters (earthquakes, forest fires, inundations, hurricanes, etc.);
2. State of the justice system, its independence and fairness;
3. Access to high-quality and high-tech medical care;
4. Efficiency of public security authorities, crime prevention;
5. Protection of economic rights, private property, and business;
6. Reliable banking system and protection of bank deposits;
7. Civic control over government decision-making and implementation of important decisions;
8. Social trust, cooperation, and mutual aid;
9. Effectiveness of inter-ethnic conflict prevention, support for the development of ethnic cultures;
10. Quality of social protection services;
11. Independence and objectivity of mass-media;
12. Efficiency of laws, impartial functioning of the legal system;
13. Access to freedoms of assembly and associations;
14. Access to public transport, good roads, and well-developed transport infrastructure;
15. Quality and safety of food products;
16. Environmental activity, control over water and air pollution.

Dependent variables consisted of indicators of social exclusion (30 indicators) (Table 1) containing statements rated on 4-point scales, where 1 point indicated the option "It is rather not about me," and 4 points—"I deal with this kind of things all the time."

## 3 Results

The comparative analysis revealed many significant differences in the level of social exclusion of the population depending on the subjective feeling of security and evaluations of objective conditions. Further description will be given only for the most important indicators with proven significance.

**Table 1** The Mann–Whitney U-tests results[a]

| Indicators of social exclusion | Evaluations of objective conditions of security (mean rank) | | Subjective feelings of security (mean rank) | |
|---|---|---|---|---|
| | Group 0 (low) | Group 1 (high) | Group 0 (feel unsafe) | Group 1 (feel safe) |
| I have trouble with mandatory payments (public utility charges, loans, etc.) | **305.7** | 280.1 | **333.2** | 243.5 |
| I am always worried about my financial situation | **316.1** | 272.6 | **305.4** | 262.9 |
| Membership in sports or other clubs is too expensive | 251.4 | **308.2** | **297.2** | 268.3 |
| I am always badly (rudely, disrespectfully) treated in public institutions | 256.0 | **307.9** | **311.4** | 259.2 |
| There is a lack of cohesion among my neighbors | **318.0** | 273.9 | **301.9** | 268.2 |
| I am not feeling safe in the neighborhood | 270.3 | **302.2** | **295.6** | 274.9 |
| I have nobody to talk about my personal issues (there are few such people around me) | 245.7 | **316.3** | **314.1** | 260.4 |
| My social contacts are impeded because of health issues | 272.0 | **300.4** | **295.9** | 272.0 |
| Commercial organizations (banks, insurance companies) refuse to provide services to me | – | – | **340.0** | 238.4 |
| I have been mistakenly denied the provision of benefits and services | – | – | **299.8** | 269.9 |
| I often have conflicts with my neighbors, they are causing inconvenience to me | – | – | **311.9** | 261.4 |
| I feel unsafe or anxious when staying home alone | – | – | **316.2** | 256.9 |
| I have payments arrears or outstanding bank loans | 264.0 | **303.5** | – | – |
| I want to change my place of residence in the next two years | 257.3 | **308.8** | – | – |
| I have long been without housing (searching for convenient accommodation) | 267.6 | **302.0** | – | – |
| I have been a victim of crime in the last five years | 263.1 | **306.3** | – | – |
| I feel abandoned or excluded from the society | 276.0 | **298.9** | – | – |
| I have no possibility to relax and have fun or I do it rarely | 273.6 | **299.5** | – | – |

(continued)

**Table 1** (continued)

| Indicators of social exclusion | Evaluations of objective conditions of security (mean rank) | | Subjective feelings of security (mean rank) | |
|---|---|---|---|---|
| | Group 0 (low) | Group 1 (high) | Group 0 (feel unsafe) | Group 1 (feel safe) |
| I do not feel supported by friends and relatives | 273.4 | **299.6** | – | – |
| I have lied and broken rules (laws) to receive benefits (services, help, etc.) | 266.3 | **303.0** | – | – |
| I always forget or do not have time to pay taxes or fines and send the meter readings | 273.9 | **298.6** | – | – |
| I earn money in different ways, including gambling, begging, selling banned substances | 271.3 | **298.5** | – | – |
| I live on a minimum wage and it suits me fine | 271.6 | **298.3** | – | – |
| I am unemployed and live off my relatives, that is fine with me | 269.3 | **300.4** | – | – |
| I wanted to get a loan, but the bank turned down my loan application | **309.8** | 259.3 | – | – |

[a] *Note* Higher average scores are shown in bold characters. *Source* Compiled by the authors

Table 1 shows the average ranks of the indicators of social exclusion of population of the Altai Krai in the groups determined by evaluations of objective conditions and subjective feelings of security.

It was found that eight parameters of exclusion depend on both independent factors. Thus, difficulties with mandatory payments (loans or public utility charges) are associated with the lack of security (the mean rank is 333.2) and unsatisfactory objective conditions of security in the region (the man rank is 305.7). For the same categories of population, it is also typical to be constantly anxious about the financial situation and feel a lack of solidarity in the neighborhood.

People whose evaluations of social security conditions are rather high cannot afford membership in sports or other clubs (the mean rank is 308.2). The same point of view is relevant for people who do not feel safe (the mean rank is 297.2). Residents badly treated in the public institutions usually have higher evaluations of objective conditions (the mean rank is 307.9) but lower level of security feelings (the mean rank is 311.4). The same categories of population feel unsafe among their neighbors, have problems to find someone to talk about their personal issues, and attribute problems with social contacts to health issues.

Let us analyze the differences found only between groups differentiated by the subjective feeling of security. Respondents feeling unsafe more often reported about refusal of services by commercial organizations (the mean rank is 340.0), unjustified or unreasonable refusal of benefits or facilities (the mean rank is 299.8), and conflicts with neighbors (the mean rank is 311.9). This group often feels unsafe and anxious when staying home alone (the mean rank is 316.2), compared to people from Group 1.

Differences obtained in comparing the groups with high and low estimates of objective conditions of security in the Altai Krai were as follows. Respondents who gave lower evaluations of social security conditions were more often denied credit and loans from banks (the mean rank is 309.8). In turn, those with high evaluations reported the following:

1. They had payments arrears or outstanding bank loans (the mean rank is 303.5);
2. They would like to change their place of residence (the mean rank is 308.8);
3. They had long been without housing (the mean rank is 302.0);
4. They had been victims of crime during the last five years (the mean rank is 306.3);
5. They felt abandoned and excluded from society (the mean rank is 298.9);
6. They had few possibilities for distraction (the mean rank is 299.5);
7. They were not supported by relatives and friends (the mean rank is 299.6).

Besides, people satisfied with the objective conditions of security in the region had more often broken rules and laws for some profit (the mean rank is 303.0) and forgotten to pay taxes or fines, send the data of meters (the mean rank is 298.6). Among people from this category of population, it is habitual to earn money through illegal ways (the mean rank is 298.6), live on the minimum wage (the mean rank is 298.3), or stay unemployed and live off family and intimate friends (the mean rank is 300.4).

## 4 Discussion

The majority of researches on social exclusion are focused on considering the issues of groups under greater risk of social exclusion (older persons, women, discriminated minorities, persons with disabilities, etc.). These works are of a rather descriptive nature. They tend to draw attention and raise awareness among decision and policy makers and providers of social services (Maximova et al. 2017). Discussions about factors affecting the extent of social exclusion of the adult population, including those related to the social environment, conditions of socialization, behavior, social experience, social interactions, and social inheritance are rather rare. There are still many difficulties in finding a working definition of social exclusion due to its strong dependence on economic variables (e.g., occupational status or professional belonging). The question about social markers of exclusion is also opened. Considering the existing uncertainty in the social exclusion area, the lack of valid methods of evaluation of different aspects of social exclusion, and the insufficient amount of data for secondary analysis, it is feasible to explore a wide range of attributes acting as indicators of social exclusion.

We applied an extended approach to the analysis of social exclusion, covering different aspects of its manifestations (not only economic, but also social and psychological ones) concerning the state of social relations, connection with the social environment, psycho-emotional parameters, and social practices, whose essential role in measuring social exclusion was explored in our previous research (Maximova et al. 2017). The emphasis was placed on the link between social exclusion and subjective evaluations of security, which shows that unsafe social conditions are often associated with exclusive practices. Even when objective circumstances are not truly dangerous, subjective feelings of security or, rather, its absence, almost always determined by socio-economic conditions, can indicate social exclusion from receiving qualitative social services. Meanwhile, another finding of our research is that social exclusion is a complex phenomenon that is difficult to deal with. Sometimes, direct signs of exclusion are indicative of individual choice strategies, where social exclusion is conscious escape directed against the existing social normative system. Such strategies would receive further deep investigation in our future research.

## 5 Conclusion

Our research allowed comparing the parameters of social exclusion of the population of the Altai Krai depending on subjective and objective evaluations of social security. Moreover, we outlined three vectors in the determination of social security. The predominance of some vectors in the evaluation of the population creates a unique configuration of peculiarities of social exclusion in the region.

The first vector was determined by coherent evaluations of objective security conditions and the subjective feelings of population. High evaluations of these parameters were related to the positive trends in social well-being and the absence of social exclusion.

The second vector was determined only by the factor of subjective evaluations. High positive values were associated with the absence or rare manifestations of social exclusion, and vice versa.

The third vector was marked with somewhat paradoxical findings. It was found that high estimations of the objective conditions of social security in the region were inherent to those who were, by virtue of their way of life and thinking, excluded from the social mainstream, discriminated and suffered from a lack of resources, but often—willingly, used illegal ways to earn a living.

**Acknowledgements** The publication was performed as a part of the state task of the Ministry of Science and Higher Education of the Russian Federation FZMW-2020-0001 "Human capital, migration, and security: Transformation in the new migration environment in Central Asia."

## References


Alekhnovitch SO (2001) Mechanisms of realization of the regional security in the paradigm of the federative state. Dissertation of Candidate of Political Sciences, Moscow State University, Moscow, Russia

Federal State Statistic Service (Rosstat) (2020) Structure of resident population (as of January 1, 2020) by gender and age groups. https://fedstat.ru/indicator/43219

Hentz JJ (2010) National and human security in Sub-Sahara Africa. Orbis 54(4):630–643

King G, Murray CGL (2001–2002) Rethinking human security. Polit Sci Q 116(4):585–610

Krause K (2006) Towards a practical human security agenda. Geneva Centre for the Democratic Control of Armed Forces, Geneva, Switzerland

Luhmann N (2007) Social systems. Essays about general theory. Nauka, St. Petersburg, Russia

Maximova SG, Noyanzina OE, Omelchenko DA (2017) Model of social exclusion of persons of elderly age in Siberian regions. Uspekhi Gerontologii 30(4):579–586

McCormack T (2011) The limits to emancipation in the human security framework. In: Chandler D, Hynek N (eds) Critical perspectives on human security cycle. Routledge, London, UK, pp 99–113

Serebryannikov V, Khlopiev A (1996) The social security of Russia. Institute of Socio-Political Research (ISPR FCTAS RAS), Moscow, Russia

Yanitsky ON (2003) Sociology of risk: key ideas. Universe of Russia 1:3–35

# Social Aspects in the Structure of Civil Aviation Transport Security

**Lev N. Elisov, Nikolay I. Ovchenkov, and Igor B. Guberman**

**Abstract** Security is one of the most important criteria for the effectiveness of the operating activities of civil aviation. The field of transport security, which in civil aviation is called aviation security, is associated with solving the issues and problems of protecting objects or subjects of aviation from dangerous and harmful factors of the industry and uncontrolled influence on its activities. Any violations of the functioning of the security system can lead to emergencies, including the death of people. Modern aviation security systems are based on analyzing security threats and adopting appropriate measures to eliminate them. This fact is primarily related to identifying hazards and managing risk factors to ensure the functioning of the aviation system in the established parameters of an acceptable security level. Simultaneously, the focus on the security of people is based only on the study of security threats; it does not consider related problems, such as the protected contingent as a social system. This research is the first attempt in the scientific field of civil aviation to study aviation security issues, taking into account social aspects. The study aims to identify the areas of security implementation that are particularly important for civil aviation.

**Keywords** Social security · Personal security · Transport security · Social aspects of security management

## 1 Introduction

Currently, we regard the issue of ensuring the security of civil aviation facilities as one of the most crucial problems. Therefore, we propose an authentic, modern approach to understanding the functional and essential air transport security components. The

L. N. Elisov (✉) · I. B. Guberman
State Research Institute of Civil Aviation, Moscow, Russia
e-mail: lev.el@list.ru

N. I. Ovchenkov
Yaroslavl State University Named After P. G. Demidov, Yaroslavl, Russia
e-mail: ovchenkov@electronika.ru



S. G. Maximova et al. (eds.), *Advances in Natural, Human-Made, and Coupled Human-Natural Systems Research*, Lecture Notes in Networks and Systems 234,
https://doi.org/10.1007/978-3-030-75483-9_52

analytical study of functionally new terminology that was not used in the study of air transport security before (but directly related to it) serves as the basis for this approach, such as (1) *indivisible security*, (2) *personal security*, and (3) *unified security*. We consider these terms from the standpoint of sociology, which represents the target subject of the aviation security implementation as a social category.

In modern society, security is an essential condition for functioning; it is a criterion for ensuring the operation of complex systems. The conditions of its provision depend on fulfilling professional duties, the comfort of an individual and various groups of the population; employees and service personnel.

"Security differs from other concepts in its ambiguity; it has the characteristic of a low level of risk, giving the subject (including both the individual and the society) a sense of security. Security also determines the stability of the functioning of other entities, objects, or their systems. In this state, any complex system does not experience any influence of external and internal factors leading to the deterioration of this system. Security provides the best opportunities for the operation and development of this system" (Elisov et al. 2020).

Security in the field of air transport and the transport sector in general is critical. Insufficient level or lack of security dramatically reduces the demand for transport services and threatens the existence and development of the industry.

Aviation security is based on identifying threats, analyzing them, and considering the impact mechanisms that eliminate threats and their consequences. In this case, there is a contradiction between the goal and the means to achieve this goal. When forming a security system, all aspects related to threats are studied, and the security subject (the human component) is practically not considered. The functioning of a complex system consisting of passengers, employees, and support personnel largely determines the requirements for the security system. Hence, we face the task of (1) studying the human component as a security subject from the point of view of sociology; (2) identifying behavioral aspects and links with security; and (3) formalizing management procedures.

## 2 Materials and Methods

During the study, we examine the definition of the term *personal security* stated in the National Security Strategy of the Russian Federation. This term presupposes the freedom of individuals and the enforcement of their rights, as well as conditions under which the right to life and the physical integrity of the individual are guaranteed.

There is a widespread practice of using the concepts of *unity* and *indivisibility* in international law to implement national security.

Using this terminology from the areas of social and philosophical sciences in civil aviation and studying the terms *indivisible security*, *unified security*, and *personal security* allows us to introduce a new concept: *integrated security*. Integrated air transport security is a state of the air transport system in which the necessary (or acceptable) level of personal security is guaranteed, provided that the entire range of aviation services and works related to them are implemented (Elisov and Ovchenkov 2017).

The presented definition identifies *personality* as the main component; it is considered an element of the subject of aviation security, which is a complex system consisting of passengers, employees, and support personnel. In this case, it is necessary to distinguish two interrelated complex systems: aviation security personnel and security system.

Ensuring an acceptable security level of the *aviation security personnel* system is possible if the requirements put forward by the system itself are met. For this purpose, it is necessary to study the system as a social category.

Sociological research is a system of logically consistent procedures (methodological and practical) united by a common goal.

Sociological research of personnel regarding civil aviation security provided us with information about the probable functioning of the system and its elements in the context of the target function of the object of the aviation transport system; it establishes functional links for the interaction of personnel and the security system. Based on this information, the requirements for the aviation security system are formed. The functioning of the personnel as a social system in the conditions of emergencies, when there is a threat to the security of the individual, is of particular interest.

Social systems have some peculiarities, such as the complexity of the system of *aviation security personnel.*

Human activity within the social system is an objective process following its laws, since the social environment in which individuals implement their activities is created by themselves.

Sociological research focuses on the study of the dynamics of social processes and phenomena. Social systems in the process of their development are in a dynamically non-equilibrium state. Static states are excluded for them.

Social laws, being trends, cannot be presented in a formalized way. For such systems, one can use classification, clusterization, systematization, structurization, or typification; however, there are specific difficulties in proving the reliability of the results obtained in this case.

Assessing the impact of individual characteristics of a person is a complex task that is difficult to formalize; it has many uncertain solutions, for which heuristic procedures are used.

The exchange of information in social systems occurs in the form of a text (document) or a dialogue (conversation).

The sociological research mainly focuses on the qualitative side of the studied processes.

There are no generally accepted paradigms in studying social systems and objects in the absence of *a single true theory*.

The direction of social research associated with analyzing the patterns of functioning of small groups and their interaction both within and outside the group is crucial for civil aviation.

The study aims to identify some areas of security implementation that are essential for civil aviation.

The research has the following goals:

- Identifying the types of interaction with the security system;
- Determining the methodological techniques for studying security;
- Formalizing the mathematical apparatus of the theory of boundary value problems for setting priorities in ensuring security.

## 3 Results

The central maxim in conducting a theoretical sociological study of personnel on the part of the integrated system when we adopt the concept of *integrated security* is presented in solving the following issues:

1. The scope of application includes the following research subjects:
    - Transport and infrastructure facilities;
    - Ensuring the security of subjects in the event of risks and threats;
    - Using various tools and methodologies to maintain security in the structure and functionality that distinguish the *aviation personnel* system. Since *integrated security* consists of continuous connections, there may be a question about the actual correspondence of the term through the domain of intermediate results based on the speculation of the environment reflecting these connections. At this stage, as a categorization of the studied environment, it is possible to use the apparatus of the theory of the studied field. We equate the definition of *personal security* with the *security field*. In this case, it is logical to introduce new terms, such as *field of threats* and *field of security*. As a result of the interaction of the emerging triad of speculative fields, *integrated security* arises. We regard the definition of *integrated security* as logical and reasonable, since the very definition of the term *security* is conditional. The studies that we conducted in the field of security are also relevant for theoretical concepts (Ovchenkov and Averin 2020).
2. As a result of the research, we conclude that the security level and the validity of identifying the essence of this issue are the methodological methods of understanding the concept of security. The solution to the problem of the optimal level of *integrated security* does not end there. In the future, one should correctly construct and formalize the model of evaluation actions.

We regard the mathematical apparatus of the theory of boundary value problems, which uses systems of partial differential equations, as the most accessible and acceptable method for the formalization of all other methods of field theory. The theory of boundary value problems offers many methods for solving differential equations. Besides, this theory allows one to choose the appropriate method of mathematical modeling. This step is necessary to determine the norms and conditions of implementation in various subject areas. Partial derivatives for differential equations *use* the boundary value problem only in certain domains. In this case, the additional constraint condition must be met in different (boundary or edge) conditions at the considered boundary points (Price and Forrest 2016; Snyman and Wilke 2018).

$n$—the $n$-order for the equation of the boundary value problem is presented in the following form (1):

$$\mathrm{Lin}(y) = f(x),\ \mathrm{U}\mu(y) = \gamma\mu, \quad \mu = 1, 2, \ldots, n \tag{1}$$

where:

$$\mathrm{Lin}(y) = \sum_{v=0}^{n} f_v(x) y^{(v)}$$

The boundary value (mixed) problem for the hyperbolic type of equation is the problem of determining the function $u(x, t) \in C^2(Q_\infty) \cap C^1\left(\overline{Q}_\infty\right)$, which would correspond to the following Eq. (2):

$$\rho\frac{\partial^2 u}{\partial t^2} = \mathrm{div}(p \text{ grad } u) - qu + F(x, t), \quad (x, t) \in Q_\infty \tag{2}$$

primary condition (3):

$$u_{|t=0} = u_0(x), \ \frac{\partial u}{\partial t_{|t=0}} = u_1(x), \quad x \in \overline{G} \tag{3}$$

and the boundary condition (4):

$$\alpha u + \beta\frac{\partial u}{\partial n}|s = 0 \tag{4}$$

Problem for parabolic type equation boundary value (mixed) is the definition of the function $u(x, t) \in C^2(Q_\infty) \cap C^1\left(\overline{Q}_\infty\right)$, $\mathrm{grad}_x u \in C\left(\overline{Q}_\infty\right)$, which corresponds to the following Eq. (5):

$$\rho\frac{\partial u}{\partial t} = \mathrm{div}(p \text{ grad } u) - qu + F(x, t), \quad (x, t) \in Q_\infty \tag{5}$$

primary condition (6):

$$u_{t=0} = u_0(x), \quad x \in \overline{G} \tag{6}$$

and the boundary condition (7):

$$\alpha u_0 + \beta\frac{\partial u}{\partial n} = v(x, t), \quad (x, t) \in S \times [0, \infty) \tag{7}$$

To determine the elliptic type of the equation, we consider the following boundary value problems (Yadav et al. 2015):

- Laplace equation in three dimensions (8):

$$\Delta = \frac{\partial^2 f}{\partial x^2} + \frac{\partial^2 f}{\partial y^2} + \frac{\partial^2 f}{\partial z^2} = 0 \tag{8}$$

- Interior Dirichlet problem: to determine the harmonic in the considered area $G$ function $u \in C(\overline{G})$, which takes the given continuous values $u_{\overline{0}}$ at the boundary $S$.
- Exterior Dirichlet problem: determine the harmonic in the considered area $G_1$ function $u \in C(\overline{G}_1)$, which takes the given continuous values $u_0^+$ and becomes zero at infinity.
- Interior Neumann problem: to determine the harmonic in the considered area $G$ function $u \in C(\overline{G})$, which has a given continuous regular normal derivative $u_{\overline{1}}$ on $S$.
- Exterior Neumann problem: to determine the harmonic in the considered area $G_1$ function $u \in C(\overline{G}_1)$, which has a given and continuous regular normal derivative $u_1^+$ on $S$, equated to zero at infinity.

Identical boundary value problems are formed for the Poisson equation (Elisov et al. 2018; Elisov and Ovchenkov 2017) (9):

$$\Delta_u = -f \tag{9}$$

The statistical apparatus of the theory of boundary value problems comprises a list of mathematical models necessary for the visual implementation of modeling various physical processes in the field theory.

Let us consider the variants of analytical approaches to solving boundary value problems and present them in the form of the following differential Eq. (10):

$$C(\varphi) = p \quad \text{in area } \Omega \tag{10}$$

where:

$C$: linear differential operator.

The solution of the equation must satisfy the conditions of the boundary value problem (11):

$$D(\varphi) = r \quad \text{at the boundary } E \tag{11}$$

In most cases, when solving boundary value problems, the presented analysis methods are limited to the scope of use. In this situation, one should use numerical methods. At the current development stage of mathematical modeling, grid and neural network methods are considered the most effective for solving boundary value problems (Aggarwal 2018; Bruce and Bruce 2017; Elisov et al. 2018; Raschka and Mirjalili 2019).

Taking into account the information above, we can formulate the main directions of interaction between two complex systems: personnel and security systems.

The first direction is related to the system of relations between the aggregate of people forming a cluster of civil aviation passengers preparing for the flight. There are two possible types of interaction with the security system: (1) passengers act as consumers of the services offered by the security system and related to ensuring their security; (2) passengers involved in the flight preparation procedure perform particular actions that accidentally or intentionally violate the rules of behavior in the specified conditions and become a threat to themselves and others.

In the first case, two situations should be considered in a complex system of personnel. The first situation is formed in normal mode conditions, when comfortable conditions of people's behavior should be provided. The sociological study aims to identify the behavioral reactions of personnel in the standard conditions of flight preparation and, on this basis, formulate the requirements for the security system. The second situation is regarded as critical, characterized by dangerous factors, up to the act of unlawful interference. A catastrophic situation is possible as a result of realized threats to the life and health of people. In this case, a sociological study is associated with obtaining the behavioral characteristics of personnel in a critical situation and developing appropriate security system requirements.

In the second case, passengers are considered as a source of threat. This situation requires a more in-depth sociological study that examines the behavioral aspects of the activity and the structural, organizational, psychological, and personal characteristics of the personnel (passengers). As a result, procedures and algorithms should be formed that exclude the development of a critical situation to the extent of a disaster.

The second direction defines the procedures for interaction within the cluster, which unites employees and service personnel in professional activity. Performing professional activity, the personnel acts strictly within the established and approved algorithms. However, for various reasons, deviations from these algorithms may occur, which sometimes can lead to a dangerous situation (i.e., the personnel here acts as a source of threat to the security of the considered object). A sociological study of personnel as a source of security threats can reveal the reasons for certain personnel actions and allow one to take appropriate security measures.

Consequently, the social aspects of personnel activities in the form of some quantitative assessments (expert assessments) are included in the partial differential equations describing the situation in a complex system of personnel, taking into account security threats (boundary value problem). The solution of the boundary value problem is obtained in the format of the specific population distribution of threats over the topology of the security object, which is considered when forming the security system of the security object.

To confirm the results obtained, we conducted an experimental simulation of hypothetical security threats in the form of a boundary value problem using the Russian

Matrix Laboratory [MATLAB] system (Gorbachenko 2011; Sweet 2008). The simulation results confirmed the possibility of solving the problem by the proposed method.

In all cases of conducting sociological research, one must take into account the following factors. The interaction of personnel in any cluster of the system is based on a conflict, which often has a social form. In most cases, conflict theory does not provide an accurate solution to the conflict. The decision is usually made based on decision theory. In this case, a compromise solution to the conflict is obtained on the basis of heuristic procedures.

## 4 Discussion

In a way, the research is groundbreaking, since it is the first attempt in civil aviation to combine security and sociology issues. The presented results indicate a positive effect; nevertheless, it is accompanied by new questions and problems (Geron 2017; Khan et al. 2020).

There is no single true theory in sociology; therefore, there is no single methodology for conducting sociological research and precise research methods due to the absence of a single security theory (Beck 2006).

The proposed method for modeling security threats, including personnel ones, in the form of a boundary value problem is ambiguous (Yadav et al. 2015). The formalization of the problem in partial differential equations requires precise initial data (Kuhn 1970). It is impossible to obtain an accurate formal description of threats until the hypothetical threat becomes real. At the same time, the study of hypothetical threats is crucial for security (Merton 1996). Sociological research also gives a qualitative result. The question arises about the compatibility of approximate source data and exact mathematical expressions. In this case, one should use heuristic procedures. The result of modeling in such conditions is possible. However, the question of acceptable accuracy remains; it has to be solved in each specific case separately.

The results obtained allow us to assert that the symbiosis of sociology and security gives a certain effect in managing aviation security of civil aviation facilities.

## 5 Conclusion

We identified some areas that are particularly significant for civil aviation.

As a source of security threats, personnel is practically not studied, although the human factor in the industry is given sufficient attention. In this regard, it is necessary to formulate sociological tasks and determine the ways and methods of their solution.

Currently, the problem of ways to automate security procedures is critical; nonetheless, the issue of the ratio of the human and technical components is quite difficult. Sociology as science can have a significant impact here.

Aviation security systems are gradually becoming information-based. They use information technology platforms, such as (1) Electronika Security Manager (ESM), a software platform; (2) Physical Security Information Management (PSIM), a software category; and (3) Data Mining (intelligent data analysis). Sociological research can find its place within these systems.

As of today, we continue our research in this area.

## References

Aggarwal CC (2018) Neural networks and deep learning. Springer, Cham, Switzerland

Beck U (2006) Cosmopolitan vision. Polity Press, Cambridge, UK

Bruce P, Bruce A (2017) Practical statistics for data scientists: 50 essential concepts. O'Reilly, Sebastopol, CA

Elisov LN, Ovchenkov NI (2017) Integral security of air transport. Nauchnyi Vestnik MGTU GA 20(6):36–44

Elisov LN, Gorbachenko VI, Zhukov MV (2018) Learning radial basis function networks with the trust region method for boundary problems. Autom Remote Control 79(9):1621–1629

Elisov LN, Ovchenkov NI, Filippov VL (2020) Modeling of potential threats to the security of civil aviation facilities. J Crit Rev 7(18):2620–2627

Geron A (2017) Hands-on machine learning with scikit-learn and tensorflow: concepts, tools, and techniques to build intelligent systems. O'Reilly, Sebastopol, CA

Gorbachenko VI (2011) Computational linear algebra with examples in MATLAB. BHV-Petersburg, Saint Petersburg, Russia

Khan N, Milner D, Marr B (2020) Introduction to people analytics: a practical guide to data-driven HR. Kogan Page, London, UK

Kuhn TS (1970) The structure of scientific revolutions, 2nd edn. University of Chicago Press, Chicago, IL

Merton R (1996) On social structure and science. University of Chicago Press, Chicago, IL

Ovchenkov NI, Averin DV (2020) The problem of the human factor in the field of aviation security and some approaches to its solution. Nauchnyi Vestnik MGTU GA 30:107–116

Price J, Forrest J (2016) Practical aviation security: predicting and preventing future threats. Butterworth-Heinemann, Oxford, UK

Raschka S, Mirjalili V (2019) Python machine learning: machine learning and deep learning with python, scikit-learn, and tensorflow, 2nd edn. Packt Publishing, Birmingham, UK

Snyman JA, Wilke DN (2018) Practical mathematical optimization. Basic optimization theory and gradient-based algorithms. Springer, New York, NY

Sweet KM (2008) Aviation and airport security: terrorism and safety concerns. CRC Press, Boca Raton, FL

Yadav N, Yadav A, Kumar M (2015) An introduction to neural network methods for differential equations. Springer, Cham, Switzerland

# The Problem of Duality of Professional Activity of Airport Aviation Security Personnel

**Dmitry V. Averin, Igor B. Guberman, and Lev N. Elisov**

**Abstract** The anthropogenic factor largely determines the problem of civil aviation security. Transport security is a multidimensional concept that includes various security factors provided by aviation personnel in the relevant areas of professional activity. The activity of personnel in the field of airport security is an essential factor. An acceptable level of airport aviation security can be achieved if the level of the professional activity of the security personnel meets the established requirements. The system of requirements is a set of algorithms and procedures for personnel activities, which is mandatory in strict compliance with the requirements. However, the physical nature of the anthropogenic factor, which is under strict control, does not guarantee absolute compliance with the requirements. Consequently, the professional activities of aviation personnel are not immune to accidental or unintentional errors. This fact leads to the duality of the professional activity of the personnel, which provides an acceptable security level and admits cases of reducing this level in critical situations to a catastrophic one. This issue is crucial for civil aviation today. It requires studying many issues, such as (1) the content of the concept of *anthropogenic factor*, (2) the identification of the concept of *unlawful interference of personnel*, (3) the structure of the professional activity of aviation personnel, (4) a competence-based model of the negative impact of personnel, and (5) the algorithm for correcting professional activity. The methodology is based on structural and logical modeling. The results are presented in the form of recommendations for minimizing the negative impact of personnel, ensuring an acceptable level of aviation security.

**Keywords** Transport security · Anthropogenic factor · Aviation security personnel · Duality of the professional activity of personnel

D. V. Averin · I. B. Guberman · L. N. Elisov (✉)
State Research Institute of Civil Aviation, Moscow, Russia
e-mail: lev.el@list.ru

D. V. Averin
e-mail: averin@gosniiga.ru



S. G. Maximova et al. (eds.), *Advances in Natural, Human-Made, and Coupled Human-Natural Systems Research*, Lecture Notes in Networks and Systems 234,
https://doi.org/10.1007/978-3-030-75483-9_53

# 1 Introduction

Security is a multifaceted concept that characterizes security and a low level of risk to a person, society, or other entities, objects, or systems; it is a state of a complex system when external and internal factors do not lead to deterioration of the system or the impossibility of its functioning and development (Elisov et al. 2020).

Security is the main criterion of civil aviation activity, which determines the advantages of this mode of transport. Achieving an acceptable level of transport security is a rather complicated and multifactorial task, which is solved by many aviation specialists using modern computer technology. The task of ensuring airport aviation security, which protects objects from unlawful interference in their activities, is of particular importance. The professional activity of aviation personnel is crucial in solving this problem.

Ensuring aviation security is implemented in strict compliance with the requirements for each procedure approved within the established procedure. In other words, it is a set of unified algorithms, each of which contains an exhaustive list of elements of the professional activity of personnel sufficient to achieve the target function of ensuring security. Any deviation from the algorithm can be considered unlawful interference, which, as a last resort, can lead to disastrous consequences.

On the one hand, aviation personnel, fulfilling their target function, ensure security and try to achieve an acceptable aviation security level. On the other hand, personnel, being an anthropogenic component in the ergatic aviation security management system, due to human physiology, may deviate from the established algorithm of activity and not necessarily deliberately. This fact leads to the paradox of duality, when personnel simultaneously perform positive and negative functions. In the end, the ratio of positive and negative aspects in the professional activity of personnel ensures an acceptable level of airport aviation security (Sweet 2008).

To maintain an acceptable level of aviation security, one faces the task of studying the structure of the personnel's professional activity and developing a methodology for reducing the negative impact of personnel on security procedures.

Therefore, the study aims to develop methods and means to ensure an acceptable level of aviation security of the transport infrastructure in the conditions of unlawful interference of aviation personnel in the airport's production procedures. We can achieve this goal by (1) studying the concepts of *anthropogenic factor* and *unlawful interference of personnel* to form a new approach to their identification; (2) studying the structure of the professional activity of aviation personnel to design a competence-based model of the negative impact of personnel; (3) developing an algorithm for managing unlawful interference of personnel in the airport's production activities.

## 2 Materials and Methods

We introduce a new concept: the threat of personnel, which is understood as the inadequacy of the professional readiness of the operator and the parameters involved in the security system, determined by the maximum level of psychophysiological parameters of the individual that allow negative events.

Monitoring and investigation of this threat involve (1) the use of mathematical and heuristic methods with their specific adaptations and limitations; (2) subsequent measurement; and (3) analysis in a quantitative format, which simplifies the decision-making process for human resources management.

The first issue we should deal with is the structure of the anthropogenic factor, which is determined by the results of professional activities. Besides, we are interested in negative manifestations or potential threats without considering common situations in production activities.

Figure 1 shows the structure of the professional activity of aviation security personnel.

High dynamics are determined by the *detection–reflection–elimination* formula with a fairly strict time frame.

A large amount of information is associated with the processing of signals from a large number of sensors, which are necessary to register the danger in the form of an abnormal signal, regardless of the form, methods, means, and coordinates of the point of application of the threat (Bruce and Bruce 2017).

The lack of time is associated with the high dynamics of professional activity.

The neuropsychiatric load is caused by a strict framework for performing individual operations according to the specified requirements.

The limitations on correcting deviations are related to the complexity of these procedures, which require adjustments to the software and hardware components of the aviation security system (Geron 2017).

The goal of the professional activity of the personnel (security) implies a high level of responsibility. Otherwise, the activities of civil aviation are meaningless.

At any level of professional training of aviation security personnel, their work does not exclude errors, the reasons for which are specific to aviation security.

The first reason is **the uncertainty of information**. Security is defined by two basic concepts: (1) threats and means of countering them; (2) personnel as the most crucial component. Countering threats is the most effective if they are anticipated. However, the real threat appears unexpectedly. In other cases, we can talk about the potential threats, when one cannot describe their parameters. Everything related to personnel (the anthropogenic component) does not have an exact mathematical description, and the parameters of processes and objects, in this case, are determined by special methods. Hence, the problems solved in the field of security are poorly formalized; they are solved using a heuristic approach, which does not exclude errors.

The second reason is **suboptimal technology** existing due to the information uncertainty (Price and Forrest 2016).

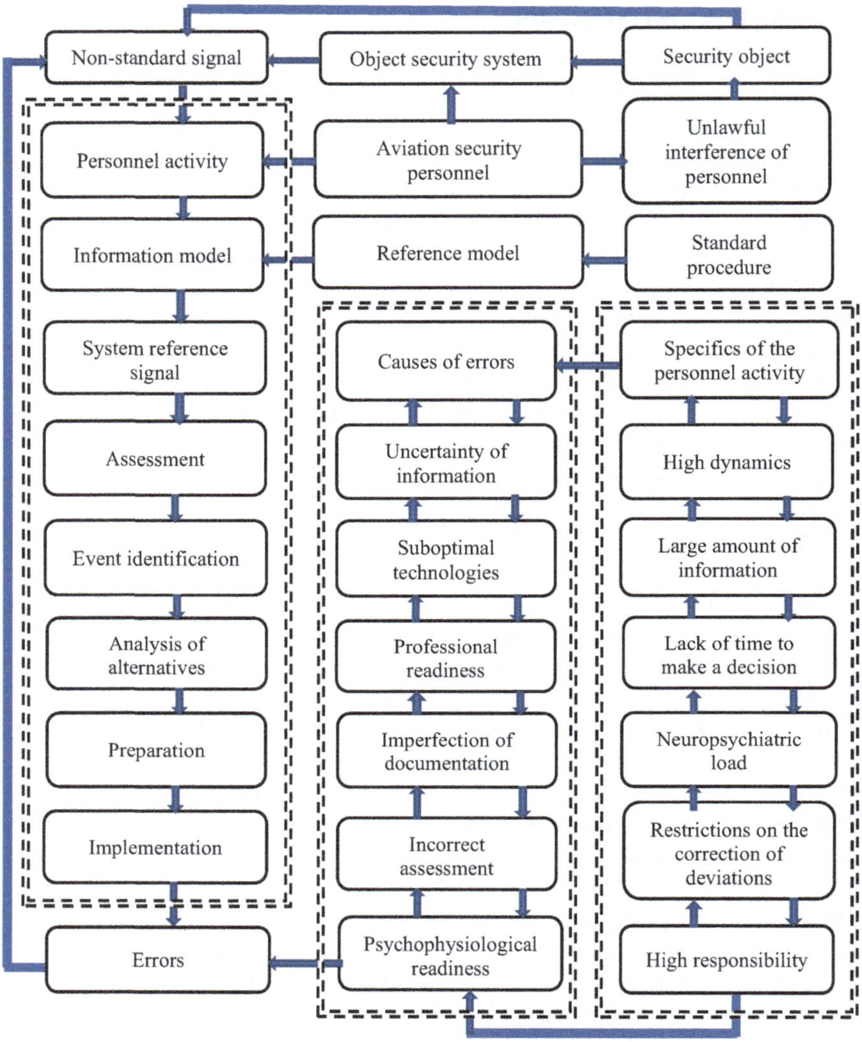


**Fig. 1** The structure of the professional activities of the aviation security personnel. *Source* Compiled by the authors

The third reason concerns **professional readiness**. We distinguish between two concepts: the level of professional training of personnel and the level of professional readiness. Both components should be at an acceptable level. Nevertheless, the level of professional readiness manifests itself in emergencies. In other words, it is the level of readiness to perform one's professional functions at a given moment. Due to some impacts, an acceptable level of training does not guarantee an acceptable level of personnel readiness.

The fourth reason is related to **the imperfection of documentation** resulting from the uncertainty of information and suboptimal technology.

The fifth reason is **the incorrect assessment** carried out using expert evaluation methods and a heuristic approach to describing processes and procedures.

The last reason is connected with **psychophysiological readiness**, which is an integral part of professional readiness. Ultimately, the specifics of professional activity and the causes of errors determine the level of unlawful interference of the aviation security personnel.

The aviation security system focuses on performing its functions according to the formula: *detection–reflection–elimination.* In normal mode, without any security threats, the system prevents hazards by using predefined algorithms. In the event of an abnormal signal, the activity of the aviation security personnel changes significantly, while some component is highlighted, which is an unlawful (illegal) interference of the personnel. Unlawful interference is determined by the specifics of the personnel's activities, which distinguish this activity from all others in the field of security.

The activities of the aviation security personnel are carried out in full compliance with the standard operating procedures [SOP], which reflect the entire set of requirements, the implementation of which guarantees the achievement of an acceptable security level. The procedure (algorithm) of personnel activity is built on this basis (Fig. 2).

In all cases of deviations from the SOP, the non-standard signal appears, indicating a particular situation in the object security system, which can grow into a real security threat. Analyzing this signal, the aviation personnel form an information model in mind to eliminate the deviations that have occurred.

The information model is compared with the reference model developed based on the SOP for various object security procedures.

Some reference signal of the system is generated taking into account the characteristics of the situation. The aviation security personnel (1) assess the situation to identify the event, (2) analyze possible alternatives to the activity, (3) make an appropriate decision, and (4) implements it. None of these stages excludes errors in the personnel activity that determine the content of unlawful interference.

There are many causes of error. However, almost all of them, except for errors related to the psychological and physiological parameters of a person, are determined by the insufficient level of professional readiness of aviation security personnel to perform their functions (i.e., level of training).

State educational standards used by all educational organizations of the country, including civil aviation educational institutions, are developed in a competence-based format.

Competence is a situational category, since it reflects and characterizes the situation in which the error occurs. Functional competence is reflected in two states. On the one hand, it includes a set of requirements for knowledge, skills, and abilities. On the other hand, it is considered the result achieved in meeting these requirements. Consequently, competencies become a useful tool for assessing unlawful personnel interference in aviation security procedures (Khan et al. 2020; Snyman and Wilke 2018).

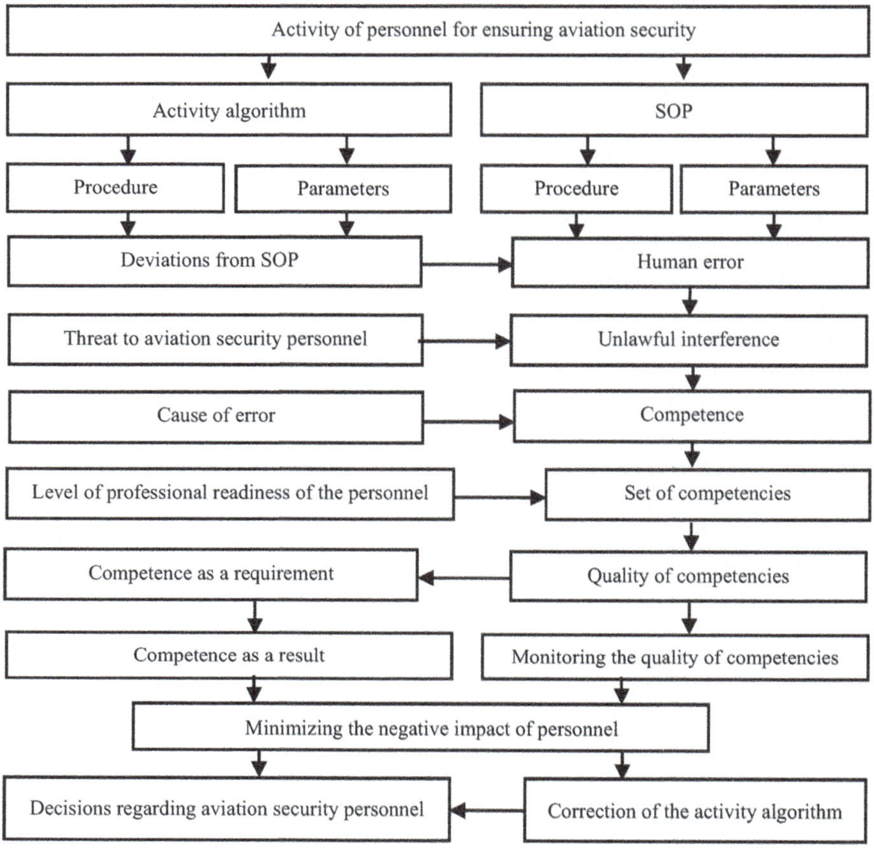


**Fig. 2** The competence-based model of the negative impact of personnel. *Source* Compiled by the authors

To do this, one should be able to assess competencies. As a result, competence is assessed by comparing requirements and characteristics (i.e., quality). The system of monitoring competence quality is developed as an expert system. The results of monitoring and analyzing the causes of errors reveal the degree of unauthorized personnel interference and determine the procedure for minimizing the negative impact of personnel. This step can correct the algorithm of the professional activity of the personnel to bring it closer to the SOP parameters and make some organizational decisions regarding a particular specialist whose actions led to the error.

The negative impact of aviation personnel on the security system and security management procedures should be distinguished from personal and technological factors as components of the anthropogenic factor. Within the framework of the personal factor, we examined negative manifestations associated with the limited psychophysiological capabilities of a human operator (specialist). At the same time, the content of the production activity of a particular specialist plays a critical role.

Such issues can be solved within the personnel training system. They are associated with the optimization of the set of competencies and the harmonization of the personnel activity model, and the content of professional activities (Fig. 3).

The technical and technological factor is associated with the inadequacy of the parameters of the technical and human components. The main problem here is that the parameters of the technical subsystem do not always correspond to the psychophysiological parameters of the human operator, including ergonomic issues.

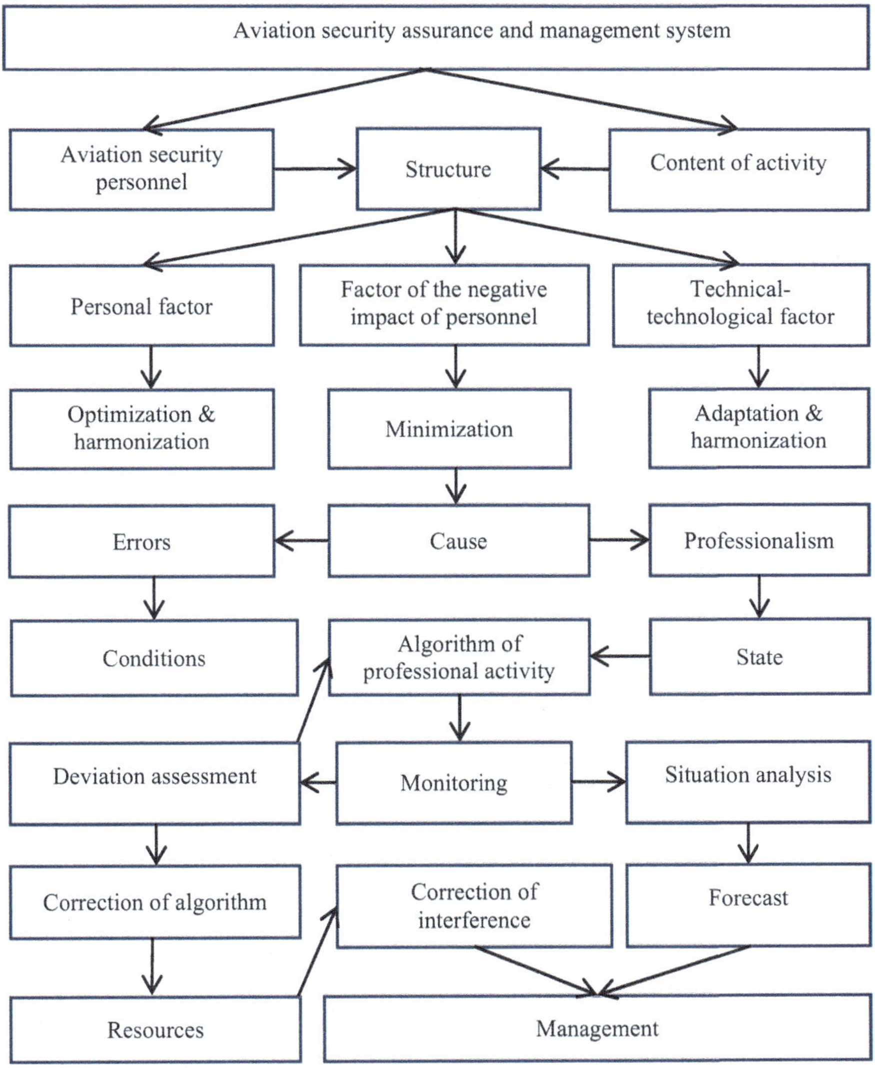


**Fig. 3** A structural and logical model for managing the negative impact of personnel. *Source* Compiled by the authors

Outside the study, the factor of the negative impact of personnel remains. Such activities often deliberately violate the algorithm of activities due to which incidents arise, which are regarded as an act of unlawful interference, posing a security threat.

Therefore, a completely new situation arises, when it is necessary to minimize the anthropogenic factor and the negative impact of the aviation personnel on the professional activity. This impact is considered a threat to the object security.

The reasons that determine the factor of the negative impact of personnel are related to (1) errors in the broad understanding of this term, (2) the conditions of professional activity, (3) the level of professional training of personnel, and (4) the state in which the specialist works. All these reasons are united by the inconsistency, deviations, and discrepancies with the activity algorithm. In aviation security, all professional activities are strictly regulated and legalized in the form of SOP, which implies compliance with the established requirements.

The aviation personnel, performing their professional function is placed in a specific situation, the parameters of which are determined by the SOP parameters and the parameters of the algorithm of the professional activity of a particular specialist. Security threat occurs if there are discrepancies between these parameters. In this case, to minimize the negative impact of personnel, it is necessary to monitor the situation, assess the discrepancies, and analyze them in terms of the measures. Minimizing the negative impact is possible by correcting the SOP algorithm and adjusting the specialist's activity procedure. In both cases, there is a need for certain resources, which minimize the negative impact of personnel, which, in turn, becomes manageable.

A personnel threat is not always represented as a potential event. Often, these threats manifest themselves in the form of negative results of the activities of an aviation specialist, which can even be a terrorist act. Such cases have already been registered in Russia. Scientific research of personnel threats aims to create and implement a special methodology for dealing with such threats according to the formula *detection–reflection–elimination*. Personnel threats are included in the set of potential threats to the transport infrastructure on an equal basis.

Personnel threats as threats to airport security are implemented in unlawful interference in the procedures for ensuring aviation security. In professional activities, aviation security personnel carry out specific actions not provided for by their functional duties and are not included in their plan. These actions may occur because of personnel errors or for some other reason. Nevertheless, in any case, they lead to deviations from certain SOP actions. In this case, aviation security personnel may be considered as a subject of illegal activity.

We consider threats to personnel from the point of view of human imperfection and poor adaptation of personnel characteristics to the parameters of technical means of protection. The requirements for personnel may exceed the level of psychophysiological capabilities of the operator working with such tools. Such threats are quite severe and numerous, given that aviation security personnel include (1) aviation security personnel, (2) passengers, (3) employees, or (4) third-party organizations. These threats require a comprehensive study based on a complex model that involves aviation personnel and considers them a single complex system.

## 3 Results

The aviation personnel, performing their professional function, are in a particular situation, the parameters of which are determined by the SOP parameters and the parameters of the real algorithm of the professional activity of a particular specialist. A security risk arises if there are discrepancies between these parameters (Fig. 2).

The proposed concept avoids the idea of the Black Sea fleet as a specific factor and considers it as a threat to the object's security from unlawful interference in the activities of this object related to the person. In other words, we suggest investigating the negative impact of aviation personnel on the procedures for ensuring and managing aviation security. It is necessary to minimize the part of the Black Sea fleet that poses a threat to professional activities and represents unlawful interference. Simultaneously, there are appropriate structural and logical models and methods of their research that allow us to identify the negative impact of aviation personnel. The research results, presented in some quantitative equivalent, can be considered a parameter for controlling the negative impact level. Therefore, the threat of the Black Sea fleet becomes manageable (Averin and Ovchenkov 2020).

Strictly speaking, CF is still not a factor but a much more complex and multifaceted category that belongs to integrated systems and has a target function that cannot be mathematically described. We introduce a new concept of *personnel threat* as a state of the inadequacy of the professional readiness of the operator and the parameters involved in the security system, determined by the maximum level of professional parameters of the individual that allows the occurrence of adverse events. There is an entirely new situation when it is necessary not to minimize the anthropogenic factor but to regulate the negative impact of the personnel of aviation organizations on professional activities. In this regard, this impact is considered as a threat to the security of the object. Moreover, the approaches to solving the problem and the research methods change (Ovchenkov and Averin 2020).

## 4 Discussion

A personnel threat is not always represented as a potential event. Often, these threats manifest themselves in the form of some negative results of the activities of an aviation specialist, which can even be a terrorist act. In professional activities, aviation security personnel carry out specific actions not provided for by their functional duties and are not included in their plan. These actions may occur because of personnel errors or for some other reason. Nevertheless, in any case, they lead to deviations from certain SOP actions. In this case, the aviation security personnel can be considered as the subject of illegal activity. The result of such activities may occur in the form of algorithmic interference, which refers to the change in the standard algorithm of the operation, its options, or incompetent interference caused by an insufficient level of professional competence (Ovchenkov and Averin 2020).

Threats from personnel engaged in professional activities to ensure aviation security are considered as part of the overall set of threats to airport security, as one of the types of threats. In fact, this opinion is not entirely true. The intruder model determines external threats and their types. Based on this model, appropriate counteraction measures were developed for each type of threat, including the activities of personnel and a set of appropriate technical means to protect the object. Personnel threats are considered a side effect of the professional activities of aviation security personnel. Therefore, it is impossible to single out a single threat or to classify threats in any way. For the same reason, it is unreal to study personnel threats as potential ones, since one cannot develop an apparatus for countering each threat in advance. This is the main difficulty in studying personnel threats, which prevents the use of mathematical models and other classical research methods (Elisov et al. 2018; Aggarwal 2018). Hence, the reduction of the negative impact of personnel on airport security is assumed based on the management of unlawful interference according to the criterion *personnel threat,* using the method of instrumental correction and the method of aviation personnel management.

## 5 Conclusion

The issue of risk modeling remains unsolved, since threats are hypothetical and cannot be formalized. With a very high degree of approximation, the diffusion equation can be proposed as a mathematical model of risk propagation through the topology of the object. In this case, having received approximate solutions, one should apply specific heuristic procedures that refine the solution based on linguistic representations. In this case, the individual results of solving the simulation problem should be analyzed together. Moreover, one should make a decision using some heuristic algorithms within the decision theory. Only a heuristic procedure can give a qualitative result, far from the real picture but sufficient to ensure aviation security.

## References

Aggarwal CC (2018) Neural networks and deep learning. Springer, Cham, Switzerland

Averin DV, Ovchenkov NI (2020) Concept, models, and methods of aviation security management based on the human factor criterion. Sci Bull State Sci Res Inst Civ Aviat 31:108–118

Bruce P, Bruce F (2017) Practical statistics for data scientists: 50 essential concepts. O'Reilly, Sebastopol, CA

Elisov LN, Gorbachenko VI, Zhukov MV (2018) Learning radial basis function networks with the trust region method for boundary problems. Autom Remote Control 79(9):1621–1629

Elisov LN, Ovchenkov NI, Filippov VL (2020) Modeling of potential threats to civil aviation sites security. J Crit Rev 7(18):2620–2627

Geron A (2017) Hands-on machine learning with scikit-learn and tensor flow: concepts, tools, and techniques to build intelligent systems. O'Reilly, Sebastopol, CA

Khan N, Milner D, Marr B (2020) Introduction to people analytics: a practical guide to data-driven HR. Kogan Page Publishers, London, UK

Ovchenkov NI, Averin DV (2020) The problem of the human factor in the field of aviation security and some approaches to its solution. Sci Bull State Sci Res Inst Civ Aviat 30(341):107–116

Price J, Forrest J (2016) Practical aviation security: predicting and preventing future threats. Butterworth-Heinemann, Oxford, UK

Snyman JA, Wilke DN (2018) Practical mathematical optimization. Basic optimization theory and gradient-based algorithms. Springer, Cham, Switzerland

Sweet KM (2008) Aviation and airport security: terrorism and safety concerns. CRC Press, Boca Raton, FL

# Modeling Emergency Modes of a Nuclear Reactor: Minimization of the Amount of Information in Training Problems

**Viacheslav S. Okunev**

**Abstract** Even in the information age, the *educational space* is limited by the time of the educational process and self-training of the university student. The study of modern information technologies takes time, limiting the study of specific software for solving practical applied problems related to the student's major. For example, it takes a year or two years to acquire independent work skills with high-precision neutron physics calculation programs at enterprises and research centers. This situation is unacceptable for universities. During eight hours of laboratory research, I gained some experience in superficial training in such software. The introduction of modern application software in the educational process of higher technical or natural education institutions requires minimizing the amount of source data. In modeling emergency modes of operation of a nuclear reactor, this is possible with the preliminary preparation of data sets using other programs. Also, I used another well-known method, which involves combining all possible emergency modes into a small number of groups. As the student learns, if this is provided for in the curriculum, they can study more complex computer programs that can independently prepare data for modeling emergency modes, including the most dangerous ones, such as anticipated transients without scram and their combinations. This approach has been successfully used in recent decades after severe accidents at nuclear power plants, when the problem of safety came to the fore.

**Keywords** Software · Emergency modes modeling · Nuclear reactor · Anticipated transients without scram · Nuclear power plant

V. S. Okunev (✉)
Bauman Moscow State Technical University, Moscow, Russia
e-mail: okunevvs@bmstu.ru



S. G. Maximova et al. (eds.), *Advances in Natural, Human-Made, and Coupled Human-Natural Systems Research*, Lecture Notes in Networks and Systems 234,
https://doi.org/10.1007/978-3-030-75483-9_54

# 1 Introduction

## *1.1 Relevance of the Problem*

Modeling of emergency operating modes of new-generation nuclear reactors must be carried out at the initial design stage, when the reactor layout is not entirely determined. For this, the author has developed software. The Fast Reactor Inherent Safety Study [FRISS] code allows one to simulate emergency conditions in fast reactors. The first version of the program was created specifically for the educational process shortly after the publication of the results of unique experiments on full-scale modeling of anticipated transients without scram [ATWS] on the Experimental Breeder Reactor-II [EBR-II] (Feldman Mohr et al. 1987; Golden et al. 1987; Lahm et al. 1987; Lehto et al. 1987; Messich et al. 1987; Mohr et al. 1987). After almost 30 years of development and repeated modernization of the FRISS program, its use in the educational process and scientific research, I did not change the name of the code and did not provide it with a version number, as is customary. Kuzmin and Okunev (1997) present a detailed description of one of the first versions of FRISS (1D) with examples of calculations. Since then, the FRISS mathematical model (FRISS-2D) has been significantly improved.

EBR-II was a sodium-cooled fast reactor that operated from 1964 to 1994 at Argonne National Laboratory, Idaho, USA (Argonne National Laboratory [ANL] 2020; Till and Chang 2012; Westfall 2004). In 1969, the reactor was brought to full capacity. EBR-II is a small reactor (with a core volume of 72 L) for an experimental nuclear power plant [NPP] providing heat and electricity to the Argonne National Laboratory. The thermal power of the reactor is 62.5 MW, the electric power is 20 MW. Metallic uranium fuel with enrichment of up to 67% was used. The studies were carried out on metallic fuels doped with zirconium (10–12% by weight) (Meneghetti and Kucera 1990) and other materials for future fast reactors (Van Tuyle et al. 1990). From 1984 to 1986, the first series of ATWS simulation experiments was conducted. The modes loss of flow without scram [LOF WS] and loss of heat sink without scram [LOHS WS] were considered. The results of the experiments were published in 1987.

Mathematical models and the corresponding software for modeling emergency operating modes of nuclear reactors are introduced into the educational process of several higher technical and natural science educational institutions. The first version of the FRISS program was developed in 1990. Early versions of the FRISS program are used at the National Research Nuclear University Moscow Engineering Physics Institute (NRNU MEPhI, Moscow, Russia), Nizhny Novgorod State Technical University (Nizhny Novgorod, Russia), Bauman Moscow State Technical University (Bauman MSTU, Moscow, Russia), Institute of Physics and Power Engineering ("SSC RF—IPPE" JSC, Obninsk, Russia).

On the one hand, the era of information technology, which humanity has entered in recent decades, has dramatically facilitated the solution of some applied technical problems; on the other hand, it has complicated their solution. For example, the development of a *user-friendly interface* for some fairly *old* and well-established computer programs for accurate calculations, such as Monte Carlo N-Particle [MCNP] (Briesmeister 1997; Oak Ridge National Laboratory 2000), led to a significant investment of time in the practical study of these programs. Sometimes the time spent is so great that it considerably complicates and sometimes excludes introducing a particular program code into the educational process. Moreover, the cost of such applications is usually more than ten times higher than the cost of the calculation program itself.

### *1.2 Research Goal*

The main research goal is developing effective mathematical models and computer programs for modeling emergency modes of new-generation fast reactors. The simplicity of the models and the user-friendly interface allows one to use the developed software to be used in the educational process of university departments that train experts in the design of nuclear reactors.

## 2 Materials and Methods

### *2.1 ATWS Modeling*

As a rule, the design of nuclear reactors for various purposes is based on (1) a broad experimental base, (2) the development of appropriate mathematical and software tools (including the development and improvement of mathematical methods and computer programs), and (3) the implementation of computational and optimization studies.

The current stage of nuclear engineering is characterized by developing concepts for improving the safety of reactors based on brand new scientific and technical solutions (International Atomic Energy Agency [IAEA] 2013; Orlov et al. 1992). At the same time, the key role is assigned to creating software for modeling the processes occurring in a new-generation reactor and optimizing the parameters of the NPP, reactor, and equipment. Experimental modeling uses subject models (Bushuev 2008; Kazansky and Matusevich 1984), while theoretical modeling uses abstract ones (Gorlach and Shakhov 2018; Slabnov 2020; Tarasik 2004). Accurate modeling is possible only in the so-called *field tests* or full-scale experiments. For example, experiments in power reactors are prohibited by regulatory documents.

Mathematical modeling plays a crucial role in modeling emergency reactor modes, accompanied by the failure of emergency protection (ATWS). Experimental modeling of ATWS is possible only on low-power research reactors; it is associated with severe accident risk. The ATWS modeling experiments on EBR-II were unique and are still of great practical importance.

Mathematical modeling of ATWS is based on the results of experiments conducted on the following facilities:

- EBR-II, Argonne National Laboratory, Idaho, USA (ANL 2020; Feldman et al. 1987; Golden et al. 1987; Lahm et al. 1987; Lehto et al. 1987; Messich et al. 1987; Mohr et al. 1987; Westfall 2004),
- BN-600, the power reactor of the third power unit of the Beloyarsk NPP, Zarechny, Sverdlovsk Region, Russia (Beloyarsk NPP 1983; IAEA 2013),
- BOR-60 (fast experimental reactor), NIIAR multipurpose research reactor, Dimitrovgrad, Russia (IAEA 2017, 2018),
- Rapsodie Experimental Reactor (Nuclear Power Research Center in Cadarache, France), the first French fast neutron reactor (IAEA 2013),
- 400 MW Fast Flux Test Facility [FFTF] mixed oxide fuel research reactor for developing new fuel types, materials, and equipment for fast reactors, Hanford, USA (Abbotts 2004; Garland 2000),
- BFS, a complex of fast physical stands, Obninsk, Kaluga Region, Russia (IAEA 2018).

All these installations require applying a sodium refrigerant. The BOR-60 provides an experimental loop with a lead coolant. At the BFS stands, one can use lead and any other liquid metal coolant.

The results of the experiments obtained during the study of the emergency conditions at all objects listed above formed the basis of the FRISS mathematical model.

I developed a series of FRISS codes for the operational modeling of emergency modes (including ATWS) and their various combinations, taking into account the non-simultaneous start. Codes can work either as part of other software systems or autonomously.

The FRISS-2D model assumes that a cylindrical reactor consists of several homogenized radial and axial zones. The calculations are carried out in $r$–$z$ geometry (Okunev 2019).

Emergency processes, including ATWS, are modeled in the point kinetics approximation, considering the feedback on average temperatures. Dependence of time $t$ on the average fuel temperatures $T_f(t)$ and coolant $T_{cool}(t)$ is estimated in each zone of the reactor, assuming that at any point of time, there are steady-state temperatures according to the laws specific for each of the considered emergencies of the change of inlet temperature of coolant $T_{in}(t)$, the external impact $\delta_p(t)$ on reactivity (introduced reactivity), and coolant flow rate $G(t)$. As a result, there are $t_w$, $t_{cool}$, $t_{cl}$, and $t_f$ time points, when the maximum power and the corresponding temperatures (coolant, cladding, and fuel) are maximum.

Maximum temperatures (the maximum is taken at $z$) are the following:

$$T_{cool}{}^{max}(r, t_{cool}) = T_{in}(t_{cool}) + w(t_{cool})/G(t_{cool}) \max\left[\Delta T_{cool}{}^{(0)}(r, z)\right],$$

$$T_{\mathrm{f,cl}}{}^{\max}(r, t_{\mathrm{f,cl}}) = T_{\mathrm{in}}(t_{\mathrm{f,cl}}) + \max\left[\left[\begin{array}{l}\Delta T_{\mathrm{cool}}{}^{(0)}(r,z)w(t_{\mathrm{f,cl}})/G(t_{\mathrm{f,cl}}) \\ + \Delta T_{\mathrm{f,cl}}{}^{(0)}(r,z)\end{array}\right]w(t_{\mathrm{f,cl}})\right]$$

where:

$w$: power;

$G$: flow rating.

The approximate mathematical model of fast reactor dynamics used for operational safety assessment or in optimization studies (which I implemented in *FRISS* series codes) is based on solving three groups of equations:

- System of point kinetics equations with a six-group description of delayed neutrons (for power and reactivity coupling),
- Non-stationary thermal conductivity equation, which is the thermal balance equation, and the coolant energy equation (to determine the reactor power proportion and changes in fuel and coolant temperatures),
- Reactivity equilibrium conditions (relation of temperature changes to reactivity).

The equations of the first group are as follows:

$$\mathrm{d}w/\mathrm{d}t = (\rho - \beta)w/t^* = \sum_{s=1}^{S} C_s\lambda_s; \quad \mathrm{d}C_s/\mathrm{d}t = \beta_s w/t^* - C_s\lambda_s,$$

where:

$w = w(t)$: the function describing the change of specific volume energy release in time (transient reactor power carried to nominal);

$w(0) \equiv w_0 = 1$: initial reactor power;

$t$: time;

$\rho$: reactivity;

$\beta$: effective delayed neutron fraction;

$\beta = \sum_{s=1}^{S} : \beta_s$;

$S$: the number of groups of delayed neutrons (typically $S = 6$);

$\beta_s$: the delayed neutron fraction of the $s$-th group;

$t^*$: prompt generation lifetime;

$C_s$: the concentration of the precursor nuclei;

$\Lambda_s$: the delayed neutron precursor decay constant of the $s$-th group ($s = 1, 2,\ldots, S$).

Equations of the second group are the following:

$$\mathrm{d}T_{\mathrm{f}}/\mathrm{d}t = (T_{\mathrm{cool}} - T_{\mathrm{f}})/t_0 + Aw, \tag{1}$$

where:

$T_{\mathrm{f}}$: average (by volume) fuel temperature;

$T_{\mathrm{cool}}$: average coolant temperature;

$t_0$: fuel element time constant (exponent constant describing the change of input parameters: $T_{\mathrm{cool}}$ or $w$);

$t_0 = 0.25 R_t d_f C_p^{(f)}$;

$R_t$: thermal impedance;

$d_f$: diameter of fuel pellet;

$C_p^{(f)}$: specific thermal capacity of fuel at constant pressure;

$A = W_0/(V_f C_p^{(f)} \gamma_f w_0)$;

$W_0$: reactor thermal power (nominal);

$V_f$: fuel volume in the core;

$\gamma_f$: fuel density.

Equation (1) corresponds to a one-dimensional parabolic (or quasi-stationary) approximation. The non-stationary differential thermal conductivity equation in partial derivatives can be replaced by an equivalent ordinary differential equation concerning the average body temperature in volume. In precise transformations, the latter equation will have an infinite order, that is, include, in addition to the average temperature and its first derivative, an infinite series of members with derivatives of increasing order from the average fuel temperature, as well as from the temperature coolant and the power over time. Equation (1) is a first-order approximation to this exact equation. The error of Eq. (1) increases with increasing perturbation rate and bio-criterion. In the worst case (instant spasmodic indignation and Bi $\rightarrow \infty$), its error does not exceed 20% of the maximum temperature deviations (Kuznetsov 1987). In practice, the error of Eq. (1) is always significantly less than 20%.

In more precise calculations, one can use an approximate second-order thermal equation:

$$\begin{aligned} &\mathrm{d}^2 T_f/\mathrm{d}t^2 + \mathrm{d}T_f/\mathrm{d}t(8\mathrm{Bi} + 24) f + 48\mathrm{Bi}(T_f - T_{\mathrm{in}}) f^2 = \\ &(6\mathrm{Bi} + 24)\delta_{\mathrm{cl}} t f^2 w + \delta_{\mathrm{cl}} t f \mathrm{d}w/\mathrm{d}t + 2\mathrm{Bi} f \mathrm{d}T_{\mathrm{cool}}/\mathrm{d}t/, \end{aligned} \qquad (2)$$

where:

$f = \chi/R^2 = \mathrm{Fo}/t$;

$\chi$: coefficient of thermal conductivity;

$R$: radius of the cylinder (fuel rod);

Fo: Fourier criterion;

Bi: Bio criterion;

$\mathrm{Bi} = \alpha R/\lambda_T$;

$\alpha$: coefficient of heat transfer to the coolant;

$\lambda_f$: coefficient of fuel thermal conductivity;

$\delta_{\mathrm{cl}}$: thickness of the fuel element cladding.

The maximum error of Eq. (2) does not exceed 1% of design temperature deviations (Kuznetsov 1987).

In Eqs. (1) or (2), the coolant temperature cannot be set independently: it is a function of the fuel temperature. The heat carrier energy equation describes the relationship between $T_f$ and $T_{\mathrm{cool}}$. In the approximations of the parabolic description of fuel temperature by fuel element radius and the linear description of coolant temperature by fuel element height, the coolant energy equation is as follows:

$$(T_f - T_{cool})\varepsilon/t_0 = dT_{cool}/dt + 2G(T_{cool} - T_{in})/t_t, \tag{3}$$

where:

$\varepsilon$: relation of the thermal capacity of fuel and the coolant (on one fuel element);

$G$: primary flow rate, carried to nominal;

$t_t$: time of transportation of the coolant through a core at rated flow.

Usually thermal capacity of coolant in the core is much less than the thermal capacity of fuel. In this case, $dT_{cool}/dt$ may be omitted from the right side of Eq. (3).

The reactivity of the system can be represented by the sum of the external impact (perturbation) $\delta_\rho$ (due, for example, to the movement of the control rod) and the internal feedback reactivity effect, taking into account the contribution of each temperature effect:

$$\rho = \sum_j \xi_j \left(T_j - T_j^{(0)}\right),$$

where:

$\xi_j$: reactivity coefficient of the corresponding ($j$) effect;

$(T_j - T_j^{(0)})$: change in the average temperature of the corresponding core component.

Reactivity balance condition is as follows:

$$\rho = \delta\rho + \xi_f\left(T_f - T_f^{(0)}\right) + \xi_{cool}\left(T_{cool} - T_{cool}^{(0)}\right) + \xi_{in}\left(T_{in} - T_{in}^{(0)}\right),$$

where:

$\xi_f$, $\xi_{cool}$, $\xi_{in}$: temperature coefficients of reactivity on fuel, coolant, and coolant temperature at the core inlet.

By definition, $\xi_j = \partial\rho/\partial T_j$, where $j = \{f, cool, in\}$. The calculated model implemented in FRISS codes allows for considering instantaneous and delayed reactivity effects.

In the general case, the coefficient $\xi_{cool}$ takes into account the following effects: (1) expansion of the coolant and structural materials of the core; (2) expansion of the coolant and structural materials in the side reflector; (3) expansion of the control rod assembly; and (4) expansion of the support manifold. The coefficient $\xi_f$ considers (1) expansion of the fuel in the core, and (2) Doppler coefficient of reactivity (taking into account the broadening of resonances depending on the interaction cross-sections of neutron energy). The coefficient $\xi_{in}$ takes into consideration the change of the coolant temperature at the entrance to the core.

FRISS calculations correspond well with experiments on EBR-II and other reactors (Rapsodie, BN-600, BOR-60, FFTR, BFS), with calculations carried out by other scientists in different countries of the world on various software systems (IAEA 2000).

## 2.2 Minimization of Initial Data Arrays and Practical Use in the Educational Process

The main task when using codes in the educational process is to minimize the data, the use of which allows one to make the necessary decision.

Modeling of emergency modes requires preliminary analysis of large volumes of heterogeneous information. At the initial stage of the reactor design, neither the emergency conditions themselves nor their development scenarios were defined.

There are two standard methods for minimizing the amount of data. The first method involves combining all the emergency conditions into a small number of groups. It is known that all possible emergencies in fast reactors with liquid–metal cooling can be initiated by perturbations of the reactivity $\delta\rho$ (introduction of positive reactivity), the coolant flow rate $\delta G$, inlet coolant temperature $\delta T_{\text{in}}$, and drainage of the core or its part. For this reason, in mathematical modeling problems, all emergency modes are combined into a small number of groups depending on the perturbation $\delta\rho$, $\delta G$, $\delta T_{\text{in}}$ that initiates them. Let us dwell on the drainage of the core. The failure of emergency protection accompanies the most dangerous modes; therefore, they should be considered in the first place. Positive reactivity input ($\delta\rho > 0$) corresponds to the transient overpower without scram [TOP WS] mode; $\delta G > 0$ corresponds to the overcooling accident without scram [OVC WS] mode; $\delta G < 0$ corresponds to the LOF WS mode; $\delta T_{\text{in}} > 0$ corresponds to the LOHS WS mode; $\delta T_{\text{in}} < 0$ corresponds to the OVC WS mode; drainage of the core corresponds to the loss of coolant accident without scram [LOCA WS] (Feldman et al. 1987; Golden et al. 1987; Lahm et al. 1987; Lehto et al. 1987; Meneghetti and Kucera 1990; Messich et al. 1987; Mohr et al. 1987; Van Tuyle et al. 1990).

The student–researcher sets the perturbation that initiates the emergency mode (the law of introducing perturbations, the amplitude value). It is possible to study combinations of several perturbations (emergency modes).

The second method lies in the preliminary preparation of the source data arrays. It is advisable to prepare data on various types of reactors, which students can use when modeling emergency conditions, in advance. For example, for fast reactors with liquid–metal cooling, files are prepared containing the primary layout parameters of BN reactors of different capacities, such as (BN-600, -800, BNM), BREST (BREST-OD-300, BREST-600, -1200, -2400), or SVBR (SVBR-10, -50, -100). More complex software systems are used to prepare the source data. These can be settlement codes, settlement and optimization codes, or codes of precision calculations. The study of such codes is possible only when completing a diploma project (work) or a master's final qualification work.

Depending on the student's qualification (course, major), some data from the source data can be prepared independently. This fact applies, for example, to temperature reactivity coefficients. They can be calculated using the Winfrith Improved Multigroup Scheme [WIMS] program (AEA Technology plc 1998; Okunev and Lisitsyn 2011), which is widely used at some *Rosatom* enterprises. WIMS is unique for its relatively high accuracy (although the program does not formally claim to be a "high precision" one). The program implements the solution of the Peierls equation; that is, it does not use the diffusion approximation. The neutron-physics calculation is carried out in a multi-group approximation with the number of groups 69 or 172 (AEA Technology plc 1998; Aldama et al. 2003; Okunev and Lisitsyn 2011). Group approximations are based on the assumption of separation of variables describing the spatial and energy dependence of the neutron flux density. This situation is true for the steady-state spectrum zone: away from the boundaries and sources (at a distance of two or three relaxation lengths). This approximation is well applicable for calculating medium- and high-power reactors. WIMS uses its microconstant library consisting of 69 or 172 groups. The entire range of possible changes in the kinetic energy of neutrons in the reactor (from 0 to 10.5 meV) is divided into 69 or 172 sub-bands. The program allows one to evaluate the change in the nuclide composition of the reactor over time (the processes of fuel burnout, poisoning, and slagging of the thermal neutron reactor).

In 2011, a training manual was prepared that allows students to independently study the WIMS program in the shortest possible time (Okunev and Lisitsyn 2011). To carry out the simplest but relatively accurate calculations, the student needs to spend no more than 1 h studying the program. The manual provides examples and source data files for calculations (Okunev and Lisitsyn 2011). One can spend more than one working day preparing the source data file oneself, which is unacceptable for most students, for example, when doing coursework.

We should note that the WIMS program has a great independent value. It is used to calculate an infinite array of fuel rods, fuel assemblies, or reactor channels. For example, an essential task in designing nuclear reactors with thermal neutrons is to study the dependence of the infinite multiplication factor on the distance between the fuel rods (or fuel channels). One can use the program when designing reactors of various types. The accuracy of the calculation of fast reactors corresponds to the accuracy of *engineering* calculations. In terms of the accuracy of calculations of thermal neutron reactors, the WIMS program is close to the precision codes.

## 3 Results

The research resulted in the development of software for optimal design and safety analysis of new-generation fast reactors.

I developed several versions of the FRISS program. The simplest of them is intended for operational assessment of the reactor behavior (assessment of safety functional) in any emergencies, including ATWS.

Moreover, I created a version of detailed modeling of emergencies (determining the dependence of changes in the characteristics of the reactor, including the safety functional, on time).

These two versions of the FRISS program are used in the educational process.

Besides, I elaborated a version of the FRISS program for use in optimization software systems. This version allows one to calculate the sensitivity coefficients of all functionals (including safety ones) to all source data considered in the problem. This version of the program, together with the optimization system, is used by senior students when completing a diploma project or a master's thesis.

There is an internal classification of each version of the FRISS program according to the dimension of the problem being solved (0-D, 1-D, 2-D).

## 4 Discussion

The use of adequate and straightforward mathematical models, as well as software with a user-friendly interface, allows one to study in detail complex combinations of emergency modes (including ATWS), which usually takes much time.

Easy-to-use high-speed software allows one to analyze the emergency operating modes of fast reactors of various types. The FRISS program allows for calculating the sensitivity coefficients of the maximum temperatures of the fuel, coolant, cladding of fuel rods, and maximum reactor power to all initial data, including the parameters of disturbances that trigger emergency modes. Knowing these coefficients allows one to evaluate the significance of various input data regarding the impact on reactor safety.

Furthermore, the program allows one to simulate complex combinations of emergency modes with a simultaneous and non-simultaneous start.

Furthermore, the program allows one to simulate complex combinations of emergency modes with a simultaneous and non-simultaneous start.

Using the FRISS-2D code, it is possible to determine the self-defense zones of the designed reactor. Using these areas, one can determine the role of reactivity coefficients in complex combinations of emergency modes, including ATWS.

Safety sets (zones) are sets of maximum permissible perturbations of reactivity, flow, pressure, the inlet temperature of the coolant, their rates of change (taking into account the possible relative delay of specific processes when they are applied), in which no protective equipment is required. Thus, in fast reactors with liquid–metal cooling, the zones of maximum permissible changes $\delta\rho$, $\delta G$, and $\delta T_{\text{in}}$ should be considered as accidents safety sets, in which the main safety barriers remain functional; in other words, the constraints on the functional characterize the behavior of the reactor in emergency modes (safety functional). Safety features include (1) maximum core component temperatures, (2) reactor power, or (3) pressure. The implementation of restrictions for such functionals simulates the trouble-free operation of the reactor. As a rule, the boundaries of security sets are determined by the maximum permissible safety functional values.

Based on the analysis of safety sets, various practical tasks can be solved.

It should be expected that the use of the software that I developed will allow students to acquire the skills of modeling emergency modes, understand, and study the physical processes occurring in the reactor in emergency modes. This situation is especially critical in the development of fundamentally new reactors, for example, fast reactors with heavy liquid metal coolant or those with new, accident-resistant metal-ceramic fuel.

## 5 Conclusion

Experimental ATWS modeling is expensive and challenging; it can damage or even destroy the reactor, posing a severe danger to society and the environment. For this reason, ATWS mathematical modeling is of paramount importance.

The introduction of software systems in the educational process requires either a significant simplification of the mathematical model or the preliminary preparation of some source data for other programs. Simplification of the model is not always acceptable, because in some cases, it hinders an adequate assessment of the characteristics of emergency modes. The second method, which involves the preliminary preparation of some data necessary for calculation, is implemented in practice. The data is entered (automatically) in advance into the source data files for ATWS calculations. This approach allows one to significantly reduce the amount of information in modeling emergency modes of a nuclear reactor.

## References


Abbotts J (2004) The long, slow death of the fast flux reactor. Bull At Sci 60(5):56–62

AEA Technology plc (1998) WIMS—a general purpose code for reactor core analysis. https://docplayer.net/7684247-Introduction-the-answers-software-service-wims-1998-aea-technology-plc.html

Aldama DL, Leszczynski F, Trkov A (2003) WIMS-D library update. Final report of a coordinated research project. International Atomic Energy Agency, Vienna, Austria

Argonne National Laboratory [ANL] (2020) Reactors designed by Argonne National Laboratory. Fast reactor technology. EBR-II (Experimental Breeder Reactor-II). Nuclear Engineering Division: think, discover, explore, innovate. https://www.ne.anl.gov/About/reactors/frt.shtml

Beloyarsk Nuclear Power Plant named after I. V. Kurchatova (Beloyarsk NPP) (1983) Booklet. Vneshtorgizdat, Moscow, USSR

Briesmeister JF (ed) (1997) MCNP—a general Monte Carlo N-particle transport code (Version 4B) LA-12625. Manual. Los Alamos National Laboratory, Los Alamos, NM

Bushuev AV (2008) Experimental reactor physics. National Research Nuclear University Moscow Engineering Physics Institute, Moscow, Russia

Feldman FE, Mohr D, Chang LK, Planchon HP, Dean EM (1987) EBR-II unprotected loss-of heat sink predictions and preliminary test results. Nucl Eng Des 101(1):57–66. https://doi.org/10.1016/0029-5493(87)90150-6

Garland MA (2000) Medical isotope production in the fast flux test facility and the high flux isotope reactor and isotope production software users guides. Pacific Northwest National Lab, Richland, WA

Golden GH, Planchon HP, Sackett JI, Singer RM (1987) Evolution of thermal-hydraulics testing in EBR-II. Nucl Eng Des 101(1):3–12. https://doi.org/10.1016/0029-5493(87)90145-2

Gorlach BA, Shakhov VG (2018) Mathematical modeling. Model building and numerical implementation. Lan, Saint Petersburg, Russia

International Atomic Energy Agency [IAEA] (2000) Series IAEA-TECDOC-1139. Transient and accident analysis of a BN-800 type LMFR with near-zero void effect. International Atomic Energy Agency, Vienna, Austria

International Atomic Energy Agency [IAEA] (2013) Series IAEA-TECDOC-1691. Status of fast reactor research and technology development. International Atomic Energy Agency, Vienna, Austria

International Atomic Energy Agency [IAEA] (2017) Nuclear energy series no. NP-T-5.8. Research reactors for the development of materials and fuels for innovative nuclear energy systems. International Atomic Energy Agency, Vienna, Austria

International Atomic Energy Agency [IAEA] (2018) Nuclear energy series, NP-T-1.15. Experimental facilities in support of liquid metal cooled fast neutron systems. International Atomic Energy Agency, Vienna, Austria

Kazansky YA, Matusevich ES (1984) Experimental methods of reactor physics. Energoatomizdat, Moscow, USSR

Kuzmin AM, Okunev VS (1997) Modeling and analysis of emergency situations in fast neutron reactors. Moscow Engineering Physics Institute, Moscow, Russia

Kuznetsov IA (1987) Emergency and transient processes in fast reactors. Energoatomizdat, Moscow, USSR

Lahm CE, Koenig JF, Betten PR, Lehto WK, Bottcher JH (1987) Driven fuel qualification for loss-of-flow and loss-of-heat sink tests without scram. Nucl Eng Des 101(1):25–34. https://doi.org/10.1016/0029-5493(87)90147-6

Lehto WK, Fryer RM, Dean EM, Chang LK, Koenig JF (1987) Safety analysis for the loss-of-flow and loss-of heat sink without scram tests in EBR-II. Nucl Eng Des 101(1):35–44. https://doi.org/10.1016/0029-5493(87)90148-8

Meneghetti D, Kucera DA (1990) Reactivity feedback components of a homogeneous U10Zr-fueled 900 MW (thermal) liquid–metal reactor. Nucl Technol 91(2):139–145

Messich NC, Better PR, Booty WF, Christensen LJ, Fryer RM (1987) Modification of EBR-II plant to conduct loss-of-flow-without-scram tests. Nucl Eng Des 101(1):13–24. https://doi.org/10.1016/0029-5493(87)90146-4

Mohr D, Chang LK, Feldman FE, Planchon HP, Betten PR (1987) Loss-of-primary-flow-without-scram tests: pretest predictions and preliminary results. Nucl Eng Des 101(1):45–56. https://doi.org/10.1016/0029-5493(87)90149-X

Oak Ridge National Laboratory (2000) RSICC computer code collection.mcnp-4c: Monte Carlo N-particle transport code system. Oak Ridge National Laboratory, Los Alamos, NM

Okunev VS (2019) Designing of new generation of the nuclear reactors. AIP Conf Proc 2195(1):020012. https://doi.org/10.1063/1.5140112

Okunev VS, Lisitsyn IS (2011) Neutron-physical calculation of the lattice of a nuclear reactor based on the gas-kinetic theory of transfer. Textbook. N. E. Bauman Moscow State Technical University, Moscow, Russia

Orlov VV, Avrorin EN, Adamov EO, Vasiliev AP, Vedikhov EP, Vertman AA et al (1992) Unconventional NPP concepts with natural safety (new nuclear technology for large-scale nuclear energy of the next stage). At Energy 72(4):317–329

Slabnov VD (2020) Numerical methods. Textbook. Lan, Saint Petersburg, Russia

Tarasik VP (2004) Mathematical modeling of technical systems. Textbook. Design-PRO, Minsk, Belarus

Till CE, Chang YI (2012) Plentiful energy: the story of the integral fast reactor. The complex history of a simple reactor technology, with emphasis on its scientific bases for non-specialists. Createspace Independent Publishing Platform

Van Tuyle GJ, Kroeger P, Slovik GC, Chan BC (1990) Examining the inherent safety of PRISM, SAFR, and the MHTGR. Nucl Technol 91(2):185–202

Westfall C (2004) Vision and reality: the EBR-II story. Nucl News 47(2):25–27

# Modeling Legal Certainty for Sustainable Development of the Russian Judicial System

**Irina V. Rekhtina, Mikhail A. Bolovnev, and Tamara V. Yakusheva**

**Abstract** For the first time, we investigate the legal certainty of civil proceedings in Russia as an independent model of the judicial system, which formation indicates a special type of legal regulation of judicial activity and establishes a platform for the sustainable functioning of the entire Russian judicial system. The sustainability of the judicial and procedural components of the judicial system of a particular country depends on the model of legal certainty transmitted at the national level. In this regard, we examine the features and types of the most popular models of legal certainty of civil proceedings that have been formed in global procedural practice: Civil Law and Common Law. Besides, we analyze their specific and typical features, considering the national model of legal certainty separately. The study of the specifics of the Russian model of legal certainty shaped in modern civil and arbitration proceedings allows us to conclude that it has a mixed nature, indicating the transformational development stage of the Russian judicial system. We prove that the model of legal certainty in the current civil proceedings of Russia can be characterized as a transitional model of legal certainty, combining the features of models of civil proceedings of Civil Law (statutory) and models of civil proceedings of Common Law (dynamic), as well as elements of the socialist type of civil proceedings. This circumstance determines the regulation of procedural relations and acts as a criterion for the efficiency of judicial protection of violated rights and legally protected interests. We believe that there are essential problems, caused partly by the transitional nature of the existing legal certainty model of civil proceedings in Russia, which require the close attention of science, practice, and corresponding decisions. The finally formed legal certainty model marks the completion of transformational processes in the judicial system and a starting point for a new development stage of Russia.

**Keywords** Justice · Legal certainty · Civil process · Legal family

I. V. Rekhtina · M. A. Bolovnev · T. V. Yakusheva (✉)
Altai State University, Barnaul, Russia
e-mail: tamara.yakysheva@mail.ru



S. G. Maximova et al. (eds.), *Advances in Natural, Human-Made, and Coupled Human-Natural Systems Research*, Lecture Notes in Networks and Systems 234,
https://doi.org/10.1007/978-3-030-75483-9_55

## 1 Introduction

For the first time, we consider legal certainty through the prism of features that shapes an independent model of legal certainty, which, being formalized at the national level, creates a platform for the sustainable development and functioning of the Russian judicial system. The finally formed model of legal certainty is a signal of the completion of the transformational processes of the judicial system and a starting point for a new stage, which the Russian Federation has yet to enter. The finalized model of legal certainty is a sign of the completion of the transformation processes of the judicial system and a starting point for a new stage Russia has yet to enter. Depending on the model of legal certainty transmitted at the national level, the sustainability of the judicial and procedural components of the judicial system of a particular country depends on the perception of the categories of (1) fair trial, (2) legal certainty, (3) finality of judicial acts res judicata, (4) accessibility of justice, (5) exhaustibility of remedies, and other factors related to the right to a fair trial (Article 6 of the Convention for the Protection of Human Rights and Fundamental Freedoms) (Council of Europe 1950; De Salvia 2004).

In the legal science of the late 20th–early twenty-first centuries, the ECtHR has established itself in Russian judicial practice and legal doctrine since the ratification of the Convention for the Protection of Human Rights and Fundamental Freedoms. Scientists conducted several comprehensive studies and noted a kind of transition period in Russian law, characterized by the stage of adaptation and revival of some previously lost institutions: (1) appeals (Borisova 2005); (2) new approaches in the system of judicial supervision (Terekhova 2007); (3) the system of principles of civil procedure law (Voronov 2009); (4) evidence law (Reshetnikova 1997); (5) the system of execution of judicial acts (Isaenkova 2007; Uletova 2002); (6) efficiency and accessibility of justice (Tsihotsky 1997); (7) unification and systematization of procedural legislation (Gromoshina 2010).

Initially, during the transformation of the legal regulation of procedural relations, the legislators did not think about the choice of the model of legal certainty of civil proceedings, which they planned to form, and the advantages of the existing models. Perhaps, such a task was not set at all. Improving the current procedural legislation, the innovations proposed by the scientific doctrine were considered. However, the specific concept of legal certainty, which includes the model of legal certainty of civil proceedings in Russia, was not developed.

The legislators face the task of (1) harmonization, (2) achieving external unification, (3) bringing national legislation into line with European legal norms and the Convention for the Protection of Human Rights and Fundamental Freedoms, which was ratified in Russia without any delay. However, this goal was never achieved. This fact is confirmed by the practice of considering complaints filed against Russia by the ECtHR. This is partly due to the lack of a developed model of legal certainty in civil proceedings and the lack of procedure for conducting such investigations at the government level, despite the requirements of society and the judicial system.

Many basic tasks are conceptually defined. First of all, the correlation between national treatment and the convention requirements require clarification. The question of whether Russia will implement the decisions of the ECtHR, which comply with the norms of the Convention, but contradict the constitutional principles of Russia and the internal regime, is controversial. The initial neglect of this aspect led to further conflicts and problems. The second issue concerns the legal force, the degree of obligation, and the procedure for applying the precedent decisions of the ECtHR. It is necessary to determine at what level of the hierarchy these acts are placed in terms of legal force, the degree of their binding nature. The third problem relates to the mechanism of application and enforcement of decisions of the ECtHR. It is essential to specify which acts can and should be applied in particular cases. In addition, it is crucial to determine how to implement the decisions of the ECtHR to exclude or minimize the failure of Russia to comply with its international obligations.

Combining the above factors identified problems in the compliance of civil proceedings with the criterion of legal certainty and updated the study on forming an optimal model of legal certainty that meets the modern requirements of the government, society, law, and the national judicial system.

## 2 Materials and Methods

The study aims to identify how a specific legal certainty model is related to the type of legal family and determined by it while reflecting the specifics of national law. We also seek to determine the model of legal certainty of Russian civil proceedings that has been formed at this stage, its advantages and disadvantages, and its impact on further development.

The study uses a system-structural approach to the object of research, which, in turn, is determined using the following methods:

- The dialectical method allows us to study the origin, formation, and development of the legal family of Russian law;
- Historical, contributing to the understanding of the specifics of the functioning of specific mechanisms and procedures of the national judicial system;
- Formal-legal methods, which are used to analyze the norms of the current Russian legislation, fixing and regulating the features of the national model of legal certainty;
- The logical method enables us to (1) identify both positive and negative elements of various models of legal certainty, (2) assess their effectiveness and the need for reception, (3) expediency, (4) legislative gaps and contradictions at the national level, provoking conflicts and hindering the formation of a modern model of legal certainty;
- The legal method allows us to study the features of various legal certainty models of legal proceedings, both in Russian and foreign legislation, to compare them

and identify the most effective elements that can later be introduced in Russian practice and adapted to national characteristics.

Within the framework of this methodology, we analyze foreign scholarship (Liebesny 1981; Merryman and Clark 1978; Wiegand 1991) and Russian and foreign theoretical bases and procedural legislation (Osakwe 2008; Saidov 2007).

Also, we use a set of sociological methods (synthesis and systematization). The study of the typical features of the legal certainty model allows us to systematize and classify them according to the type of legal family into Civil and Common legal models. The analysis allows us to consider the overall model of legal certainty as a system and examine the components of the model. This step facilitates the identification of positive and negative aspects of the legal certainty model. Besides, we identify specific features of the judicial system and legal proceedings at the national level. As a result, we draw a conclusion about the transitional nature of the legal certainty model of the proceedings. Furthermore, we provide modeling of the prospects for the development of the Russian judicial system in terms of forming a final specific model of the legal certainty of civil proceedings to develop other legislative decisions and law enforcement recommendations.

## 3 Results

The legal certainty model as a system of interrelated elements of legal certainty is formed in legal systems by choosing the optimal combination of legal certainty with other categories of law. Therefore, the requirement of the legality of the balance of private and public interests by the appeal system according to the criteria for reviewing final court decisions is provided by a set of legislative, enforcement, and doctrinal means to the extent that the appeal system provides a specific model of the legal framework.

The analysis of the manifestation of legal certainty in different countries allows us to conclude that the type of legal family is the main factor in choosing the legal certainty model that a particular country forms at the national level. This fact allows us to distinguish several legal certainty models with specifics determined by a particular type of legal system.

Besides, we revealed that differences in the establishment of a regime of legal certainty in Common and Civil Law are sufficient to justify the existence of two independent models of legal certainty in civil proceedings: (1) models of legal certainty in civil proceedings and civil law systems; (2) models of legal certainty in civil proceedings in the Common Law.

The first model of legal definiteness (Civil Law, Romano-Germanic, Civil or Continental Law) is characterized by the following features: (1) the aim of legal doctrine to formulate the concept of legal certainty, definition, and scientific concept of legal certainty; (2) legal certainty is formed through the normative (legislative) consolidation of individual elements; (3) it is addressed primarily to the court and the judiciary, which highest instance ensures legal certainty and unity of judicial practice.

The second model of legal certainty in the Anglo-American system has the following features: (1) the legal doctrine does not seek to develop a single definition of legal certainty; (2) it is replete with numerous concepts concerning the means and mechanisms for ensuring legal certainty, aimed at their proper use in the process of legal certainty; (3) the regime of legal certainty is formed mainly by creating precedents of legal certainty, which are addressed to the court and judicial authorities.

We can say that the legal certainty model of the modern civil process in Russia is transitional. This model combines the features of two legal certainty models in the civil process: (1) model of civil process of civil law (statutory) and (2) model of civil process (dynamic), as well as elements of the socialist type of civil process that undermine the legal certainty of civil process.

The study shows that in the transition from the Russian model of legal certainty to the legal certainty model of Civil Law, the legal doctrine seeks to formulate the concept of legal certainty and develop a scientific concept of legal certainty addressed to the court and the judiciary. However, the highest authority does not always provide legal certainty and unity of judicial practice; the transformation of elements is carried out both in legislation and developing guidelines.

The research reveals that the model of legal certainty in Russian civil proceedings includes the following elements: (1) the regime of legal certainty is partially formed based on judicial acts adopted at the level of the Supreme Court of the Russian Federation (second cassation); (2) the absence of a single universal concept of legal certainty in the scholarship; (3) independent indication and use by the subjects of procedural relations of mechanisms ensuring legal certainty (res judicata, estoppel).

We found that the costs (*outstrips* of the socialist period in the national model of legal certainty) include (1) a multi-authoritarian system of appeal and duplication of instances; (2) subjective, discretionary powers of officials, on which the revision of the uncertainty of the grounds for revoking final court decisions depends; and (3) the transformation of extraordinary cassation and supervisory proceedings into a normal method of verification with authority to verify the legality and validity of judicial acts.

## 4 Discussion

The modern model of legal certainty of Russian civil proceedings is formed under the influence of historical moments. Until 1917, taking into account the provisions of the Civil Procedure Charter and the establishment of judicial institutions in 1864 (Chistyakov 1991), the model of legal certainty of Civil Law, characteristic of the Romano-German states, was fixed and developed in pre-revolutionary Russia.

In 1917, the Romano-German legal system was radically transformed into a socialist legal system denying some traditional principles of Civil Law, the principles of socialist legality, objective truth, and the active role of the court were hyperbolized. The authorities introduced investigative and discretionary elements, and the Soviet supervisory proceedings developed in an unlimited period.

In 1991, the transformation of the political course and the beginning of judicial reform towards the democratization of the judicial process began. RSFSR adopted the concept of judicial reform, which established the following provisions: (1) justice, (2) formation of an independent judiciary, (3) guarantees of respect for fundamental human rights and freedoms in the judicial process and many other points. Later, these provisions were enshrined in the Constitution of the Russian Federation of 1993 (Russian Federation 1993), which regulates the activities of the courts and the Russian judicial system at the government level.

After the recognition of the jurisdiction of the European Court of Human Rights in 1998, these provisions were taken into account in the development of the following procedural codes: (1) the Arbitration Procedural Code of the Russian Federation of 2002 (Russian Federation 2002a), the Civil Procedural Code of the Russian Federation of 2002 (Russian Federation 2002b), in which the legislator attempted to achieve compliance with European standards of justice and the principle of legal certainty within the framework of a more extensive right – the right to a fair trial.

Such unification was characterized by the adaptation of existing procedural mechanisms to the European standards of justice enshrined in the norms of the Convention. This document was fragmented, unsystematic, and often contradictory (Yarkov 2011). No regular, consistent, or comprehensive changes were made. Unification and differentiation of civil proceedings were not taken into account, which predetermined the problems of Russian integration into the European legal field (Rekhtina et al. 2019).

## 5 Conclusion

In current conditions, it is possible to talk about a transitive model of legal certainty in civil proceedings in Russia since the archaic elements of the former socialist period remain. Such rudimentary provisions as (1) a multi-authoritarian system of appeal; (2) bifurcation and duplication of instances; (3) subjective, discretionary powers of officials, on which the review depends; (4) uncertainty of the grounds for revoking final court decisions, which introduce an imbalance and hinder the completion of the model formation.

It is premature to speak that a legislative model of legal certainty has been formed in the Russian Federation since appropriate, effective mechanisms have not been introduced and are not functioning in civil proceedings. We cannot but say that this model, which is traditional for Russia, is the most suitable for civil proceedings in current conditions.

The scientific community faces the task of developing a model of legal certainty, considering the standard of legal certainty enshrined at the convention level, and proposing a concept of systemic changes in the current civil procedure legislation to ensure legal certainty and guarantee the protection of rights of subjects of procedural relations from arbitrary revocation of final court decisions.

## References


Borisova EA (2005) Verification of judicial acts in civil cases. Gorodets, Moscow, Russia

Chistyakov OI (ed) (1991) Russian legislation of 10th–20th centuries. In: Judicial reform, vol 8. Yuridicheskaya Literatura, Moscow, Russia

Council of Europe (1950) Convention for the protection of human rights and fundamental freedoms (November 4, 1950 ETS No. 005). Rome, Italy

De Salvia M (2004) Precedents of the European Court of human rights: guidelines on judicial practice related to the European convention for the protection of human rights and fundamental freedoms: jurisprudence from 1960 to 2002 (Zhukova AA, Pashkova GA trans). Law Center Press, Saint-Petersburg, Russia

General doctrine of the state and the law of transition. Yurlitinform, Moscow, Russia

Gromoshina NA (2010) Differentiation and unification in civil proceedings. Dis. of Doc. of Legal Sci. Kutafin Moscow State Law University, Moscow, Russia

Isaenkova OV (2007) Executive law in the Russian federation: peculiarities of formation and development prospects. Yurlitinform, Moscow, Russia

Liebesny HJ (1981) Foreign legal systems: a comparative analysis. George Washington University, Washington, D.C.

Merryman JH, Clark DS (1978) Comparative law: Western European and Latin American legal systems. Bobbs-Merrill Company, New York, NY

On the unification of norms regulating the proceedings under supervision in civil proceedings. J Russ Law 10(118):65–71

Osakwe K (2008) Comparative jurisprudence: schematic commentary. Jurist, Moscow, Russia

Rekhtina IV, Bolovnev MA, Prasolova IA, Kalashnik NI, Kazantseva OL (2019) Problems of Russia's integration into the legal framework of Europe. KnE Soc Sci 3(16):63–75. https://knepublishing.com/index.php/KnE-Social/article/view/4474

Reshetnikova IV (1997) Law of evidence in civil proceedings. Dis. of Doc. of Legal Sci., The Liberal Arts University, Ekaterinburg, Russia

Russian Federation (1993) Constitution of the Russian Federation (December 12, 1993, edited on July 1, 2020). Moscow, Russia. http://www.consultant.ru/document/cons_doc_LAW_28399/

Russian Federation (2002a) Arbitration procedural code of the Russian Federation (July 24, 2002a No. 95-FZ, edited on December 8, 2020). Moscow, Russia. http://www.consultant.ru/document/cons_doc_LAW_37800/

Russian Federation (2002b) The civil procedural code of the Russian Federation (November 14, 2002b No. 138-FZ, edited by December 8, 2020). Moscow, Russia. http://www.consultant.ru/document/cons_doc_LAW_39570/

Saidov AKh (2007) Comparative jurisprudence. The main legal systems of modernity. Jurist, Moscow, Russia

Terekhova LA (2007) System of review of judicial acts in the mechanism of judicial protection. Wolters Kluwer, Moscow, Russia

Tsihotsky AV (1997) Theoretical problems of justice efficiency in civil cases. Nauka, Novosibirsk, Russia

Uletova GD (2002) Practice of applying legislation on enforcement proceedings. Legislation 9:45–53

Voronov AF (2009) Principles of civil proceedings: the past, present, and future. Gorodets, Moscow, Russia
Wiegand W (1991) The reception of American law in Europe. Am J Comp Law 39(2):229–248
Yarkov VV (2011) Novels of the RF CPC: new wine in old furs? Yuridicheskaya Gazeta 1–2. https://justicemaker.ru/view-article.php?id=3&art=2517

# Status and Dynamics of the Level of Domestic Corruption in the Border Region of Russia

**Anastasya S. Spirina**

**Abstract** The necessity of carrying out investigations into domestic corruption in the border region are determined by the high significance of the negative impact of corruption on the state and public security of the country. In the Altai region, as in other regions of Russia, corruption is a threat to the security of individuals, society, and the state. Therefore, addressing such a threat is an important challenge for the whole region. The main purpose of this paper is to determine the level of domestic corruption practices in the Altai region, as well as to analyze the dynamic characteristics of domestic corruption practices, relying on the data obtained from a number of sociological surveys conducted in the Altai region in 2018–19. The analysis presented in the paper reveals the extent of corruption in the Altai region. The author analyzes the relationship between such indicators as the territorial location, corruption coverage, changes in the level of corruption, and the way in which corruption situations are addressed. The concepts of pervasiveness, demand, and risk of corruption are clearly defined and separated. Corruption indicators are examined with respect to (a) the overall population of respondents and (b) different situations. The analysis of indicators from different cases allows to identify the most corrupt areas.

**Keywords** Domestic corruption · Pervasiveness of domestic corruption · Dynamics · Border region · Demand for corruption · Corruption risk

## 1 Introduction

The concept of corruption is very contradictory and ambiguous in nature. The significance of corruption can differ among representatives of different social groups, depending on the context of their perception. Identification of the corrupt subjects is essential for sociological analysis. Currently, the opinions on the origin of the subjects of corruption differ. Corruption can be considered a form of social relations

A. S. Spirina (✉)
Altai State University, Barnaul, Russia
e-mail: natali93_08@mail.ru

 
S. G. Maximova et al. (eds.), *Advances in Natural, Human-Made, and Coupled Human-Natural Systems Research*, Lecture Notes in Networks and Systems 234,
https://doi.org/10.1007/978-3-030-75483-9_56

because several actors generally take part in it. The actor of corruption tends to act as an official (a group of officials) or as a private person (a group of persons). The violation of formal rules of conduct by a public person in exchange for remuneration in a transaction is an act of corruption.

There are two possible reasons for an illegal transaction. In the first case, a private person claims for a benefit that belongs to them. However, a person faces challenges in processing papers and decides to give a bribe to accelerate the process. In the second case, an individual gives a bribe in order to get something that is considered illegal (Khan 2008). While considering this definition, it can be seen that in the first and second cases, the result of the interaction is the receipt of benefits by the two parties, with a variation of the objective. It may seem that such an interaction brings a positive result. However, the involvement of participants in corruption is considered a negative phenomenon in most cases. From the moral perspective, corruption is interpreted as a cynical misusing of laws and misunderstanding of social life norms and principles such as honesty, decency, and modesty. The result is evident—illegal enrichment of someone at the expense of others, the discrediting of official power and solidifying the power of corrupt officials, weakening healthy moral principles, and establishing immoral norms (Okhotskiy 2009).

Although at the individual level, in some cases, corruption can be presented positively, in fact, "everything is bad," not only at the individual level but also at the social level and the level of individual communities and groups.

The negative impact of corruption can be aggravated by social inequality, the difficulty in receiving free public services (educational, medical, etc.), which are important indicators of a person's standard of living. Additionally, corruption worsens the quality of services provided and turns free services granted by the government into paid by citizens, leads to tax evasion, non-efficiency of social programs, and the discrediting of social institutions that are created specifically for them. Corruption transfers resources from the masses of society to the elites and mainly from the poor to the rich. Corruption is rooted in the unequal distribution of resources in society and, in the future, tends to acquire institutional characteristics (Ardelyanova 2014).

Corruption has been experienced for a relatively long period. Currently, there is a mutual influence of corruption and the prevailing culture, structure of organizational and managerial activities, economic and political development, poverty, value orientations of citizens, "loopholes" in laws and their effectiveness, judicial and investigative practice, mechanisms of control and counteraction, etc. (Shakhtakhanov and Dmitrieva 2019).

## 2 Materials and Methods

The empirical basis for analyzing corruption practices among citizens results from sociological surveys among urban and rural citizens of Altai Krai aged 18–90 years conducted in 2018 ($n = 1{,}110$) and 2019 ($n = 1{,}259$). The subject of the study is the state and dynamics of petty corruption in Altai Krai. The research goal is to characterize petty corruption practices in Altai Krai in the course of history.

The analysis of the state and dynamics of petty corruption in Altai Krai was carried out based on the following indicators:

- The pervasiveness of petty corruption in Altai Krai;
- The degree of personal participation of citizens in corrupt practices;
- Involvement of citizens in petty corruption.

Corruption practices were analyzed utilizing 13 different situations of the interaction of citizens with municipal authorities:

- Receiving free medical care services in a clinic or a hospital;
- Interaction with pre-school institutions—day-care provision for young children or provision of day-care services;
- Secondary school—entering the desired school and successfully graduating from it;
- University—entering, transferring from one university to another, passing exams and tests, etc.;
- Pensions—registration, recounting, etc.;
- Social payments: registration of rights, recounting, etc.;
- Solving issues regarding avoiding mandatory military service or issues related to it;
- Acquiring a job or a promotion;
- Issues with a land plot—acquisition or registration of the right to it;
- Obtaining, registering, or privatizing residential space;
- Settling a matter in court;
- Police—request for help and protection;
- Traffic police—acquiring rights, violating rules, etc.).

The quantitative data were processed utilizing the statistical processing software SPSS Statistics 23.0, including frequency analysis and independence tests for contingency tables.

## 3 Results

The pervasiveness of corruption was measured using the question: "In general, have you ever found yourself in a situation where you knew, expected or felt that an informal reward, a bribe, was necessary to solve a problem—whether you gave it or not?" The value of this indicator in 2019 amounted to 16.2%, which is almost two times lower than in 2018, which represented 30.1% for that period. It was peculiar that the proportion of those who found challenging to answer mentioned question remained at the level of 10% (10.1 and 11.3%, respectively). A statistically significant relationship was found between the level of corruption pervasiveness and place of residence ($\chi^2 = 0.193$, $p \leq 0.001$)—corruption pervasiveness among the urban

**Table 1** Assessment by the population of changes in corruption, %

| | Increased | | Remained unchanged | | Decreased | | Difficult to answer | |
|---|---|---|---|---|---|---|---|---|
| | 2018 | 2019 | 2018 | 2019 | | 2019 | 2018 | 2019 |
| In the city, village | 22.7 | 17.2 | 42.4 | 45.0 | 12.7 | 10.0 | 22.2 | 27.8 |
| In the region | 32.0 | 21.3 | 38.6 | 39.2 | 9.2 | 9.1 | 20.2 | 30.5 |
| In Russia | 49.3 | 33.5 | 27.8 | 35.9 | 8.9 | 8.0 | 13.9 | 22.6 |

*Source* Compiled by the authors

citizens in 2018 was higher (33.5%) than among the rural citizens (24.6%). In 2019, the pervasiveness of corruption between urban and rural residents was almost at the same level (13.2 and 17.5%, respectively), which could be explained by the specifics of the 2019 sample, where the percentage of rural residents prevailed over urban residents (71.9 and 28.1%).

In the research framework, the respondents were also asked to evaluate petty corruption changes over the past two years at the federal, regional, and local levels. Table 1 presents the obtained data.

Evaluations of petty corruption changes at the country level in 2018 showed that citizens noticed an increase in corruption over the previous two years (49.3%). Simultaneously, almost a quarter of citizens (27.8%) noted the absence of changes. In 2019, the opinion of citizens was divided between the options "no change" and "increased corruption" (35.9 and 33.5%, respectively). In comparing the dynamics of perceptions of Altai Krai citizens about changes in petty corruption at the federal level, a decrease was noted in the share of people who believed that petty corruption grew and increased in the share of people who noted the absence of changes.

There also was a decrease at the regional level in 2019 in the share of citizens who had noticed an increase in petty corruption in Altai Krai over the previous two years. The proportion of citizens reporting the lack of change remained at the same level. However, there was a trend of an increase in the proportion of those who believed that corruption was rising, especially at the local level. Although that trend was more prevalent among rural citizens of Altai Krai, urban residents were more likely to express their opinion on high corruption.

There was a statistically reliable difference between opinions on changes in petty corruption at the local level. It differed between the urban and rural citizens ($\chi^2 = 0.231$, $\chi^2 = 0.128$, $p \leq 0.001$). The urban citizens tended to look at the situation from a more negative perspective than the rural citizens. Among urban citizens, almost a quarter of citizens (25.1 and 27.3%) reported increased corruption in their city (18.9 and 22% among rural). Only 7.2–7% of urban residents noted a decrease in corruption, in contrast to 22–16.6% among rural residents.

The research determined the level of demand for corruption, which was considered to be the proportion of cases in which the citizens gave bribes in corruption situations. The mentioned indicator showed citizens' willingness to resort to corruption to solve their problems or as an inability to resist to corruption pressure from the officials. The total sample of demand for corruption in 2018 was 30.8%. In other words, among those who once had found themselves in a corruption situation, almost one of the three eventually decided to give a bribe. In 2019, the percentage of citizens who gave the bribe amounted to 21%, which allowed us to point out the reduction in demand for corruption.

Then the demand for petty corruption was analyzed in the context of thirteen possible situations, describing how citizens solve their daily problems. The demand for corruption among citizens of Altai Krai was evaluated over the past two years. In 2018, the maximum value of demand for petty corruption amounted to 13.2% in those cases when a citizen had interactions with traffic police. In 2019, there was a decrease in the value of the leading indicator to 5.8%. Traffic police ranked first in the situation of the level of demand for petty corruption despite the declining index. Table 2 presents the obtained data.

We also noticed the general tendency of “narrowing” the demand for corruption among Altai Krai citizens. The 2019 results made it possible to identify six areas in which more than 5% of citizens gave bribes in the previous two years. Those areas include courts, primary and secondary education institutions, military service, universities, pre-schools, and traffic police. Traffic police retained its top position in 2018.

For some situations, statistically significant differences were revealed between urban and rural citizens in ways of solving the corruption situation. In the case of medical care services ($\chi^2 = 0.132, p \leq 0.001$), the demand for corruption was higher among urban citizens in contrast to rural citizens. In the case of traffic police ($\chi^2 = 0.206, p \leq 0.001$), the demand for corruption was on the same level. It meant that urban and rural residents were equally inclined to bribe to solve the traffic police

**Table 2** Demand for corruption, %

| The situation | 2018 | 2019 |
|---|---|---|
| Preschools | 5.1 | 4.7 |
| Social payments | 3.0 | 2.4 |
| Job | 2.3 | 2.0 |
| Getting free medical care | 4.5 | 3.8 |
| School | 7.4 | 4.0 |
| University | 5.7 | 1.0 |
| Solving problems related to conscription | 6.8 | 1.7 |
| Land plot | 3.4 | 1.0 |
| Housing area | 4.4 | 1.5 |
| Court | 7.8 | 2.3 |
| Traffic police | 13.2 | 5.8 |
| Pensions | 1.5 | 2.8 |
| Police | 2.4 | 3.2 |

*Source* Compiled by the authors

**Table 3** Corruption risk, %

| The situation | 2018 | 2019 |
|---|---|---|
| Preschools | 24.5 | 27.1 |
| Social payments | 15.0 | 20.4 |
| Job | 20.5 | 18.5 |
| Getting free medical care | 19.0 | 18.7 |
| School | 21.2 | 18.0 |
| University | 27.4 | 14.6 |
| Solving problems related to conscription | 23.9 | 16.5 |
| Land plot | 21.8 | 15.3 |
| Housing area | 21.8 | 26.3 |
| Court | 27.6 | 21.0 |
| Traffic police | 40.8 | 29.3 |
| Pensions | 14.2 | 26.1 |
| Police | 14.0 | 24.5 |

*Source* Compiled by the authors

issue. However, urban citizens were more inclined to not give a bribe in a corruption situation in contrast to rural citizens.

Another considered indicator in the study was the risk of corruption, which represented the proportion of cases when citizens were involved in a corruption situation during an interaction with public authorities. Table 3 presents the obtained data.

We analyzed the dynamics of the level of risk of petty corruption in the context of "narrowing" of the index spread. In 2018, the level of petty corruption risk was in the range of 14.0–40.8%. In 2019—14.6–29.3%. Traffic police were back in the lead position with a slight decrease in indicators. There was a decrease in the risk of corruption in such areas as work, school, university, military services, land plot, and court. The risk of corruption increased in such areas as pre-school institutions, social benefits, housing, pensions, and the police. The risk of corruption is established at a unified level in the situations regarding the medical care service.

Comparison of changes in the level of demand for corruption and the risk of corruption enabled us to divide all areas into two subgroups. The first subgroup was described as a "weakening" group, which has included areas for applying to the school, court, military service, land, and university—the level of demand for corruption and the level of risk of corruption decreased in these areas. The second subgroup—the "risk group"—included the police and pension issues. Areas of this group were characterized by increased demand for corruption and increased risk of corruption.

## 4 Discussion

The pervasiveness of corruption is seen as the proportion of citizens who have involved themselves in the corruption situation at least once in their life regardless of the outcome—have they paid a bribe or not. The pervasiveness of corruption in Altai Krai in 2018 and 2019 can be described as "relatively low." It is possible to note "decreases" in the scope of corruption in Altai Krai by 2019. Besides, the level of corruption pervasiveness among urban citizens is higher than among rural citizens, according to 2018 data. There has been a decline in the level of preparedness for corruption action among Altai Krai citizens.

Data on the perception of corruption at the federal, regional, and local levels have shown that the most significant increase in corruption has been perceived among citizens at the federal level. However, the citizens have no change in corruption level in both years at regional and local levels. However, it is essential to note the continued negative point of view of urban residents on the deterioration of the situation with corruption at the city level. The traffic police were the most corrupt area according to urban and rural citizens. In this situation, in addition to high indicators, statistically significant differences between urban and rural citizens in ways of solving the corruption situation were also revealed.

## 5 Conclusion

The latent nature of petty corruption directly impacts the difficulty of determining the level of corruption. General ideas about the level of petty corruption are based on various indicators. In this study, the state and dynamics of petty corruption were carried out by determining the pervasiveness of petty corruption in Altai Krai, the degree of personal participation of citizens in corruption practices, and citizens' inclusion in petty corruption. Differences were identified in citizens' assessment of corruption changes at the federal, regional, and local levels. Apart from that, according to some indicators, statistically significant differences were revealed depending on the territorial residence of citizens.

**Acknowledgements** The reported study was funded by RFBR, project No. 19-31-27001: "Social trust and corruption: problems of interdependence and transformation in contemporary conditions of regional society" (2019–2021).

## References

Ardelyanova YA (2014) Corruption, trust and inequality in modern societies. Mosc State Univ Bull. Ser 18. Sociol Polit Sci 2:212–219

Khan M (2008) Corruption and governance. Prognozis 1:125

Okhotskiy EV (2009) Socio-legal nature and the main signs of corruption. State Audit Law Econ 3:25–36

Shakhtakhanov EL, Dmitrieva EN (2019) Corruption as a social phenomenon and directions for improving the fight against corruption in the public service system. Stolitsa Nauki 6(11):158–167

# Systemic Challenges and Changes in Education Systems in Russia and Around the Globe

# Legal and Economic Regulation of Athletes' Remuneration in the Context of Sustainable Development of the Region

Olga A. Shavandina and Ekaterina Y. Kovalenko

**Abstract** The article is devoted to the development of legal and economic regulation of remuneration of athletes in order to determine the factors that affect their wages. In addition, the article analyzes the most discussed theoretical issues of legal regulation of labor relations in sports to attract attention from the state and the business community. In the context of sustainable development of the region and taking into account the problems of legal regulation of remuneration of professional athletes, the authors conclude that matrix modeling will determine the significance of individual indicators (indicators, achievements) of athletes and their impact on the formation of the premium part of wages. Providing information in matrix form will ensure its confidentiality.

**Keywords** Legal regulation of labor relations · Athletes' remuneration · Individual indicators of athletes · Matrix model of the formation premium payment

## 1 Introduction

The sustainable development of the Russian Federation as a whole and each region, in particular, presupposes the development of physical culture and sports as one of the important directions. This, in turn, contributes to increasing the population of the country, improving the citizens' standard of living, creating comfortable conditions for their living, opportunities for self-realization, and disclosure of each person's talent. The Strategy for the development of physical culture and sports in the Russian Federation for the period up to 2020, approved by order of the Government of the Russian Federation of August 7, 2009, No. 1101-R, notes that the concept of long-term socio-economic policy determines the role of physical culture and sports in the development of human potential in Russia.

This document declares the necessity of creating conditions for the population of the country to regularly engage in sports and lead a healthy lifestyle.

O. A. Shavandina (✉) · E. Y. Kovalenko
Altai State University, Barnaul, Russia
e-mail: shao07@rambler.ru



S. G. Maximova et al. (eds.), *Advances in Natural, Human-Made, and Coupled Human-Natural Systems Research*, Lecture Notes in Networks and Systems 234,
https://doi.org/10.1007/978-3-030-75483-9_57

Despite the improvement of the main indicators, the level of development of physical culture and sports does not correspond to the overall positive socio-economic transformations in the Russian Federation. At the same time, government expenses on physical education and sports are a cost-effective investment in the development of human potential and improving the quality of life of Russian citizens. Thus, the sphere of physical culture and sport faces global challenges, requiring the modern approach of solution.

According to Paragraph 3 of the Order of the President of the Russian Federation May 7, 2018, No. 204 *On National Goals and Strategic Objectives of the Development of the Russian Federation for the Period up to 2024*, the Government of the Russian Federation, when developing a national program of demographic development, should proceed from the fact that by 2024 it is necessary to ensure achieving the following goals and targets:

- Increasing the healthy life expectancy to 67 years;
- Increasing the total birth rate to 1.7;
- Increasing the proportion of citizens who lead a healthy lifestyle;
- Increasing to 55% the proportion of citizens who regularly engage in physical culture and sports.

Besides, the Government of the Russian Federation should concentrate on solving the following issues:

- Forming a system of motivating citizens to a healthy lifestyle, including a healthy diet and giving up bad habits;
- Creating conditions for all categories and groups of the population to engage in physical culture and sports through increasing the level of provision of the population with sports facilities, as well as preparing a sports reserve.

In recent years, there has been an improvement in the main indicators of the development of physical culture and sports due to the implementation of the Strategy for the development of physical culture and sports in the Altai Krai for the period until 2020, approved by order of the Administration of the Altai Krai of May 4, 2011 No. 100. Sports facilities are being built within the federal and regional target programs, and additional investments are being attracted to improve the infrastructure of the sport industry in the province. However, the pace and level of development of physical culture and sports, especially in municipalities, is noticeably lower than the overall positive trends in the region.

Thus, the most important factors that require increased attention from the scientific community are the human resourcing for the sports industry, motivation, and remuneration of professional athletes, and coaches. We suppose that these problems require deep analysis, legal regulation, and financial support for professional sports.

## 2 Materials and Methods

The main condition of labor relations between an employer and an employee is the salary, which is established by an employment contract under the payment systems of the organization. The athletes' remuneration depends on many factors, such as:

- Kind of sport;
- Amount of funding for a sports organization;
- Results achieved during training and competition;
- Terms of the employment contract concluded with the athlete, etc.

The professional athlete's renumeration includes the official salary, compensatory, and incentive payments. Bonuses (premiums) or incentives are an integral part of payments, and the amount is much higher than the salary and compensation payments (Alekseev 2012, 2018). The salary structure of athletes often does not have elements that would reduce its final value. Only some sports organizations apply a system of penalties.

We consider that the salary of the athletes should reflect all aspects of their professional activity. To highlight this problem, we use indicators based on the individual achievements of the athlete. The names of the proposed indicators and their reasons are presented in Table 1.

It is difficult for a sports organization to determine the significance of the presented in the indicator table. However, one can answer this question by making a matrix model that allows using indicators when a formation of the bonus part of the salary of each athlete sports organizations. Since the size of the official salary of the professional athlete is much lower than incentive payments, then the adequate reflection in

**Table 1** Indicators of individual achievements of an athlete

| Indicator | Reason for indicator |
|---|---|
| Training system effectiveness | Training methodology;<br>Competition system rationality;<br>Progressiveness of technology and tactics;<br>Material and technical support;<br>Scientific and methodological support;<br>Medical and biological support;<br>Information support |
| Sports achievements | Individual talent of an athlete;<br>Training level for sports achievement;<br>Socio-economic support |
| Professional ethics | Ethic relations in sports;<br>Corporate honor compliance |

*Source* Matveev (2019), Yagodin (2016)

the matrix model of individual achievements (in indicators) encourages athletes to maximize their bonus part of the salary.

Using the selected indicators, the matrix model can be represented as follows:

$$B = I \times FS, \tag{1}$$

where:

*B*—matrix column that defines the size of the premium payment. $(n \times 1)$,
*I*—matrix of individual indicators (indicators), size $(n \times m)$,
*FS*—matrix column of financial evaluation of indicators, size $(m \times 1)$,
*n*—number of athletes or groups,
*m*—number of indicators.

The matrix of individual indicators 1 can be formed on the basis of statistical data recorded by athletes-instructors, coaches, medical, and other employees of sports clubs (organizations) during the professional activity of athletes. This data is a part of reports that are submitted to higher-level organizations, leagues, federations, etc. The reliability of the indicators used in the model makes it apodictic.

Thus, the matrix model can reflect the individual characteristics of professional activities of athletes who are members of a sports organization. The presentation form ensures the confidentiality of information, the level of which can be determined for different groups of users.

## 3 Results

The empirical level of scientific knowledge includes the description of the observation results, namely, statistical data, and experiment. Using economic and mathematical modeling, we can evaluate the indicators of individual achievements of an athlete to form the premium part of the athlete's salary.

A large number of sports organizations operate on the territory of the Altai Krai. In terms of the experiment, we use the individual characteristics of the athletes of the university volleyball club. This club plays in the Highest League "A", has repeatedly been the winner, and prize-winner of the Games of the people of Siberia. The personal composition of the club includes:

- Head coach (the Director of the club);
- Senior coach;
- Seven stickers;
- Three blockers;
- Two passers;
- Two liberos.

We group the individual characteristics of athletes by their roles, which are only 4 (strikers, blockers, passers, and liberos), thus $n = \overline{1.4}$ determining the number of rows of the *B* and *I* matrices.

To form the matrix, we use data from "University" club athletes, which form the basis of a statistical report for higher-level organizations and financial partners. Quantitatively, the indicators "effectiveness of the training system" and "sports achievements" take positive values. The peculiarity of the indicator "professional ethics" is that it can have a negative rating. This is due to the inclusion of "penalties" in this indicator as an element of "moral attitude in sport." Penalties may include the following items:

- Overweight;
- Delay for training or a competition/game;
- Absence of training without a valid reason;
- Using mobile communications during training (competitions, games);
- Using prohibited drugs, alcohol;
- Misconduct with a coach, teammate, or opponents;
- Misconduct outside of the training and competition period, etc.

Each position of penalty has a negative rating in points and, as a result, reduces the amount of bonus payments to the athlete. This list can be developed and documented by the sports organization independently, and the labor contract of the athlete must have a reference to the administrative document, which approves the list of penalties and quantitative-monetary assessment of each position. The penalties for the module may exceed the positive component of the indicator "professional ethics," and, therefore, the final value of this indicator may take a negative value.

In Table 1, the authors have identified three indicators, each of which can be detailed. For example, all the positions of the penalties listed above can be defined as indicators of the second level.

In this study, we combine quantitative estimations of individual characteristics into three groups. The first group is the indicators of athletes that determine their achievements during the training (training camps, etc.) and correspond to the indicator "effectiveness of the training system." In the second group, we sum up the quantitative characteristics of athletes that they achieved during tournaments (competitions, games, etc.). This group of indicators corresponds to the "sports achievements" indicator. The reasons for the third group (indicator) "professional ethics" include quantitative assessment in points of penalties and moral activity, which is based on integrity, hard work, discipline, and other positive moral qualities, such as:

- Following the norms of behavior based on moral principles;
- Respect for the opponent;
- Adequate attitude to the competition rules and judges' decisions;
- Denial of doping and any artificial stimulation;
- Refusal to use dirty methods, tricks that prevent the opponent from tuning in to perform an exercise;
- Self-control of the athlete (the ability to restrain their emotions, to adequately perceive any outcome of the match);
- Traditions, noble behavior, etc. (Yagodin 2016).

**Table 2** General and average individual indicators of athletes

| Role of an athlete | | Training system effectiveness | | Sports achievements | | Professional ethics | |
|---|---|---|---|---|---|---|---|
| Role | Amount, ppl | Total points | Average score | Total points | Average score | Total points | Average score |
| Strikers | 7 | 563 | 80.4 | 845 | 120.7 | −85 | −12.1 |
| Passers | 2 | 193 | 96.5 | 219 | 109.5 | −29 | −14.5 |
| Blockers | 3 | 272 | 90.7 | 312 | 104.0 | −34 | −11.3 |
| Libero | 2 | 160 | 80.0 | 259 | 129.5 | −17 | −8.5 |
| Total | 14 | 1,188 | | 1,635 | | −165 | |

The number of indicators allows determining the second parameter of the size of the matrix of individual indicators $I$, that is the number of columns $m = \overline{1.3}$.

Let us imagine matrix $I$ in an expanded table form, using the data of the "University" club athletes, which are provided to partners and higher-level organizations in the form of reports (Table 2).

For all role groups, the "professional ethics" indicator has a negative value, since it reflects only penalties. The largest modulus indicator characterizes the "strikers," whose number of seven people is a half of the team. The average number of penalties for the first "striker" is just over 12 points, while the first "libero" has 3.6 points less. The highest average penalty rate of 14.5 points characterizes "passers." This value is six points or 70.5% higher than the lowest value in the third group of indicators.

Analyzing the sports performance of club members, it is worth noting that the indicator "sports achievements" is the most significant. The average rating of each athlete for this criterion exceeds 100 points, and the length of the range is 25.5 points or 24% of the lowest value. Individual indicators for "performance of the training system" are lower than the previous indicator ("sports achievements"). For athletes with the roles of "passers" and "blockers," the difference is only 13.5 and 14.7%, respectively, while for "strikers" and "libero," the percentage gap increases to 50–60%.

The next matrix model element is the financial component (FS). In general, the FS matrix column of financial evaluation of indicators can be represented by:

$$FS = \begin{pmatrix} S_1 \\ S_2 \\ S_3 \end{pmatrix} \tag{2}$$

The first and second indicators $S_1$ is $S_2$ are the cost of the first point of the indicators "effectiveness of the training system" and "sports achievements," respectively, and $S_3$ is the cost of the first point of penalties. Thus, the matrix model presents the following:

$$B = I \times FS = \begin{pmatrix} 563\ 845\ -85 \\ 193\ 219\ -29 \\ 272\ 312\ -34 \\ 160\ 259\ -17 \end{pmatrix} \times \begin{pmatrix} S_1 \\ S_2 \\ S_3 \end{pmatrix} \quad (3)$$

The result of multiplying two matrices is a column matrix $B$:

$$B = \begin{pmatrix} 563 \cdot S_1 + 845 \cdot S_2 - 85 \cdot S_3 \\ 193 \cdot S_1 + 219 \cdot S_2 - 29 \cdot S_3 \\ 272 \cdot S_1 + 312 \cdot S_2 - 34 \cdot S_3 \\ 160 \cdot S_1 + 259 \cdot S_2 - 17 \cdot S_3 \end{pmatrix} \quad (4)$$

The economic meaning of the B-matrix is determined by the economic content of each element. The first element, a line, defines the amount of bonus payments to all athletes who perform in the role of "strikers" (7 people). The number of athletes in this group is the largest, which determines an adequate amount of financial resources. The next largest bonus is the line number three, which establishes the correspondence between the athletes whose role is "blocking" (3 people), and the amount of their bonus payments. Lines 2 and 4 of matrix $B$ have the lowest financial rating. Indicators of these lines determine the amount of one-time incentive (bonus) payouts responsibly to "passers" and "libero." Representatives of these roles are the smallest (2 people each), which is a limitation when forming their premium fund.

To estimate the average bonuses per athlete, we use a matrix model in which the $I_{\text{cp}}$ matrix will contain the average individual performance of athletes.

$$B_{cp} = I_{cp} \times FS = \begin{pmatrix} 80.4\ 120.7\ -12.1 \\ 96.5\ 109.5\ -14.5 \\ 90.7\ 104.0\ -11.3 \\ 80.0\ 129.5\ -08.5 \end{pmatrix} \times \begin{pmatrix} S_1 \\ S_2 \\ S_3 \end{pmatrix} \quad (5)$$

As a result of multiplication, the matrix will take the form:

$$B_{cp} = \begin{pmatrix} 80.4 \cdot S_1 + 120.7 \cdot S_2 - 12.1 \cdot S_3 \\ 96.5 \cdot S_1 + 109.5 \cdot S_2 - 14.5 \cdot S_3 \\ 90.7 \cdot S_1 + 104.0 \cdot S_2 - 11.3 \cdot S_3 \\ 80.0 \cdot S_1 + 129.5 \cdot S_2 - 08.5 \cdot S_3 \end{pmatrix} \quad (6)$$

Indicators of the HRV matrix allow us to conclude that bonus payments will be the highest for athletes who are "libero" (Line 4), they have a high average score for the most significant indicator "sports achievements" and the lowest modulus value of the average score for "penalties." Athletes, whose role is "passers" (Line 2), have the highest number of penalty points on average for the first person, which causes a proportional reduction in the bonus. In the middle of the range of one-time incentive payments are bonuses for athletes with the roles of "strikers" (Line 1) and "blockers" (Line 3). A more detailed analysis can be performed if the financial evaluation of the indicators is determined, that are $S_1$, $S_2$ and $S_3$ will take specific values in conventional monetary units.

Thus, when forming bonus payments, the matrix model allows you to consider the professional achievements of athletes, their individual moral behavior, and corporate responsibility. Despite the possibility of using personalized (individual) indicators, the modeling process can provide the necessary level of confidentiality.

## 4 Discussion

One of the main objectives of the Strategy for the development of physical culture and sports in the Russian Federation is the need to improve the financial support of physical culture and sports activities. In this regard, it is relevant to consider the features of legal and economic regulation of remuneration of athletes. In the Labor code of the Russian Federation of December 30, 2001 (Labor Code) contains General rules governing the labor relations of all categories of employees and special rules establishing features of legal regulation of the work of athletes and coaches (Chapter 54.1 of the Labor Code Chapter 54.1 *Peculiarities of the Regulations of the Labor of Sportsmen and Trainers* by Federal law of February 28, 2008 No. 13-FL).

According to Article 135 of the Labor Code of the Russian Federation, the employee's salary is established by an employment contract by the current employer's payment systems. Collective agreements, local regulations under labor legislation, and other legal acts containing labor law norms regulate remuneration systems, including the following aspects: (1) amounts of tariff rates; (2) salaries (official salaries); (3) additional payments; (4) allowances of a compensatory nature, including for work in conditions that deviate from normal; (5) system of additional payments and allowances of a stimulating nature; (6) bonus system.

The terms of remuneration determined by the collective agreement, agreements, and local regulations may not be worsened in comparison with those established by labor legislation. Deductions from the employee's salary by Article 137 of the Labor Code of the Russian Federation are made only in cases provided for by this Code and other federal laws.

As for special rules, the provisions of Chapter 54.1 of the Labor Code of the Russian Federation regulate labor relations with employees whose labor function is to prepare for sports competitions and participate in sports competitions in certain types of sports (athletes). Also, it refers to employees whose labor function is to conduct training events with athletes and manage the competitive activities of athletes to achieve sports results (coaches). People defined by Chaps. 3 and 4 of Article 20 of the Labor Code of the Russian Federation may act as employers. The exception of employers are individuals who are not individual entrepreneurs. In general, the regulation of the activities of employees of sports organizations is established by labor legislation and other regulatory legal acts containing labor norms, collective agreements, and local rules adopted by employers under the requirements of Article 8 of the Labor Code of the Russian Federation in accordance with the rules approved by Russian sports federations, and the conclusion of the elected body of the primary trade union organization. Features of regulation of work of athletes, coaches, which under Article 252 Labor Code of the Russian Federation can be set exclusively by the Labor Code of the Russian Federation. Cases and procedures for establishing such features in other acts containing Labor law norms shall be determined by Chapter 54.1 of the Labor Code. The working hours of athletes and coaches, their involvement in overtime work, work at night, on weekends, and non-working holidays, as well as the specifics of remuneration of athletes and coaches for the specified period, are regulated by collective agreements, agreements, and local regulations, what is stipulated in article 348.1 of the labor code of the Russian Federation.

On the legal literature pages, many topical legal regulation issues of athletes' labor relations and their remuneration are discussed. In particular, it is noted that the labor legislation allows diapositives in the sphere of regulating the remuneration of athletes and coaches, providing an opportunity in local acts of corporate organizations to establish the specifics of regulating the remuneration of labor, considering the interests of all participants in labor relations (Batusova and Belitskaya 2012).

Aguzarov, considering the issue of remuneration of athletes and coaches and its legal regulation, points to the need to improve the combination of government and contractual regulation of remuneration for their work, which should be based on fixing a certain framework for each level of regulation. We suggest setting a salary cap for athletes and coaches, considering the norms approved by all-Russian sports federations, and adding Articles to the Labor Code of the Russian Federation that provide for specific remuneration for athletes and coaches (Aguzarov 2013).

Vavilkin and Tashiyan, studying the features of the labor contract of a professional athlete, note that the issues concerning the differences in the implementation of their sports activities by professional athletes based on either a civil contract or an employment contract are still a matter of scientific discussion (Barshai et al. 2014; Kurennoy 2014; Ponkina 2013). Besides, the authors point out that in sports, an employment contract that formalizes the labor relationship between an athlete and an employer dominates. They suggest including some limiting conditions in the content of employment contracts with professional athletes, for example, the following provisions:

1. Defining the duty of professional athletes to maintain their physical and psychological condition at the necessary level;
2. Restricting a professional athlete in the use of specific products, beverages, medicines, etc. in their free time, including restrictions on certain forms of private life;
3. Concerning the restriction of personal rights of athletes due to their performance of functions under an employment contract;
4. Defining the obligations of a professional athlete to comply with the rules and guidelines established by the sports organization (employer);

5. Defining the obligation of a professional athlete to undergo certain medical examinations and take care of their health (Vavilkin and Tashchiyan 2017).

Investigating the issue of applying a fine in practice as a disciplinary penalty against athletes, Popov points out the illegality of such application and suggests introducing the concept of "sports contract" into the legislation within the framework of the Institute of sports Contracting (Popov 2019).

Other authors pay attention to the term "sudden death of an athlete," regulate the procedure for mandatory insurance of professional athletes, propose to introduce the terms "sports season," "sports passport of an athlete" into labor legislation, as well as significantly expand guarantees and compensation for athletes and coaches in terms of remuneration, working hours, and termination of employment contracts (Makarova and Paliy 2016).

In Shevchenko's works, the analysis of foreign labor legislation regulating the work of professional athletes is carried out. In particular, it is indicated that in a number of countries, the labor of professional athletes is specifically allocated by the legislator and is regulated either in the labor legislation or in a separate law on sports (Shevchenko 2015). The author also pays attention to the prospects and directions of the development of international legal support for sports and labor relations in professional sports. In particular, regarding the concept of international standards of work in professional sports in terms of payment points to the necessity of securing provisions defining minimum social guarantees determined by the state and subject to recognition and support of sports organizations sports-labor and hierarchical relationships with professional athletes, including minimum standards of work, leisure, security, remuneration and social security of professional athletes (Shevchenko 2014). It is worth noting that the theoretical issues of legal regulation of labor relations in sports are quite relevant and debatable. Currently, some problems can be partially considered in the Strategy for the development of physical culture and sport in the Russian Federation for the period up to 2030, which is being developed and should be approved by the end of this year. Other problems identified by the scientific community and requiring early solutions can be implemented through special or local regulations.

## 5 Conclusion

Thus, the legal regulation of remuneration for professional athletes and coaches requires close attention from the state. According to the analysis of scientific papers (Alekseev 2012, 2018; Vavilkin and Tashchiyan 2017; Shevchenko 2014), the practice of law enforcement reveals gaps in the legal regulation of relevant relations.

When solving the problem of remuneration of employees of a sports organization, the matrix modeling technique can be considered, which allows considering individual achievements in the formation of bonus payments.

The validity of the matrix model makes it necessary to make changes to the local acts of sports clubs (federations, leagues, etc.). The section on bonuses of the Regulations on remuneration of employees of a sports organization (Federation, League, Committee, etc.) should reflect the cost of each point by type of indicators and contain methodological recommendations for the formation of a B-model to accrue one-time incentive payments.

In conclusion, we note the main features of the matrix model, which are that:

- Modeling is based on individual statistical data that is provided to higher-level organizations (leagues, federations, committees, etc.) and financial partners;
- Number of indicators can be expanded by detailing them with second-level indicators;
- Grouping by the role of athletes can be personalized by the personal roster of a team, club, or sports organization;
- Model can use indicators that allow you to reduce the amount of bonus payments when forming them;
- Flexibility of the model is provided by the financial component, which may depend on the amount of inflation, financial capabilities of the sports organization, and other objective factors;
- Calculation algorithm does not require the development and implementation of special software.

Matrix modeling, while maintaining the confidentiality of information, allows you to consider personalized indicators in the formation of one-time incentive payments to professional athletes and makes it possible to assess the impact of individual indicators on the value of the premium bonus.

Thus, the matrix model, as an economic component, can be considered as an element of the system of legal regulation of athletes' labor.

**Acknowledgements** The authors express their gratitude to the Russian Foundation for Basic Research for stimulating theoretical research of the most pressing problems in various branches of fundamental science. The authors are grateful to I. F. Voronkov, the head coach of the volleyball club, for providing information about the problems that the coaching staff and members of the club have to face in their professional activities.

## References

Aguzarov AM (2013) Legal regulation of salaries of athletes and coaches. Gaps Russ Legis 3:97–100

Alekseev SV (2012) Labor relations in sports. Features of remuneration of athletes, coaches, and other workers in the industry. Sport: Econ, Law, Manag 3:4–11

Alekseev SV (2018) Sports law. UNITY-DANA: Zakon i Pravo, Moscow, Russia

Barshai VM, Tashchiyan AA, Simavokyan KS (2014) Civil law regulation of relations in the field of physical culture and sports. In: Collection of works of the all-Russian scientific and practical conference. Rostov-on-Don, Russia

Batusova ES, Belitskaya IY (2012) Actual problems of labor legislation in the context of economic modernization. Yusticinform, Moscow, Russia
Kurennoy AM (2014) Globalization and its impact on the legal regulation of the labor of professional athletes: state and prospects. Sport 3:36–37
Makarova VA, Paliy OI (2016) Features of the legal regulation of the work of athletes. Polythematic Netw Online Sci J Kuban State Agrar Univ 01(115):798–818
Matveev LP (2019) General theory of sports and its applied aspects. Sport, Moscow, Russia
Ponkina AI (2013) Sports autonomy: theoretical and legal research. Spark, Moscow, Russia
Popov GA (2019) Fine as a disciplinary sanction against athletes: legality and prospects. Electron Suppl Russ Juridical J 2
Shevchenko OI (2014) Prospects and development trends international legal support sports labor relations in professional sports. RUDN J Law 3:198–207
Shevchenko OI (2015) Features of the legal regulation of the labor of professional athletes and coaches in foreign countries and Russian legislation. Cour Kutafin Mosc State Law Univ 4:111
Vavilkin DS, Tashchiyan AA (2017) Features of the employment contract of a professional athlete. Yurist-Pravoved 3(82):200–206
Yagodin VV (2016) Fundamentals of sports ethics. Izdatelstvo Uralskiy Universitet, Yekaterinburg, Russia

# Contemporary Postgraduate Studies at a Flagship University

**Oxana N. Tikhonova**

**Abstract** In this paper, the analyses of contemporary postgraduate studies effectiveness has been conducted in the context of a flagship university, in connection with key changes aimed at improving the training of scientific and scientific-pedagogical personnel in the framework of the national project "Science". Strategic objectives of postgraduate school development at a flagship university have been outlined based on the results of monitoring the postgraduate students' activities. This publication is aimed at analyzing the real state of contemporary postgraduate studies under conditions of a flagship university to set up strategic objectives. The methodological fundamentals of analyses are based on the motivational program-target management, according to which, the management function of planning is anticipated by a system analysis. We studied the effectiveness of postgraduate studies using the survey method, which was carried out anonymously, according to a questionnaire developed by the author, which includes closed and open questions aimed at researching the motivation for entering postgraduate school, the degree of postgraduate students' satisfaction with the educational process, and the productivity of their research activities. The survey elicited a high percentage of postgraduate students involved in research activities, active participation in grant and project activities, and high publication activity. The survey data allow us to determine the strategic tasks for the university in terms of the postgraduate studies development.

**Keywords** Postgraduate student · Postgraduate studies · Strategic objectives under flagship university conditions

## 1 Introduction

Currently, training scientific and scientific-pedagogical personnel within candidate's programs is characterized by a significant decrease in the number of dissertations defended, which is associated with the inclusion of candidate's programs into the

---

O. N. Tikhonova (✉)
Altai State University, Barnaul, Russia
e-mail: tihonova.oxana@yandex.ru

 
S. G. Maximova et al. (eds.), *Advances in Natural, Human-Made, and Coupled Human-Natural Systems Research*, Lecture Notes in Networks and Systems 234,
https://doi.org/10.1007/978-3-030-75483-9_58

higher education system regulated by the Federal Law *On Education in the Russian Federatio*n of December 29, 2012, No. 273-FZ (Russian Federation 2012) and entered into force in 2014. As a result of the reform, the academic workload for students training within the candidate's program has increased. More than that, the requirements for intermediate and final certification, as well as the system of final state certification, have changed for students receiving a candidate's degree. As part of implementing the National Project *Science* in 2020, the State Duma of the Russian Federation has adopted the draft law *On Amendments to Certain Legislative Acts of the Russian Federation Regarding the Training of Scientific and Pedagogical Personnel Within the Candidate's Programs* (General Committee of the Presidential Council of the Russian Federation for Strategic Development and National Projects 2018). This draft law, developed by the Ministry of Education and Science of Russia, introduces fundamental changes to improve scientific and pedagogical personnel training. The federal requirements underlying the project aim at students training within the candidate's program to defend their dissertations and earn the candidate's degree. However, these goals are intermediate in the strategy of students, and strong potential motivation can be provided by training potential managers of significant projects in relevant areas. A competitive person creating intellectual property in a high-tech business should result from the candidate's program. Therefore, postgraduate training should include a narrow circle of specialists immersed in scientific activities and should no longer be the next stage of training for people with predominantly extrinsic motivation. Stricter requirements should exclude the possibility of training a wide range of applicants, which will drastically reduce the number of students training within the candidate's program.

A system analysis of the real state of the candidate's programs under the conditions of the flagship university has allowed us to determine the strategic objectives for developing the candidate's programs and understand the level of motivation of candidate's students and their involvement in the training process and research activities.

The paper aims to analyze the effectiveness of modern candidate's programs under the context of the flagship university and set strategic goals.

## 2 Materials and Methods

The motivational target-oriented program approach has formed the methodological basis of the study (Shalaev 2006).

I have used the following research methods to conduct the system analysis:

- Analysis of the reported documentation (Shalaev 2006).
- Survey based on the technique that I have developed and containing closed and open questions aimed to study the following issues:

  1. Motivation for joining the candidate's program;
  2. Features of the educational process and emerging difficulties of learning;

3. Degree of satisfaction with training and interaction with the dissertation advisor;
4. Degree of involvement in research activities and the implementation of grants and projects;
5. Publication activity;
6. Secondary employment;
7. Plans for professional activities.

- Anonymous questioning.
- Processing and analysis of statistical data.

## 3 Results

According to the analysis of documents of the enrollment campaign for 2019, 81 applicants have joined the candidate's program (54 full-time and 27 part-time students). These indicators exceed those in 2018 by 20% (16 applicants). Moreover, the number of candidate's students enrolled in full-time education has increased by 23% (19 applicants).

Under the agreement on providing paid educational services, 59 applicants have been accepted for training (32 full-time and 27 part-time students). The number of candidate's students studying for a fee has increased by 31% (18 students) compared to 2018. At the same time, full-time education predominates and demonstrates an increase of 36% (21 students).

The number of foreigners wishing to train within the candidate's program at the Altai State University and accepted to the first year of study has amounted to 18 applicants, which corresponds to 2018.

As of December 31, 2019, the number of candidate's students amounted to 246 people (163 full-time and 83 part-time students).

In 2019, there were 58 educational programs, included in 16 integrated groups of training areas, for training scientific and pedagogical personnel within the higher education programs, particularly candidate's programs.

Full-time and part-time first-year candidate's students of the 2019–2020 academic year have participated in the survey. The number of respondents has amounted to 81 students (54 full-time and 27 part-time students). Among them, there are 38 women and 43 men, as well as 18 foreign citizens.

When studying the motivation for joining the candidate's program, I did not limit the respondents to the number of answer options. The survey has found that the majority of the first-year candidate's students (70%) aim to defend a candidate dissertation. One-half of the respondents (50%) are interested in scientific work and desire to increase their educational level. About one-third of the students (34%) are guided by recognizing the prestige of the degree diploma and focusing on building a successful career as a university teacher. For 24% of the respondents, training within the candidate's program is the next stage of the study. Another 10% of the students consider the possibility of deferring military service.

I should note that 93% of the candidate's students are satisfied with the educational process. Almost everyone (97%) is well-informed about the curriculum and acquainted with the terms of study, mainly due to browsing the university website or contacting the dean's office (48%).

The structure of the candidate's education program meets the expectations of the candidate's students. The survey has found that 69% of them assess it as optimal, 17% find it complex and overly voluminous, and 14% consider the candidate's program easy.

Some candidate's students (31%) note the need for elective courses aimed at the in-depth study of professional disciplines (legal psychology, English, and information technology).

The candidate's students (76%) are more satisfied with the lecture halls and libraries of the university, as well as the laboratory equipment necessary for conducting experiments on the scientific research topic. However, they note the absence of a computing cluster for processing large amounts of data and the need for technologization of the library.

At the same time, 83% of the candidate's students are satisfied with computer classrooms and an electronic educational environment that allows them to perform independent work effectively. However, they note the need to update technical means.

In a significant part of cases (66% of the students), the dissertation advisors have formulated the dissertation research topic considering their scientific interests and experience. Another part of the students (28%) has formulated the research topic independently based on their interests. As for an individual plan, 79% of the candidate's students have made it together with dissertation advisors. Most students (90%) are engaged in research with pleasure, while 10% are reluctant and force themselves. More than one-half of the candidate's students (52%) note that their research interest has changed since entering the candidate's education program: the respondents report an increase in interest (42%) or a partial loss of interest (10%). Most students (93%) training within the candidate's program work closely with their advisors, whereas 7% complain that their advisors are busy.

Almost one-fourth of the respondents (24%) meet with their dissertation advisors every week, 38% of the students meet the advisors once in two weeks, 29% note that they meet with their advisors once a month, and 9% of the candidate's students meet their dissertation advisors every three months.

More than a half of the respondents (66%) emphasize the high efficiency of their work with the dissertation advisors since they latter take an active part in research providing theoretical, methodological, and organizational assistance when preparing research and teaching within the candidate's program.

About one-third of the candidate's students (28%) are satisfied with solving individual questions and problems. Simultaneously, 6% of the respondents are undecided in assessing the effectiveness of dissertation advisors in joint research work.

An insignificant part of the students (7%) notes an increase in the effectiveness of interaction with the dissertation advisors at the scheduled meetings. I should note that 17% of the respondents consider the distance format of joint work as a tool to increase this effectiveness.

The scientific library renders substantial assistance in conducting research work by the candidate's students by providing access to foreign and domestic databases.

Furthermore, there is a system of intramural competitions for grants and projects (in which 40% of the candidate's students participate), as well as an academic mobility program (which provides young scientists with foreign business trips and participation in conferences of different levels), but constant live communication with advisors remains a priority for 81% of the respondents.

The candidate's students consider the following options as the most effective for organizing work on scientific research:

- Combining scientific research with working outside the university, which coincides with the research topic;
- Combining the implementation of scientific research with participating in the research projects of faculties or departments under the guidance of the dissertation advisors (or jointly with them);
- Combining the implementation of scientific research with working in a scientific and educational laboratory on a similar topic.

By the end of the first year of study, almost all candidate's students (97%) have published their papers in journals reviewed by the Higher Attestation Commission of the Russian Federation.

More than one-half of the respondents (57%) consider their prospects and plans to continue their scientific and teaching activities at Altai State University as possible. About one-third of the candidate's students (33%) work in other organizations. Almost one-fourth of the respondents (24%) find it difficult to answer.

The respondents note that it is important for the university to take decisive steps in opening dissertation councils in promising scientific areas making training within the candidate's program even more interesting and productive.

At the present stage, young scientists of Altai State University are actively engaged in active research in 16 accredited areas for training scientific and pedagogical personnel within the candidate's programs.

Currently, three councils for the defense of doctor's and candidate's dissertations are successfully working at the Altai State University.

In 2019, 32 candidate's students received diplomas with the qualification "*Researcher. Educator-Researcher*," and four candidate's students defended their candidate's dissertations after graduation.

## 4 Discussion

At the present stage, according to Karmazinsky, any university aims to create a stream of external investments to support the research of candidate's students. He considers training within the candidate's programs as a source for creating intellectual property and intangible assets, as well as a basis for creating an innovative business.

In this case, the following competence indicators are necessary for candidate's students (Karmazinsky 2012):

- Ability to independently provide resources for research and development;
- Creation of intellectual property;
- Possession of methods and techniques of conceptual thinking;
- Prediction of trends for developing scientific directions for building models, objects, and technologies, and transforming intangible assets into tangible ones;
- Ability to present scientific activities through PR campaigns;
- Ability to collaborate for implementing ideas and developments;
- Scientific and technical activities interconnected with *infrastructure* activities, including those at the international level.

In the Altai State University, the Center for Developing Technological Entrepreneurship, Technology Transfer, and Intellectual Property Management, as well as Student Business Incubator, largely provide this relationship.

## 5 Conclusion

Thus, developing the training within the candidate's programs under the context of the flagship university pursues a set of goals and objectives.

One of these goals is a rebranding of promising areas for candidate's programs, as well as positioning the most *potent* scientific topics and specialties mastered within the candidate's programs, and promoting and popularizing scientific achievements on the Internet and in print media.

Expanding cooperation with leading foreign universities on the implementation of joint training within the candidate's programs and the number of existing partner universities through the system of open scientific and practical seminars (webinars) for candidate's students (including those with the participation of representatives of the Russian Academy of Sciences and leading foreign universities) is another goal pursued by the flagman university.

More than that, developing candidate's programs within the flagship university aims to strengthen the research work of candidate's students through the disciplines that form professional competencies (writing papers, preparing and forming grants in the scientific field, etc.) and introduce a system for international dissemination of research results for candidate's and doctor's students.

Involving candidate's and doctor's students into the activities of scientific laboratories and university centers based on project activities, as well as providing support for the research activities of candidate's and doctor's students via academic mobility, is also necessary for developing the candidate's programs.

Besides, the system analysis has shown that the candidate's and doctor's education sector necessitates the development of a model for effective training within the candidate's and doctor's programs using qualimetric standards and rules that allow regularly monitoring the quality of training candidate's and doctor's students (including evaluating the effectiveness of candidate's dissertation advisors according to the requirements of each education year, as well as the effectiveness of training within the candidate's and doctor's programs according to the established criteria).

## References

General Committee of the Presidential Council of the Russian Federation on Strategic Development, and National Projects (2018) Passport of the national project "Science" (December 24, 2018 No. 16). Moscow, Russia

Karmazinsky AN (2012) Postgraduate school: problems and solutions. Mod Probl Sci Educ 6:323

Russian Federation (2012) Federal law "On Education in the Russian Federation" (December 29, 2012, No. 273, as amended has come into force since July 1, 2020). Moscow, Russia

Shalaev IK (2006) Program-targeted management psychology. Barnaul State Pedagogical University, Barnaul, Russia

# Strategic Directions for Continuing Professional Development of Medical University Instructors in Russia

**Irina Y. Plotnikova, Alexandr A. Filozop, Svetlana Yu. Berleva, Anna V. Podoprigora, and Ilya V. Stepanov**

**Abstract** Within the paper, we consider the degree of scientific development of strategic directions in the sphere of continuing professional development (CPD) of educators at medical universities and identify the principal approaches towards the pedagogical aspect of continuing professional education at Russian medical universities. Bibliometric analysis of pedagogical courses from the journals included in the National Bibliographic Database of Scientific Publications (Russian Science Citation Index [RSCI]) is a key method to investigate this problem. The study has found a low level of publication activity of Russian authors and critical state of the pedagogical component in the CPD system at medical universities in Russia. Publications included in the bibliometric analysis cover a wide range of issues related to the following topics: (1) professional competencies of the academic staff of medical universities; (2) application of innovative approaches in the postgraduate training of medical educators; (3) patterns providing CPD for medical educators. However, publications in the above-listed areas have appeared to be the views of the authors on the problem with no factual theoretical and methodological basis that neglect, to a greater extent, global experience in coping with CPD challenges of medical educators. Provided CPD patterns for medical educators have an exclusively local vector of application and have not been tested for effectiveness.

**Keywords** Continuous professional development · Additional professional education · Medical educators · Patterns of continuing professional education

I. Y. Plotnikova · A. A. Filozop · S. Yu. Berleva (✉) · A. V. Podoprigora · I. V. Stepanov
N.N. Burdenko Voronezh State Medical University, Voronezh, Russia
e-mail: romazzzhka@mail.ru

I. V. Stepanov
e-mail: ivstepanov@vrngmu.ru



S. G. Maximova et al. (eds.), *Advances in Natural, Human-Made, and Coupled Human-Natural Systems Research*, Lecture Notes in Networks and Systems 234,
https://doi.org/10.1007/978-3-030-75483-9_59

## 1 Introduction

The need for significant changes and transformations of the education system in Russia arose at the turn of the 20th–21st centuries. This was due to several factors. First of all, after the collapse of the USSR in 1991, the crisis that involved the economic, legal, and socio-cultural spheres of the post-Soviet society, lasting for a decade, contributed to the destabilization of the Russian education system and its gradual destruction. Many scientists left the country, state universities did not receive adequate funding, payments for research and educational activity were low and untimely, the social status of the teacher was devaluated in the public perception—all these factors inevitably contributed to a decrease in the quality of higher education, including medical universities.

Second, at the beginning of the twenty-first century, due to the intensification and dynamics of the developing information society and new challenges to humanity in various spheres of life, the need to continually develop novel knowledge, skills, and competencies was globally accepted to be a must. These competencies would help persons navigate the complex dynamics of current processes and make them adaptive, creative, and competitive in the ever-changing social space. By the 90s years of the twentieth century, the shift from the paradigm *Education for all Life* to the paradigm *Education over a Lifetime* has been completed. In the first decade of the twenty-first century, high-quality lifelong learning became firmly established as one of the significant conditions for the successful development of an individual and modern society, being the key to solving the global problems of humanity in the third millennium (Grudeva and Dimitrova 2018).

*National Doctrine of Education in the Russian Federation,* approved by the Government of the Russian Federation in 2000, has become the first strategic guideline in reforming Russian education. It has established the priority of education in public policy and reflected major perspectives of its development until 2025. Creating conditions for continuing professional education throughout the life of a person has become one of the domains in reforming the education system in Russia (Government of the Russian Federation 2000).

In 2003, the Russian Federation signed the declaration at the meeting of European Ministers of Education in Berlin and joined the Bologna Process.

In 2012, the State Duma of the Russian Federation adopted the new Federal Law No. 273-FZ *On Education in the Russian Federation*, which established the legal, organizational, and economic foundations of education in the Russian Federation, basic principles of the state policy of the Russian Federation in the field of education, general rules for the functioning of the education system and the implementation of educational activities, and determined the legal status of participants in the sphere of education (Russian Federation 2012).

In 2017, the Government of the Russian Federation approved the state program of the Russian Federation *Development of Education* for the period 2018–2025, where the strategic objectives for reforming higher education were as follows (Government of the Russian Federation 2017):

- At least 70% of the academic staff in every educational institution of higher education that provides training for the basic sectors of the economy and the social sphere must continuously update their professional knowledge and competencies based on the current achievements of science and technology, modern professional requirements, and perspectives of their professional sphere by 2024 (Government of the Russian Federation 2017);
- A system of continuous professional development of the academic staff in partnership with organizations of the real economic sector should be formed by 2024 (Government of the Russian Federation 2017).

In the documents on the long-term socio-economic projected growth of the Russian Federation for the period until 2030, developed by the Ministry of Economic Development of the Russian Federation in 2018, it is noted that modernization of the education system, which is the basis for the dynamic economic growth and social development of society, is a pre-requisite for forming the innovative economy. Modernization of professional higher education is impossible without a high-quality update of the teaching staff, which involves the restructuring of the advanced training system for the academic staff and essential measures to improve the quality of management and further training of administrators of the professional education system (Ministry of Economic Development of the Russian Federation 2013).

Thus, at present, the *National Doctrine of Education in the Russian Federation*, agreements of the Bologna Process, the Federal Law *On Education in the Russian Federation*, and the state program of the Russian Federation *Development of Education* until 2025 appear to be the conceptual basis for developing strategic directions in ensuring the continuous professional development of the teaching staff of medical universities in Russia.

## 2 Materials and Methods

The paper aims to investigate the degree of scientific development of strategic directions in the sphere of continuing professional development (CPD) of educators at medical universities and identify the principal approaches towards the pedagogical aspect of continuing professional education at medical universities in Russia. In order to achieve the research goal, we have performed a bibliometric analysis of pedagogical courses published in the journals included in the National Bibliographic Database of Scientific Publications (Russian Science Citation Index [RSCI]), being the largest scientific electronic library in Russia (eLIBRARY.RU).

We have selected publications according to the following criteria:

- Publications should have been published in peer-reviewed scientific journals;
- Publications should have been published within 2016–2019;
- Publications should include keywords for search in eLIBRARY.RU, such as *continuing professional development of medical educators/academic staff at*

*medical universities* and *continuing the pedagogical education of medical educators/academic staff at medical universities.*

Considering the given criteria, we have selected 29 papers highlighting strategic directions for ensuring the continuous professional development of educators at medical universities for bibliometric analysis (six papers published in 2016, seven papers published in 2017, eight papers published in 2018, eight papers published in 2019).

## 3 Results

The results obtained have demonstrated low publication activity of Russian scientists on the CPD problem of educators at medical universities. This fact has resulted from the lack of interest in this issue among the Russian scientific and pedagogical community and the critical state of the system of additional professional education of educators at medical universities. The permanent reform of the education system in Russia lasting for two decades has not currently resulted in developing uniform standards for the pedagogical training of educators at medical universities and clear criteria for assessing their professional competence, as well as adopting a strategy to provide CPD over a lifetime. This challenge to develop the state system of continuing professional growth of academic staff is set in the state program of the Russian Federation *Development of Education* (Government of the Russian Federation 2017) and will have to be solved only in 2024.

The European project *Tempus IV 159328-TEMPUS-1-2009-1-FR-TEMPUS-SMHES Life-Long-Learning Framework for Medical University Teaching Staff* (2010–2012), attended by six Russian medical schools and European universities from France, Germany, Luxembourg, and Italy, has had a significant impact on determining the strategy for providing the continuity of education of the medical teaching staff at the local level (Novikov et al. 2012). In this regard, publications in the RSCI on CPD of educators at medical universities have reflected primarily the experience and perspectives of further improving the system of advanced training of academic staff at the local level (in particular medical universities).

Bibliometric analysis has revealed the most publication activity on the studied issues:

- Six publications in Kursk (Kursk State Medical University);
- Six publications in Saint Petersburg (four publications in Saint Petersburg State Pediatric Medical University and two publications in North-Western State Medical University named after I. I. Mechnikov);
- Four publications in Krasnoyarsk (Krasnoyarsk State Medical University named after Professor V. F. Voino-Yasenetsky).

Here is the full list of publications and affiliations of the authors:

- Krasnoyarsk State Medical University named after Professor V. F. Voino-Yasenetsky (Krasnoyarsk, Russia)—four publications;
- North-Western State Medical University named after I. I. Mechnikov (Saint Petersburg, Russia)—two publications;
- Saint Petersburg State Pediatric Medical University (Saint Petersburg, Russia)—four publications;
- Omsk State Medical University (Omsk, Russia)—two publications;
- I. M. Sechenov First Moscow State Medical University (Moscow, Russia)—one publication;
- Rostov State Medical University (Rostov-on-Don, Russia)—one publication;
- Pirogov Russian National Research Medical University (Moscow, Russia)—two publications;
- Kursk State Medical University (Kursk, Russia)—six publications;
- Voronezh State Medical University named after N. N. Burdenko (Voronezh, Russia)—one publication;
- Volgograd State Medical University (Volgograd, Russia)—two publications;
- Smolensk State Medical University (Smolensk, Russia)—one publication;
- Tyumen State Medical University (Tyumen, Russia)—one publication;
- Ryazan State Medical University named after academician I. P. Pavlov (Ryazan, Russia)—two publications;
- Yaroslavl State Medical University (Yaroslavl, Russia)—one publication.

Issue-related bibliometric analysis of the selected publications has revealed several research areas in the CPD problem of educators at medical universities (Table 1).

The first section, *Professional and Pedagogical Activity Aspects of the Academic Staff at Medical Universities*, has included publications investigating the structural (gnostic, design, constructive, organizational, and communicative) components of the professional and pedagogical activity of the academic staff of medical universities. Theoretical analysis has included various viewpoints on the concepts, such as *competence*, *competencies of medical educators*, and *competence-based approach*

**Table 1** Issue-related publications on the CPD problem of educators at medical universities

| No | Issue | Number of publications |
|---|---|---|
| 1 | Professional and pedagogical activity aspects of the academic staff at medical universities | 2 |
| 2 | Using e-learning, distance learning, and media-oriented educational approaches for developing the professionalism of educators at medical universities | 3 |
| 3 | Professional culture and professional competencies of educators at medical universities | 7 |
| 4 | Designing a CPD model for educators at medical universities | 17 |

*Source* Compiled by the authors

*considering professional standard* in the CPD organization of educators at medical universities. (Lonskaya and Malyutina 2016; Saharov 2018).

The second section, *Using E-learning, Distance Learning, and Media-oriented Educational Approaches for Developing the Professionalism of Educators at Medical Universities*, has included publications on the following issues:

- Modern patterns of media-oriented education, actively used for the professional development of educators at medical universities;
- Options and limitations of distance learning technologies in the advanced training of educators at medical universities, as well as information and communication technology (ICT) competence of educators at medical universities as an integral part of their pedagogical skills (Getman and Kotenko 2017).

The third section, *Professional Culture and Professional Competencies of Educators at Medical Universities*, has included publications discussing the key competencies of educators at medical universities and offering different viewpoints on the concept *of the professional culture of a modern teacher* (Pahomova and Rodikov 2016).

The fourth section, *Designing a CPD Model for Educators at Medical Universities*, has included most publications (17 papers). Publications have highlighted the CPD patterns of educators at medical universities evaluated at medical universities where the authors work. Despite the diversity of the presented patterns, one thing is in common that the authors have offered their solutions to one of the critical challenges of organizational and methodological nature in the CPD sphere, namely, the integration of formal and non-formal education and self-development of educators at medical universities.

## 4 Discussion

Formulating the following fundamental principles for providing CPD of educators at medical universities is recognized as innovative for the Russian system of additional professional education:

- Principle of autonomy of learners;
- Principle of professionally-oriented learning;
- Principle of problem-based learning;
- Principle of innovation-oriented learning (Gavrilyuk et al. 2017; Mandrikov et al. 2018; Markovina et al. 2017).

However, in this section, there have been no publications related to the study of the effectiveness of the CPD patterns proposed for educators at medical universities.

## 5 Conclusion

Bibliometric analysis of publications from the National Bibliographic Database of Scientific Publications of Russia, published over the three past years and highlighting strategic CPD directions for educators at medical universities, has allowed reporting the current state of the problem at Russian medical universities.

The results obtained have demonstrated low publication activity of Russian authors, which evidences the critical state of the system of additional pedagogical professional education of educators at medical universities in the Russian Federation. Issue-related bibliometric analysis has revealed four main research areas regarding the CPD problem of educators at medical universities. They cover a wide range of issues, such as the following: (1) professional competencies of academic staff at medical universities; (2) use of innovative forms of postgraduate training of educators; (3) CPD organization patterns for educators at medical universities.

However, publications in all areas are only the views of the authors on the problem with no factual theoretical and methodological basis that neglect global experience in coping with CPD challenges of medical educators (the average number of sources in the *References* is 13 items, and only 10% of the sources are foreign publications on the studied issue). The patterns proposed for providing CPD of educators at medical universities have an exclusively local vector of application and have not been tested for effectiveness.

Ultimately, we should note that at present, there is a situation in the Russian system of additional professional education of medical educators where strategic federal documents announcing innovative changes in the educational system neither normatively nor organizationally regulate a single approach in a real academic environment to ensure CPD of the academic staff at medical universities.

## References

Gavrilyuk OA, Nikulina SY, Avdeeva EA, Artyuhov IP (2017) Organization of the system of personal and professional development of a medical university teacher: a problem and solutions. J Novosib State Pedagog Univ 7(3):19–36

Getman NA, Kotenko EN (2017) Media-oriented educational options of the Department of Pedagogics and Psychology in the system of additional professional education of a medical university. Int J Exp Educ 8:22–27

Government of the Russian Federation (2000) Resolution of the Government of the Russian Federation "National doctrine of education in the Russian Federation" (October 4, 2000, No. 751). Moscow, Russia

Government of the Russian Federation (2017) State program of the Russian Federation "Development of Education" for the period 2018–2025 (December 26, 2017, No. 1642). Moscow, Russia

Grudeva MI, Dimitrova YS (2018) Strategic directions for improving the professionalism of teachers of higher educational institutions. Educ Self-Dev 13(3):59–65

Lonskaya LV, Malyutina TV (2016) The content of the components of professional and pedagogical activity of a teacher at a medical university. Mod Probl Sci Educ 6. http://www.science-education.ru/ru/article/view?id=25691

Mandrikov VB, Artyukhina AI, Chumakov VI (2018) Continuing pedagogical development of teachers at medical universities: problems and perspectives. Med Educ Univ Sci 2(12):55–58

Markovina IY, Gavrilyuk OA, McFarland J (2017) International continuing education program for teachers of English in medical universities: a new model. Med Educ Univ Sci 1(9):60–64

Ministry of Economic Development of the Russian Federation (2013) Forecast of the long-term socio-economic development of the Russian Federation for the period until 2030, March 2013. Moscow, Russia

Novikov AI, Vodolazskiy NB, Tvorogova ND (2012) Modern approaches to lifelong learning in the profession of teachers at medical universities: an analytical review, 1st edn. Poligrafichesky Tsentr KAN, Omsk, Russia

Pahomova RA, Rodikov MV (2016) Professional culture of a teacher at a medical university. J Mod High Technol 6(1):179–182

Russian Federation (2012) Federal law "On Education in the Russian Federation" (December 29, 2012, No. 273-FZ). Moscow, Russia

Saharov SP (2018) Competence of a medical university teacher taking into account a professional standard. Eur Soc Sci J 11:303–308

# Interaction of Lecturers and Students in Information and Educational Environments of the University

**Svetlana A. Burilkina, Nelli G. Suprun, and Aleksey V. Tomarov**

**Abstract** In the current conditions of higher education development, the issues of modeling the information educational environment of a higher education institution come to the fore. The research aims to study the interaction between teachers and students in the educational process when solving the problem of students' absenteeism and academic debts, difficulties that hinder the active interaction between teachers and students in a multifunctional digital educational space. The research uses the method of surveying teachers and students. The authors analyze (1) the content, structure, and dynamics of changes in the educational space and (2) the relationship with the components and indicators of information culture of educational activity subjects in the educational process and counseling work. The authors identify problems associated with the insufficient use of current information technologies by teachers in educational and counseling processes. The urgent task of today's university teachers is to develop a method and methodology for teaching a new generation of students in the information space. Nowadays, the education faces the task of preparing students for professional activities in the information environment. This causes certain difficulties. Problems of interaction between teachers and students in the e-learning process are associated with the high intensity of teachers' work and insufficient use of other online learning tools—webinars, web conferences, etc. The research shows that teachers and curators need to adapt to new conditions and professional and pedagogical activity requirements. The identified problems of the educational space must be considered when establishing an effective interaction between students, teachers, and university executives.

**Keywords** Information and educational environment · Educational organization · Information culture · Educational work · Curator · Student · Teacher · Interaction of educational activity subjects

S. A. Burilkina (✉) · N. G. Suprun · A. V. Tomarov
Nosov Magnitogorsk State Technical University, Magnitogorsk, Russia
e-mail: svetlanaburilkina@yandex.ru



S. G. Maximova et al. (eds.), *Advances in Natural, Human-Made, and Coupled Human-Natural Systems Research*, Lecture Notes in Networks and Systems 234,
https://doi.org/10.1007/978-3-030-75483-9_60

## 1 Introduction

The information educational environment of universities requires a search for new tools, forms, and methods of teaching specific disciplines.

The electronic information and educational environment of the university provide access to curricula, work programs of disciplines, practices, electronic educational resources, and publications of electronic library systems. University teachers need to keep statistics and record the progress of the educational process, the results of the midterm assessment, and mastering the program. Conducting classes and assessing learning outcomes must be presented to all educational process participants in electronic form. The implementation of these activities is carried out using the educational portal, where the teacher and the student interact. Sometimes teachers and curators have difficulties when communicating with students. This happens due to the academic debts of students, absenteeism, and the problems of teachers that arise in pedagogical interaction with students. The work of a teacher is assessed by (1) the number of indebted students, (2) the level of activity on the educational portal, and (3) the results of qualitative and quantitative performance. Consequently, the amount of work is increasing due to the need to combine a high teaching load, methodological work, and intensive research and publication activities.

The paper analyzes the problem of interaction between teachers and students in the information and educational environment of the university. This system exists separately and can be diagnosed, predicted, simulated, and designed depending on the students' needs.

## 2 Materials and Methods

The methodological base of the research is the provision of philosophy, sociology, psychology, and pedagogy. They reveal the general scientific category of "interaction" as a reflection of the relationship between the individual and society. The interaction of a teacher and a student is considered as a phenomenon that integrates the aspects of a persons' value assimilation of the surrounding reality, educational, and counseling processes in the information environment of the university.

The authors rely on concepts that reveal the development features of the information society (Bell 1986). In particular, they rely on the increasing role of information processes in the social institutions of current society. Bell, in his work "The social framework of the information society," substantiated the importance and relevance of codified theoretical knowledge for the implementation of educational innovation (Bell 1986).

Such scientists as E. K. Henner, E. G. Belyakova, L. K. Geikhman, V. V. Kulish, N. A. Matveeva, N. V. Nazarov, and N. F. Radionova worked within the theory of pedagogical interaction, communication, and technologies of their implementation. The works of D. Masoumi, I. S. Batrakov, V. G. Gladkikh, V. A. Novikov, N. K.

Sergeev, E. I. Tikhomirov, etc., are devoted to the research problems of current university education and educational systems (Masoumi 2015; Mrochko et al. 2019; Novikova 2008).

The research object is the interaction of the teaching staff and students of Nosov Magnitogorsk State Technical University in the educational process of the information and educational environment of the university.

The research consists of a survey of 400 students of various fields and 120 teachers. It analyses the questionnaire results conducted in April–May 2019 by the Laboratory of Social and Psychological and Pedagogical Research of the Social Work, Psychological and Pedagogical Education Department. The research aims to study the interaction of teachers and students in the educational process of the university. The authors analyze the problem of students' academic debts, absenteeism, and teachers' difficulties arising in pedagogical interaction with students.

## 3 Results

One of the research objectives is to study the activities of teachers and curators to support the students. Curators of student groups are the most important subjects of the educational process. They are both a source of information for students and teachers of academic disciplines. According to the survey data, among the surveyed teachers, 74 are curators. Worth noticing that some curators have more than one group—from two to four. Curators carry out educational work, form a socially active civil position of students and a conscientious attitude towards learning, and inform students on all issues of the educational process.

According to the survey, curators see their groups quite often:

- Once a week—20 people;
- Twice a month—36 people;
- When necessary—8 people;
- Monthly—4 people;
- 3–4 times per month—4 people.

However, the frequency of group meetings does not affect the attendance and academic performance of students. Those students who do not attend classes are also absent from the curatorial hours. If a student does not attend classes, they miss information on organizational issues and issues of educational activities. Therefore, it is necessary to convey important information using modern tools.

Teachers and curators have difficulties informing students. The research data shows that teachers mainly interact with students in academic and educational processes (Table 1).

According to the survey, the curators keep in touch with the group:

- On curated hours—60 people;
- By phone—48 people;

**Table 1** Interaction of the teacher-curator with the group

| Options | Number of respondents | % of the respondents' number |
|---|---|---|
| On curated hours | 60 | 81.1 |
| By phone | 48 | 64.9 |
| Personal communication | 48 | 64.9 |
| At lectures, practices | 38 | 51.4 |
| Through the educational portal | 26 | 35.1 |
| Through the social networks | 14 | 18.9 |
| Respondents in total | 74 | 316.2 |

*Source* Compiled by the authors

- At lectures, practices—38 people;
- Through the educational portal—26 people;
- Through the social networks—14 people.

Most of the curators (64 people) have student phone numbers. Forty-six curators have students' contact info. Only half of the curators have their parents' phone numbers.

It should be noted that there is a curators' office on the educational portal. However, curators are more accustomed to talking to a student during a class, curated hours, or by phone. Besides, most teachers (91.6%) prefer to use e-mail, the educational portal—48.3%. Just over one-third of the curators did not want to use social media when interacting with students.

Teachers note that it is easy to work in groups with stronger students, good attendance, less working students, and senior groups. The 30 curators find it easy to work in any group. However, there are groups where it is difficult to work. Sixty-four teachers skipped the question about difficulties. Learning difficulties arise because educational motivation among students is not always the leading one.

The authors state that 100 teachers out of 120 do not want to be a curator. Those teachers who are already curators do not want such a position. The curator needs to bring important information to the students, help to solve their problems, establish the microclimate in the group, be respectful towards the students, reward them for achievements, control discipline, and attract students to participate in cultural events. Moreover, these "activities" take most of the teachers' time but are not considered when evaluating the teachers' work.

The intensity of the teacher's work increases, and they must adapt to new conditions and requirements of accreditation indicators, choose priority areas of activity. They have to combine teaching, scientific, methodological work, work on career guidance, curatorial activities, and conduct contract-based work. If we range the activity types of a teacher by importance, then the main activity is teaching—it takes a lot of time to prepare for classes and develop teaching materials. The second is a scientific activity, in the third—is the conclusion of economic contracts, in the fourth or fifth place is an educational activity.

The current information society is characterized by the spread of the principles of formality and rationality when quantitative assessments prevail over qualitative ones.

Quantitative indicators in higher educational institutions have a significant influence when passing a competitive selection for a position. An increase in university competitiveness depends on the effective performance of teachers. This allows universities to carry out research, conduct publishing activities, and develop educational technologies. The teaching activity is also assessed quantitatively by the number of indebted students, activity on the educational portal, qualitative and quantitative academic performance. Consequently, the amount of work increases due to the need to combine a high pedagogical load, methodological work, and intensive research and publication activities. The project "Modern Digital Educational Environment in the Russian Federation" leads to a new format of competition between universities and teachers throughout Russia. Besides, it reduces the teachers who are not ready to engage in advanced training, including through the integration of classroom and e-learning technologies (Noskova et al. 2016).

Currently, some students require constant social, psychological, and pedagogical support and control because they do not cope on their own, show immaturity of behavior, infantilism, inability to make well-considered decisions, do not want to take responsibility for the events in their lives (Burilkina and Suprun 2015). The highest motivation to "gain knowledge" and acquire a profession is observed among freshmen. Students perceive higher education as a new stage of education. First and second-year students learn basic disciplines that aim to master general cultural and professional competencies and not always give complete knowledge about the future profession.

Students like different disciplines because:

- They understand the material—47%;
- The teacher explains the material in an accessible form—44%;
- Materials are posted on the educational portal—43%;
- Student likes the teacher—39%;
- One can always contact a teacher—34%;
- The teacher always fascinates with interesting material—29%;
- The discipline is not difficult—25%;
- Deadlines for assignments are specified—25%;
- There is an opportunity to retake missed classes—11%.

Not all expectations that exist upon admission are met. Because of this, the motivation for gaining knowledge is significantly reduced. Gradually getting acquainted with the higher education system, at the end of the first and second year, the student acquires the ability to evaluate reality critically. If students have no interest in learning, they start skipping classes and cannot pass the first final examination. Students are superficially familiar with their future profession. The second or third year introduces the professional cycle disciplines and production practices at enterprises. There, students get to know better the nature of their future work. If students do not like their future profession, then the number of absenteeism increases, academic performance decreases, which may ultimately result in expel.

The survey revealed what students think about the consequences of missing classes. According to students, the most likely consequences that can occur if classes are missed are: (1) a phone call from the principal—52%, (2) a conversation with a curator—31%, (3) a conversation in the principals' office—25%, and (4) a phone call from a curator—13%. Only a fifth of the students is not afraid to miss classes because they believe that nothing wrong will happen and there will be no consequences.

According to almost half of the surveyed students, the curator does not call them because of absence. The parents of half of the students also do not receive calls or were invited to the office. The majority of students have never been called to the department (60%) nor received a call from the deputy director of academic and educational work (56%). Most often, students ask classmates about their studies (Fig. 1).

The information environment involves feedback. Therefore, the student's classmates will answer questions about their studies faster than teachers, using social networks and various messengers.

The teachers' responses about the consequences of student absence slightly differ from the students' opinion. If a student skips classes, then teachers:

- Talk to them—94.4%;
- Ask the monitor of the group—88.9%;
- Call the student—77.5%;
- Call the curator—62.9%;

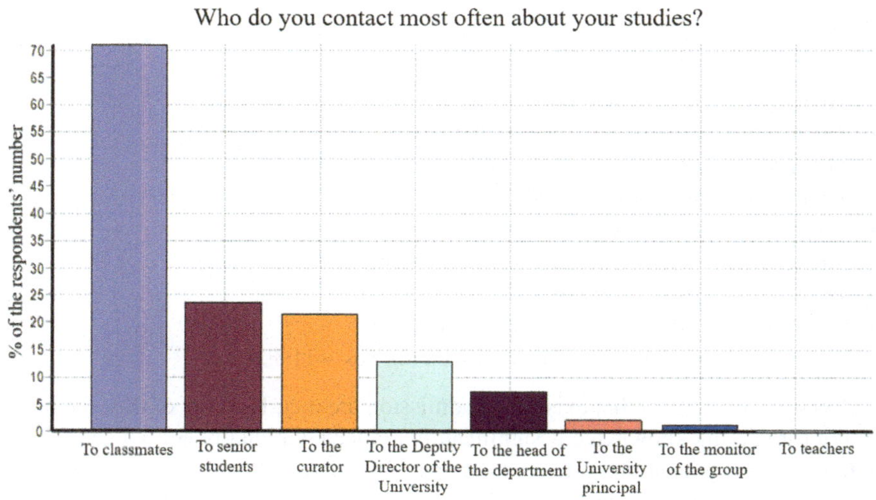


**Fig. 1** Who students reach out to about their studies. *Source* Compiled by the authors

- Inform parents—61.1%;
- Contact a student through a social network—29%.

According to the research, both teachers and students adhere to traditional and conservative methods of interaction.

Therefore, the essential task of a university teacher is to develop a method and methodology for teaching a new generation of students in the information space. It is also necessary to make more active use of Internet options to develop and support the educational process. The electronic system Moodle contains such elements as a chat, forum, a student work review for teachers, and internal mail. It allows to exchange the information between the teacher and students quickly. Both need to actively use not only the Moodle e-learning system but also other online learning tools. Teachers can make webinars with students and teachers, web conferences that expand the geography of professionally oriented communication and go beyond the educational process of their university. For example, online meetings, online lectures, and video lessons that complement the real educational process and help students to study a complete material at a convenient time.

## 4 Discussion

For most students, the use of information technology is a common activity in life. On the other hand, teachers cannot always use the tools of these technologies to interact with students more effectively. The efficiency of the interaction in the educational process is possible only when the contact "teacher–student" is established. Current information technologies allow using different tools (Yandex, Vkontakte, Viber, WhatsApp, Skype, Moodle, etc.). However, not all teachers are confident users of these information technologies. According to the research, teachers most often use e-mail and an educational portal. Sometimes they have technical capabilities but avoid using modern tools of conveying information. Teachers use the ones that they can use confidently. Besides, the teacher is loaded with different types of work. Besides teaching (the leading one), they carry out scientific, methodological, self-supporting, and educational activities. Using information technologies, teachers note high work input in the design of disciplines at the initial stage. Moreover, teachers spend a lot of time working with the educational portal.

The activity of teachers is evaluated according to various criteria—for example, (1) the teachers' activity, (2) the dynamic of activities on the educational portal, (3) work with indebt students. The quantitative and qualitative performance of students is also considered. Teachers combine all activities to meet performance. They are forced to master new information technologies to carry out the educational process and educational work more effectively, including the preservation of the student body.

Some students require constant social, psychological and pedagogical support, and control, as they have little motivation to study. Not all students adapt to the learning process immediately. Therefore, as active participants in the educational process, both students and teachers are required to be able to respond to innovations and apply different options for interaction quickly.

## 5 Conclusion

The information and educational environment of the university has its own specific technological, psychological and pedagogical features. The analysis of the interaction between teachers and students in the information educational environment of the university shows that its efficiency depends not only on the ability of teachers to use electronic resources but also on students' ability to adapt and their desire to study. Teachers note high work input in the design of disciplines at the initial stage. Besides, it takes a lot of time to work with the educational portal. The good efficiency of the interaction in the educational process is possible only when the interaction "teacher–student" is established. It requires selecting information technologies and tools, and the definition of the digital education space content. The personnel should also be competent for technological and informational support, functioning and development, and resource provision. In conclusion, the authors state that the tasks and resources of modeling the information and educational environment of the university require further research.

## References


Bell D (1986) Social framework of the information society (Nikulicheva YV Trans) In: Gurevich PS (ed) New technocratic wave in the West. Progress, Moscow, Russia, pp 330–342

Burilkina SA, Suprun NG (2015) Dynamics of social well-being of students (based on the materials of a sociological study in Magnitogorsk). Hum, Socio-Econ Soc Sci 11–1:90–92

Masoumi D (2015) Preschool teachers' use of ICTs: towards a typology of practice. Contemp Issues Early Child 16(1):5–17. https://doi.org/10.1177/2F1463949114566753

Mrochko LV, Mrochko OG, Yakovchuk TG (2019) Electronic information and educational environment of the university: monitoring of student achievements (based on research experience). Econ Socio-Humanitarian Res 1(21):149–154

Noskova T, Pavlova T, Yakovleva O, Smyrnova-Trybulska E, Morze N (2016) Modern education quality requirements and information technologies in academic teachers' activities. Int J Contin Eng Educ Life-Long Learn (IJCEELL) 26(4):434–459. https://doi.org/10.1504/IJCEELL.2016.080968

Novikova VA (2008) Formation of the university's information space: directions and prospects. Inform Educ 1:116–117

# New Vectors in Continuing Education

**Tatiana Yu. Krotenko, Fanis F. Sharipov, Maria A. Dyakonova, Yurii A. Brodov, and Valentina G. Loktionova**

**Abstract** Sociocultural, ecological, and economic circumstances of the constantly changing world insistently call for the emergence of new directions and research subjects in the educational sphere. This article aims to identify promising vectors of the research for the space of continuing education. The authors applied the following methods of research: (1) analysis of combinations of terms, which cover philosophical, socio-cultural, economic, and educational meanings; (2) analysis of the authors' pedagogical expertise in formal, informal, and spontaneous education; (3) online expert survey performed during the quarantine; (4) content analysis of respondents' extensive answers to the questions. As a result, the article indicated new previously not obvious sources of setting transdisciplinary methodological base that can emerge at the interface of different spheres of knowledge, in practice focused attempts to implement the ideas of transdisciplinarity. In the research, the authors diagnose innovative methodological approaches available in science and contemporary philosophy and use them in considering continuing cognition as a vital feature of a present-day person. The article discusses the potential hidden in paradoxical combinations at first glance and at the intersections of knowledge spheres, which seemed to be quite independent before. This is important for the solution of the task of sustaining anthropic qualities and becomes clear in the course of unfocused development of

T. Yu. Krotenko (✉) · F. F. Sharipov · M. A. Dyakonova
State University of Management, Moscow, Russia
e-mail: krotenkotatiana@rambler.ru

F. F. Sharipov
e-mail: fanissh@rambler.ru

M. A. Dyakonova
e-mail: marie.d@mail.ru

Y. A. Brodov
Moscow State Technological University "STANKIN", Moscow, Russia

V. G. Loktionova
Pyatigorsk Medical and Pharmaceutical Institute, branch of the Volgograd State Medical University, Pyatigorsk, Russia
e-mail: vglokt@mail.ru



S. G. Maximova et al. (eds.), *Advances in Natural, Human-Made, and Coupled Human-Natural Systems Research*, Lecture Notes in Networks and Systems 234,
https://doi.org/10.1007/978-3-030-75483-9_61

tuition systems, integrating human nature and artificial intelligence. The scientific novelty of the research is contained in the conclusions about the research methods, which are most suitable for the generation of new pedagogic and psychological knowledge of continuing education.

**Keywords** Continuing education · Educational space · Vectors of development of education · Methods of research of education · Cognitive processes · Transdisciplinarity · Clustering in education

## 1 Introduction

After this technical term appeared in the UNESCO reference, reporting, and regulatory documents continuing education starts acting as a special subject of scientific research. The idea of continuing cognition as a separate concept has been actively elaborated with the first steps to develop inclusive education. The idea of continuing education becomes a motto of the post-war epoch in the 40–50 s of the twentieth century. A kind of expressed hope of people for life in the society where they could use their potential for the benefit of peace, friendship, and progress.

Starting from the 1960s of the twentieth century the central topics are successively concentrating on developing the following initiatives:

- Compulsory education for grown-up citizens;
- Tuition of people in difficult life situation;
- Professional education responding to the social and technological stage of the state;
- Additional training maintaining the proficiency required from workers;
- Economic models that provide permanent access to learning environments etc.

We see a gradual shift of priorities from guaranteed education via respective institutes of the multilevel system towards the problems of creating motivation for lifelong cognition, the commitment of the individual to continuing education, fostering values of self-education.

We witness how real life of the developing society and pedagogic reality strive for post non-classical ideals and acquire respective features. Strategies, values, and models of behavior of the education entities are changing (Rozin 2020). However, education is still reflected with the use of classical pedagogical thinking tools: the concept of endless learning is poorly backed up with the information on the nature of the processes of cognition and is often limited to pragmatic interpretations of the Olympic motto "Faster, Higher, Stronger!" We could forecast the emergence of a new methodology of studying continuing educational space at the juncture of different fields of knowledge in practice-oriented attempts to realize transdisciplinarity ideas.

The pursuit of finding fresh impulses in education research faces the necessity to answer the question "what and why should be researched in the phenomenon of continuing education." Another issue related to the importance of identifying what new methodological sources are available in science and present-day philosophy is what potential is hidden in seemingly paradoxical combinations and intersections of knowledge sectors that looked quite independent before. Another significant question concerns the methods of the research, which most accurately suit the task of generating new pedagogic-psychological knowledge about continuing education (Kolesnikova 2017).

Our ever-changing world faces new socio-cultural, ecological, economic circumstances, which insistently require the emergence of new vectors and research subjects in education.

## 2 Materials and Methods

As we said in the introduction, the research problems of continuing education are clear in the UN and UNESCO reporting and reference documents. It is also expressed, in the materials of the national and foreign educational institutions, on the official Internet sites, in the articles and reviews of the last three years as well as in the monographs on the questions of lifelong education and forming in people readiness for continuing learning (Onosov et al. 2019).

In addition to the above sources, we also used the following methods:

- Analysis of combinations of terms, which contain philosophical, socio-cultural, economic and educational meanings;
- Analysis of the authors' pedagogical expertise in formal, informal and spontaneous education;
- Online expert survey carried out during the quarantine;
- Content analysis of the respondents' extensive answers to the questions.

In April 2020, 75 highly qualified specialists in education, professional authors of scientific and educational content were engaged in the local survey "Current Potential of Continuing" made the members of the department of International Industrial Business and the department of Theory and Organization of Management of the State University of Management (Moscow, Russia) and answered the following questions:

- Please, name the subjects of study of the phenomenon of continuing education, which are meaningful for you. Please, explain your preferences.
- In your view what theoretical and conceptual sources are available in science and philosophy at the interface of different knowledge areas?
- Please, identify the most suitable methods for generation of the present-day scientific and practical pedagogical knowledge in the situation of avalanche of changes in the world.

## 3 Results

Sociocultural, economic, ecological, technological circumstances of the last decades are producing new features of educational space, broaden understanding of the world, and change vision of interpersonal roles in the processes of the transfer of knowledge. Globalization, informational openness, vague boarders between the real and the virtual, multilevel and variable means of obtaining and processing information, profit vectored areas of life, the behavior transformation of subjects of tuition—all give birth to new and unknown before qualities and the necessity of new directions and subjects of research in the field of education (Kozlova, 2019).

Practically all (97%) of the participants of our survey identified in similar wordings the same issue. We have an obvious deficit of information about the arguments of personal choices of directions, formats, and modes of tuition (Manpreet & Balwant, 2019). We lack sufficient knowledge about the culture of this or that social stratum which participates in educational processes, about specific features of life activity of the students and tutors caused by the change of educational standards (especially when educational formats change radically, which happened during pandemic and self-isolation). Such notions like "continuity," "transdisciplinarity," "development," "education," and "consistency" require an entire understanding.

Identifications of important today subjects of study emerge at the maintained borders between formal, informal and non-formal education as well as at its intentional "delusion" during the research. Formal direction discusses goals, tasks, actions and meanings. Informal and spontaneous directions are part of the texture of life and, by definition, broaden the space of consciousness, set the area of the "closest development," educate and inform (Kicherova, Zyuban & Muslimova, 2020). The interlocking of these entirely different lines produces variability, individual educational paths. 78% of the respondents found this moment important.

It is worth considering (and they are just interesting) the answers of the experts to the second question about the theoretical and conceptual sources which one can find today in science, philosophy at the juncture of fields of knowledge (Kurennoy, 2020).

In many philosophical systems, life and cognition are equated. The philosophy and psychology of existentialism are particularly suitable for the continuing discussion of education. They see any bit of behavior as an act of cognition, and such links as "cognition and life," "study and life," "self-realization and life" find their philosophical and psychological interpretation. Education in the field of existential meanings is understood as an enlargement of means of self-actualization, as an attributive component of creative development and building of one's self. Here find attractive study issues of existential branching of self-realization, an increase of the number of its forms using keeping continuing studentship actual and established not only by the student's institutional status but also by way of living. Such research locus is interesting for one-third of the experts (33%). Tuition monitored in its continuity as a lifelong process generates interest in the tutor and the student's modes of life, fixed attention to their liaison as two subjects of the process of education, which integrity is impossible without such cooperation (Uszynska-Jarmoc, Zak-Skalimowska & Kunat, 2019).

In view of 43% of the respondents, the idea of continuity echoes with the basic principles of humanistic psychology: the centrality of personality, integrity of human nature, mindfulness of the experience, and striving for learning "lessons of life," self-realization, and self-actualization. Here long-term observations over the same educational processes and correlation of their socially significant achievements, personality development paths with the chosen or enforced educational routes would be extremely useful.

The main source of positive, promising, opportunity-giving perception of Genesis is contained in continuing education. Positivism psychology discusses exactly this: mobilization of the resources for life in miseries and misfortunes, changing life attitudes and existential accents; it is about the motivations, strivings of man for the continuous overcoming of the inner barriers and ongoing review of his past achievements and fulfilled tasks. Scientific arguments in this direction are connected with continuing education for 37% of the respondents.

The desire of maximum, good quality comprehension of a person to find response in the current anthropological research (once again at the interface of philosophy and psychology) generation of understanding, reflection, cognitive, and communicative capabilities, in respect, to the sensitivity of this or that stage of maturing to the above categories. Sixty-three percent of the experts actively discuss transforming a contemporary person's anthropological features, which imprint the individual's cognitive capabilities. Here we mean fast-maturing, the appearance of thinking peculiarities, perception, emotions not known in children before-slow aging, enlargement of active time, and useful life activity (Delal & Oner, 2020).

Many manifestations and reactions unknown before are caused to a certain extend by the digital transformation. Current information technologies change computers' technical performance and affect a person's structures of cognition and thinking and not always positively to anthropological psychologists and physiologists (Kuzu, 2020). An individual subjected to continuous stress caused by the consequences of global digitalization has to live and accept stoically that his attention is dispersed. It is just impossible to concentrate it on one pending task. However, as stressed by the cognitive psychologists IQ (intelligence quotient) of an average computer user has significantly grown in the last years. Developing speed of reaction, peripheral vision, increasing IQ, an individual puts at risk the development of the emotional-volitional and needs motivating spheres. These constraints certainly affect forming a child as a subject of cognition and, moreover, at the initial stages of continuing education. The solution to these quite literally serious problems draws the attention of 87% of the experts who participated in our survey.

Pedagogy does not have enough knowledge about cognition, which is continuously accumulated by natural science. To be more exact, pedagogy undertakes subtle and, until now, weak attempts of their interpretation. We mean brain functions, capacities, special features of operation of cerebral hemispheres, sensory, etc. Here, 58% of respondents maintain that best practices in cognitive and gender psychology may help, for example, studies of the forms and modes of organization of available experience, the correlation between physiology and thinking processes, subconscious nature of many cognitive acts, accumulation and synergetic action of cognitive programs, etc. Meanwhile, it is a legitimate question how different tuition variants influence the creation of such programs with some degree of probability. To the experts, view these and other questions can be discussed at the juncture of dynamic and synergetic approaches. The individuals determine themself the importance of educational events that in every separate case guarantee the success of remembering, fast and good quality decision, and the effectiveness of the next step. The mechanisms of self-organization switch on (Bermous, 2020). The significance of the independent choice of modes and forms of tuition and individual cognition path increases. In continuing education, the cognition path may be a vertical, parallel, or horizontal line with the elements of virtuality and openness. The emergence of temporal and spatial dimensions creates a sort of chronotype.

Living and acting simultaneously in the fields of information and education people rediscover themselves. In the context of pedagogical impact, here comes the necessity to diagnose temporal features of the subject of tuition. It may be the need of evaluation of the relevancy of certain educational step, readiness to present the level of training, diagnose the stage of development of the individual from the point of view of the sensitivity to next volume of knowledge, evaluation of motivation and striving for another "reboot" and resumption of the tuition. Forty-five percent of the respondents say that special value is acquired by the research works on meaning making, purposefulness of actions, intentionality of consciousness, emotional evaluation of the events on the educational path and the correlation of actual level of cognition and experience.

Transdisciplinary and cross-curricular approaches have a special importance because, in addition to comprehensive interdisciplinary knowledge, they provide students the knowledge of the relationship between communication and information sciences to create humanitarian technologies preventing manipulation of consciousness.

Another meaningful focus of research is the problem of maintaining human capital in education, the problem of mutation, and distortion of the "self" caused by the interaction of natural and artificial intelligence. Dismissal of a new generation, most often unintentional, from learning values accumulated by man is caused as it looks by the rapid relocation of these values in databases where a machine, not a human being, is a carrier and keeper. The computer is still not keen enough to sense the emotional attitude of the communication partner. It cannot respond with empathy to what has been seen, heard, or felt in the sphere of art. Here arises a critical issue of sociocultural consistency (Kamnev, 2020).

The amount of education in view of our young contemporaries more and more often associated with personal success, career development, comfortable and streamlined living, and social control over it but seldom with personal movement forward and upward. These two things do not exclude each other, their priority—that is what important. Along with this, a reduction of importance of tolerance, compassion, delicacy to the motivation of behavior of the other has been observed. Erich Fromm's maxima “to be, not to seem” becomes more and more actual in the situation of the simulated educational activity (Fromm, 1976).

There is another moment in which continuous education should not overlook the opinion of the experts. We see an increase in the number of people not engaged in the real economy. They are the elder generation or the youth, people poorly adapted to the realities of the fifths or sixths technological mode, people who are out of the order of the days, people having probably unique expertise which can be hard but necessarily used—reasonably and effectively (Sergeeva et al., 2019). It is well known that the problem of superfluous people and the problem of premature mortality are connected. Quite often, people start resuming responsibility for the existence in a social action, which is significant not only for them. Exactly the above categories of the citizens need continuing education more than anybody else and opportunities to find and apply its potential. There is a need for a respective system of diagnosis, motivation, retraining courses, and engagement in lifelong education.

Research works at a juncture of pedagogy, economy, and management are of extraordinary importance (Salehan et al., 2018). Sixty-three percent of the respondents bring this up in their extensive answers. Education is considered as a sphere of capitalized knowledge, as a space of the transformation of knowledge into intellectual potential. Meanwhile, there is an understanding that the increase of the variety of modes of educational work and institutions responsible of the continuing education should not become an uncontrolled process. Here we should rely only on the proven facts and the results of good quality forecasts since education contends with the lives of particular individuals and with responsible investments in the overall prospect of society and culture.

## 4 Discussion

Discussing the issue of openness and limitlessness of the space of information and the paradoxical reluctance of human beings to use its endless capacities, one can find interesting research directions at the juncture of ecology and psychology, ergonomics and sociology, axiology, and cognitive science. The present-day general school has been taken as a bright example of limited options of behavior. Its future was discussed by 78% of the respondents. Here comes a problem of finding tools for purposeful designing of educational situations and environments where lack of variation would be impossible.

Global coherence and communication through media became typical makers of the area of the digital revolution. Two-thirds of the experts (66%) note that such links like “information—knowledge,” “knowledge—education,” “information—tuition” worth good quality research for the generation of structured information-educational space and simultaneous self-education of students.

Resources of the informational approach are priceless here. In respect to continuing education, it allows discussing the questions of accessibility, the limitlessness of uncontrolled flows of information, distortion of personality’s framework of values, and skewed perception of the external world. In such a case, the decision appears to be in studying the possibilities of effective and cautious consumption of the content, thoughtfully designed search of information, purposeful use of network resources to remove existential uncertainties, and improve quality of life (Janelli, 2018).

As a matter of fact, in this field of research, the mechanisms which help to retrieve really needed data from informational chaos are unexplored. The emergence of the so-called "informational pedagogy" looks like an important movement towards understanding the transfer of information and its assimilation in tuition. Of course, here we need to apply such concepts like an "interactive performance," "information technology," etc.

## 5 Conclusion

Many problems addressed in our research are recognized and discussed—and this is extremely valuable—by the educators themselves, who regularly, in the line of duty, generate educational content and transmit it to students. The experts indicate an obvious scarcity of methodological bases for the solution to these problems. In the areas of contacts of practical interests and diverse research directions, there emerge new impulses for studies of the sphere of education.

Trying to answer the questions set at the beginning of the article, we can just say that only the time and combined efforts of the professionals may tell what extend pedagogy can solve them. According to the institutional theory developed by economists, innovation follows the improvement of the institutes, which are responsible for the generation of new knowledge. Science lays vectors, directions of movement where basic science and applied research can merge. Points of intersection of scientific interests and needs of practice are the birthplace of clusters accumulating the potential of innovation. To authors, view clustering could also be useful in the field of continuing education.

Education has an important mission, which cannot be shattered by technological revolutions, ecological disasters, or transformations of the economy. This is the space where unavoidable contradictions between man and society should be settled consciously, effectively, and meaningfully. Addressing the education with multi-directional research and practical interests, people safeguard themself since both social and biological life exists due to the translating of knowledge and reproduction of positive modes of thinking. Education is dedicated to the importance of going on all lifelong meetings of man, society, and culture.

## References


Bermous AG (2020) To the issue of scientific research programming in continuous education (field approach). Lifelong Educ: 21st Century 1(29):2–19

Delal H, Oner D (2020) Developing middle school students' computational thinking skills using unplugged computing activities. Inform Educ 19(1):1–13

Fromm E (1976) To have or to be? Harper & Row, New York, NY

Janelli M (2018) E-learning in theory, practice, and research. Educ Studies Mosc 4:81–98

Kamnev VM (2020) On concept of media education. Vopr Filos 3:9–12

Kicherova MN, Zyuban EV, Muslimova EO (2020) Non-formal and informal learning: international recognition practices. Educ Studies Mosc 1:126–158

Kolesnikova IA (2017) On retrospective reflection of the individual educational way. Lifelong Educ: 21st Century 4(20):31–49

Kozlova TA (2019) Modern philosophy of education and modern philosophical anthropology: common problems and ways of interaction. Lifelong Educ: 21st Century 3(27):18–27

Kurennoy VA (2020) Philosophy of liberal education: the principles. Educ Stud Mosc 1:8–39

Kuzu OH (2020) Digital transformation in higher education: a case study on strategic plans. High Educ Russ 29(3):9–23

Manpreet K, Balwant S (2019) Teachers' well-being: an overlooked aspect of teacher development. Educ Self-Dev 14(3):25–33

Onosov AA, Tumanov SV, Savina NE (2019) Integration of foreign learning platforms into the national education system: sociological expertise of the situation. Lifelong Educ: 21st Century 1(25):2–17

Rozin VM (2020) Reflection of the bases of interdisciplinary study of sociality. Vopr Filos 1:64–73

Salehan M, Kim DJ, Lee JN (2018) Are there any relationships between technology and cultural values? a country-level trend study of the association between information communication technology and cultural values. Inf & Manag 55(6):725–745

Sergeeva M, Shilova V, Evdokimova A, Arseneva N, Degtyareva V, Zuykov A (2019) Future specialists' socialization in the context of competence approach. Rev Prax Educ 15(34):571–583

Uszynska-Jarmoc J, Zak-Skalimowska M, Kunat B (2019) Creativity and need for cognition in students of pedagogy. Educ Self-Dev 14(3):48–61

# Children Socialization in the Information Age

**Evdokiya P. Yakovleva and Nadezhda A. Efremova**

**Abstract** The paper focuses on the actual problem of the socialization of children in the information age. The research covers the current problems of socialization of children are considered. The paper indicates the current situation in Russia based on statistical data for the last five years. The priority goals for the socialization of children in digitalization and outlined ways are to solve the problems of the asocial behavior of children in the information age. The main purpose is to study the process of socialization of children in the information age. The bulk of theoretical and empirical studies on the socialization of children is devoted to studying individual problems and therefore has a pronounced fragmentary character. The methodological basis of the study is a synergistic approach and a method of non-included observation. The source study base is the works of leading modern Russian and foreign researchers in the field of social philosophy, pedagogy, and psychology. One can apply it in the development of regional programs for the implementation of social protection. The conclusions and materials can serve as the basis for special university courses and an additional source in the study of the relevant sections of social and humanitarian disciplines. As a result of the study, we conclude that socialization is a complex multifactorial phenomenon. In our opinion, it is an interdisciplinary approach with its synergistic methodology that can investigate such a complex, relevant, and multifactorial process as the socialization of children and adolescents in the information age.

**Keywords** Socialization · Digitalization · Asocial behavior · Deviant behavior · Teenager · Information age

E. P. Yakovleva · N. A. Efremova (✉)
North-East Federal University Named After M.K. Ammosova, Yakutsk, Russia
e-mail: Enadin1979@mail.ru

E. P. Yakovleva
e-mail: yakov_eva@mail.ru



S. G. Maximova et al. (eds.), *Advances in Natural, Human-Made, and Coupled Human-Natural Systems Research*, Lecture Notes in Networks and Systems 234,
https://doi.org/10.1007/978-3-030-75483-9_62

## 1 Introduction

In the context of digital transformation, current society can no longer solve the accumulated problems in the field of socialization and prevention of children's deviant behavior using the old algorithm. The relevance of the study of the process of socialization of children in the information age is due to the global changes taking place in the modern world. Understanding and recognizing the problems of socialization of children in the context of the transformation of society has an existential value. Today, this phenomenon is the most important topical scientific problem of an interdisciplinary approach in children's upbringing and education in the information age.

The formation of a destructive oriented deviant lifestyle, death rate increase, disability, the development of various children's health pathologies, and increased crime among teenagers actualizes the need to find ways to form a responsible attitude towards socialization issues in the younger generation, as well as to improve the system of prevention of deviant behavior forms at various levels among teenagers from the countryside (Dorfman, 2015).

The most considerable problems of children's socialization at present are the following:

- Virtual autism;
- Conformism;
- Aggression;
- Network addiction;
- Decreased socialization skills;
- Atrophy of cognitive skills;
- Increase in depression and deviant behavior in adolescents.

Increasingly, we come across the problem that instead of a deep study of the richest heritage of Russian and world culture, children are fond of entertaining forms of leisure, virtual games that do not contribute to the growth of spirituality.

Children's socialization is one of the central psychological and pedagogical problems. The social significance of the problem becomes more relevant in the context of the digitalization of education and many spheres of life. Currently, there is a noticeable increase in children's activity, though not always constructively directed.

The most common example of adopting a normative character by deviation is the orientation toward an innovative culture as the leading type of orientation among young people. An orientation toward novelty in combination with another widespread type of culture—the culture of moral anomie, suggests that young people to:

- Choose favor of innovative social practices,
- Not to be guided by certain norms (Zubok & Sorokin, 2020).

## 2 Materials and Methods

The synergetic approach serves as a rationale for explaining the socialization processes, changes in consciousness, and deviant behavior of children in the age of digital transformation (Arshinov, 2001; Bransky & Pozharsky, 2004; Knyazeva, 2003; Kurdyumov & Knyazeva, 2005; Haken, 1980). The problem of children's socialization in the information age is the focus of this study. The new social reality reveals many examples of the manifestation of the self-organization mechanism under instability conditions in society. However, the article aims not to describe the socialization of children but rather at approaching the level of theoretical observation and analysis.

The study of asocial behavior through the prism of a synergistic approach provides an objective assessment of the level of children's social adaptation in the information age, identifying factors that destroy stability and harmony in the digital world, and outline ways to solve these problems. The potential of synergetics is enormous in developing such important methodological problems as the role of the individual in history and the alternativeness of historical development.

## 3 Results

As an actual social phenomenon, the socialization of children is always secondary, a derivative of violations of one or more aspects of the process of social adaptation—the child, environment, conditions, and their interaction. Under the actual asocial behavior of children, one should understand a decrease in cognitive skills and the lack of interaction of a child with the environment in the conditions presented to a child by the individual social environment.

It is known that the Internet is the main educator of the younger generation. With the proliferation of social networks on the Internet, the phenomenon of "publicizing" private life has emerged. To increase their ratings, teens are filming violent scenes and posting them on their pages. Moreover, teenagers are directing bloody reality shows (Pozdnyakova, 2017).

In recent decades, many innovations have been adopted to improve the quality of upbringing and education. All conditions are created for the development of intelligence, creativity, talents, and interests of children. The achievements of many children's teams, art groups in literature, music, and sports are pleasing. However, the number of children with antisocial behavior is increasing. Currently, we note cynicism, indifference, depression, and an increase in deviant behavior. Considering the current situation, we assume that the effectiveness of children's socialization will be achieved in the presence of the following psychological and pedagogical conditions:

- Creating special social services or crisis centers for children and centers for helping children, where they could receive the necessary assistance. Such centers or services may have psychological counseling and rehabilitation centers;

- Providing multidisciplinary children's social service that consider the whole range of problems of a socio-psychological nature which appeared recently. With such a social protection service, special "social services" can be opened for children who find themselves in a difficult life situation, where they can receive psychotherapeutic and counseling assistance;
- Creating psychological services with the functions of psychodiagnostic, psychohygiene, and correction, with counseling centers for teachers, parents, and children of different ages in cities and regional centers;
- Practicing the "golden rule"—"each person is unique and talented." It should be put into the basis of the pedagogical activity of every teacher and educator with any, even the most "difficult and socially neglected" children;
- Providing the cooperation of formal youth structures with informal teenage associations;
- Organizing a health program, medical examination of children, the development and implementation of preventive and curative pedagogy in boarding schools, especially in the northern and arctic regions, are relevant. Providing the necessary measures for the social protection of sick, weak children and people with disabilities.

In this regard, one cannot ignore the urgent problem of the indigenous people of the Arctic and the whole world, who have the problem of survival in a new social reality. This problem is so actual at present that drastic measures are required not only to prevent the asocial children's behavior but also to improve the level and quality of life radically.

All factors are essential to improve the level and quality of life in the Arctic regions. Natural and climatic living conditions in the Arctic are difficult and extreme. Changes in environmental and climatic parameters are caused by various consequences that need to be studied and predicted. Many researchers note in the Arctic that "climatic and environmental conditions not only affect human health but can lead to a higher level than in other areas, criminal, antisocial, self-destructive behavior (alcoholism, drug addiction, suicide). This is confirmed by the situation typical for a number of regions of the North and the Far East, the Kemerovo region, and Magnitogorsk" (Mudrik, 2006).

The research data confirm that the natural and climatic conditions of the Arctic for the health of the population are much more complicated than in other regions. In the Arctic conditions, human health is influenced by many factors that do not affect the human body in central Russia. It is necessary to adapt to frequent geomagnetic disturbances, sharp fluctuations in atmospheric pressure, low air humidity, and other equally significant natural factors that significantly impact the health of children.

Natural and climatic features of the environment form a unique way of life. The above is reflected in all spheres of life of the Arctic peoples, including the quality of life, especially children's health. At present, children's health requires close attention, especially in the extreme conditions of the Arctic region, where the level and quality of life require additional measures to improve the current state of affairs.

In this regard, one cannot but recall the major theoretician's works under discussion—V. A. Robbek. He noted that in the difficult period of the country's transition to new social and economic relations, the indigenous peoples of the Far North regions, including small numbers, found themselves on the verge of a catastrophic situation manifested in negative ecological and demographic, sociocultural processes. These include the destruction of the traditional economy, way of life, deterioration of health, and loss of the gene pool (Robbeck, 2011).

Thus, the analysis of various forms of asocial behavior shows that social deviations of various etiologies in the world are aggravated and aggravated. The most acute and common forms of deviant behavior are such as:

- Alcohol;
- Drugs, and other psychoactive substances;
- Loneliness;
- Depression;
- Suicide;
- Various forms of violence;
- Aggression;
- Alienation (Table 1).

The untimely attention of social institutions to this problem, to the analysis of the factors influencing this process, to developing state programs for managing this process leads to increased instability. Children's instability and antisocial behavior are manifested in alienation, increased virtual addiction, increased depression and deviant behavior, decreased emotional intelligence, loss of communication skills, and an increase in the aggressive information environment (Fig. 1).

The study of this problem is distinguished by a close interweaving of theoretical with practical problems, which are currently one of the most pressing and acute of children's socialization in the information age.

Any conflict of different etiology can cause psychological stress in a child, incompatible with effective socialization. Thus, the analysis shows the following:

**Table 1** The incidence of mental disorders and behavioral disorders associated with the use of psychoactive substances in children throughout the country over the past five years, according to the ministry of health of the russian federation

| Дети с диагнозом | Age | Years | | | | |
|---|---|---|---|---|---|---|
| | | 2014 | 2015 | 2016 | 2017 | 2018 |
| Alcohol and alcoholic psychosis | 0–14 | 19 | 16 | 15 | 14 | 7 |
| | 15–17 | 444 | 359 | 308 | 235 | 173 |
| Drugs | 0–14 | 29 | 34 | 32 | 23 | 21 |
| | 15–17 | 662 | 882 | 802 | 671 | 546 |

*Source* (Federal State Statistics Service, n.d.)

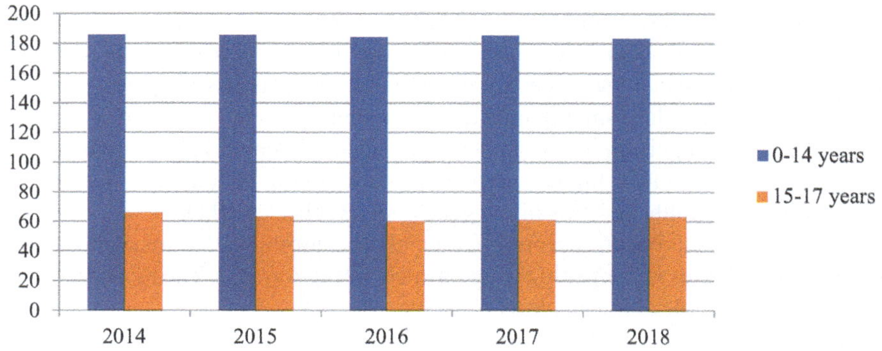


**Fig. 1** The incidence of children's mental and behavioral disorders throughout Russia. *Source* (Federal State Statistics Service, n.d.)

- There is a stable connection between the picture of the world that determines the child's life and the complex of behavioral reactions to stress;
- Any conflict and alienation causes psychological discomfort and stress in the child;
- The severity of stress reflects the degree of the child's sensitivity level;
- The occurrence of stress means that this model of the child's behavior can be used for a short time until a new one is developed that more fully meets own needs;
- In a situation where children are forced to use an adaptive model of behavior that does not satisfy them, psychological defense mechanisms come into play. As a rule, this leads to prolonged stress, which corresponds to inhuman socialization.

## 4 Discussion

The problem of deviant behavior and socialization of children in Russia has been discussed at large discussion platforms, scientific conferences, and pedagogical forums by leading scientists (Dorfman, 2015; Zubok & Sorokin, 2020; Kon, 1988; Mudrik, 2006; Pozdnyakova, 2017; Trushkina, 2018). Foreign researchers have written about the influence of civilization on peoples living in Russia and the Arctic (Bruce, 2000; Keskitalo, 2007; Nuttall, 1998; Young, 1986; Kefeli, 2013).

We note with gratitude that deep research in the field of philosophy by N.A. Berdyaev (Berdyaev, 1999), pedagogical research by A. V. Mudrik (Mudrik, 2006), M. E. Pozdnyakova (Pozdnyakov, 2017), A. S. Trushkina (Trushkina, 2018), psychological research by E. P. Belinskaya (Belinskaya, 2002), A. N. Ayanyan and T. D. Martsinovskaya (Ayanyan & Martsinkovskaya, 2016), fundamental research W. A. Robbeck (Robbeck, 2011).

In N.A. Berdyaev's work, we emphasize the complicated relationship of man with the information society. We try to reveal the meaning of human existence and spiritual creativity (Berdyaev, 1999). A.V. Mudrik connects this problem with economic security in Russia (Mudrik, 2006). In the work of M.E. Pozdnyakova, a deep analysis of the problems of modern youth, as well as the information environment, which has become an integral part of our lives, is made (Pozdnyakova, 2017). In E.P. Belinskaya's article, a person is considered in the world of mass communication; that is, characterization is given from a psychological point of view (Belinskaya, 2002). A. N. Ayanyan and T. D. Martsinkovskaya argue that the territory where children live affects the ratio and content of sociocultural and personal identity and the process of socialization itself in general (Ayanyan & Martsinkovskaya, 2016). V. A. Robbeck makes efforts to protect against the negative impact of globalization processes (Robbeck, 2011).

## 5 Conclusion

Thus, we conclude that the priority goals of children's socialization in the context of digitalization can be the following:

- Inclusion of the progressive achievements of human culture and civilization, the formation of a holistic and harmonious view of the world;
- Development of communication and creative skills, the ability of the permanent learning and adapting to new changing conditions;
- Readiness for dialogue, interaction, and cooperation in the context of digital transformation;
- Stimulating the need for self-development, self-improvement, and the formation of social skills.

Consequently, in the information age, the flexibility of adaptive strategies and behaviors is a prerequisite for successful socialization.

The study affords to make the following conclusions:

- Need to create the special social services or crisis centers for children, the so-called social security service centers for helping children, in which they could receive the necessary, timely assistance;
- Children's social service should be multidisciplinary and consider the whole range of difficulties and problems that have appeared recently. With such a social security service, it is possible to open special "social clinics" for children;
- Psychological services should be created in cities and Arctic regional centers with the functions of psychodiagnostics, psychohygiene, and correction, with consulting centers for teachers and parents;
- Prevention of asocial behavior, constructive cooperation of formal youth structures with informal associations;
- Urgent implementation requires a health program, medical examination of children, development, and therapeutic pedagogy implementation in boarding schools, especially in the North and Arctic regions.

In conclusion, we note that living in harmony with nature and people is the basic foundation of life. The socialization of a person to any extreme conditions forms a unique personality. Berdyaev noted that people cannot realize the fullness of their own life while self-absorption. A person is not only a social being and cannot wholly belong to society. People must defend their originality, independence, spiritual freedom, and fulfill their vocation in society (Berdyaev, 1999). Surviving, despite the extreme conditions and challenges of modern civilization, means affirming the strength, resilience, and uniqueness of destiny.

**Acknowledgements** The research was carried out without financial support based on the Department of Philosophy and the Department of Stylistics of the Yakut Language and Russian-Yakut Translation of the North-Eastern Federal University named after M.K. Ammosov.

## References

Arshinov VI (2001) Synergetics as a phenomenon of post-non-classical science. IFRAN, Moscow, Russia

Ayanyan AN, Martsinkovskaya TD (2016) Socialization of adolescents in the information space. Psychol Res 9(46):8

Belinskaya EP (2002) Man in the information world. In: Andreeva GM, Dontsova AI (eds) Social psychology in the modern world. Aspect Press, Moscow, Russia, pp 203–220

Berdyaev NA (1999) About people, their freedom, and spirituality. Flinta, Moscow, Russia

Bransky VP, Pozharsky SP (2004) Globalization and synergetic historicism. Polytekhnika, St. Petersburg, Russia

Bruce LW (2000) The conquest of a continent: siberia and the russians. Cornell University Press, New York, NY

Dorfman IA (2015) The manifestation of deviant forms of behavior among rural adolescents: state and trends. Mod Sci Probl Educ 4:1–3

Federal State Statistics Service (n.d.) *Official website*. Retrieved from https://gks.ru

Haken G (1980) Synergetics. Mir, Moscow, Russia

Kefeli IF (2013) Circumpolar civilization in the geo-civilizational structure of the global world. Partnerstvo Civilizacii 4:12–13

Keskitalo CE (2007) Development of the arctic as an international region. Coop Confl: J Nord Int Stud Assoc 42:187–205

Knyazeva EN (2003) Balancing on the edge of chaos as a way of creative renewal. In: Avanesova GA (ed) Synergetic Paradigm. Pearson and society in conditions of instability. Progress-Traditsiya, Moscow, Russia, pp. 123–140

Knyazeva EN, Kurdyumov SP (2005) Foundations of synergetics. Synergetic worldview. KomKniga, Moscow, Russia

Kon IS (1988) Scientific and technical revolution and problems of youth socialization. Znanie, Moscow, Russia

Mudrik AV (2006) Socialization yesterday and today. Moscow Psychological and Social Institute, Moscow, Russia

Nuttall M (1998) Protecting the arctic: indigenous peoples and cultural survival. Harwood Academic, Netherlands, Amsterdam

Pozdnyakova ME (2017) Features of deviant behavior of current russian youth: on the way to criminalization and desocialization. Rossiya Reformiruyuschayasya 15:516–548

Robbeck VA (2011) Sustainable development of the peoples of the north of russia. basic and applied research. Nauka, Novosibirsk, Russia

Trushkina AS (2018) Prevention of deviant behavior in adolescents. In: Ispulova SN, Mametyeva OS (eds) Science and practice: continuity, traditions, and innovations. Omega Sains, Ufa, Russia. pp. 168–172

Young O (1986) The age of the arctic. Foreign Policy 61:160–179

Zubok Yu A, Sorokin OV (2020) Social regulation of deviant behavior of young people in a changing reality. In: Rostovskaya TK (ed) State youth policy: national projects 2019–2024 in the social development of youth. Perspektiva, Moscow, Russia. pp. 15–20

# Teaching Efficiency in the Conditions of Active Development of IT Technologies

Olga V. Ternovskaya, Aleksei N. Ivlev, Ekaterina Yu. Ternovskaya, Mihail N. Podoprikhin, and Sergey A. Ivanov

**Abstract** The paper reveals the need for an automated assessment of teaching effectiveness in training courses, which arises in connection with the development of information and communication technologies and their use as innovative forms of education. The study reveals that pedagogical and psychological factors are considered separately when determining the teaching effectiveness. They influence the process of assimilation of educational materials when using electronic information and educational technologies. The paper analyzes new methods for assessing teaching effectiveness using digital technologies. We offer an innovative assessment system that considers both pedagogical (qualitative) and quantitative criteria. Accordingly, we determine the influence of the quantitative characteristics (study duration and the number of students in the group) on teaching and learning success for an individual and an entire study group. The results obtained using the methodology considered in this work can serve as a basis for adjusting training courses: (1) changing the training course duration, (2) the number of students in the group, or (3) the content of the material presented. It will improve the efficiency of the educational process. We tested the technique in the educational process of the university when studying the course of graphic disciplines. It can be useful to study other disciplines at the university and educational institutions of all levels.

**Keywords** Informatization of education · Information technology · Teaching efficiency

## 1 Introduction

Like other spheres of human life, current education depends on the level of technological progress, the development of information and communication technologies. The use of current technical teaching aids provides the expansion of education possibilities at all stages. Information technologies allow moving to a new level of

O. V. Ternovskaya (✉) · A. N. Ivlev · E. Yu. Ternovskaya · M. N. Podoprikhin · S. A. Ivanov
Voronezh State Technical University, Voronezh, Russia
e-mail: olgaternovskay@yandex.ru



S. G. Maximova et al. (eds.), *Advances in Natural, Human-Made, and Coupled Human-Natural Systems Research*, Lecture Notes in Networks and Systems 234,
https://doi.org/10.1007/978-3-030-75483-9_63

obtaining and controlling knowledge, adjusting the training effectiveness. Informatization of any educational process is the optimal use of information technologies that ensure the implementation of the psychological and pedagogical goals of training and education. The development of computer tools contributes to the development of the self-cognitive, personality-oriented activity of the student.

An increasing number of educational institutions use electronic and information technologies in the educational process. This is manifested in the active use of various multimedia tools, presentations, and open lessons from the Internet. There is a growing contingent of teachers and instructors who are developing their electronic materials for classroom teaching. These electronic resources can be used to help student's self-study learning for lagging. They can also be used for a more in-depth study of the educational material for interested learners, students, and listeners.

It becomes customary when teachers use the network for consultation, testing, and remote interaction with users engaged in general educational activities. It helps to increase the flexibility of education. The student determines the sequence and duration of the study of the educational material. The learning process depends on the abilities, capabilities, and needs of the student completely. Training transfers to the level of knowledge acquisition and not to the issuance of educational information.

Current teaching using information technology contributes to the rapid generation and knowledge transfer. The learning process acquires mobility. Today, it is possible to use innovative teaching materials, standards, and technologies faster and to assess the acquired knowledge faster according to clearly established criteria, regardless of the specific educational institution.

However, we consider that the development of educational web resources is not structured. Individual innovative educators are currently performing it. Electronic textbooks, reference books, and methodological literature for practical and laboratory classes are under development. Besides, their development directly depends on the development of the carriers of electronic programs, the qualifications of teachers working in this area, and their level of computer and Internet technology proficiency.

Quality is a correspondence with the standard. In education, diagnostics of the quality of mastering educational materials is interpreted as efficiency. Efficiency is the level of mastery of the taught knowledge, competencies according to the curriculum, and educational standards. Teaching efficiency is the degree of success of the pedagogical process in achieving educational goals. It must be precisely defined by the educational process requirements. The trainees' results determine the efficiency at optimal training costs (time, resource, etc.) and guaranteed achievement of training and education goals. It is possible to increase the effectiveness of the educational process by changing the pedagogical or psychological terms.

Pedagogical conditions are related to the organizational side of the educational process, by the use of the following methods:

- Various approaches to the training sessions organization;
- Different principles of studying educational materials;
- Educational and methodological complexes with innovative teaching technologies;

- Multimedia, interactive information systems;
- Training programs with different specifics;
- Innovative principles of structuring educational materials, etc.

The psychological terms that ensure the training effectiveness based on the individual personality traits of trainees include the following:

- Processes of development of different forms of thinking;
- Individual abilities of trainees;
- Difference in the types of thinking of trainees;
- Differences in needs and motivation for learning, etc.

The work aims to assess the effectiveness of the educational course, considering its quantitative (course duration, number of trainees) and qualitative (degree of assimilation of the material) characteristics.

## 2 Materials and Methods

We used 30 groups of students (from 10 to 15 people) who studied the discipline "Descriptive geometry and engineering graphics," as the initial data analysis.

The duration of the training course in groups was from 36 to 104 h by the curriculum.

One should determine the duration of the training, and the number of trainees in the development of an educational course must get the required results of mastering the material offered within this course.

It is preferable to use A. N. Ivlev, O. V. Ternovskaya, and I. V. Logunova (Ivlev, Ternovskaya & Logunova, 2019) technique, supplemented by the method of expert estimates and the method of regression analysis. There is a structure of the teaching efficiency assessment methodology (Fig. 1).

Determination of the duration of the training and the number of trainees is based on the analysis of the teaching efficiency coefficients of the group $K_{gr}$ and the coefficient of teaching efficiency $K_{ef}$.

The initial data for calculating the teaching efficiency coefficient are:

- Number of trainees;
- Maximum possible number of trainees in the group;
- Duration of the training course;
- Maximum of time for studying the course;
- Effectiveness coefficient of teaching group.

The learning efficiency coefficient of the group $K_{gr}$ is a generalized result of the students' progress. Within this study, the effective coefficient of group training was determined based on the test results. We tested students for the whole course. The learning efficiency coefficient of the group $K_{gr}$ was determined as the arithmetic means of the knowledge assimilation coefficients of each of the trainees.

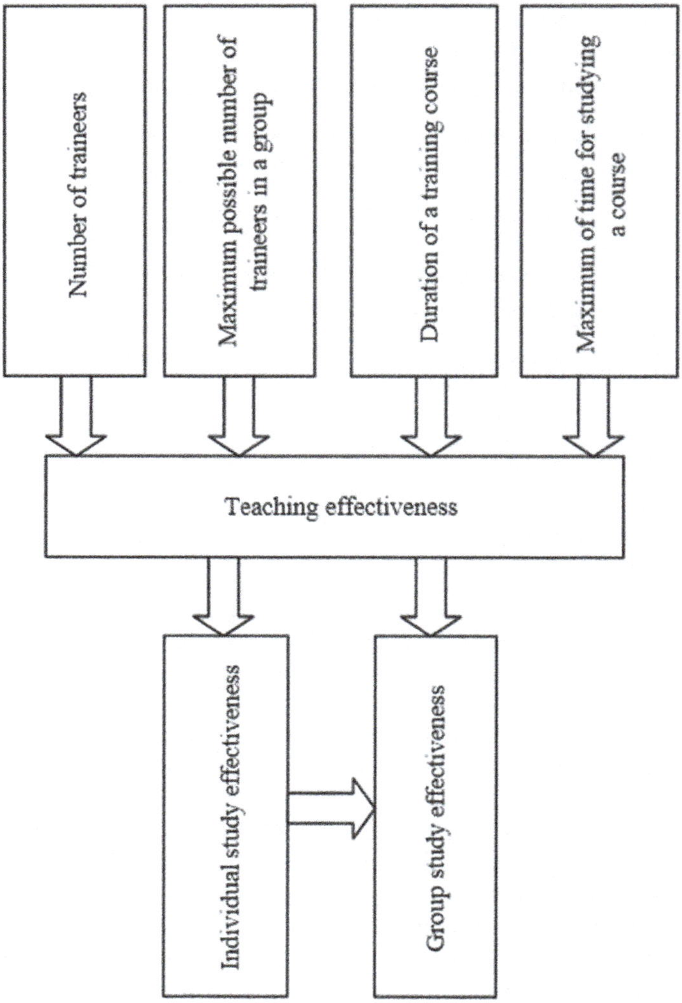


**Fig. 1** The structure of the teaching effectiveness assessment methodology. *Source* Compiled by the authors

The teaching efficiency ratio is calculated according to the formula:

$$K_{ef} = \frac{pT_{max}K_{gr}}{p_{max}T} \quad (1)$$

where:

$p$—number of trainees.

$T_{max}$—maximum of time devoted to studying the course.

$K_{gr}$—effectiveness of group learning.

$p_{max}$—maximum possible number of trainees in the group.
$T$—duration of the training course.

## 3 Results

To determine the relationship between the number of trainees and the effectiveness of group training and the duration of the training, and the group training effectiveness, we constructed the corresponding scattered diagrams and determined the correlation coefficients (Fig. 2).

The correlation coefficient for the sample: number of trainees ($p$)—learning efficiency of the group ($K_{gr}$) $r_1 = 0.962$. It indicates a close connection between the features.

Correlation coefficient for the sample: training duration ($T$)—group training efficiency ($K_{gr}$) $r_2 = 0.996$. It also indicates a close relationship between the features.

The regression equation, which determines the relation between the number of trainees and the group training effectiveness, has the following form:

$$K_{gr} = 1.2902 - 0.047p \quad (2)$$

The regression equation that determines the relation between the duration of training and a group training effectiveness is the following:

$$K_{gr} = 0.4874 + 0.0032T \quad (3)$$

According to formula (1), using regression Eqs. (2) and (3), the teaching efficiency coefficient Kef was calculated. The calculation results are presented in Table 1.

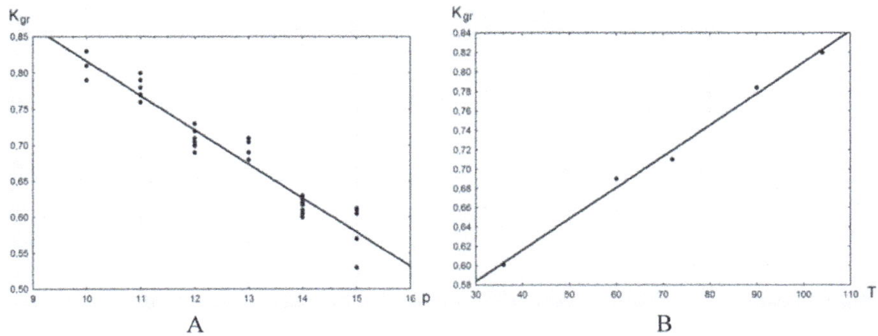


**Fig. 2** Scatter diagrams: **a**—scatter diagram of the sample by the number of trainees ($p$) – group training effectiveness ($K_{gr}$); **b**—by the duration of training ($T$)—group training effectiveness ($K_{gr}$). *Source* Compiled by the authors

**Table 1** Dependence of the training efficiency coefficient, on the number of trainees, duration of training and group training efficiency

| Number of trainees *p*, ppl | Duration of training *T*, hr | Group training efficiency $K_{gr}$ | Teaching efficiency coefficient $K_{ef}$ |
|---|---|---|---|
| 10 | 102.8 | 0.816 | 0.551 |
| 11 | 87.9 | 0.769 | 0.667 |
| 12 | 73.1 | 0.721 | 0.821 |
| 13 | 58.3 | 0.674 | 1.042 |
| 14 | 43.5 | 0.627 | 1.398 |
| 15 | 28.7 | 0.579 | 2.099 |

*Source* Compiled by the authors

Analysis of the table shows that an increase in the teaching efficiency coefficient is observed with a simultaneous increase in the number of students in a group, a decrease in the training duration, and an increase in group training effectiveness.

Teaching group depends on the number of trainees in the group and the duration of the training course. Moreover, an increase in the students' number in a group leads to decreased group teaching effectiveness. On the contrary, an increase in training duration leads to an increase in group training effectiveness.

Thus, an increase in the teaching efficiency coefficient is associated with an increase in quantitative criteria, and at the same time, with a qualitative criteria deterioration.

To find the optimal value of the teaching efficiency coefficient, one needs to establish the teaching efficiency (quality criterion) at the required level, which is determined by expert assessments. In the case under consideration, such a level can be $K_{gr} = 0.769$, which will correspond to training $T = 87.9$ academic hours, the students' number in the group $p = 11$ people. In this case, the teaching efficiency coefficient will be $K_{ef} = 0.667$.

## 4 Discussion

Evaluating the effectiveness of teaching remains the most important problem in education. Scientists devoted many works to this problem. Some works focus on trainees' psychological characteristics, which affect the effectiveness of the assimilation of educational material, regardless of the pedagogical conditions of training. Other studies determine students' psychological characteristics and their influence on the learning process at the stage of assimilating information and forming knowledge. One notes that mastering the theoretical and practical material directly in the classroom is higher than distance learning (Balashova, 2011). The patterns identified in traditional forms of education have a different manifestation in modern electronic teaching aids in the educational process. Scientists have experimentally established that when reading an e-book, fatigue and emotional tension in learners are higher than reading on paper (Vyatleva, Kurgansky, Kuchma & Teksheva, 2012).

Currently, many scientists are solving the problem of assessing the educational process. Besides, they offered some methods for determining the effectiveness of teaching and learning. D. Kirkpatrick's method (Kirkpatrick, 1998) is one of the most well-known foreign methods at present. Russian scientists also offer many different methods for determining efficiency by the level of assimilation, by the strength of the knowledge gained, and by the time spent on this process (Makarov, 2016). Also, there are techniques for evaluating nonparametric criteria (Nasledov, 2008). Scientists have also identified psychological and didactic conditions for the effectiveness of information assimilation by students in e-learning (Gnedykh, 2015). However, to automate the process of assessing educational institutions' knowledge, it is necessary to develop a system that implements this task in practice. Unfortunately, currently, only one development is known that corresponds to our direction of research (Khromushina et al., 2002). The disadvantage of this design is that it does not consider the number of trainees participating in the survey. Also, the structure and peculiarities of the training course are not considered, according to which testing is carried out and the effectiveness of the entire educational process is determined.

The acquired knowledge is the result of the cognitive activity of the subject. In etymology, knowledge is a person's beliefs, which are based on facts, rational beliefs and correspond to real situations in a particular field of study. Knowledge is the result of a person's knowledge, information accumulated by him in the process of learning. Knowledge can be verified empirically or practically, for example, through testing.

According to the point-rating system of the exam and test (regardless of the form of their conduct), knowledge assessment does not always really reflect the level of organization of the educational process. Assessment is the teacher's subjective opinion (Halfond, 2004; Gubar & Feigin, 2005). An integrated approach that combines taxonomic, thesaurus, and qualimetric approaches can be used to increase the objectivity of assessing the quality of students' knowledge (Snigiryova & Grishanova, 2014). The taxonomic approach allows us to assess the hierarchy of the system of goals and results, moving from knowledge to activity. Thesaurus approach evaluates the completeness and systematization of knowledge. Qualimetric approach evaluates the change in knowledge quality. One can assess the knowledge quality using several assessment methods simultaneously. This allows to individualize the assessment (Ulanovskaya et al., 2013). Scientists are studying the methodology for assessing the quality of students' knowledge. However, they have not yet come to a consensus on the definition of a universal method for diagnosing the quality of assimilation of educational material and determining the effectiveness of the educational process.

Increasing the efficiency of educational material digestion depends on didactic and pedagogical conditions to optimize the educational process. Besides, the assimilation of the material depends on the trainees' psychological, individual, and personal characteristics. An indisputable fact is that information technology allows to automate the routine work of issuing theoretical materials. Also, information technologies help the current check of the assimilation level of the acquired knowledge and the formation of ordinary skills among trainees. As a result, the teacher is freed up to implement creative, scientific aspects in teaching.

We have to admit that only one expert system has been developed to automate the knowledge assessment process. It is comparable to that proposed by the authors but has some disadvantages. These shortcomings do not make it possible to consider some indicators important for the educational process. These indicators include the number of test-takers and consideration of the peculiarities of the training course, which determines the effectiveness of teaching.

## 5 Conclusion

The use of the considered methodology for assessing the effectiveness of an educational course determines the teaching effectiveness, considering the qualitative and quantitative characteristics of the course of study. The use of testing allows assessing each student's success and then determining the indicator of the effectiveness of training for the entire group of students. Improving teaching efficiency can be achieved by adjusting the content of the training course or training time and the students' number.

The methodology for assessing the effectiveness of the educational course is universal. This allows us to recommend its use for assessing the effectiveness of an educational course in any academic discipline in any educational organization.

## References

Balashova YuV (2011) Cognitive and personal characteristics of full-time and distance learning students (dissertation of a candidate of psychological sciences). Sholokhov Moscow State University for the Humanities, Moscow, Russia

Gnedykh DS (2015) The effectiveness of the assimilation of educational information by students in the context of e-learning (Dissertation of the Candidate of Psychological Sciences). Saint Petersburg State University, Saint Petersburg, Russia

Gubar SA, Feigin AV (2005) On the assessment of the quality of knowledge. Probl High Educ 1:52–53

Halfond JA (2004) Grade inflation is not a victimless crime. *The Christian Science Monitor*. Retrieved from https://web.archive.org/web/20121012112315/, http://www.csmonitor.com/2004/0503/p09s01-coop.html.

Ivlev AN, Ternovskaya OV, Logunova IV (2019) Evaluation of personnel training efficiency in modern conditions: applied aspect. Int J Manag Theory Pract 12:31–37

Khromushina AM, Korosteleva OA, Rufov VE (2002). Expert assessment of teaching efficiency (Patent of the Russian Federation No. RU 25615 U1). Rospatent, Moscow, Russia

Kirkpatrick DL (1998) Evaluating training programs: the four levels, 2nd edn. Berrett-Koehler Publishers, San Francisco, CA

Makarov VM (2016) Evaluation of the effectiveness of the study of various methods of teaching students. Int J Appl Fundam Res 12(4):709–712

Nasledov AD (2008) Mathematical methods of psychological research. analysis and interpretation of data. Rech, Saint Petersburg, Russia
Snigiryova TA, Grishanova IA (2014) The integrated approach application at the quality assessment structures of the trained knowledge. Fundam Res 11–6:1382–1385
Ulanovskaya KA, Kalyuzhnov E Yu, Antropov IV (2013) Individualization of knowledge quality assessment. Izv Volgogr State Tech Univ 12(2 (105)):130–133
Vyatleva OA, Kurgansky AM, Kuchma VR, Teksheva LM (2012) Features of the perception of information from an electronic device for reading (reader). Probl Sch Univ Med Health 1:39–46

# Selecting ICT in Alternative Training

**Alla F. Matuszak, Olga Yu. Pavlova, Elizaveta V. Kalugina, Olga V. Mukhametshina, and Rinat I. Kusarbaev**

**Abstract** The research goal is to investigate pedagogical students' preferences of ICT in alternative (distance) education. The primary research issue concerns selecting the most helpful ICT for temporary distance education of pedagogical students. It splits into more specific issues concerning (1) the ranking of ICT in temporary distance education in different groups of students, (2) gadgets mostly applied in distance education, (3) technologies preferred by students and teachers for learning new information, (4) and technologies for testing knowledge and skills. We apply the case study method and method of ranking in the research. Internet survey serves as a data collection technique. The results show that full-time and part-time students select the same three essential technologies for alternative education, but their selection of ICT differs from teachers' preferences. Besides, we have registered some gender differences in ICT selection. The results deviate from other scholars' research as students' preferences of ICT in this paper depend on the distance of education. The research results may be of interest to educators engaged in teacher training.

**Keywords** Alternative education · Teacher training · ICT

A. F. Matuszak · O. Yu. Pavlova · E. V. Kalugina · O. V. Mukhametshina · R. I. Kusarbaev (✉)
South Ural State Humanitarian Pedagogical University, Chelyabinsk, Russia
e-mail: rinkus@inbox.ru

A. F. Matuszak
e-mail: lilac0@yandex.ru

O. Yu. Pavlova
e-mail: pavlovaou@cspu.ru

E. V. Kalugina
e-mail: kaluginaev@cspu.ru

O. V. Mukhametshina
e-mail: muhametshina2010@mail.ru



S. G. Maximova et al. (eds.), *Advances in Natural, Human-Made, and Coupled Human-Natural Systems Research*, Lecture Notes in Networks and Systems 234,
https://doi.org/10.1007/978-3-030-75483-9_64

# 1 Introduction

Contemporary education faces challenges regarding the necessity to provide distance education. Earlier Information and Communication Technologies (ICT) were complementary to the education process for full-time and part-time students and were part of blended education. Today the whole education system is distance-based and also based on ICT.

The task of digitalizing all spheres of life, including education, has been set earlier in the Priority project in the field of education Modern digital educational environment in the Russian Federation (Government of Russian Federation 2016). Education is rapidly changing, so is teacher training. Today scholars even admit the necessity to create e-didactics (Andryukhina, Sadovnikova et al. 2020; Tchoshanov 2013). Pedagogical graduates and post-graduates 2020 complete educational programs using ICT tools instead of interacting with professors directly.

Nowadays, teacher training and other levels and types of education largely depend on ICT. The first challenge faced by future teachers is to choose particular gadgets and technologies while studying at university. The second challenge is to prepare undergraduates to base future school students teaching on ICT. There is a confirmation that one-third of pre-school children can communicate in an electronic environment (Batenova et al. 2019). Therefore, future teachers need to know how to organize educational process applying ICT in order to show their competence in pedagogics. They should select ICT for their own education in new conditions to provide for that.

Previous research stated that both teachers and students were ready to implement ICT in different education models, for example, in different types of blended education (Pletyago et al. 2019). Scholars have studied aspects of the choice of ICT in training. The new generation of students is called "Digital Natives," consequently their conception of education might be based on Massive Open Online Courses (Avetisyan & Gevorkyan 2018); e-learning and digital content such as digital textbooks, online platforms (Tyukavkin 2019); wiki and blogs (Kruchinina & Svetkina 2016); online learning (Klimenskikh et al. 2019); mLearning (Valeeva et al. 2019). Scholars have also studied the choice of gadgets by non-pedagogical students and have found out the prevalence of mobile phone applications in the educational process. The researchers conclude that "in order to achieve a high-quality standard of higher education, a prerequisite is the introduction of electronic gadgets into the educational process" (Kabanova & Vetrova 2019).

The paper aims to find out *What ICT is most helpful for temporary distance education for pedagogical students*. The research question can be split into several specific ones:

1. What is the ranking of ICT in temporary distance education in different groups of students?
2. What gadgets are mostly applied in distance education?
3. What technologies are of greater use in learning new information from students' point of view?

4. What technologies are of greater use in testing knowledge and skills from students' point of view?
5. What technologies are of greater use in learning the new information from teachers' point of view?
6. What technologies are of greater use in testing knowledge and skills from teachers' point of view?

Different groups of students include full-time and part-time pedagogical students. Part-time undergraduates are adjusted to distance education, while full-time undergraduates are accustomed to everyday face-to-face communication with professors. In addition, it would make sense to observe gender differences in the selection of ICT.

## 2 Materials and Methods

We used several methods and techniques to study the selection of ICT—the method of case study provided for selecting a pedagogical university. Generally, the case study method applies when needed to "observe effects in real contexts" (Cohen et al. 2007). The chosen write up of the case study is the narrative report. The real context of the study is the studied university. It is a typical higher education institution licensed by the government that meets the Ministry of Education requirements. It is situated in a prominent central city of a region (population over million people). The university teaches full-time and part-time students using undergraduate, graduate, and doctoral pedagogical training programs. Furthermore, it is well equipped with ITC tools. Nonetheless, the university has no own platform for distance education.

The method of ranking has allowed us to show the students' preferences of particular types of ICT. Data collection techniques include Internet surveys. We have interviewed pedagogical students to obtain the answers to the research questions 1–4. Interviewing teachers have allowed us to obtain the answers to the research questions 5–6.

The list of technologies for research question 1 included mobile applications, e-learning, mLearning, blogs, clips, podcasts, videoconferences, educational films, and self-shot films. The list of gadgets for selection consisted of computers, netbooks/notebooks, cameras, smartphones, tablets, TV sets, etc. The list of knowledge mastering technologies was as follows: Internet materials, mobile applications, e-learning, mLearning, blogs, clips, podcasts, videoconferences, educational films, and self-shot films. The list of knowledge testing technologies included e-mail, mobile applications, e-learning, videoconferences, self-shot films, etc. The highest rank selection was attributed to one point.

The sample consisted of full-time pedagogical students studying using distance methods temporarily ($n = 57$) and part-time pedagogical students studying using distance methods permanently ($n = 88$), a total of $n = 145$. The research results were analyzed by studying the relationship between the choice of students and the form of receiving education. In the process of research, we observed some gender differences. Additional data have been obtained from interviewing teachers to compare the results (n = 9).

## 3 Results

The results for research question 1 (What is the ranking of ICT in temporary distance education in different groups of students?) show that both full-time and part-time students select three types of technologies most used in their distant learning: mobile applications, videoconferences, and educational films. However, their ranking for the two groups of students varies. Videoconferences are mostly applied for full-time students ($\sum = 156$), then they mention mobile applications like Viber and WhatsApp ($\sum = 186$), and the third place is occupied by educational films ($\sum = 203$). For part-time students, the ranks of the technologies are mobile applications ($\sum = 260$), videoconferences ($\sum = 351$), and educational films ($\sum = 356$).

Since the research is conducted among pedagogical students, the majority of respondents are women. Only four respondents are men. Their answers to the first research question differ from those obtained from women. Men also rank first mobile applications ($\sum = 9$), but the second and the third places are given to e-learning ($\sum = 12$) and all kinds of mLearning ($\sum = 14$).

The second research question shows the preferences of gadgets for distance education. For both groups of respondents,' the preferences are the same. The first place is occupied by netbooks/notebooks: full-time students ($\sum = 114$), part-time students ($\sum = 238$). Then both groups of respondents name computers: full-time students ($\sum = 150$), part-time students ($\sum = 274$). The third selected option is smartphones: full-time students ($\sum = 179$), part-time students ($\sum = 279$).

Men choose the same preferred gadgets as women, but the most chosen one is indicated to be a computer ($\sum = 4$). The difference is not significant.

The answers to the third research question show the technologies preferred by respondents to master new information. Full-time students indicate videoconferences as the best way to learn new material ($\sum = 180$). The second and third place is given to the information from the Internet ($\sum = 201$) and educational films ($\sum = 225$). Part-time students also choose three technologies, but one of them is different from the one chosen by full-time students; the other two are attributed to different places. Hence, the first one is Materials from the Internet ($\sum = 284$), the second—videoconferences ($\sum = 291$), and the third—mobile applications ($\sum = 246$).

The selected technologies to master the material by men include Internet sources ($\sum = 14$), mobile applications ($\sum = 15$), and e-learning ($\sum = 18$).

The fourth research question concerns the technologies for testing knowledge and skills. The answers for the fourth research question show that full-time students choose, primarily, e-mail ($\sum = 123$), then videoconferences ($\sum = 165$), and then mobile applications ($\sum = 147$). Part-time students prefer, primarily, e-mail ($\sum =$

110), then mobile applications ($\sum = 119$), and then videoconferences ($\sum = 298$). There is no difference in the selection of technologies by men and women in this research question.

The next research question concerns teachers' opinions of the technologies used in learning information. The teachers' opinions do not coincide with the students' opinions. The teachers give the first place to mobile applications ($\sum = 14$). Teachers prefer to give oral instructions and to record them in Viber or WhatsApp. Then teachers point out videoconferences ($\sum = 18$) by implementing which they can conduct regular classes and explain the new material. Apart from that, teachers rely on the information obtained independently from the Internet ($\sum = 21$).

The answers for the last research question show that teachers prefer to test students' competencies using mobile applications ($\sum = 12$), e-mail ($\sum = 20$), and videoconferences ($\sum = 22$). The inventory of the technologies coincides with the one pointed out by both full-time and part-time students.

## 4 Discussion

The answers for the first research question differ in two groups of respondents mainly due to the fact that part-time and full-time students are used to different manners of cooperating with the teacher. Part-time students mostly do their work independently and need the means for taking a credit or an exam—they only need to be consulted. Thus, they prefer mobile applications. While full-time students are used to classes conducted by their university teachers, so they rank videoconferences first. As other scholars' researches show, in regular educational environment students anticipate their education to be aided by ICT and 76% of them choose blended education (Pletyago et al. 2019). Full-time students tend to apply "face-to-face" model of blended education, while part-time students choose other options. During our previous 2019 researched we investigated full-time undergraduates' preferences in selecting technologies for mastering foreign languages. The students' preference was Internet (Matuszak & Uvarina 2019). Smartphones occupied the fifth place. The mentioned research shows how students' technological choices change in the conditions of distance education.

Gender differences were also involved in such choices. Though the sample of men could hardly be considered representative, some differences were observed. The selection of e-learning and all kinds of mLearning pointed to the fact that men tended to be more independent from teachers in their learning.

The results that were showing gadget choices deviated from the ones obtained by other researchers. In the research (Kabanova & Vetrova 2019), it was described that "almost every student (95%) pointed out that a mobile phone was the most effective electronic gadget that could be used in the educational process" and that "the overwhelming majority of respondents (98%) indicated that they often used an electronic gadget in the educational process and only 2% of the students did not use it in their studies at all." The deviations of the current research can be explained by the following: in this research, students find themselves in the situation of distance education, while in the article by E. E. Kabanova & E. A. Vetrova, day department learning is investigated.

The answers to the third research question show differences in the selected technologies. Full-time students' choice is videoconferences, as they allow teachers to teach students as if they are located in the classroom. That is why this preference by the full-time undergraduates can be expected. The other two chosen options (materials from the Internet and educational films) are usually complementary to the teaching process in the day department. Consequently, these answers can also be anticipated. Part-time students decide to rely on materials from the Internet, as they usually do while studying in the extra-mural department. The third chosen technology is unexpected (mobile applications), but the new learning conditions can explain this result. Mobile applications allow students to ask teachers any questions at any time, which explains its appeal to students.

The difference in selecting technologies to learn the new material is in men's choice of e-learning. The selection of this technology by very few respondents may also result from the fact that the technology is not spread in the selected university.

The answers to the fourth research question show that full-time undergraduates prefer to test knowledge via e-mail, which gives opportunities to send both oral and written tasks. The teacher can quickly correct written tasks. Videoconferences are also selected, as they give the opportunity to be asked during the regular class. The third option (applications) also allows one to send oral and written tasks, but it does not allow one to check papers quickly.

Part-time students selected options they are used to e-mail, applications (these two means of testing their learning results are for distance learners). Videoconferences that allow one to give oral answers are mostly for regular classes. Part-time students are used to studying without using them. The absence of difference in selecting testing technologies by men and women might result from the fact that students mostly agree to teachers' choice.

In analyzing teachers' response for research questions 5 and 6 an unexpected result was only partial coincidence of opinions of teachers and students. Full-time students do not expect teachers to offer mobile application means as explanations of the new material. However, they are ready to use educational films via the Internet. They also expect their teachers to conduct more classes through videoconferences. Part-time students expect teachers to give them more free hand in searching materials on the Internet. They rank third teachers' instructions via Viber and WhatsApp. It concludes that university teachers should ask their students for their opinions before choosing the technology of temporary distance education.

The teachers' response for the sixth research question shows that both students and teachers choose the same knowledge testing technologies. Differences are found in their ranking. While teachers consider applications to be the most convenient type of knowledge testing technology, students voted for an e-mail. The differences are not very noticeable and do not hinder the process of competency control.

## 5 Conclusion

The selection of ICT depends on the conditions of learning. It has become a rule that regular modern classes generally use blended education. They are mostly accompanied by mLearning. Alternative education in the form of temporary distance education requires a shift from ICT selected in blended education. Alternative education is characterized by ample use of e-mails and videoconferences.

There have been noted differences in full-time and part-time pedagogical students' ICT preferences in learning new information. Part-time students select technologies that provide them with more independence in learning.

It has also been observed that full-time students in a stable environment and in the conditions of temporary distance education select different ICT and tools.

Teachers and students prefer different technologies for explaining the new material. Teachers use mobile applications for recording their classes, while full-time students prefer to have "eye contact" with them while being consulted and, thus, choose videoconferences. Videoconferences also allow one to ask questions "in the moment."

There have been noticed some differences in the choice of ICT by men and women undergraduates. Men tend to try technologies with which they have never worked during regular classes, for instance, e-learning.

The learning conditions are rapidly changing. Pedagogical students have to be able to apply different ICT and analyze how useful they might be in different educational tasks to be ready to meet the new challenges and participate in alternative education. Participation in the research of this type assists in considering the impact of ICT on self-education and future teaching.

**Acknowledgments** This article was financially supported by the Mordovian State Pedagogical Institute Named After M. E. Evsevyev under the contract to implement research work dated 01.06.2020, No. 16-305 on the topic "Organization of Alternative Education in a Socio-Cultural Context."

## References

Andryukhina LM, Sadovnikova NO, Utkina SN, Mirza-ahmedov AM (2020) Digitalization of professional education: prospects and invisible barriers. Educ Sci J 3(22): 116-147. https://doi.org/10.17853/1994-5639-2020-3-116-147

Avetisyan PS, Gevorkyan NM (2018) Human capital development in a digital economy: challenges the education system is facing. Bull Russ-Armen Univ 2:122–128

Batenova Ju V, Dolgova VI, Emelyanova IE, Ivanova I Yu, Pikuleva LK, Filippova OG (2019) Readiness for informational interaction of subjects of the educational process. Perspekt Nauk I Obraz 41(5): 34-46. https://doi.org/10.32744/pse.2019.5.3

Cohen L, Manion L, Morrison K (2007) Research methods in education, 6th edn. Routledge Tailor and Francis Group, London and New York, UK and NY

Government of the Russian Federation (2016) Priority project in the field of education "Modern digital educational environment in the Russian Federation" (October 25, 2016 No. 9). Moscow, Russia

Kabanova EE, Vetrova EA (2019) The use of modern electronic gadgets in the educational process of the university. Eur J Contemp Educ 8(3):524-533. https://doi.org/10.13187/ejced.2019.3.524

Klimenskikh MV, Lebedeva Ju V, Maltsev AV, Savelyev VV (2019) Psychological factors of online-learning efficiency of students. Perspekt Nauk I Obraz 42(6):312-321. https://doi.org/10.32744/pse.2019.6.26

Kruchinina GA, Svetkina EG (2016) The use of innovative forms of training in the development of health-saving competencies of University students. Vestnik of Lobachevsky State University of Nizhni Novgorod. Series: Social Sciences, 3(43): 155-162

Matuszak AF, Uvarina NV (2019) ICT in teaching undergraduate students. Balt MaritE Forum 1:528–533

Pletyago TY, Ostapenko AS, Antonova SN (2019) Pedagogical models of blended learning: on the experience of Russian and foreign practice of design and implementation. Educ Sci J 5(21): 113-130. https://doi.org/10.17853/1994-5639-2019-5-113-130

Tchoshanov M (2013) Engineering of learning: Conceptualizing e-Didactics. UNESCO Institute for Information Technologies in Education, Moscow, Russia

Tyukavkin NM (2019) Digitalization of educational process in universities. Expert: Theory Pract 1(1): 35-41

Valeeva NG, Pavlova EB, Zakirova Yu L (2019) M-learning in teaching ESP: case study of ecology students. Eur J Contemp Educ 8(4): 920-930. https://doi.org/10.13187/ejced.2019.4.920

# Readiness to Continuous Mathematical Self-Education Among Pedagogical University Student

**Lubov A. Odintsova, Larisa M. Bronnikova, and Oksana Yu. Grigoreva**

**Abstract** There is a shift in modern higher education in emphasis on preparing students for self-education, self-development, and innovation activities. It is substantiated by the government and social procurement, the rapid rate of obsolescence of information and the rapid emergence of a new information flow in the modern world, and the need for young professionals who obtain modern knowledge and use modern technology and also are mobile. The research goal is to theoretically substantiate and construct the system of readiness of pedagogical university students to implement an ongoing mathematical self-education. The research bases on methodological principles of competence, system, and activity approach in education, the results of research in the field of the organization of ongoing education, the leading provisions of the methodology of self-education, self-analysis in the educational process, the theory of harmonization of theoretical and practical training of future teachers in teaching mathematics. We use the following methods: (1) analysis, (2) synthesis, (3) abstraction, (4) concretization, (5) generalization and systematization of scholarly experience, (6) modeling, and (7) pedagogical experiment. Based on outlined principles (systematic, activity-based, relationships of all types of independent work, the harmonization of fundamental theoretical and practical components of training of the future professional in the field of mathematical education, self-analysis, and control), a system of formation of readiness of pedagogical university students to implement ongoing mathematical self-education, which comprises the research results. The structure and content components (target-, content-, activity-, management-, and efficiency-based) are described. We conduct a pilot test of the functioning efficiency of all components of the proposed system. Approbation of the results is presented at regional and international conferences.

L. A. Odintsova · L. M. Bronnikova · O. Yu. Grigoreva (✉)
Altai State Pedagogical University, Barnaul, Russia
e-mail: oxanagray@mail.ru

L. A. Odintsova
e-mail: lubo.odintsova@yandex.ru

L. M. Bronnikova
e-mail: bronnikova_laris@mail.ru



S. G. Maximova et al. (eds.), *Advances in Natural, Human-Made, and Coupled Human-Natural Systems Research*, Lecture Notes in Networks and Systems 234,
https://doi.org/10.1007/978-3-030-75483-9_65

**Keywords** Ongoing education · Self-education · Mathematical education · Competence-based approach · Readiness for ongoing mathematical self-education

## 1 Introduction

Modernization of current national higher education occurs in the conditions of emerging market relations in all activities under the influence of such factors as the rapid growth of the information flow, which is intended to be assimilated by students, and the need to update outdated information. The inclusion of Russia into The European Higher Education Area resulted in the necessity of taking into account the European standards of its content and implementation, reducing the share of classrooms in general labor intensity of educational programs, the emergence of new educational paradigms, technologies, methods, and forms of organization of the educational process.

The demand for competent, mobile, capable of self-education and prone to innovation activities professionals in the labor market reflects in the legal documents that regulate the professional training in the field of mathematical and ongoing education (USSR State Committee for Public Education 1989, pp. 9–20; Ministry of Education and Science of RF 2018). The idea of ongoing education, which originated in the works of Aristotle, Plato, and Confucius, was given an impetus for development at the end of the last century and as a pedagogical concept that had remained relevant since. The research of Russian scientists aimed at studying the experience of the formation of ongoing education in various countries allowed us to identify the modern features of ongoing education, to set the tasks of developing the structure, content, technologies, methods, means, and forms of the functioning of the national system of ongoing education, to clarify the role of universities in the structure of ongoing education (Medvedev 2012; Lobanov et al. 2018).

The concept of "self-education" is firmly established in pedagogical science, but it is not interpreted unambiguously (more often as a cognitive activity for self-improvement and as a self-creating, self-educational activity). A retrospective analysis of the development and clarification of this concept is carried out in the works of L. G. Titova, V. V. Yudina (Lobanov et al. 2018), Medvedeva (2012), Karpova (Karpova & Shabalkova 2015). Skatkin first identified the creative nature of the teacher's self-educational activity as the basis for the teacher's self-education (Lobanov et al. 2018). In aspect of the conducted research, the experience of organizing a prolonged assessment of students' educational activities, based on regular monitoring of the progressive development towards the assimilation of educational material, the formation of students' self-assessment skills by involving them in the process of developing criteria for evaluating results, is of interest (Saribeyli 2018).

In management theory, the phrase "education management" is usually considered from the point of view of implementing managerial functions by the organizers of education (Lobanov et al., 2018). From our perspective, the educational process should be managed at the level of teacher and student. To date, the question of the formation and organization of the student's management of their educational activities remains insufficiently studied. Therefore, the literature review shows that the holistic approach to the formation of the university students' readiness for the implementation of ongoing self-education is not sufficiently developed, namely, from the point of view of the relationship of its content, organizational-activity-based, managerial, and efficiency-based components.

The research goal is the theoretical substantiation and construction of the "System of formation of readiness of pedagogical university students to implement ongoing mathematical self-education," which reflects the integral nature of this process and provides for its implementation.

## 2 Materials and Methods

The research theoretically bases itself on (1) the primary provisions of the competence, systemic and activity-based approaches in education, which are reflected in the works of I. A. Zimnyaya (Zimnyaya, 2013), A. V. Khutorskoy (Khutorskoy, 2017), etc.; (2) the leading provisions of the self-education methodology which are pointed out in the works of L. V. Karpova, I. F. Medvedev (Karpova & Shabalkova, 2015; Medvedev, 2012), (3) the research results of A. Ya. Khinchin (Khinchin, 2006) in the field of identifying the essence of formalism and its overcoming by the students' knowledge; (4) the research results in the field of professional self-analysis development by P. N. Osipov (Osipov & Sedankina, 2006), L. A. Odintsova, and L. M. Bronnikova (Odintsova & Bronnikova, 2020); (5) the primary provisions of G. I. Sarantsev's research (Sarantsev, 2016) in the field of substantiation and implementation of harmonization of theoretical and practical training of future teachers in teaching mathematics.

We must reveal the conceptual construct of the study. Based on the analysis of the development of the concept of ongoing education, we choose the interpretation of L. G. Titova and V. V. Yudin as the basis for the study. "Ongoing education is a purposeful process of personal and professional development throughout the life of a person" (Lobanov et al., 2018). The mentioned authors, while revealing the meaning of the ongoing education system, include the following in it: (1) the network of educational institutions (public and private), (2) forms of teaching and self-education, (3) levels of education; (4) philosophical foundations and moral characteristics, which perform the task of preservation and assertion of Russian national culture; (5) education regulatory framework; (6) hierarchy, which allows one to optimize the management, learning and transferring of experience of formation and functioning of ongoing education (Lobanov et al., 2018).

The interpretation of ongoing education mentioned above implies that a professional in the field of mathematical education should:

(1) Be able to be up to speed in continually changing needs of society and government for professionals in the appropriate area, their competence requirements,

and programs of advanced training and retraining (offered by institutions of postgraduate education);
(2) Have the ability to build an individual (in accordance with their inclinations and needs) trajectory of personal, professional, and creative development.

The professional should have self-education skills in order to implement the mentioned provisions. Based on the views of I. F. Medvedev (2012), self-education implies systematic, purposeful research-oriented cognitive activity aimed at achieving specific personal or socially significant educational goals, managed by the individual themselves. The features of self-education include as follows:

(1) Developed internal motivation;
(2) High level of activity, initiative, and personal integrity;
(3) Purposefulness and the voluntary acquisition of new knowledge and types of activities;
(4) Organized and systematic nature of this activity;
(5) Independence and selective nature of cognition.

The allocation of cognitive activity independence as an essential feature of self-education does not lead to the identification of independent work with self-education. It is considered as one of the forms of self-educational activity.

The research uses the following methods: (1) analysis, (2) synthesis, (3) abstraction, (4) concretization, (5) generalization and systematization of the scholarly experience, (6) modeling, and (7) pedagogical experiment.

## 3 Results

While relying on an established theoretical platform of ongoing research, we put forward the critical idea of the conversion process of training future professionals in the sphere of mathematical education in a pedagogical university—the inclusion of formation of readiness of high school graduates to implement ongoing mathematical postgraduate education systemic character, providing the relationship and direction of all components of the training.

Touching upon the readiness of pedagogical university students to implement ongoing mathematical self-education, fundamentally, we signify the formation of professional competence, readiness for self-analysis, critical self-assessment of the assimilation of knowledge and professional experience, self-organization of activities to improve them, and develop pedagogical creativity. The essential characteristics of such a competence are closely related to the characteristics of information, educational, cognitive, research, and communication competencies. However, suppose the last four competencies are formed in the educational process under the teacher's direct control. In that case, the first should be established in the process of students' self-educational activity, where the interaction of the teacher and the students is mediated through a system of tasks for independent work, guidance and recommendations for their implementation self-control, and mutual control over the results.

After analyzing the set of competencies recommended for mastering by future professionals in the field of mathematical education (major "Pedagogical Education," specialty "Mathematics"), we highlight the following characteristics of readiness for the implementation of ongoing mathematical self-education:

(1) Operating and freely navigating in the legal documentation which regulates the activities of a mathematics teacher at various education levels (general secondary, secondary vocational, and higher education);
(2) Mastering techniques of searching for the necessary literature and electronic resources in mathematics, pedagogy, and mathematics teaching methods;
(3) Mastering techniques of working on the study of mathematical information necessary for professional activity;
(4) Knowing, being able to implement and possessing the methods of articulating a scientific issue, techniques, and means of finding rational ways to solve it, and analyzing the results obtained;
(5) Reasonably combining theoretical knowledge in mathematics and its application to the substantiation of statements and the solution of practical problems;
(6) Mastering methods and technologies of organizing the educational process that meets the requirements of the implemented educational paradigms and the achievements of modern pedagogical science;
(7) Mastering techniques of choosing methods and technologies for organizing educational activities that are relevant to the content of the taught educational material;
(8) Knowing, being able to implement and possessing the methods of articulating a scientific issue, techniques, and means of finding rational ways to solve it, and analyzing the results obtained;
(9) Mastering techniques of organizing self-analytical and managerial activities in the process of performing independent work on the study of theoretical material and pedagogical experience.

The outlined characteristics serve as the basis for the formation of pedagogical university students' readiness to implement ongoing mathematical self-education in the process of studying mathematics. Considering the description of the "System of formation of readiness of pedagogical university students to implement ongoing mathematical self-education," first of all, we define the principles of its construction and further functioning. Based on the scholarly experience in organizing self-educational activities of future mathematics teachers, we believe that the following provisions can be chosen as such principles:

(1) System-based, meaning the specified object to build must be approached systematically—the content, forms, methods, means, organization of interaction of educational process subjects must meet the general objectives, and results are monitored by specifically developed criteria;
(2) Activity-based. Constructing an allocated competence by including students in specific activities corresponding to the contents of the studied training material and the formation of their own abilities to perform versatile educational

actions (finding necessary information, its comprehension, absorption, application to the issue solution, the development of new scientific methods of cognition, methods of reasoning claims, methods of communication, presentation of results, etc.), which constitute the readiness of graduates for inclusion in ongoing postgraduate education;

(3) The interconnection of all types of independent work and the content of independent work offered for extracurricular work should be a continuation of the classroom work, should deepen it, and systematize students' activities. It may include tasks that are more complex in content and more time-consuming. The mentioned condition should be taken into account in forming the system since it focuses on ongoing postgraduate self-educational activities designed to help graduates consciously choose areas and forms of professional development in specialized postgraduate educational institutions. The analysis of non-contact independent work allows a teacher to adjust the organization of contact independent work;

(4) Harmonization of the fundamental theoretical and practical components of the training of the future professionals in the field of mathematical education. The formation of self-educational activities based on a rational combination of mastering fundamental scientific knowledge, methods of mathematical activity, and professional experience allows a graduate to understand and evaluate the latest achievements in mathematical and pedagogical science, new technologies, methods, and means of teaching mathematics and make the most rational choice;

(5) Self-analysis-based. It means that each subject task for independent work should provide for the implementation of self-analytical actions that will help the student learn to evaluate and regulate the course of their reasoning critically;

(6) Students' self-management of their self-educational activities. The principle focuses on the inclusion in the formed task system for self-educational activities of tasks for self-checking the results of mastering theoretical questions (specifying the definitions of concepts, using different information media in their formulation, highlighting the stages of proving a theorem or solving a problem, the argumentation of each stage), identifying mistakes made by a student, selecting and solving problems in order to eliminate them. It allows students to master their self-analysis skills of the results of their activities, planning, self-organization, self-control, self-regulation of activities, and their results.

The formulated principles of constructing the "System of formation of readiness of pedagogical university students to implement ongoing mathematical self-education" determine its structure, consisting of components: target-, content-, activity-, management-, and efficiency-based.

The following is the description of the system components.

The target contains the general goal—the formation of students' readiness to implement ongoing mathematical self-education and specific sub-goals aimed at forming students' internal motivation for self-education. It also includes skills of

(1) searching and working with various information media, (2) working with mathematical information; (3) using general scientific, mathematical, and logical methods of constructing argumentative statements; (4) communication skills, (5) managerial skills and experience in self-educational activities, (6) development of independence, activity, and initiative; and (7) fostering purposefulness and perseverance. Being a system-forming component, it determines the functions and content of the others.

The content component is a system of educational tasks that serve as a means of organizing students' self-educational activities aimed at mastering scientific knowledge on the subject, about self-educational activities, and mastering the techniques of their use in practice, which will ensure the acquisition and development of professional experience. The mentioned system is divided into the following groups:

(1) Tasks involving the disclosure of the essence of general scientific methods of cognition, logical methods of constructing reasoning, and developing skills of their practical use. Such methods are thoroughly substantiated in the process of performing the following types of tasks:

    (a) Based on the available data, making an assumption about the existence of a pattern and proving or disproving it;
    (b) Identifying the scope and content of the concept, defining the concept by its content, giving examples and counterexamples to the concept, establishing the dependence of changes in the scope of the concept (extensions, abbreviations) on its content, comparing the contents of different concepts, and determining the relations between the concepts (subordination, co-ordination, opposite features, contradictions, etc.);
    (c) Finding out the equivalence of the given definitions, formulating a definition that is equivalent to the given one;
    (d) Indicating from the proposed definitions several interrelated concepts, namely, which definition follows from the other, etc.

(2) Tasks that serve as a means of forming ways of working with mathematical concepts and statements. This group includes tasks for independent statement proof and independent problem solutions. In addition to proving the theorem (solving the problem), the student is offered:

    (a) To specify which known mathematical propositions are used in the proof (solution);
    (b) To point out at what stage is it used;
    (c) To determine why each of them can be applied in the conditions of the theorem being proved or the problem being solved.
    (d) The tasks in which one is asked to think about what the conclusion of the theorem will be if one or more conditions are removed or replaced by others are proven to be rather useful. Similar work with definitions of mathematical concepts is also useful.

(3) Tasks that involve students' creativity—setting up new tasks (reverse, opposite, similar, etc.), selecting tasks on a topic, preparing reports, reviews, projects,

cases, and portfolios. Readiness for research activities in the process of self-education contributes to the performance of tasks of this type:

(a) Reformulating the problem (theorem);
(b) Formulating for a given statement (theorem, problem) the reverse or opposite one and finding out their validity, generalizing a theorem, formulating and proving a theorem by analogy;
(c) Creating and solving a problem based on the proposed data and for the application of specific theoretical facts (theorems, attributes, properties, etc.);
(d) Solving the problem (proving the theorem) in several ways, choosing the most rational way.

(4) Applied tasks that involve developing the motivational and personal component of readiness for self-education since they demonstrate the application of relevant subject knowledge in life situations and increase interest in mathematics. The solution of practical problems is carried out using the method of mathematical modeling, which allows one to reveal the value and power of the mathematical language and research methods.
(5) Tasks of a self-analytical nature that contribute to forming such personal qualities as creativity, initiative, focus on cooperation, co-creation, self-confidence, a tendency to introspection, and the ability to anticipate and predict the results of future professional activities. Here are examples of questions of a self-analytical nature related to the tasks that are asked after the solution: What difficulties did one have in the process of completing the task? What mistakes were made? What should be done to prevent making such mistakes? What kind of help is required?
(6) For the formation of managerial skills, it is advisable to include the following tasks in the system:

(a) Analyzing the actual situation, including the determination of the level of knowledge formation and methods of activity for working on a fragment of educational material;
(b) Activity planning, which involves the development of a reasonably structured sequence of actions to achieve the set goal;
(c) Organizing activities—implementing a planned sequence of actions in the system of independent work of students;
(d) Implementing self-control over the results obtained during the implementation of this sequence of actions;
(e) Organizing correctional work (repetition of the theoretical foundations of the tasks, correction of errors).

The activity-based component consists of a set of forms, methods, and means of implementing the developed system of educational tasks to form students' readiness to implement ongoing mathematical self-education. The primary form of its organization should be independent work performed individually, in pairs, in groups, which corresponds to the professional and life situations that the student faces after graduation. The methods of working on the tasks are supposed to use general scientific methods of cognition (observation, comparison, analysis, synthesis, induction, deduction, modeling (including mathematical modeling), abstraction, an ascent from the abstract to the concrete, formalization, and hypothetical method), as well as logical laws in the argumentation of statements and the construction of conclusions. While studying at the university, the student must learn the essence of all these methods, master the techniques of their use in professional activities, actively engaging in the implementation of the system of tasks for self-educational activities. From these tasks, the teacher forms a series of tasks for the current independent work, tasks for thematic control sections, final semester sections that can be of a search, creative, research, or practice-oriented nature.

The management component is designed to master the techniques of implementing all the components of the management cycle. It contains recommendations and memos for including students in the process of managing educational activities at all stages of its implementation (starting with the analysis of initial achievements, planning, organizing activities on the study of a fragment of educational material, monitoring the progress and results of this activity, and ending with the analysis of the results of control, evaluation, correctional work, analysis of its results and transition to the study of a new fragment of educational material). The work on illustrating the performance of managerial functions in the classroom and the issuance of tasks for their independent implementation will allow one to form knowledge and skills to implement self-management of educational activities. The skills of self-management in their educational activities are developed in the process of independent work through the implementation of managerial tasks.

The efficiency-based component is created to guide the formation of pedagogical university students' readiness to implement ongoing mathematical self-education. It contains three criteria: personal, cognitive, and activity-based, which are specified by the system of indicators:

(a) Personal—indicators of the formation of purposefulness, perseverance, independence, active nature, initiative;
(b) Cognitive—indicators of the formation of the knowledge system—evident characteristics of concepts and scientific facts, scientific methods, conditions, and boundaries of their application;
(c) Activity-based—the requirements of mastering include methods of searching for scientific literature and electronic resources in mathematics and methods of teaching it, methods of working on the study of necessary information; methods of setting a scientific issue, methods, and means of finding rational ways to solve it; analysis of the obtained results, methods and technologies the educational process organization, methods of choosing approaches and technologies of organizing educational activities that are relative to the content of the educational material, methods of organizing self-analytical and managerial activities in the process of performing independent work on the study of theoretical material and pedagogical experience.

## 4 Discussion

The theoretical substantiation of the structure, content, and purpose of the components of the "System of formation of readiness of pedagogical university students to implement ongoing mathematical self-education" are discussed and approved at the meeting of Educational Research Laboratory "Didactic Development of Ongoing Physics and Mathematical Education." The system has passed a pilot test in the framework of one university in the teaching of mathematical analysis in the first year. Summing up the results of the organized annual monitoring of the course of the pedagogical experiment shows a positive trend in the formation of students' readiness to implement mathematical self-education. The advantages of the developed didactic support of the content block of the system were particularly evident in the organization of distance educational activities during the pandemic. Approbation of intermediate research results is presented at seminars and conferences at the regional and international levels.

## 5 Conclusion

The current study constructs the pedagogical "System of formation of readiness of pedagogical university students to implement ongoing mathematical self-education" based on the created methodological apparatus. The conducted experimental verification of its functioning (based on the developed system of criteria and indicators) in the educational process in the specialty "Mathematical Analysis" allows us to state the following:

(1) The use of the proposed system built based on systematic, activity-based, and competence-based approaches contributes to achieving the integrity of the process of preparing students for the implementation of ongoing mathematical self-education;
(2) Strict adherence to the principles of reflection and harmonization of the fundamental theoretical and practical components of the training of a future specialist in the field of mathematical education contribute to the development of skills: to argue their actions, to reflect on their own achievements, and to gradually master the managerial skills of analysis, planning, and regulation of their activities;
(3) The revealed positive dynamics of the level of competence formation and readiness of pedagogical university students to implement mathematical self-education indicates the effectiveness of the developed pedagogical system to form readiness for ongoing mathematical self-education.

## References


Karpova OL, Shabalkova TN (2015) Conceptual characteristics of actualization of self-educational activity of university students. Theory Pract Phys Cult 5:8–10

Khinchin A Ya (2006) Pedagogical articles: Questions of teaching mathematics. Fight against methodological cliches (2nd ed) In: Gnedenko BV (ed). KomKniga, Moscow, Russia

Khutorskoy AV (2017) Methodological grounds for applying competence-based approach to education construction. High Educ Russ 12:85–91

Lobanov NA, Titova LG, Yudin VV (ed) (2018) Ongoing education: Methodology, technology, management. Editorial and Publishing Department of the Yaroslavl State Pedagogical University named after K. D. Ushinsky, Yaroslavl, Russia

Medvedev IF (2012) The concept of self-education: basic concepts and structure. Educ Sci J 2:32–42. https://doi.org/10.17853/1994-5639-2012-2-32-42

Ministry of Education and Science of Russian Federation (2018) Order of the ministry of education and science of the russian federation "on approval of the federal state educational standard of higher education—bachelor's degree majoring in 44.03.05 pedagogical education (with Two Specialties) (February 22, 2018 No. 125). Moscow, Russia

Odintsova LA, Bronnikova LM (2020) Self-analytical tasks for organizing independent work of pedagogical university students in mathematics as a means of preventing formalism in knowledge and methods of activity. Modern Problems of Science and Education. Surgery 2. Retrieved from https://science-education.ru/ru/article/view?id=29630

Osipov PN, Sedankina TE (2006) Development of professional self-analysis in the course of the educational process. Bull Kazan Technol Univ 6:357–362

Sarantsev GI (2016) Harmonization of professional training of the bachelor majoring in "pedagogical education." Integr Educ 20(2):211–219

Saribeyli FR (2018) Theoretical and practical aspects of student self-assessment. Educ Sci J 20(6):183–194. https://doi.org/10.17853/1994-5639-2018-6-183-194

USSR State Committee for Public Education. (1989). The concept of ongoing education. Bulletin of the USSR State Committee for Public Education. Series: Higher and Secondary Special Education (March 18, 1989 No. 7). Moscow, Russia

Zimnyaya IA (2013) Competence in the context of the competence-based approach in education. Sci Notes Natl Soc Appl Linguist 4(4):16–31

# Digital Competence in University Students: Insights from Various Training Programs

**Zoia M. Kondrashova and Nikolai N. Solokhin**

**Abstract** The paper describes the current approaches to the formation of digital competence of a contemporary person. We compare the approaches to the definition of digital competence. Besides, we identify and substantiate the indicators of achieving students' digital competence. More than that, we provide examples of indicators of achieving students' digital competence depending on their majors. The paper analyzes the definition of the concept of blended education. We reveal that various scholars consider blended education at the level of the concept of modern education, learning technology, and teaching method. The basics of modern didactics allowed us to conclude that blended education is a teaching method, along with full-time, part-time, and distance education. The consideration of blended education from the standpoint of a teaching method has become the basis for building a structural and functional model of the formation of a person's digital competence. The research illustrates component features of the structural–functional model of the formation of the students' digital competence, including the target-, content-, organization-, and efficiency-based components. The goals, objectives, and principles of building the model are disclosed within the target-based component. We consider the principle of proactive training as the primary one. It consists in building a teaching process aimed at creating conditions for the training of future professionals who will be in demand in a dynamically changing labor market and have competencies of the future professions.

**Keywords** Digital competence · Structural–functional model · Blended education

Z. M. Kondrashova (✉)
Southern Federal University, Rostov-on-Don, Russia
e-mail: zm.66@mail.ru

N. N. Solokhin
Don State Technical University, Rostov-on-Don, Russia
e-mail: nik2007.72@mail.ru

 
S. G. Maximova et al. (eds.), *Advances in Natural, Human-Made, and Coupled Human-Natural Systems Research*, Lecture Notes in Networks and Systems 234,
https://doi.org/10.1007/978-3-030-75483-9_66

# 1 Introduction

The research topicality is determined by the development strategy of the modern information society. The pace of technological development, creation, processing, and dissemination of information has significantly exceeded most people's ability to master and apply knowledge. The shift of emphasis in the perception of the surrounding world, especially in relevance to the Internet, from scientific, educational, and cultural to entertainment-informational, has formed a new perception model–the so-called "mosaic thinking," which implies a fleeting perception of information by the masses. Due to the factors mentioned above, we should highlight the directions of forming the information knowledge space. It includes (1) implementing educational projects aimed at providing access to knowledge, achievements of modern science and culture; (2) providing facilities for scientific and technical creativity, including the creation of areas for self-actualization of educational and scientific organization representatives; (3) using and developing various educational technologies, including distance education, e-learning during the implementation of educational programs; (4) developing and implementing partnership programs of educational institutions of higher education and high-tech organizations, including the improvement of educational programs; (5) creating and developing citizens' legal consciousness and their responsible attitude towards the use of information technologies, including consumer culture. The aforementioned is reflected in the accelerated introduction of digital technologies in the economy and social sphere.

The activities of modern educational organizations at various levels aim to create a modern and secure digital educational environment that ensures high quality and accessibility of education of all types and levels. It also ensures the modernization of vocational education, including through the introduction of flexible educational programs. The possibility of creating flexible educational programs is based on the principles and models of blended education as a condition for the formation of students' digital competencies.

The research goal is to (1) develop the indicators of the formation of students' digital competence, (2) clarify the concept of blended education, (3) construct a versatile model for the formation of digital competence of students of various majors in the process of studying different disciplines using blended education.

# 2 Materials and Methods

The theoretical research methods include the following:

- Analysis of scientific literature on social and psychological-pedagogical issues (Kozlova 2019; Strekalova 2019; Fomina 2014) related to the formation of digital competence in students of various universities (Khramtsova and Mayboroda 2019), the process of building the educational environment of a university on a digital basis, its impact on students' personalities (Dneprovskaya 2018);

- Analysis of the Federal State Educational Standard for Higher Education in the areas of pedagogical education and construction; study and analysis of university programs (Southern Federal University, Don State Technical University (Rostov-on-Don, Russia)), textbooks, books covering specific aspects of the use of digital technologies in the educational process;
- Generalization and systematization of the data obtained as a result of the analysis, identification of the general directions of building a digital educational environment, identification of the features of implementing blended education at a university;
- The ascent from the abstract to the concrete in considering the functionality of digital products and their specific application in the construction of educational programs of individual universities;
- The system method identifies the system-forming elements of the model of the formation of undergraduate students' digital competence.

Among the empirical research methods, there is a comparison of individual approaches to building a digital educational environment in Russian and foreign universities.

## 3 Results

We should differentiate the concept of information competence from the concept of digital competence. Information competence is associated with analyzing, processing, and selecting various information types using modern technical means. Digital competence includes methods and techniques of using and creating digital resources for various spheres of human life. In other words, we associate digital competence with the ability to select, use, and create digital products for solving various tasks. Digital competence is a broader concept than information competence. Digital competence involves the possession of digital resources at various levels–at the user's, creator's, and instructor's. A user is a person who successfully uses digital products in various areas of their activities. A creator is a person who generates different types of digital products. An instructor is a person who builds technology for using digital products for various purposes. In this regard, we can determine the indicators of achieving the students' digital competence:

- They use digital products in various areas of their activities;
- They search, select, systematize, generalize, and critically analyze digital products;
- They create a digital product package with regard to its functionality to achieve individual goals and results in various areas of activities;
- They create digital products for use in various areas of activities.

The achievement of the results mentioned above is possible within the framework of various majors in the higher education system. Within the framework of a specific major, the result of achieving students' digital competence is formulated in accordance with a specific area of activities. For example, if they select the pedagogical major, then the results are as follows:

- They use digital products in pedagogical activities;
- They search, select, systematize, generalize, and critically analyze digital products taking into account the peculiarities of pedagogical activities;
- They create a digital product package with regard to its functionality to achieve specific goals and results in pedagogical activities;
- They create digital products for use in pedagogical activities.

The categories of "familiarity," "being able," and "having the skills" should be formulated, taking into consideration the specific major and subject area.

For example, if the major is "Pedagogical Education" and the specialty is "Primary Education" (Southern Federal University (Rostov-on-Don, Russia)) of the discipline "Methods of Teaching Mathematics" or "Theory and Technologies of Mathematical Education," then the phraseology is as follows:

- They use digital products in pedagogical activities; in other words, they are familiar with digital resources that can be used in teaching mathematics to students, construct a math lesson using digital resources, and model a math lesson using digital resources. This approach allows us to determine the evaluation tools that reveal the competence indicator level formation in these disciplines. In this context, the category "familiarity" is evaluated by means of a colloquium, the category "being able"–by means of an interview, the category "having the skills"–by means of pedagogical case-problems.

In the previous example, we used disciplines from the part of the curriculum formed by educational relations participants. Next, we consider an example of a discipline from the mandatory part of the curriculum.

We take into consideration one of the majors of a technical university, "Construction" (Don State Technical University, (Rostov-on-Don, Russia)), the specialty "Building Design," and discipline "Mathematics" from the mandatory part of the curriculum. Indicators of achievement of students' digital competence of this university are as follows:

- They use digital products in professional activities;
- They search, select, systematize, generalize, and critically analyze digital products concerning professional activities;
- They search, select, systematize, generalize, and critically analyze digital products taking into account the peculiarities of professional activities;
- They create digital products for use in professional activities.

We formulate the categories of "familiarity," "being able," and "having the skills," taking into consideration the specific specialty "Building Design" and the subject area "Mathematics." For this discipline, it is appropriate to consider the indicator "They search, select, systematize, generalize, and critically analyze digital products with regard to professional activities." They are familiar with the package of digital products that allow one to deepen, refine, and expand the knowledge of individual sections of the calculus course. They are able to find, select, and organize digital products considering their educational needs in the field of calculus. They have the skills of generalization and critical analysis of digital products according to the achievements in the course of calculus. The assessment tool selection can be as follows: the category "familiarity" is assessed by means of control work in individual sections of calculus, the category of "being able"—by means of an online test. The category "having the skills" is assessed by conducting an interview concerning the individual sections of calculus.

The presented mechanism for constructing university students' digital competence achievements in various majors and specialties is versatile in nature. It should be noted that each indicator is achieved through several disciplines of a specific specialty.

The formulated indicators of students' digital competence and the categories of "familiarity," "being able," and "having the skills" determine the direction of change in the educational space of the university in the direction of digitalization. Nowadays, the digitalization of the educational space provides for the university to have at least its own electronic educational and information environment, studios for creating online educational courses, teachers who are ready to work with various digital products and are ready to create online educational courses within their disciplines. Not all modern universities have the mentioned capabilities (Dneprovskaya 2018). Only the first attempts to organize a digital educational environment have been outlined. Changing the educational environment of the university in the direction of digitalization will allow one, along with such forms of education as intra-mural, extra-mural, intra-extramural (or evening classes), to use blended education.

Moreover, intra-mural and extra-mural forms of education are reflected in the primary educational program. However, blended education is exclusively the choice of the teacher, who implements a separate discipline within the primary educational program in the major and the specialty. Therefore, given the need to create a digital educational environment of the university, it is essential, along with such forms of education as intra-mural, extra-mural in the primary educational program, to add blended education. It is also crucial to create curricula for blended education and organize students' recruitment for a blended mode of study.

The concept of blended education requires clarification. We understand blended education as a mode of study, along with such forms as intra-mural, extra-mural, and distance education. The difference in the use of each of these forms is determined by the percentage of contact work with a teacher in the curriculum. The analysis of the curricula of various universities allowed us to identify approximate indicators of contact work with a teacher. Approximate contact work indicators with a teacher by mode of study are as follows: intra-mural–50%, extra-mural–20%, blended–approx. 30%, and distance–from 5 to 10% of contact work or complete absence of contact work with a teacher. The student has the ability to choose the mode of study. The goals of choosing a particular mode of study can be completely different from the student's ability to master the disciplines independently to the need to study and work simultaneously. Digital products are a means of implementing each of the modes of study. The availability of education tools, their quality, and their level depend on the capabilities of an individual university.

The layout of the structural and functional model of the formation of digital competence of students in the higher education system includes four main components: target, content, organizational, and effective.

*The target component.*

The target is to develop the ability to select, use, and create digital products for solving various tasks in various fields of activity.

Tasks:

- To analyze the possibilities of using digital products in various fields of activity;
- To substantiate the criteria for the search, selection, systematization, generalization, and critical analysis of digital products;
- To disclose the conditions for the formation of a package of digital products, taking into account their functionality to achieve specific goals and results in various fields of activity;
- To identify mechanisms for creating digital products for use in various fields of activity.

Principles:

- The principle of proactive training is the construction of a learning process aimed at creating conditions for the training of a future professional in demand in a dynamically changing labor market, possessing the competencies of the future professions;
- The principle of theory and practice interconnection is a theoretical substantiation and practical use of digital products within the professional needs of an individual and the realities of life;
- The principle of students' consciousness, proactivity, and independence.

*The content component* includes disciplines and modules included in the educational program for training future professionals in various fields and specialties. The educational program should be provided with digital products of various functionality as much as possible. It implies digital products of various functionality—digital educational platforms, information environments, digital resources that allow the teacher to create online training courses, a systematic package of training webinars, mass online courses suitable for the successful implementation of the educational program.

*Organizational component.*

The organizational component includes building the learning process using various modes of study—intra-mural, extra-mural, and blended and distance education.

We determine the percentage of the number of hours for contact work with the teacher for each mode of study in the educational program. Contact work with the teacher is represented by a series of lectures and practical and laboratory classes. The non-contact work of the student is represented by the forms of organization of independent activity. The student's independent activity should be structured within the educational program. It should contain such structural elements as (1) the content (section of the work program of the discipline, a specific topic), (2) the number of hours allocated for independent activities, (3) a list of tasks for each topic, (4) a list of digital products required for each task, (5) evaluation criteria for each task, and (6) a description of the form of presentation of the task.

*The efficiency component.*

The efficiency of the achievement of individual indicators and the competence as a whole is evaluated using various evaluation tools. The construction of the evaluation tool should depend on the form of implementation of the educational program. The identification of learning outcomes occurs through direct contact with the teacher in the case of intra- or extra-mural modes of study. In the case of blended or distance education, the identification of learning outcomes occurs with the maximum use of the capabilities of digital products. The last statement needs to be clarified. The theoretical material for distance and blended education is tested using online testing. In this case, it is possible to use various test methods. It is crucial to use testing at various stages of studying the discipline (module). It is possible to use educational tests that reveal gaps in the student's knowledge of a particular topic and contain links to theoretical sections to eliminate gaps. Practical tasks are evaluated using peer review. It is recommended to develop criteria according to which a student can organize an expert assessment of another student's work in the group for the successful organization of such an assessment.

The presented model of the structural and functional model of the formation of students' digital competence in the higher education system is versatile since the characteristics of each component will have its own characteristics when built within a specific major and specialty of students' education.

## 4 Discussion

The issue of the formation of students' digital competencies in the system of higher and secondary vocational education is reflected in the researches of E. V. Vasilyeva, E. L. Vartanova, A. V. Vyrkovsky, E. A. Dolgikh, M. V. Lyashenko, T. Yu. Mayboroda, M. I. Makeenko, T. A. Pershina, V. N. Pulyaeva, S. S. Smirnov, N. G. Khramtsova, V. A. Yudina et al.

Digital competencies include students' ability to use information technologies, information and communication networks, and digital equipment, and apply skills of working with application software (Vasilyeva, Pulyaeva and Yudina 2018; Dolgikh and Pershina 2019). They also include IT-literacy, information hygiene, and analytical skills (Vartanova et al. 2017). The research on the subject identifies specific indicators of competence in the modern educational paradigm (Khramtsova and Mayboroda 2019), among which there is the creation of electronic presentations using special software, a spreadsheet, a text editor, and the skills of working with digital equipment and application software. Lyashenko M. V. describes four types of students' digital competence that depend on the sphere of life—information, communication, functional, and consumer (Lyashenko 2019). The scholar substantiates the need to improve students' digital competence while associating it with the issue of self-development of the graduate.

The necessity to build a new educational environment is substantiated in the research of Bondareva G. A., Dadalko V. A., Kozlova N. Sh., Petrova N. P., Solovkina E. D., Strelkova N. B. et al.

Dadalko V. A. and Solovkina E. D. state the importance of building a new educational ecosystem that "will consist not only of schools, technical schools, universities, but also of fundamentally new elements such as global online platforms, various urban educational formats, or practicing communities" (Dadalko and Solovkina 2018). The authors support the expansion of educational structures and the enrichment of educational formats. A slightly different view of the issue is presented in the study of Kozlova N. Sh. (2019), which associates the solution of the issue with the development and implementation of an electronic information environment in the educational process of an educational institution. The author reveals the most evident advantages of such an innovation. The paper by Petrova N. P. and Bondareva G. A., devoted to digitalization and digital technologies in education, reveals an ambiguous approach to the concept of the digital educational environment and provides examples of the formation of a digital educational environment by individual researchers (Petrova and Bondareva 2019). Strekalova N. B. (2019) identifies and substantiates the risks of informatization and possible education system changes. The identified risks allow one to outline the ways of education digitalization and the conditions for forming students' digital competencies.

The issue of using blended education in higher education is reflected in the research of Antonov S. N., Kravchenko G. V., Ostapenko A. S., Pletyago T. Yu., Fomina A. S. et al. The scholars use a series of terms such as "blended education," "distance education," and "e-learning" in their works. From the scholars' perspective, blended education is a means of implementing e-learning, and distance education are synonymous with the concept of e-learning. Fomina A. S. identifies the aspects of blended education. They are institutional, organizational, and pedagogical. These aspects characterize the strategy of a university in the field of e-learning, the organization of educational activities, and methodological support of the educational process (Fomina 2014). A slightly different view of the issue of using blended education in higher education is presented in the materials of Kravchenko G. V. (Kravchenko 2014). The scholar considers blended education as a distance education course format, in which active learning methods are embedded, and as a model for using information and educational resources using elements of asynchronous and synchronous distance education. Antonova S. N., Ostapenko A. S., Pletyago T. Yu. approach the issue of blended education from the standpoint of the pedagogical model. Their research analyzes the practice of using this type of education in national and international universities (Pletyago et al. 2019).

The analysis of source material depicts no unified approach to the definition of digital competencies. There are no characteristics of indicators of the formation of digital competence. There are only general attitudes or a list of individual skills. It is necessary to identify, substantiate, and systematize the list of learning outcomes necessary for a modern person in changing society for improving the quality of the formed competencies. Each competence should be represented by a list of indicators, the measurement of which will allow one to draw a conclusion about the formation of the particular competence in a modern person. Besides, it is crucial to demonstrate that any subject area contributes to forming the individual's digital competence.

The didactic essence of the issue is crucial in studying the construction of blended education. The basics of modern didactics clarify the concept of blended education from the point of view of existing educational paradigms and concepts and allow one to rethink such concepts as traditional education, e-learning, computer learning, distance education, and online learning.

## 5 Conclusion

The study of the issue of the formation of digital competence of university students allows one to draw the conclusions enlisted further.

The following indicators characterize students' digital competence—it (1) uses digital products in various fields of activity; (2) searches, selects, systematizes, summarizes, and critically analyzes digital products; (3) forms a package of digital products with their functional capabilities for achieving specific goals and results in various fields of activity; and (4) creates digital products for using in various fields of activity.

The formation of students' digital competence is possible within any major and specialty using a specific structural–functional model. The structural–functional model is based on the choice of the mode of study—intra-mural, extra-mural, blended, and distance education. A feature of each mode is the number of hours in the curriculum allocated for contact work with the teacher and independent activity of the student. Each mode of study is accompanied by a package of digital products formed or developed by teachers of a particular university. Various assessment tools are used to check the results of achievement, the nature of which varies depending on the mode of study.

## References

Dadalko VA, Solovkina ED (2018) Competencies for digital economy and transformation of the educational system on the sixth economic paradigm. Natl Interes: Priorities Secur 14(5):913–926. https://doi.org/10.24891/ni.14.5.913

Dneprovskaya NV (2018) Assessment of the readiness of the russian higher education for the digital economy. Stat Econ 4:16–28

Dolgikh EA, Pershina TA (2019) Statistical study of students' digital competencies. E-Manag 2(3):64–72. https://doi.org/10.26425/2658-3445-2019-3-64-72

Fomina AS (2014) Blended education at the university: institutional, organizational-technological, and pedagogical aspects. Theory Pract Soc Dev 21:272–279

Khramtsova NG, Mayboroda T Yu (2019) Approaches to the development of digital competencies of law students. Sci Educ Perspect 1(37): 80-93

Kozlova N Sh (2019) Digital technologies in education. Bull Maykop State Technol Univ 1: 85-93

Kravchenko GV (2014) The use of the blended education model in the higher education system. Izv Altai State Univ 2–1(82):22–25

Lyashenko MV (2019) Self-development of digital competence of the students at college. Bulletin of the South Ural State University. Series Education. Educational Sciences 11(3): 74–94

Petrova NP, Bondareva GA (2019) Digitalization and digital technologies in education. World Sci, Cult, Educ 5(78):353–355

Pletyago T Yu, Ostapenko AS, Antonova SN (2019) Pedagogical models of blended education in higher education: generalization of the experience of russian and foreign practice. Educ Sci 21(5): 112-129 https://doi.org/10.17853/1994-5639-2019-5-113-130

Strekalova NB (2019) Risks of introducing digital technologies to education. Samara Univ Bulletin Hist, Pedagog, Philol 25(2): 84–88. DOI: https://doi.org/10.18287/2542-0445-2019-25-2-84-88

Vartanova EL, Vyrkovsky AV, Smirnov SS, Makeenko MI (2017) Russian media industry: The digital future. MediaMir, Moscow, Russia

Vasilyeva EV, Pulyaeva VN, Yudina VA (2018) Digital competence development of state civil servants in the russian federation. Bus Inform 4(46):28–42

# Additional Education in Digital Technologies as Competitiveness in the Labor Market

**Alexandra S. Ermolaeva**

**Abstract** Over the past 20 years, the spread of digital technologies has accelerated and continues to evolve. In this regard, technological change is one factor that can affect the labor market, particularly the demand for competitive specialists with digital skills. The paper examines the problems of obtaining additional education in digital and information technologies to ensure the competitiveness of individuals. The author analyzes the survey results conducted by the "NAFI Research Centre," with 1600 participants from 52 regions of Russia. The author uses empirical (analysis and generalization of the obtained results) and theoretical (survey and analytical research) research methods. During the research, the factors that influence the readiness to receive additional education are identified. The research analysis shows that working Russians are aware of the risk of possible job loss and, consequently, loss of personal competitiveness in the labor market due to the high rates of digital technology development and a lack of knowledge in this area. It is substantiated that obtaining additional education in digital technologies is not only a matter of meeting the needs of the population in digital skills but also one of the factors ensuring the competitiveness of an individual in the labor market. The author concludes that the interaction of employers, educational organizations, and the government in understanding labor market trends will make it possible to form the necessary expected digital skills among employees, motivating them to receive appropriate additional education.

**Keywords** Additional education · Digital technologies · Digital literacy · Digital skills · Competence · Competitive advantages · Competitiveness of an individual · Labor market

---

A. S. Ermolaeva (✉)
Financial University under the Government of the Russian Federation, Moscow, Russia
e-mail: asermolaeva@fa.ru



S. G. Maximova et al. (eds.), *Advances in Natural, Human-Made, and Coupled Human-Natural Systems Research*, Lecture Notes in Networks and Systems 234,
https://doi.org/10.1007/978-3-030-75483-9_67

## 1 Introduction

The development of digital technologies and their spreading speed that stimulates various innovations, contribute to (1) the creation of an information space considering the population needs and (2) the emergence of new business models in many sectors of the economy that aim to increase labor productivity, quality of goods and services. Since 2019, the project "Personnel for the Digital Economy" within the national program "Digital Economy of the Russian Federation" is developing (Government of Russian Federation, 2017). Its main priority is to improve the education system and provide the digital economy with competitive personnel with a high level of relevant skills and competencies. That is why the current research aims to study and identify the problems of obtaining additional education in digital and information technologies as one of the factors that ensure the competitiveness of an individual.

According to the American economist, M. Porter, competitiveness has the ability to develop continuously. First, one needs to achieve a competitive advantage by changing the competition basis, and then maintain ones' advantage by constantly improving the product, production methods, and other factors so quickly that competitors cannot catch up and overtake them. The improvement and renewal are ongoing processes that allow one to create competitive advantages (Porter, 2007). However, the fundamental characteristic of competitiveness is its multilevel character. In current conditions, the issue of the competitiveness of products and services, individual companies, countries, regions, etc., are under research (Fatkhutdinov, 2004). Therefore, it is appropriate to assume that the multilevel organization (Nemtsev, 2011) of competitiveness can be supplemented by the competitiveness of the individual (Fig. 1).

The competitiveness of an individual is also determined by the ability to constantly develop and improve—professionals in various fields have to remain competitive in the labor market, which requires a high level of professionalism, competence, and knowledge.

Additional education allows professionals to adapt to the changing requirements of the labor market in the current conditions of digitalization. In particular, the first level of ensuring individual competitiveness forms competitiveness factors that generate knowledge, competencies, and professionalism. The next level creates competitive advantages based on the use of additional education. At the same time, the transition from one level to another requires the use of various forms, methods, and technologies of additional education (retraining, advanced training, internships, etc.). Scholars understand additional education as a special organized process focused on the initial freedom of the students' choice of field and time for studying the curriculum, various education activities, and the formation of the necessary conditions for the implementation of the student's existing potential in their work (Rukina, 2009). Employees obtain additional education because of the labor market conditions, which requires not only the availability of additional education (Akhmedov & Shatalov, 2017) but also the willingness to constantly develop and improve the level of their professional competence and expand the range of knowledge and skills.

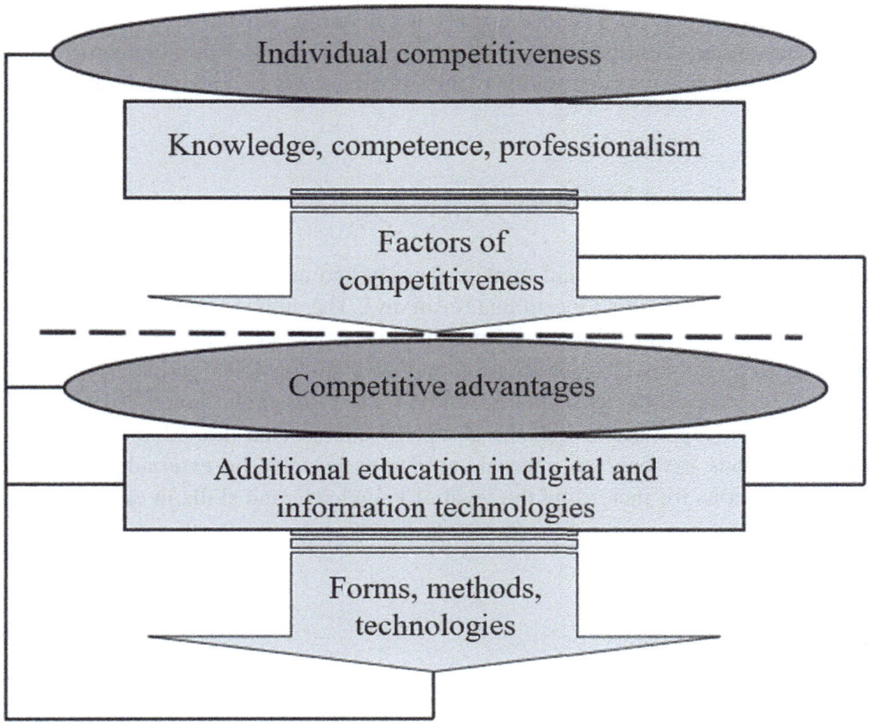


**Fig. 1** Model of ensuring the competitiveness of the individual. *Source* Compiled by the authors

Competition among professionals in the rapidly changing labor market makes it necessary to solve the problems of increasing competence arising in connection with new technologies, especially against the background of the growing trend towards automation. Over the past decade, companies from different industries digitized their operations and processes using artificial intelligence to optimize business processes (Trofimov, 2019).

As innovative technologies develop, new professions emerge (Pachler & Turvey, 2018). It creates a demand for employees with digital skills. The author understands digital skills as skills necessary to use digital technologies (Grand-Clement, 2017). According to statistics, 40% of the population will have general digital skills by 2024 (Kolykhmatov, 2018). It makes sense that digital technologies permeate not only employment (digital skills) but also everyday life (digital literacy). The author focuses on the analysis of digital technical skills, while "soft skills" (Posypkina, 2018), such as (1) digital literacy, (2) information retrieval and computer skills, (3) information literacy and related abilities (Stopar & Bartol, 2019) are also significant and are the subject of further research.

Thus, in the context of an increasingly deeper digital transformation in the current world, the change of necessary skills (including programming skills, data analysis, design technologies, etc.) is increasingly accelerating and requires employees to possess the digital skills that are significant in terms of increasing the competitiveness of an individual. It leads to saturation of the labor market with competitive specialists.

## 2 Materials and Methods

The research uses a theoretical method (survey and analytical research) based on empirical research results (questionnaire survey). The author analyzes the results of a survey to identify the attitude of Russian citizens to receive additional education in digital technologies. The survey was conducted among 1,600 people aged 18 and over in 52 regions of the Russian Federation (NAFI Research Center, 2019).

The author hypothesizes that the competitiveness of an individual in the labor market depends not only on internal aspirations but also on external factors that create conditions for increasing the level of knowledge and skills in current digital technologies.

## 3 Results

The author describes the obtained results according to the criteria identified above.

***Data on readiness to obtain additional education***. The survey revealed indicators that, according to the personal experience of respondents, have the greatest impact on their readiness to obtain additional education in digital technologies, namely:

- Diploma or certificate as confirmation of acquired knowledge;
- Motivation;
- Self-discipline;
- Financial interest;
- Need for lifelong learning;
- Need to study at a young age to obtain additional education.

The revealed indicators characterize certain groups of factors for obtaining additional education. Analysis of the respondents' answers made it possible to rank the factors according to the degree of their influence on the willingness to obtain additional education. As seen in Fig. 2, groups of social needs and internal motivation factors have a decisive influence on the readiness for additional education.

The leading position is taken by a group of factors of social need, which determines the need for continuous learning. It dictates the special importance of lifelong education, which becomes essential in the context of digitalization. One of the main tasks of education is to train specialists seeking continuous self-education and self-development (Zainullina, 2015). Indeed, considering the increased use of rapidly changing digital technologies, people need to acquire new skills and knowledge in the workplace, which becomes an ongoing process throughout ones' life. This allows

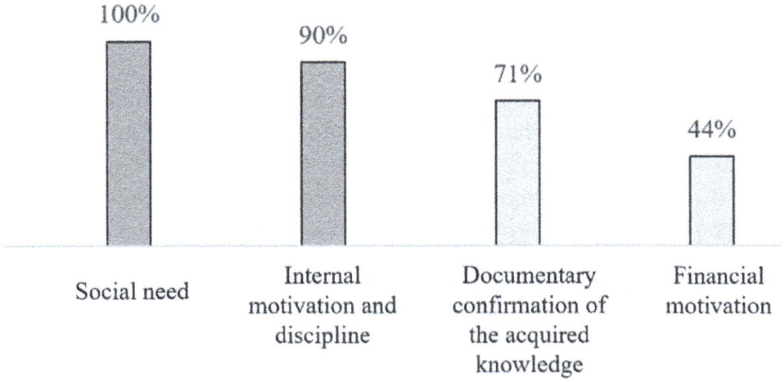


**Fig. 2** Groups of factors that affect the readiness to obtain additional education. *Source* Compiled by the authors

us to keep updated on innovations and new developments in digital and information technology. Internal motivation and discipline group of factors is also significant and affect the readiness to obtain additional education. Lifelong learning indeed is impossible without strong moral motivations (Gabdullina, 2012). At the same time, there is a problem of a person's internal motivation for additional training, which correlates with external causes. They give rise to the critical need to improve the level of qualifications. To increase the internal personal motivation for learning, some scientists offer an individual approach. The true needs of an individual are revealed during the study. Therefore, the training program can be adjusted to understand the goals and objectives of the educational process and meet the individual needs of students (Cherkasova, 2014).

***Data on the experience of obtaining additional education***. According to scientists, the digital skills of employees in various fields should be updated every three years (Reis et al., 2011). The survey results show that in the past three years, the majority of the population did not improve their level of knowledge through additional education (Fig. 3).

As seen in Fig. 3, 79% of respondents did not receive additional education over the past three years, 21% improved their qualifications through additional education.

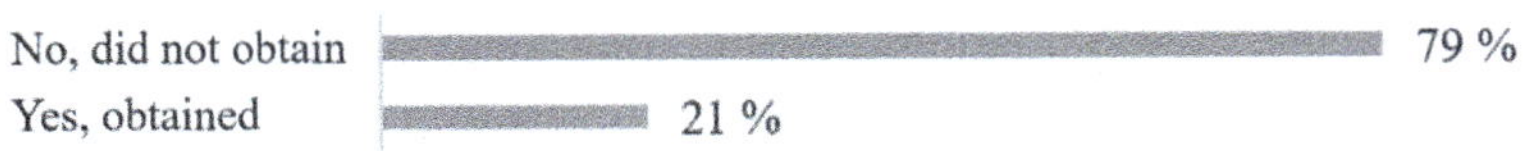


**Fig. 3** Results of the questionnaire about the experience of obtaining additional education. *Source* (NAFI Research Centre, 2019)

## 4 Discussion

The research revealed that more than 80% of Russians want to obtain additional education, but only slightly more than 20% realize this desire. The author believes that the reason is the inconsistency of interaction between employers, educational organizations, and the government. Strengthening the interaction, on the one part, from the perspective of educational organizations (or the education system) and the employer, and from the point of understanding labor market trends on the other part, will clearly define and formulate the need for the necessary digital skills among employees. Building more coherent relationships between educational organizations and employers lets educators know the specific problems employers face and vice versa. In the third part, in the digital world, the great role of the government is to ensure the accessibility of education and the provision of highly qualified personnel, that both employers and employees have the opportunity to take advantage of the new job opportunities that are opening up.

As for the liability of each mentioned subject of labor relations, the employer is responsible for improving its employees' skills, where a well-developed motivation system plays an important role. A crucial element of motivation is to stimulate an employee to obtain additional education in digital technologies. As digital skills needs are constantly changing, it is essential to motivate workers to meet new requirements and retrain. Thus, using an effective work organization and management (particularly based on a well-developed motivation system), employers will be able to use the digital skills of their employees fully. Eventually, it will increase the competitiveness of the organization without removing the individual responsibility of the employee to increase their qualifications and renew knowledge constantly.

The development of the so-called integration of educational processes contributes to the involvement of employers in the development of educational policy and education quality standards. Thanks to them, the rapidly changing requirements of the labor market are considered more correctly (Bautin & Mychka, 2018). In these conditions, it becomes necessary to update existing skills and abilities (Yusupova & Pozdeeva, 2018): various special knowledge, pro-active attitude, knowledge of foreign languages, digital and information technologies, etc. In sum, this suggests that education strives for constant advanced learning and retraining of a professional to adapt to new conditions and changes (Cibizova, 2011).

At the same time, a broader responsibility falls on the government to increase the relevance, efficiency, and accessibility of education, create conditions for improving the quality of additional education, and provide highly qualified personnel. The role of the education and training system is to assess and anticipate changing needs and skills to adapt to the proposed educational programs quickly.

In this regard, additional training of digital skills is one factor that increases the competitiveness of individuals in the labor market, which is graphically shown in Fig. 4.

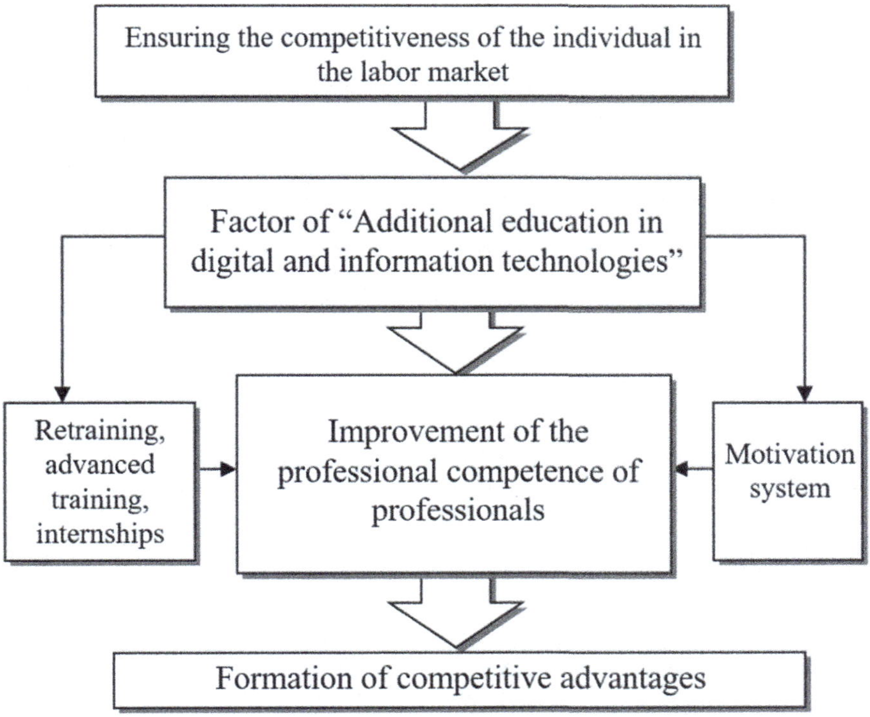


**Fig. 4** Characteristics of "Additional education in digital and information technologies" factor. *Source* Compiled by the authors

## 5 Conclusion

The authors' hypothesis is confirmed. The competitiveness of an individual in the labor market is determined not only by the internal willingness and desire to obtain additional education in digital technologies but also by the coherence of external factors—employers, educational organizations, and the government.

With the increasing use of rapidly changing technology, there is also a social need to acquire new skills in current digital technologies for work. During the research, the author found that obtaining additional education is becoming one of the necessary conditions for an employee to achieve a more competitive position in the labor market. The research made it possible to characterize the factor of increasing individual competitiveness in the labor market—an additional education in digital technologies.

The interaction of educational organizations, employers, and the government in understanding labor market trends will create the necessary digital skills among the population and stimulate it to receive additional education.

## References

Akhmedov AE, Shatalov MA (2017) Problems of the development of higher education in the context of the internationalization of educational services. In: Igolkin SL (ed) Actual problems of the development of vertical integration of the education system, science, and business: economic, legal, and social aspects. Voronezh Institute of Economics and Law, Voronezh, Russia, pp. 30-34

Bautin VM, Mychka S Yu (2018) Factors of competitiveness of higher education institutions in the system of market relations. In: Igolkin SL (ed) Actual problems of the development of vertical integration of the education system, science, and business: economic, legal, and social aspects. Voronezh Institute of Economics and Law, Voronezh, Russia, pp. 34-40

Cherkasova OV (2014) To the question of the motivation of students to receive additional professional education. Koncept: Sci Methodol E-Mag S25: 36-40

Cibizova T Yu (2011) On the problems of training highly qualified specialists in the system of continuing professional education. Sci Educ: Sci Ed MSTU NE Bauman 10: 54

Fatkhutdinov RA (2004) Managing the competitiveness of the organization. Eksmo, Moscow, Russia

Gabdullina EN (2012) Continuing professional education as a factor of preparing the competitiveness of an individual. Alm Contemp Sci Educ 3:36–38

Government of Russian Federation (2017) Order of the government of the Russian federation "on approval of the program 'digital economy of the russian federation'" (July 28, 2017 No. 1632-r). Moscow, Russia

Grand-Clement S (2017) Digital learning. education and skills in the digital age. Retrieved from https://www.rand.org/content/dam/rand/pubs/conf_proceedings/CF300/CF369/RAND_CF369.pdf

Kolykhmatov VI (2018) Main directions of additional professional education development in conditions of the digital economy. Sci Notes Name Univ P F Lesgaft 10(164): 132-136

NAFI Research Centre (2019) Ready to study: Rssussians want to learn new things about the digital environment. Retrieved from https://nafi.ru/analytics/uchitsya-uchitsya-i-eshche-raz-uchitsya-rossiyane-khotyat-uznavat-novoe-o-tsifrovoy-sfere/?fbclid=IwAR0JYX5P7744rbiRSCEv4_zD8CFTNTSST5naeeCHgYvV3Om_MKmsHuOXGaI

Nemtsev VN (2011) Research of the risk management problems of an innovative enterprise. Ankil, Moscow, Russia

Pachler N, Turvey K (2018) Looking back, moving forward: Impact and measurability of the use of educational technology. In: Voogt J, Knezek G, Christensen R, Lai KW (eds) Second handbook of information technology in primary and secondary education. Springer International Handbooks of Education, New York, NY, pp 1–19

Porter ME (2007) Competitive strategy: techniques for analyzing industries and competitors (N. Minervin, Trans.), 3rd edn. Alpina business books, Moscow, Russia

Posypkina A (2018) Analysts evaluated the level of digital literacy among Russians. Retrieved from https://www.rbc.ru/technology_and_media/20/06/2018/5b29331c9a79477930b03101

Reis J, Amorim M, Melão N, Matos P (2011) Digital transformation: a literature review and guidelines for future research. In: Rocha Á, Adeli H, Reis LP, Costanzo S (eds) Proceedings of WorldCIST'18: *World Conference on Information Systems and Technologies. Advances in Intelligent Systems and Computing*. Springer, Budva, Montenegro, pp. 411-421

Rukina NM (2009) Additional education of adult as a factor in the professional development of personality. Bull Samara Sci Cent Russ Acad Sci 11(4–5):1137–1142

Stopar K, Bartol T (2019) Digital competencies, computer skills and information literacy in secondary education: Mapping and visualization of trends and concepts. Scientometrics 118(2):479–498

Trofimov VV (2019) Artificial intelligence in the digital economy. Bull St Petersburg State Univ Econ 4(118). Retrieved from https://roscongress.org/materials/iskusstvennyy-intellekt-v-tsifrovoy-ekonomike/

Yusupova SE, Pozdeeva SN (2018) Education of digital economy. Econ Syst Manag 108. Retrieved from http://uecs.ru/uecs-108-1082018/item/4786

Zainullina LN (2015) Information technology in the activation of students' independent work. Educ Self-Dev 4(46):75–79

# Distant and Traditional Training Formats in Natural-Science Disciplines

**Tatiana K. Kurilenko, Elena A. Chanchaeva, Sergey S. Sidorov, Marina N. Lyovkina, and Elena N. Polnikova**

**Abstract** Initially developed for a specific contingent, distance learning is replacing the traditional format. Many people wonder if it is a worthy alternative to become exceptional or one of the education system types. The forced transition to distance learning format made it possible to compare distance and traditional formats based on the results of teaching natural sciences. A comparative analysis was conducted in March–May 2020 on the base of the Gorno-Altaisk State University (Gorno-Altaysk, Russia). The comparison results showed that in distance learning, it is difficult for a teacher to assess the quality of mastering the material according to traditional criteria when passing a lecture course. When performing laboratory work in a distance format, the student becomes an observer, not an experimenter. There is no element of developing practical skills and analysis of the obtained results. The assessment of students' knowledge in a distance format is flexible and convenient for all participants in the educational process. To increase the relevance of test results in the distance format, it is necessary to expand the fund of assessment tools. In traditional learning, the teacher acts as a curator responsible for the level of knowledge of students. In the distance format, the teacher is an intermediary between the educational material and the student's independent work. Final control in a distance format does not allow to assess the knowledge and level of students' competencies objectively. The means and methods of distance learning are an effective addition to traditional education in the natural sciences. However, they cannot replace all elements of full-time education.

**Keywords** Distance learning format · Traditional learning · Students · Natural science disciplines

---

T. K. Kurilenko · E. A. Chanchaeva · S. S. Sidorov (✉) · M. N. Lyovkina · E. N. Polnikova
Gorno-Altaisk State University, Gorno-Altaisk, Russia
e-mail: sidorovss10@mail.ru



S. G. Maximova et al. (eds.), *Advances in Natural, Human-Made, and Coupled Human-Natural Systems Research*, Lecture Notes in Networks and Systems 234,
https://doi.org/10.1007/978-3-030-75483-9_68

## 1 Introduction

Distance learning Initially focused only on a specific contingent (military, people with disabilities, or living in remote regions) (Simonson et al., 2011). The currently improved distant form of interaction between participants in the educational process became an element that complements the traditional forms of full-time education (Zimina et al., 2019). However, no one knows whether this alternative form of education is worthy enough to become an exclusive form of education, regardless of external conditions.

According to the authors (Zhang et al., 2019), traditional methods of assessing student knowledge are becoming increasingly irrelevant in the era of increasing digital learning. The results of a survey among students and teachers of medical universities showed the recognition of the advantages of distance assessment (Zimina et al., 2019) and the need to preserve the elements of contact learning (Bordovskiy et al., 2020).

Natural science teaching recognizes the effectiveness of virtual interactive laboratories. In conditions of poor material equipment of universities, they can act as an analog of real laboratories (Devyatkin & Khasanova, 2018; Sharipova et al., 2020). Currently, the possibility of introducing distance learning in various fields of the education system is widely discussed (Kegeyan, 2016; Sharipova et al., 2020; E. Shcherbakova & T. Shcherbakova, 2020; Skrypnikova, 2020; Sozontova, 2020).

The research aims to compare distance and traditional learning formats of natural sciences.

## 2 Materials and Methods

The current research methodology was based on the theoretical provisions on distance learning by D. Keegan (1996). The normative base is Article 15 of the Federal Law *The Network Form of Implementing Educational Programs* of December 2, 2019 N 403-FZ, which provides the possibilities, using the conditions of implementing the educational programs in the network form (Russian Federation, 2019).

To assess the effectiveness of traditional and distance learning formats, the authors used an empirical comparison method. Learning is based on the interaction of students with the teacher. Therefore, they used the comparison criteria applicable to the activities of students and teachers. According to these criteria, the authors compared the advantages and disadvantages of traditional and distance learning formats based on their own practical experience. The students' activity was compared with the cognitive work, demonstration of motivation, and material assimilation quality. Among the teachers, the authors compared their ability to convey the material in an accessible form and to control the quality of material assimilation at the learning stage. Besides, the authors compared the objective assessment of knowledge and skills at the stage of the final control and the ability to implement educational tasks.

The research used (1) a theoretical method for analyzing the practical experience of foreign and domestic researchers and (2) a method for comparing the results of teaching activities in the distance and traditional learning formats. Their comparison was made based on teaching natural sciences in March–May 2020 at the Gorno-Altaisk State University (Gorno-Altaysk, Barnaul). During this period, students were taught only in a distance format.

The transition to a distance format requires certain conditions, namely:

- Readiness of the university infrastructure;
- Digitization of the educational process;
- Introduction of amendments to existing and development of new local regulations;
- Readiness of teachers (level of proficiency in distance learning technologies, processing of the content of the courses, and psychological readiness for full asynchronous communication);
- Readiness of students (availability of functional equipment, psychological readiness (possibly to a lesser extent than teachers) for asynchronous communication).

The distant interaction of teachers and students at the Gorno-Altaisk State University is organized in the Modular Object-Oriented Dynamic Learning Environment (Moodle) (GNU GPL license). Work in this system was tested by all teachers and students and actively used for two years as an addition to the traditional teaching format. For a comparative analysis, a list of used educational technologies was compiled by type of training (lectures, practical exercises) and control.

## 3 Results

One of the education tasks is to provide a student with a sufficient level of knowledge and form their competencies. In traditional education, students who do not meet these requirements go through a system to improve their performance. They use the opportunity to work with the teacher individually (completing missed laboratory work, consulting on topics that caused difficulties, and interviewing). Such a system, mandatory in the education system of the last century, is currently not relevant.

In the context of the forced transition to a distance learning format, the university introduced a feedback system for students through the Deans' Office. The Dean's Office provides feedback by asking students about the conditions of distance learning. Students have the opportunity to regulate the examination rules (remove the time limit, use additional attempts when performing current examinations, etc.). For comparison, during the control work in the classroom, students must rely on their own memory and basic knowledge since all information sources are excluded, and time is strictly regulated.

Figure 1 shows that the lectures in a traditional learning format consider a set of indicators characterizing a student as an active participant in the educational process.

When listening to lectures, the students are already able to show such personal qualities as (1) discipline, (2) interest in the field related to their future professional activity, and (3) research interest in the scientific field.

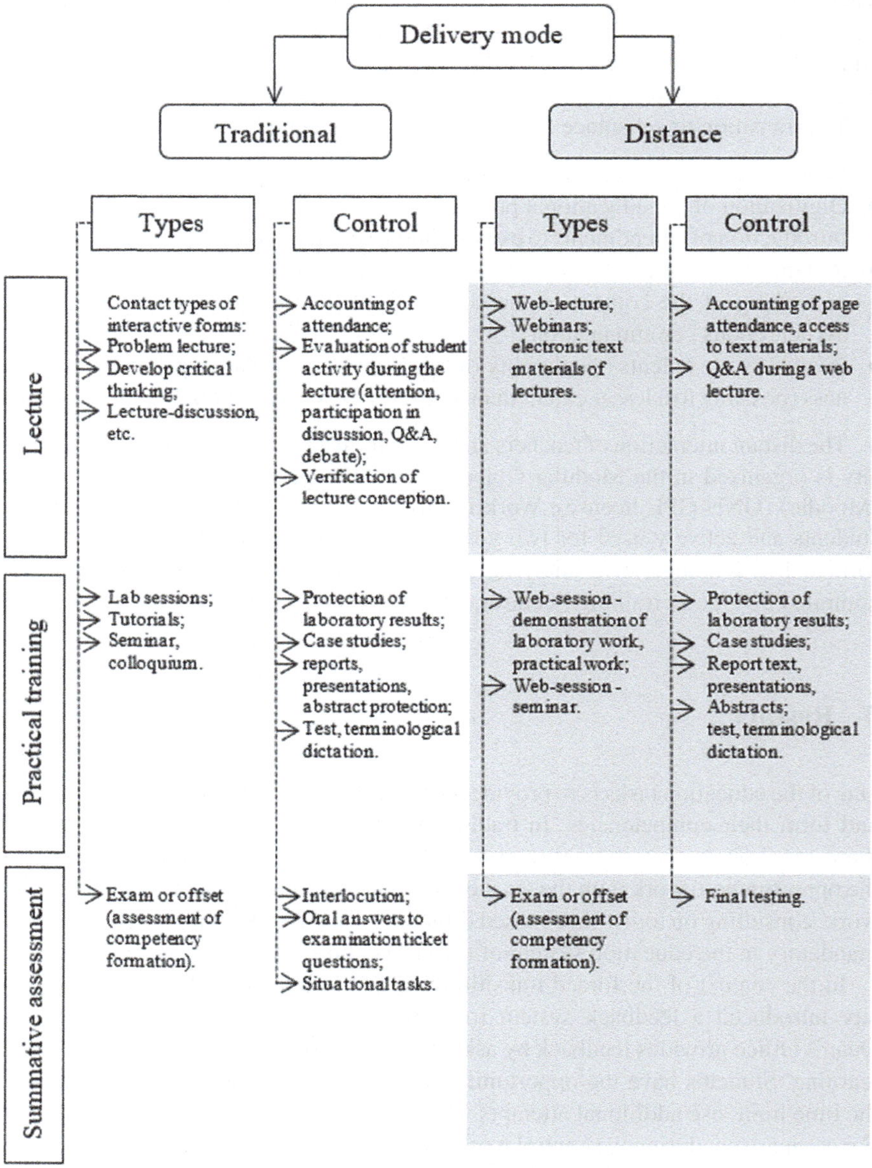


**Fig. 1** Content of lessons in the distance and traditional learning. *Source* Compiled by the authors

The students are obliged to take down the lecture material. This way, they use various perception and memory types (visual, auditory, and motor).

Observing students' attention, the lecturer uses techniques to eliminate exhaustion, involve students in discussion, debate, and reflection. Consequently, listening to lectures is a crucial stage in traditional teaching. This stage allows the teacher to track the quality of assimilation of the studied material. An educational element is mandatory in lectures.

Students are free to choose the time, place, and duration of listening or reading the lecture material with a distance learning format. Students can ask the teacher, assess the quality of the material presentation (openly or anonymously). At the same time, it is difficult for the teacher to assess at this stage whether students perceived the material efficiently. The possibilities of education personalization are essential; they allow tracking the students' academic activity by referring to the proposed information sources.

A compulsory type of practical learning in natural science disciplines is laboratory work, which forms practical skills. Often, laboratory work requires special equipment (reagents, biomaterial samples, etc.). In a distance learning format, this type of work is carried out only in the form of broadcasting trials, experiments, or practical work. Interactive lab programs allow students to perform programmed experimental work step-by-step virtually. However, the result of such work is obvious, and the experimenter has no doubts about obtaining a different result. To consolidate the studied material, students draw up the work according to the following plan: (1) a description of the methodology of the work results, (2) answering control questions, and (3) the formulation of conclusions. Thus, when distantly performing laboratory work, direct perception is replaced by the perception of abstract phenomena and images. At the same time, students do not play the role of an experimenter, but an observer, which cannot but affect the learning quality.

In real conditions, students practically work on the experiment, observing the necessary conditions and following the instructions. The obtained result may differ from the expected one, and then it requires analysis. For example, at what stage an error was made that could affect the result distortion. Besides, in the conditions of joint practical work, the teacher can implement educational tasks.

A mandatory element in the compared learning formats is the current monitoring of students' progress. The control types used in traditional and distant work in both cases allows to assess the quality of acquired knowledge fully. However, in the distance format, to assess students' knowledge, the teacher needs to create an assessment tools fund in such a volume that it will exclude copying the answers to stereotyped questions among students. The current student control in the distance format is highly flexible; students use various sources of information.

In a distance format, students are facing the task of self-replenishment of knowledge using innovative technologies, which only highly motivated students can do. Thus, in traditional teaching, the teacher acts as a curator responsible for students' level of knowledge. Moreover, with distance learning, the teacher is an intermediary between their own developed material for classes and the student's independent work.

In the traditional format, for the final control of the discipline, the teachers use an interview, oral exam, and situational tasks. These classical forms of control reveal the level of knowledge, practical skills, and abilities of students. They allow to assess the development of the competencies of the attested student. In the distance format, final testing is mostly used, which is convenient for all participants of the education process. The student's overall performance during the semester is compared with the final grade. There may be a dissonance between the student's academic performance and the result of the final test. An interview, oral exam, and solving situational problems allow the teacher to assess the students' level of competence objectively. Thus, the remote format of the final control, as an exceptional form, does not allow an objective assessment of the student's level of training.

## 4 Discussion

Distance learning means and methods find many positive responses when they are complementary to traditional learning (Dronova, 2018; Fedina et al., 2017; Morozov & Tereshchenko, 2020; Sharipova et al., 2020; E. Shcherbakova & T. Shcherbakova, 2020; Sozontova, 2020; Zimina et al., 2019). Analysts admit that in current realities, the education system is not ready for a full transition to a distance format. Therefore, today we use only the distance learning model (Skrypnikova, 2020). Limiting exclusively to the framework of the distance learning format is a forced measure, which reveals, with all recognized advantages, its obvious disadvantages over traditional teaching. 46% of the surveyed students believe that their education became less effective after switching to a distance format. 34% note the lack of face-to-face discussions with teachers. For 29%, it is difficult to concentrate when studying the material on their own. Almost a third of students from low-income groups of the population do not have sufficient functional techniques (Klyagin et al., 2020).

According to the authors, distance learning does not provide a student with a high level of knowledge (Boykova, 2020; Dronova, 2018; Palatova, 2020). At the stage of information perception, with the dominance of information educational technologies, there is a cognitive distortion effect (Baiguzhin et al., 2019). Self-education, which is the main component in the distance format, requires high motivation and abilities (Boykova, 2020; Sozontova, 2020). For motivated and gifted students, distance methods and tools expand the opportunities to realize their potential. For an average in abilities and motivation, a struggling student, without the supervising function of a teacher, the distance format turns into a formal educational environment (Abakumova et al., 2019). The problem of self-education is especially acute among first-year students. Student collectivism is a special form of interaction that is effectively formed through joint contact activities. Based on the already established synchronous interaction of students and teachers, organizing independent work is less problematic than among a psychologically unformed group of first-year students. They have no experience of working together with a new team. In the context of distance learning, it becomes more difficult to fulfill educational tasks that are effectively implemented during joint training sessions, additional educational activities, and collective leisure activities (Kuryan & Voronina, 2019; E. Shcherbakova & T. Shcherbakova, 2020).

## 5 Conclusions

According to the given comparison criteria, the efficiency evaluation of traditional and distance forms of learning natural science disciplines revealed the advantages of the traditional teaching format for lectures, laboratory work, and the final assessment of knowledge and skills. The use of the distance learning format only reduces the quality of assimilation of the material by most students. This complicates the objectivity of the assessment of knowledge and does not allow the full implementation of educational tasks. The means and methods of distance learning are an effective addition to the traditional one in the natural sciences. However, they cannot replace all elements of traditional learning. The further improvement of digital technologies will allow comparing the next stage of traditional and distance learning.

## References

Abakumova IV, Bakaeva IA, Grishina AV, Dyakova EV (2019) Active learning technologies in distance education of gifted students. Int J Cogn Res Sci, Eng Educ 7(1):85–94

Baiguzhin PA, Shibkova DZ, Aizman RI (2019) Factors affecting psychophysiological processes of information perception within the context of education informatization. Sci Educ Today 9(5): 48–70. http://dx.doi.org/https://doi.org/10.15293/2658-6762.1905.04

Bordovskiy PG, Petrenko EV, Stradina MS (2020) Teaching biomedical disciplines based on distance education technologies at physical education and sports university. Teor I Pracktika Fiz Kult 4:36–37

Boykova AV (2020) Some problems of transition to distance training in a military higher education institution under conditions of pandemia. Int J HumIties Nat Sci 4–2(43):29–31

Devyatkin EM, Khasanova SL (2018) Interactive tools of e-learning and distance learning of natural sciences. Mod Probl Sci Educ 6. Retrieved from http://www.science-education.ru/ru/article/view?id=28256

Dronova EN (2018) Technologies of remote training in higher school: experience and difficulties of use. Prepod XXI Vek 3–1:26–34

Fedina NV, Burmykina IV, Zvezda LM, Pikalova OS, Skudnev DM, Voronin IV (2017) Use of distance learning technologies in the course of implementing educational programs in preschool education. Eurasia J Math, Sci Technol Educ 13(11): 7561–7571. https://doi.org/10.12973/ejmste/80095

Keegan D (1996) Foundations of distance education, 3rd edn. Routledge, London, UK

Kegeyan SE (2016) Distance learning: Its advantages and disadvantages. Int J Prof Sci 1:71–74

Klyagin AV, Abalmasova ES, Gorev KV, Gruzdev IA, Egorov AA, Zakharova US, Shibanov EYu (2020) First weeks storm: how higher education entered into the reality of pandemic. National Research University Higher School of Economics, Moscow, Russia

Kuryan ML, Voronina EA (2019) Students and faculty interaction outside the classroom: perception and actual experience. Sci Educ Today 9(3):42–57

Morozov AV, Tereshchenko AY (2020) Distance education technologies and their legal regulation. Obrazovanie i Pravo 3:262–267

Palatova DA (2020) Distance education: a replacement for traditional forms of education. In: Pedagogicheskie Masterskie. The independent non-profit educational organization the Interregional Center of Innovative Technologies in Education, Perm, Russia, pp 76–79

Federation R (2019) Federal law "on amendments to the federal law 'on education in the Russian federation' and certain legislative acts of the Russian federation" (December 2, 2019 N 403-FZ). Russia, Moscow

Sharipova FS, Kungirotova AI, Jakubova M (2020) Implementation of systems and models of distance education in the teaching of natural sciences. Young Sci 16(306):331–333

Shcherbakova EV, Shcherbakova TN (2020) Pros and cons of distance education in higher education. Rev Pedagog Res 2(3):21–28

Simonson M, SmaldinoSh, Albright M, Zvacek S (eds) (2011) Teaching and learning at a distance: foundations of distance education. Prentice-Hall, Upper Saddle River, NJ

Skrypnikova NN (2020) The future of education: total distance learning or total abandonment of it. Prof Obraz I Rynok Tr 2: 58–59. https://doi.org/10.24411/2307-4264-2020-10213

Sozontova EA (2020) On evaluating the effectiveness of using distance technologies in higher education in the study of mathematics. Mod Probl Sci Educ 2. Retrieved from http://science-education.ru/ru/article/view?id=29721

Zhang J, Burgos D, Dawson S (2019) Advancing open, flexible, and distance learning through learning analytics. Distance Educ 40(3):303–308. https://doi.org/10.1080/01587919.2019.1656151

Zimina VA, Zhilenkova TI, Stuf IY, Kozlov AV, Syasina TV, Bolshakova GD, Kachanova EV (2019) Distance learning implementation problems in medical university (Moodle platform). Int Res J 12(90): 93–95. https://doi.org/10.23670/IRJ.2019.90.12.064

# Developing Executive Functions of Cognitive Abilities in the Information Culture of Preschool Children

**Yulia V. Batenova, Irina E. Emelyanova, Oksana G. Filippova, and Marina N. Tereshchenko**

**Abstract** The research goal is to identify quantitative and qualitative predictors of knowledge in the development of information culture of preschool children. The research results may be applied as a basis for preschool education professionals' activities in preparing educational information preschool children must acquire to achieve high educational results. The goal mentioned above is achieved using a relevant set of methods—theoretical analysis, textological methods for extracting knowledge, pedagogical experiment, expert assessment, and mathematical and statistical processing of pedagogical experiment results. The listed methods allow us to define the requirements for educational information intended for preschool children in the process of information culture formation. The possibility of applying the proposed solutions is demonstrated on the example of two texts during a psychological and pedagogical experiment, which allows its authors to conclude that it is advisable to use a universal method for constructing educational information to achieve high learning outcomes in older preschool children during information culture formation, particularly, cognitive component formation.

**Keywords** Information culture · Digital technologies · Educational information · Knowledge · Volume · Quantitative and qualitative predictors · Preschool age

Y. V. Batenova (✉)
South Ural State University, Chelyabinsk, Russia
e-mail: batenovauv@cspu.ru

Y. V. Batenova · I. E. Emelyanova · O. G. Filippova · M. N. Tereshchenko
South Ural State Humanitarian Pedagogical University, Chelyabinsk, Russia
e-mail: emelyanovaie@cspu.ru

O. G. Filippova
e-mail: mishanovaog@cspu.ru

 
S. G. Maximova et al. (eds.), *Advances in Natural, Human-Made, and Coupled Human-Natural Systems Research*, Lecture Notes in Networks and Systems 234,
https://doi.org/10.1007/978-3-030-75483-9_69

# 1 Introduction

The increasing rates of knowledge acquisition in the education system determine the need for more efficient educational practices than the traditional ones. The distinctive feature of these practices is the integration consensus of modern education approaches (Anisimovich et al., 2012; Bova & Leshchanov, 2017; Ekhlakov & Malakhovskaya, 2018).

Moreover, the innovative research educational strategies should be developed within the framework of relational systems which would take into account the nature of relations between subjects of the educational process, environment, physical, psychological and social processes able to have both positive and negative impact on different aspects of children's development and education (Kravtsov & Korostelev, 2015).

For more than two past decades, many researchers have emphasized the need to develop new skills of the 21 century, particularly critical thinking, the ability to use an appropriate toolkit needed to solve difficult problems, search, analyze, synthesize, and use the acquired knowledge in various circumstances, etc. (Ayusheeva & Dikikh, 2017; Baranyuk et al., 2016; Ismailova et al., 2018; Podruzhkina & Fedorov, 2017).

The development of the skills mentioned above requires a different kind of education not regarded as a process involving information perceived and used "as it is" acquisition and transfer, but as those matching child's learning and development needs. Such education is considered part of the information environment, productive learning strategy, social and individual opportunities, a multi-level system of pedagogical support, and the preservation of a child's physical and mental health (Batenova, 2019b).

Developing preschool children's information culture, considered within the framework of its cognitive component, as the child's ability to perform various operations with information, to model and design information objects, can be represented as a set of exercises aimed at increasing the level of their cognitive functioning (for example, the ability to remember specific tasks).

The importance of the selected component is defined by its close relationship with other significant areas of a preschooler's life (socio-communicational, educational, etc.).

## *1.1 Theoretical Basis for the Research Problem*

Many researchers pay attention to preschool education, mainly to the issues of identifying the degree of their preparedness to demonstrate cognitive competencies in the process of cognition of the surrounding world, which involves identifying potential predictors (Collins-Thompson et al., 2011) that define the boundaries of the child's capabilities. One of these predictors includes the preschool child's cognitive abilities, which one can use in predicting the amount (quantitative indicator) and content (qualitative indicator) of information that they can absorb within the education process (Averyanova et al., 2017; Chupakhin, 2016).

Cognitive strategies of pedagogical intervention are of particular importance in the context of this research, as well as the information educational technologies and external factors affecting the development of the child. External factors (including the educational situation) can significantly impact their cognitive development without taking into account the individual characteristics of the development of a preschooler in general. For instance, it includes general cognitive abilities and executive functions, including cognitive abilities of higher level—operational memory, attention, cognitive variability, and planning skills. Moreover, according to the results of several studies (Linebarger & Walker, 2005), the use of modern information technologies integrated into the educational process requires more intensive comprehension than reading books or watching TV. Considering the importance of developing cognitive abilities for future intellectual achievements, we can notice a growing interest in understanding the mechanisms that can be used in numerous external factors to influence cognitive outcomes (Batenova, 2019a).

The research interest aims at solving two issues—the impact of information technology on the cognitive development of preschoolers and recommendations on the development of useful information and educational programs and applications in the context of rapidly changing media ecology. The mentioned interest is related to the need of developing an effective program for the formation of the information culture of preschool children, in particular, in the issue of the influence of information and communication technologies designed for them to master the skills of function performing (for example, sequencing, as a set of methods to establish the sequence of information received; conditioning, as behavioral development skills by modifying stimulus/response associations, etc.).

Function performing of the preschool child's cognitive abilities is often formed at the limit of emotional capabilities, as the line between real and virtual life is "blurred," which determines the unpredictability of the events displayed on the screen and, accordingly, the strength and way of responding to them. In turn, the mentioned factors can cause stress, rapid fatigue, and, as a result, a decrease in their ability to interact in the best way possible during the learning process.

In the light of the foregoing, information technologies in all their diversity are capable of producing high cognitive results only in the state of the comfort of a preschooler interacting with them.

To achieve the comfort of a preschooler, one should establish several psychological and pedagogical conditions that increase preschoolers' enthusiasm in interaction with information technology. These conditions may include the optimal level of pedagogical support and the quality of information content (acting characters, emotional richness of events, etc.) offered to children, and hence ensuring the para-social interaction—socially conditioned interactions between a child and a character. The importance of achieving para-social relationships by creating an intelligent character prototype, integrated into the information context of the message, is in the significant facilitation of the perception of learning information in the context of digital technology by the preschooler, which is achieved by using the "add one" rule that consists in sequentially adding another event to the existing ones.

Another psychological and pedagogical condition essential for the formation of executive functions of cognitive abilities in preschoolers is introducing an 'exergame' in the educational process (Oh & Yang, 2010) (namely, games that include physical exercises and other forms of physical activity). Generally, they are included after information interaction with students, and the necessary messages are presented through digital technology. Manipulating the level of physical activity and cognitive involvement during the implementation of education helps achieve high cognitive abilities in a preschooler. It is due to the need for selective attention to specific game actions while suppressing others, which acts as a means of increasing the efficiency of the learning process and affects the ability to sequentially reproduce the information received in a "digital broadcast" situation. It also contributes to developing the ability to perform the task in a new context in contrast to the one in which it has been provided earlier (analysis and synthesis of the information received).

The indicated form of development of the cognitive abilities in a preschooler can be considered from the point of view of the biomechanics of controlling a computer mouse while performing educational tasks that require pointing and clicking motor actions. Manual dynamics, such as tracking the mouse movements and the reach of a specific target on the touch screen (i.e., using your fingers to move from one position on the touch screen to another), helps preschoolers develop cognitive control and its essential processes in the context of solving learning issues. Thus, the quality of the developed information tool interfaces due to the need to enable fine motor skills of hands can positively affect the overall performance of cognitive tasks. Apart from that, studying the reaction time necessary to complete the movement or its spatial trajectory allows us to study another crucial biomechanical aspect of preschoolers' cognitive development in the digitalization era.

Based on the factors mentioned above, the research aims to establish quantitative and qualitative predictors of knowledge in the process of developing executive functions of cognitive abilities in the formation of the information culture in preschool children.

### *1.2 Research Hypothesis*

The hypothesis of the study is the assumption that the regular interaction of preschoolers with information technologies, together with the use of special educational information applications, can significantly affect the growth of indicators reflecting the level of performing functions of preschoolers. They, in turn, will find a positive response to the development of their cognitive abilities in general—to (1) preservation of the content of the information received, (2) the manifestation of selective attention to the proposed content information message, (3) the use of spatial thinking, (4) fluency in reading and understanding, and (5) the formation of the ability to argue, namely, working memory and cognitive flexibility.

## 2 Materials and Methods

The research methods set to achieve the research goal include (1) theoretical analysis of the research results presented in literary sources, (2) textological methods for extracting knowledge, (3) pedagogical experiment, (4) expert assessment method, and (4) mathematical and statistical processing of the results of a pedagogical experiment.

### *2.1 Research Organization*

It is necessary to use a mathematical apparatus based on the determination of semantic distances between information units presented in the form of basic concepts of a subject area in metric space in order to establish the amount of knowledge contained in semantic networks (an information model of a subject area) and offered to a preschooler to learn. A text with a predicative structure with related concepts will be used in the course of this study as an information unit of semantic networks. In this case, the measurement of the acquired volume of knowledge by a preschooler is carried out by calculating the number of predicates perceived by them contained in the text. Representation of the semantic network (S) in the form of an ordered set of concepts (n1, n2, … nn), the predicate of which is expressed by the semantic distance (L), the formula for calculating the amount of knowledge mastered by a preschooler will look like this:

$$S = \sum L(n_1, n_2,\ n_n) \tag{1}$$

where:

S—scope of knowledge (quantitative characteristic of knowledge).

L—the distance between concepts (qualitative characteristic of knowledge).

n1, n2, nn—concepts in the text.

The calculations of the values of L in the used text material presented below are carried out based on a tree of parsing the sentences included in it and in the first text demonstrate a robust, meaningful connection between all concepts included in the text, and in the second—its complete absence.

Its increase characterizes the process of acquiring knowledge. During the education, it is possible to distinguish the stage of providing a preschooler with a fragment of domain knowledge (provided knowledge) and the stage of its assimilation (acquired knowledge), which can be expressed through the formula:

$$A\ (S_1,\ S_2) = K(S_1 S_2) - K(S_1) \tag{2}$$

where:

$A\ (S_1, S_2)$—the amount of acquired knowledge.

K ($S_2$)—the amount of provided knowledge.

Thus, the values obtained when performing mathematical calculations by formula (2) determine the degree of increase in the scope of knowledge of a preschooler due to the acquisition of new knowledge (Ayusheeva & Dikikh, 2017).

To calculate the scope of knowledge acquired by preschoolers, one provides a practice text, which differs from other types of texts by a predicative structure, characterized by the consistency of reasoning with an unambiguous interpretation, which makes it possible to equate it to a semantic network (Averkin eta al., 2019). The following are distinguished as features of the educational text: (1) each subsequent fragment of the text does not have shared knowledge with the previous ones, (2) the connection of fragments of the text is carried out by creating new predicates with the concepts formerly introduced into the previous fragment. The research uses two texts, including several sentences; the concepts and relations of the semantic network are highlighted: Text I "Snow White made breakfast for the Dwarves in the morning ($S_1$). She cooked everything very tasty ($S_2$). Snow White poured tea with milk for them, baked pancakes with condensed milk and jam ($S_3$). But she never allowed them to sit at the table with unwashed hands ($S_4$). The gnomes got used to it and began to wash their hands before eating without a reminder ($S_5$)."

Text II "Every morning, Snow White offered breakfast to the Dwarves ($S_1$). The food she prepared was always delicious ($S_2$). There was tea with milk and pancakes with condensed milk and jam on the table ($S_3$). Only those who washed their hands thoroughly could have breakfast ($S_4$). Everyone followed this rule with no reminders($S_5$)."

Within the framework of the pedagogical experiment, these texts are displayed in the text user interface of a personal computer (PC) for children of preschool age to independently familiarize themselves with it, which involves reading them once. Simultaneously, the interval between reading texts is at least 24 h.

After that, as part of the conversation with the expert, preschoolers are asked to reproduce the material they have read as close as possible to the original, while the use by the preschooler of all the predicates included in it (A) and the sequence of presentation of events (SP) are essential.

The textological method used to determine quantitative and qualitative predictors of knowledge in the process of forming the information culture of preschool children is based on a statement that has found its scientific justification in previous studies. In these studies, the perception of the text is a phased formation of understanding of information, which includes (1) contextual meanings of words included in the sentence (language—the first level), (2) the meaning of the text (semantic—the second level), (3) the characters and their behavior (personality—the third level), and (4) copyright relations (reflective—the fourth level).

Evaluation of the results obtained during the organization of the pedagogical experiment is carried out by calculating A according to formulas (1) and (2) the number of predicates translated into points (five predicates—five points, four predicates—four points, three predicates—three points, two predicates—two points, and one predicate—one point) and expert assessment for the sequence of presentation of the material read by the preschooler on a five-point scale. In accordance with it, five points represent full compliance with the sequence of presentation of the material by the preschooler; four points—one mistake made; three points—two mistakes made; two points—three mistakes made; one point—complete violation of the sequence of presentation.

## 3 Results

The pedagogical experiment is attended by children of preschool age (six-seven years) in the number of 24 people (11 boys, 13 girls) belonging to the primary medical group. The experts are teachers of a preschool institution in the number of 3 people with at least 20 years of professional experience. The duration of the pedagogical experiment is two days (November 19–20, 2019). The results obtained during the organization of the pedagogical experiment are depicted in Table 1 below.

The results of the pedagogical experiment, subjected to comparative analysis, of the primary and secondary statistical processing of the experimental data displayed in Table 1, make it possible to draw many conclusions. The use of two texts, one of which has a predicative structure built taking into account the semantic network of the subject area, make it possible to conclude by comparison that the scope of knowledge (quantitative indicator) of preschoolers depends on the features of the provision of educational information, in particular, on the introduction to new predicates which are concepts in the previous fragment of the text. Besides, the increase in the quality indicators of preschoolers' knowledge, expressed through the observance by preschoolers of a given sequence of sentences that are part of the proposed texts, should be noted. The difference in the reproduction of the predicates perceived by preschoolers contained in the texts is 1.29 points, and the sequence−1.83 points.

## 4 Discussion

The obtained data allow us to outline a particular algorithm for building up work with preschool children in the context of education digitalization. In the process of working with the educational text, the child acquires knowledge, develops such cognitive skills as critical thinking, search, analysis, and synthesis of consumed information, its evaluation, and integration of new information into the existing body of knowledge. New requirements arise in this regard concerning the construction of educational information that considers the needs and capabilities of a child. Thus, by developing cognitive abilities through the presentation of specially organized educational texts, we lay the foundations of information culture. Similar results are obtained in previous relevant studies, however, in relation to children of adolescent age. They indicate that educational information applications can be used to improve indicators that reflect the level of performance and cognitive abilities (Podruzhkina & Fedorov, 2017). We currently observe research interest in this topic among Russian

**Table 1** Results of quantitative and qualitative predictors of knowledge of preschool children in the course of pedagogical experiment

| Participant | Text I | | Σ, points | Text II | | Σ, points |
|---|---|---|---|---|---|---|
| | A, points | SP, points | | A, points | SP, points | |
| 1 | 2 | 3 | 4 | 5 | 6 | 7 |
| P. M | 5 | 5 | 10 | 3 | 2 | 5 |
| A. K | 4 | 4 | 8 | 2 | 1 | 3 |
| M. G | 4 | 4 | 8 | 2 | 2 | 4 |
| E. N | 3 | 4 | 7 | 2 | 2 | 4 |
| A. P | 4 | 4 | 8 | 3 | 2 | 5 |
| M. K | 4 | 4 | 8 | 2 | 2 | 4 |
| C. Sch | 5 | 4 | 9 | 3 | 2 | 5 |
| M. A | 4 | 4 | 8 | 3 | 2 | 5 |
| N. G | 3 | 4 | 7 | 3 | 2 | 5 |
| P. V | 5 | 4 | 9 | 4 | 2 | 6 |
| M. M | 5 | 4 | 9 | 3 | 2 | 5 |
| H. Yu | 4 | 4 | 8 | 2 | 1 | 3 |
| K. S | 3 | 3 | 6 | 1 | 2 | 3 |
| V. V | 4 | 4 | 8 | 3 | 2 | 5 |
| V. D | 4 | 4 | 8 | 3 | 2 | 5 |
| S. E | 5 | 5 | 10 | 4 | 3 | 7 |
| L. D | 4 | 4 | 8 | 3 | 3 | 6 |
| N. N | 4 | 4 | 8 | 3 | 2 | 5 |
| K. A | 4 | 4 | 8 | 3 | 3 | 6 |
| E. T | 4 | 4 | 8 | 3 | 3 | 6 |
| V. L | 4 | 4 | 8 | 3 | 3 | 6 |
| S. E | 4 | 4 | 8 | 3 | 2 | 5 |
| D. A | 5 | 4 | 9 | 4 | 3 | 7 |
| V. R | 5 | 4 | 9 | 4 | 3 | 7 |
| $\overline{X}$ | 4.17 | 4.04 | 8.20 | 2.88 | 2.21 | 5.08 |

*Source* Compiled by the authors

scholars, while some of their researches are more theoretical and present an average empirical significance.

The present study results solve only a part of the issue of forming the information culture in preschoolers, determining quantitative measures and qualitative requirements for knowledge that must be comprehended and learned by a preschooler. However, in our opinion, the presented material should form the basis of the activities of professionals in the preschool education system in preparing educational information offered to preschoolers that will consequently ensure high results in their education.

## 5 Conclusion

In conclusion, according to the results obtained during the experiment, it can be summarized that the frequency of information presented to preschoolers increases the amount of knowledge they acquire, and a decrease in the frequency of predicates in the text significantly reduces not only quantitative but also qualitative indicators of assimilation of the proposed educational material. Summarizing the facts mentioned above, we can conclude that it is necessary to observe the predictive structure of the text and the depth of the traced links between the concepts used in the information message. Undoubtedly, one should not ignore preschoolers' individual abilities in establishing mental connections between predictors by making inferences with a specific number of initial judgments in them. In addition, the volume and quality of learning material are determined by the scope of knowledge that is new to preschoolers since, with their addition, fewer links are formed in relation to the total number of newly educated ones.

**Acknowledgements** RFBR funded the following study according to the research project No. 18-013-00743 A "Establishing the Basics of Information Literacy for Preschoolers."

## References

Anisimovich KV, Druzhkin K Ju, Minlos FR, Petrova MA, Selegey VP, Zuev KA (2012) Syntactic and semantic parser based on ABBYY Compreno linguistic technologies. Comput Linguist Intellect Technol 11(18):91-103

Averkin AN, Sobolev SV, Vorontsov AO (2019) Comparison of various techniques for analyzing emotions to solve the task of visualizing the mood index. Soft Meas Calc 11(24):30–34

Averyanova MV, Asakura EA, Kireev VS (2017) Algorithm for categorizing texts based on a semantic network of concepts of medical and musical subject areas. Theory Practice Innov 7(19):13–20

Ayusheeva, NN, Dikikh A Yu (2017) Technology for constructing a semantic network of a scientific text based on an analysis of the ontology of a subject area. In: Knowledge – Ontologies – Theories (ZONT-2017). Sobolev Institute of Mathematics Siberian Branch of the Russian Academy of Sciences, Novosibirsk, Russia. pp 38–46

Baranyuk VV, Bogoradnikova AV, Smirnova OS (2016) Definition of the semantic content of a subject domain based on the formation of a Thesaurus. Int J Open Inf Technol 4(9):74–79

Batenova YuV (2019a) Intensification of the digital experience of preschool children: cognitive consequences. Mosc Univ Psychol Bull 4:4–20

Batenova Yu V (2019b) Developing the way of thinking in preschool children in the conditions of computer games. Commun Comput Inf Sci 1038:611-621

Bova VV, Leshchanov DV (2017) Semantic search of knowledge in the environment of functioning of interdisciplinary information systems based on the ontological approach. Bull South Fed University Tech Sci 7(192): 79–90

Chupakhin IA (2016) Formalization of the knowledge base about the subject area in the form of a semantic network. In: Modern Trends in the Development of Science and Production. West Siberian Research Center, Belgorod, Russia, pp 154–158

Collins-Thompson K, Bennett P, White R, Chica S, Sontag D (2011) Personalizing web search results by reading level. In: Proceedings from ICIKM: the 20th ACM International Conference on Information and Knowledge Management, pp 403–412. https://doi.org/10.1145/2063576.2063639

Ekhlakov Yu P, Malakhovskaya EK (2018) Semantic network for the formation of the content of a text communication message for promoting mobile applications in the consumer market. Bull Dagestan State Tech University Tech Sci 45(1): 129–138

Ismailova L Yu, Kosikov SV, Kholodov VA (2018) Development and implementation of a system for editing metadata objects to coordinate methods for considering a subject area. Inf Math Technol Sci Management. 1(9): 16-25

Kravtsov DV, Korostelev DA (2015) Algorithm for isolating a given subject area from wikipedia in the form of a categorized semantic network of concepts. Latest Achiev Sci Educ: Domest Foreign Exp 2:149–151

Linebarger DL, Walker D (2005) Infants' and toddlers' television viewing and language outcomes. Am Behav Sci 48(5):624–645

Oh Y, Yang S (2010) Defining exergames & exergaming. Retrieved from https://www.researchgate.net/publication/230794344_Defining_exergames_exergaming

Podruzhkina TA, Fedorov D Yu (2017) Approach to building a trajectory of learning on the basis of semantic knowledge networks. In: Bachevsky SV (eds) Actual Problems of Information and Telecommunications in Science and Education (APINO 2017). The Bonch-Bruevich St. Petersburg State University of Telecommunications, Saint Petersburg, Russia, pp 306 310

# Digital Technologies in Career Guidance for Preschool and School-Age Children

**Elena L. Perchenko, Tatyana V. Gudina, Pavel S. Vakhrameev, and Vladimir V. Ivanov**

**Abstract** The paper presents one of the ways to solve a significant social problem, such as early professional self-determination of children of different age groups using diagnostic tools to obtain a digital footprint formed in the process and as a result of career guidance work in a mobile application. Developing information and communication technologies based on gamification principles makes it possible to use gaming techniques in non-gaming issues, such as professional self-determination based on personal preferences and choices with obtaining feedback. The proposed diagnostic program will allow users to form an idea of a predisposition to one of the types of professions (human–human, human-nature, human-technology, human-symbolic system, or human-artistic image). More than that, the program can also act as a tool for the work of a teacher-psychologist, a vocational guidance counselor, or a parent. Besides, children can use this program themselves when independently choosing a learning path. The development of a mobile application in a digital platform system based on gamification principles makes it possible to work at different stages of age development. Simultaneously, children are subjects of their development. Using gaming technologies will allow actualizing the mental activity of the personality of children and contributing to the development of professional heuristic and creative thinking. Thus, relying on meaningful experience, students will be able to choose a training profile and labor self-determination responsibly.

**Keywords** Gamification · Career guidance · Diagnostics of personal self-determination · Mobile application

E. L. Perchenko (✉) · T. V. Gudina · P. S. Vakhrameev · V. V. Ivanov
Cherepovets State University, Cherepovets, Russia
e-mail: pel.mail@mail.ru

P. S. Vakhrameev
e-mail: vakhrameev@rntc.chsu.ru

V. V. Ivanov
e-mail: vvivanov1@chsu.ru



S. G. Maximova et al. (eds.), *Advances in Natural, Human-Made, and Coupled Human-Natural Systems Research*, Lecture Notes in Networks and Systems 234,
https://doi.org/10.1007/978-3-030-75483-9_70

## 1 Introduction

Information technologies are an integral attribute in the life of a modern child and a factor that changes the development situation in the direction of education, self-education, and independent design of professional self-determination, as well as in the direction of addictions and health disorders. Using interactive services and mobile applications, capable of solving educational and career guidance problems, is one of the conditions for forming the readiness of children for a responsible approach to professional choice as the basis for socially useful activities that consider their interests and abilities. Children of Generation Z (digital generation) receive information from digital sources and the Internet. In addition, they focus on such means of communication as smartphones and gadgets.

The market economy, as well as increased competition in the labor market, places high demands on future specialists (e.g., the ability to think independently and creatively and the readiness for unpredictable situations), which determines the need and awareness of the approach to the professional self-determination of children.

When using modern gaming and computer technologies, children will be able to deepen and expand their knowledge based on personal preferences, interests, and inclinations, gain additional knowledge in a particular direction of activity according to their developmental trajectory, and receive the ability to focus on their future profession.

The paper aims to develop diagnostic tools for obtaining a digital footprint that children of different age groups form in the process and as a result of professional self-determination in a mobile application.

Within the study, we have solved a set of the following tasks:

- Form a package of monitoring methods to determine the propensities for one of the types of professions and obtain a digital footprint in a mobile application;
- Fix quantitative and qualitative indicators of involvement of children for each type of profession (human–human, human–nature, human–technology, human–symbolic system, and human–artistic image);
- Represent a *molecule* of professions in the form of a digital footprint based on the experiment results.

In recent years, the attitude towards the professional self-determination of children has changed significantly, which makes professional orientation multi-stage and determines the continuity of education for children of different age groups, beginning with over-fives: choosing the direction of education, checking the correctness of choice, and choosing a profession as a result.

## 2 Materials and Methods

The strategic tasks of the country include employment and psychological and pedagogical support of the young generation solved with consideration of the state of the regional labor market, traditions formed in the region, and available resources. In this regard, there is an objective need for the scientific development of an effective mechanism, such as a mobile application in the digital platform system, which implies early career guidance based on gamification principles with the use of modern vocational guidance techniques. Such an application enables, on the one hand, children to independently pass the test and determine their tendency to one of the types of professions (Klimov 2003) and, on the other hand, teachers-psychologists and vocational guidance counselors to work with students even more effectively.

Solving the tasks set is based on the interdisciplinary approach by specialists of different profiles who design a modern mobile application, scientifically grounded in essence, and applied in content, as one of the tools of a digital platform that allows recording the number of visits and a qualitative indicator of the involvement of children in one of the types of professions (human–human, human-nature, human-technology, human-symbolic system, or human-artistic image). Empirical and interpretive methods allow extracting reliable information about the professional preferences of children, determining the semantics of the results, identifying patterns, and formulating conclusions about the meaning of the data obtained. The forecasting method allows performing a proactive synthesis of an indicative definition of the *image of oneself* as a future professional compared to the *image* of the desired characteristics of the prospective profession.

Now, we present an illustrative type of experiment within the pilot study, in which 30 children participated. The application includes the following methods:

- *Differential-diagnostic questionnaire* (DDQ) developed by E. A. Klimov (2003) to select a type of professions according to the classification by the author;
- Card game *Profesyans: Who? What? Where?* (PWWW), developed by N. S. Pryazhnikov (2005) based on the ideas of E. A. Klimov about the formula of professions.

## 3 Results

The proposed application, which can be used on smartphones and tablets, provides a system for administering roles: child, teacher-psychologist, and administrator. Students receive access to play activities in the form of tests and vocational guidance techniques, depending on their age. Teachers have access to the results of passing the game achievements of children. Administrators are connected to the entire system, including adding new game tasks. After the authorization and briefing procedure, users choose a test game option (Griffiths and Griffiths 2016).

For preschool and primary school children, the DDQ questionnaire was modified to use color cards representing options for professional activities and having their *weight* in the entire array of values. In each of the 20 pairs of proposed activities, children choose one type of profession according to their interests. Election data are saved on the server and opened in the personal accounts of users in the form of a diagram. The results of preferences testify to the propensity of children to work with the corresponding objects of labor (professional preferences are based on the classification of types of professions by E. A. Klimov (2003).

We have selected the PWWW card game for teenagers based on their reflexive capabilities and interest in professional self-determination. Therefore, we should dwell on it in more detail. The main task of the inherently activating methodology is to inform teenagers about the modern labor market, awaken their interest in the world of professions and the desire to take responsibility for their choice, and make them active subjects of self-determination.

The general logic of the proposed methodology is based on the choice and construction of the *image of oneself* as a future professional compared to the *image* of the desired characteristics of future labor. The subject makes nine choices out of 27 cards of professionally important qualities (*Qualities*), five out of 27 cards of typical labor actions (*What to do?*), and five out of 27 cards of typical workplace organizations (*Where to do?*).

For transferring the PWWW card game to a digital version, we have developed the following algorithm. There are three datasets with 27 values each. They are interpreted as tuples consisting of strings and two sets of integer values. Teenagers choose nine cards from each array of values, which are interesting to them, and have their integer *weight* and connections with cards of another group in the form of an integer code concept. After children finish choosing the cards of interest and completing the implementation of the methodology, the program looks for relationships in each group and checks for matches (i. e., subjects make a choice, and the program identifies the correlation and the same integer code value).

The data on passing each test become available only to the user and a teacher to whom the student is assigned (if necessary). The subsequent passing of tests and execution of tasks replenish the result statistics for every user. For completing assignments, virtual rewards are provided. They are displayed in the personal account of students and act as additional motivation. For teachers, they serve as an assessment of the involvement of students.

As a result, we get a fixation of the qualitative and quantitative indicators of the involvement of children in a particular type of professions according to their choice and the idea of the desired future of a professional (human–human, human-nature, human-technology, human-symbolic system, or human-artistic image).

Figure 1 presents an example of choosing necessary professionally important qualities (*Qualities* [Q]) combined with labor actions (*What to do?* [What]) and typical workplace organizations (*Where to do?* [Where]) (Egor, 15 years old).

The combination of the selected *What* and *Where* correlates with the selected *Q* since specific labor actions *What* and the place of work *Where* assume the use of the quality *Q* and contribute to its development. This procedure distinguishes nine

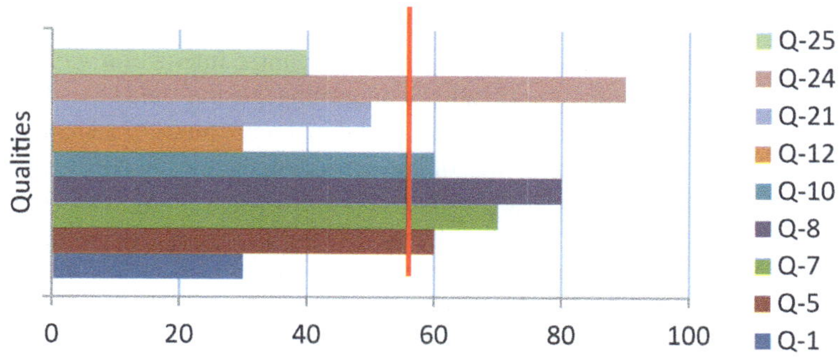


**Fig. 1** An example of choosing necessary professionally important qualities (%). *Source* Compiled by the authors

selected qualities *Q*, the most relevant for the subject when the total accumulation of *What* and *Where* represents the largest indicator (more than 60%). Searching for the presence of matches allows the program to identify correlations and represent the molecule of professions in the form of a digital footprint based on the experiment results (Fig. 2).

When addressing the analysis of the diagram and the diagnostic results, we can find the maximum combination of choices of the subject in a group of qualities that can be attributed to the qualities that underlie personal development. This suggests that the development of activity, initiative, will, and adherence to principles while maintaining dignity and decency accompany the self-awareness and social maturation of a teenager. Choosing *Q-7*, *Q-8*, and *Q-10* qualities confirms this fact.

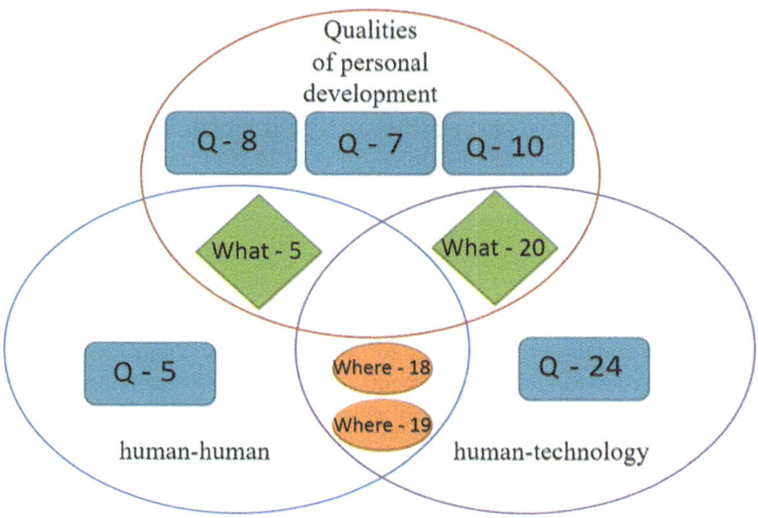


**Fig. 2** *The molecule* of professions. *Source* Compiled by the authors

The other two maximally presented choices of the subject are equivalent. Choosing *Q-5* (ability to communicate) and *What-5* (lead, organize interaction) confirms the *human–human* type of professions. It allows suggesting that the child has a propensity for a profession where it is necessary to build and maintain contacts with people and possible to realize oneself in the sphere of law enforcement or military service, which is confirmed by choice of *Where-18* (law enforcement agencies) and *Where-19* (army and navy).

Choosing *Q-24* (endurance and efficiency, the ability to work with constant loads and mobilize one's strength for a long time) corresponds to the choice of *What-20* (to quickly navigate (think) in situations of stress and risk) and determines the *human-technology* type of professions. The subsequent reflexive analysis contributes to the awareness of the vector of changes by the teenager, which the human-technology type of professions requires, namely, a high level of development of visual-figurative thinking, spatial representations, technical awareness, and ingenuity.

The activating effect of this experiment allows the child to leave a particular profession and focus on several specialties by making an independent choice. In cases when the program does not match, and the subject does not show a tendency to a particular type of profession and vivid professionalization, there is an identification of *problem areas* requiring additional advice from a psychologist and a vocational guidance counselor.

## 4 Discussion

Despite a significant number of works on labor and professional choice of children (Klimov 2003; Polyakov 1999; Pryazhnikova and Pryazhnikov 2007; Pryazhnikov 2005; Sasova 2001; Chistyakova 2005), there is a change in the labor market at present. Therefore, a request is being formed to search for new technologies to promote the employment of citizens of working age (Usov 2020).

Developing scientific and practical tools, such as a mobile application in a digital platform system based on the principles of gamification and designed to record the digital footprint of each participant in the game time, will reveal professionally important and personal qualities that contribute to effective career guidance and gaining knowledge about the prospects of one's personal development in future professional activities.

The mobile application is written in the *Swift* programming language for the *iOS* operating system and the *Java* language for *Android* and developed using *Xcode* and *Android Studio* programming environments, respectively (Pryazhnikov 2005).

The proposed mobile application will allow doing the following:

- Choosing a game scenario that requires active switching of attention in real time;
- Evaluating the statistics of achievements using game tools such as virtual awards, points, statuses, and achievements;

- Recreating a game experience that contributes to the emotional and personal involvement of a child;
- Ensuring inter-user interaction of all subjects of game participation (child, specialist (psychologist), and administrator) (Zotova 2005).

The educational profile predetermines the professional choice of students, and the success of high school education, as well as the preparation of students for transiting to the next educational level and for future professional activities in general, depends, to a large extent, on this choice. Profile education of a modern school is a challenge of time since teenagers choose their future profession already in the 7–8 grade when they are not ready for this yet. The more precise the self-determination, the more likely it is that the society will get a good specialist-professional. Many Russian scholars analyze the features of the development of the modern school in Russia in recent decades and the pre-profile and profile problems as well (Afanasyeva and Nemova 2004; Denisenko and Kosenko 2017; Zotova 2005; Kapustnyak et al. 2004; Novikova et al. 2007; Pinskiy 2004; Chistyakova 2005).

However, the problem of choosing a training profile is not yet properly implemented and is often done by parents or under their influence (traditions, family business, *fashion* for a profession), as children do not see and do not represent the whole picture of the world, the palette of possible professions, and market trends.

The content and forms of vocational guidance work should correspond to the age characteristics of students and contribute to personal development. At various levels of education, we see the priorities of this work as follows:

- In older preschool age, in connection with the establishment of a worldview and increased cognitive activity, it is necessary to form a primary idea of the world of professions, their role in society and the life of each person, as well as interest in work and a positive attitude towards different types of work.
- At primary school age, intellectualization, awareness, and arbitrariness of all mental processes occur, making it possible to diagnose the inclinations of children and determine additional activities according to their interests.
- In the teenage years, social and professional self-determination is based on personal reflection and actual interest in choosing a profession. The development of a sense of adulthood and self-awareness in teenagers aims to form the ideal of personality through the awareness of their characteristics. At this age, career guidance contributes to studying the abilities and motives of activity, the definition of pre-profile and profile education, and reflections on their professional future (Dik and Khutorskoy 2003).

Developing a mobile application in the digital platform system allows working at different stages of age development using scientific and practical tools as diagnostic tools for obtaining a digital footprint formed in the process and as a result of career guidance work.

# 5 Conclusion

In general, the developed mobile application is one of the tools of the digital platform, which allows recording the choices of participants in the form of a digital footprint, keeping quantitative statistics on visits to the game lessons for each type of profession, and showing the involvement of children in a specific professional area according to the proposed assessment methods.

Working with the application will contribute to the early career guidance of children of different age groups and possible areas of work based on their personal preferences and choices in gamification with receiving feedback. The proposed diagnostic program will allow users to get an idea of the predisposition to one of the five groups of professions (human–human, human-nature, human-technology, human-symbolic system, or human-artistic image) or use the information for personal self-development and making the right decisions in selecting a learning path.

**Acknowledgements** We have prepared the paper with the support of the Russian Foundation for Basic Research (Project no. 19-29-07313 *Gamification in the Formation of a Digital Footprint as a Model of Modern Career Guidance and Professional Self-Determination of Children of Different Age Groups in the System of Lifelong Education*).

# References

Afanasyeva TP, Nemova NV (2004) Supporting the activities of educational institutions of municipal methodical service. APK i PRO, Moscow, Russia

Chistyakova SN (2005) The problem of self-determination of high school students when choosing a training profile. Pedagogika 1:19–26

Denisenko S Yu, Kosenko PV (2017) How to choose a future profession for high school students. Obrazovanie i Vospitanie 3–1(13):18-21

Dik Yu I, Khutorskoy AV (eds) (2003) Profile education in the context of modernization of school education. Institut Obshchego Srednego Obrazovaniya Rossiyskoy Akademii Obrazovaniya, Moscow, Russia

Griffiths D, Griffiths D (2016) Head first: Android development. Piter, Saint Petersburg, Russia

Kapustnyak AG, Roslova LO, Rybchenkova LM (2004) New forms of state (final) attestation of ninth-year students. APK i PRO, Moscow, Russia

Klimov EA (2003) Psychology of a professional: Selected psychological works. Moscow Psychological and Social Institute, MODEK, Voronezh, Moscow, Russia

Kuznetsov AA, Filatova LO (2004) New basic curriculum as a foundation for implementing profile education in high school. APK i PRO, Moscow, Russia

Novikova TG, Prutchenkov AS, Pinskaya MA (2007) Methodological recommendations for supporting work on the portfolio of secondary school students. Profilnaya Shkola 1:13–21

Pinskiy AA (ed) (2004) Pre-profile preparation: Beginning of the experiment. Alyans-Press, Moscow, Russia

Polyakov VA (1999) Professional self-determination of young people. Pedagogika 5:33-37

Pryazhnikov NS (2005) Career guidance at school: Games, exercises, and questionnaires (8–11 grades). Vaco, Moscow, Russia

Pryazhnikova E Yu, Pryazhnikov NS (2007) Reflections on the professional self-determination of young people. Bulletin of Practical Psychology of Education 3(12):42 45

Sasova IA (2001) Socio-economic education and upbringing of students. Narodnoe Obrazovanie 10:165–170

Usov VA (2020) Swift: Basics of developing applications for iOS, iPadOS, and macOS. Piter, Saint Petersburg, Russia

Zotova NN (2005) On the developmental functions of computer support in counseling on choosing a profession. Mosc Univ Psychol Bull 2:70–76

# Influence of the COVID-19 Pandemic on Information and Educational Environments

**Maxim A. Khokhryakov and Tatyana U. Vilkova**

**Abstract** The paper aims to study the impact of the COVID-19 pandemic on the development of the information and educational environment of the university. The research relevance is due to the urgent nature of the transition of universities to the use of information and educational environments. This must be done in full for implementing educational programs, as well as intermediate and final certification, in a remote form. In the paper, we have analyzed changes in the regulatory legal acts of universities from the top ten ranking of digital activities in the Russian Federation in connection with the measures taken to counter the coronavirus infection. More than that, we have surveyed teachers and students of the Kutafin Moscow State Law University (Moscow, Russia) to identify the difficulties teachers and students faced when switching to a distance-learning format. The study has made it possible to conclude about the positive impact of the COVID-19 pandemic on developing the information and educational environment of the leading universities of the Russian Federation. At the same time, we have identified problems manifesting themselves in a sharp and unprepared transition to a complete distance learning format using the information and educational environment of the university. Among these difficulties, we should note the lack of digital literacy of teachers, the digital divide, and problems with software and hardware. When eliminating the identified shortcomings, many students and teachers admit learning certain disciplines and courses in a distance form in the future using the information and educational environment of the university.

**Keywords** Information and educational environment · Distance learning · Online learning · Pandemic · COVID-19

M. A. Khokhryakov · T. U. Vilkova (✉)
Kutafin Moscow State Law University (MSAL), Moscow, Russia
e-mail: tvilkova@yandex.ru



S. G. Maximova et al. (eds.), *Advances in Natural, Human-Made, and Coupled Human-Natural Systems Research*, Lecture Notes in Networks and Systems 234,
https://doi.org/10.1007/978-3-030-75483-9_71

## 1 Introduction

The modern world is a world of high technology. It increasingly penetrates all spheres of public life. Even industries traditionally considered as work of exclusively human consciousness are rapidly digitalized. For example, the transition to an electronic criminal case is performed in jurisprudence (Vilkova and Maslennikova 2019). In addition, the possibility of using artificial intelligence (Sushina and Sobenin 2020) and big data technology (Tabolina and Tabolin 2020) is investigated.

Developing and introducing digital technologies into all spheres of public life is a key to the successful and competitive development of the country in the twenty-first century. The Decree of the President of the Russian Federation *On the National Goals and Strategic Objectives of the Development of the Russian Federation for the Period Until 2024* of May 7, 2018 (Presidential Executive Office 2018), set the task for the Government of the Russian Federation to achieve the ambitious national development goals of the country. Among them, there are the tasks of joining Russia to the five largest economies in the world, the transition to economic growth rates higher than the world ones, and some others. This program is implemented through a number of national projects, including in the areas such as health care, education, demography, culture, science, etc. Simultaneously, digitalization is declared as a tool for new opportunities in all areas.

In particular, the passport of the National Project *Education* (National Project Education 2020; Presidential Executive Office 2018) provides that by the end of 2024, 20% of students will master certain courses and disciplines (modules), including online. This will be implemented using the resources of educational organizations, including universities, ensuring that the quality of training of students is consistent with the world level.

All leading universities of the Russian Federation work on creating an information and educational environment but with varying degrees of readiness, fullness, and use in the educational process.

The COVID-19 pandemic had a significant impact on the timing and methods for digitalizing education.

Since March 16, 2020, according to the order of the Ministry of Science and Higher Education of the Russian Federation (2020), all Russian universities have been recommended to organize training for students of all forms and levels of education using distance learning technologies. Besides, the degree of technical and methodological readiness of the information and educational environment has not affected this.

Currently, researchers actively study the impact of the coronavirus pandemic on using digital technologies in the educational process in different countries.

The scientific literature analyzes how the resources specially created for learning and other mobile and electronic tools not pursuing such goals initially influence students. For example, in 2019, positive experimental results were obtained from applying *Instagram* in education for increasing the level of knowledge of high school students (Stojanovic et al. 2019), as well as positive opportunities of *DIGITgame*

(Bonora et al. 2019). At the same time, there has been a negative impact of social networks, such as *Facebook*, on the learning process of students (Jere et al. 2018).

In general, a positive attitude of students at humanitarian universities has been revealed towards e-education, even though such studies are still few (Al-Naqbi and Alshannag 2018).

The COVID-19 pandemic has led to the need for an urgent unplanned transition to total distance synchronous and asynchronous education, and, if this is not possible, to the suspension of classes. Active development of distance courses has begun even in medical schools, where online learning is difficult due to the large proportion of clinical practice among students (Ruthberg et al. 2020; Longhurst et al. 2020).

The experience of the last months of such training has revealed the advantages and disadvantages of distance learning. In particular, some scholars have found that videoconferencing software increases the cognitive load of a teaching session (Kwon et al. 2020).

Currently, some researchers (Almaiah et al. 2020) study the factors affecting the e-learning system and challenges, which the e-learning system faced during the COVID-19 pandemic (e.g., issues of change management, technical issues of the e-learning system, and financial support). Another group of scholars studies behavioral, cognitive, and emotional reactions of students to the forced conditions of distance learning (Dushkevych et al. 2020; Moawad 2020).

It is established that the possibility of conducting crisis online education and its effectiveness depend on the field of education. In medical schools, such training is almost impossible and should be suspended (Bowden 2020). Nevertheless, teaching foreign languages has suffered much less. Planned online (language) learning is as effective as in-person learning (Gacs et al. 2020).

The proposals for creating high-functioning professional networks online have been substantiated to improve the quality of digital online education during the crisis period (Knight 2020). More than that, online assessment in higher education is developed to ensure lifelong learning in a pandemic (Cuaton 2020; Garcia-Penalvo et al. 2020).

In this study, we have considered the state of information and educational environments of the leading universities in Russia until March 2020, whether they were ready for the transition to a full-distance mode of operation, and how they have changed under the influence of the COVID-19 pandemic.

## 2 Materials and Methods

In order to solve this problem, we have analyzed statistical data on the activities of ten universities leading in the digital educational process according to the *YuraytStatistics* service (Yurayt 2020).

The rating has been compiled on the basis of digital footprints of various activities of students and teachers on the Russian educational platform *Yurayt*. It shows how much time (in days) a particular university has spent for its educational process on the platform since the beginning of the 2019–2020 academic year.

The educational behavior of 275,859 students and 37,986 teachers from 3,210 organizations of Russian-language higher education in Russia and the Near Abroad has been analyzed in compiling the rating.

The indicator integrates various types of activities on the *Yurayt* educational platform:

- Reading, abstracting, and discussing educational texts;
- Watching educational videos and performing video assignments;
- Listening to audio tracks;
- Passing the adaptive formative testing;
- Educational interaction between students and teachers.

For the research objectivity, we have analyzed the activities of universities occupying the first, second, eighth, and ninth lines in the ranking.

We have used a normative approach. It has manifested itself in the analysis of Russian legislation in the field of digitalization of educational activities and local regulations of some universities that regulate the content, structure, and activities of the electronic educational environment. Additionally, we have used regulations that were in force before the pandemic and those adopted in the period from March to June 2020 (Financial University under the Government of the Russian Federation 2016; High School of Economics 2019; Ministry of Science and Higher Education of the Russian Federation 2020; Kutafin Moscow State Law University 2020; Russian Presidential Academy of National Economy and Public Administration under the President of the Russian Federation 2017).

Besides, we have used the questionnaire method to receive answers from 80 teachers and 65 students at the Kutafin Moscow State Law University (MSAL) (Moscow, Russia) to the questions about education using the electronic educational environment.

Furthermore, we have used general logical methods: analysis and synthesis, induction and deduction, abstraction and ascent from the abstract to the concrete, etc.

## 3 Results

The Russian Presidential Academy of National Economy and Public Administration under the President of the Russian Federation (RANEPA) occupies the first place. The regulation on using e-learning and distance learning technologies in the implementation of educational programs in RANEPA (Moscow, Russia) was adopted on September 22, 2017 (Russian Presidential Academy of National Economy and Public Administration under the President of the Russian Federation 2017). The regulation did not allow the implementation of educational programs using only e-learning and distance learning technologies. The final state certification was also not provided using these technologies.

However, from March 19, 2020, all forms of training and certification began to be conducted using a distance learning system to implement measures to counter the coronavirus infection. Simultaneously, instructions and regulations were developed for using distance learning technologies in the context of measures taken to counter the spread of the coronavirus infection.

For the Financial University under the Government of the Russian Federation, which ranks second in the ranking of the digital activity of universities, the regulation for maintaining the information and educational portal of the Financial University was adopted on December 27, 2016 (Financial University under the Government of the Russian Federation 2016). The information and educational portal is an electronic information resource that provides registered users with authorized access to information, as well as educational and methodological materials.

Until March 17, 2020, the information and educational environment of the university has been used for distance learning only in the extramural bachelor's and master's programs.

The regulation on electronic educational resources adopted on January 17, 2020, is in force at the MSAL (eighth in the ranking) (Kutafin Moscow State Law University 2020). This regulation does not provide using the information and educational environment for organizing training in higher education programs.

Despite this fact, since March 2020, all forms of classes and attestations, including the final state certification, have been performed remotely using the information and educational environment.

The Academic Council of the High School of Economics approved the procedure for using e-learning and distance learning technologies during the implementation of educational programs at the High School of Economics (Moscow, Russia) (ninth line in the top ten) on February 1, 2019 (High School of Economics 2019).

Electronic information and educational environment (LMS) and sites of educational disciplines created in it with educational materials placed in digital form are the main means of electronic and distance education. In addition, electronic and distance learning tools include software and hardware for organizing webinars, video conferencing equipment, and massive open online courses created at the university or in other organizations.

Thus, all universities have created the information and educational environment and developed it in one or another form. However, this development has occurred with varying degrees of using e-learning and distance learning technologies in implementing educational programs or their parts.

A survey of 163 teachers and 163 students at the MSAL conducted in June 2020 has also shown that the information and educational environment of the university was not practically used before the spread of the coronavirus and the forced transition to online and distance learning.

Until March 2020, 41.3% of the surveyed students did not know whether an educational information environment and a learning management system (LMS platform) were created.

As for the surveyed teachers, 29.6% of them noted that until March 2020, there was practically no work on this platform. Only 11.1% of the teachers answered that working on the platform was an obligatory part of the educational process.

At the same time, until March 2020, 81% of the surveyed students did not work in the information and educational system. Only 14.8% of the teachers answered that they worked in the system.

## 4 Discussion

Admittedly, COVID-19 has had and continues to have a significant multifaceted impact on the transformation of educational activities. Additionally, the coronavirus affects its funding, and the effectiveness of online learning has yet to be evaluated.

Teachers everywhere had to use new methods of online learning. Simultaneously, teachers often did not have the necessary experience for this. They also had to develop new curriculum materials to cope with the changing environment, new assessment rules for students, including ensuring the validity of exam grades, and even guidelines for creating a professional appearance when teaching from home (Dwivedi et al. 2020; Guangul et al. 2020; Yang 2020). The pandemic has highlighted the lack of technical equipment of many participants (universities, teachers, and students), and the digital divide between generations (Colpitts et al. 2020) and different population groups (Bhagat and Kim 2020). Problems related to the communicative, developmental, upbringing, socializing, and even physiological components of education have received actualization (Lubkov et al. 2020). Changes in university funding for various reasons are the consequences of the pandemic for education. For example, in China, government spending on higher education will soon decline as the focus will be on sustaining the economy (Yang 2020). In Australia, universities with a large proportion of international students feel a decrease in the proportion of such students today due to border closures and a corresponding decrease in income (Ross 2020). In Canada, teachers are expected to decline due to declining tuition revenues (Metcalfe 2020).

The study has shown that many of these phenomena concerning introducing digital technologies into the educational process and forced online learning are also inherent in Russian higher education institutions.

We have analyzed the use of digital technologies in the educational process before and during the coronavirus pandemic in leading Russian universities. The analysis has allowed us to assert that until March 2020, not a single university implemented educational programs in full remotely using the information and educational environment.

The Financial University under the Government of the Russian Federation is an exception as it allows implementing bachelor's and master's programs, including the final state certification, remotely, in extramural form.

Simultaneously, universities that began to develop and use the information and educational environments earlier than others (e.g., RANEPA in 2017 and Financial

University in 2016), as expected, ranked higher in the ranking of universities leading in the digital educational process than universities that began such developments later (e.g., MSAL in 2019, High School of Economics in 2019).

The coronavirus pandemic has forced universities to experiment with distance learning and make full use of the information and educational environment. At the same time, leading universities have faced a non-trivial task. The education sector has undergone major transformations, but it is still unclear whether they are successful. We have identified what difficulties universities, teachers, and students face in the transition to full distance learning.

The conducted survey of teachers and students at the MSAL has shown that the main reasons for the difficulties in the transition to online education include the following provisions.

*Low digital literacy of teachers.* Teaching online requires more than just an understanding of modern technology from educators. The problem of using new, previously unfamiliar technologies that arises during the session suddenly becomes difficult and can lead to the failure of the entire session. This is especially true when there are 20 or 30 students on the other side of the screen. Solving this problem necessitates training teachers quickly and effectively. This should be preferably done using the same tools teachers will use in their work with their students later. We have asked teachers about the difficulties they encountered while working in the information and educational environment of the university. A quarter of the teachers surveyed (25%) noted a lack of understanding of the capabilities of various platforms and resources for one or another form of remote work and a lack of digital literacy.

At the same time, we have asked students how they could improve using the information and educational environment of the university and the learning management system (LMS platform). One-tenth of the students (10%) answered about the need to improve the digital literacy of teachers. When asked about the form of conducting classes remotely, 70% of the teachers questioned answered that they asked students to send practical works, answers to questions, and completed tests by e-mail. Almost the same number of students surveyed (77.8%) confirmed these data when answering this question. This fact indicates that teachers do not understand the capabilities of the information and educational environment of the university and the skills to work with it.

*Digital divide.* We tend to think that everyone has a computer and a smartphone these days. But in fact, not everyone has a stable and high-speed Internet connection. In addition, the data plan may end, and older computers may not support the modern programs needed to install for training. We should consider this fact if we do not want to restrict access to training for the entire layers of students. More than that, students who easily use all social network functionality often cannot complete such simple tasks as attaching a file to an email or saving a file with a different name. The problem of connecting to the Internet is leading among the difficulties teachers and students face while working in the information and educational environment of the university (44.4% of the students and 38.3% of the teachers surveyed).

*Software.* For a perfect online lesson, teachers and students need to have access not only to the tools for creating video and sharing a screen. It is preferable for teachers to see the faces of their students, who, in turn, can raise their hands and temporarily give teachers control over their computers in some cases. Almost a quarter of the teachers surveyed (23.3%) and almost a half of the students (42.9%) experienced problems of using software tools when working in the information and educational environment.

*Technical support means.* Teachers who conduct online classes should have stereo headsets, cameras, powerful computers with good graphics, and large monitors. This is necessary for teachers to be able to simultaneously open a presentation and see images of students, a chat window, and text files. The survey has shown that 17.5% of the students and 15% of the teachers faced the lack of the necessary technical support.

Despite the difficulties that have arisen, both sides of the educational process rate the quality of distance learning quite highly. One-half of the teachers surveyed (50%) believe that teaching is fully implemented without compromising the quality of the educational process. About one-third of the teachers (34.4%) note that teaching is implemented in full but with a decrease in the quality of the educational process. Interestingly, the students are more critical. Almost one half of the students surveyed note that teaching is implemented in full but with a decrease in quality. Simultaneously, 39.7% of the students believe that teaching is fully implemented without compromising the quality of the educational process. Despite this fact, only 37.5% of the teachers allow the completely transferring of particular disciplines and courses into a distance learning format using the information and educational environment of the university and the learning management system (LMS platform). However, 54% of the students already consider themselves ready to study certain disciplines and courses in a complete distance format using the information and educational environment of the university. Simultaneously, the students and teachers believe that hybrid (mixed) training will be the most effective form of work (70.3% and 63.5% of the respondents, respectively). It implies using all possible forms of organizing communication, namely online (using the information and educational environment of the university) and offline.

The research performed was limited in time. We have conducted it during the first wave of the spread of the coronavirus infection and have not considered the later experience. The study has concerned only the issues of using the information and educational environment of the university. In addition, the study was also limited to the circle of people who participated in the survey. Moreover, the study has covered only teachers and students at only one university.

Studying the impact of the COVID-19 pandemic on the development of the information and educational environment of the university during the second wave of the COVID-19 pandemic is another potential research area. It will also be necessary to study the impact of the pandemic on other aspects of higher education. For further research, it is important to perform a comparative analysis of the changes that have occurred over the entire 2020 year, identify unresolved and growing problems, and substantiate proposals for their solution.

## 5 Conclusion

As a research result, we can conclude that the presence and development of the information and educational environment of the university is not a forced measure during the COVID-19 pandemic. The need to develop a modern university, society, technology, new challenges, and solve the tasks set within the national projects objectively determines this measure. Unfortunately, until March 2020, the process of creating and using information and educational environments of universities was mostly not active enough and often formal. An urgent and abrupt transition to distance learning has accelerated the formation and expanded use of such environments. However, this transition has revealed the current technical and methodological shortcomings. Distance learning with using the information and educational environment of the university has become a reality. Its further development and increasing efficiency necessitate consideration of the existing experience and the elimination of the identified shortcomings.

Based on the analysis of statistics and the study, we can conclude that a hybrid (mixed) approach to teaching humanitarian disciplines is promising. In these disciplines, distance learning in the information and communication environment of the university is organically combined with traditional teaching methods. Some courses can be converted to a complete distance learning format. Nevertheless, when taking into account the identified shortcomings, we find the complete transition to distance learning using the information and educational environment of the university impossible at the moment and consider it as an emergency measure in a pandemic.

**Acknowledgements** We have prepared this paper within the state assignment No. 075-00293-20-02 *Transformation of Russian Law in the Face of Great Challenges: Theoretical and Applied Foundations* (Topic No. FSMW-2020-0030) of May 25, 2020.

## References

Almaiah MA, Al-Khasawneh A, Althunibat A (2020) Exploring the critical challenges and factors influencing the e-learning system usage during the COVID-19 pandemic. Educ Inf Technol 25:5261–5280. https://doi.org/10.1007/s10639-020-10219-y

Al-Naqbi AK, Alshannag Q (2018) The status of education for sustainable development and sustainability knowledge, attitudes, and behaviors of UAE University students. Int J Sustain High Educ 19(3):566–588. https://doi.org/10.1108/IJSHE-06-2017-0091

Bhagat S, Kim DJ (2020) Higher education amidst COVID-19: challenges and silver lining. Inf Syst Manag 37(4):366–371. https://doi.org/10.1080/10580530.2020.1824040

Bonora L, Martelli F, Marchi V (2019) An amazing way to learn STEM concepts developing sustainable cities idea in the citizens of the future: the methodology of erasmus+ project DIGITgame (digital improvement by game in smart city projecting). In: Proceedings of the 10th international conference on e-education, e-business, e-management and e-learning. Association for Computing Machinery, Tokyo, Japan, pp 18–22. https://doi.org/10.1145/3306500.3306536

Bowden S (2020) COVID-19: Canadian medical students ready to step up. Univ Tor Med Journal Microbiome 97(3):8–11

Colpitts BDF, Smith MD, McCurrach DP (2020) Enhancing the digital capacity of EFL programs in the age of COVID-19: the ecological perspective in Japanese higher education. Interactive technology and smart education. https://www.emerald.com/insight/content/doi/10.1108/ITSE-08-2020-0123/full/html

Cuaton GP (2020) Philippine higher education institutions in the time of COVID-19 Pandemic. Revista Romaneasca Pentru Educatie Multidimensionala, 12(1Sup2):61–70. https://doi.org/10.18662/rrem/12.1sup2/247

Dushkevych M, Barabashchuk H, Hutsuliak N (2020) Peculiarities of student distance learning in emergency situation condition. Revista Romaneasca Pentru Educatie Multidimensionala, 12(1Sup2):71–77. https://doi.org/10.18662/rrem/12.1sup2/248

Dwivedi YK, Hughes DL, Coombs C, Constantiou I, Duan Y, Edwards JS, Upadhyay N (2020) Impact of COVID-19 pandemic on information management research and practice: Transforming education, work, and life. Int J Inf Manage 55:102211. https://doi.org/10.1016/j.ijinfomgt.2020.102211

Financial University under the Government of the Russian Federation (2016) Order of the rector "on approval of the regulations for the maintenance of the information and educational portal of the financial university under the government of the Russian Federation" (December 27, 2016, No. 2530/o). Moscow, Russia

Gacs A, Goertler S, Spasova S (2020) Planned online language education versus crisis-prompted online language teaching: Lessons for the future. Foreign Lang Ann 53(2):380–392. https://doi.org/10.1111/flan.12460

Garcia-Penalvo FJ, Corell A, Abella-Garcia V, Grande M (2020) Online assessment in higher education in the time of COVID-19. Education in the Knowledge Society, 21. https://revistas.usal.es/index.php/eks/article/view/eks2020211 2/22274

Guangul FM, Suhail AH, Khalit MI, Khidhir BA (2020) Challenges of remote assessment in higher education in the context of COVID-19: a case study of Middle East College. Educ Assess Eval Account 32:519–535. https://doi.org/10.1007/s11092-020-09340-w

High School of Economics (2019) Order of the rector of the national research university high school of economics "on the implementation of the procedure for the use of e-learning, distance educational technologies in the implementation of educational programs at the national research university high school of economics" (February 12, 2019, No. 6.18.1–01/1202–05). Moscow, Russia. https://www.hse.ru/docs/242875929.html

Jere NR, Boikhutso T, Maoneke PB (2018) Social media for enhanced e-education at Namibian schools. Inf Commun Technol Adv Intell Syst Comput 625:129–137. https://doi.org/10.1007/978-981-10-5508-9_12

Knight SWP (2020) Establishing professional online communities for world language educators. Foreign Lang Ann 53(2):298–305. https://doi.org/10.1111/flan.12458

Kutafin Moscow State Law University (2020) Order of the rector of the Kutafin Moscow State Law University "on approval of the regulation on electronic educational resources in the Kutafin Moscow State Law University (Moscow State Law Academy)" (January 17, 2020, No. 17). Moscow, Russia

Kwon R, Zhang ML, VandenBussche CJ (2020) Considerations for remote learning in pathology during COVID-19 social distancing. Cancer Cytopathol 128(9):642–647. https://doi.org/10.1002/cncy.22289

Longhurst GJ, Stone DM, Dulohery K, Scully D, Campbell T, Smith CF (2020) Strength, weakness, opportunity, threat (SWOT) analysis of the adaptations to anatomical education in the United Kingdom and the Republic of Ireland in response to the covid-19 pandemic. Anat Sci Educ 13(3):301–311. https://doi.org/10.1002/ase.1967

Lubkov AV, Gordienko OV, Sokolova AA (2020) A humanitarian approach to the digitization of education. Educ Self-Dev 15(3):89–96. https://doi.org/10.26907/esd15.3.08

Metcalfe AS (2020) Visualizing the COVID-19 pandemic response in Canadian higher education: an extended photo essay. Studies in Higher Education. https://www.tandfonline.com/doi/full/10.1080/03075079.2020.1843151

Ministry of Science and Higher Education of the Russian Federation (2020) Order "on the organization of educational activities in organizations implementing educational programs of higher education and relevant additional professional programs in the context of preventing the spread of the new coronavirus infection on the territory of the Russian Federation" (March 14, 2020, No. 397). Moscow, Russia

Moawad RA (2020) Online learning during the COVID-19 pandemic and academic stress in university students. Revista Romaneasca Pentru Educatie Multidimensionala, 12(1Sup2):100–107. https://doi.org/10.18662/rrem/12.1sup2/252

National project Education (2020). http://edu.gov.ru/national-project/

Presidential Executive Office (2018) Decree "on the national goals and strategic objectives of the development of the Russian Federation for the period until 2024" (May 7, 2018, No. 204). Moscow, Russia. http://www.kremlin.ru/acts/bank/43027

Ross J (2020) Economic ramifications of the COVID-19 pandemic for higher education: a circuit breaker in Australian universities' business model? High Educ Res Dev 39(7):1351–1356. https://doi.org/10.1080/07294360.2020.1825350

Russian Presidential Academy of National Economy and Public Administration under the President of the Russian Federation (2017) Order of the rector of the Russian Presidential Academy of National Economy and Public Administration under the President of the Russian Federation "on approval of the regulation on the use of e-learning, distance learning technologies in the implementation of educational programs" (September 22, 2017, No. 01–6230). Moscow, Russia. https://www.ranepa.ru/images/docs/prikazy-ranhigs/Pologenie_EO_DOT.pdf

Ruthberg JS, Quereshy HA, Sh, Ahmadmehrabi, Trudeau S, Chaudry E, Hair B, Mowry SE (2020) A multimodal multi-institutional solution to remote medical student education for otolaryngology during COVID-19. Otolaryngology-Head and Neck Surgery 163(4):707–709. https://doi.org/10.1177/0194599820933599

Stojanovic D, Bogdanovic Z, Despotovic-Zrakic M, Naumovic T, Radenkovic M (2019) An approach to using Instagram in secondary education. ICVL 2019: the 14th international conference on virtual learning. University of Bucharest Publishing House, Bucharest, Romania, pp 247–253

Sushina TE, Sobenin AA (2020) Artificial intelligence in the criminal justice system: leading trends and possibilities. Proceedings of the 6th international conference on social, economic, and academic leadership (ICSEAL-6–2019). Atlantis Press, Prague, Czech Republic, pp 432–437. https://doi.org/10.2991/assehr.k.200526.062

Tabolina K, Tabolin V (2020) Prosecutor's supervision in criminal proceedings in the context of the leading development of digital relations. Proceedings of the sixth international conference on social, economic, and academic leadership (ICSEAL-6–2019). Atlantis Press, Prague, Czech Republic, pp 376–381. https://doi.org/10.2991/assehr.k.200526.055

Vilkova TY, Maslennikova LN (2019) Legitimacy and unification in criminal proceedings: from procedural document forms to the electronic criminal case. Perm Univ Herald Juridical Sci, 46:728–751. https://doi.org/10.17072/1995-4190-2019-46-728-751

Yang R (2020) China's higher education during the COVID-19 pandemic: some preliminary observations. High Educ Res Dev 39(7):1317–1321. https://doi.org/10.1080/07294360.2020.1824212

Yurayt (2020) YuraytStatistics: national and regional statistics of the digital activity of universities and colleges. http://urait.ru/info/stat

# Advances in Cultural Traditions and Innovation, Development Barriers, and Social Stability

# Transformations of Cultural Environments in Small Towns

**Tamara A. Semilet**

**Abstract** The research uses the methodological and heuristic power of the concept of sustainable development to analyze and evaluate proposed and implemented projects for transforming the cultural environment of Russian small towns. I aim to verify whether the gnoseological and ontological aspects of the small-town renovation project, which dominates at the governmental level, correspond to the goals and objectives of sustainable development. In this regard, I seek to find the optimal transformation model. The analysis of theoretical and empirical studies revealed that the prevailing project, both in theory and in practice, aimed at preventing small towns from further degradation, dilapidation, and disrepair by turning them into tourist destinations and changing their cultural environment is far from optimal and does not meet the goals of sustainable development. As a result, I came to the conclusion that the project of harmonizing the cultural environment of small towns in ecological, social, and humanitarian aspects can become an effective alternative to the project of their *touristification* and turn small towns into self-efficient, prestigious, and comfortable places.

**Keywords** Cultural environment · Small town · Sustainable development · Tourism · City of dreams

## 1 Introduction

The relevance of the paper has ontological and gnoseological grounds. The deplorable state of Russian small towns is an ontological ground. The need to develop a sound program for the revival of small towns and the search for theoretical and methodological foundations for transforming the cultural environment of small towns are the gnoseological ground.

T. A. Semilet (✉)
Altai State University, Barnaul, Russia
e-mail: 7let@mail.ru

 
S. G. Maximova et al. (eds.), *Advances in Natural, Human-Made, and Coupled Human-Natural Systems Research*, Lecture Notes in Networks and Systems 234,
https://doi.org/10.1007/978-3-030-75483-9_72

Small towns are an integral part of the Russian social space. They equalize the population density, balance the uncontrolled growth of large cities and the over-concentration of people in megacities. Moreover, the small towns promote the development of small and medium-sized businesses. Such spaces are devoid of such sins of a civilized person as (1) overpopulation, (2) break with traditions, (3) a generation gap, and (4) psychological stresses.

Small towns are studied from different angles. However, the scholarship have some common grounds, namely, (1) the recognition of their decadent state, (2) lack of funding sources, (3) depopulation, (4) deterioration of infrastructure, (5) stagnation of cultural life, and (6) the need to find solutions to these problems. There is an opinion that the crisis of small towns, which makes them disappear, passed the following stages: (1) industry shutdown, (2) loss of jobs and income, (3) rising unemployment rate, (4) a gradual depreciation of infrastructure, (5) decline in the level of services due to low wages, (6) depopulation, and (7) depression among people (Nikonorov 2016). Analysts are unanimous in the necessity of renovation of small towns and transformation of their cultural environment.

Numerous programs and projects are aimed at the sustainable development of small towns, defined in a narrow sense as "capable of staying at the same level for a long time". The purpose of this development remains unclear.

In a broad sense, the concept of sustainable development eliminates this disadvantage and provides criteria of assessing the validity of transforming the cultural environment of small towns.

The aim of this research is to analyze both proposed and implemented projects for transforming the cultural environment of small towns in terms of sustainable development.

## 2 Materials and Methods

The theoretical basis of the study comprises the works of Russian and foreign scholars, namely, Harshbarger (2012); Pirogov (2011); Schegolkov and Metelkina (2017); Tilak (2003); Nikonova et al. (2016); Furschik (2019); Agenda 21 (United Nations [UN] 1992) adopted by more than 178 countries at the United Nations Conference on Environment and Development [UNCED], held in Rio de Janeiro, Brazil, from 3 to 14 June 1992; the report of the World Commission on Environment and Development Our Common Future (UN 1987); Decree of the Government of the Russian Federation No. 762 of June 28, 1996 "On the Federal comprehensive program of development of the small and middle towns of the Russian Federation in the context of the economic reform" (Russian Federation 1996).

In the research, we used the following methods: (1) systemic structural analysis, (2) structural–functional analysis, (3) axiological approach, (4) projective approach, (5) explication method.

## 3 Results

The heuristic perspectives of the concept of sustainable development are manifested, inter alia, through the formulation of axiological research angle of the projects of transforming the cultural environment of small towns, assessing them from the point of view of viability, longevity, and ability to harmonize ecological, economic, social, and humanitarian areas of life and ensure comfortable conditions for their residents.

Business-oriented approach based on deficiency needs and philosophy of survival dominates the ways of transforming the cultural environment of small towns.

The reorientation of transforming the cultural environment of small towns to tourism as a leisure industry is regarded as the main tool of its renovation strategy and therefore dominates over the anthropocentric subject-oriented axiological approach, which puts in the first place the problem of the value of the environment and its ability to improve the world view, quality of life and positive cultural identification of people.

The reliance on tourism solves the issues of small towns, namely (1) capital inflow, (2) creating and promoting a vivid brand of the town, (3) constructing resorts and recreation places for people, (4) revitalizing urban public spaces, (5) building renovation, (6) creating new jobs, good roads, parking lots, gas stations, cafes, drugstores, and sports grounds.

Projects for the transformation of the cultural environment of small towns, both proposed and implemented, disregard the main points of the concept of sustainable development, such as (1) balanced and conflict-free development of the urban conglomerate and (2) focus on a comfortable life for residents. Empirical studies have shown that residents are socially disintegrated, deprived of the processes of transformation of towns and tourist activities, urban centers are gentrified.

Methodologically, it would be advisable to supplement the *from past to present* approach with the *from present to future* one, where small towns are an alternative to megacities. Small towns show the type of settlement tuned to a post-industrial deurbanized society living in harmony with nature and characterized by (1) cleanliness, (2) low population density, and (3) extensive use of digital technologies that allow one to work from anywhere, from an *electronic cottage*. It reflects an understanding of sustainable development in a broad sense as a harmonious and balanced process of economic and social change, in which the exploitation of natural resources, investment, scientific, technological, and personal development are interdependent and expand both present and future opportunities to meet human needs.

## 4 Discussion

Discussing the problem of transforming the cultural environment of Russian small towns (towns with less than 50,000 residents), most scholars agree that small towns are in a depressive tailspin and need renovation. Many scholars refer to the Decree of the Government of the Russian Federation No. 762 of June 28, 1996 *On the Federal comprehensive program of development of the small and middle towns of the Russian Federation in the context of the economic reform* (Russian Federation 1996), which states that in the 1960s, 1970s, and 1980s the problems of small towns were repeatedly addressed by the government with a particular focus on centralized and subsidized mechanisms. Simultaneously, small towns were funded from leftovers, "which was one of the main reasons of the depressive state of many small and medium-sized towns" (Russian Federation 1996). Y. Velichko notes, "In the USSR, the government aimed to develop large cities where industry was mainly concentrated. Both the cultural and social environments of small towns was financed "from the leftovers." Small towns were forgotten as soon as the reforms began" (Velichko 2005). G. N. Bazhenova points out that today small towns attract the attention of scholars and politicians due to the "growing crisis of large-scale forms in all areas of social life" (Bazhenova 1998). Russian architect Andrey Bokov warns that, "small towns are getting extinct in the shadow of large cities and the main task is to soothe these degradation trends" (Bokov 2016).

Many participants in the discussion of the problem of small towns in Russia admit that small towns are an important component of the Russian socio-cultural space. Moreover, they (1) make it more homogenous, (2) prevent the growth of population density in megacities, and (3) serve as accommodation for villagers moving to urban areas, mitigating their *cultural shock* when adapting to urban life.

There are some pessimistic points of view on this topic such as, "we must fight degradation before we stubbornly go further" (Bokov 2016); and minimalistic strategies to support the survival of small towns by obtaining subventions from the budget and accepting migrants. Nevertheless, some scholars suggest optimistic projects for turning small towns into centers of the recreation and hospitality industries, where tourism is seen as the main catalyst for the revival of small towns. Furthermore, this optimistic opinion dominates the discussion about strategies for transforming the cultural environment for the *evaporating towns*.

Proponents of the optimistic path regard such reorientation as "one of the most promising means" (Schegolkov and Metelkina 2017) or even as the "construction of a dream" (Harshbarger 2012), which will ensure the sustainable (in terms of self-reproduction) functioning and development of small towns. The abundance of resources that small towns have gives reason for optimism. Consequently, "since small towns have some advantages over industrially developed regions, namely (1) favorable environmental conditions, (2) undeveloped territories, and (3) opportunities for infrastructure projects, small towns are a strategic resource of Russia (Oborin et al. 2017).

When implementing such a project for the development and construction of the cultural environment of the town, the components of the environment turn into the following objects of the tourism industry:

1. Natural resources given by nature "for granted" (rivers, lakes, forests, mountains, cliffs, waterfalls, caves, etc.);
2. Human-made landscapes (buildings, towers, temples, monuments);

3. Cultural and historical sites (battlefields, memorial sites, movie locations, museums, art galleries, ateliers, folk crafts characteristic of the region, ethnic attractions, fishing villages, stud farms);
4. Infrastructure facilities (hotels, campsites, guest houses, cafes, parking lots, grocery stores, drugstores, souvenir shops, tourist information centers);
5. Events (*Cucumber Day* in Suzdal, the old car festival *Samokhod* in Myshkin, the aviation festival *the Sky Fair* in Kungur, *Carrot Fest* in Dmitrov);
6. The image of the town (Aslanov and Perminova 2018), the brand of the town (Andreeva 2016), the unique pivot the town would spin around (Pirogov 2011) which would form a certain concept of the town and become the basis of local identity of people.

We regard this scenario of cultural environment transformation in terms of the concept of sustainable development.

In a narrow sense, sustainable development is development that is "to ensure that it meets the needs of the present without compromising the ability of future generations to meet their own needs" (UN 1987), "economic development that is conducted without depletion of natural resources" (Lexico 2020). It is the biocentric and ecocentric approach to understanding the term, which emphasizes an ecological perspective based on the two principles: (1) "abstain from doing harm" and (2) "preserve". "Today, this concept is the most common and is often called "the worldwide model of the future of civilization,"—the magazine *Sustainable business* writes (Sustainable business 2020).

In a broad sense, sustainable development refers to a sustainable development strategy designed with an economic, social, and ecological perspective, based on the document entitled *Agenda 21* adopted by more than 178 countries at the UNCED held in Rio de Janeiro, Brazil, from 3 to 14 June 1992 (UN 1992).

The broader meaning of the term "sustainable development" is extremely important for our analysis. Along with the ecological component (environmental protection) and the economic component (sustainable manufacturing, production, and consumption models), it includes (1) an anthropological component (creating conditions for satisfying the pyramid of human needs, from physiological ones to self-actualization), (2) a social component (conflict-free coexistence of different social groups) (Belousov 2013), and (3) cultural component (high quality of the cultural environment as the requirement for a comfortable life).

A broad understanding of sustainable development implies the co-evolution of man and nature within the dialectic of "humanization of nature and naturalization of humans" (Tilak 2003). More than that, it gives particular importance to the harmonious relationship between people and a decent life for everybody. This understanding of sustainable development also gives special significance to long-term, harmonious, and conflict-free coexistence of people.

This position is the basis for the analysis and assessment of proposed and implemented projects for transforming the cultural environment of small towns.

On the one hand, sole reliance on tourism solves the issues of small towns, namely (1) capital inflow, (2) creation and promotion of a vivid brand of the town,

(3) building resorts and recreation places for people, (4) revitalization of public spaces, (5) building renovation, (6) creation of new jobs, good roads, parking lots, gas stations, cafes, drugstores, sports fields.

On the other hand, the cultural environment is not only an environment of artefacts but also a "continuum of relationship" where residents are connected to each other, with "urban artefacts, place of residence, landscape, their personal space, significant events, etc.—to everything that concerns the relationships" (Pirogov 2011). Moreover, such an environment determines whether everyday life is comfortable or not. Empirical studies have revealed numerous negative social and humanitarian effects of urban environment transformation for the needs of the tourism industry.

The first effect is the alienation of urban residents from tourist activities and the transformation of the cultural environment of their towns, manifested in the attitude of arriving tourists as invaders, trampling the territory and leaving behind trash (buses, private cars, picnics, outdoor workouts) (Nikonova et al. 2016). Also, alienation manifests itself in the filling of jobs with workers from other places, over whom local residents do not have hiring priorities (Schegolkov and Metelkina 2017). The third manifestation of alienation lies in the fact that local residents are alienated from the money of tourist; they cannot rent guest houses and bed-and-breakfasts to tourists, sell souvenirs and other local goods (Valkova and Shabalina 2019).

The second effect is social disintegration. Being involved in the hospitality industry without a proper level of organization, the local community is disintegrating. As I. Zadorin points out, "some experts say that the town Myshkin is beginning to resemble some resort towns in the South of Russia, where everybody is chasing newly arrived tourists to rent them a room, earn money on it in high season" (Aslanov and Perminova 2018).

The third effect is manifested in the gentrification and touristification (a combination of two words—gentrification and tourism) (Touristification, www) of territories, when local residents are regarded as a source of attraction for tourists, which should be transformed through decrees and bans (e.g., one cannot plank one's old (and ramshackle) house, install double glazed windows or change anything). Local residents often *relocate* from urban centers to clear a tourist area that is not visited by local residents due to prices that are too high for them but affordable for tourists (Andreeva 2016). Therefore, one can observe social inequality.

From the analysis of the current situation, I can conclude that the transformation of the cultural environment of small towns in Russia only for the needs of tourism, based on deficiency needs, the philosophy of survival, and objectification of nature and residents, which exclude harmonious development and prevents small towns from approaching the dream of a comfortable life, is not a solution to the problem of their sustainable development.

## 5 Conclusion

The concept of sustainable development provides both criteria for assessing the implemented projects to transform the cultural environment of small towns, as well as a view to the future and guidelines. The expert opinion of Moisey Furschik, the managing partner of the company *Financial and organizational consulting* can illustrate this phenomenon. He predicts that in developed countries, large cities and small towns will swap their roles. "Small towns will become the most desirable and expensive places for life due to their cleaner environment, lower population density, and friendly nature. They will shape homogeneous and comfortable communities" (Furschik 2019). Social areas (education, medicine, restaurants, entertainment) will be served by people, not machines, which will be regarded as a high-quality mark. Furschik adds that to eliminate the problem of their remoteness, small towns will have fast transport links to the nearest megacities. Most services will be provided remotely, which will make life in a small town not less civilized than life in a metrolopolis.

## References


Andreeva N (2016) Sociologist Alisa Maksimova on how the influx of tourists is changing the small town. https://www.poehali.tv/adventures/9196

Aslanov I, Perminova M (2018) How to develop small towns? Interview with the sociologist Igor Zadorin. https://fomlabs.ru/material/kak-razvivat-malye-goroda

Bazhenova GN (1998) Cultural environment of a small historical town as a socio-philosophical problem (Dissertation of Candidate of Philosophical Sciences). Russian Presidential Academy of Public Administration, Moscow, Russia. http://cheloveknauka.com/kulturnaya-sreda-malogo-istoricheskogo-goroda-kak-sotsialno-filosofskaya-problema#ixzz6I66XYW9V

Belousov KY (2013) The current evolution stage of the concept of sustainable development and the formation of the corporate sustainability paradigm. Probl Mod Econ 1(45):47–50

Bokov A (2016) Revitalization of Russian single-industry towns in current conditions. Report at the Saint Petersburg International Cultural Forum

Furschik M (2019) How 21st century technologies are changing the face of cities. https://www.if24.ru/budushhee-gorodov-cherez-20-let/

Harshbarger B (2012) Small town tourism: building the dreams. Rural Res Rep 22(6):1–8

Lexico (2020) Sustainable development. https://www.lexico.com/definition/sustainable_development

Nikonorov SM (2016) Revitalization of Russian monotowns in current conditions. https://istina.msu.ru/conferences/presentations/38612654/

Nikonova TV, Vesloguzova MV, Murtazina GR (2016) The role of local communities in the processes involving the intangible cultural heritage in tourism (based on the Bolgar historical and archaeological complex, Tatarstan). Serv Tour: Curr ChallS 10(3):124–130

Oborin MS, Sheresheva MY, Ivanov NA (2017) Problems and prospects of development of tourist-recreational potential of small cities. Tyumen State Univ Herald Soc , Econ, Law Res, 3(3):169–192

Pirogov SV (2011) City as cultural phenomenon: cognitive approach. Tomsk State Univ J Cult Stud Art Hist 2:31–37

Russian Federation (1996) Resolution "on approval of the federal comprehensive program for the development of small and medium-sized cities of the Russian Federation in the context of economic reform" (June 28, 1996 No. 762). Moscow, Russia

Schegolkov YY, Metelkina PV (2017) Development of tourism in small towns of Russia. World New Econ 11(1):86–91

Sustainable business (2020) Sustainable development: Concept, principles, goals. http://csrjournal.com/ustojchivoe-razvitie-koncepciya-principy-celi

Tilak JB (ed) (2003) Education, society, and development: national and international perspectives. APH Publishing Corporation, New Delhi, India

United Nations (1987) Report of the world commission on environment and development "our common future". Oxford University Press, Oxford, UK. http://www.un-documents.net/wced-ocf.htm

United Nations [UN] (1992) Agenda 21. (June 3–14, 1992). Rio de Janerio, Brazil

UrbanNext (2020) Touristification: how tourism and gentrification are changing the cities around the world. https://urbannext.net/touristification-how-tourism-and-gentrification-are-changing-the-cities-around-the-world/

Valkova TM, Shabalina NV (2019) Tourism in small towns: Zaraysk. Bull Mosc RegN State University Ser: Nat Sci, 2:144–145.

Velichko YV (2005) The phenomenon of the cultural environment of small towns in Western Siberia (Dissertation of Candidate of Philosophical Sciences). Derzhavin Tambov State University, Tambov, Russia. http://www.dslib.net/teorja-kultury/fenomen-kulturnoj-sredy-malyh-gorodov-zapadnoj-sibiri.html

# Traditional Legal Consciousness in Sustainable Development: The Case of China

**Rafail V. Nasirov and Aleksey V. Ivanov**

**Abstract** The Modern world is at the crossroad of enormous changes in all spheres of life. The era of Western dominance, has been lasting for several centuries, seemed to be finished along with the dissemination of European Enlightenment. The project of Modernity implied secularization of public life reflects the theory of natural human rights and principles of the law state. However, the slogan of the rule of law turned not only legalization, but the etatization of social life, the destruction of traditional forms of social organization and communication. Under these conditions, the crisis of Western society spreads to the East and humankind tries to find other variants of normative-value system construction. The purpose of the article was the determination of *Fa* (formal law) in value-normative system of traditional Chinese philosophy. We revealed a holistic system of main categories of ancient Chinese philosophy looking at the ideas of Taoism, Confucianism, and legalism. We substantiated the protective purpose of *Fa* in the public life. We assumed that the emphasis in the regulatory system of Chinese society determined the values and patterns of behavior that focus on the sublime beginning of humanity and virtue, explained the preservation of Chinese civilization and its prospective development.

**Keywords** Traditional legal consciousness · Sustainable development · China · Ancient Chinese legal consciousness · Statute · Fa · Tao · De · Zhen · Li

## 1 Introduction

A legal expert has to consider the modus of social existence with formal law and, take into account the relationship of law with other non-legal phenomena of public life. Scientist use to come up with his own unique interpretation of human life development. A common feature of a traditional worldview is related to a formal law in social norms' system, as the media is not absolute, but the relative values. This formal official law tends to manifest in different cultures, civilizations, their scales,

R. V. Nasirov (✉) · A. V. Ivanov
Altai State University, Barnaul, Russia
e-mail: nasirov.rafail@yandex.ru



S. G. Maximova et al. (eds.), *Advances in Natural, Human-Made, and Coupled Human-Natural Systems Research*, Lecture Notes in Networks and Systems 234,
https://doi.org/10.1007/978-3-030-75483-9_73

directions, and forms. In general, the traditional world was the syncretic and holistic; therefore, it is not so easy to select the legal category in value-normative system of traditional society.

## 2 Materials and Methods

Content article is based on the works of ancient Chinese thinkers (Hin-Shun 1994; Lao-Tzu 2017). On the theoretical level, the traditional legal consciousness of China was developed by foreign and Soviet researcher (Granet 2008; Kovler 2002; Titarenko et al. 2006; Martynenko 2015).

The subject of the article is the traditional legal consciousness of China reflecting the concept of *Fa* (law). Our basic research methods were historical and comparative-legal. We considered that it is impossible to determine the authentic meaning of *Fa* (law) concept without determining its relationship with other categories of traditional Chinese philosophy—Tao, Te, Zhen, and Li. A systematic method has allowed to reveal the hierarchical structure of concepts and values of ancient Chinese philosophy and to determine the place of *Fa* (law). The study allowed to combine the modernization and sustainable development with traditional Chinese values and norms.

## 3 Results

A sign of civilization (along with sedentary lifestyle, producing economy, social stratification, public authorities, writing) is a formal law. As a rule, to identify the legal component in the traditional legal consciousness is difficult for the reason that corresponding worldview wore syncretic in nature and were based on mythological and religious basis. But a fertile ground to study this question is the ancient Chinese worldview, in a complex system of concepts which clearly distinguish the term, similar to modern law—*Fa.* But before giving him the definition, as stated above, it is necessary to have a general idea about the outlook of the ancient Chinese.

In the traditional worldview of the primary principle of life is monism, and accordingly a beginning of the perception and cognition of the world is the belief in its unity. It would seem that this thesis contradicts the history of religious beliefs, where polytheism was preceded by monotheism. But historians of religions find that in polytheistic beliefs there is the idea of the first principle, according to which the many gods are secondary, subordinate to the general laws of being.

But if the world is one, then how to explain the diversity of its manifestations and a complex structure? By archaic ideas of Ancient China was preceded by some primordial Whole way which was the egg, which was carried, apparently, a value of "not yet born". But more fundamental was the eternal *Dao*, which is impossible to describe and to identify clearly; the presence of this first principle of being is obvious, but it could not be adequately and fully reflected in the consciousness of man, as the man himself and his inner world are only manifestations of the Tao. In Chap. 42 of "Tao Te Ching" says: "The Way [Tao] begets one. One produces Two. Two produces Three. And Three begets the myriad things. All things carry Yin and embrace Yang. All the penetrating breath leads to agreement" (Lao-Tzu 2017). Note that the judgment is built in a spiral is disclosed as a Single embodied in "the dark things," and at the end of the phrase, so that this process of complication and differentiation of the world does not become a straight line and not left in a meaningless (bad) infinity, to determine the source of preserving the unity of the universe—"the all-pervading breath," providing "consent" of things. Breathing is always a unity of opposites, therefore, metaphors and so on in relation to the *Tao* paired beginnings of life, *Yin* and *Yang*, can be breath and exhale.

Traditional Chinese worldview can be considered as one of factors of maintenance of cultural and historical identity of Chinese society, its ability to modernize while maintaining the fundamental ideological frameworks and social institutions.

## 4 Discussion

*Tao* determines the development and harmony of all creation, including society and human consciousness. The feature of Chinese culture is that in a series of concepts of *Tao*, *Yin* and *Yang*, *Zhen*, *De*, *Li* and others can distinguish the concept of *Fa*, which is close to the modern term "law" in the legal sense of the word. Initially, the hieroglyph—"fa" had a common meaning: "sample," "ideal" of being; in Chap. 25, "Tao Te Ching" says: "the Person sample [the law, the ideal of] Earth. The ground sample the Sky. The sky is the sample—Path [Tao]. And the Way a sample that itself in itself" (Lao-Tzu 2017). This character contains two images—flowing water, whose smooth surface symbolized fairness and justice, and the unicorn, which, according to legend, were able to identify the perpetrator and to punish him (Chinese hieroglyphes: law 2009). Given that the term *Fa* had a sense of deontological (binding) rules of conduct, it has received a narrower meaning of a norm has external sanctions. *Fa* began to oppose *De*—virtue that characterizes the internal state of the person which, in turn, is the result of reflection in the human soul of the world law of harmony of *the Tao*. The researchers draw attention to the fact that the opposition of the "*Fa—De*" matches the dichotomy of law and grace in the christian tradition (Titarenko et al. 2006).

It is also necessary to consider the relationship of *Fa* with this important category of ancient Chinese thinking as *Li*—propriety, ethics, ritual, ceremony, setting (Titarenko et al. 2006). The researchers point out: "the Category sent in the letter character *Li* practically, it is impossible to translate into European languages, he picked up exact linguistic analogue" (Martynenko 2015). *Li* similar to *Fa* in that it contains requirements for external human behavior. Confucius says: "To overcome oneself and the appeal to Li is humanity (Zhen)… should not look at inappropriate Li, you should not listen to inappropriate Li, you should not speak inappropriate Li"

(Confucius 2008). But *Li* it is unacceptable to equate with modern understanding of etiquette, as to the rituals and acts of courtesy are required of *Xin*—sincerity and compliance with the external act of behavior and inner experiences; (The Teacher said "If a man has no love for mankind, how he can observe the ritual?" (Confucius 2008)). On the other hand, the term *Fa* was to evoke violence and to get closer to such notions as *Li*—force, violence and *Xing*—punishment (Titarenko et al. 2006). Finally, as a term with legal meaning is called character *Lu*, which is usually translated as "law" in the sense of a legal act, statute (Kovler 2002).

The question arises: "How ancient Chinese sages explained the need for such drastic means as *Fa, Li, Xin*?" The traditional worldview in general were not typical for the flat, respectively, idealism and utopianism; admitted that in public life along with a good basis function and negative, that cannot be eliminated, but achievable goal to minimize them, hold within certain limits and transformation. Confucius is not talking about filling the whole public life of *Zhen*, i.e., humanity and morality, and about reaching the "right balance of love and hate" (Confucius 2008). The ancient Chinese sages, without recourse to sophisticated philosophical reflection, realized complex dialectic of good and evil, which reflects the ratio of the fundamental elements of being the Yang and the Yin, the process of interaction, of mutual transitions, rebirths; in "Book of changes" said: "The Yin, the Yang is called Dao" (Titarenko et al. 2006). This kind of judgments did not justify the relativity of moral evaluations, but, required the person relevant intellectual, spiritual and emotional focus. It will become clear that this was reflected in the categories of Yin and Yang, if you refer to the famous modern philosopher Jean Baudrillard, who often draws attention to the reversibility (invertibility) of good and evil, and advises: "Evil should not deny: it is necessary to join the game, they need to play, it is necessary to beat" (Baudrillard 2006). Fearing that this advice will be accepted in the spirit of relativism, the philosopher stresses that the idea is "obliged to remain humanistic, attentive to man and thus to open the reversibility of good and evil, humane and inhumane" (Baudrillard 2006).

Along with open manifestations of social evil that can and must respond to the relevant public (official and legal) resources, there are a variety of mediated forms of evil affecting the very atmosphere of public life and its fundamental basis. The attitude of the *Fa* as a reaction to external and discrete manifestations of evil, but not as a means of establishing social peace explains the evaluation of Confucius the act of a young man who denounced his father stole the sheep; the wise man sadly sighed and said: "Truly honest man is not. The father will hide the crime of the son, the son will hide the crime of the father—that means honesty (Confucius 2008). It is clear that Confucius does not justify theft, but the threat of destruction of the family is more dangerous and fundamental as an informal, outside of the actions of the Fa family relationships largely supported in public life at the beginning of *Zhen* and *De*. Confucius cautiously accepted the modern principle of "rule of law" that recognizes the possibility of legalization all social relations: "Today is reverence to the parents called their content. But people also contain dogs and horses. If parents are not respected, what will be different the attitude towards them from relationship to dogs and horses?" (Confucius 2008).

Thus, in Ancient China, the traditional consciousness of the protective function of law is manifested clearly; in General, the *Fa* was perceived as "forced evil". In aphoristic form Laozi is identified as: "...When you grow up laws and orders, the number of thieves and robbers" (Lao-Tzu 2017). The representative of Western culture might argue that the laws are good, their purpose is to create conditions to counter and combat the troublemakers. But the sage has in mind that a healthy society most citizens and without legal requirements will not become robbers in force the existing inner belief; in that human behavior is manifested in the world of *Tao*. Therefore, not only the offender, but also the carrier of the marginal behaviour, not commit crimes only out of fear of punishment, not to the same degree (one internally and externally and the other internally) not just violate or willing to violate state law and ignore the requirements of *Tao* and bring in all the universe of grains of mess of chaos. This setting is expressed in the famous aphorism of Seneca: "The Intent to commit a crime is a crime!", and the approval of F. M. Dostoevsky: "I do not want a society where I could not do evil, and so it was that I could do all evil, but I didn't want to do it myself."

Ancient Chinese civilization shows that the focus on the lofty principles of social existence is quite compatible with the rationalism and pragmatism of thinking. Together with lofty categories of *Tao*, *De*, *Zhen*, and the other, the emphasis was on the formal side of social life. M. Granet writes: "ultimately they conceptualize the universe as if it was regulated by the Protocol, and debug it in the manner of the ceremony—the goal to which, in their assurances they seek. Their morals, their physics, their logic is just different sides of the effective knowledge, which is the etiquette" (Granet 2008). However, this has not led to the emergence of a worldview that would be based on the principles of scientism, and legal consciousness did not become formalized and statist. "The idea of the law was not strongly developed, and that in the future, the observation of nature was given empirical, and the organization of society was built on compromises. And yet the notion of the Right or rather the concept about the Sample, allowing the Chinese to keep flexible and understanding of the Procedure, not encouraged them to imagine over the world people the world of transcendental realities. Their wisdom, all imbued with a sense of nature, the strongly humanistic" (Granet 2008).

It is worth noting that in Chinese history is known and the period of the Qin Empire (III-II centuries BC), in which the basis of social life determined the rational-utilitarian beginning of the school *Fa-Chia* (legalists) with the principle of *Fa-Zhi* (rule by law). Even two thousand years prior to the publication in Western Europe of the treatise N. Machiavelli's "The Prince" was founded as a political doctrine based on the principles of misanthropy and consistent pragmatism. In "Shang Jun Shu" ("Book of the ruler of Shan state") was written: "If you manage people as virtuous, the inevitable turmoil and the country will perish; if we manage people as evil, it always maintained perfect order and the country achieves power"; "Punishment begets power, power begets power, power begets greatness and majesty, inspiring awe produces virtue. So virtue has its origin from punishment (Hin-Shun 1994). The views of the representatives of this school of ancient Chinese thought is hard to accept as true, but it is necessary to understand them. It is important to consider that the school and the practice of *Fa-Jia* originated in the crisis of ancient Chinese society, when needed emergency funds for the end of the civil war. Bright representative of the legists Feng Fei wrote: "The Confucians culture undermine the laws daredevils [use of weapons] to cross the prohibition and the rulers [believe] that another conform to the ritual; and occur because of the troubles" (Hin-Shun 1994). Ideas and practical advice of the legists were implemented, and the society out of the crisis, but in such periods of history and there is a risk of mixing absolute and relative values and principles when the necessary conditions of social existence—law and order—begin to be perceived as the highest good and an end in itself.

The truth is reflected in the fact that the Qin Empire lasted only a few decades, and after the civil war, established a regime of the Han Empire (II BC–III century), which officially became recognized as the teachings of Confucius. The sage stated: "to lead the people by laws and to maintain order with punishments, the people will tend to shy away [from punishment], and will not feel shame. If you lead the people through virtue and maintain order with the help of the ritual, the people will know shame, and he corrected" (Hin-Shun 1994).

Of course, within the Han Empire on the basis of the postulates of the doctrine of the *Fa-Jia* emerged a developed system of bureaucracy and formal law. In the formation of the centralized state is inevitable and dangerous tendency of strengthening the public-began formal public life was neutralized by the fact that between the lofty principles of *De* and *Zhen*, on the one hand, and legal principles of the *Fa* and *Xin*, on the other hand, was the principle of *Li* (ritual). Figuratively speaking, *Li* assumed "kick" of etatization and legalization public life, which subsequently led to a well-known formalization of this crucial element of the value-normative system of Chinese society. It can be argued that due to this "victim" in the end, there was no mixing of absolute and relative principles of social existence. Orientation to the principles of *De* and *Zhen* was not completely abolished, which was one of the reasons for the surprising vitality of this culture, a connoisseur of which they write: "The Chinese civilization has always found the inner resources of resilience and renewal. The most important prerequisite for this is its high adaptability, focus on self-improvement and learning from other cultures" (Titarenko et al. 2006).

## 5 Conclusion

China again becomes a major center of the global world. Speaking the biblical words, "everything is the way it was". But, what are the values and attitudes this ancient civilization will distribute and protect in the world politics? Did the relatively short (towards three millennia of Chinese history) colonialism change the code of civilization in this country?

With that said it is worth to point out the uniqueness of Western civilization, which about three centuries ago went on a very peculiar path of development with almost total legalization and, accordingly, the etatization of social life. Deep knowledge of the history of Chinese society and the characteristics of its legal consciousness allows to determine the post-Communist stage of China development compared to the USSR. Up to now, the dominance of the West World presents not only in economy and politics, but also in the ideological sphere. We don't know the answer to the question: "Is there any algorithm of sustainable and equitable development, in which modernization is not matched with the abolition and preservation of traditional values and institutions?" In this regard, the achieving of economic, political and ideological worldview of self-sufficiency is the main Chinese challenge.

## References

Baudrillard J (2006) Passwords. From fragment to fragment. U-Factoria, Ekaterinburg, Russia
Confucius (2008) Lunhui. Sayings. Eksmo, Moscow, Russia
Granet M (2008) Chinese thought from Confucius and Lao Tzu. Algorithm, Moscow, Russia
Hin-Shun J (ed) (1994) Ancient Chinese philosophy, vols 1–2. PrinT, Moscow, Russia
Kovler S (2002) Anthropology of law. NORMA, Moscow, Russia
Lao-Tzu (2017) The book of the way (Tao Te Ching). AST, Moscow, Russia
Martynenko NP (2015) "Li" and "Fa"—polar category of Chinese social philosophy. The Moscow University Herald. Series 7 Philosophy 1:51–63
Titarenko ML, Kobzev AI, Lukyanov AE (eds) (2006) Spiritual culture of China: Encyclopedia, vol 1 "Philosophy". Vostochnaya literature, Moscow, Russia
The Epoch Times (2009) Chinese hieroglyphes: law. http://www.epochtimes.com.ua/ru/kitayskie-ieroglify/kitayskie-ieroglify-zakon-88264

# State Confessional Policies with Respect to Western Siberian Muslim and Christian Communities in Second Half of the 1970s—Early 1980s

**Petr K. Dashkovskiy, Natalia S. Dvoryanchikova, and Elena A. Shershneva**

**Abstract** The paper analyzes the influence of state confessional policy on Muslim and Christian communities' position and activities in Western Siberia in the second half of the 1970s—early 1980s. During the specified period, the state-confessional relations were modernization in the Soviet Union in general and in Western Siberia in particular. Using legal documents and archival data, the authors examine the regulation practice of religious community registration in the region. During the studied period also resumes the registration process of the All-Union of the Evangelical Christians-Baptists, Mennonites, and Lutherans communities. The authorities tried to control the opposing to legalization communities of Pentecostal and All-Union of the Evangelical Christians-Baptists Council. The study unveils the peculiarities of Muslim and Christian communities, namely, the community registration process, economic, social, and demographic changes within communities, religious education of children, and church ministers' education. It concludes that there is a change in the state confessional policy towards religious organizations—the communities resume the registration process, and Muslims, the All-Union of the Evangelical Christians-Baptists Council, Baptists, Pentecostals, Lutherans, and Seventh-Day Adventists revive their activities.

**Keywords** Religious communities · State confessional policy · USSR · Western Siberia

## 1 Introduction

The relationship history of the USSR government and religion is a crucial topic of historical science. The research and analysis of the relationship between the Soviet

P. K. Dashkovskiy · N. S. Dvoryanchikova (✉) · E. A. Shershneva
Altay State University, Barnaul, Russia
e-mail: natali.dvoryanchikova@mail.ru

P. K. Dashkovskiy
e-mail: dashkovskiy@fpn.asu.ru



S. G. Maximova et al. (eds.), *Advances in Natural, Human-Made, and Coupled Human-Natural Systems Research*, Lecture Notes in Networks and Systems 234,
https://doi.org/10.1007/978-3-030-75483-9_74

State and religious communities will help avoid problems in the government's confessional policy. It also contributes to the building up of an interaction model between various confessions and government. Thus, it is necessary to analyze changes in the confessional situation in the 1970s–1980s in Western Siberia. The position of religious communities in the Soviet period is relatively well covered in the modern historical science works of Vasileva (2001); Odintsov (2012); Nikolskaya (2009). However, scholars primarily examine nationwide tendencies in government-confessional relations. The scholar has not studied regional problems between government institutions and religious communities, including Siberian communities, thoroughly. Specifically, the position and activities of Muslim and Protestant communities in Altai Krai and Novosibirsk region are not exact, although specific points are highlighted by Zherebyatyeva (2009); Dashkovsky and Dvoryanchikova (2019); Gorbatova (2008); Siebert and Dashkovsky (2015).

## 2 Materials and Methods

Archive materials and legal documents provide a means of conducting the most objective analysis of Muslim and Christian communities and their position in the USSR confessional policy in the second half of the 1970s–early 1980s. Archive sources are represented by the business correspondence of the authorized Council for Religious Affairs in Altai Krai, as well as in Novosibirsk region with Soviets, district and city committees, and religious associations; statistical data; administrative materials; legislative normative legal acts of state and party bodies (State Archives of Altai Krai. Fund 1692. List 1. File 136. p. 1; State Archives of Altai Krai. Fund 1692. List 1. File 144. p. 4, etc.). We use materials from the State Archives of the Novosibirsk Region and Altai Krai (State Archives of the Novosibirsk Region. Fund 1418. List 1. File 230. p. 16; State Archives of the Novosibirsk Region. Fund 1418. List 1. File 213. p. 8, etc.). The historicism method and its principles allow us to consider the regulating process of religious communities in various social spheres. The research methods include (1) comparative-historical, (2) historical-systematic, and (3) retrospective. The comparative-historical method allows us to analyze the position of religious organizations in the studied period. The historical-systematic method makes it possible to generalize the status of Muslim, Orthodox, Old Believers, Baptist, Lutheran, Pentecostal, and several other Western Siberia religious communities and recreate a unified system adhered to by the Soviet leadership. The retrospective method recreates a holistic view of the past. Besides, the retrospective method makes it possible to analyze tendencies in the believers' relationships from the community to community, and, also, their interaction with authorities.

## 3 Results

In the 1970s, the severity of the confrontation between the Soviet and Western blocs began to decrease. That process was commonly called in history the "relaxation of international tension". The USSR and the USA were involved in local conflicts in different parts of the world and needed a break. Besides, L. I. Brezhnev did not seek reckless actions outside of the "traditional" Soviet sphere of influence. During the "relaxation" period, the confrontation intensity between the two systems decreased, and the cooperation in various fields, on the contrary, expanded. The signing of the Helsinki Final Act of the Conference on Security and Co-operation in Europe in August 1975 was significant for the confessional policy. The most significant part of the Conference was dedicated to respect for human rights. "Relaxation" evidently affected the government-confessional relations in the mid-1970s USSR. Specifically, the Decree of the Presidium of the Supreme Soviet of the RSFSR of June 23, 1975 (Presidium of Supreme Soviet of Russian SFSR, 1975) confirmed the provision mentioned above. The Decree amended the Soviet legislation on religious cults. As a result, the capacity of religious cults expanded. However, this decision remained "nominal".

In 1977, the fact of developing a new Constitution gave a certain hope to change the situation for the believers' part of the population. Its Bill was published for public discussion in early 1977 (Odintsov 2012). The adoption of the new Constitution of the USSR softened the attitude towards religious organizations in the country. Article 52 of the Constitution enshrined the believers' right to freedom of religion (Constitution 1977). It contained the following wording, "Citizens of the USSR are guaranteed freedom of conscience that is the right to profess or not to profess any religion, and to conduct religious worship or atheistic propaganda. Incitement of hostility or hatred on religious grounds is prohibited. In the USSR, the church is separated from the government and the school from the church. Although Article 52 of the Constitution proclaimed guarantees of freedom of conscience, the situation did not change—the administrative-command method of implementing that right continued to dominate and did not suit believers or non-believers (atheists) (Klimenko 2007)".

The signing of the agreements mentioned above and the adoption of the new Constitution had a severe impact on the activities of religious communities, including the religious situation development in Western Siberia. Since 1977, the registration process of the All-Union Council of Evangelical Christians-Baptists, Russian Orthodox Church, Seventh Day Adventists, and Old Believers communities resumed in Altai Krai and Novosibirsk region.

Only one Muslim community (in Novosibirsk) legally operated in the Novosibirsk region. Apart from that, two more communities operated without registration (Islam at the End of the World 2015). Separate groups of believers were also located in the villages of Yurt-Ora, Kazanka, Yurt-Akbalyk, and others (Fund 1418. List 1. File 230. p. 16). The registered community was holding worship services once a week on Fridays and religious holidays. By the end of 1970, the Muslim community numbered 700 people. Ethnically, the believers were predominantly Tatars, and only a small number were Kazakhs. Primarily older people attended mosque services. Simultaneously, young adults also attended mosque services in small numbers on religious holidays. From 1978 through 1982, such rites as naming, wedding, and funeral were regularly held in the community. It was noteworthy that, in comparison with the previous period, the income of the Novosibirsk Muslim community increased. In 1979, the income amounted to 15,526 rubles, 17,200 rubles in 1980, 10,500 rubles in 1981, and 10,300 rubles in 1982 (Fund 1418. List 1. File 251. p. 9).

The Russian Orthodox Church in Western Siberia ranked first in believers' number. In 1976, there were only three registered Orthodox churches in Altai Krai—in Barnaul, Biysk, and Rubtsovsk. During religious holidays, 2,000–3,000 people attended the church in Barnaul, 400 to 500—in Biysk, 250–300—in Rubtsovsk (Fund 1692. List 1. File 136. p. 1). In the Novosibirsk region, there also were three Orthodox churches. In 1977, the number of clergies in the Novosibirsk region increased and amounted to 20 officially active clergies. Age and sex composition of the clergy in the region were as follows: from 30 to 50 years—13 people; from 50 to 60 years—two people, from 70 to 85 years—3 people, from 20 to 30 years—two people (Fund 1418. List 1. File 213. p. 8).

It was peculiar that in 1978 the work of the authorized Council for Religious Affairs in the south of Western Siberia was based on the requirements of the new Constitution of the USSR on freedom of conscience. After the open publication of religious and church law, believers from many villages began to raise questions more persistently about registering religious associations in the Western Siberia south. As a result, the city executive committees of Slavgorod and Kamensk and the regional executive committees were forced to register and open two more Orthodox communities. It defused the tension of believers who have been writing statements to various authorities since 1960—when the communities were in an illegal position. The illegal prayer meetings operated at the time. It emerged that an illegal Orthodox prayer house operated in Slavgorod for several years and performed religious rites. The authorized Commissioner for Religious Affairs in Altai Krai, A. P. Sedeshov noted, that the religious communities' registration also had negative tendencies, since believers in other settlements (Aleisk, Gorno-Altaisk, Troitsk, Talmenka) also began to apply for registration persistently. In 1979, believers of Aleisk used the same approach to obtain registration. The City Executive Committee eventually was forced to compromise: believers were permitted to gather for prayers in a private residence and occasionally invite a priest from Barnaul to perform the confession and communion rituals. However, believers did not accept this option. The Commission for Religious Affairs noted that the official clergy "have a hand in it". In their conversations, believers adamantly emphasized the patriotism of the Russian Orthodox Church, its contribution to the victory in World War II and the "tank column" in particular, and also its contribution to the Peace Fund development. As of January 1, 1979, there were five registered Orthodox communities in Altai Krai (Fund 1692. List 1. File 144. p. 4).

The income of the Russian Orthodox Church in Western Siberia also increased in the studied period. For example, the income of Saint Seraphim of Sarov Church in the Novosibirsk region amounted to as follows: 35,000 rubles in 1976, 41,000 rubles in 1977, and 108,000 rubles in 1979–1980. In 1982, the community received 21,000 rubles in six months. The most income came from religious object sales. On top of that, there was an increase of believers' number in the Novosibirsk region. While in 1979, the Orthodox communities consisted of 1,495 people, it changed to 2,286 people already by 1980 and 2,339 by 1981 (Fund 1418. List 1. File 252. p. 23).

It is worth mentioning that there were also Orthodox communities in the Gorno-Altai Autonomous Oblast (a part of Altai Krai at that time); however, none of them were registered. In March 1979, a group of Orthodox believers in Gorno-Altaysk wanted to legalize their activities. In this regard, the believer E. V. Zueva and the Assumption Cathedral parson in Biysk Archimandrite Yermogen held a conversation with the City Executive Committee on the community registration issue of the Russian Orthodox Church in Gorno-Altaysk. The number of believers in Gorno-Altaysk amounted to 50 people. The believers gathered in the E. V. Zueva's residence for worship since there was no church or priest. The City Executive Committee representatives visited one of these meetings and realized the necessity of registration. The local authorities did not refuse to do it but asked to collect and submit the relevant documents to the Gorno-Altaysk City Executive Committee (Fund 1692. List 1. File 139. p. 33).

The confessional situation in the studied period was ambiguous. On the one hand, religious sentiments had faded in some areas. On the other—for the most part, we could notice the growth of spiritual life in the region. In early 1980 there was a noticeable peak of interest of Altai Krai people towards the religion. In one report, the Commissioner for Religious Affairs declared that formerly people were afraid to attend church and perform rites. Nevertheless, in the 1980s, there was absolute freedom in this regard (Fund 1692. List 1. File 287. p. 63).

In contrast to the Orthodox Church, there was a notable decrease in rites in the Old Believers community. In 1979, for example, the Novosibirsk Old Believer community decreased the number of baptisms. Only seven people were baptized (up to three years old—three, from three to seven years old—one, from eight to 16 years old—one, and over 16 years old—two) (Fund 1418. List 1. File 228. p. 5). In Altai Krai, two Old Believer communities were legally operating in Barnaul in the mid-1980s (Bezpopovtsy ("priestless ones") of Pomorian Old-Orthodox Church and Popovtsy ("priested ones") of Belokrinitskaya Hierarchy). The Old Believers of the registered Belokrinitskaya Hierarchy community hoped to revive religious life through registration and priests, but their hopes were not fulfilled. It was due to the long-time inability of the church to find a priest; thus, one was appointed from Novosibirsk. The issue arose—the priest could not come often. The Commissioner for Religious Affairs also noted that the Old Believers listened to radio broadcasts, watched television, and were generally loyal to the Soviet state (Fund 1692. List 1. File 162. p. 7).

In the 1970s–1980s, the Novosibirsk Catholic community intensified its actions. Believers repeatedly raised the issue of registering the Catholic community. However, the community activists gradually moved away; thus, the association was not registered. At the end of 1970, the authorities registered the Catholic community of Novosibirsk with 16 members (Fund 1418. List 1. File 223. p.11). The Novosibirsk Catholic Church was registered only in 1982. By 1983, the number of members of the community significantly raised to 240 people. During this year, the community performed 30 baptisms and 16 weddings (Fund 1418. List 1. File 255. p. 21). Two Catholic communities functioned legally in Altai Krai. Catholic priest, I. A. Svidnitsky, rather often came to the community from Novosibirsk to conduct divine services and perform rites (State Archives of Altai Krai. Fund 1692. List 1. File 162. p. 12).

Protestant denominations also operated in the USSR in general and in Western Siberia in particular. As of January 1, 1979, over 3,000 Evangelical Christians-Baptists communities were identified, with a total number of 259,000 believers. More than 2,500 communities were members of the All-Union Council of Evangelical Christians-Baptists or supported it. The communities included about 237,000 believers of the evangelical direction (Evangelical Christians-Baptists, Pentecostals, and Mennonite Brethren). Most of them were located in the Ukrainian Soviet Socialist Republic (1209 members) and RSFSR (699 members) (Fast 2007).

By the mid-1970s, four communities and two branches of the All-Union Council of Evangelical Christians-Baptists were officially operating in Altai Krai. By 1977, the number of communities grew to seven (Fund 1692. List 1. File 162. pp. 147). As a result, the authorities tried to legalize the All-Union Council of Evangelical Christians-Baptists activities. The All-Union Council of Evangelical Christians-Baptists had the largest number of registered communities (Dvoryanchikova 2018). The number of believers in registered communities was 664, and 111 in unregistered. Those communities were located primarily in the cities Barnaul, Biysk, and Rubtsovsk. Barnaul had the largest community, which consisted of 274 members. In addition, 129 members lived in Blagoveshchenka village and 94—in Glyaden village. Apart from seven registered communities, there were another six that acted without registration. Despite the desire of some communities to act legally, local authorities still found reasons to refuse to register. Undoubtedly, there were other cases of registration denial in regional history. For example, that was the case with the Baptists of the Rodino village that operated since 1960. The community was registered by the District Executive Committee and then authorized in 1975 and defined as the Glyaden branch of the All-Union Council of Evangelical Christians-Baptists of the Blagoveshchensky District of Altai Krai. The community obtained its independent registration only in 1982 (Fund 1692. List 1. File 218. pp. 10).

In 1983, there had been a decrease in the All-Union Council of Evangelical Christians-Baptists community members. The initially small number of believers In the Aleksandrovka village dropped from 14 to 11 people. In Slavgorodskoye village, the number of believers decreased from 20 to 15 people; also, there was no priest. In Talmenka, the number slightly dropped from 34 to 32 people. The increase from 758 to 800 people occurred primarily due to German believers who lived in Glyaden and Blagoveshchenka villages—the number of believers in mentioned villages increased from 123 to 137 people and from 120 to 160 people, respectively. In spite of the registration process, regional authorities, however, implemented active atheistic propaganda. That propaganda was not as malign as it was during the Khrushchev Thaw. The major event in church life was the interregional meeting of Western Siberia clergies on January 26, 1985. The agenda included the following: (1) the condition of churches in Western Siberia, (2) lack of regional clergies meetings, (3) poor priest education, and (4) discussion of separate church-related issues (particularly in Altai Krai). Additionally, K. P. Borodinov was elected as the senior presbyter of Western Siberia at the meeting (Fund 1418. List 1. File 269. pp. 1).

Altai Krai also comprised 26 communities of the All-Union Council of Evangelical Christians-Baptists Churches. All communities operated without registration. In total, there were about 1242 followers in the region. The largest community of the Council in the Novosibirsk region was located in Novosibirsk—it included 190 believers. Followers gathered in believers' private residences. In 1981, the activity of the Council in the Novosibirsk region became more active—210 people gathered for the service on the Whitsun Day feast in the Novosibirsk community (Fund 1418. List 1. File 242. pp. 8).

Passing down the religious knowledge to children was a pivotal moment for the Council. Followers of the Council organized a camp for children in a private residence in the Kochenyovsky District of the Novosibirsk region. For several years they taught children religion. In 1984, the camp closed down, and the Novosibirsk community leaders K. K. Krekker, V. A. Orlov, and V. E. Zherebnenko were convicted and imprisoned. Nevertheless, young adults of the Council continued to gather there during the summer for recreation (Fund 1418. List 1. File 271. pp. 11).

It is crucial to mention that the Council followers actively interacted with believers from other regions. In 1981, the Novosibirsk Council follower was transporting religious literature from the Kazakh Soviet Socialist Republic and got detained on the railway in the Semipalatinsk region. However, Novosibirsk was a transshipment base for the literature distribution in Siberia and the Far East communities (Fund 1418. List 1. File 242. pp. 21).

Apart from the mentioned above Protestant communities in Altai Krai, there were also 16 Lutheran communities with 360 believers. None of the communities were registered. Ethnically, the believers were mostly Germans (Fund 1692. List 1. File 210. pp. 8–10). In 1978, there was one registered Lutheran community in the Novosibirsk region with 210 people (16% of men and 84% of women). In addition to that, seven communities operated illegally. According to the official figures, the total number of Lutherans in this region was about 380.

In 1977, three small Seventh-Day Adventist communities operated in Altai Krai. None of them operated legally. Seventh-Day Adventists opposed the registration for a long time. At the same time, there was no consensus on the number of communities. Three communities operated in the Novosibirsk region. In August 1977, the Berdsk Seventh-Day Adventist community that operated for a long time received official registration. The community was not large as well and consisted of only 19 believers. In 1978, the community held a large Spring Festival with the participation of 59 people (Fund 1418. List 1. File 225. pp. 4).

Some religious communities took a negative stance on registration. In 1977, there were 111 Pentecostal communities in Altai Krai. Their unique feature was the division into small groups, which gave communities several advantages, for example, mobility, closeness, and secrecy. Therefore, the control and suppression of their activities by the local authorities were difficult. In 1983, five Pentecostal communities operated illegally in Altai Krai and acted against the legalization of their activities. Only three Pentecostal communities operated illegally in the Novosibirsk region. The Novosibirsk Pentecostal community was the largest one—60–70 people attended the prayer house. There were quite remarkable data on the Pentecostal community of the Tatar district of the Novosibirsk region. The authorized Commissioner for Religious Affairs in Altai Krai noted that Pentecostals carried out missionary activities and recruited new adepts. In 1980, there was also a ten-member Pentecostal community in Bolotnoye. The presbyter of the community, F. R. Ilyin, was also the leader of the Pentecostal Association of Siberia and the Far East (Fund 1418. List 1. File 235. p. 16).

Another branch of Western Siberia Protestantism was represented by two Mennonite community trends: Mennonite Brethren and Church Mennonites. In 1977, there were about 216 believers in Altai Krai. Mennonite Brethren, with a total number of 174 people, operated in the Blagoveshchensky and Kulundinsky districts. Church Mennonite communities were located in the Slavgorodsky and Khabarsky districts of the region. Mennonite Brethren were close to the Council of Churches of Evangelical Christians-Baptists stance on religious legislation. Thus, several Mennonite communities with the Council for Religious Affairs recommendations were accounted as the Council of Church followers. It is essential to consider that in the second half of the 1980s, there were officially operating Mennonite Brethren communities in the villages of Nikolaevka, Protasovo, and Grishkovka. In the Novosibirsk region, three Mennonite communities also operated legally. In 1978, the Novosibirsk community consisted of 127 members, the Berdsk one—of 8 members. There were 53 members of the Neudachino village of the Novosibirsk region (Tatarsky district). All mentioned above religious communities consisted ethnically of Germans (Fund 1418. List 1. File 223. p. 9).

## 4 Discussion

It was significant that religious organizations began to engage in a more "open" dialogue with foreign churches and pastors, although government structures, including KGB, undoubtedly controlled that process. On April 2, 1983, Eparchy governer and Novosibirsk and Barnaul Archbishop Gedeon held a reception in honor of the United States ecumenical religious leaders. The delegation included a Baptist Church pastor, Dr. A. Haraszti, the consultant to Dr. Billy Graham's Evangelical community in Eastern Europe, and the Free Evangelical Church pastor, Blair Carlson. The foreign guests were accompanied by the Council representative from Moscow, I. M. Orlov. Nonetheless, religious leaders could not meet without the supervision of the Religious Affairs commissioner. Therefore, the Novosibirsk Region Council for Religious Affairs, A. S. Nikolaev, attended the event. The leaders and representatives of the Novosibirsk Catholic, Lutheran, Baptist, Mennonite communities were also invited to the reception. Religious leaders discussed (1) the problematic world issues, (2) the trip of Patriarch Pimento America, and (3) the necessity to end the arms race. The authorized Commissioner also spoke about the Novosibirsk development, constitutional rights, and confession of any religion. Billy Graham stated that Americans must change the opinion towards the Soviet people (Fund 1418. List 1. File 256. pp. 1–8).

## 5 Conclusion

With the adoption in 1977 of the new USSR Constitution, there was a gradual change in government-confessional relations, both in the USSR and Western Siberia. Since 1977 there had begun the process of legalizing the activities of communities. Seven communities and two branches of the All-Union Council of Evangelical Christians-Baptists officially operated in Altai Krai. One of the registered communities was the Mennonites. However, it received registration under the All-Union Council of Evangelical Christian Baptists' charter. More than that, the following communities were registered in Altai Krai: Orthodox—two, Catholic—one, Evangelical Christian Baptists—two, and Seventh Day Adventists—one. At the beginning of the 1980s, 28 religious organizations began to operate legally in the Novosibirsk region. The income of the Orthodox and Muslim communities and the number and education level of priests in the Orthodox Church parishes increased. Several communities of Lutherans, Seventh-Day Adventists and Mennonites of the Western Siberia south began to operate legally. At the same time, Pentecostal communities and the Council continued to operate in Western Siberia, opposed the official legalization, and both were not ready for constructive interaction with government power institutions.

**Acknowledgements** The work was supported by the Russian Foundation for Basic Research within the implementation of the grants "Influence of Ethnocultural and Political Processes on the Religious Situation in Siberia and Adjacent Regions of Mongolia in the 19th—Early 21st Centuries" (Project no. 20-59-44004) and "Socio-Cultural Adaptation of Orthodox and Non-Orthodox Communities in Southern and Western Siberia in the Context of Modernization Processes in Russia During the Imperial and Soviet Periods" (Project no. 20-39-70010).

## References

Borozdin SS, Bortnikova YA, Garifullin IB, Zinnatullina GI, Pachezhertsev NI, Silantyev RA, Yarkov AP (2015) Islam at the end of the world. The history of Islam in Western Siberia: the transformation of the ummah in the XIX–XXI centuries, vol 3. Tyumen State University, Moscow, Russia

Dashkovsky PK, Dvoryanchikova NS (2019) The influence of state-confessional politics on the situation of religious communities in Altai in the second half of the 1960s—early 1980s. In: Dashkovsky PK (ed) Ethnoreligious processes in the cross-border space of Western Siberia, Kazakhstan and Mongolia in the context of state-confessional policy in the XX – the beginning of the XXI centuries, vol 2. Altai State University, Barnaul, Russia, pp 165–178

Dashkovsky PK, Zibert NP (2015) Features of registration of religious communities in the Altai Territory in the mid-1950s—Early 1980s. Nations Relig Eurasia 8(8):230–241

Dvoryanchikova NS (2018) Activity of Protestant communities in the Altai territory in the mid-1970s–1980s. Izv Altai State Univ 5(103):35–39

Fast AA (2009) The Soviet state, religion, and church. 1917–1990. Documents and materials. IPP Altai, Barnaul, Russia

Gorbatov AV (2008) The state and religious organizations in Siberia in 1940-s–1960-s. Tomsk State Univ J, Tomsk, Russia

Klimenko EN (2007) The relationship between the state and religious associations in the Russian Federation: constitutional and legal aspects (Dissertation of Candidate of Law Sciences). Russian Academy of Justice, Moscow, Russia

Nikolskaya TK (2009) Russian Protestantism and state power 1905–1991. Publishing House of the European University, Saint Petersburg, Russia

Odintsov MI (2012) You will gain power when the Holy Spirit comes upon you... History of the Pentecostal churches in Russia in the 19th–20th centuries. Russian Religious Research Association, Saint-Petersburg, Russia

Presidium of the Supreme Soviet of the Russian SFSR (1975) Decree "on amendments and additions to the resolution of the All-Russian central administrative committee and quality estimation service of the RSFSR of April 8, 1929 "on religious associations" (June 23, 1975). SFSR, Moscow

State Archives of Altai Territory. Council for Religious Affairs under the Council of Ministers of the USSR for the Altai Territory (Fund 1692a, List 1, File 136)

State Archives of Altai Territory. Council for religious affairs under the council of ministers of the USSR for the Altai Territory (Fund 1692b, List 1, File 144)

State Archives of Altai Territory. Council for religious affairs under the council of ministers of the USSR for the Altai Territory (Fund 1692a, List 1, File 139)

State Archives of Altai Territory. Council for religious affairs under the council of ministers of the USSR for the Altai Territory (Fund 1692b, List 1, File 287)

State Archives of Altai Territory. Council for religious affairs under the council of ministers of the USSR for the Altai Territory (Fund 1692c, List 1, File 162)

State Archives of Altai Territory. Council for religious affairs under the council of ministers of the USSR for the Altai Territory (Fund 1692d, List 1, File 218)

State Archives of Altai Territory. Council for religious affairs under the council of ministers of the USSR for the Altai Territory (Fund 1692c, List 1, File 210)

State Archives of the Novosibirsk Region (1927–2000) Council for relations with religious and charitable organizations [Management documentation]. References, information on the activities of religious associations in the city of Berdsk, Bagansky, Barabinsky, Bolotninsky, Vengerovsky, Dovlensky, Iskitimsky, Karasuksky, Kochenevsky, Krasnozersky, Kuibyshevsky, Novosibirsk rural, Northern, Tatar, Toguchinsky, Chulymsky districts of the NSO (Fund 1418, List 1, File 252)

State Archives of the Novosibirsk Region (1977) Council for relations with religious and charitable organizations [Management documentation]. Financial statements of the Novosibirsk Diocesan Administration and the Ascension Cathedral of Novosibirsk for 1977 (Fund 1418, List 1, File 213, p. 14)

State Archives of the Novosibirsk Region. (1978a). Council for relations with religious and charitable organizations [Management documentation]. Information on the number of religious societies of German nationality operating in the Novosibirsk Region and on the contribution of funds by the Diocesan Administration to the Peace Fund (Fund 1418, List 1, File 223, p. 10)

State Archives of the Novosibirsk Region (1978b) Council for relations with religious and charitable organizations [Management documentation]. References, information about the activities of religious associations in the city of Berdsk, Bagansky, Barabinsky, Bolotninsky, Dovlensky, Iskitimsky, Karasuksky, Kochenevsky, Kochkovsky, Kuibyshevsky, Suzunsky, Ubinsky, Ust-Tarsky, Chulymsky districts of the Novosibirsk region for 1978b (Fund 1418, List 1, File 225, p. 60)

State Archives of the Novosibirsk Region (1979a) Council for relations with religious and charitable organizations [Management documentation]. References, information, correspondence about the activities of religious associations in Bagansky, Barabinsky, Dovolensky, Karasuksky, Kargatsky, Kolyvansky, Kochenevsky, Krasnozersky, Kuibyshevsky, Kyshtovsky, Maslyaninsky, Tatarsky, Toguchinsky, Ubinsky, Cherepanovsky, Chulymozerny regions of Novosibirsk region (Fund 1418, List 1, File 230, p. 88)

State Archives of the Novosibirsk Region (1979b) Council for relations with religious and charitable organizations [Management documentation]. Financial reports of St. Nicholas Church from Novolugovoye, Seventh-day Adventist Church in Novosibirsk, St. Nicholas Church in Bolotnoye, Novosibirsk Communities: Old Believers, Church Mennonites of Lutheran Confession for 1979b (Fund 1418, List 1, File 228, p. 21)

State Archives of the Novosibirsk Region (1980) Council for relations with religious and charitable organizations [Management documentation]. References, information, correspondence about the activities of religious associations in Bagansky, Barabinsky, Bolotninsky, Dovolensky, Iskitimsky, Kolyvansky, Kochenevsky, Kochkovsky, Maslyaninsky, Moshkovsky, Novosibirsk rural, Tatarsky, Toguchinsky, Ubinsky, Ust-Tarsky, Chistoozerny regions of the Novosibirsk region 1980 (Fund 1418, List 1, File 235, p. 53)

State Archives of the Novosibirsk Region (1981) Council for relations with religious and charitable organizations [Management documentation]. References, information on the activities in the city of Novosibirsk and the Novosibirsk region of religious associations of supporters of the Council of Churches of Evangelical Christians-Baptists and conclusions on the distributed literature (Fund 1418, List 1, File 242, p. 51)

State Archives of the Novosibirsk Region (1982) Council for relations with religious and charitable organizations [Management documentation]. References, information, correspondence about the activities of religious associations in Dzerzhinsky, Zheleznodorozhny, Zaeltsovsky, Kalininsky, Leninsky, Pervomaisky, Sovetsky, Central districts of Novosibirsk (Fund 1418, List 1, File 251, p. 60)

State Archives of the Novosibirsk Region (1983a) Council for relations with religious and charitable organizations [Management documentation]. Financial reports of St. Nicholas Church from Novolugovoye, Novosibirsk communities: Lutheran, Church Mennonites, Muslim, Catholic, Old Believers for 1983a (Fund 1418, List 1, File 255, p. 29)

State Archives of the Novosibirsk Region (1983b) Council for relations with religious and charitable organizations [Management documentation]. Reports on the reception in the Novosibirsk diocese in honor of religious leaders from the USA, the trip of Archbishop Gideon to the GDR, his Christmas message to God-loving lyricists, honest monastics and laity of the Novosibirsk diocese (Fund 1418, List 1, File 256, p. 17)

State Archives of the Novosibirsk Region (1985a) Council for relations with religious and charitable organizations [Management documentation]. Minutes of the interregional meeting of clergymen of Western Siberia, meetings of members of the Presidium of the regional peace committee with the clergy. Report on the meeting of the Commissioner: and the heads of religious associations in Novosibirsk and the region with representatives of the religious communities of West Berlin (Fund 1418, List 1, File 269, p. 35)

State Archives of the Novosibirsk Region (1985b) Council for relations with religious and charitable organizations [Management documentation]. References, information, correspondence on the activities of religious associations in Bolotninsky, Iskitimsky, Kochenevsky, Tatar, Chistoozerny districts of the region (Fund 1418, List 1, File 271, p. 23)

USSR (1977) The constitution (fundamental law) of the Union of Soviet Socialist Republics (October 7, 1977). USSR, Moscow

Vasileva OY (2001) The Russian orthodox church in the politics of the Soviet State in 1943–1948. RAS, Moscow, Russia

Zherebyatieva EE (2009) The legal status of protestant organizations and atheistic work in the Altai territory (1945–1980). Izv Altai State Univ 4–1:79–82

# The Image of Russia and Intercultural Interactions in Views of the Population in Western and Central Mongolia

**Petr K. Dashkovskiy, Elena A. Shershneva, and Navaanzoch Cedav**

**Abstract** The paper focuses on the results of ethno-sociological research carried out in Mongolia within the international grant implementation. Scholars from Altai State University, Khovd State University, and National University of Mongolia aimed to study a set of cross-cultural communication aspects among the population of Western and Central Mongolia. Despite the ethnic tolerance of the population of Mongolia, many respondents noted that it is the integration with representatives of other ethnic groups that can lead to the loss of ethnic uniqueness of Mongolia society. Almost 60% of the respondents informed that they have no difficulty communicating with the Russian ethnos representatives. At the same time, more than 80% surveyed emphasized that communication with the Russian people evokes mostly positive emotions. During the study, it was found that in general, the population of Mongolia has a positive image of Russia, which builds good-neighborly relations with different countries, including Mongolia. Moreover, it is noted that to a great extent, the positive image of Russia is associated with a long relations history between the two countries, especially during the Soviet period. In addition, some respondents of the younger generation, particularly those between the ages of 22 and 35, indicated that they had difficulty determining their attitude towards interaction with the Russian people.

**Keywords** Intercultural communication · Mongolia · Russia · Inter-ethnic relations · Tolerance · Cooperation · Image of the country

P. K. Dashkovskiy (✉) · E. A. Shershneva
Altai State University, Barnaul, Russia
e-mail: dashkovskiy@fpn.asu.ru; D2703@yandex.ru

N. Cedav
National University of Mongolia, Ikh Ulaanbaatar, Mongolia



S. G. Maximova et al. (eds.), *Advances in Natural, Human-Made, and Coupled Human-Natural Systems Research*, Lecture Notes in Networks and Systems 234,
https://doi.org/10.1007/978-3-030-75483-9_75

## 1 Introduction

The long-lasting neighborhood and cooperation of Russia and Mongolia have led to the formation of a particular image among the population of Mongolia about their northern neighbors. The changes taking place in the economic, political, and cultural life of Mongolian people since the 1990s have interest for political figures and scholars. For more than ten years, the Altai State University, the Khovda State University, and the National University of Mongolia with P. K. Dashkovskiy, N. Cedav, B. Nyamdorzha have been studying ethnic, religious, and social processes in different parts of Mongolia within the implementation of international Russian-Mongolian grants. Some results of these studies have already been partially put into scientific discourse (Dashkovskiy 2015; Dashkovskiy and Cedav 2017; Munkhbat 2018; Tsogzolmaa 2014; Cedav 2017). The new study aims to conduct sociological surveys of the population in Central and Western Mongolia in order to establish the attitude of Mongolian people towards representatives of various ethnic and religious communities and assess the attitude of Mongolian people towards Russia as one of the strategic partners.

## 2 Materials and Methods

In developing the questionnaire for sociological research, the authors relied on methodological developments of Maksimova (2012, pp. 57–93) and Drobizheva (2013, pp. 446–471). In 2016, as part of the study in Mongolia, the authors interviewed 164 people. The respondents surveyed during the research were divided into four age groups: (1) 16–21 years old, (2) 22–35 years old, (3) 36–45 years old, and (4) 46–60 years old. Simultaneously, most of the respondents were residents of regional centers (94 people, 57%) and small villages (31 people, 19%). Such age distribution came from the task to look at the potential perception features of other culture members in different age groups of the Mongolian population.

The increased interest in the processes of national identification made it possible to assess the ethnic composition of the respondents. In recent years, the desire of Mongolian people to preserve their national uniqueness can be illustrated by their ethnic self-identification. The majority of people clearly identify their ethnic group, emphasizing a specific subethnic group, which is demonstrated by Fig. 1.

## 3 Results

The Mongolian People's Republic entered a new stage of political and economic development in the 1990s. It was accompanied by (1) the democratization of society, (2) the revision of foreign policy guidelines, and (3) the search for new strategic

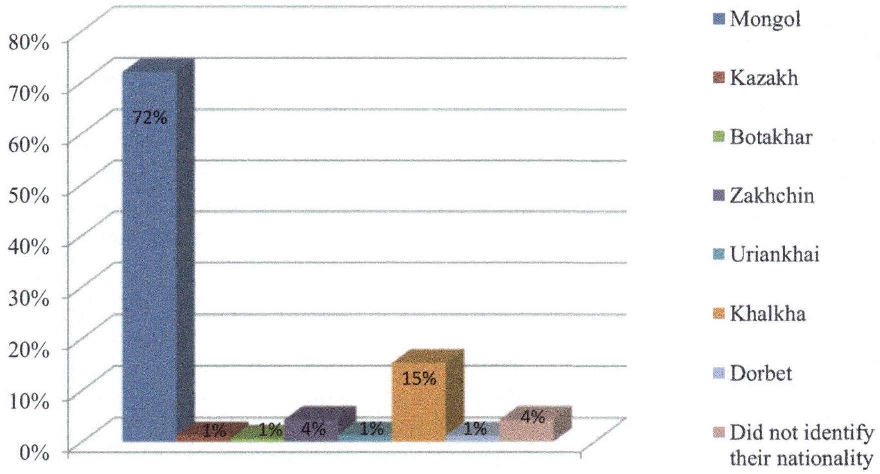


**Fig. 1** The ethnicity of the population of Central and Western Mongolia. *Source* Compiled by the authors

partners. However, considering a certain extent of multiethnic and multireligious diversity in Mongolia, the authors pointed out the particular importance of interethnic relations in the fundamental modernization of the country. In the course of the study, the authors asked the respondents if they could easily find contact with representatives of other nationalities. Figure 2 presents the results of the responses.

Figure 2 indicates that (1) 46% of the respondents can easily find contact with representatives of another ethnic group, (2) 30% of the respondents indicate that

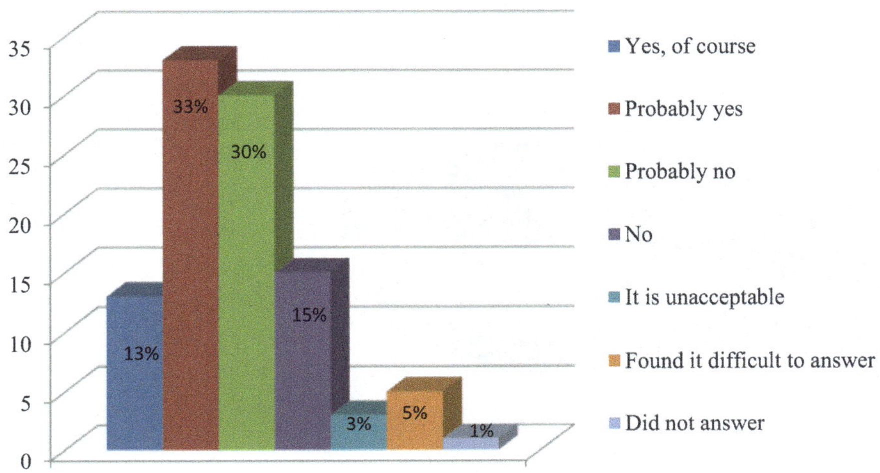


**Fig. 2** Answers of the respondents from Mongolia to whether they can easily find contact with representatives of different nationalities. *Source* Compiled by the authors

it is probably difficult for them to do this, (3) 15% of the respondents unequivocally respond that it was difficult for them, and (4) 3% of the respondents noted that, in general, it is not acceptable for them to communicate with representatives of other ethnic groups. The relatively low percentage of respondents' positive responses regarding their readiness to interact with various ethnic group representatives is probably associated with the following circumstances. According to the opinion of a significant part of the population of the country, it is the integration with representatives of different cultural groups that can lead to the loss of the ethnic uniqueness of Mongolian society.

The attitude towards national identity preservation is also observed when answering the question about different nationality representatives in a family. The results were the following: (1) 9% of the respondents (14 people) responded positively to this question; (2) 63% (104 people) indicated that there were no representatives of another ethnic group in their family, (3) 28% (46 people) of the respondents found it difficult to answer the question asked. At the same time, 46% of the respondents (75 people) could not agree to marry a representative of a different nationality, 22% (36 people) found it difficult to answer, and only 32% surveyed (53 people) could not mind seeing a representative of a different cultural tradition in their family.

In the process of building interethnic relations, special attention was paid to specific ethnic groups with which relations were most friendly. According to the conducted survey, the most attractive ethnic group for the Mongolian population were Russians, with whom there were the least of contradictions. More than half of the respondents (59%) confirmed that they had no difficulties in relations with the Russian ethnos representatives. It is also peculiar that 12–16% of respondents indicated that they easily contacted the Kazakhs, Tuvans, and Chinese (peoples who also live like Russians in the regions bordering on Mongolia). Figure 3 reflects the mentioned indication more clearly.

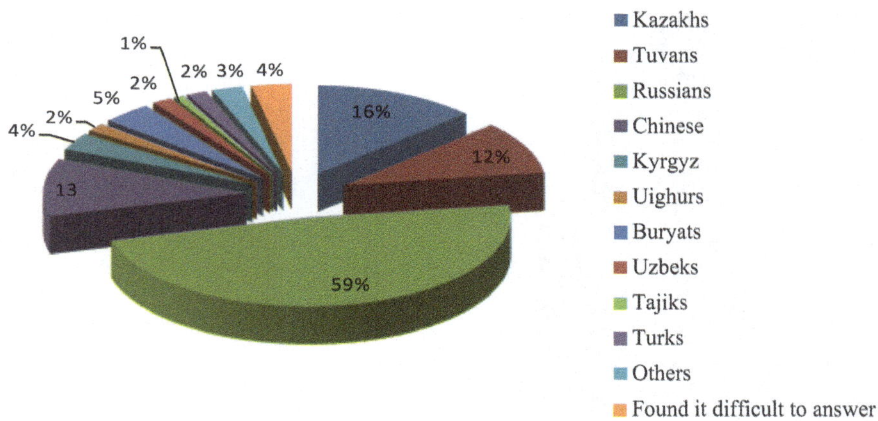


**Fig. 3** Peoples with whom the respondents from Mongolia experience no difficulties in relations. *Source* Compiled by the authors

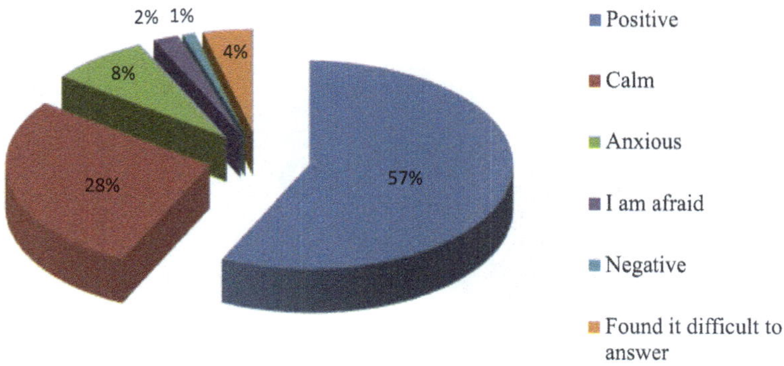


**Fig. 4** Answers of the respondents from Mongolia to how they feel when communicating with the Russian people. *Source* Compiled by the authors

In that case, it was necessary to pay attention to the fact that to clarify the degree of the respondents' positive or negative attitude to the Russian ethnos; the authors asked the question about the feelings the respondents have during communication with Russian people. Figure 4 presents the results of the responses.

Figure 4 depicted that, in fact, 85% of the respondents noted that communication with the Russian people evokes a positive and calm feeling in them, and only 11% of the respondents expressed anxiety or fear concerning communication with representatives of the Russian ethnos. The authors emphasized that many researchers noted a positive attitude towards the Russian population in Mongolia even at the beginning of the twentieth century. That stance was present primarily due to the lack of territorial claims of Russia to Mongolia and the provision of political and economic support (Boykova 2014). During the Soviet period, the USSR provided significant economic assistance and political support. In general, it had a positive effect on the image of the country. At the present stage, a friendly attitude towards the Russian people in Mongolia is primarily associated with the new policy of Russia, which aims at supporting its compatriots abroad. Also, the alignment of a new political agenda in Russian-Mongolian diplomatic relations since the beginning of the twentieth century consisting in mutually beneficial cooperation, has become noticeable. Therefore, the visit of the President of the Russian Federation to the Mongolian People's Republic in 2000 was of great importance. An agreement on receiving the right by the Russian-speaking population to purchase housing in Mongolia was reached during this visit. Additionally, the restoration of the parish of the Russian Orthodox Church began in the capital of Mongolia, Ulaanbaatar (Mikhalev 2012). Simultaneously, there is a trade increase, arrangement and reequipment of customs terminals, and activities aimed at acquaintance with the Russian and Mongolian cultures in both countries.

During the study, the authors pay special attention to the attitude of the Mongolian population to the population of the Russian border regions (Buryatia, Tuva, Altai, and the Irkutsk region). Figure 5 presents the survey results on the mentioned issue.

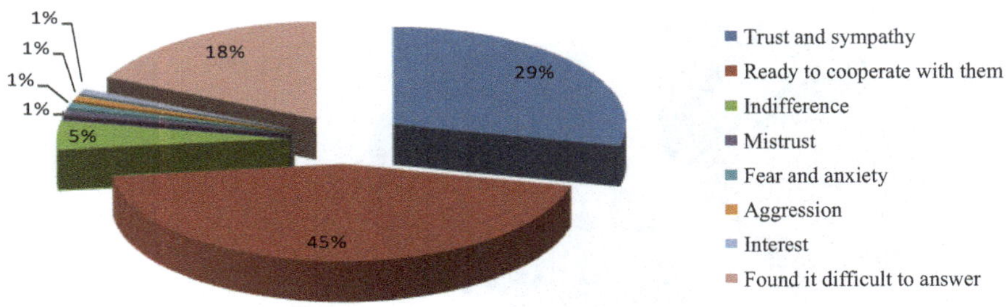


**Fig. 5** The attitude of the respondents of Mongolia to the population of the Russian border regions. *Source* Compiled by the authors

Figure 5 shows that most respondents feel trust and fondness for their Russian neighbors and are ready to cooperate with them. Although, 18% of the respondents found it difficult to answer the question. A considerable percentage of the respondents who were not ready to clearly define their position in relation to their Russian neighbors were noted among Western Mongolia residents aged from 22 to 35 years. Among the 29 people interviewed in this age group, eight people (28%) were undecided about their opinion on this issue.

Unique relations have developed throughout the long history of cross-border cooperation of Mongolia with Russia. Particular links have been established on the borders of Mongolia and Buryatia. In the post-Soviet period, contact with Mongolia in the border zone is more complicated and sometimes acquires an illegal character (Nanzatov and Sodnompilova 2012). Also, demographic and economic asymmetries are observed in the border areas. In accordance with some researchers, it is the shared history and low living standards on both sides of the border that unite the Buryat and Mongolian populations (Bazarova and Batbuyan 2013).

Generally, despite arising difficulties and misunderstandings in the border areas, the population of Russia and Mongolia develops fairly balanced relations. The peoples inhabiting the border area see each other as self-sufficient ethnic groups with their own culture and traditions (Oidup and Kylgydai 2012). The policy pursued by both countries also contributes to establishing such neighborly relations (Bazarov 2016). In this respect, the Treaty "On the Regime of the Mongolia-Russia State Border" between Russia and Mongolia signed in 2006 is of great importance. The Treaty makes it possible to build cross-border cooperation on the current legal basis. In the subsequent period, the constituent entities of the Russian Federation located in the Siberian Federal District and the regional authorities of the Mongolian leagues signed additional agreements (Rodionov 2009). Currently, a visa-free regime when visiting these countries has been introduced for citizens of Russia and Mongolia.

It should be emphasized that at the present stage, cooperation with Russia is multifaceted. Meanwhile, cooperation is actively developing in various Russian and Mongolian border regions (Kimeev et al. 2012). However, historically speaking, there

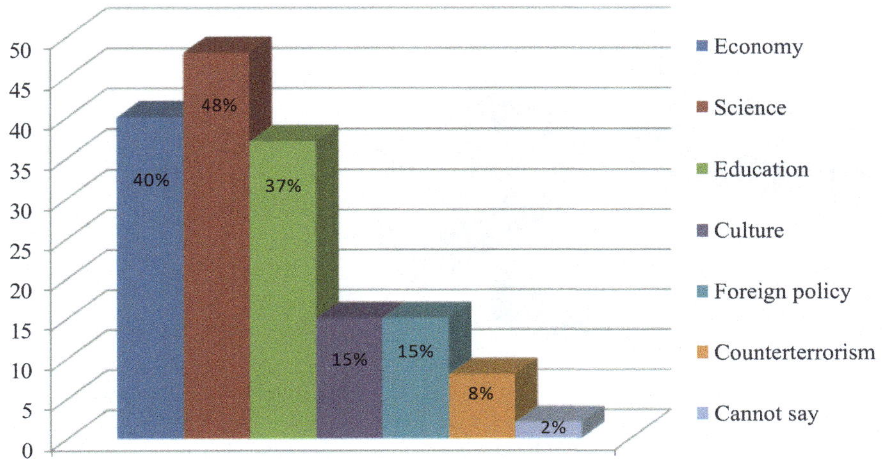


**Fig. 6** Areas in which relations with Russia should be developed according to the respondents from Mongolia. *Source* Compiled by the authors

has always been a predominance of contacts in the economic sphere. The respondents were also asked to determine in which area they saw the most promising cooperation with Russia. Figure 6 demonstrates the results of the answers.

Figure 6 depicts that most of the respondents see cooperation with Russia in three areas: economy, science, and education. Since the Soviet period, the USSR and then the Russian Federation have been the primary economic partners of Mongolia. Those close contacts were caused by similar economic systems and the allied political relations between Russia and Mongolia (Rodionov 2009).

The authors pointed out that among the cooperation spheres, the respondents listed science in the first place. That position was expressed by 48% of surveyed. It was due to the fact that it was in the Soviet years that scientific and cultural cooperation between the two countries was a channel for strengthening allied relations. Russia contributed to creating a scientific and technical base and an education system in Mongolia. However, since the late 1980s–early 1990s (twentieth century), a Russian-Mongolian relations crisis had caused a negative attitude towards Russian influence on Mongolian culture. One of the slogans for national identity preservation was the idea of returning to the Mongolian script (Rodionov 2009). A revival of warm, neighborly relations began only in the 2000s, including in the field of cultural and scientific interaction. Researchers accentuate the importance of learning the Russian language for the further economic development of the region (Badmaev 2011).

Within the study, the authors asked the respondents to rate the level of Russian-Mongolian relations. Figure 7 shows the results of the responses to this issue.

Figure 7 demonstrates that a significant part of the respondents (72%) assessed the level of Russian-Mongolian relations as decent, high, and very high.

The image of Russia, developed among the Mongolian population over the long history of cooperation between the countries, facilitates this kind of relationship. To

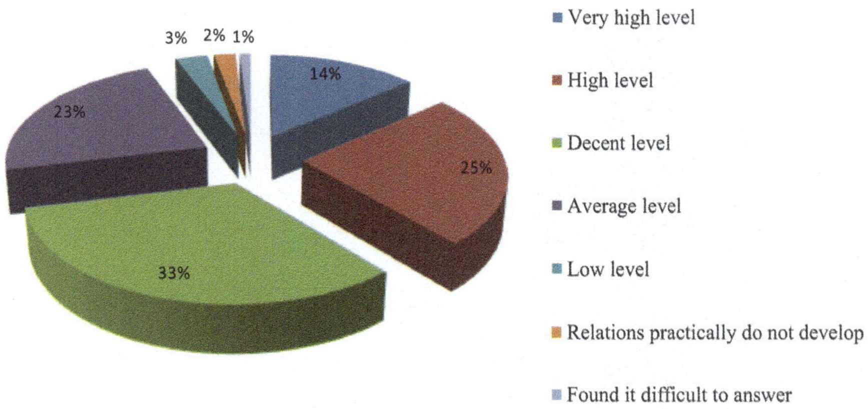


**Fig. 7** The current level of Russian-Mongolian relations. *Source* Compiled by the authors

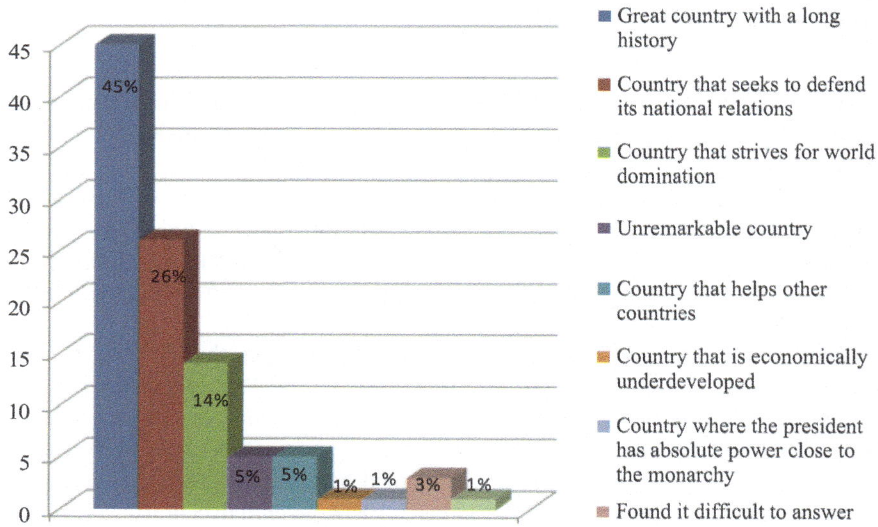


**Fig. 8** Image of Russia among the Mongolian population. *Source* Compiled by the authors

detail the image of Russia, the authors asked the respondents to additionally answer the question about the associations that Russia evoked among them.

Figure 8 shows that the population of Mongolia sees Russia as a relatively strong partner. Furthermore, 45% of surveyed consider Russia a great country with a long history. Since Mongolia has close relations with its northern neighbors, the authors asked the respondents about the Russian influence on the development of Mongolia. Figure 9 presents the results of the responses to this question.

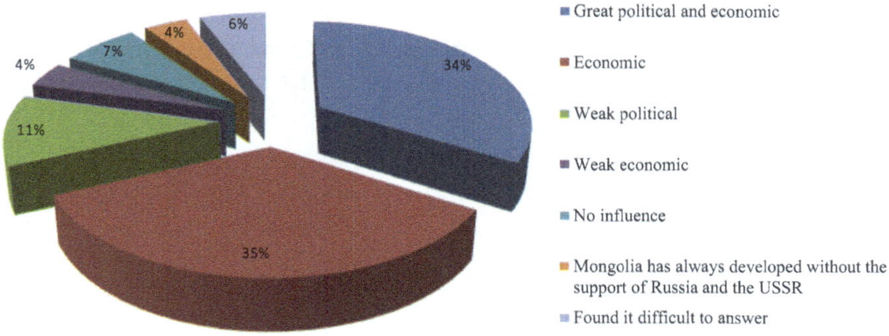


**Fig. 9** Respondents' answers to the question about the influence on the development of Mongolia that Russia had in the 20th and early 21st centuries. *Source* Compiled by the authors

## 4 Discussion

The study shows that 34% of the respondents (55 people) believe that Russia has a tremendous economic and political influence on Mongolia. Meanwhile, 35% (57 people) believe that Russia has only an economic impact on their country. At the same time, 51% of surveyed Western Mongolia residents (37 people) notice Russian influence on both the political and economic life of Mongolia. In addition, 47% of the respondents from Central Mongolia (43 people) believe that Russia has only an economic impact.

At present, Russia remains the leading strategic partner for Mongolia (Badmaev 2011). It is confirmed by the leadership policy in both countries and the data of the survey. Figure 10 reflects the results of the responses to the question about strategic partners of Mongolia.

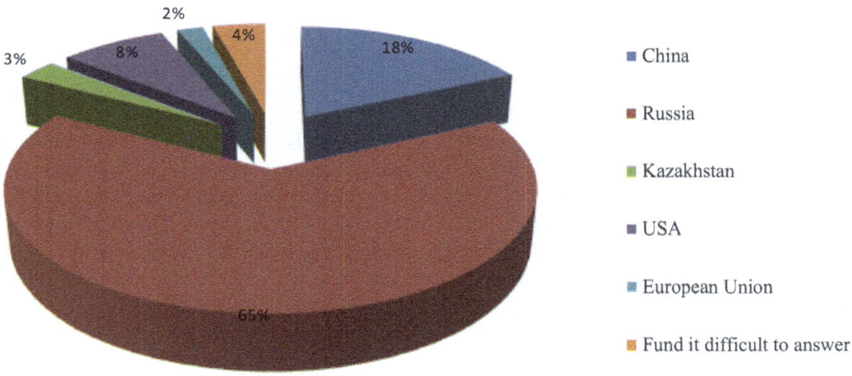


**Fig. 10** Countries which the respondents from Mongolia named as the leading strategic partner of their country. *Source* Compiled by the authors

Simultaneously, the authors consider that 35% of the respondents aged from 22 to 35 in Western Mongolia consider both Russia and China as the primary strategic partners. Older people, especially those who are over 61, still consider Russia as their primary ally.

## 5 Conclusion

Consequently, a rather respectful and warm attitude towards Russia and Russian people has developed in the view of the Mongolian population employing the close contacts between Russia and Mongolia over a long historical period. According to the opinion of Mongolian people, the Russian Federation can remain the primary strategic partner of their country. The developing diplomatic relations contribute to the formation of the image of Russia in the perception of the Mongolian population as a strong neighbor capable of coming to the rescue. Simultaneously, a positive moment in such relations for the Mongolian population is that Russia does not seek to change and even influence their national identity.

**Acknowledgements** The paper was prepared within the implementation of the grant of the Russian Foundation for Basic Research, and the Ministry of Culture, Education, Science, and Sports of Mongolia on the topic *Ethno-religious and social processes as factors of intercultural interaction of the population in the Altai, Tuva, and Mongolia transboundary space: History and modernity* (project No. 19-59-44002).

## References

Badmaev AZ (2011) The image of Russia and the Russian language in Mongolia: from the Soviet legacy to the present. Bull Buryat State Univ 8:127–134

Bazarov VB (2016) Russian-Mongolian relations in the post-Soviet period: cooperation deepening. Vlast 6:81–85

Bazarova AG, Batbuyan B (2013) The standard of living of adjacent border areas of Buryatia and Mongolia. Geogr Nat Resour 2:132–138

Boykova EV (2014) Russian military researchers of Mongolia (the second half of the 19th–the early 20th centuries). Institute of Oriental Studies of Russian Academy of Sciences, Moscow, Russia

Cedav N (2017) Some problems of studying the ethno-confessional situation in Mongolia. Nations Relig Eurasia 3–4(12–13):128–134

Dashkovskiy P (2015) Ethnic and religious processes in Western Mongolia (based on social research). Procedia—Social Behav Sci, 185:109–116

Dashkovskiy PK, Cedav N (2017) Ethno-confessional processes in the Mongolian Altai. In: Dashkovskiy PK (ed) The religious landscape of Western Siberia and adjacent regions of Central Asia, vol 3. Altai State University, Barnaul, Russia, pp 122–134

Drobizheva LM (ed) (2013) Civil, ethnic, and religious identity: yesterday, today, tomorrow. ROSSPEN, Moscow, Russia

Kimeev VM, Terentev VI, Akulova AS, Shirin YV, Lkhagvasuren B (2012) Essays on Western Mongolia. Kemerovo State University, Kemerovo, Russia

Maksimova SG (ed) (2012) Social security—social risks: theoretical and empirical models of deviation reproduction. Altai State University, Barnaul, Russia

Mikhalev AV (2012) "Russian question" in the system of foreign policy relations of Mongolia in the 20th century. Vlast 7:161–164

Munkhbat D (2018) Pilgrimage of the Mongols: tradition and modernity. Nations Relig Eurasia 2(15):79–84

Nanzatov BZ, Sodnompilova MM (2012) Border zone: actual problems of everyday life (on the example of border regions of Buryatia). Vlast 3:165–168

Oidup TM, Kylgydai AC (2012) Features of interethnic links of the population in the Tuva-Mongolian border area. Sociol Stud 6:136–140

Rodionov VA (2009) Russia and Mongolia: a new model of relations at the beginning of the 20th century. Buryatia Scientific Center of the Siberian Branch of Russian Academy of Sciences, Ulan-Ude, Russia

Tsogzolmaa N (2018) The spread of new religious movements and their impact on the education system of Mongolia. Nations Relig Eurasia 1(14):117–124

# State Regulation of the Process of Creating Muslim Parishes of Western Siberia

**Petr K. Dashkovskiy, Elena A. Shershneva, and Natalia S. Dvoryanchikova**

**Abstract** The paper focuses on the problem of state regulation of the foundation process of Islamic communities on Western Siberia territory since the latter half of the 19th to the beginning of the twentieth century. Analysis of archive materials identifies the main problems the Muslim population of Western Siberia faced in constructing religious structures and creating independent communities. The conducted historical and religious research contributed to the analysis of the principles of building state-confessional policy towards the Muslim population of the country. The introduction of previously unpublished archive materials, as well as the analysis of legislative sources, appraises the role of the state and its institutions in the formation of Islamic communities on the territory of Western Siberia since the latter half of the 19th to the beginning of the twentieth century. At the same time, the identification of several problems faced by the Muslim population when opening communities is now possible. First of all, it is the financial insolvency of Muslim communities. As part of the policy of the country, the construction of the religious building and its further maintenance was to be carried out at the expense of the Islamic people. Not every Muslim community could whip around the required sum at the initial stage for the construction of the mosque. Second, even the moneyness did not guarantee the community to obtain permission to build a cult facility. Based on the legislation of the Russian Empire, a restriction was introduced on the number of community members. Thus, the community could not be organized and receive official status if it had less than 200 male people. As a result, Moslems were completely subordinate to the provincial authorities. Besides, Orenburg Magometan Spiritual Assembly did not contribute to the construction of other communities, the Empire territory.

**Keywords** Islam · Muslims · Mosques · Muslim communities · Siberia · Russian empire

P. K. Dashkovskiy (✉) · E. A. Shershneva · N. S. Dvoryanchikova
Altai State University, Barnaul, Russia
e-mail: dashkovskiy@fpn.asu.ru



S. G. Maximova et al. (eds.), *Advances in Natural, Human-Made, and Coupled Human-Natural Systems Research*, Lecture Notes in Networks and Systems 234,
https://doi.org/10.1007/978-3-030-75483-9_76

## 1 Introduction

According to Muslim traditions, the mosque is a particularly respected place where believers can communicate with God and solve the most important issues in the life of the community. Muslims never spare money for constructing and maintaining religious buildings. Following the established tradition, Muslims living on the territory of the Russian Empire have always strived to create independent communities that state authorities could recognize. Within the legislation of the Russian Empire, constructing a mosque from a purely personal matter of the community has become an event regulated and controlled by the institutions of state power. Since the eighteenth century, the norms for constructing religious buildings for Muslims have acquired a legislative nature (Zagidullin 2007).

## 2 Materials and Methods

Archival materials and legal acts dated from the second half of the nineteenth century to the beginning of the twentieth century allow analyzing the role of state confessional policy that influenced the formation of Muslim communities in Western Siberia at that time. Archival sources contain circular orders and petitions of Muslims addressed to the mufti of the Orenburg Muslim Spiritual Assembly (OMSA) and various provincial and government bodies (Russian State Historical Archive (CSHA), Central State Historical Archive of the Republic of Bashkortostan (CSHA RB), Historical Archive of the Omsk Region (HA OR)). Legislative acts regulating the norms and principles for constructing and organizing Muslim communities of the Russian Empire are of particular importance. As part of preparing this paper, we used the CSHA, CSHA RB, and HA OR materials (CSHA funds 821, 1265 & 1286; CSHA RB fund 295; HA OR fund 67). The principles of historicism used in the paper allowed examining the position of non-Orthodox religious communities through the prism of state confessional policy towards Islam in the Russian Empire. We also used problem-chronological and retrospective methods. The problem-chronological method made it possible to trace the principles of state policy related to the formation of Muslim communities on the territory of the Russian Empire. Moreover, this method allowed establishing the features of legal regulation in organizing Muslim communities. The retrospective method provided an opportunity to identify the reasons for the increased activity of Muslims in constructing religious buildings.

## 3 Results

The state, aimed to control the construction of Muslim religious buildings, formalized the construction norms in the legislation and assigned the responsibility for monitoring the construction to the OMSA. This institution, established in 1788, monitored Muslims who lived on the territory of the Russian Empire. As a state institution, it was obliged to monitor the construction of religious buildings and the repair of already existing similar facilities. At the same time, it could not resolve these issues alone and had to coordinate its decisions with the regional authorities (Monich and Manonina 2015).

Constructing religious buildings was especially acute in the regions with a pronounced poly-confessional composition of the population. For example, Muslims have repeatedly faced the problem of obtaining permission to build a mosque in Western Siberia. Despite the fact that the OMSA was the central spiritual body of Muslims, it did not contribute to solving this problem. On the contrary, the cases when the religious authorities supported the provincial authorities and interfered in constructing the mosque were quite common. Insufficient material resources allowing building a mosque and maintaining a religious building and clergy were often pointed as the reasons for refusal (CSHA RB fund 295 list 3 file 5094 pages 1–2; CSHA fund 821 list 8 file 677 pages 10–17).

Russian legislation of that period fully regulated the procedure for obtaining permission to build a mosque. The decision to construct a religious building could be made at the general meeting, where two-third of the male population had to vote to form a community. In addition, after the Muslim men substantiated their decision and the readiness to maintain a mosque and clergy, they sent the Public sentence to the OMSA. After the OMSA gave a positive decision on the petition submitted, it was sent for approval to the provincial government. The provincial authorities also approved the mosque project, which should not contradict the norms approved by the legislation of the Russian Empire (Starostin 2019; Zagidullin 2005).

We should note that the standard mosque project was developed and legislated in the Russian Empire in the first half of the nineteenth century (Bakiev 2008; Full collection of laws of the Russian Empire 1830). However, the Muslims who fully assumed the responsibility of establishing and maintaining religious knowledge felt the need to satisfy their confessional architecture ideas. The Decree issued in 1862 and allowing Muslims to construct mosques according to their designs approved by the construction department of the provincial government satisfied this need (Zagidullin 2005; Pirozhkova 2010).

The lack of mosques on the territory of Western Siberia became especially acute in the second half of the nineteenth century. In Siberia, the number of the Muslim population in need of worship increased due to the state reforms performed in this period. However, the approved standards for constructing mosques did not contribute to constructing religious buildings in every settlement where the Muslim population lived (Azamatov 1999; Arapov 2001). Thus, by the beginning of the twentieth century, there were 8,539 Muslim men living on the territory of the Tomsk province and only 48 religious mosques (CSHA RB fund 295 list 2 file 8 page 835 back side 884). This ratio made it possible to judge the full observance of the norms of Russian legislation in the field of religious construction on the territory of the Tomsk province. Nevertheless, the location of the religious facilities in the province was uneven.

We should note that the introduction of legislative norms regulating the number of people in communities led to the remoteness of mosques from the places of residence of Muslims in Western Siberia. In this regard, in the second half of the nineteenth century, Muslims began to request separating communities from the existing ones (CSHA RB fund 295 list 3 file 6021; CSHA RB fund 295 list 3 file 6930 page 12 back side 12). They asked for creating new communities and dividing the existing ones (CSHA RB fund 295 list 6 file 585 pages 1–11; CSHA RB fund 295 list 6 file 1849 pages 1, 8). Such requests were usually due to the remoteness of mosques or prayer houses from the place of residence of a community. Besides, such an arrangement of religious buildings made it difficult to perform religious rites. Despite the evidence of this problem, the process of obtaining permission to build a mosque or organize a new community was rather complicated and lengthy. Such appeals required clarification of the reasons, confirmation of the impossibility of visiting an existing mosque, and clarification of the fact of non-violation of the law. This approach to resolving the issue was since the norm established by law on the presence of at least 200 men who needed a community was violated in some cases of separating communities (CSHA fund 821 list 8 file 678 pages 1–9).

We should emphasize that in addition to these difficulties, the Muslim population faced one more problem when constructing mosques. Particularly, it was the most optimal inclusion of a religious building in the urban space. The Building Code established norms for constructing religious buildings of different faiths (Zagidullin 2005; Bernyukevich 2018). Thus, the erected mosque should not have attracted attention. Muslims tried to consider this when applying for permission to build a mosque, indicating that this object would not harm the Orthodox population (Starostin 2019).

The financial viability of the community played an essential role in constructing religious buildings. All community members tried to participate in the construction of a mosque since they considered it a particularly charitable deed. Besides, constructing and maintaining religious buildings of the Muslim faith could be performed only at the expense of the community. This fact found its confirmation in constructing a stone mosque in Tomsk (Markova 2008). At the same time, we should note that not every community could begin its building even with permission received. In these cases, the wealthiest Muslims took all the building costs. For instance, in 1859, a Muslim from Central Asia who lived in Petropavlovsk submitted a petition to the OMSA to allow building a prayer house because his fellow believers, having received permission to build a mosque, could not exercise this right due to the financial lack. However, this process was not a simple community decision. It was accompanied by a long correspondence between the provincial authorities and the OMSA, which ended with a resolution by the Governor-General of Western Siberia to prohibit constructing a prayer house. This decision was explained by the fact that Muslims received the right to build a mosque but not a prayer house. In this regard, it was indicated that constructing a religious facility of a different type would postpone constructing a mosque for an indefinite period (CSHA RB fund 295 list 3 file 3270). The Muslims of the Siberian Linear Cossack Army faced a similar problem as they were refused to begin constructing a mosque due to the lack of the necessary amount of money for its construction (HA OR fund 67 list 1 file 729 page 59–61 back side 61).

We should emphasize that the provincial authorities monitored the financial solvency of the community very closely and often hindered the construction of a mosque, motivating their refusal with a lack of funds. Charitable work did not always solve the financial problem. Thus, in 1905, a petition was sent to the Department of Religious Affairs of Foreign Confessions to obtain permission to collect donations for constructing a mosque and a school in Barnaul. The Barnaul community had a prayer house and a school acquired on credit. In this regard, the Tomsk Governor refused permission to collect donations for constructing a mosque, as the existing building, in his opinion, satisfied the needs of the Muslim community members. The debt of the community for the building purchased earlier testified to its financial insolvency (CSHA fund 821 list 8 file 678 pages 14–17).

In some cases, the collection of the required amount for construction was performed by leasing land plots (Gibadullina 2009). Mosques also had their waqfs (i.e., the immovable and movable property received from Muslims as donations that went to the needs of the Muslim community). The independent maintenance of mosques led to many disputes among Muslims. In 1874, a telegram was sent to the Department of Religious Affairs of Foreign Confessions from a Kirghiz (Kazakh) living in Omsk. It contained information that he had built a mosque in the Bababagashevskaya county. However, after its construction, a conflict arose with relatives over the plot of land on which it was built. The military governor of the Akmola region issued an order to destroy the mosque. Due to the termination of the dispute in 1875, the order of the military governor was asked to be left unfulfilled until the arrival of N. G. Kaznakov, the West Siberian Governor-General (CSHA fund 821list 8 file 677 pages 1–9).

Russian legislation regulated the construction of mosques and the position of previously erected religious facilities. Within the Russian legal framework, a system of office work was developed for regulating the requests of Muslims for constructing and repairing religious buildings. Nevertheless, no funds were allocated for repairing religious buildings, even those previously erected at the initiative of the Russian government. That is why the mosque, erected in 1827 in Omsk, as well as the embassy house, constructed in the second half of the nineteenth century, were decided to transfer to the possession of the local Muslim community for its full maintenance. The provincial authorities explained this decision by the fact that the treasury did not include funds for repairing these buildings. In addition, they noted that the state needed them no longer. More than that, the authorities indicated that since Islam belonged to the tolerant confessions in the Russian Empire, Muslims had no legal basis for counting on financial support from the state (CSHA fund 1286 list 23 file 1329 pages 1–6; CSHA fund 1265 list 12 file 75; RGIA fund 821list 8 file 661 pages 1–3).

Even though Islam belonged only to a tolerant religion, the confession followers repeatedly applied for financial support to the communities. This behavior of Muslims can be to a certain extent explained by their significant number in the Russian Empire. However, the imperial authorities indicated that no funds were provided to support Muslim communities in the country, and the communities should bear all expenses for maintaining Muslim facilities themselves (CSHA fund 821 list 8 file 678 pages 70–72 back side 72; CSHA fund 821 list 8 file 678 pages 45–46 back side 46).

This kind of difficulty in constructing mosques led to the fact that by the beginning of the twentieth century, they sometimes acquired an illegal character. The imperial government tried to fight this kind of buildings in every possible way. As a result, those guilty of building a mosque without permission or building it in an improper place were punished within the framework provided by the legislation of the Russian Empire (Code of Laws 1912). However, even such legal measures did not stop Muslim communities from solving the problem of constructing religious buildings.

## 4 Discussion

In recent years, Russian science has paid close attention to the problem of forming and functioning of Muslim communities on the territory of the Russian Empire. Many scientists, such as I. K. Zagidullin, D. D. Azamatov, D. Yu. Arapov, etc., analyzed the legal framework for the formation of Muslim communities and the functioning of mosques. The authors paid particular attention to the principles of forming the Muslim ummah and the sources of its funding. In the works of modern researchers (Bakieva 2008; Bernyukevich 2018), a Muslim mosque is viewed as an architectural structure and a social institution. Due to the formation and legislative approval of the communities, Muslims received an opportunity to build relationships with state authorities. Some researchers (Monich and Manonina 2015; Bakieva 2008; Starostin 2019) paid special attention to the regional specifics in forming communities. Works of a regional nature form an idea of the problems of Muslim communities that arise during the construction of religious buildings. Within the study, we have obtained new results of state regulation of the process of creating Muslim communities in Western Siberia from the second half of the 19th to the beginning of the twentieth centuries.

## 5 Conclusion

Thus, in the Russian Empire, as a poly-confessional and poly-ethnic state, the problem of controlling the religious life of the peoples living on its territory was very acute. The imperial authorities paid special attention to the problem of organizing Muslim communities and their control. The OMSA, created at the end of the eighteenth century, was a state body in its essence supposed to help monitor the activities of the Muslim communities. However, this control was purely ideological and did not involve financial support for the Muslim communities. Constructing and maintaining a mosque and clergy was entrusted entirely to the community itself. At the same time, the provincial authorities, guided by the current legislation, often impeded the creation of Muslim communities. The OMSA also did not provide the desired support to Muslim communities. Besides, it is important to note that even the availability of funds did not always speed the process of permitting the construction of a religious object. Correspondence on establishing communities and constructing religious buildings could take several years, which sometimes led to the illegal construction of mosques in the regions.

**Acknowledgements** The work was supported by the Russian Science Foundation within the implementation of the grant "Religion and Power: Historical Experience of State Regulation of the Activities of Religious Communities in Western Siberia and Adjacent Regions of Kazakhstan in the 19th–20th Centuries" (Project no. 19-18-00023).

## References

Arapov D (ed) (2001) Islam in the Russian Empire (legislative acts, descriptions, statistics). Academic Center, Moscow, Russia
Azamatov DD (1999) The Orenburg Mohammedan Spiritual Assembly at the end of the 18th–19th centuries. Gilem, Ufa, Russia
Bakieva GT (2008) Russian state and Siberian Tatars in the 17th—early 20th centuries: confessional aspects. Izv Altai State Univ 4:7–15
Bernukevich TV (2018) Political aspects of the construction of religious buildings in Russia in the 19th—early 21st centuries. Bull ZabGU 24(2):47–58
Central State Historical Archives [CSHA]. Fund 1265. List 12. File 75.
CSHA RB. Fund 295. List 3. File 3270.
CSHA RB. Fund 295. List 2 File 8.
CSHA RB. Fund 295. List 3. File 6021.
CSHA RB. Fund 295. List 3. File 6930.
CSHA RB. Fund 295. List 6. File 1849.
CSHA RB. Fund 295. List 6. File 585.
CSHA. Fund 1286. List 23. File 1329.
CSHA. Fund 821. List 8. File 661.
CSHA. Fund 821. List 8. File 677.
CSHA. Fund 821. List 8. File 678.
Gibadullina EM (2009) Mechanisms for financing the construction and maintenance of rural mosques in Samara Province in the second half of XIX—Early XX century. In: Zagidullin IK (ed), Sources of Islamic institutions in the Russian Empire. Collection of articles. Kazan, Russia: Institute of History named after I. K. Zagidullin; Sh. Mardjani Institute of History, Academy of Sciences of the Republic of Tajikistan, pp. 37–43
HA OR. Fund 67. List 1. File 729.
Historical Archive of the Omsk region [HA OR]. Fund 67. List 1. File 729.
Markova MF (2008) Muslim schools in Tomsk Province at the turn of the XIX–XX centuries. In: Zinoviev VP, Tuchkov AG (eds) Proceedings of the Tomsk Regional museum of Local Lore: collection of articles. Publishing house "Veter", Tomsk, Russia, pp 268–275
Monich GI, Manonina TN (2015) Architecture of mosques in Tomsk province in the late 19—early 20th centuries. Vestnik of Tomsk State University of Architecture and building. J Constr Arch 4:49–61

Pirozhkova IG (2010) Normative regulation of cult construction in the Russian Empire. Tomsk State Univ J 1(81):298–305

Russian Empire (1830) Complete collection of laws of the Russian Empire. St. Petersburg, Russian Empire: type. 2nd division of the property. E.I.V. Chancellery, Collection II, T. IV

Russian Empire (1912) Code of laws of the Russian Empire, vol XII. Russian Book Fellowship "Deyatel", Saint-Petersburg, Russian Empire

Starostin AN (2019) Formation of Muslim communities in mining settlements of the Urals and Siberia in the late XIX—early XX centuries. News Ural State Min Univ 2(54):173–180

Zagidullin IK (2005) Islamic cult architecture in the imperial legal space (19—Early 20 centuries). Vestn Samara State Univ 4(38):60–65

Zagidullin IK (2007) Islamic institutes in Russian empire: mosques in European Russia and Siberia. Tatarstan, Kazan, Russia

# The Involvement of Youth in Sociopolitical Life as One of the Conditions for the Development of Border Areas: The Case of the Novosibirsk Region

**Aleksey V. Sharapov**

**Abstract** The Novosibirsk Region is a border area with a high economic development level and political competition, compared to other Siberian regions. This study provides a detailed analysis of the information about *young candidates* for the Legislative Assembly of the Novosibirsk Region who participated in election campaigns during 2005–2015. A *young candidate* is a person with the passive electoral right under the age of 30. Currently, many Russian scholars consider young people to be politically passive. The official statistical data analysis demonstrated that the number of candidates (including the young ones) fielded by political parties grows with each electoral cycle, while the number of young self-nominated candidates decreases. This is mainly caused by difficulties in collecting signatures for non-party candidates—man candidates drop out of the elections at this stage. However, in the last three electoral cycles, the number of young people elected as deputies was stable: 3–5 people from different political parties (in single-mandate constituencies and on party lists). At the same time, the number of vacant deputy mandates decreased from 98 to 76. Among the parliamentary parties of the Novosibirsk Region, the Liberal Democratic Party of Russia attracted the greatest number of young people as candidates for deputies in single-mandate constituencies and on the party lists for the last three electoral cycles. The lowest percentage of *young candidates* was attracted by the United Russia party (the ruling party of Russia). Such staffing policy may be explained by the presence of many well-known and credible candidates in the party's ranks. Other parties are experiencing staff shortages, resorting to attracting *young candidates* in some electoral districts during election campaigns.

**Keywords** Novosibirsk region · Political parties · Electoral cycle · Young candidates

A. V. Sharapov (✉)
Altai State University, Barnaul, Russia
e-mail: asharapov1981@gmail.com



S. G. Maximova et al. (eds.), *Advances in Natural, Human-Made, and Coupled Human-Natural Systems Research*, Lecture Notes in Networks and Systems 234,
https://doi.org/10.1007/978-3-030-75483-9_77

# 1 Introduction

The Novosibirsk Region shares borders with the Republic of Kazakhstan, Altai Krai, Omsk Region, Tomsk Region, and Kemerovo Region. The city of Novosibirsk is the capital of the Siberian Federal District. Therefore, all social, economic, and political processes in the region are closely related to the federal level. However, the outmigration of young people in the region is increasing. One of the effective means for keeping younger generations in the home regions is involving them in regional politics and providing them with a chance to enter government and municipal service. Young people's awareness of the need for their skills and talents in the region is a significant factor in restraining their migratory moods.

This problem is of great significance for the Novosibirsk Region. In 2019, the region faced a natural population decrease (−5,569 people) but a more significant increase in immigration (+10,337 people) (Federal State Statistics Service 2020).

The study of the electoral activity of youth in Russia is becoming an increasingly pressing issue. Sociologists and political scientists explore it as part of regional research. The empirical part of this research at its core possesses the fairly extensive theoretical base associated with psychological, social, and ideological reasons for youth participation in socio-political activities. The following authors represent the regional aspect (the Novosibirsk Region) of the study of young people's political and social activity.

O. V. Lagutin, E. O. Negrov, and O. V. Popova studied middle-class youth in various regions of the Russian Federation and identified three main political models: *proactive*, *cooperative*, and *loyalist* models. In addition, they studied the characteristics of political consciousness of the Russian youth and found that it is not radically different from that of the adult population. The negative reaction of the local population to the federal government is accompanied by a more positive reaction to local government and political structures. In other cases, positive attitudes toward the authorities may be associated with the public trust in them. Federal authorities usually have more support than regional and local ones.

According to some scholars, there is a small opportunity for protest civic activity growth; however, young people have no urge to protest. At the same time, nearly a quarter of young people have no interest in socio-political activities, and more than 7% are ready to use immoral and illegal ways to defend their interests (Lagutin and Negrov 2019; Popova and Lagutin 2019).

T. A. Aseeva and Ya. Yu. Shashkova came to similar conclusions (Shashkova and Aseeva 2019). They studied the electoral behavior of the Altai Krai youth in the electoral cycle of 2016–2018 and discovered that electoral absenteeism was dominant. Young people deem elections an important element of the political system. However, they are concerned about the forms of its implementation: (1) the "transparency" of the elections, (2) the existence of barriers in pursuing a political career, and (3) the virtual ban on the socio-political initiative. Presently, there are no effective means for overcoming youth absenteeism. It has to do with the low level of youth confidence in most socio-political institutions and the unwillingness of the government and other political forces to correct this.

Yu. O. Novikova and F. G. Suleymanova conducted a mass survey on the attitudes of rural youth to youth policies in the Novosibirsk Region. The scholars found out that young people are insufficiently involved in the development of the home region. Despite the great opportunities for rural youth to participate in youth policy programs, their social activity remains low. The underlying reasons include: (1) poor awareness of events, (2) low funding of youth policy, and (3) low spending on children's recreation and health improvement (1,825 RUB per person of 14–30 years) (Novikova and Suleymanova 2014).

The works mentioned above are based on opinion polls and surveys. To date, there has been no political assessment of the regional electoral activity of young people and their participation in the election campaigns. Moreover, there have been no attempts to analyze quantitative data on the involvement of young people in this process. In this study, I aim to evaluate the role of youth in the long-term formation of regional representative bodies. Moreover, the study aims to evaluate the attractiveness of the Novosibirsk Region as a place for realizing the potential of young people. The presented research methodology may be used to assess the political activity of young people in other Siberian regions.

## 2 Materials and Methods

The leading political parties of the Novosibirsk Region actively use youth themes to attract young people as candidates for Legislative Assembly in the electoral period. This hypothesis is proved by the empirical method. The necessary statistics were obtained from the official website of the Election Commission of the Novosibirsk Region (Election Commission of the Novosibirsk Region 2020). The website contains information about deputy candidates for the past 15 years.

## 3 Results

In 2005, 98 deputies of the Novosibirsk Regional Council were elected via a mixed system for the first time—49 deputies were elected in single-mandate constituencies and 49 on party lists of regional party branches. The study of the information about *young candidates* for the Novosibirsk Regional Council, nominated on party lists and single-member constituencies, provided the following data. Of the 35 young single-mandate candidates, 34 were men; 11 were nominated by political parties, 24 were self-nominated. Only one young man became a deputy in a single-mandate constituency. However, this demographic group has a significant weight in single-mandate constituencies (about 14% of 249 people). Within the mixed system, *young candidates* were nominated by the following political parties:

- *Union of Right Forces*: 14 people,
- *Liberal Democratic Party of Russia* (LDPR): 10 people,
- *United Russia*: 1 person;
- *Communist Party of Russian Federation* (CPRF): 2 people;
- *Agricultural Party of Russia*: 1 person;
- *Rodina* (Motherland) *Party*: 13 people.

Out of 379 party candidates, 60 people (16%) were candidates under the age of 30; this is significantly more than in single-mandate constituencies. Two young men received deputy mandates. We note that there were no young people on the party list of *Fair and Just Russia* in that electoral cycle (Election Commission of the Novosibirsk Region 2020).

The next elections were held in 2010. These were the regular elections to the Legislative Assembly of the Novosibirsk Region. Its deputy composition was reduced to 76 deputies: 38 deputies under a mixed system and 38 deputies for single-mandate constituencies. There were 42 *young candidates* in single-mandate constituencies: 18—self-nominated, 24—political party candidates; 39 of those were men, 3 were women. Two young men won the election race and became deputies. In total, there were 230 *young candidates* in single-mandate constituencies in the election campaign. So, the proportion of young people of the total number of candidates was about 18%. Within the mixed election system, *young candidates* were represented by the following parties:

- *United Russia*: 14 people;
- *CPRF*: 13 people;
- *LDPR*: 22 people;
- *Fair and Just Russia*: 34 people;
- *Patriots of Russia*: 27 people;

Three people received deputy seats: two representatives of *United Russia* and one of *Fair and Just Russia*. The number of young participants in the election process decreased because the list of *Patriots of Russia* was not registered. Only four parties were included on the ballot on election day. Despite this, 114 out of 571 candidates were young people under the age of 30 (20%) (Election Commission of the Novosibirsk Region 2020).

In the 2015 elections, the Legislative Assembly of the Novosibirsk Region underwent two changes: (1) party lists were divided into general and territorial parts, and (2) the number of political parties increased.

In the new socio-political conditions of 2011–2016 (growth of protest moods), youth activity changed. In 2015, there were 1,670 people nominated on party lists, 472 of whom were young people (28%). Such a great number of candidates may be explained by the significant interest of small parties in the regional elections, in which a major part of the party list is represented by young people. Unfortunately, most of the small parties were unable to register for the elections. Among the registered party lists, the *Civic Platform Party* had the largest number of young people in their ranks: 68 people; in single-mandate constituencies 21 people (14%) out of 216 candidates.

**Table 1** Nomination of young candidates to the legislative assembly of the Novosibirsk region

| Political parties/self-nominees | Young candidates in single-mandate constituencies | | | Young candidates on party lists | | |
|---|---|---|---|---|---|---|
| | 2005 | 2010 | 2015 | 2005 | 2010 | 2015 |
| United Russia | 0 | 0 | 2 | 1 | 14 | 12 |
| CPRF | 1 | 3 | 1 | 2 | 13 | 7 |
| LDPR | 5 | 9 | 8 | 10 | 17 | 25 |
| Yabloko | | | 4 | | | 25 |
| Union of right forces | 2 | | | 14 | | |
| Fair and just Russia | 0 | 5 | 2 | 0 | 34 | 17 |
| Agricultural Party of Russia | 0 | | | 1 | | |
| Patriots of Russia | 0 | 6 | | 0 | 27 | 32 |
| Communists of Russia | | | | | | 31 |
| Rodina | 2 | | 0 | 13 | | 8 |
| Narodnaya Volya | 0 | | | 6 | | |
| The social democratic party of Russia | | | | 13 | | |
| Eurasian Union Party | 1 | | | 0 | | |
| People's Alliance Party | | | | | | 81 |
| Women of Russia | | | | | | 18 |
| Democratic party of Russia | | | | | | 75 |
| Communist party of social justice | | | | | | 20 |
| Civic platform | | | | | | 68 |
| Green Party | | | 0 | | | 28 |
| Republican party of Russia—people's freedom party | | | | | | 22 |
| Self-nominated | | | | 24 | 18 | 4 |

*Source* Compiled by the authors

Three people were elected on party lists and one in single-mandate constituencies (the *CPRF* candidate, A.S. Burmistrov, refused the mandate) (Table 1). The empty columns in Table 1 indicate that the party either did not participate in the elections or had no *young candidates*.

## 4 Discussion

The analysis of the electoral activity of *young candidates* revealed a number of regularities. The last three electoral cycles have shown a steady increase in the number of *young candidates* nominated under the mixed electoral system. This positive trend is related to an increase in the number of party lists. Despite the changes in the Novosibirsk Region Legislative Assembly membership and the electoral law, the number of elected young deputies has remained approximately the same (3 in 2005; 5 in 2010; 4 in 2015). However, not all youth representatives included in party lists have a real chance to become a deputy. In parliamentary parties, young people usually occupy a bottom position in their lists, which significantly lowers their chances to be elected. In small and little-known parties, young people make up a significant part of candidates but have no chance to win the election either. This means that political parties of the Novosibirsk Region mainly attract young people to bolster their reserves. This study shows an increase in the number of young people who actively participate in election campaigns.

## 5 Conclusion

As a result of monitoring data of the election results posted on the Election Commission of the Novosibirsk Region website, I managed to reveal several patterns. The region features high political competition. Along with the parliamentary parties, one or two small parties expect the distribution of mandates both on party lists and in single-mandate constituencies. Due to the abolition of the electoral deposit (which means that candidates must collect a certain number of signatures), the participation of young self-nominated candidates in elections decreased considerably. *CPRF* and *United Russia* parties are the most conservatives in their youth policies. *United Russia* has considerable human resources, which allows them to choose the strongest candidates belonging to the age group of 40–60 (higher management of enterprises, heads of public authorities, and other recognizable personas). *CPRF* bets on more mature candidates due to the great number of elderly supporters. *LDPR* and non-parliamentary parties do their best to attract young people as candidates. These parties suffer from the lack of candidates the most. For the last 15 years, young people have been elected both on party lists and in single-mandate constituencies. Political parties effectively fulfill one of their main functions—the human resources function.

## References

Election Commission of the Novosibirsk Region (2020) Official website. http://www.novosibirsk.izbirkom.ru/

Federal State Statistics Service (2020) Socio-economic situation of federal districts. https://gks.ru/folder/11109/document/13260

Lagutin OV, Negrov EO (2019) Political behavior models of the youth middle class of modern agglomerations: the case of the main northwest regions and southern Siberia. Vestnik RUDN 2:157–174

Novikova YO, Suleymanova FG (2014) The social base of state youth policy in rural areas. RegN: Econ Sociol, 4:153–167

Popova OV, Lagutin OV (2019) Political mood of youth: loyalty or protest? Bull Peoples' Friendsh Univ Russ 4:599–619

Shashkova YY, Aseeva TA (2019) Youth in the regional electoral processes of the 2010s: trends and prospects (The case of South-Western Siberia regions). Polit Sci, 1:186–205

# Regional Factors in the Formation of Religious Identity Among the Students of East-Kazakhstan (Based on Results of Sociological Research)

**Petr K. Dashkovskiy, Aljbina S. Zhanbosinova, and Eleonora O. Stolyarova**

**Abstract** In the context of globalization, the problem of religious identity has become a focus of interdisciplinary research. The interest is due to the general destabilizing phenomena associated with non-traditional religious movements. The post-Soviet space, including Kazakhstan, has not escaped the activation and politicization of ethno-confessional processes. There is a hypothesis that permanent residence and primary socialization do not affect the religious identity formation in Kazakhstan. In order to confirm or disprove this hypothesis, we conducted a pilot study. The study participants were three groups of university students from the East-Kazakhstan region. The paper presents the results of the first stage of our research to identify similarities and dissimilarities in students' religious identity in the Eastern and Southern regions of Kazakhstan. The study showed that regional differences in socio-economic and cultural-historical nature are possible factors influencing the religious identity of youth.

**Keywords** Students · Youth · Identity · Religion · Kazakhstan

## 1 Introduction

The identity issue became the subject of focused research in the second half of the twentieth century. This problem was addressed by such prominent scientists as Erikson (1968), Bourdieu (1977), Bauman (2001), and others. Although the concept research has been conducted mainly since the 2000s, Russian psychologists and sociologists took an interest in various aspects of identities, including ethnic and religious ones, as early as the 1980s. It was reflected in the works of V. V. Lapkin, I. S. Semenenko, V. A. Tishkov, L. N. Drobizheva, L. A. Fadeeva, and

P. K. Dashkovskiy (✉)
Altai State University, Barnaul, Russia
e-mail: dashkovskiy@fpn.asu.ru

A. S. Zhanbosinova · E. O. Stolyarova
East-Kazakhstan State University, Ust-Kamenogorsk, Kazakhstan



S. G. Maximova et al. (eds.), *Advances in Natural, Human-Made, and Coupled Human-Natural Systems Research*, Lecture Notes in Networks and Systems 234,
https://doi.org/10.1007/978-3-030-75483-9_78

other scholars (Drobizheva 2013; Mchedlov 2007; Panov et al. 2012). As a result, today, we recognize the multiplicity of identity models and the multilevel nature of their manifestation. It is also essential to identify the peculiarities of various identity models.

In recent decades, religious identity has been actively studied in both Russian and Kazakh science. It is due to the intensification of intercultural contacts in the context of globalization (which can be assessed as a positive trend) and the growth of religious radicalism and extremism (Laumulin 2012). The term *identity* has many different interpretations. According to the researchers, the identity is formed and maintained during human life (Galchuk 2017, p. 45). The issues of religious identity are intensively studied in the cross-disciplinary interaction of sociology, history, anthropology, and other sciences.

The concept of religious identity is ambiguous. On the one hand, certain religious practices are serving as a reflection of faith, and on the other, it is a specific sense of belonging to a particular group. Scientists believe that individual religious identity is defined as subjective involvement in a group with the following structure: *individual (subject)—belonging—group*. In other words, it is a subjective experience of an individual belonging to a religious community (Bondarenko and Leonova 2010, p. 286). The problem is debatable due to the correlation between the concepts of *religious identity* and *religiosity*, when respondents identify themselves from two positions: *believer* and *religious affiliation* (Fedorova 2016, pp. 168–169). Some authors consider religious identity a social difference, along with gender, social classes, and ethnicity (Soleimani 2017).

These processes are especially relevant for the post-Soviet space. In the '90 s of the twentieth century, public interest in religious issues increased dramatically. It was a response to the collapse of the communist ideology. The number of people identifying themselves as followers of a particular faith has increased manifold. It is also necessary to mention the emergence of new religious movements that were not traditional for those regions. There was a politicization of the ethno-confessional space. It represented a threat to social consolidation. At the same time, religious beliefs were given the right to exist on an equal basis with secular ones (Ryzhova 2015). Today, youth have already passed primary socialization when religiosity has become an integral part of society. Secondly, since the 2000s, there has been a significant increase in the birth rate in the Central Asian region. It means that the new generation occupies an important place in society.

The factors mentioned above are related to Kazakhstan and the East-Kazakhstan region in particular. We should agree with the opinion that "the interest in youth religiosity is caused by the fact that this social group will determine the religious and ideological climate in the near future" (Arinin and Petrosyan 2016).

According to many studies, one of the main factors influencing religious identity is ethnic identity. However, it is essential to keep in mind that specific factors may change over time. What was important for one or several generations may manifest itself differently in the new conditions. In this context, the process of the religious identity formation among the youth, especially students, are of particular interest for this study.

There are two approaches to understanding the social role of the youth as a particular socio-demographic group. On the one hand, the youth is the part of society that is most vulnerable to negative influence. That is why young people are seen as a threat to social stability. On the other hand, it is young people who are expected to renew and modernize society. The religious aspect of the worldview can significantly impact the civilizational choice of the future Kazakh society.

The total number of students in Kazakhstan at the beginning of the 2017–2018 academic year was 496,209; 269,649 or 54.3% were women. Compared to the last academic year, the number of students in the current academic year increased by 4% (Higher education institutions of the Republic of Kazakhstan at the beginning of the 2017/2018 academic year 2019). The total number of young people aged 15 to 29 years, according to 2017 data, was 4,090,905 people. Consequently, the students accounted for 12.1% (Committee on Statistics under the Ministry of National Economy of the Republic of Kazakhstan 2017, p. 22). However, in this case, absolute or relative indicators are not that crucial. The role of the educated and professionally qualified youth in society is much more critical. The change in students' status in modern Kazakhstan is caused by a higher level of social responsibility assigned to them in industrial and innovative development.

## 2 Materials and Methods

To verify this hypothesis, in 2018, we conducted a study. Its participants were the students from universities in the East-Kazakhstan region. The research objective is to confirm or disprove the hypothesis that the process of the religious identity formation among Kazakh youth is determined by regional affiliation. Perhaps, soon, it is the regional affiliation that will have the most significant impact on the religious identity formation of the youth and the rest of Kazakhstan.

Three groups represented the survey participants. The main difference between them was the place of residence. The first group was represented by the students from Ust-Kamenogorsk (the regional and industrial center of the East-Kazakhstan region) and surrounding areas. The second group consisted of students from Semey (the second largest city in the region) and the surrounding areas. As mentioned above, Ust-Kamenogorsk and Semey have always had significant economic and socio-cultural differences. The third group included students studying in Ust-Kamenogorsk under the Serpin program. They lived in the southern regions of Kazakhstan. The Serpin educational program is designed to compensate for the large differences in population density in the southern and eastern regions of Kazakhstan. It aims to reallocate students between the overloaded southern and eastern universities. These educational institutions have enrollment problems due to the specific demographic and geographical location; namely, the aging population and educational migration outside Kazakhstan (the east-Kazakhstan region borders the Russian Federation).

The first group (Ust-Kamenogorsk) consisted of 58 people; 45 (77.6%) were Kazakhs. The second group (Semey) included 50 people, 38 of them (76.0%) were

Kazakhs. 61 people represented the third group (Serpin), 59 of them (96.7%) were Kazakhs. A total of 169 respondents were interviewed; 84.0% of them were Kazakhs. The analysis will focus on representatives of the Kazakh ethnic group, most of the population, and young people in particular. Since the vast majority of the population of Kazakhstan are Kazakhs, the analysis will focus on representatives of this ethnic group, and young people in particular. We compared the survey results of ethnically homogeneous groups that differ in socio-cultural and socio-economic conditions of personality formation. It allowed us to raise the question of the significance of the regional factor in the religious identity formation among Kazakh youth.

The study involved three groups of students (almost equal in number) who were different in their cultural traditions and life experience. This approach was used to determine the features of the religious identity formation in the student community. The results and preliminary conclusions will be presented below.

## 3 Results

In modern society, no age group can claim internal social homogeneity. In the context of the development of market relations, such factors as education, income inequality, permanent residence, and access to information resources lead to a specific social differentiation in the youth environment. It is logical to assume that such differentiation can also manifest itself in the religious identity formation.

The religious identity formation among young people differs significantly from that among the representatives of other ages. On the one hand, young people are more susceptible, flexible, and have little life experience. Therefore, the fundamental values and patterns of everyday behavior of young people are often irrational and unconscious. They just passively replicate the values and patterns typical of their environment. To a large extent, this applies to the religious element in the worldview of students. On the other hand, active existence in the modern information space, the open mind quickly adapting to global standards, and patterns of behavior also influence the religious identity of young people in general and students. Kazakhstan is truly diverse in socio-cultural and economic aspects. Therefore, the factors of permanent residence and primary socialization have a significant impact on the religious identity formation.

The concepts of *faith* and *religion* are often considered synonymous in collective consciousness. This confusion can lead to problems with one's religious identity, as well as the conflicts or misunderstandings in social relations. At the same time, it is important to emphasize that faith is an almost constant deep emotional feeling, while religiosity is more focused on external social manifestations (Kolyakina 2007). We can conclude that faith belongs to the realm of ethics and is purely individual; religiosity belongs to morality and has clearly defined social characteristics. Of course, faith and religion can coexist, but this is not a necessary condition.

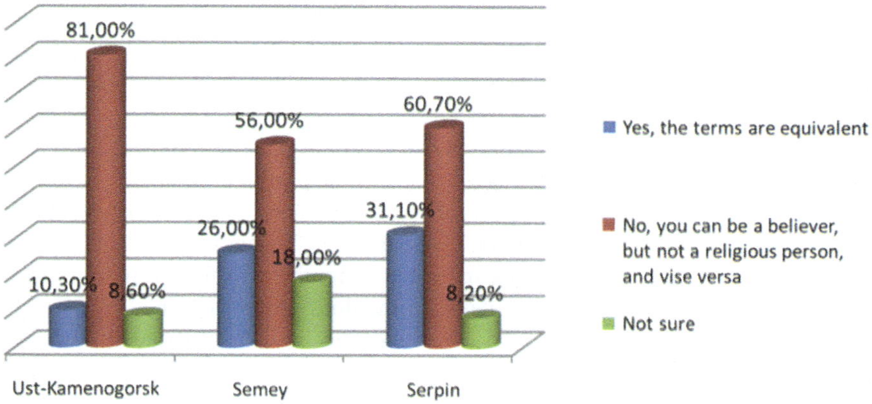


**Fig. 1** Do you think being a religious person and a believer are two different things? *Source* Compiled by the authors

Data comparison obtained during the survey of students from different regions reveals significant discrepancies in the answers to the question: Do you think being a religious person and a believer are two different things? (Fig. 1).

In Ust-Kamenogorsk, the number of students distinguished between the concepts of faith and religion was significantly higher than in other regions. It is noteworthy that in Semey, the number of respondents who found it difficult to answer was twice as high as in other groups.

The next question confirms that the differences we have found are not accidental. The number of students who identified themselves as believers, but not religious people is 15% higher in Ust-Kamenogorsk than in Semey and the southern regions of Kazakhstan.

A significant number of students from Semey and the southern regions of Kazakhstan, who took part in the survey (more than 40%), consider themselves believers and religious people at the same time. In Ust-Kamenogorsk, this indicator is more than twice as low. The following data show how this statement correlates with the behavior in society.

We revealed the number of those who identified themselves as believers and religious people and those for whom religion is a way of life: the first group (Ust-Kamenogorsk) accounted for 17.2 and 7.0%, the second group (Semey)—44.0 and 26.0%, the third group (students of the Serpin program)—43.4 and 31.0%, respectively. It is easy to notice considerable discrepancies.

Most often, respondents in all three groups lead an ordinary life, only celebrating religious holidays: minor fluctuations from 37.9 to 40.4%. We see a contradiction: the concept of religiosity rarely relates to a strictly defined lifestyle.

We could assume that such a lax following the canonical way of life results from the multi-confessional environment. It may relate to the need to adapt to the social environment. To some extent, this assumption is relevant to the first and second groups: almost half of the respondents (53.4 and 44.9%, respectively) indicated that they were surrounded by people whose views on religion differed. However, this factor cannot be considered determining for the respondents of the third group, two-thirds of which indicated that there were no people with different religious views in their inner circle.

More than half of the respondents' parents had the same religious beliefs: from 57.9% (Ust-Kamenogorsk) to 62.7% (students of the Serpin program). It is reasonable since parents are the primary agents of socialization and have the most significant influence on identity formation, including religious identity. Surprisingly, a significant number of survey participants indicated differences in their religious views with their parents. It is especially crucial for the third group (Serpin)—37.3%. The southern regions of Kazakhstan were characterized by a more respectful attitude to the older generation in general and their parents. In these circumstances, going one's own way of choosing religious beliefs is a brave decision or even a challenge. Perhaps here we see the gap between the Soviet and post-Soviet generations.

At the same time, when analyzing the answers to the question "Do you think that family and nationality should determine the choice of religion: for example, if you were born in Kazakhstan, you should be a Muslim, if you were born in Russia, then you should be an Orthodox?" (Fig. 2); we see that most respondents in the first and third groups and more than half of the second group were convinced that family and nationality should not determine religious beliefs. However, it was the respondents' national identity that almost ultimately determined their choice of religion. Therefore, there is an absolute contradiction. Students believed: that was not how it was supposed to be, but in their case, it was precisely so.

The age at which respondents assumed their religious self-determination occurred also indicates a significant impact of the family (Fig. 3).

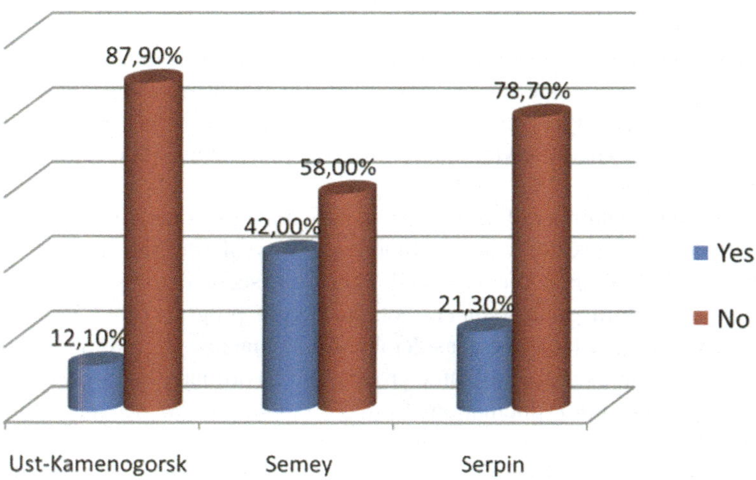


**Fig. 2** Do you think that family and nationality should determine the choice of religion? *Source* Compiled by the authors

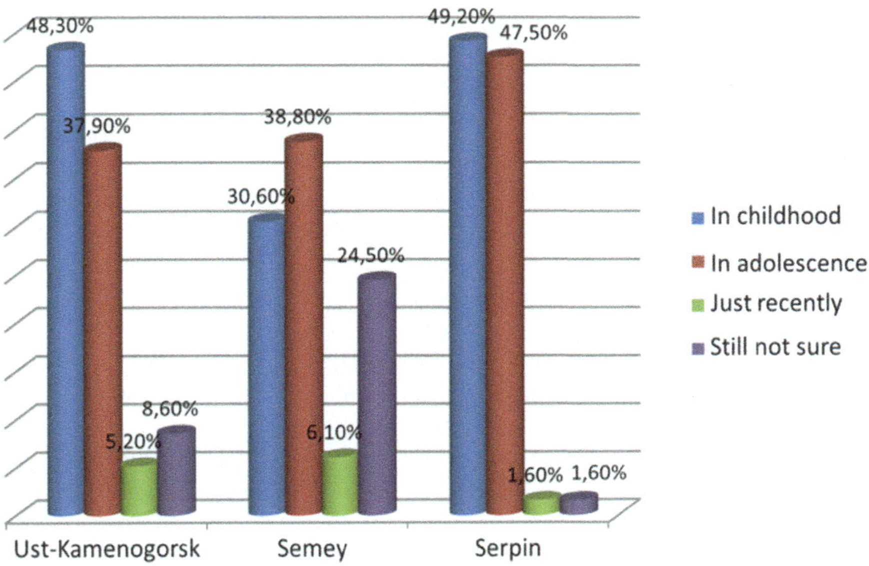


**Fig. 3** At what age did you realize your attitude to religion? *Source* Compiled by the authors

Among the third group (Serpin), the number of those who decided only recently or have not yet decided is minimal: 1.6%, respectively. In the first group (Ust-Kamenogorsk), the number of students who were not sure about their attitude to religion was already 8.6%, in the second group (Semey)—24.5%.

However, not only the family (parents) played an important role in the religious identity formation. Some hint of other causes can be found when analyzing the answers to the question about the factors that had the greatest impact on the attitude of respondents to religion: 26% of the second group (Semey) named life experience (tragic or crisis situations); in the first group this indicator was 24.1%, in the third group it was even less—16.7%. It is noteworthy that the first group respondents (Ust-Kamenogorsk), rather than the third group (Serpin), were more likely to name the family as the main factor of influence.

Kazakhstan's society is truly multicultural, multi-ethnic, and multi-confessional. The coming years will show to what extent young people from different regions of the country will engage in constructive and civilized interaction with the representatives of other cultures and religions. However, even today, we can assume that religious beliefs (even if they are only implied and formal) are determining for many people in the perception of another person. In this case, one should pay attention to Table 1. It contains the survey results regarding the rating of qualities that respondents consider when building relationships with other people.

It is easy to notice that the respondents of the second and third groups placed religion in the first positions of qualities that are important for them in other people. At the same time, the respondents of the first group ranked religion only in ninth place out of eleven. The contrast is even more dramatic because the entire survey focused

**Table 1** The rating of factors the respondents value in other people (based on the survey)

| Ust-Kamenogorsk | | Semey | | Serpin | |
|---|---|---|---|---|---|
| 1 | Personal qualities | 1 | Personal qualities | 1 | *Religion* |
| 2 | Interests and hobbies | 2 | *Religion* | 2 | Personal qualities |
| 3 | Age | 3 | Family, background | 3 | Interests and hobbies |
| 4 | Gender | 4 | Interests and hobbies | 4 | Family, background |
| 5 | Nationality | 5 | Age | 5 | Nationality |
| 6 | Job | 6 | Nationality | 6 | Age |
| 7 | Place of residence | 7 | Citizenship | 7 | Gender |
| 8 | Family, background | 8 | Gender | 8 | Citizenship |
| 9 | *Religion* | 9 | Job | 9 | Place of residence |
| 10 | Citizenship | 10 | Place of residence | 10 | Job |
| 11 | Wealth | 11 | Wealth | 11 | Wealth |

*Source* Compiled by the authors

on issues of faith and religiosity. In other words, this result cannot be considered random. Despite this, in all three groups, the personal qualities of the other person were prioritized, while nationality in all three groups was only in the middle of the list. Thus, the respondents of the second and third groups considered the other person's religious beliefs more important than their nationality, while the respondents of the first group had the opposite view.

Another important question arises: do respondents rate themselves according to the same criteria as others? How does the attribution of other people relate to the self-perception? Are we dealing with the so-called *double standards*? If this is the case, we can identify another symptom of an identity crisis. To answer these questions, let us refer to Table 2.

As we can see from Table 2, religion, as an element of self-presentation of the individual, remains almost in the same positions as the assessment of another person (Ust-Kamenogorsk—eighth and ninth place, respectively, Semey—third and second place, Serpin—second and first place). It is worth paying attention to another fact that applies only to the second group (Semey). Evaluating another person, these respondents put nationality only in the sixth position, while describing themselves this characteristic suddenly appeared on the 2nd place, right after the personal qualities.

We stated above that ethnic and religious identity are closely related. However, we can assume that nationality as an attribute of personality plays a vital role in some regions. We assume that there are still unexplored, purely regional factors that affect

**Table 2** Rating of qualities that respondents use to characterize themselves

| Ust-Kamenogorsk | | Semey | | Serpin | |
|---|---|---|---|---|---|
| 1 | Personal qualities | 1 | Personal qualities | 1 | Personal qualities |
| 2 | Interests and hobbies | 2 | Nationality | 2 | *Religion* |
| 3 | Family, background | 3 | *Religion* | 3 | Interests and hobbies |
| 4 | Job | 4 | Gender | 4 | Nationality |
| 5 | Gender | 5 | Interests and hobbies | 5 | Family, background |
| 6 | Age | 6 | Age | 6 | Citizenship |
| 7 | Nationality | 7 | Job | 7 | Gender |
| 8 | *Religion* | 8 | Family, background | 8 | Job |
| 9 | Place of residence | 9 | Place of residence | 9 | Age |
| 10 | Wealth | 10 | Citizenship | 10 | Place of residence |
| 11 | Citizenship | 11 | Wealth | 11 | Wealth |

*Source* Compiled by the authors

the religious identity of young people. To check this assumption, we compared the results obtained in three groups from representatives of only one ethnic group: the Kazakhs.

Table 3 shows the answers to the first set of questions connected with the inner world, everyday life, and primary socialization. For greater comparability of results, only the most frequent answers are presented.

From the presented data, we can already conclude that young residents of the South consider religious self-attribution far more critical for everyday life than their peers in the East of the Republic do. The data presented in Table 4 clearly show the same differences between regions. Representatives of the same ethnic group, the same age, and the same social status (students) give a different picture. Therefore, we still have reason to assume that additional regional factors affect the students' religious identity formation (and probably not only them).

Table 4 shows the answers to the second set of questions connected with the religious identity manifestation in the external social, political environment.

The presented data shows that socio-political views related to the place of religion in Kazakhstan society and the state are much closer among the representatives of the second and third groups of respondents (Semey and Serpin, respectively). Although geographically and administratively, they are separated by a considerable distance. At the same time, the respondents of the first group (Ust-Kamenogorsk) demonstrate a unique attitude that differs from other survey participants.

**Table 3** Answers to the questions given by representatives of the Kazakh ethnic group only (Part 1)

| | Ust-Kamenogorsk (%) | Semey (%) | Serpin (%) |
|---|---|---|---|
| Do you think being a religious person and a believer are two different things? | | | |
| No, you can be a believer, but not religious, and vice versa | 77.8 | 52.6 | 61.0 |
| Do you consider yourself a believer but not a religious person? | 54.5 | 50.0 | 31.0 |
| How does religion influence your life? | | | |
| I celebrate religious holidays, but the rest of the time I lead an ordinary life | 45.5 | 42.1 | 39.3 |
| Are there people among your family and friends who have religious beliefs different from yours? | | | |
| No | 51.1 | 56.8 | 67.8 |
| Do your parents share the same religious beliefs? | | | |
| Yes, they brought me to religion | 59.1 | 60.6 | 63.2 |
| Do you think that family and nationality should determine the choice of religion: for example, if you were born in Kazakhstan, you should be a Muslim; if you were born in Russia, then you should be an Orthodox? | | | |
| No, the choice of religion should not depend on this | 84.4 | 57.9 | 78.0 |
| At what age did you realize your attitude to religion? | | | |
| My attitude to religion was shaped in my childhood | 53.3 | 37.8 | 49.2 |
| What has had the most significant impact on your attitude to religion? | | | |
| Family, relatives | 62.2 | 42.1 | 46.6 |

*Source* Compiled by the authors

**Table 4** Answers to the questions given by representatives of the Kazakh ethnic group only (Part 2)

| | Ust-Kamenogorsk (%) | Semey (%) | Serpin (%) |
|---|---|---|---|
| Do you think Kazakhstan should be a secular state? | | | |
| Yes | 82.2 | 73.0 | 74.6 |
| Do you need laws in Kazakhstan that protect the feelings of believers? | | | |
| Yes | 53.3 | 36.8 | 44.8 |
| Should the state pay special attention to religious people? | | | |
| No, the state should treat all citizens equally, regardless of their religious beliefs | 82.2 | 73.7 | 76.3 |
| Are you interested in other cultures and religions? | | | |
| Yes, I am interested in it, and I try to learn more about other cultures and religions | 53.3 | 36.8 | 32.2 |

*Source* Compiled by the authors

**Table 5** The place of religion and ethnicity in the system of assessing other people from Kazakh students' views

| | Ust-Kamenogorsk (%) | Semey (%) | Serpin (%) |
|---|---|---|---|
| What is the most crucial fact about another person? (arrange the choices in order of importance: the most important thing comes first, and the least essential thing comes last) | | | |
| Religion | | | |
| | 18.4 | 35.3 | 38.1 |
| Nationality | | | |
| | 8.1 | 26.7 | 31.3 |

*Source* Compiled by the authors

**Table 6** The place of religion and ethnicity in the system of self-attribution of the respondents

| | Ust-Kamenogorsk (%) | Semey (%) | Serpin (%) |
|---|---|---|---|
| If you had to define yourself in five words, what characteristics would you choose? | | | |
| Religion | | | |
| | 35.6 | 54.3 | 57.9 |
| Nationality | | | |
| | 46.7 | 65.7 | 50.9 |

*Source* Compiled by the authors

We should also note that Kazakh students pay attention to religion and ethnicity, evaluating other people. The results of the survey concerning this aspect are shown in Table 5. It shows the proportion of respondents who put religion and ethnicity first.

Table 5 shows that the respondents of the second and third groups show much more similarities between themselves than with the first group.

At the final stage of the study, we need to compare the place of religion and ethnicity in the respondents' self-attribution system. In this case, we considered the proportion of respondents who put these characteristics in the first place. The results obtained are presented in Table 6.

## 4 Conclusion

Summarizing the results, we can draw several preliminary conclusions.

1. Attempts to solve the identity crisis using the traditional *ethno-confessional* scheme generate external, superficial, optional forms of religiosity, devoid of deep moral content.

2. These contradictions are especially relevant to young people due to the natural marginality of their social status. This process is complicated by cultural, historical, and socio-economic differences between the regions of Kazakhstan.
3. We can assume that the socialization process of young people goes through the development and determination of the social environment. Completely different mentalities collide. Against the background of uncertainty in the minds and the lack of common ideologies, a new mindset is being formed in the younger generation.
4. Several factors are affecting the religious identity of young people that have not yet been studied. These factors include regional socio-economic and cultural-historical differences.

It should be noted that these conclusions need to be confirmed or improved with additional data. As for the processes of resolving the crisis of religious and other identities, they should be covered in additional research.

We used to think that ethnicity and religion are intricately connected. However, students of the Ust-Kamenogorsk region show some significant differences from students of the Semey region and southern regions of Kazakhstan. The differences relate to the conditions of religious socialization, as well as the role of religion in their lives and the life of the Kazakh society in general. The survey results reveal new directions in the study the religious identity formation among Kazakh youth. Several factors are affecting the religious identity of young people that have not yet been studied. These factors include regional socio-economic and cultural-historical differences.

## References

Arinin E, Petrosyan D (2016) The peculiarities of students' religiosity. Sociol Stud 6:71–77

Bauman Z (2001) The individualized society. Polity, London, UK

Bondarenko O, Leonova M (2010) Religious identity: an explication of the concept. Humlties Soc Sci 6:285–291

Bourdieu P (1977) Outline of a theory of practice. Cambridge University Press, Cambridge, UK

Drobizheva LM (ed) (2013) Civil, ethnic, and religious identity: yesterday, today, and tomorrow. ROSSPEN, Moscow, Russia

Erikson E (1968) Identity, youth, and crisis. W. W. Norton Company, New York, NY

Fedorova MV (2016) Religious identity in the conceptual space of religious sciences. Manuscript 4–2(66):168–171

Galchuk DS (2017) The concept of "personal identity." Buryat State Univ Bull 5:44–51

Higher education institutions of the Republic of Kazakhstan at the beginning of the 2017/2018 academic year (2019) https://www.zakon.kz/4900658-vysshie-uchebnye-zavedeniya-respubliki.html

Kalyakina EA (2007) The definition of the concept of "religiosity". Bull Kostroma State Univ Ser: Pedagogy Psychology Socio-Kinet, 13(3):51–55

Laumulin M (2012) The religious situation and the threat of religious extremism in Central Asia. CentL Asia Cauc 15(1):60–79

Mchedlov MP (ed) (2007) Faith, ethnicity, nation. The religious component of ethnic consciousness. Cultural Revolution Publishing, Moscow, Russia

Ministry of National Economy of the Republic of Kazakhstan Committee on Statistics (2017) Demographic yearbook of Kazakhstan: statistical compendium. Ministry of National Economy of the Republic of Kazakhstan Committee on Statistics, Astana, Kazakhstan

Panov PV, Sulimova KA, Fadeeva LA (eds) (2012) The struggle for identity and new communication institutions. ROSSPEN, Moscow, Russia

Ryzhova SV (2015) Ethno-confessional identity and formation of religious, political orientations. Sociol Stud 10:136–144

Soleimani S (2017) The concept of religious identity. Soc: Sociol Psychol, Pedagog 4:30–32

# State Registration Practices of Christian Communities in Western Siberia in the Mid-1960s and 1970s

**Petr K. Dashkovskiy and Natalia S. Dvoryanchikova**

**Abstract** The article studies the peculiarities of Christian communities' status within the state-confessional policy of the USSR in the mid-1960s—mid-1970s on the territory of Western Siberia. Based on legal documents and archive data, the regulations regarding religious communities in the region were studied. During the country leadership of L. I. Brezhnev, there was a departure from the previous policy of N. S. Khrushchev in state-confessional relations. During the studied period, the process of registration of religious communities was resumed. At first, the authorities mainly legalized Protestant communities, which was later expressed by the predominance of officially active Protestant associations in Western Siberia. Those were the first cases of communities being granted legal status since the *thaw* period. However, despite this, the state continues to actively adhere to atheistic policies to eradicate religious thoughts from the minds of Soviet citizens.

**Keywords** Church-state relations · The Christian community · Atheistic propaganda · Soviet union · Western Siberia

## 1 Introduction

The history of church-state relations has been one of the most prevalent Russian historiography topics in recent years. This trend is due to the critical role of religious organizations in Russian citizens' lives and the growing influence of the church on society. A key role in the religious communities' status in the USSR was occupied by official (state) registration, which enabled them to create religious associations and perform religious rites. By the mid-1960s, the legislative acts regulating the activities of religious communities included: (1) the January 23, 1918 Decree of the Council of People's Commissars of the Russian Soviet Federative Socialist Republic on Separation of Church from State and School from Church, (2) the Decree of the All-Russian Central Executive Committee and the Council of People's Commissars of the Russian

P. K. Dashkovskiy · N. S. Dvoryanchikova (✉)
Altay State University, Barnaul, Russia
e-mail: natali.dvoryanchikova@mail.ru

 
S. G. Maximova et al. (eds.), *Advances in Natural, Human-Made, and Coupled Human-Natural Systems Research*, Lecture Notes in Networks and Systems 234,
https://doi.org/10.1007/978-3-030-75483-9_79

Soviet Federative Socialist Republic of April 8, 1929, on Religious Associations, (3) the 1936 Constitution of the Soviet Union. State-confessional policy in the USSR was also determined by decrees and resolutions of the Council of Ministers of the USSR on religion-related issues. Besides, there were special instructions and explanations of the Council for Religious Affairs under the Council of Ministers of the USSR, which clarified and specified the provisions of the essential documents (Fasikhova 2002).

With the 1964 L. I. Brezhnev's arrival to the USSR leading position, the state-confessional policy started to change. Many Protestant movements found themselves in the most challenging situation, although the Soviet government made certain liberalizations in religious life. Thus, on September 30, 1965, The Presidium of the Supreme Soviet adopted a Decree on lifting restrictions on special settlements for members of the *Jehovah's Witnesses, True Orthodox Christians, Innokentievites, and Adventists-reformists* sects, as well as their family members (Odintsov 2002a, b). According to scientific research, the legalization of religious communities resumed in the late 1960s (Odintsov 2002a, b).

## 2 Materials and Methods

Legal documents and archive materials provide the most objective analysis of the regulation practices of the state registration of Christian communities in Western Siberia in the context of the state-confessional policy of the USSR in the mid-1960s and mid-1970s. Archive sources include (1) business correspondence of the authorized Council for Religious Affairs in Altai Krai and Novosibirsk region with Councils of People's Deputies, district committees, city committees, and religious associations; (2) statistical data, (3) administrative materials, (4) legislative regulations of state bodies and bodies of the party. We primarily used the State Archive of Novosibirsk region and the State Archive of Altai Krai while writing the article. The methodological basis of the research is the principles of the scientific study of the system of state-confessional relations in Russia, developed by M. I. Odintsov, O. Yu. Vasilyeva and other scientists. Also, we used the principle of historicism, which allowed us to study the process of religious community registration in the system of state-confessional policy at the regional level in the mid-1960s–mid-1970s. The study was conducted using a historical-systematic method that allows us to generalize the status of Christian communities in Western Siberia in state-confessional relations.

## 3 Results

In 1965, a new body for implementing the state's religious policy was created—the Council for Religious Affairs Under the Council of Ministers of the USSR. It combined the abolished Council for the Affairs of the Russian Orthodox Church and the Council for Religious Cults, which dealt with other confessions (Geraskin 2008). The newly created body in charge of state-confessional relations existed until 1991. The main goal of the new institution was *a consistent implementation of the policy of the Soviet state in relation to religions and monitoring compliance with legislation on religious cults* (Novikov 1985). The main tasks of the Council were to monitor compliance with the Constitution of the USSR and the Decree of the Council of People's Commissars of the Russian Soviet Federative Socialist Republic of January 20, 1918, on Separation of Church from State and School from Church. The functions of the Council also included making decisions (1) on the registration and deregistration of religious associations, (2) opening and closing prayer buildings and houses, (3) giving mandatory instructions on the elimination of their violations, (4) raising the issue of prosecuting people, who violate the law on cults to disciplinary, administrative or criminal liability (Molodov 2015).

Locally, the Commissioner was the Head of the Council for Religious Affairs. The Regional Executive Committees of Deputy Councils of Workers had the right to appoint or dismiss from the mentioned position. The functions of the commissioners of the Council for Religious Affairs were: (1) to monitor religious associations, (2) to monitor the number and population of communities, (3) to interpret legislation on religious cults, (4) to make annual reports on the religious status in Altai Krai and Novosibirsk region, (4) to help Commissions for Assistance to comply with legislation on cults in carrying out atheistic propaganda. In addition, the commissioners' functions included an examination of requests from religious communities regarding opening religious buildings and registration of clergy and religious communities.

The specifics of state registration regulation of religious communities are relatively well reflected in the regional aspect in archive documents. Thus, in Altai Krai in 1966, there were only four registered religious associations: three Orthodox churches and one community of the All-Union Council of Evangelical Christian Baptists. By the mid-1960s, there were nine officially registered religious organizations on the territory of Novosibirsk region: three Orthodox churches (one church did not operate but was not deregistered), three communities of the All-Union Council of Evangelical Christian Baptists, one Jewish, one Muslim and one Old Believer community (Fund 1418. List 1. File 80. Page 16). It should be noted that the profits of the Russian Orthodox Church increased slightly during the period under study. In 1964, the total church profit in Altai Krai had estimated 372,000 rubles and in 1965—445,800 rubles (Fund 1692. List 1. File 74).

It is crucial to note that local authorities tried to restrain the registration of religious communities during the studied period. In 1967, the community of the Russian Orthodox Church in Kamen-na-Obi, Altai Krai, sent the registration request to the Council for Religious Affairs, as did the community in Slavgorod (Fund 1692. List

1. File 76). Significantly, these requests were not complied with by the Executive Committees. Besides, during the studied period, the number of religious rites performed by the Russian Orthodox Church communities located in Barnaul, Biysk, and Rubtsovsk increased. In the Orthodox Church of Rubtsovsk, there was an increase in the number of baptism rites by 23%, and the number of the last rites increased by 47%. According to the Commissioner of the Council for Religious Affairs in Altai Krai, the main reason for the increase in religious rites was the small number of civil holidays. It should be stated that, in reality, the increase was caused by the activization of Orthodox believers (Fund 1692. List 1. File 76).

An important issue in the study of religious organizations' status is the study of clergy—its number and age structure. In this regard, it is necessary to address religious education since spiritual enlightenment is one of the main aspects of the life of the church. One of the most critical problems for the church during the studied period was the insufficient number of clergy. In 1971, there were 7,274 churches in the Soviet Union. At the same time, 6,234 priests and 618 deacons were registered. If one compared the number of parishes with the number of clergies, also having taken into account that several priests who served in the Cathedral and some parish churches, one could see that by the end of the 1960s, there was a real shortage of clergy, especially in the regions, including Western Siberia. In this regard, many parishes, which did not have their abbots, were served concurrently by the clergy of neighboring churches. Besides, natural demographic processes occurred. Specifically, the older generation of priests was dying, and due to the reduction in the number of theological school students, it was not replaced by graduates of seminaries (Tsypin 1997).

It is essential to point out that during the studied period, Orthodox communities were also located in the Gorno-Altai Autonomous Region, which was part of Altai Krai at that time. After the final closure of churches in the first two decades of Soviet rule, religious life in this part of the Altai, as in many other areas, proceeded under underground and semi-underground conditions. According to the former rector of Gorno-Altaisk State University, Yu. V. Tabakaev, in the 1950s–1960s in the part of the city on the other side of the river, where he lived, there was a small building in the neighborhood, which the locals called the "monastery house." Several nuns lived here, presumably originally from a shut down in the 1920s Ulalinsky Nikolsky Convent (Kreydun 2010). Most groups of believers lived in places where churches had previously operated. For example, an Orthodox church operated in the village of Mayma until 1938. With the closure of this church, religious life began to diminish, although the believers continued to live in the village for a long time. In 1929, in the village of Shebalino of the specified Autonomous Region, a church was destroyed. However, despite religious persecution, there were still people who retained their faith through the following decades. Among them were M. N. Chabaevskaya, A. S. Tcheltugasheva, E. S. Kurbatova. In the 1950s and 1960s, these women traveled to Biysk on weekends and holidays to visit one of the few active churches (the Uspensky Cathedral) in the South of Western Siberia (*History of the Altai Spiritual Mission*).

It is important to note that the emerging changes in state and religious policy were to some extent caused by the signing of the Helsinki Final Act at the Conference on Security and Co-operation in Europe in August 1975, which contains provisions on respect for human rights and fundamental freedoms, including freedom of conscience, religion, and belief. The mid-1970s were marked by a particular positive shift in the relationships of the ruling party and the Soviet state to religious issues. It signified that human rights, including freedom of religion, was at the heart of the historical dispute between the capitalist and socialist systems. There was a growing prisoner of conscience of the USSR defense campaign in the West, which was against the religious associations' and believers' rights violations.

In addition to the Russian Orthodox Church communities, there were communities of Old Believers on the territory Altai Krai. Old Believers lived in Barnaul, Biysk, and in Smolensky and Zalesovsky districts of Altai Krai. The community of the Barnaul Beglopopovtsy Old Believers of Belokrinitskaya Hierarchy requested for church opening and the legalization of activities. However, in his response, the Commissioner pointed out the inexpedience of such activities. The main reason was that the local authorities were afraid of the activization of the Old Believers' communities (Fund 1692. List 1. File 76). During the studied period, one community of Old Believers (Belokrinitskaya Hierarchy) and one unregistered community of Old Believers (Bespopovtsy) legally operated on the territory of Novosibirsk. In 1965, the Old Believers community (Belokrinitskaya Hierarchy) consisted of 287 people (57 men, 230 women). On Sundays, about 150 people attended the meetings, and on religious holidays, believers from Novosibirsk region, Altai Krai, and Kemerovo region traveled to the community. In total, the meetings were attended by about 300–400 people (Fund 1418. List 1. File 80. Page 16).

The most massive Protestant movement in terms of the number of registered communities in Western Siberia was the All-Union Council of Evangelical Christian Baptists. In 1965, there were three registered groups of Baptist followers of the All-Union Council of Evangelical Christian Baptists in Novosibirsk region. The largest community of 1,415 people was located in Novosibirsk. The gender composition of the community members was: (1) men—270 people, and (2) women—1,145. By occupation, believers were divided in the following way: (1) workers and employees—485 people, (2) pensioners—248, and (3) housewives—682. The age group of up to 30 years was 100 people, from 30 to 40—205 people, from 40 to 50—156 people, from 50 to 60—251 people, and over 60—703 people. Russians, Germans, and Ukrainians dominated the ethnic composition of the community. The community held prayer meetings on Wednesdays, Saturdays, and two times on Sundays. At the services themselves, 300–400 people were directly present, and on holidays 700–800 (Fund 1418. List 1. File 80. Page 16). It is crucial that in 1973 in Novosibirsk region, there were already 12 registered communities of Evangelical Christian Baptists, and three of those were registered. By the beginning of 1974, there were already more than 1,700 believers in the communities. The vast majority of believers—81%—were women, while men made up only 19%. According to their social status, believers were divided in the following way: (1) workers—26%, (2) employees—3%, (3) collective farmers—1%, (4) pensioners—29%, and (5) homemakers—41% (Fund 1418. List 1. File 179). By ethnicity, the majority of believers were: (1) Russians—78%, (2) Germans—15%, and (3) Ukrainians. Other ethnic groups made up 3%.

In 1972 The Novosibirsk Community of the All-Union Council of Evangelical Christian Baptists was actively engaged in missionary work, and the Commissioner for Religious Affairs noted violations of the legislation on cults in the region. Thus, members of the community traveled to the cities of the South of Western Siberia—Bolotnoye, Tatarsk, Kamen-na-Obi, and Mariinsk. Besides, the believers were engaged in charity activities by providing financial help to other religious associations (e.g., they organized a fundraiser for church construction in Leninsk-Kuznetsky). The community also invited believers from other cities to water baptism on July 2, 1972, without the permission of the Executive Committee. The Executive Committee mentioned that if the community continued to commit illegal actions, the question could be raised regarding the withdrawal of the community members from the executive body and the deregistration of ministers of religion культа (Fund 1418. List 1. File 172). In 1975, the registered community of the All-Union Council of Evangelical Christian Baptists in Novosibirsk already consisted of 1,300 believers. It is important to note that in 1976 another community of the All-Union Council of Evangelical Christian Baptists in the village of Kozhurla in Novosibirsk region received official registration. It is peculiar that this religious community had been active since 1935, but only in 1976 it was registered (Fund 1418. List 1. File 208).

An ambiguous situation with the status of Protestant communities was also observed in Altai Krai. Thus, in 1967, there were 12 communities of the All-Union Council of Evangelical Christian Baptists in Altai Krai, but only one was registered in Barnaul. The activities of those communities were renewed during the studied period. In particular, there was an increase in profit in the registered Barnaul community due to believers' donations. In 1969, two communities of the All-Union Council of Evangelical Christian Baptists were registered, located in Rubtsovsk and the village of Glyaden. In 1971, there was also an increase due to the registration of two branches of the All-Union Council of Evangelical Christian Baptists. Thus and so, the Evstratovskaya group of Baptists as a branch of the village community and the association in the working village of Volchikha as a branch of the Rubtsovsk organization. It is also noteworthy that in 1973 the Slavgorod Baptist community was registered. Thus, by the mid-1970s, four communities and two branches of the All-Union Council of Evangelical Christian Baptists operated on the territory of Altai Krai (Fund 1692. List 1. File 127). Also, unregistered communities of the All-Union Council of Evangelical Christian Baptists were located in Kulunda, Talmenka, Biysk, Gorno-Altaisk, and other localities of Altai Krai (Fund 1692. List 1. File 107). By 1971, the All-Union Council of Evangelical Christian Baptists was comprised of 990 believers (Fund 1692. List 1. File 115).

There were six unregistered communities of the Council of Churches of Evangelical Christian Baptists, comprised of about 500 people. The most numerous were the Barnaul and Slavgorod communities. By the beginning of the 1970s, there was an increase in the number of believers in the communities, which brought the number of believers to 577. In 1973, in Novosibirsk region, there were two communities and two groups of the Council of Churches of Evangelical Christian Baptists. In total, there were about 180–190 people in the communities. In 1973, 18 members of the Council of Churches of Evangelical Christian Baptists were imposed administrative liability. The Commissioner of the Council for Religious Affairs in Novosibirsk region concluded from the analysis of the sermons that the believers were *in hostile positions* to the Soviet government (Fund 1418. List 1. File 228). It should be noted that one of the communities applied for registration in 1966 and 1969. In 1970, the court confiscated a house that was illegally purchased for religious practices. After that, two executives (Miroshnichenko and Orlov) were sentenced to two years in prison each for attempts to cause riots regarding the confiscation of the house. After that, the community began to convert a residential building into a prayer house in 1970. However, the Executive Committee issued a warning concerning the illegality of such actions, but the believers did not stop this process. As a result, the house extension was eventually demolished. In 1972, the community reapplied for registration, which the Executive Committee reviewed, and the believers were interviewed about the legislation on cults. The community did not recognize the Soviet legislation, and in 1973 the registration of the community was considered inappropriate (Fund 1418. List 1. File 188).

Besides, three Mennonite communities operated legally in Novosibirsk region. The largest was the Novosibirsk community, which by 1974 was comprised of 120 believers. It was dominated by women—77%, while men made up only 23%. In 1973, the registered community of Church Mennonites in Berdsk dissociated due to its leader's death. The Commissioner for Religious Affairs refrained from raising the issue of this group's deregistration. By the end of 1970, the Berdsk community resumed its activities (Fund 1418. List 1. File 19). On the territory of Altai Krai, there were seven Mennonite associations with a total number of 832 people (Fund 1692. List 1. File 115). The communities were located in the Blagoveshchensky, Tabunsky, and Slavgorodsky districts. It is also noteworthy that the communities maintained contact with the Council of Churches of Evangelical Christian Baptists' followers. There were also cases of community mergers.

For many years, Seventh Day Adventist communities operated in Altai Krai. With a total number of 140 people, they were located in Barnaul, Biysk, and Rubtsovsk. There were three Adventist communities in Novosibirsk region. Adventists-reformists lived in the village of Alexandrovka in the Kolyvansky District and the village of Kazachy in the Tatarsky District and had no connection to each other. In some Adventist families, people refused to serve in the military or receive legal documents for religious reasons. In connection with illegal activities, the Commissioner of the Council for Religious Affairs sent a submission to the regional Prosecutor with a request to bring them to criminal responsibility (Fund 1418. List 1. File 154).

It is important to note that since the mid-1960s, many Pentecostal communities have joined the All-Union Council of Evangelical Christian Baptists. However, communities often did not agree with Baptists' positions on compulsory military service and the state registration of associations. Therefore, in 1969, initially as an experiment, the Council for Religious Affairs allowed Pentecostal communities to register on an autonomous basis, meaning without joining any particular union. This possibility applied to both unregistered communities and communities that were part of the All-Union Council of Evangelical Christian Baptists. However, there was also a prefix—upon the condition of abandoning *fanatical rites*. The latter immediately pushed the number of associations away from this path. This practice of registration was used not in the every republic of the USSR.

In Novosibirsk region, there were three unregistered Pentecostal societies, which consisted of about 65 people. Pentecostals applied to join the Novosibirsk registered community of the All-Union Council of Evangelical Christian Baptists, but their own registration was out of the question. It is peculiar that the cases of transition to the All-Union Council were numerable. Thus and so, one of the Pentecostal communities (village of Andreevka, Kuybyshevsky District) after its leader's A. I. Mazur's leave, merged with the All-Union Council of Evangelical Christian Baptists. However, there were still communities that did not want to register. Thus, the Novosibirsk and Bolotninskaya groups of Pentecostals refused to register and, in this regard were imposed an administrative liability in 1971 (Fund 1418. List 1. File 154). A similar situation occurred with the community of Evangelical Christians of the Tatarsky District of Novosibirsk region. The Pentecostal community was headed by the accountant of the mechanized leskhoz S. L. Donkaev. The community itself was opposed to registration (Fund 1418. List 1. File 230). It is significant that Pentecostals maintained contact with other communities. Therefore, in 1974, P. Yusina came to the community from Alma-Ata and tried to communicate with them. According to the Commissioner, the leaders of the Pentecostals kept in touch with each other but observed secrecy, which made it difficult to take any measures against them (Fund 1418. List 1. File 192). In Altai Krai, there were two communities of Evangelical Christians, with a total of 50 believers. According to A. V. Gorbatov, the lack of administrative center and unity within the organization hindered its wider spread and dynamic development (Gorbatov 2008).

It is important to note that by the end of the 1960s, there was a decrease in the number of followers of *Jehovah's Witnesses* in Siberia due to travel to the Western regions of the country and Central Asia. Migration flows were associated with the adoption of the Decree on Elimination of Restrictions on Special Settlement of Members of the Sects *Jehovah's Witnesses*, *True Orthodox Christians*, *Innokentievites*, *Adventists-reformists*, as well as of their family members by the Presidium of the Supreme Soviet on September 30, 1965. According to this confidential decree, *Jehovah's Witnesses*, as well as members of other anti-Soviet sects, were exempt from administrative supervision and could leave the places of special settlement (Odintsov, 2002a, b). *Jehovah's Witnesses* in Novosibirsk region mainly operated in Krasnozyorsky and Chulymsky districts. Besides, some believers lived in Novosibirsk. Those groups were created in the early 1960s by people who came for construction work from the Zakarpattia region of the Ukrainian SSR (Fund 1418. List 1. File 154). In the Krasnozyorsky district, there were two groups of *Jehovah's Witnesses* consisting of 11 people. Local authorities were wary of the believers' behavior. Thus, according to the Commissioner, the antisocial nature of believers was manifested in the fact that they did not participate in elections, refused to serve in the army, did not belong to a trade union, and their children did not join the Young Pioneers. In addition, in the fall of 1971, in the area of the Novosibirsk Airport, there were cases of distribution of Jehovah's pamphlets *Two Ways to Heaven* and others through the local residents' mailboxes (Fund 1418. List 1. File 154).

Similar data on the decline in the number of believers are typical for Catholic communities. So, if in 1970 in the Catholic communities of Novosibirsk region there were 150 believers, then by the middle of 1975, there were only 42. Some believers emigrated to West Germany (21 people), some went to the cities of Central Asia and other regions of the country, and some simply stopped attending prayer meetings. Novosibirsk believers raised the issue of state registration, but many activists left the region. Due to these circumstances, the community was not registered. The Catholic community of Novosibirsk was registered and given consent to purchase the building for prayer purposes only on August 9, 1982 (Fund 1418. List 1. File 255. List 1).

Based on the analysis of archive materials, one can notice that by the mid-1970s, there was a noticeable process of activization of Lutheran communities. The number of all Lutherans in Altai Krai was 371 (Fund 1692. List 1. File 119). There were seven Lutheran communities in Novosibirsk region, but only the Novosibirsk community operated legally, and the rest were noticed. Small Lutheran groups also existed in the village of Oktyabrsky, Karasuksky district, the village of Krasny Yar, and Ordynsky district in the number of 3 to 10 people. The pastor of the registered community was G. F. Ikkert, born in 1920. He was a construction worker by occupation. In total, the Novosibirsk community consisted of 201 people. In terms of gender, it was dominated by women—77%. There were 23% of men. In 1970, 62 baptism rites, 48 last rites, and 30 confirmations were performed in the registered community. The following year, the number of rites decreased slightly—41 baptism rites, three wedding ceremonies, 37 last rites, and 15 confirmations. Among those who were confirmed, 26% were boys, girls—74%. In 1970, the community held 131 prayer meetings. The profits of the community increased to a certain extent from 1970 to 1971. Thus and so, if in 1970 the profits amounted to 1,852 rubles, then in 1971—2,326 rubles (Fund 1418. List 1. File 154). Generally, it should be noted that since 1970 there has been an impressive reduction in the number of Lutherans. Thus, by the beginning of 1970, there were 440 people in the communities, and by 1974 there were 270. The reduction in the number of Lutherans occurred mainly due to the collapse of rural associations and emigration to West Germany (Fund 1418. List 1. File 192. List 1).

## 4 Discussion

The state-confessional policy of the 1960s and 1970s was primarily directed toward several Protestant communities. The number of antisect literature published in the country overwhelmed the *educational* needs. The life of believers was depicted as gloomy. They were associated with fanatical rites and even child sacrifices. It should be noted that during the studied period, the main emphasis in antireligious propaganda was placed on the atheistic propaganda among children of Protestant believers through the school and higher education system. In 1971, the Ministry of Education of the RSFSR adopted a document named *A Letter on Strengthening the Atheistic Education of Secondary School Students*. Thus, in the Pervomayskaya school in Novosibirsk region, traditional antireligious evening gatherings were held annually, and the *Science and Religion* journal was advised for a subscription. Also, the school regularly held homeroom periods on antireligious topics. The primary school in the village of Neudachino actively worked on atheistic education. The school Director spoke about the experience of such work at one of the meetings in Tatarsk, which was attended by secretaries of city Committees, teachers, and leaders of the *Knowledge* society. Besides, teachers actively worked with parents of religious students at parent-teacher meetings and conducted conversations on atheistic topics. It is worth noting that in the plans of schools in Novosibirsk region, the section *Atheistic Education, Religion and Law*, and *About Miracles, Superstitions, and Rituals* was introduced. Teachers were also assigned to work individually with Protestant believers. Thus, in the report of the Soviet District Executive Committee of Novosibirsk, they noted: *in each school, two people were selected who supervised atheistic education, they worked with the children of believers*. Authority representatives themselves noted the intolerance to *sectarians* from the party and state officials. In 1965, the Commissioner of the Council for Religious Cults in Leningrad region, N. M. Vasiliev wrote, "Among the frontline, and often leading employees of the party-Soviet apparatus, under the influence of the press or other factors, it is considered that the faithful of the Orthodox Church are one thing, other confessions are also tolerated, but sectarians are some kind of incomprehensible ugliness, and, despite a conscientious attitude to work, discipline, religious sectarians as opposed to other believers are placed in the category of people who do not deserve full political trust" (Nikolskaya 2003). Thus, during the studied period, Pentecostal, Baptist (adherents of the Council of Churches of Evangelical Christian Baptists), and several other communities were under attack.

## 5 Conclusion

In conclusion, we should point out that since the mid-1960s, the state-confessional policy has begun to change. It is confirmed by the Resolution of the Presidium of the Supreme Soviet of the USSR on Certain Facts of Violation of Socialist Legality Concerning Believers, according to which many court decisions against believers were canceled. During the studied period, revoking the registration of religious communities, which was common during the *thaw* period, stops. However, it is worth noting that most of the closed religious associations during the office of N. S. Khrushchev, including in Altai Krai and Novosibirsk region, were not restored to their rights. During the studied period, the communities of the Council of Churches of Evangelical Christian Baptists and associations of Evangelical Christian Baptists were subject to persecution. Many members of these communities were arrested and sent into exile. In this regard, the Council of Churches of Evangelical Christian Baptists' adherents began to operate under the radar. It was reflected, in particular, by the fact that supporters of the Council of Churches often changed places of worship services, places for teaching the younger generation religious basics. In addition, there is a revival of activity in the communities of the Council of Churches, which was expressed in establishing contacts with the believers of the Union Republics.

**Acknowledgements** This research was carried out within the framework of the Grant *Religion and Power: the Historical Experience of State Regulation of Religious Communities in Western Siberia and Neighboring Regions of Kazakhstan in the 19th–20th centuries (project No. 19-18-00023).*

## References

Fasikhova MN (2002) The policy of the Soviet state towards religious associations in the 1960s–1980s in Tatarstan: historical and political analysis (Dissertation of Candidate of Historical Sciences). Kazan Federal University, Kazan, Russia

Geraskin YuV (2008) Organization and development of institution of authorized agent for control of Russian Orthodox Church under USSR Council of Ministry. Izv Altai State Univ 4:45–51

Gorbatov AV (2008) Government and religious organizations of Siberia in the 1940s–1960s. TSPU, Tomsk, Russia

History of the Altai Ecclesiastical Mission. https://eparhia-gorniyaltay.ru/?page_id=1023

Kreydun YA (2010) Churches of Gorno-Altaisk and its suburbs in the 20th–early 21st centuries. ASU, Barnaul, Russia

Molodov OB (2015) Comissioners of the council for religious affairs and realization of the state religious policy in the Russian North, 1960s–1980s. Perm Univ Her 1(28):242

Novikov MP (ed) (1985) Atheistic Dictionary. Politizdat, Moscow, Russia

Nikolskaya TK (2003) Russian Protestantism and state power in 1905–1991 (Dissertation of Candidate of Historical Sciences). European University at St. Petersburg, Saint Petersburg, Russia

Odintsov MI (2002a) Russian Orthodox Church in the 20th century: history, relationships with state and society. CIOE, Moscow, Russia

Odintsov MI (2002b) The Council of Ministers of the USSR decides: "to exile forever!"—collection of documents and materials about Jehovah's Witnesses in the Soviet Union (1951–1985). Russian association of scholars in religion, Moscow, Russia

State Archives of Altai Krai. (Fund 1692a "Council for Religious Affairs under the Council of Ministers of the USSR (1943–1991)". List 1. File 76). Barnaul, Russia

State Archives of Altai Krai. (Fund 1692b. "Council for Religious Affairs under the Council of Ministers of the USSR (1943–1991)". List 1. File 127). Barnaul, Russia

State Archives of Altai Krai. (Fund 1692c. "Council for Religious Affairs under the Council of Ministers of the USSR (1943–1991)". List 1. File 107). Barnaul, Russia

State Archives of Altai Krai. (Fund 1692d. "Council for Religious Affairs under the Council of Ministers of the USSR (1943–1991)". List 1. File 115). Barnaul, Russia.

State Archives of Altai Krai. (Fund 1692e "Council for Religious Affairs under the Council of Ministers of the USSR (1943–1991)". List 1. File 119). Barnaul, Russia

State Archives of Altai Krai. (Fund 1692 "Council for Religious Affairs under the Council of Ministers of the USSR (1943–1991)". List 1. File 74). Barnaul, Russia

State Archives of Novosibirsk Region. (Fund 1418a "Council for Relations with Religious and Charity Organizations of the Novosibirsk Region Administration, Novosibirsk, 1943–present time". List 1. File 80. Page 16). Novosibirsk, Russia

State Archives of Novosibirsk Region. (Fund 1418b "Council for Relations with Religious and Charity Organizations of the Novosibirsk Region Administration, Novosibirsk, 1943–present time". List 1. File 179). Novosibirsk, Russia

State Archives of Novosibirsk Region. (Fund 1418c "Council for Relations with Religious and Charity Organizations of the Novosibirsk Region Administration, Novosibirsk, 1943–present time". List 1. File 172). Novosibirsk, Russia

State Archives of Novosibirsk Region. (Fund 1418d "Council for Relations with Religious and Charity Organizations of the Novosibirsk Region Administration, Novosibirsk, 1943–present time". List 1. File 208). Novosibirsk, Russia

State Archives of Novosibirsk Region. (Fund 1418e "Council for Relations with Religious and Charity Organizations of the Novosibirsk Region Administration, Novosibirsk, 1943–present time". List 1. File 228). Novosibirsk, Russia

State Archives of Novosibirsk Region. (Fund 1418f "Council for Relations with Religious and Charity Organizations of the Novosibirsk Region Administration, Novosibirsk, 1943–present time". List 1. File 188). Novosibirsk, Russia

State Archives of Novosibirsk Region. (Fund 1418g "Council for Relations with Religious and Charity Organizations of the Novosibirsk Region Administration, Novosibirsk, 1943–present time". List 1. File 192). Novosibirsk, Russia

State Archives of Novosibirsk Region. (Fund 1418h "Council for Relations with Religious and Charity Organizations of the Novosibirsk Region Administration, Novosibirsk, 1943–present time". List 1. File 255. List 1). Novosibirsk, Russia

State Archives of Novosibirsk Region. (Fund 1418a "Council for Relations with Religious and Charity Organizations of the Novosibirsk Region Administration, Novosibirsk, 1943–present time". List 1. File 19). Novosibirsk, Russia

State Archives of Novosibirsk Region. (Fund 1418b "Council for Relations with Religious and Charity Organizations of the Novosibirsk Region Administration, Novosibirsk, 1943–present time". List 1. File 154). Novosibirsk, Russia

State Archives of Novosibirsk Region. (Fund 1418c "Council for Relations with Religious and Charity Organizations of the Novosibirsk Region Administration, Novosibirsk, 1943–present time. List 1. File 230). Novosibirsk, Russia

State Archives of Novosibirsk Region. (Fund 1418d "Council for Relations with Religious and Charity Organizations of the Novosibirsk Region Administration, Novosibirsk, 1943–present time". List 1. File 192. List 1). Novosibirsk, Russia

Tsypin V (1997) History of the Russian church, 1917–1997. The Valaam Monastery, Moscow, Russia

# Inter-Religion Relationships in Christian Communes of South Western Siberia in the 1920s–1930s

**Natalya P. Zibert**

**Abstract** In this study, we analyzed the religious policies of the Soviet government and its influence on the relations between different Christian communes of South Western Siberia in the 1920s–1930s. The analysis of archive documents has shown that the government systematically oppressed Tikhon communes, while Renovationist communes, supported by the Bolsheviks, were able to proselytize in various regions of the country, including South Western Siberia. But the general population was quite conservative and opposed the Renovationist church, continuing to visit official churches. This fact and the release of Patriarch Tikhon from house arrest in 1923 led to the increased severity of the Soviet anti-religious campaign, depriving Renovationist communes of their privileges and the equalization of all religious communes.

**Keywords** Religious policies · Renovationism · Russian Orthodox Church · Soviet period · Siberia

## 1 Introduction

Researching the Renovationist communes in USSR, their interactions with other communes and the government is one of the trending topics in Russian historical sciences. This paper examined the activities and relations of Christian communes in the 1920s–1930s in South Western Siberia. This territory was always ethnically and religiously diverse. Therefore, researching inter-commune relations allows one to fully understand the religious environment in the region at that time.

N. P. Zibert (✉)
Altai State University, Barnaul, Russia
e-mail: zibert.natalya18@mail.ru



S. G. Maximova et al. (eds.), *Advances in Natural, Human-Made, and Coupled Human-Natural Systems Research*, Lecture Notes in Networks and Systems 234,
https://doi.org/10.1007/978-3-030-75483-9_80

## 2 Materials and Methods

The material for this study was provided by the State Archive of Altai Krai and the State Archive of the Novosibirsk Region. I analyzed the data on: (1) the activities of the executive committee [Ispolkom] in the Altai Governorate; (2) the correspondence of Altai eparchy and the higher levels of church and state government bodies, church councils, and priests; (3) priest reports on various issues. The methodology of this paper is based on the principles of determinism, objectivity, historicism, and systematicity. The main methods of research are the retrospective method and the historical-comparative method. The historical-comparative method allowed me to compare the religious communes in South Western Siberia. The retrospective method allowed for tracing the patterns in the interaction of religious communes between each other and the government.

## 3 Results

After the revolution of 1917, Russian society changed considerably, including its religious side. The "Declaration of the Rights of the Peoples of Russia" effectively canceled all national and religious privileges. The government signed a number of laws aimed at confiscating the financial resources of the churches, and primarily of the Orthodox Church. The government also made several attempts to introduce a schism into the Russian Orthodox Church. In 1918, Vladimir (Putiata) founded the "Free Folk Church" in Penza. In 1919, the government formed the "Ispolkom of the Clergy on the Basis of the Communist System". In 1920, Protoiereus Koblov founded the "Puritan Party of Revolutionary Clergy and Laity" in Saratov. In 1921, the "Renovationist Church" appeared in Saratov. But the lack of support from the state in the first post-revolutionary years prevented these organizations from playing any significant role in society. However, the ideas of Christian socialism were still popular among the clergy. This created the conditions for church reforms that would soon follow.

In the first post-revolutionary years, the Orthodox communes were persecuted and repressed the most. Various Protestant communes did not have as much wealth and, therefore, were more favored by the state. Since 1918, the Protestant commune started selling agricultural products (All-Union Council of Evangelical Christians-Baptists 1989). In winter 1919, the government issued the decree "On religious exemption from conscription", which allowed for a complete or partial draft exemption for people whose religion prohibits military service Administrative Directorate of Council of People's Commissars of the USSR 1943). Thus, draft dodgers considerably increased the number of people in the Protestant communes (Administrative Directorate of Council of People's Commissars of the USSR 1943).

To prevent this, in 1923, the government signed the "Civil Procedure Code of the Russian Soviet Federative Socialist Republic" that contained the chapter "On religious exemption from military service". Pursuant to the new rules, a person could be exempted only according to the verdict of several religious and secular commissions. These commissions aimed to establish whether a person actually follows a religion that prohibits military service. The increase in the population of the Protestant communes happened not only because of the draft dodgers, but also due to the overall friendliness of the communes, mutual economic help, and their isolation from the government (Savin 1999).

But the increase in Protestant communes caused local governments and the State Political Directorate [GPU] to take measures against the increasing spread of Protestantism (Veis 2007). This forced many Protestant groups to either cease their activities or go underground (Savin 2004).

In 1922, on behalf of L. D. Trotsky, the government developed a new religious policy aimed at the internal dissolution of religion. The officials believed that continuous replacement of religion by communistic values would cause the Russian Orthodox Church to vanish. This idea was popular among the party activists. For example, the works of E. M. Yaroslavsky frequently mention the priests that want to join the party and form a cluster of communist priests. G. E. Zinovyev and A. V. Lunacharsky believed that a church, with proper training, can help in propagating communist ideas. However, P. A. Krasikov, I. I. Skvortsov-Stepanov, and V. I. Lenin were against such methods. Lenin believed that "if for some people the statement 'socialism is a religion' is a form of transitioning from religion to socialism, for others it is vice versa" (Nezhivykh 2012). But when Lenin fell ill and retreated from government affairs, Trotsky was free to force his policies. He effectively split the Russian Orthodox Church into two: the Renovationist church (loyal to the Bolsheviks) and the Tikhon Church. (Petrov 2001). The Soviet government used the Renovationist movement not only to cause a schism (Churkin 2012) but also to confiscate Orthodox Church property. Supported by the Bolsheviks, the Renovationist movement quickly gained influence in most regions of the country, including South Western Siberia. Sometime later, a group of priests, including Krasnitsky and Vvedensky, received the right to govern church activities from the Patriarch (Nezhivykh 2012). In May 1922, newspapers announced the creation of a High Church Administration, which was to perform the functions of the church management (Churkin 2012). Then, on the so-called "constituent assembly" (synonymous, but unrelated to the "All Russian Constituent Assembly"), the organization of the "Living Church" was created, which now governed the Patriarch's residence and executive office. Moreover, the "constituent assembly" approved the "Program of Church Reforms, outlined by the clergy and laypeople of the "Living Church". This program rejected the "scholastic perversions" of Christianity and returned to the Evangelical faith. They aimed to develop the idea of God being a source of truth, love, and mercy, and the human being the pinnacle of God's creation. The liturgy was to discard all rituals that can be traced back to pagan faiths and make divine services more accessible and understandable (Golovushkin 2014).

The Central Committee of the Bolshevik Party was somewhat cordial to the Renovationists. The Renovationists met little to no resistance from the local governments and were able to proselytize in Siberian villages easily (Fund 2, List 3, File 37). In Biysk, the local ispolkom deposed the Archbishop Innocent (Sokolov) and gave the post to Ioann Zavadsky, a Renovationist priest (Fund 536, List 1, File 233). The newspapers regularly wrote about court cases against the priests who remained loyal to Patriarch Tikhon. The press actively called Tikhon and his supporters the enemies of the revolution (Cherkazyanova 1994).

In Siberia, the Renovationist movement started from Tomsk. In this city, the Renovationist church started issuing "The Siberian Church" magazine, propagandizing Evangelical reforms in the church. The Renovationist church was the most active in Tomsk, Krasnoyarsk, and Novonikolaevsk (now "Novosibirsk") eparchies (Cherkazyanova 1994). Most of the churches in Novonikolaevsk Governorate were under the Renovationist rule. In some cases, a church could have two communes: both Tikhon and Renovationist ones (Shabunin). The situation in the other regions of the country was similar. For example, in the Astrakhan eparchy, the GPU arrested the priests that refused to submit to the Renovationist bishop Anatoly. Most of the churches were transferred to the Renovationists (Skolota 2008). Renovationism was actively supported in the Northern Caucasus. By the beginning of 1926, there were 876 Renovationist parishes in the region (Pospelovsky 1996). By October of the same year, their number rose to 960 (Pantyukhin 2014).

In 1922, bishop Victor of Tomsk eparchy was accused of resisting the confiscation of church property and arrested. In June 1922, a Renovationist "Siberian Church Administration" was founded, led by P. F. Blinov, a priest (Shilkina 2011). In September 1922, bishops Sophronius and Gavriil promoted Ioann Zavadovsky, a priest, to the bishophood. He would later lead the Barnaul Renovationist Church. In October 1922, P. F. Blinov was too promoted to bishophood on an All-Siberian Assembly of the "Living Church". The arrest of Sophronius at the end of 1922 allowed the Revationist church to consolidate their positions in Novonikolaevsk, which then became the center of the Siberian Renovationist Metropolis (Gavrilov 2006).

In 1924, Patriarch Tikhon created a separate Novonikolaevsk Eparchy and gave it to Archimandrite Nikifor (Astashevsky). The arrival of Nikifor in Novonikolaevsk caused mass return from Renovationism to the Orthodox Siberian Metropolis (Ovchinnikov 2015). M. Lashevich, the chairman of the Siberian Revolutionary Committee, said that the government was cordial to the Orthodox Church but stressed out the need to reunite the two groups. When the Orthodox Siberian Metropolis refused to unite with the Renovationists, the local government arrested bishops Sophronius and Innocent on suspected support of Patriarch Tikhon (Fund 1, List 1, File 759). This caused several churches of Novonikolaevsk Governorate to come under the influence of the Renovationists: Alexander Nevsky Cathedral, the Prophet Daniel Church, the Church of Our Lady of Kazan, St. Nicholas Chapel. In the Church of the Resurrection and the Church of the Protection of the Holy Virgin, the communes of Renovationists and Tikhon supporters co-existed (Shabunin). Such a situation was common in other regions of the country. For example, in the Kolpino church, which was under the control of Renovationists, Tikhon supporters conducted their services in one of the parakklesions (Skolota 2008). The state financially supported the Renovationist churches. There were tens of uncensored Renovationist magazines in the country, while the Tikhon church published none.

Some clergy of Omsk joined the Renovationist Church, not knowing the true intentions of the organization. Some joined it to avoid being arrested and exiled. One could split the clergy into three groups (Cherkazyanova 1994). Those who joined out of their own volition; 2. Those who were pressured by the situation or their superiors; 3. Those who remained loyal to Patriarch Tikhon. The Bolsheviks did not shy from blackmailing Tikhon priests (Govorova 2006). The priests that refused to join the Renovationist Church were often accused of resisting the confiscation of church property or counter-revolution and then exiled to labor camps (Nezhivykh 2012). For example, Tobolsk protoiereus V. Khlynov was sentenced to 5 years in the Central-Asian Labor Camp. He died in the camp (Kapkov 2018). Omsk protoiereus V. Infantyev was arrested and executed by shooting for his anti-Renovationist work. Irkutsk archbishop Anatoly and Metropolitan Veniamin of Petrograd were executed for "hiding church property from confiscation". Omsk priest P. Olersky was arrested multiple times for refusing to join the Renovationist Church. In 1922, for refusing to relinquish church property, including eight silver icon plating, Omsk priest Lebedyansky was sentenced to 1.5 years of prison (Cherkazyanova 1994). In 1923 in Barnaul, a group of priests was arrested for hiding church property.

One must note that the clergy was initially interested in the church reforms, including the modernization and democratization of church services. The legitimacy of the Renovationist "Holy Synod" was confirmed by the Ecumenical Patriarchate of Constantinople, which boosted the popularity of the Renovationism further. The Renovationist Metropolitan of Siberia, P. F. Blinov, stated: "...all Siberian parishes have their own Renovationist allies" (Sokolov). Some priests that refused to join the Renovationism had to leave their parishes. For example, the priest of Sredne-Krasilovskoe village, Fedor Zavyalov, was forced to move to the villages of Shadrinskoye and Buranovskoye of Altai Governorate. Ioann Pulkhrov, who refused to join the Renovationist Church, conducted services in the Pokrov Temple in Barnaul, despite being prohibited from doing so (Fund 218, List 1, File 218). However, the Renovationist Church was not popular among the laypeople. The laity mainly opposed all changes in church services and dogmas. The population refused to follow the Gregorian calendar (also called "New Style"). Thus, most Renovationist parishes conducted no holiday services (Borisova 2016). The population especially opposed the new ability of the clergy to marry a second time, the closure of monasteries, and the "theory of class struggle with Christian ideas." The disunity of the Renovationist movement also inhibited its popularity (Cherkazyanova 1994). This was also common in the neighboring regions. In the Trans-Baikal Governorate, the priests refused to conduct the "Old Church" services. The population remained distrustful and suspicious of the Renovationists (Grigoryeva 2015).

The Renovationists also had strained relations with the communes of other religions. For example, the higher ranks of the Renovationist Church refused to maintain a dialogue with the Council of Evangelical Christians-Baptists, despite the efforts of the latter. Sometime later, the newspaper "Izvestiya" featured a Renovationist article accusing the Baptists of counter-revolution. At the same time, the Baptists noted the "danger of flirting with the people alien to the Christian spirit" (Fund 2, List 4, File 24). The Renovationists were also condemned by the Old Believers. In their "Appeal to the deacons, hierarchs, honest deacons and all Orthodox Christians of the Old Orthodox Church", they pointed their non-involvement with the "Living Church" and the fact that they always lived by the old dogmas (Cherkazyanova 1994).

When Patriarch Tikhon returned from his house arrest in 1923, he called on the clergy and laity to support Soviet authorities. This increased the number of clergymen that openly supported the Patriarch. In 1925, the Renovationist parishes started returning to the Tikhon Church. From January to May 1925, fifty Renovationist communes returned to the Tikhon Orthodox Church (Fund 2, List 6, File 209). Renovationist clergymen that lost the respect of their parishes frequently became irritable and turned to drinking, which drew the laity even further away (Fund 2, List 6, File 209). The priests that returned to the Tikhon Church gave different explanations for their apostasy. For example, the deacon of the Pavlovsk village cited that he had to follow the senior priest of the church (Fund 218, List 1, File 225). The deacon of Vasilyev Church in Solonovki village, Slavgorod district, publically repented and acknowledged the falsehood of the Renovationist beliefs. The priest of the Church of St. Michael the Archangel in Rogozikha village did the same. The believers of Nerchinsk and the village Kostino-Logovskoe wrote a letter of repentance and asked to be brought under the Biysk bishop Nikita. In Novokrayushikno, the congregation forced their priest to return to the Tikhon Church. Nikolai Kostin, a priest, in his report to the bishop Nikodim, states that, in 1923, he joined the Renovationist Church because of the printed propaganda. In March 1923, when he arrived at Barnaul and saw the truth, he announced his leave from the "Living Church" in the letter to the Altai Eparchial Council (Fund 218, List 1, File 218).

The "Second Particular Assembly of Russian Church" (also known as "First Renovationist Assembly") was described by Tikhon supporters as "a non-orthodox assembly of schismatic bishops, clerics, and laypeople". Communicating with the Renovationist priests was announced to be a grave sin. Children baptized by the Renovationist priests were commonly re-baptized (Fund 2, List 4, File 24).

Renovationist priests started serving in Tikhon parishes, unauthorized. For example, in 1925, Dmitry Unzhakov arrived from Biysk to serve in the temple of Manzherok village without proper authorization. Soon, Biysk Eparchial Church Council removed him from service (Fund 2, List 6, File 209).

In 1925, the hierarchs of the Renovationist Church wrote to bishop Nikodim and asked him to choose the delegates for the "Third Particular Assembly of Russian Church", which would include both the Renovationists and Tikhon supporters. One of the tasks of the Assembly was to prepare for the upcoming Ecumenical council (Fund 218, List 1, File 225). Bishop Nikodim promised to answer by July the 5th of 1925, after convening with his colleagues and getting appropriate permissions.

Barnaul clergy was skeptical of uniting with the Renovationists. For example, Valentin Popov, the priest of the Holy Trinity Church, thought that mending the schism is only possible if the schismatics return to the Tikhon Church. The priest of

Znamenskaya Church, Pavel Pobedonostsev, found irony in the fact that the schismatics were the first to propose mending the schism (Fund 218, List 1, File 225). On October the 1st of 1925, the Third Assembly took place in the Cathedral of Christ the Saviour, Moscow. The Renovationists announced that they would not further reform the divine services and dogmas. By 1930, the confrontations have significantly subsided, mainly because of the Soviet anti-religious measures (Nezhivykh 2015).

One should note that the relationship between the church and the state in the 1920s–1930s was strained, despite the Patriarchial attempts to ameliorate them.

## 4 Discussion

Most secondary literature on the Renovationist movement researches either all-Russian tendencies and key points (Kirichenko 2011; Mazyrin 2019; Zakharova 2013), or the tendencies in the smaller subdivisions of South Western Siberia: Novosibirsk Region, Tyumen Krai, Altai Krai, etc. (Nikulin 2011; Mavlyutova 2008; Mazyrin 2019; Dronova 2012). Currently, the main issue is the disunity of opinions on religious life in the 1920s.

Some scholars say that the Renovationist had genuine concerns about modernizing and renovating the Church. Other researchers believe that the schism was mainly caused by the intra-religion power struggle. The Tikhon supporters in the 1920s stated that the main cause of the schism was governmental repressions.

## 5 Conclusion

In conclusion, one must say that all attempts of the Soviet government to abolish religion failed. Despite the fact that Renovationist schism divided the clergy, most laypeople preferred the traditional Orthodox Church. Other non-Orthodox confessions also avoided the Renovationists movement, seeing them as allies of the state. The dis-unity of the movement and frequent misconducts of the priests considerably harmed their popularity among laypeople. All these facts caused the Renovationist Church to lose its privileged position—The soviet government started persecuting it with all other denominations.

**Acknowledgements** The work was supported by the Russian Science Foundation within the implementation of the project "Religion and Power: Historical Experience of State Regulation of the Activities of Religious Communities in Western Siberia and Adjacent Regions of Kazakhstan in the 19th–20th Centuries" (No. 19-18-00023).

## References


Administrative Directorate of Council of People's Commissars of the USSR (1943) Collection of governmental decrees and orders for 1919. Administrative Directorate of Council of People's Commissars of the Soviet Union, Moscow, Russia

All-Union Council of Evangelical Christians-Baptists (1989) History of the evangelical Christians-Baptists in the USSR. All-Union Council of Evangelical Christians-Baptists, Moscow, Russia

Borisova VV (2016) Renovationist schism in Toblosk eparchy. Orthodoxy Science Educ 2(2):21–31

Cherkazyanova IV (1994) Renovationism in Siberia. Bull Omsk State MusM Hist RegNal Stud 3:165–173

Churkin VF (2012) Ideocracy: the history of its formation in the USSR. Sci Bull Belgorod State University Ser: History Polit Sci, 7:244

Dronova VV (2012) On the question of church and state relations in Tobolsk eparchy in 1917—late 1930s. Bull Altai State Univ 4–1(76):89–92

Firsov SL (2014) Power and fire: Church and the Soviet state: 1918—early 1940s: historical essays. St.Tikhon's Orthodox University, Moscow, Russia

Gavrilov PY (2006) Novosibirsk eparchy: history and modern days. Novinvest plyus, Novosibirsk, Russia

Golovushkin DA (2014) Russian renovationist movement in 1922–1923: reformation or church revolution? Bull Irkutsk State University Ser "Politology Relig Studies", 8:232–240

Govorova IV (2006) Confiscation of church property in 1922 in the context of religious policy. Bull Samara Natl Res University Ser: Hist Hist Sci 3:99–103

Grigoryeva ES (2015) Trans-Baikal eparchy in the 1920s: relations of renovationism and traditional orthodoxy. In: Berdnikova TB, Shapieva AV (eds) Youth Scientific Spring—2015. Transbaikal State University, Chita, Russia, pp 122–124

Kapkov KG (2018) The last confessor of Emperor Nicholas II and his family: Tobolsk protoiereus Vladimir Khlynov. Based on archive documents. Tonchu, Tobolsk, Russia

Kirichenko OV (2011) Anti-renovationist activities of orthodox nuns in the 1930s. State, Relig, Church Russ Abroad 29(3–4):414–418

Mavlyutova ZS (2008) Renovationist movement in Tyumen Krai in 1920s. Bull Archaeol, Anthropol, Ethnogr 8:115–120

Mazyrin AE (2019) Soviet renovationism: church phenomenon or a tool of state security? State, Relig, Church Russ Abroad, 1/2(37):226–248

Nezhivykh NA (2012) The religious new economic policy. The religious policy of the state in the 1930s (on Western Siberia). Omsk State University, Omsk, Russia

Nezhivykh NA (2015) Some aspects of the "Gregorian schism." Bull Int Sci Conf 5(9):7–10

Nikulin AA (2011) Renovationist schism in Western Siberia in the 1920s–1930s. Bull St-Petersburg University Hist 2(2):85–90

Ovchinnikov VA (2015) Periodization of Russian Orthodox Church history in Siberia in Soviet times (1917–1991). Bull Kemerovo State University Ser: Hist, Hist Sci, 1(61):72–80

Pantyukhin AM (2014) Renovationist movement in 1920–1940 (on Stavropol and Terek). AGRUS, Stavropol, Russia

Petrov SG (2001) Renovationist schism in Siberia. HumIties Sib 4:79–82

Ponomarev S. Formation of alternative civil service in Russia. http://polit.ru/article/2004/04/08/civil_service/

Pospelovsky DV (1996) Russian Orthodox Church in XX cent. Direct-Media, Moscow, Russia

Savin AI (1999) On the question of the governmental policies towards sects in the early 1920s in Siberia. In: Shishkin VI (ed) Problems in the history of local governance in Siberia in the late XVI—early XX centuries. Novosibirsk, Russia, pp 274–278

Savin AI (2004) Soviet state and Evangelical churches of Siberia in 1920–1941: documents and materials. Posokh, Novosibirsk, Russia

Shabunin EA. Renovationism in Siberia. http://www.orthedu.ru/kraeved/1661-10.html

Shilkina MV (2011) Approaches to researching "Renovationism" in religious and secular literature. Bull Russ State Univ Humanities Ser: Hist Philology Culturology Orient Stud 17:174–180
Skolota RV (2008) Essence and regional features of the Renovationist schism in the Astrakhan region (1920–1925). Bull Novgorod State University Ser: Hist Hist Sci 49:29–32
Sokolov RA. Russian Orthodox Church in Altai in the period of repressions. https://ruskline.ru/analitika/2007/11/07/russkaya_pravoslavnaya_cerkov_v_period_gonenij/
State Archives of the Altai Krai. (Fund 2 "Altai Governorate committee of Russian Communist Party (Bolsheviks) (1917–1925)", List 3, File 37). Barnaul, Russia
State Archives of the Altai Krai. (Fund 2 "Altai Governorate committee of Russian Communist Party (Bolsheviks) (1917–1925)", List 4, File 24). Barnaul, Russia
State Archives of the Altai Krai. (Fund 2 "Altai Governorate committee of Russian Communist Party (Bolsheviks) (1917–1925)", List 6, File 209). Barnaul, Russia
State Archives of the Altai Krai. (Fund 218 "Novoaltaisk local council of Zmeinogorsk district of Tomsk Governorate, village Novo-Altaiskoe (unkn.—1919a)", List 1, File 225). Barnaul, Russia
State Archives of the Altai Krai. (Fund 218 "Novoaltaisk local council of Zmeinogorsk district of Tomsk Governorate, village Novo-Altaiskoe (unkn.—1919b)", List 1, File 218). Barnaul, Russia.
State Archives of the Altai Krai. (Fund 536 "Barnaul eparchial office of the traditional Orthodox Church (1920–1937)", List 1, File 233). Barnaul, Russia
State Archives of the Novosibirsk Region (n.d.)(Fund 1 "Siberian Revolutionary Committee. 1919–1925 (from 18 of September 1919—in Chelyabinsk, from November 1919—in Omsk, from June 1921—In Novonikolaevsk)", List 1, File 759). Novosibirsk, Russia
Veis AI (2007) Some of the history of the evangelization of Siberia. Sib Fields 1(58):25–28
Zakharova LB (2013) Renovationist schism in the Russian Orthodox Church in the 1920s. Bull High Education Volga Region Humlties, 1(25):52–58

# Altai "Stonemasons" as a Religious Movement in the Altai-Asian Frontier in Patriotic History of the 19th—Early 21st Centuries

**Kirill S. Matytsin, Petr K. Dashkovskiy, and Elena A. Shershneva**

**Abstract** Russian historiographers have different views on the Altai ethnographic group called "stonemasons." They lived in the Altai area in the eighteenth century. Most of them were old believers called Bespopovtsy. Their confessional affiliation and attitude towards the authorities and the church provoked repressive actions from the government. Consequently, it contributed to the development of new territories outside the Russian Empire and led to interaction with various ethnic groups living on the Altai-Asian Frontier. Due to the research studies on the Imperial, Soviet and Russian periods, we found out the change of views on the given problems. As a result, the novelty of the research lies in the fact that the Altai "stonemasons" were not only an ethnographic group formed in the foothills of Altai and existed until the end of the eighteenth century but also a popular religious movement existing throughout the entire period of the Russian Empire.

**Keywords** Altai "stonemasons" · Altai · Old believers · Historiography · Altai-Asian frontier

## 1 Introduction

Religious movements as an alternative to official religions require a comprehensive and detailed study. As of today, the researchers are paying much attention to this phenomenon. It remains one of the main issues in religious studies. In our opinion, the premises for the formation of many religious movements and their ideological basis are intricately connected with historical science.

K. S. Matytsin (✉) · P. K. Dashkovskiy · E. A. Shershneva
Altai State University, Barnaul, Russia
e-mail: cyril.matytsin@gmail.com

P. K. Dashkovskiy
e-mail: dashkovskiy@fpn.asu.ru

E. A. Shershneva
e-mail: D2703@yandex.ru

 
S. G. Maximova et al. (eds.), *Advances in Natural, Human-Made, and Coupled Human-Natural Systems Research*, Lecture Notes in Networks and Systems 234,
https://doi.org/10.1007/978-3-030-75483-9_81

The first alternative religious movements in Russia appeared in the Middle Ages. The state and the church considered them heretical (Kazakova and Lurie 1955). Later, the development of the religious movements was facilitated by the reforms of Russian Orthodox Church in the seventeenth century. They caused indignation and discontent among some parts of the population. As a result, the most radical peasants became the leaders of popular religious movements. They wanted freedom from state oppression. In our opinion, the Altai "stonemasons" is one of such movements. Initially, this ethnographic group was included fugitive peasants, artisans, and soldiers. Over time, due to the search for Belovodye (a sacred place for the old believers) it expanded its influence in the Altai territory, and then transformed into a popular movement, affecting other regions of the Russian Empire as well.

## 2 Materials and Methods

To understand the peculiarities of this popular religious movement, we should analyze how the topic was studied and research conclusions. It is essential to determine the approaches, the methods, and the evolution of the authors' opinions. Consequently, the historiographical approach to the given research seems to be the most relevant.

The proportion of the historiographical studies in relation to the volume of the published historical works on this topic is currently small. Since the Altai "stonemasons" study is a narrowly focused topic, it requires a particular approach to historiographical research.

The source base of the study includes a wide range of historiographical and historical materials. All identified sources are presented in two groups: (1) monographs, papers, theses, conference materials; (2) periodical press materials. In the meantime, the groups mentioned above are classified into two types of historical knowledge, according to the phenomenological concept of the source studies: (1) socially-oriented historical description; (2) scientific research.

The research methodology and the methods are based on an interdisciplinary approach. Consequently, we can highlight its multifaceted nature. It is also worth noting that studying the dynamics of the processes relates to historical science, culture, and social relations. There are many current modern historical and conceptual models, but we use the cultural-civilizational approach in this research. With a deep understanding of the terms "civilization" and "culture," we can reveal individual nations and states" peculiarities and those of regional groups. Consequently, the Altai "stonemasons" study considers ethnographic, religious, socio-political, and cultural-historical aspects. The civilizational approach allows us to investigate historical processes and, at the same time, consider objective (socio-cultural) and subjective (personal) factors relating to the coverage of phenomena and facts.

The principles of scientific objectivity and historicism are essential for this research. They allow us to reveal historiographical patterns of various scientific knowledge stages shaped under the influence of science and the development of society.

In the study, we use the problem-chronological and retrospective methods as well as the method of actualization. The method of periodization of the historical process allowed us to organize the research. It also enabled us to classify the periods in the study of the Altai "stonemasons." This method allowed us to define the main trends of scientific thought development in changing historiographical situations of Russia.

The study employs analysis and comparative research. The latter method allows us to compare the historiographical content on the same topic and to reveal the degree of analogies, inconsistencies, and additions. Using synthesis and classification methods, we summarized historical data, aspects, and concepts, related to the research topic. Then we classified them to represent historical and historiographical dynamics.

Our research also involves the source-based approach in the study of ethnographic works and archival materials.

## 3 Results

A comprehensive historiographical study of the Altai "stonemasons" has not been conducted so far. Nevertheless, it should be noted that the first attempts to make a historiographical analysis of the origin, formation, and worldview of the Altai "stonemasons" were made in the nineteenth century.

The analysis of the works of Imperial period was closely related to the society rather than scientific goals. Throughout the Imperial period, there were a small number of historical works on this topic. That is why they can be found mainly in ethnographic studies. In this case, we can identify two types of analysis of the predecessors' work. In the first case, the ethnographer of the Imperial period, while analyzing the studies of other authors, did not subject them to critical analysis, but used them as references or confirmation of his assumptions. Consequently, when considering the Altai "stonemasons," A. A. Printz referred to S. I. Gulaev, M. V. Shvetsov, P. S. Palace, K. F. Ledebur, and B. G. Gerasimov referred to N. M. Yadrintsev (Printz 1867; Shvetsova 1898; Gerasimov 1911). The second type includes criticism of the previous works. It should be noted that the critical view of the problem was not always observed here as well. For example, N. M. Yadrintsev, criticizing A. A. Printz, who claimed that the Uimon Valley was inhabited from the Bukhtarma River, presented his vision of the problem instead of detailed consideration of the findings (Yadrintsev 1886). The ethnographic works of the 19th—early twenty-first centuries also contain objective assessments. The project of A. E. Novoselov is an example of such work. During the trip to the settlements of "Polish" old believers, he found no confirmation of the view of M. V. Shvetsova on the problem of forming an ethnographic group of the Altai "stonemasons" (Novoselov 1913). The last example can testify to a certain scientific leap in ethnographic research at the end of the Imperial period.

Among the few historical works of the Imperial period, it is worth mentioning E. F. Shmurlo. In addition to the archival and court cases, his work also involved the ethnographic notes and essays. E. F. Shmurlo, relying on historiographical sources, disagrees with ethnographers who considered the Altai "stonemasons" independent. In particular, the historian subjected N. M. Yadrintsev's work to critical analysis (Shmurlo 1898).

During the Soviet period, the historiographical approach has changed largely due to the foundation of USSR Academy of Sciences. Published monographs and papers began including historiographic materials, mentioning the Altai "stonemasons." It is necessary to single out the works of K. V. Chistov, V. F. Lobanov, and N. N. Pokrovsky. K. V. Chistov gave a historical overview on the problems of the Altai "stonemasons" in Soviet times (Chistov 1962). It presented the most common opinions of pre-revolutionary researchers about the Altai "stonemasons." Besides, the research of E. E. Blomkvist and I. P. Grinkova was described positively (Blomkvist and Grinkova 1930). Dedicating a whole chapter to the Altai "stonemasons" in his monograph, N. N. Pokrovsky considered the views of researchers of Imperial and Soviet periods as well. He criticized the authors who wrote about the formation of an ethnographic group of Altai "stonemasons" in the '20s of the eighteenth century. The conclusions of N. M. Yadrintsev and A. N. Beloslyudov about the settlement of the Uimon Valley were also critically assessed. The view of K. V. Chistov on the old believers' communities called *soglasiya*, which influenced the spread of the idea of Belovodye, was also partially revised by the historian (Pokrovsky 1974). V. F. Lobanov questioned the time of creation of the old believers' text *Traveller*, a guidebook to Belovodye through the settlements of "stonemasons," proposed by K. V. Chistov (Lobanov 1980).

In the works of scholars of the Russian period, we noticed the trend of appealing to predecessors' works. It should be noted that modern scholars based their conclusions on their predecessors' opinions and adequately analyzed the source base and interpretation methods. Consequently, many views on the problems of the Altai "stonemasons" were reconsidered. Among other works, we should highlight the study of N. I. Shitova, which is related to the clothes of Uimon old believers. Using many historiographical sources, she analyzed all available information about the Uimon ethnographic group of the old believers. Due to the analysis, the researcher presented various hypotheses on the formation of this group. In this regard, N. I. Shitova suggested her version based on her ethnographic research. (Shitova 2005).

Most of the historiographical reviews are found during the coverage of the Belovodye topic. Consequently, due to the works review connected with the *Traveller*'s lists, A. A. Chuvyurov determined the geography of these texts and, as a consequence, identified the regions from which the old believers escaped to Bukhtarma and Uimon Valleys (Chuvyurov 2011). Yu. D. Rykov provides the most detailed historiographical overview in his paper. Before analyzing the new list of *Traveller*, Yu. D. Rykov provides an overview of all historiographical sources on Belovodye known at that time. Besides, he used a lot of secondary literature related to the Altai "stonemasons." The lack of information on this topic can be explained by the fact that Yu. D. Rykov considered only the ideological aspect of the old believers (Rykov 2011).

## 4 Discussion

The old believers' settlements located in the foothills of Altai became the research object during the pre-revolutionary period. At that time, travelers, ethnographers, and historians mainly studied ethnographic groups, including the Altai "stonemasons." It should be noted that travelers from Germany, such as: P. S. Pallas, K. F. Ledebur, A. A. Bunge, and K. A. Mayer; were the first to notice the old believers of Altai. The research was started on December 22, 1766, when St. Petersburg Academy of Sciences appointed P.S. Pallas as the Professor of Natural History. From 1768 to 1774, expeditions were made to the different regions of the Russian Empire. The expeditions to Asia aimed to collect information about nature, climate, fauna and flora, and mineralogical resources. Another mission was to obtain historical and ethnographic material. The expeditions were sponsored by St. Petersburg Academy of Sciences and the authorities.

In 1877, due to the establishment of the West-Siberian Department of the Imperial Russian Geographical Society, interest in the study of ethnographic groups increased. The works dedicated to the Altai "stonemasons" began to appear in the regional publications. It is important to emphasize that the researchers had different political views. Many of them were representatives of the so-called Narodnichestvo movement. This social and political movement rejected the idea of the superiority of the monarchic power over the peasantry. According to the opinion of the Narodniks, the peasantry must be free in the economic sense, and not be exploited, as it was in the Russian Empire. In this regard, the Narodniks tried to pay attention to the peasant communities, which sought independence. Consequently, the ethnographic group of the Altai "stonemasons" was singled out by the Narodniks as a group that contributed to the colonization process in Altai. The researchers also emphasized the contribution of the old believers to the economic development of the Russian Empire through maral breeding, beekeeping, animal husbandry, and trade with China. The Narodniks promoted A. P. Shchapov's theory, claiming that the old believers were primarily a political movement (Schapov 1906).

In general, the ethnoreligious group of the Altai "stonemasons" during the Imperial period was represented by two opposed political parties: conservatives and the Narodniks. The Narodniks, following the ideas of A. I. Herzen, saw an example of the peasant socialism in the Altai old believers. In this regard, the results of the ethnographic material analysis, collected during their expeditions, seem to be the idealization of real processes. This opinion is also supported by the fact that the ethnographic works were socially-oriented. On the other hand, the statements of the conservative party also do not seem objective. They sought to clarify the need for a hard line of the authorities on the old believers regarded as a sectarian movement, which denied the state system in its radical form. Nevertheless, both sides contributed to the study of Altai "stonemasons" by publishing ethnographic and historical materials on the given topic. During the Imperial period, the researchers raised issues related to the reasons for the formation of popular movements, their ideology, and their spreading.

The study of the Altai "stonemasons" in the Soviet period continued within ethnography and historical science. The historical area covered historical research considering written texts related to the Altai "stonemasons."

On the one hand, ethnographic expeditions to the settlements of the Altai were organized to collect information about this movement (their household, buildings, clothes, and art), and, on the other hand, to enrich the museum collections. During the Soviet period, expeditions were carried out in the '20s, late '70s, and early '80s of the twentieth century. In 1927–1928, S. I. Rudenko proposed E. E. Blomkvist and I. P. Grinkova to participate in the Kazakhstan expedition of the USSR Academy of Sciences to study the life of Russian population in Kazakhstan. The researchers chose the area of the Bukhtarma river. As a result, E. E. Blomkvist and I. P. Grinkova managed to get extensive information about the old believers, who represented most of the Bukhtarma district population, including the Bukhtarma "stonemasons" (Blomkvist and Grinkova 1930). The theory of A. P. Shchapov, developed in the previous Imperial period, was reflected in Soviet historical science. The "Raskolniks" and "sectarians" as popular movements were marked as a form of antifeudal protest. As a result, the Soviet historical science took as a basis the statements of K. Marx, F. Engels, and V. I. Lenin about the driving role of the class struggle in the development of antagonistic formations, and those about the everyday nature of peasants social protests in the context of the Imperial feudalism. Within this research, the first works mentioning the Siberian old believers' communities were those of A. S. Dolotov, A. A. Preobrazhensky, Yu. S. Bulygin (Dolotov 1930; Preobrazhensky 1972a, b; Bulygin 1974). This approach allowed scholars to continue viewing popular religious movements, including the Altai "stonemasons," as capitalist-oriented communities with the ideas of utopian socialism.

In the '60s of the twentieth century, the school of archaeologists created by N. N. Pokrovsky was founded in Novosibirsk. It was a significant moment in the study of the Altai "stonemasons." Its significance lies in the fact that it could bring together scholars from different branches of humanitarian knowledge. It became possible to share experience in the study of the old believers, sectarianism, and popular religious movements of the Russian Empire. In addition to the works of N. N. Pokrovsky, an outstanding contribution to the study of various aspects of life of the Altai "stonemasons" was made by other scholars as well, such as Mamsik (1987), Minenko (1979). Representatives of this school paid particular attention to the topic of the old believers' escapes. In the Soviet period, they it was perceived as antifeudal class struggle of peasants.

We should also mark the library stock work, which was carried out in the Soviet period. It gave more information about the Altai "stonemasons." Thanks to M. N. Speransky, I. M. Kudryavtsev, K. V. Chistov, V. F. Lobanov, it became possible to clarify the lists of the *Traveller*. The archaeographic expeditions have also made a significant contribution to it by thorough scientific analysis of the text of the route and its comparison with the legend of Belovodye existed among the "stonemasons." It is also important to mark the role of K. V. Chistov in the formation of a scientific understanding of the old believers' communities, which spread the legend of Belovodye (Chistov 1962).

In the Soviet period, science played one of the crucial roles. At that time, the course was taken to create a society which worldview was based on Marxism-Leninism ideology. Meanwhile, the researchers were under the control of various state and party authorities. Their works must have reflected the ideological principles of the state. These postulates discouraged the objective study of historical events. As a result, most researchers considered any oppositional movement in the Russian Empire to be a new evolutionary stage on the way to socialism. It should be noted that in the Soviet period, there were researchers who focused on the ideological aspects of the peasant movements. They found out that their religious principles could contribute to the development of capitalism. The scholars were looking for a comprehensive historical analysis of the ethnoreligious groups. However, the ideological principles of the Soviet period did not always allow it. Nevertheless, despite the influence of the ideology on science in the Soviet period, the scholars contributed to research on the ethnographic groups, including Altai "stonemasons." The archaeographic and ethnographic literature of the Soviet period was included in the source base of modern researchers.

In the Russian period, the identification of the Altai "stonemasons" was questioned due to the study of their ideology. It was suggested to consider the "stone" not only as a geographical object (a mountain) but also as a "pustyn" (a kind of the old believers' monastery) in the sense of the Altai old believers (Matytsin 2018). During the Russian period, there were practically no studies related to the research of the settlements of the Bukhtarma "stonemasons" in the Altai territory. The study by T. S. Mamsik is an exception in this sense. She found out that the "stonemasons" inhabited the Upper Priirtyshje in the '20s of the eighteenth century (Mamsik 2009).

Besides, the Russian period gave rise to the new versions of the legend of Belovodye. In particular, the theory of K. V. Chistov is being criticized and revised. One publication on the topic of Belovodye stands out among the others. It is a critical analysis conducted by N. N. Pokrovsky, A. I. Maltsev, A. A. Chuviurov. For example, A. I. Maltsev, criticizing the concept by K. V. Chistov, wrote:

> None of the pilgrims' works of the $18^{th}$–$20^{th}$ centuries mention Belovodye, and none of their manuscripts contains the lists of the *Traveller*. The idea of the existence of an entire country, where there is no Antichrist and where the "Donikon" priesthood (i.e., before the church reforms of Patriarch Nikon) has been preserved in its totality, fundamentally violates the pilgrims' teachings about the global and ultimate victory of the Antichrist and his final elimination of the "true" priesthood. The origin and the spread of the legend of Belovodye are not connected with the activity of an individual old believers' community. It existed among the Pomortsy, the Chasovennye, and the Beglopopovtsy (Maltsev 2002).

At present, despite the comprehensive analysis of the sources, the scholars still do not have an accepted view on the origin of some groups of the Altai "stonemasons" (Uimon old believers). In our opinion, this can be explained by the lack of a purposeful study on the issue.

In modern works, some researchers assess the Altai "stonemasons" protests in the Imperial period as positive, while others assess them as negative examples of interaction with the authorities. Nevertheless, it should be noted that the growing interest in the old believers at the present stage led to the reassessment of the activity of popular religious movements. For example, now they are studied as units of cultural and traditional values. The old believers are considered to be a historical and cultural phenomenon in scientific circles in Russia and abroad. The rejection of the previous views and methods, accepted in the historical science of the USSR, allowed scholars to rethink the previously published works and materials. It also enabled them to find new approaches, due to which the ideas about the Altai "stonemasons" were corrected and improved.

## 5 Conclusion

Summarizing the results of the conducted research, we can draw the following conclusions. In Russian historiography, during the Imperial, Soviet, and Russian periods, methodological changes took place in the study of the Altai "stonemasons" ethnographic group. Identifying these periods, we also compared them with each other on the following criteria: (1) methods; (2) approaches; (3) assessments in the study of the old believers and other popular movements. We can also conclude that historical assessments and interpretations relate to the changes in the political course of the state. Politics had a huge impact on science.

Analyzing the works of the researchers of the Imperial period, we can reveal the following topics related to the Altai "stonemasons": (1) the origin of the ethnographic group of the Altai "stonemasons"; (2) confessional composition; (3) household characteristics; (4) the idea of the free land called Belovodye, and its search. According to their issues, we can classify the historiographical sources into two groups: (1) the formation of the Altai "stonemasons"; (2) the ideological basis of the Altai "stonemasons."

In the Soviet period, scientific research was conducted under the influence of Marxism-Leninism. Consequently, the Altai "stonemasons" were viewed as a form of the antifeudal protest. Despite the small number of ethnographic expeditions in this period, the central studies related to the Altai "stonemasons" were conducted by analyzing historical sources.

During the Russian period in historical science, we can notice the trend to use different methodological approaches. This strategy allowed scholars to abandon the former Marxist-Leninist ideological clichés studying the problems mentioned above. Consequently, we can see a variety of opinions and assessments in science. The research is also affected by the changes in the ideology, the economic and political life of the Russian period. The idea of a confrontation between Westernism and Slavophilism is becoming vital again. The Soviet researchers' views on the state and religious policy in the Russian Empire are being reconsidered. As a result, the study of the old believers and related ethnographic groups, including the Altai "stonemasons," once again becomes scientifically relevant. Also, having considered the historiography of the three periods, we can conclude that the Altai "stonemasons" were initially regarded as an ethnographic group located in the valley of Bukhtarma. Only later did they begin to be regarded as a mass popular religious movement on the Altai-Asian frontier.

**Acknowledgements** The work was supported by the Russian Foundation for Basic Research within the implementation of the grants "Ethno-Religious and Social Processes as Factors of Cross-Cultural Interaction of the Population in the Cross-Border Area of Altai, Tuva, and Mongolia: History and Modernity" (Project No. 19-59-44002) and "Socio-Cultural Adaptation of Orthodox and Non-Orthodox Communities in Southern and Western Siberia in the Context of Modernization Processes in Russia During the Imperial and Soviet Periods" (Project No. 20-39-70010).

## References

Blomkvist EE, Grinkova IP (1930) Bukhtarma old believers. Materials of the expeditionary research commission. Academy of Sciences, Leningrad, USSR

Bulygin YuS (1974) The first peasants in Altai. Altai Book Publishing House, Barnaul, USSR

Chistov KV (1962) The legend of Belovodye. Quest Lit Folk Art 35:116–181

Chuvyurov AA (2011) The Traveler of Mark Topozersky: the geography of the existence of manuscript collections. In: Baiburin AK, Schepanskaya TB (eds) Folklore and ethnography. On the ninetieth birthday of K. V. Chistov. MAE RAS, St. Petersburg, Russia, pp 218–232

Dolotov AS (1930) Church and sectarianism in Siberia. Sibkrayizdat, Novosibirsk, USSR

Gerasimov BG (1911) In the Bukhtarma Valley (Brief historical and geographical essay with 3 tables of numbers). Notes Semipalat Banch ZSO IRGO 5:1–125

Kazakova NA, Lurie YaS (1955) Anti-feudal heretical movements in Russia of the 14th–early 16th century Moscow. Leningrad, USSR

Lobanov VF (1980) The new list of "Traveler" monk Michael. Sib Source Study Archeogr 1:208–211

Maltsev AI (2002) Belovodye. Orthodox Encycl 4:534

Mamsik TS (1987) Peasant movement in Siberia in the second quarter of the 19th century. Nauka, Novosibirsk, USSR

Mamsik TS (2009) Bukhtarma masons are the first settlers of the Upper Irtysh Region: ethno-cultural, estate, confessional composition. In: Shilovsky MV (ed) Upper Irtysh in the 17th–21st centuries. Parallel, Novosibirsk, Russia, pp 28–41

Matytsin KS (2018) On the issue of identification of Altai "masons." Proceed Young Sch Altai State Uni 15:222–224

Minenko NA (1979) Russian peasant family in Western Siberia (18th—first half of the 19th century). Nauka, Novosibirsk, USSR

Novoselov AE (1913) Report on a trip to Altai. At the old believers of Altai. Notes ZSO IRGO 1(2):1–18

Pokrovsky NN (1974) Antifeudal protest of the Ural-Siberian peasants of the Old Believers in the 18th century. Nauka, Novosibirsk, USSR

Preobrazhensky AA (1972a) Ural and Western Siberia at the end of the 16th—beginning of the 18th century. Nauka, Moscow, USSR

Preobrazhensky AA (1972b) Ural and Western Siberia at the end of the 16th—beginning of the 18th century. Nauka, Moscow, USSR

Printz AA (1867) Stonemasons, yasak peasants of the Bukhtarminsky volost of the Tomsk province and a trip to their villages and to the Bukhtarminsky Territory in 1863. Notes IRGO on Gen Geogr 1:543–582

Rykov YuD (2011) Yaroslavl list of "Traveler" by Mark Topozersky. In: Poloznev DF (ed) Book culture of the Yaroslavl Territory, 2014. IPC "Conversion"—Higher School of Business, Yaroslavl, Russia, pp 11–57

Schapov AP (1906) Zemstvo and Schism. Works 1:476–495

Shitova NI (2005) The traditional clothing of the Uimon Old Believers. Gorno-Altaisk State University, Gorno-Altaysk, Russia

Shmurlo EF (1898) Russian settlements beyond the southern Altai Range. Notes ZSO IRGO 25:1–64

Shvetsova MV (1898) "Poles" of the Zmeinogorsk district. Notes ZSO IRGO 26:1–77

Yadrintsev NM (1886) Schismatic communities on the border of China. The farmer is a diplomat and a warrior. Sib Collect Append East Outlook 1:21–47

# Peace, Justice, and Strong Institutions

# Constitutional and Legal Anomalies as a Threat to Sustainable Development of Russia

**Eugene S. Anichkin**

**Abstract** The article deals with constitutional and legal anomalies that are a serious obstacle to the effective legal regulation of public relations and sustainable legal development of modern Russia. The article analyzes such types of anomalies as: (1) the dualism of nominal constitutional law and reality, (2) interpretative and law enforcement deviations from the formal constitutional law norm, (3) normative contradictions, (4) gaps, and (5) other formal legal defects of individual sources and norms of constitutional law. Each of the anomalies is interpreted via the historical, comparative, system-structural, and formal-legal methods of the constitutional legal doctrine, current legislation, and law enforcement practice of the Constitutional Court of the Russian Federation. The author concluded that the poor knowledge of constitutional and legal anomalies, the complexity and discussion of this issue, and the practical relevance of the stated topic indicate the need for further research. As notable features of constitutional and legal anomalies, one may define: (1) the close connection with the Constitution of the Russian Federation, (2) the special scope, (3) specific content, and (4) high negative effect on the entire legal system of Russia. Further analysis of constitutional and legal anomalies would allow creating adequate measures to prevent and eliminate them, which will ensure the sustainable legal development of our country and the formation of a legal state.

**Keywords** Legal anomaly · Fictitiousness · Deviation · Contradiction · Gap defect · Efficiency · Sustainable development · Constitution

E. S. Anichkin (✉)
Altai State University, Barnaul, Russia
e-mail: rrd231@rambler.ru



S. G. Maximova et al. (eds.), *Advances in Natural, Human-Made, and Coupled Human-Natural Systems Research*, Lecture Notes in Networks and Systems 234,
https://doi.org/10.1007/978-3-030-75483-9_82

# 1 Introduction

The issues of improving legal regulation and the quality of law are of constant relevance in legal science. However, their solution is hardly possible without a balanced understanding of the factors that prevent normal legal education and effective operation of the law. Among these factors, a prominent place is occupied by legal anomalies. They are one of the many threats to the sustainable development of any country. Anomalies represent the objective and natural aspect of legal life. One aspect of legal life is the normal state of law, its principles, institutions, sources, norms, and compliance with the behavior of legal subjects, public relations, and facts. But the other unavoidable aspect of the life of the law is all sorts of deviations from its normal correct state, "which have a negative impact on the processes of formation and action of law as a regulator of public relations" (Nanba, 2014, p. 146). The author suggests calling all these deviations and inconsistencies by the collective term *legal anomaly* or *anomaly in the law*.

# 2 Materials and Methods

The methodological basis of the research is formed by a wide range of general and special methods of research. Dialectic, analysis and synthesis, induction and deduction, and abstraction were used as general scientific methods. Historical, comparative, system-structural, and formal-legal methods were used as special methods.

The historical method allowed us to study the conditions and processes in the evolution of anomalies in the constitutional law of the post-Soviet period, taking into account the changing socio-political context. Using the comparative method, the author analyzes the Russian constitutional and legal regulation on the background of previous stages of development of constitutional law and other sectoral regulations, and identifies the distinctive features of the development of anomalies in the modern constitutional law of Russia. The system-structural method was used to study the hierarchical, spatial, and subject elements of modern constitutional law to clarify the anomalies in them and their relationship to each other. The study of doctrinal, legislative, and law-enforcement sources was conducted by means of the formal legal method, which allowed forming the logic of the material presentation and the conceptual apparatus of the stated topic.

# 3 Results

The question of anomalies in law is poorly studied in legal science. A. V. Petrov, who was particularly interested in this issue, came to the conclusion that anomalies of law are "deviations in the content of law, the necessary connections of its elements and forms that are random and arbitrary for law as a purposeful subject activity..." (Petrov 2015, p. 177). Taking these definitions as a basis, the author makes the reservation that sometimes an anomaly of the law is understood in the broad sense of the word (i.e., it also includes socially useful, positive deviations in law) (Nanba et al. 2015). In this article, the subject of research is the negative anomalies that are found in various branches of law. Since constitutional law is a leading branch of Russian law, the constitutional-legal anomalies are of system-forming importance for the entire legal system of Russia (they are of scientific interest in the first place). The author believes that it is possible to identify five types of constitutional and legal anomalies. Some of them can be reproduced in other law branches, but in constitutional law they have special content due to industry specifications.

### *3.1 Dualism of Nominal Constitutional Law and Reality*

There is a formal and actual gap between the norm of constitutional law and the realities of public life. This is most clearly seen in the example of the conflict between specific norms of the Constitution of the Russian Federation (Russian Federation 1993) and informal constitutional practice. The progressivism of the constitutional norms, or, on the contrary, their conservatism gives rise to the fictitiousness of the Constitution and its isolation from real practice. For example, the development of our country at the beginning of the twenty-first century neutralized the potential of several provisions of the Russian Constitution. On the background of active changes of the state policy outdated and fictitious nature of the acquired norms about the contractual method of delimitation of competence between Federal bodies of state power and bodies of state power in the subjects of the Russian Federation (Part 3, Article 11), the residual field of reference regions (Article 73), the Republic as a State within the Russian Federation (Part 2, Article 5), on the implementation of Executive power in the Russian Federation the Government of the Russian Federation (Part 1, Article 110) and several others. Refusal of intra-federal treaties, a noticeable narrowing of the scope of the residual jurisdiction of the constituent entities of the Russian Federation, the unification of the legal status of constituent entities of the Russian Federation, the actual exercise of executive power not only by the Government of the Russian Federation but also by the President of the Russian Federation, as well as their authorized agents in the federal districts, have become the laws of the socio-political life of modern Russia, entering into a disagreement with the Constitution of the Russian Federation.

### 3.2 Interpretative and Law Enforcement Deviations from the Formal Constitutional and Legal Norm

Such deviations are most clearly seen in the example of the norms of the Constitution of the Russian Federation. They represent a difference between the literal meaning of constitutional norms, on the one hand, and their semantic development in current legislation and interpretation within the framework of the pro-enforcement practice of the Constitutional Court of the Russian Federation, on the other hand. Sometimes the norms of the Constitution get a "different interpretation" in federal constitutional laws, federal laws, decisions, and rulings of the Constitutional Court of the Russian Federation. However, the formal difference between the text of the current Constitution and laws that develops its meaning, or the legal position of the Constitutional Court of the Russian Federation does not always have an adverse legal effect. For example, through several decisions of the Constitutional Court of the Russian Federation, the number of objects of constitutional review along the lines of abstract and concrete normative control was noticeably increased compared to the provisions of Art. 125 of the Constitution of the Russian Federation and Art. 3 of the Federal Constitutional Law of July 21, 1994 "On the Constitutional Court of the Russian Federation" (Russian Federation 1994). Federal constitutional laws and the laws of the constituent entities of the Russian Federation on issues of their residual maintenance became the objects of abstract normative control. The federal constitutional laws, resolutions of the Supreme Council of the Russian Federation on the order of enactment of laws, repealed or expired laws, and resolutions of the State Duma on amnesty became the objects of specific normative control. In the given example, there is a discrepancy of a formal legal nature, but it is hardly possible to talk about violating the meaning of the norm of the Constitution.

However, the constitutional practice also demonstrates the facts of an unfortunate correlation between the provisions of the Constitution of the Russian Federation and its updated semantic "shell," that is, a negative deviation from the norm. Thus, the Federal constitutional law of December 17, 2001, "On the Procedure for Admission to the Russian Federation and the Formation of a New Constituent Entity of the Russian Federation" (Russian Federation 2001) illustrate the limitation of the literal meaning of the Constitution norm. It was adopted based on Part 2 of Article 65 of the Constitution, according to which "the admission to the Russian Federation and the creation of a new subject in it shall follow the procedure prescribed by Federal constitutional law." A literal understanding of the term "compound entity" in this context allows us to talk about such cases as the Union of the subjects of the Russian Federation, the division of the subject of the Russian Federation into two or more new subjects of the Russian Federation, the separation of one subject of the Russian Federation from the other. However, according to Part 1 of Art. 5 of the Law, "the formation of a new subject within the Russian Federation may be carried out as a result of the merger of two or more bordering subjects of the Russian Federation." In this example, the legislator identified the categories "compound entity" and "association," although the former is broader in meaning.

### 3.3 Regulatory Contradictions

The Russian constitutional law of the end of the 20th—first two decades of the twenty-first century is characterized by contradictions of general legal and industry-specific nature. In the literature, general contradictions within the legal system include the contradictions: (1) between the dynamics of social relations and the stability of the legal system, (2) between the tendency to establish uniformity of legal norms and the need to take into account industry and local characteristics, and (3) between the differentiation of legislation and its integration (Dzybova 2011). The subject of our study is the special contradictions that are characteristic of constitutional law as a branch of law. One can distinguish at least three types of such contradictions: (1) intra-textual; (2) hierarchical; (3) spatial.

(1) Intra-textual contradictions can be shown by the example of the Constitution of the Russian Federation as the main source of constitutional law (although they also occur in other normative legal acts). Thus, it is reasonable to talk about the contradictions between Chap. 1, "The Foundations of Constitutional Order," and the provisions of other chapters of the Constitution between the provisions of Chaps. 2 through 9 of the Constitution and constitutional values in Chap. 1. In the first case, this is, for example, a contradiction between the principle of equality of subjects of the Russian Federation (Part 1 of Article 5) and the right of republics to establish their own state languages (Part 1, Article 68), the right of the Autonomous region or Autonomous district to initiate the adoption of the Federal law on the Autonomous region or the corresponding Autonomous district (Part 3, Article 66). In the second case, it is reasonable to specify, for example, the contradiction between Paragraph "A" of Article 71 and Article 136: Paragraph "A" of Article 71 relates to the subject matter jurisdiction of the Russian Federation change of the Federal Constitution. In contrast, Article 136 provides that the amendments to Chaps. 3–8 of the Constitution enter into force after their approval by the legislative bodies of not less than two-thirds of subjects of Federation. In the third situation, the Constitutional Court of the Russian Federation in a number of its decisions (Anichkin and Rudt 2019) resolved the issue of overcoming the conflict between individual constitutional values reflected in Chap. 1 of the Constitution (the social state and market economy, state sovereignty and priority of international law over national law, recognition of human rights and freedoms as the highest value and security of the state).
(2) Hierarchical contradictions between normative legal acts with different legal force levels are standard for different branches of law. Constitutional law is no exception, but it has a specific content of hierarchical conflicts. Mainly, it is reduced to a contradiction between the norms of the Constitution of the Russian Federation, as an act that has the highest legal force, and the norms of other legal acts, and in the past, the provisions of treaties (Morozova and Karasev 2015, p. 28) on the division of subjects of competence and powers between Federal state authorities and state authorities of the subjects of the

Russian Federation. The rich practice of the Constitutional Court of the Russian Federation (Resolution of the Constitutional Court of the Russian Federation 2017; Resolution of the Constitutional Court of the Russian Federation 2020) shows that the norms of "unconstitutional" acts are often recognized by some extent as inconsistent with the Constitution of Russia, which entails the loss of their legal force. Similarly, within the legal system of a constituent entity of the Russian Federation, the Constitution or Charter has the highest level of legal force.

(3) The branches of law that regulate issues of joint jurisdiction of the Russian Federation and its subjects are characterized by spatial contradictions, that is, contradictions between normative legal acts of the Federal and regional levels. The Constitution (Part 5, Article 76) sets the priority of the federal laws accepted by subjects of joint jurisdiction over the legislation of constituent entities of the Russian Federation, resulting in higher legal force the first and the primacy of their application. In this sense, constitutional law has similarities with administrative, labor, family, and other branches of the law (Part 1 of Article 72 of the Constitution of the Russian Federation), but the normative contradictions themselves have a special constitutional and legal branch content. However, an important caveat should be made: at present, this type of contradiction is no longer relevant since they were eliminated, but it was an attribute of the constitutional practice of the 1990s. The key to the entire legal system of the country had numerous contradictions between the Russian Constitution and the constitutions of the individual republics within the Russian Federation (Tatarstan, Bashkortostan, Tuva, Adygea, Karelia, Yakutia, Karachay-Cherkessia) (Sarycheva 1997; Sidorenko 2000).

## *3.4 Gaps in Constitutional Law*

It seems that they can have two main manifestations: the absence of separate sources of constitutional law due to direct prescription or real needs of legal regulation of certain relations, and normative voids within a specific source of constitutional law.

For example, most federal constitutional laws and federal laws directly provided for by the Constitution of the Russian Federation have been adopted over the 26 years of its operation. However, the Federal constitutional laws on the constitutional Assembly (Part 2 of Article 135) and on changing the status of a subject of the Russian Federation (Part 5 of Article 66), as well as the Federal law on the status of the Security Council of the Russian Federation (Item "G" of Article 83) have not yet been adopted. Laws that are not directly provided for in the Federal Constitution but that follow from the meaning of certain of its provisions are also awaiting adoption. For example, in the provisions of Part 1 of Article 85 of the Constitution should be the adoption of the Federal law "On procedures for overcoming differences and resolving disputes between bodies of state power of the Russian Federation and bodies of state power of subjects of the Russian Federation, as well as between bodies of state power of subjects of the Russian Federation." Obviously, this situation indicates the incomplete implementation of certain norms of the Constitution of the Russian Federation and the resulting incompleteness of constitutional and legal regulation.

The second manifestation of the gap can be demonstrated by the example of the current Constitution of the Russian Federation. A certain disadvantage of the Constitution is both the "understatement" of its individual provisions and the absence of several standard norms for modern constitutions. So, in the provisions of Part 3 of Art. 11 of the Constitution, according to which "differentiation of subjects of conducting and powers between public authorities of the Russian Federation and bodies of state power of subjects of the Russian Federation is the Constitution, Federal and other treaties on the delimitation of subjects of competence and powers" is not mentioned Federal law, although it was and remained the main source of legal regulation of these relations. Or Article 33, according to which only citizens of the Russian Federation have the right to appeal, but not foreigners and stateless persons. It is not surprising that the Federal law "On the Procedure for Consideration of Appeals of Citizens of the Russian Federation" adopted on May 2, 2006 (Russian Federation 2006) eliminated this omission and extended its effect to foreigners and stateless persons (Part 3 of Article 1).

At the same time, it must be recognized that the gaps are eliminated by amendments to the Constitution of the Russian Federation. In the bill of the Russian Federation on the amendment to the Constitution of the Russian Federation No. 885214–7 "On Improving the Regulation of Certain Issues of the Organization and Functioning of Public power" ("https://sozd.duma.gov.ru/bill/885214-7" 2020), adopted by the State Duma on March 11, 2020, in the final version there was a norm on the principles of Russian foreign policy (Article 79.1) since such an important issue was not regulated at the constitutional level before.

## 3.5 *Other Formal and Legal Defects of Individual Sources and Norms of Constitutional Law*

The author believes that they may include ambiguity of terms and concepts used in the norms and other semantic ambiguity of the norms, excessive volume of regulation, imperfections of the structure of the normative legal act (including inconsistency of the name of the structural element and the norms contained in it), as well as provisions that allow unreasonably broad discretion of legal entities.

For example, the term "law," used in the Constitution, has at least three meanings: (1) "the totality of normative legal acts" (Article 54), (2) "any source of legal norms" (Part 1 of Article 19), (3) in its literal meaning, denoting the same name the act passed by the State Duma, approved by the Federation Council, and signed by the President of the Russian Federation (Part 3 Article 55). The semantic ambiguity of certain norms of the Constitution of the Russian Federation has repeatedly prompted authorized subjects to apply to the Constitutional Court of the Russian Federation with requests for its interpretation (Resolution of the Constitutional Court of the Russian Federation 1995; Resolution of the Constitutional Court of the Russian Federation 1998).

The Constitution also contains many multivariate framework norms that can be filled with different semantic content, depending on the current socio-political and economic situation. Under Part 3 of Article 80 of the Constitution of the Russian Federation, "...the President of the Russian Federation, in accordance with the Constitution of the Russian Federation and Federal laws, determines the main directions of domestic and foreign policy of the state." However, the question remains about the nature of such policies, ranging from democratic to authoritarian.

Over the past 10–15 years, excessive legal regulation has become a trend in the development of federal legislation on subjects of joint jurisdiction of the Russian Federation and its subjects. In particular, the Federal laws "On General Principles of Organization of Legislative (representative) and Executive Bodies of State Power of the Subjects of the Russian Federation" and "On Basic Guarantees of Electoral Rights and the Right to Participate in Referendums of Citizens of the Russian Federation," contrary to their name, are detailed regulations of the relevant relations, leaving almost no room for regional legislation. The constant increase in the volume of Federal legal regulation on joint jurisdiction subjects represents a serious abuse of law on the part of the Federation and a violation of the balance of legal support for this sphere.

The aforementioned bill on amendments to the Constitution of the Russian Federation No. 885214–7 of 2020 may serve as an illustration of unreasonably broad discretion within specific legal norms. In the provisions of three Articles (Part 4 of Article 111, Part 4 of Article 112 and Part 4 of Article 117), the possibility of dissolving the State Duma of the Russian Federation depends on the political discretion of the President of the Russian Federation, on what they consider more appropriate: to terminate the activities of the lower house of Parliament or to maintain the current composition of The State Duma. It is noteworthy that the solution to this most important political issue is not due to the formal establishment of any criteria for the choice of a particular behavior option by the President of the Russian Federation.

## 4 Discussion

The main differences in the doctrinal understanding of constitutional and legal anomalies are as follows. First of all, there are different frameworks for understanding this phenomenon. Some scientists, including the author of this article, consider anomalies in the narrow sense of the word, referring only negative deviations from the normal state of law to them, other (Nanba S. B., Pomazansky A. E., Tsomartova F. V.) define anomalies as a broad legal category that includes positive, permissible deviations from the norm. Second, the different substantive content of legal anomalies is observed. The author of the article believes that it is correct to include only formal legal and normative phenomena in the number of anomalies. At the same time, there are points of view about such components of anomalies as deviations of the current law from the legal doctrine (Erygina 2015, pp. 38–39), informal law (Tikhomirov 2005, pp. 81–82), shadow law (Baranov 2002). Third, there is an uneven assessment of anomalies. Most scientists see them as a purely negative meaning and effect, but some evaluate certain anomalies positively. For example, T. M. Pryakhina positively assesses the gaps in the Constitution of the Russian Federation (Pryakhina 2005), and V. G. Gulumyan generally denies the possibility of gaps in the text of the Constitution (Gulumyan and Konyukhova 2003, p. 31). The inadequate knowledge of constitutional and legal anomalies, the complexity and discussion of this issue, and the practical relevance of the stated topic undoubtedly indicate the need for further research.

## 5 Conclusion

The author of this study believes that the constitutional law of modern Russia is characterized by several types of anomalies, each of which has special substantive content. All of them are in one way or another connected with the Constitution of the Russian Federation as a key and constituent act in the legal system of Russia, and from that, they are of special significance. The diversity of anomaly types may vary, but their negative effect remains unchanged. They have undoubtedly negative impact on the development of constitutional law, and through it, on the entire Russian legal system. Therefore, the identification and analysis of such anomalies will allow the government to create adequate measures to prevent and eliminate them, contributing to the successful and sustainable legal development of our country.

## References

Anichkin ES, Rudt YuA (2019) Constitutional balancing in Russia: law and prospects. Right J High School Econ 5:80–100

Baranov VM (2002) Shadow law. nizhny novgorod academy of the ministry of internal affairs of russia, N. Novgorod, Russia

Constitutional Court of the Russian Federation (1995) Resolution "on the case of interpretation of part 2 of article 137 of the constitution of the Russian federation". Moscow, Russia, p 15

Constitutional Court of the Russian Federation (1998) Resolution "on the case about interpretation of provisions of part 4 of article 111 of the constitution of the Russian federation". Moscow, Russia, p 28

Constitutional Court of the Russian Federation (2017) Resolution "On the verification of constitutionality of paragraph 2 of article 40, parts 10 and 11 of article 42 of the federal law "on elections of deputies of the state duma of the federal assembly of the Russian federation", paragraphs 2 and 3 of part 1 of article 128 and section 10 of article 239 of the code of administrative procedure of the Russian federation in connection with complaints of citizens I. L. Trunov and M. V. Yurevich". Saint-Petersburg, Russia, p 11

Constitutional Court of the Russian Federation (2020) Resolution "on checking the constitutionality of parts 2 and 3 of article 13, paragraph 5 of part 5 of article 19 and part 1 of article 20 of the

federal law "on the basics of health protection of citizens in the Russian federation" in connection with the complaint of a citizen. Saint-Petersburg, Russia, p 1
Dzybova SG (2011) General laws and contradictions of the legal system of society. Russ Justice 2:50–53
Erygina VI (2015) State-legal doctrine on the representative nature of political parties: conceptual patterns and deviations in the practice of their implementation. Const Municipal Law 8:37–41
Gulumyan VG, Konyukhova IA (eds) (2003) Prospects for the development of the Russian constitution. KRZSP: The Constitution of the Russian Federation and the development of legislation in the modern period: materials of the all-Russian scientific conference, vol 1, RAP, Moscow, Russia
Morozova AS, Karasev AT (2015) Problems of improving federal relations in Russia: regularities and deviations. Const Munic Law 10:27–31
Nanba SB (2014) Pravo: models and deviations. J Russ Law 6(210):145–149
Nanba SB, Pomazansky AE, Tsomartova FV (2015) Positive deviations in law. J Russ Law 5(21):26–36
Petrov AV (2015) The anomalies of the law: concept and nature. Bull Lobachevsky Univ Nizhny Novgorod 1:170–181
Pryakhina TM (2005) On the issue of self-sufficiency of the Russian constitution. Const Munic Law 1:7–9
Russian Federation (1993) Constitution of the Russian federation. Moscow, Russia
Russian Federation (1994) Federal constitutional law "on the constitutional court of the Russian federation" (21 July 1994, No 1-FKZ). Moscow, Russia
Russian Federation (2001) Federal constitutional law "on the procedure for admission to the Russian federation and the formation of a new subject of the Russian federation within it" (17 Dec 2001 No. 6-FKZ). Moscow, Russia
Russian Federation (2006) Federal law "on the procedure for considering appeals from citizens of the Russian federation" (2 May 2006 No. 59-FZ). Moscow, Russia
Russian Federation (2020) Bill of the Russian federation on amendments to the constitution of the Russian federation" No. 885214−7 "On improving the regulation of certain issues of the organization and functioning of public power". Retrieved from https://sozd.duma.gov.ru/bill/885214-7
Sarycheva I (1997) "Sprawl" of the legal system-a threat to the constitutional security of Russia. Federalizm 4:121–142
Sidorenko EN (2000) Ministry of justice of Russia and problems of regional law-making. Yurist 5:7–10
Tikhomirov YuA (2005) Law official and informal. J Russian Law 5:80–87

# Analyzing Formal Criteria of Legitimate Restrictions of Human Rights in the Criminal Sphere in Russia and the Republic of Kazakhstan

**Valentina I. Plokhova**

**Abstract** The aim of the study is to consider the best practices of strengthening and interpreting of formal (legality, certainty, reasonable stability, predictability) criteria of in legal limitation of human rights in the criminal sphere in Russia and the Republic of Kazakhstan. We did it on the basis of comparison (comparative method of research) of constitutions and criminal codes of these countries, practices of constitutional control of the Constitutional Court of the Russian Federation and the Constitutional Council of the Republic of Kazakhstan, as well as legal literature. As a result of the study, new forms of violation of the legality criterion in Russia were identified. The doctrine and practice of control over regulatory legal acts used in the interpretation of criminal law norms have been supplemented with the following new criteria-indicators of legal restrictions on human rights in the criminal sphere: (a) criminal law is an extreme measure that can be used only when the protection of public relations by other means cannot be properly ensured; (b) all laws and regulations by which a blanket rule of the criminal code is interpreted must be published and verified accordingly. We proposed that the sustainable development program of Russia should indicate observance of human rights as the main legal basis for sustainable development, while in our opinion, the legal literature on sustainable development and national security should take into account the criteria of legitimate restriction of human rights. This study is interesting for both countries, as it promotes more adequate use of the theory of legitimate restriction of human rights in the legislative and law enforcement activities.

**Keywords** Human rights restriction · Legality · Certainty · Blanket norms · Criminal law · Norm control · Russia · Republic of Kazakhstan

V. I. Plokhova (✉)
Altay State University, Barnaul, Russia
e-mail: VPlohova@yandex.ru



S. G. Maximova et al. (eds.), *Advances in Natural, Human-Made, and Coupled Human-Natural Systems Research*, Lecture Notes in Networks and Systems 234,
https://doi.org/10.1007/978-3-030-75483-9_83

## 1 Introduction

Achieving sustainable development goals at both the national and international levels (Sakharov and Kolmar 2019) and ensuring Russia's national security and the Republic of Kazakhstan is impossible without meeting the criteria of legitimate restrictions on human rights in general and in the criminal sphere in particular. The improvement of social processes (digitization, robotization), countering new national and international threats of the twenty-first century (armed conflicts, natural disasters, the COVID-19 coronavirus pandemic, terrorism, etc.) often occurs with violations of human rights. In particular, it concerns the regime of citizen self-isolation introduced in Russia to prevent the spread of coronavirus infection COVID-19 based on decrees issued in March 2020 by the mayor of Moscow and governors (Russian Federation [RF] 2020a, b). Hasty criminalization of acts in the Code of Administrative Offences of the Russian Federation (hereinafter—the Code of Administrative Offences of the Russian Federation) and the Criminal Code of the Russian Federation (April 2020)) (Russian Federation 2020c, d) require in-depth analysis concerning compliance with the criteria of legality, certainty, legal validity, the proportionality of the act, and responsibility. Thus, the amended sanction of Part 3 of Article 236 of the Criminal Code "Violation of sanitary and epidemiological rules" does not even comply with the provisions of the General Part (Article 531) of the Criminal Code. Scientists have also found violations of citizen rights and freedoms and have noted an increasing bias towards protecting the nation and public interests to the detriment of private rights (Lukasheva 2014, pp. 29–35).

The author's research of constitutionalists on the legal category of human rights restrictions (Prikhodko 2017; Eleupova 2006), which the author has conducted over the past ten years in the criminal sphere, depicts that the majority of criteria-indicators of legal restrictions on human rights in a general, indirect form are enshrined in constitutional norms. They are also formulated in international treaties, interpreted in decisions of the European Court of Human Rights (hereinafter, the "ECtHR") and the Constitutional Court of the Russian Federation (hereinafter, the "Constitutional Court"). However, as the constitutionalists claim over several decades (Lapaeva 2013, p. 14), little attention is paid in the legal literature to the issue of general limits—criteria of lawful restriction of human rights, and even more so to particular limits in the criminal sphere.

The interpretation of many of them is controversial. Some are disputed, and some have not yet been studied (e.g., observance of the principle of legality of human rights restrictions based on a verdict, by-law, decrees of Moscow governors and mayor; resolutions of the Plenum and Presidium of the Supreme Court of the Russian Federation (hereinafter referred to as "PSC RF"); blanket rules, etc.). The ECtHR assumed the primary burden in interpreting, defining, and developing the criteria (their indicators) for legal limitation of human rights, the Constitutional Court of the Russian Federation (Chepenko 2017), and the Constitutional Court of the Republic of Kazakhstan (Amandykova and Dil 2012) in conjunction with the Supreme Courts of Russia and the Republic of Kazakhstan.

The Republic of Kazakhstan was chosen as an object of comparison with Russia because these countries have similar legal frameworks, basic principles of law, and law enforcement—respect for human rights (the human being and his rights and freedoms are the supreme value). However, the mechanism of constitutional rights protection in the Republic of Kazakhstan is partly different from Russia. Moreover, Kazakhstan has recently adopted a new Criminal Code of the Republic of Kazakhstan (No. 226 dated 03.07.2014) (hereinafter–the Criminal Code of the Republic of Kazakhstan [RK] 2014), while in Russia over the past few years, the draft Criminal Code of the Russian Federation and the Code of Administrative Offences of the Russian Federation have been under discussion. In connection with these circumstances, there is a timely demand for mutual enrichment of both countries with correct practices in the development, definition (interpretation) of criteria for legitimate limitation of human rights in the criminal sphere.

## 2 Materials and Methods

The circumstances mentioned above explain the used research methods, including the object, goal, and comparative research levels (Zinovieva 2014; Kalashnikova 2014; Yudina 2014). The paper analyzes and compares decisions of the Constitutional Council of the Republic of Kazakhstan [CC RK] 2008, 2009, 2015, 2019; Constitutional Court of the Russian Federation [CC RK] 2007, 2008a, b, 2010, 2015, 2017, 2018, 2019), constitutions, criminal codes and other normative acts of the Republic of Kazakhstan and Russia (RF 1993, 1994, 1996, 1998, 2016, 2020a, b, c, d, e; Kazakhstan 1995, 2014, 2016), results of theoretical reflection on the criteria for legal limitation of human rights by scientists of these countries (Amandykova and Dil 2012; Eleupova 2006; Kostrova 2017; Lapaeva 2013; Lukasheva 2014; Obrazhiev 2015; Pikurov 2009; Plokhova 2014, 2017, 2018; Prikhodko 2017; Sakharov and Kolmar 2019; Filippova 2017; Chepenko 2017; Shishk 2004; Yusupova 2013; Lange 1956; Lohberger 1968; Plohova 2017; Weidenbach 1965). Generalological methods of cognition are applied (Ruzavin 2015), such as system analysis (which allows the author to reveal the place of criminal law in the system of law and the relationship between constitutional and criminal norms), synthesis, induction, and deduction. The methods mentioned above make it possible to analyze and characterize the general aspects and specifics in the consolidation and interpretation by scientists and practitioners of formal criteria-indicators in the constitutions, criminal legislation, constitutional and criminal law, in the legal positions of higher judicial instances with respect to the norms of criminal law in Russia and Kazakhstan and to identify the best practices available in each country, to formulate new formal criteria-indicators of legitimate limitation of human rights in the criminal sphere.

## 3 Results

The author obtains the following results in the course of study.

There is a tendency to expand the means of restricting human rights in the practice of constitutional control of the Russian Federation. Initially, the restriction was possible only by federal constitutional law, then merely by federal law, then by a law of a constituent entity of the Russian Federation, by a court decision, by a resolution of the Government of the Russian Federation, by decrees of top officials of constituent entities of the Russian Federation, as long as such powers were granted by federal law.

The primary introductory constitutional provisions that set forth the criteria for legally restricting human rights and which are always referred to by the RF Constitutional Court when it sees unconstitutional restriction of human rights are contained in Part 3, Article 55 of the Constitution of the Russian Federation and Paragraph 1, Article 39 of the Constitution of the Republic of Kazakhstan. They list the main explicit formal criterion of legal limitation of human rights—legality. According to the Constitution of RK, human and citizen rights and freedoms can be restricted only by law. At present, Part 3 of Article 55 of the Constitution of the Russian Federation states that federal law may restrict human rights (in the previous version—federal constitutional law). However, in the legal positions of the CC RF, it is noted that since subjects of the Russian Federation have the right to limit, for example, the right of ownership by establishing regional taxes and dues, the CC RF provided that "…the laws of the constituent entities of the Russian Federation on taxes and dues are fully subject to the provisions of Article 55 (Part 3) of the Constitution of RF. However, limitation of rights on the basis of the law of a constituent entity of the Russian Federation is lawful only if the possibility of limitation of rights by the laws of the constituent entities of the Russian Federation is provided for in the federal law" (Plokhova 2014, p. 85).

Under the same condition, human rights may be restricted by a court decision under the Russian Constitution. Thus, Part 2 of Article 20 of the Constitution of the Russian Federation recognize the death penalty "… until it is abolished by federal law as an exceptional measure of punishment for particularly grave crimes against life, while granting the accused the right to have their case heard by a court with the participation of a jury." However, the restriction of liberty provided for in Article 22, Paragraph 2 of the Constitution of the Russian Federation does not contain both conditions. It is possible only by judicial decision. In this respect, Paragraph 2 of Article 22 of the Constitution of the Russian Federation does not contain both conditions.

Additionally, Paragraph 2 of Article 16 of the Constitution of the RK corresponds more to the legality criterion, which provides the possibility to restrict this right only in cases stipulated by law and by a court decision. "Arrest and detention shall be allowed only in cases stipulated by law and only with the sanction of a court with the granting of the right of appeal to the arrested person. Without a court order, a person may be detained for a period not exceeding 72 h."

The possibility to specify a blanket rule of criminal law by by-laws, including a government decree, is justified in approximately the same way as the possibility of restricting human rights by judicial decision (as long as such powers are granted by federal law).

The legislator and decrees of the highest officials of the constituent entities of the Russian Federation on preventing the spread of coronavirus infection COVID-19 are trying to bring under this scheme by introducing on 01.04.2020 (RF 2020e) amendments to Federal Law No. 68-FZ of 21.12.1994 "On Protection of the Population and Territories from Natural and Man-caused Emergency Situations" (RF 1994), broadening the definition of an emergency situation to include "the spread of a disease that poses a danger to others" (Article 1). Besides, the amendment of Article 11 of this law extended the powers of the government authorities of the constituent entities of the Russian Federation and local authorities in the field of protection of the population and territories from emergency situations (Paras. "y," "f") and introduced a conflict of laws rule. It read, "In the event that the Government of the Russian Federation establishes mandatory rules of conduct for citizens and organizations stipulated by Subparagraph "a.2" of Article 10 of this Federal Law, the rules of conduct established by the government authorities of the constituent entity of the Russian Federation in accordance with Article 10 of this Federal Law shall be the following. It follows that beforehand the March decrees of the constituent entities of the Russian Federation can hardly be called legal and they require checking for their compliance at least with the federal law.

The blanket nature of the majority of the norms of the Criminal Code of the Russian Federation and the Criminal Code of the Republic of Kazakhstan, the achievement of the criterion of certainty of the norms of criminal law in Russia with the help of clarifications contained in the decisions of the Plenum of the Supreme Court lead to non-compliance with the criterion of legality." Besides, at the present stage of development of society, legal science, and constitutional control in the Russian Federation, there are new aspects of this problem, the solution of which supplements the theory of justification of blanket dispositions of the norms of criminal law. Depending on which normative act (law or subordinate normative act) details the blanket features of the criminal code norm, one can distinguish, conventionally speaking, "legal" and "sub-legal" blanket, bearing in mind that limitation of human rights is possible only by law. Since attributes of a blanket rule may be interpreted based on regulatory and other, besides criminal, tort legislation, to the extent that it makes sense to conditionally distinguish "legal regulatory" and "legal tort" blanket, as in this aspect too the problems manifest themselves. In the case of the "legal regulatory" blanket, it is necessary to observe one of the legality criterion indicators—consistency of law norms and its main requirement: what is permitted in the regulatory legislation cannot be prohibited in criminal legislation. This rule has long been revealed and proved in the Russian legal literature (Shishko 2004) and in the decisions of the RF Constitutional Court No. 8-P dated May 27, 2008, and No. 15-P dated July 13, 2010 (RF Constitutional Court 2008b, 2010). Thus, in its Resolution No. 8-P of May 27, 2008, the Constitutional Court of the Russian Federation noted that "...there should be no such regulation, as a result of which criminal liability should be imposed for the commission of certain actions permitted by the law regulating the respective sphere of relations, that is legal in nature" (Constitutional Court of the Russian Federation 2008b). Besides, the author believes that in order to comply with the criterion of consistency of norms, the CCRF should initially indicate the following rule (legal position)—the laws included in the blanket part of the CCRF disposition should not come into force before the changes in the articles of the CCRF. The evaluation algorithm should supplement the mentioned provision. In the future, using the experience of Kazakh colleagues (RK 2016), this rule should be written in regulatory acts. However, the requirement "one term—one definition of it" in criminal law and other branches of law is not always feasible due to different boundaries of regulated and protected social relations (different legal fields).

Besides, the legal positions of the ECtHR and the CC of RF dictate additional rules for the situation, when the same term in its meaning can be interpreted both in a broad and in a narrow sense—the prerogative of the established judicial practice, but with the restriction "not to the detriment of the accused."

What is important for "legal tort" blanketing is the insistent, repeatedly voiced by the RF Constitutional Court (since 2001) legal position on the place of criminal law in the system of law in general and in relation to administrative law in particular (RF Constitutional Court 2008b, 2010, 2015, 2018, 2019). Criminal law "...by its very nature is an extreme (exclusive) means ... with the help of which the government reacts to the facts of unlawful conduct in order to protect public relations if it cannot be properly ensured only by means of legal norms of another branch", is stated in p. 5.1 of the 2008 CCRF Resolution (RF Constitutional Court 2008b).

For all considered aspects of the constitutionality of blanket rules of the criminal code, to ensure one of the indicators of legality—accessibility—the law and the by-law must be published accordingly.

An indicator of compliance with the certainty of the means of limiting human rights, its reasonable stability, and predictability of the consequences of conduct is a clear definition of the rules of the law in time. The regulation of the operation of criminal law throughout history is practically the same in the Constitutions and Criminal Codes of Russia and the Republic of Kazakhstan. The only difference is that Article 6 of the Criminal Code of the Republic of Kazakhstan interprets some provisions of the retroactive force of criminal law in more detail, while the Normative Decision of the Supreme Court of the Republic of Kazakhstan No. 15 of 22.12.2016 "On Judicial Practice in Applying Article 6 of the Criminal Code of the Republic of Kazakhstan" (RF Constitutional Court 2016) discloses other nuances of this issue, many of which are interpreted in the RF Constitutional Court decisions. However, the current development of the legal system, including in the criminal sphere, the search for new ones, or the return to the forgotten mechanisms of influence on offenders (in Russia—administrative forensics, in Kazakhstan—a criminal offense), raise new problems for the law enforcement. Mainly, concerning the rules of operation of administrative legislation adjacent to criminal law in two situations: the first arise in Russia in the case of the introduction of criminal liability for the repeated commission of an administrative offense (administrative forensics). The Message of the Constitutional Assembly of the RK (2019), expressed concern that the issues of a retroactive force of the new law (enforcement of previously imposed administrative penalties) in the transfer of administrative offenses to the category of criminal offenses have not yet been resolved (CC 2019). In several decisions of the last five years, the RF CC proposes and justifies the solution of the issue in the first situation—in the case of mitigation of liability and related issues with regard to crimes with administrative forensics. First of all, in 2015, The RF Constitutional Court was declared unconstitutional by Article 1.7 of the CAO (CC RF 2015). Second, in several decisions, the RF Constitutional Court proposes a new procedure for the administrative law to be applied in time, which is also overdue in the Republic of Kazakhstan. In its 2018 Resolution, the CC of RF clarifies, "...if the responsibility for the act has been mitigated but not eliminated, and the act itself has acquired another branch legal qualification, then its decriminalisation, while at the same time fixing the identical composition of the administrative offence, cannot be regarded as establishing a new unlawful act which is not previously punishable." Such decriminalization is mitigation of public legal liability for the commission of the relevant offense manifested by the introduction of less strict (as compared to criminal) administrative sanctions, less restriction of rights (as compared to the criminal legal institution of criminal record) in the application of administrative liability measures (CC RF 2018). It is further emphasized that if the law abolishes criminal liability for a specific act with its simultaneous transfer under the CAO of the RF, "...the federal legislator continues to regard this act as offending, but assesses the nature of its public danger in a different way." Therefore, proceeding "...from the constitutional principles of justice and equality of requirements of responsibility inevitability for the committed offence, as well as certainty, clarity, unambiguousness of legal norms and their coherence in the general system of legal regulation. It is assumed that persons who commit such acts, although before the specified date, will be brought to administrative responsibility" (Sub-Clause 4–5 Paragraph 4.1) (CC RF 2018). Such a decision is justified by the principle of inevitability of liability, the similarity of administrative and criminal liability (CC RF 2015).

In addition, administrative liability for certain offenses falls under the concept of "criminal sphere" (Clause 4) (CC RF 2017). Federal Law No. 195-F of 23.06.2016 introduced Part 2.1 in Article 1.7 of the Code of Administrative Offences of the Russian Federation, which established a general rule (for the case of mitigation and toughening of public legal liability): "In the event of simultaneous entry into force of the provisions of the law abolishing administrative liability for the act and establishing criminal liability for the same act, a person shall be subject to administrative liability on the basis of the law which was in force at the time of committing the administrative offence" (RF 2016). The author believes that this provision should become one of the indicators of the legality criterion.

The most frequently contested criterion for legally restricting human rights in Russian constitutional control practice is criminal law certainty. Definition of criminal law norms by the RF Constitutional Court is interpreted broadly to include all formal criteria of legal limitation of human rights, such as legality, accessibility, certainty, reasonable stability, and predictability. In the decisions of the Constitutional Court, it is interpreted in approximately the same way. "...The law restricting constitutional human and civil rights and freedoms must comply with the requirements of legal accuracy and predictability of consequences, i.e., its norms must be formulated with a sufficient degree of clarity and be based on understandable criteria that allow one to differentiate legitimate from illegal behavior with all certainty, excluding the possibility of arbitrary interpretation of provisions of the law" (CC RK 2008). It follows from the constitutional principle of legal equity (Article 14 of the Basic Law) since such equity can only be ensured if a legal norm is uniformly understood, interpreted, and applied (CC RK 2015). The legal certainty of the criminal law norm of the CC of the Russian Federation derives from the systemic nature of the norms of law, including within the criminal law, and, in most cases, from the interpretation of the law in the resolutions of the CC of the Russian Federation (Plohova 2017). In some cases, under the guise of forced ambiguity of criminal law norms, the contested norm of the RF Criminal Code has not been corrected for decades (Plohova 2018). The definitions of criminal law in the Republic of Kazakhstan are slightly broader than in the Russian Federation Criminal Code.

## 4 Discussion

For decades, the restrictions on human rights by by-laws have been discussed in the legal literature and by constitutional oversight bodies. Explaining the peculiarity and acceptability of the restriction of the right to confidentiality of personal deposits and savings, the 2009 Normative Resolution of the Constitutional Court of the RK notes that it "…is the prerogative of the legislator and the legislator has no right to authorize another governmental body or official to regulate by-laws the limits of restriction of confidentiality of personal deposits and savings". The issue was also raised in the messages of the RK Constitutional Court in 2017 and 2019. "…In the current legislation, there are still by-laws that regulate the most important social relations specified in Paragraph 3 of Article 61 of the Basic Law, including the mechanism for the application of certain measures of government coercion, which is unacceptable …" It also applies to legal acts containing the rights and obligations of citizens, which contribute to the establishment of signs of offenses and affect the definition of liability measures (Traffic Rules and others), which is advisable to thoroughly review of the legislation on the subject. In criminal law science, the mentioned issue is addressed in the framework of the longstanding problem of the constitutionality of blanket criminal law rules, which are in the majority in the criminal codes of both countries. For many decades, the issue has been discussed by both Russian (Shishko 2004) and Western lawyers (Lange 1956; Lohberger 1968, p. 12; Weidenbach 1965, p. 24). N. I. Pikurov's vision of the situation is rather convincing, including an analysis of the decisions of the RF Constitutional Court in 2009. He points out that the crime characteristics are defined by the RF Criminal Code, while the norms of other branches of law detail the blanket rules (Pikurov 2009, pp. 29–46). However, the incomplete resolution and complexity of some of the issues are evident. With regard to the constitutionality of the establishment by the Government of the Russian

Federation of a list and amounts of narcotics (the same practice has developed in Kazakhstan), the Constitutional Court has repeatedly emphasized that "… acting on the basis of and in compliance with federal law (Article 115, Paragraph 1 of the Constitution and Articles 2 and 3 of the Federal Constitutional Act on the Government of the Russian Federation) may not, like other executive authorities, establish grounds for criminal liability that are not provided for by federal law." The list of narcotics, psychotropic substances and their precursors and the amounts subject to control in the Russian Federation for the purposes of the articles of the Criminal Code on illicit trafficking of these items, including the list of items subject to control in the Russian Federation for the purposes of the articles of the Criminal Code of the Russian Federation on illicit trafficking of these items, Article 2 of the Federal Act on narcotics and psychotropic substances (1998) is prescribed.

However, the Constitutional Court of the Russian Federation asserts that "the federal legislator thereby did not grant the Government of the Russian Federation the power to carry out normative regulation on the issue of establishing the grounds for criminal liability…. They do not establish the criminality of an act, the punishment for it and other criminal-legal consequences; they are determined only by the Criminal Code of the Russian Federation, Article 228" (RF Constitutional Court 2007). However, the amounts of narcotics determine the differentiation of responsibility (criminal, administrative) and punishment.

The justification for restricting human rights by by-laws, as proposed in the legal literature, because they represent "…only a concretization of the legislative restriction already established," may be appropriate at the moment but requires further research and resolution. In our opinion, all laws and by-laws interpreting a blanket rule of the criminal code should be checked for compliance with constitutional norms, including compliance with the criteria of a lawful limitation of human rights, and therefore be included in the algorithm we developed to assess (Plokhova 2017) compliance of laws and law enforcement activities with the conditions of a lawful limitation of human rights in the criminal sphere (hereinafter—the algorithm). The mentioned aspect of the issue associates with the questions of legal expertise of criminal legislation, which is not the research subject.

The relationship between the branches of law in the legal literature is studied mainly by representatives of criminal law branches—the relationship between the norms of criminal law and regulatory legislation (Pikurov 2009; Shishko 2004), tax law (Filippova 2017), criminal proceedings (Kostrova 2017), administrative law, etc. Representatives of the regulatory industries, and the legislator, do not particularly consider how changes in non-criminal legislation will manifest themselves in criminal law. This aspect sometimes leads to problems not only in the application of criminal law but also to the violation of the principle of law equity, failure to meet the criteria of proportionality of the act and responsibility, legal certainty (e.g., the unified social tax in the Tax Code of the Russian Federation) In this connection, the author believes that for all branches of law (for a legislator, law enforcer, and researcher) this provision should be assessed as a criterion of legitimate restriction of human rights. The most valuable experience in this regard is that of Kazakh colleagues, particularly Paragraph 1 of the article on human rights. Article 12, paragraph 2, of the Law of the Republic of Kazakhstan "On Legal Acts" contains the following provision—"...the norms of laws in cases of their inconsistency with the norms of the codes of the Republic of Kazakhstan may be applied only after making the appropriate amendments or additions to the codes" (Kazakhstan 2016).

Due to the fact that the certainty of criminal law norms in the Russian Federation is achieved in most cases by the interpretation of their features in the resolutions of the PSC of the Russian Federation, there is a multi-year discussion on the legal nature (normality) of these resolutions (Obozhiev 2015, pp. 253–281). The legislator of the Republic of Kazakhstan in Article 1 of the Criminal Code of the Republic of Kazakhstan is unequivocal and, according to researchers-constitutionalists (Yusupova 2013), rather well decided this discussion: "... normative decrees of the Constitutional Council and the Supreme Court of the Republic of Kazakhstan are part of the criminal legislation of the Republic of Kazakhstan" (RK 2014, Para. 2, Art. 1). In addition, Article 3 of the Criminal Code of RK contains 42 definitions of some concepts contained in the code.

## 5 Conclusions

The research has shown that in both countries, research on the limits of legal restrictions on human rights in the criminal sphere is currently more topical than before. The interpretations of formal criteria for legal restrictions on human rights in the criminal sphere developed in the theory and practice of constitutional control in Russia and the Republic of Kazakhstan are mutually interesting. They are complemented by the following new aspects of the legal doctrine and algorithm. In order to comply with the legality criterion (in the broad meaning of this criterion), it is necessary: (1) to publish and check for compliance with the Constitution, including compliance with the criteria of legal limitation of human rights of all laws and by-laws by which the blanket part of the criminal code norm is interpreted; (2) to enshrine the following rule in the legal positions of the Constitutional Court of the Russian Federation—laws included in the blanket part of the disposition of the Criminal Code of the Russian Federation should not come into force until amendments are made to the articles of the Criminal Code of the Russian Federation; (3) to recognize criminal law as extreme (exceptional); and (4) to supplement the conditions for the same definition of concepts in the criminal and other branches of law to a situation where the same term in its meaning can be interpreted in both broad and narrow senses, with a provision on the prerogative of established jurisprudence, but not to the detriment of the accused.

## References

Amandykova SK, Dil AV (2012) The role of the Constitutional Council in the formation of the concept of human and citizen's rights and freedoms in the Republic of Kazakhstan. Bull Karagand Univ 1(65):9–13

Blokhova VI (2019) Detailed criteria for legally restricting human rights in the criminal sphere. Vestnik Tomsk State Univ 448:239–243

Chepenko YK (2017) Legal problems in the mechanism of constitutional provision of human and citizen's rights and freedoms (in the light of the practice of the Constitutional Court of the Russian Federation). Thesis of candidate of laws. St. Petersburg State University St. Petersburg, Russia

Constitutional Council of the Republic of Kazakhstan (2008) Normative decision of the Constitutional Council of the Republic of Kazakhstan. On verification of constitutionality of parts 1 and 4 of article 361 of the Criminal Code of the Republic of Kazakhstan on the appeal of Kapshagai city court of Almaty region (February 27, 2008 No 2). http://ksrk.gov.kz/index.php/solutions/np-ks-rk-ot-27022008-g-no2-o-proverke-konstitucionnosti-chastey-pervoy-i-chetvertoy-stati

Constitutional Council of the Republic of Kazakhstan (2009) Normative resolution of the Constitutional Council of the Republic of Kazakhstan. About official interpretation of item 2 of article 18 of the Constitution of the Republic of Kazakhstan and check on conformity to the Constitution of the Republic of Kazakhstan of the Law of the Republic of Kazakhstan. About combating legalization (laundering) of illegally-gained income and terrorism financing and the Law of the Republic of Kazakhstan. About modification and additions to some legislative acts of the Republic of Kazakhstan on combating legalization (laundering) of illegally-gained income (September 20, 2009 No. 5). http://adilet.zan.kz/rus/docs/S090000005

Constitutional Council of the Republic of Kazakhstan (2015) Normative decision of the Constitutional Council of the Republic of Kazakhstan. About the official interpretation of item 1 of article 27 of the Constitution of the Republic of Kazakhstan and about check on conformity to the Constitution of the Republic of Kazakhstan of the Law of the Republic of Kazakhstan. About protection of children from the information, causing harm to their health and development and the Law of the Republic of Kazakhstan. About modification and additions to some legislative acts of the Republic of Kazakhstan concerning protection of children from the information, causing harm to their health and development (May 18, 2015 No. 3). https://tengrinews.kz/zakon/konstitutsionnyiy_sovet_respubliki_kazahstan/konstitutsionnyiy_stroy_i_osnovyi_gosudarstvennogo_upravleniya/id-S1500000003/

Constitutional Council of the Republic of Kazakhstan (2016) Normative Resolution of the Supreme Court of the Republic of Kazakhstan. On Judicial Practice on Application of Article 6 of the Criminal Code of the Republic of Kazakhstan (December 22, 2016 No. 15). https://online.zakon.kz/document/?doc_id=33231406

Constitutional Council of the Republic of Kazakhstan (2019) Message from the Constitutional Council of the Republic of Kazakhstan. On the state of constitutional legality in the Republic of Kazakhstan (announced at the joint session of the Chambers of the Parliament of the Republic of Kazakhstan 20.06.2019). http://adilet.zan.kz/rus/docs/S1900000117

Constitutional Court of the Russian Federation (2007) Decision of the Constitutional Court of the Russian Federation on the refusal to accept for consideration a complaint from a citizen of Chugurov, Pavel Nikolaevich, on the violation of his constitutional rights under list I of the List of narcotic drugs, psychotropic substances and their precursors under control in the Russian Federation (February 8, 2007 No. 292-O-P). https://legalacts.ru/doc/opredelenie-konstitutsionnogo-suda-rf-ot-08022007-n-292-o-p-ob/

Constitutional Court of the Russian Federation (2008a) Definition of the Constitutional Court of the Russian Federation. On refusal to accept for consideration the complaints of citizens of Konstantin Aleksandrovich Amosov, Firth Arnold Mikhelevich, Dmitry Vasilyevich Yadrihinsky and Yana Viktorovna Yakovleva on violation of their constitutional rights by article 234 of the Criminal Code of the Russian Federation (February 7, 2008a No. 79-O-O). https://www.garant.ru/products/ipo/prime/doc/1686201/

Constitutional Court of the Russian Federation (2008b) The Decision of the Constitutional Court of the Russian Federation. On Case of Verification of the Constitutionality of the Position Ch. 1 Art. 188 of the Criminal Code of the Russian Federation in Connection with the Complaint of Citizen M. A. Aslamazyan (May 27, 2008b No. 8-P). http://www.ksrf.ru/ru/Decision/Pages/default.aspx

Constitutional Court of the Russian Federation (2010) The Decision of the Constitutional Court of the Russian Federation. On Case about Verification of Constitutionality of Position Ch. 1 Art. 188 of the Criminal Code of the Russian Federation, Ch. 4 article 4.5, part 1 article 16.2 and part 2 article 27.11 of the Code of the Russian Federation about administrative offences in connection with complaints of citizens V. V. Batalova, L. N. Valueva, Z. Ya. Ganievoj, O. A. Krasnoj and I. V. Epova (from July 13, 2010 No. 15-P). http://www.ksrf.ru/ru/Decision/Pages/default.aspx

Constitutional Court of the Russian Federation (2015) Resolution of the Constitutional Court of the Russian Federation. On the Case of Checking Constitutionality of Part 2 of Article 1.7 and Paragraph 2 of Article 31.7 of the Code of the Russian Federation on Administrative Offences in Connection with the Request of the Magistrate of the Court Plot No. 1 of the Vyksa Judicial District of Nizhny Novgorod Region (July 14, 2015 No. 20-P). http://www.ksrf.ru/ru/Decision/Pages/default.aspx

Constitutional Court of the Russian Federation (2017) Ruling of the Constitutional Court of the Russian Federation. On a case about check of constitutionality of provisions of article 2121 of the Criminal code of the Russian Federation in connection with the complaint of the citizen I. I. Dadin (from 10.02.2017 No. 2-P). Retrieved from http://www.ksrf.ru/ru/Decision/Pages/default.aspx.

Constitutional Court of the Russian Federation (2018) Ruling of the Constitutional Court of the Russian Federation. On the Case on Verification of the Constitutionality of Part 1 of Article 1.7 and Part 4 of Article 4.5 of the Code of the Russian Federation on Administrative Offences, Paragraph 4 of Article 1 of the Federal Law. On Amendments to the Criminal Code of the Russian Federation and the Criminal Procedural Code of the Russian Federation on Issues of Improvement of the Grounds and Procedure of Exemption from Criminal Liability and Paragraph 4 of Article 1 of the Federal Law. On Amendments to Certain Legislative Acts of the Russian Federation. http://www.ksrf.ru/ru/Decision/Pages/default.aspx.

Constitutional Court of the Russian Federation (2019) Ruling of the Constitutional Court of the Russian Federation. On the Case of Verification of the Constitutionality of Items 1 and 6 of Article 5 of the Law of the Republic of Komi. On Certain Issues of Holding Public Events in the Republic of Komi. In Connection with the Complaints of Citizens M. S. Sedova, V. Sedova., P. Tereshonkova (from 01.11.2019 No. 33-P). http://www.ksrf.ru/ru/Decision/Pages/default.aspx

Eleupova AA (2006) Implementation of the constitutional control by the Constitutional Council of the Republic of Kazakhstan. Dissertation of the candidate of legal sciences. Chelyabinsk State University, Chelyabinsk, Russia

Filippova CI (2017) Criminal liability for tax crimes: inter-sectoral communications. Thesis of thesis of Ph. Lomonosov Moscow State Law University. O. E. Kutafin Moscow State Law University, Moscow, Russia

Kalashnikova EM (2014) Methodological significance of comparative approach in social cognition. Vestnik Perm State Humanit Pedagog Univ Ser 3 Humanit Soc Sci 2:11–12

Kostrova MB (2017) Relationship between criminal law and criminal procedure law, 2nd ed., interpretation and additional. Prospekt, Moscow, Russia

Lange R (1956) Strafgezetzgeber der Schuldlcher. Juristische Zeitung, 73

Lapaeva VV (2013) Criteria for Limiting human and civil rights in the constitution of the Russian Federation. State Law 2:14–24

Lohberger LK (1968) Blanhettstrafrtht und Grundgesetz. Munhen, DDR

Lukasheva EA (2014) The 21st Century: The decline in human rights values. In: Soshnikov TA (ed) Human rights in the modern world: New challenges and difficult solutions. Moscow Publishing House, Moscow, Russia, pp 29–35

Obrazhiev KV (2015) The system of formal (legal) sources of the Russian criminal law. Yurlitinform, Moscow, Russia
Pikurov NI (2009) Qualification of crimes with blanket features. Russian Academy of Justice, Moscow, Russia
Plohova VI (2017) The Use of the Decisions of the European Court of Human Rights, the Constitutional Court of the Russian Federation when Investigating the Problems of the Criminal Law of the Russian Federation (Article One). J Sib Fed Univ. Humanit Soc Sci 10(7):1090–1100
Plokhova VI (2017) Algorithm for assessing the conformity of laws and law enforcement activities with conditions for legally restricting human rights in the criminal sphere. In: Gracheva YV et al (ed) Proceedings of the 14th international scientific and practical conference (26–27 January 2017): Criminal Law: strategy of development in the XXI century, pp 137–138. Prospekt, Moscow, Russia
Plokhova VI (2018) Forced uncertainty of criminal law norms and (or) incorrect fixation, interpretation of their features (by the example of article 138.1 of the Criminal Code of the Russian Federation). Vestnik Tomsk State Univ 433:207–215
Plokhova VI (2014) Conditions for legally restricting human rights and freedoms in the criminal sphere. Criminol J Baikal State Univ Econ Law 4:83–92
Prikhodko TV (2017) The Modern Model for Restricting Human and Civil Rights and Freedoms in the Russian Federation and the Federal Republic of Germany: Constitutional and Legal Research, PhD thesis. F. M. Dostoevsky Omsk State University, Omsk, Russia
Republic of Kazakhstan (adopted by national referendum on 30 August 1995, as amended and supplemented on 23 March 2019) (2014) Criminal Code of the Republic of Kazakhstan (No. 226-V dated 03.06.2014, with amendments and additions as of 04.05.2020). https://online.zakon.kz/document/?doc_id=31575252.
Republic of Kazakhstan (2016) Law of the Republic of Kazakhstan “On Legal Acts” (from 06.04.2016 No. 480-V, with amendments and additions as of 01.01.2020). https://online.zakon.kz/document/?doc_id=37312788
Russian Federation (1994) Federal Law. On the protection of the population and territories from natural and technogenic emergency situations (December 21, 1994, No. 68-FZ (last edition)). Moscow, Russia
Russian Federation (1996) Criminal code of the Russian Federation (from 13.06.1996 № 63-FZ, ed. from 27.12.2018). Collection of Legislation of the Russian Federation. 17.06.1996. № 25. Art 2954
Russian Federation (1998) Federal Law. On narcotic drugs and psychotropic substances (January 8, 1998 No. 3-FZ (last version)). Moscow, Russia
Russian Federation (2016) Federal Law. On amendments to the code of administrative offences of the Russian Federation (23 June 2016, No. 195-FZ (last version)). Moscow, Russia
Russian Federation (2020a) Decree of the mayor of Moscow. On introduction of the high readiness regime (March 5, 2020a No. 12-UM). Moscow, Russia
Russian Federation (2020b) Decree of the Governor of the Altai Territory. On individual measures to prevent the introduction and spread of new coronavirus infection COVID-19 (March 26, 2020b No. 42). Moscow, Russia
Russian Federation (2020c) Federal Law on amendments to the code of administrative offences of the Russian Federation. (April 1, 2020c No. 99-FZ). Moscow, Russia
Russian Federation (2020d) Federal Act on amendments to the Criminal Code of the Russian Federation and articles 31 and 151 of the Code of Criminal Procedure of the Russian Federation (April 1, 2020d No. 100-FZ). Moscow, Russia
Russian Federation (2020e) Federal Law on amendments to certain legislative acts of the Russian Federation on issues related to the prevention and liquidation of emergency situations (April 1, 2020e No. 98-FZ). Moscow, Russia
Russian Federation (as of 06.04.2011 No. 480-V, with amendments and additions as of 01.01.2020) (1993) Constitution of the Russian Federation (adopted by popular vote on 12.12.1993, with amendments introduced by the Laws of the Russian Federation of 30.12.2008 No. 6-FKZ, of

30.12.2008 No. 7-FKZ, of 05.02.2014 No. 2-FKZ, of 21.07.2014 No. 11-FKZ). Collection of Legislation of RF, 04.08.2014 No. 31, Art. 4398
Ruzavin GI (2015) Methodology of scientific cognition: textbook. Unity-Dana, Moscow, Russia. http://biblioclub.ru/index.php?page=book_red&id=115020&sr=1
Sakharov AG, Colmar OI (2019) Prospects for implementing UN sustainable development goals in Russia. Bull Int Organ 14(1):189–204
Shishko IV (2004) Economic offences: questions of legal assessment and responsibility. Legal Center Press, Saint-Petersburg, Russia
Weidenbach P (1965) Die verfassungsrechtliche Problematik der Blankettstrafgesetze. Tubingen, BRD
Yudina AV (2014) Objects, objectives and levels of comparative legal research. PFUR Newsl Ser Leg Sci 1:129–134
Yusupova GH (2013) On normative judicial practice of the Supreme Court of the Republic of Kazakhstan. Bull Inst Legis Repub Kaz 1–2(29–30):274–277
Zinovieva LE (2014) Comparatist approach: some methodological features, analytical practices and their educational potential. Vestnik Perm State Human Pedagog Uni Ser 3: Humanit Soc Sci 2:18–23

# Accusatory Prosecutorial Bias in Russian Criminal Proceedings and Contradictions with Scientific Forensic Means

**Yury P. Garmaev, Elena V. Gulina, and Vitaly V. Polyakov**

**Abstract** The paper analyzes the problems of criminal proceedings in Russia associated with the phenomenon of the so-called "accusatory bias" as a type of illegal actions of individual police officers and other law enforcement agencies. The purpose of the study is not so much practical as theoretical. The paper proposes a new approach to developing scientific provisions and applied recommendations in the framework of science and the academic discipline "forensic science," aimed at overcoming the accusatory bias. Using the comparative jurisprudence (comparison with sources from the United States), statistical, sociological, and other methods, we carried out an analysis and generalization based on the materials of criminal cases on corruption crimes, completed by convictions and acquittals of Russian courts. The study's primary results are the first formulated conclusions about the need to introduce a special principle of the formation of scientific provisions and applied recommendations in forensic science—the principle of the inadmissibility of accusatory bias. Its application by a scientist, a criminologist-developer, includes six elements, including the development and implementation of recommendations on common facts and typical situations of illegal prosecution, non-criminal acts in a particular field of activity, the prevention of typical mistakes and violations of the law, the investigation of crimes committed by law enforcement officers as manifestations of an accusatory bias.

**Keywords** Accusatory bias · Acquittal · Investigation of corruption crimes in Russia · Principles of forensic investigation methodology · Protection from unlawful accusation · Forensic methodology · Appointment of criminal proceedings · Corruption crimes

---

Y. P. Garmaev
Krasnoyarsk State Agrarian University, Krasnoyarsk, Russia

E. V. Gulina
Buryat State University Named After Dorji Banzarov, Ulan-Ude, Russia

V. V. Polyakov (✉)
Altai State University, Barnaul, Russia
e-mail: agupolyakov@gmail.com

 
S. G. Maximova et al. (eds.), *Advances in Natural, Human-Made, and Coupled Human-Natural Systems Research*, Lecture Notes in Networks and Systems 234,
https://doi.org/10.1007/978-3-030-75483-9_84

# 1 Introduction

Previous studies devoted to the problems of investigating corruption crimes in Russia (Garmaev 2018; Gulina 2019) led to the conclusion that the main shortcoming of identifying and investigating corruption, economic, and several others crimes is the circumstance that the efforts of the law enforcement agencies of the Russian Federation are more focused on combating those encroachments, the identification and proof of which does not present any particular legal and factual difficulty. Many other scientists are coming to similar conclusions. Thus, the authoritative criminologist Luneev (1995) notes that the constant demands "to intensify the fight against crime," "organizational conclusions" in the event of an increase in the number of crimes committed in the controlled territory, the resonance of criminal cases on corruption crimes, and accusations by the media of the prejudice of the police gave rise to the practice of preferential registration and investigation of non-laborious cases.

Therefore, in Russia, doctors and school and university teachers are "leading" for several decades in terms of the number of registered and investigated corruption and criminal encroachments. Furthermore, with a large quantitative margin, there are private and junior commanding staff of the traffic police, and other officials of the lower and middle level in the region (in government bodies and local governments). Numerous studies concluded that this "anti-rating" does not correspond to the modern realities of Russian reality due to the high latency of criminal corruption of more dangerous types—"business" (associated with big business), "apex" (in the upper echelons of power in regions from the federal center), and "political" (associated with political decision-making). Modern investigative and judicial practice on corruption in Russia is mainly one-episode criminal cases that are not particularly difficult. According to the composition of crimes under the Criminal Code of Russia (Russian Federation 1996), most often are acts related to giving a bribe (Article 291 of the Criminal Code of the Russian Federation) or attempted bribery, and with petty bribery (Article 291.2 of the Criminal Code of the Russian Federation). Simultaneously, the "apex" and systematic corruption violations are detected and exposed relatively rarely. They are investigated for a long time and often ineffectively. The initiation of "high-profile" criminal cases of this kind is usually accompanied by a public outcry and large-scale media coverage (Gulina 2019).

The preceding allows us to hypothesize the widespread prevalence of one of the most negative phenomena in the law enforcement and judicial spheres of Russia—"accusatory bias" in Russian.

This concept is not used in national legislation. However, in scientific research, the term and the content of the accusatory bias are the subjects of close attention. In the scientific and reference literature, it is defined as a particular direction of the activities of the investigators, interrogator, prosecutors, and court in criminal and administrative proceedings, in which these officials take the position of the accusation, ignore the arguments of the defense, and neglect the circumstances testifying in favor of the person involved in criminal or administrative responsibility. This affects the objective investigation and consideration of the case (Barabanov 2007; Koryakin 2008; Kudryavtsev 2008).

We tried to study foreign sources, in particular, North American, using the tools of comparative jurisprudence. In the scientific literature of the United States, we did not manage to find a term and definition completely similar in meaning. Simultaneously, many scientific works are devoted to more general negative phenomena—the police's illegal actions and crime (Hough et al. 2018). For example, in Baltimore, Chicago, Ferguson, and New Orleans, four independent studies were carried out aimed at specific types of misconduct associated with police misconduct and serious violations of citizens' civil rights. The results show significant similarities and differences between police misconduct across several US police departments (D'Souza et al. 2019).

American scholars offer noteworthy methods of dealing with these negative phenomena. These methods are mainly practical. Thus, in response to police misconduct, one of the studies proposed to actively use the so-called "civilian allegations" to identify and expose problem police officers (Rozema and Schanzenbach 2019). This practice would certainly be useful for Russia. However, such examples of "civil charges"—what in our country would probably be called "civil suits and initiation of criminal prosecution against police officers by citizens" are rare. We are not aware of any complex applied scientific research on this topic.

The study proposes theoretical approaches and methods aimed at changing the methodology of scientific forensic research and teaching forensic science to future and current lawyers. Similar, but not identical, studies were conducted in the United States. For example, the "police conscientiousness theory" is opposed to the negative pattern of police misconduct (Kutnjak Ivković et al. 2018).

## 2 Materials and Methods

The methodological basis of this study is the conceptual approach of Baev (2002, 2011), Khalikov (2019), Nazarov (2015), and other Russian scientists on the need to overcome the trend of accusatory bias in criminal proceedings and the science of forensic science. We used comparative jurisprudence methods, formal-logical, system-structural, statistical (to summarize the materials of judicial practice), and sociological methods (when conducting questionnaires and interviews).

Statistical and other data on combating crime in Russia were used as empirical material (Gulina 2019). The materials of 215 criminal cases on corruption and other crimes were studied, which ended with convictions, acquittals, and termination of criminal prosecution. We also studied 27 materials on complaints from citizens to law enforcement agencies and courts about the illegality of the actions of law enforcement agencies.

A questionnaire and interviews were conducted. The method of expert assessment was applied in relation to 84 investigators, 65 employees of bodies—subjects of the operational search activity, 27 employees of the prosecutor's office, and 340 citizens representing different segments of the population of five constituent entities of Russia (Irkutsk and Novosibirsk regions, Transbaikal and Primorsky Krais, and the Republic of Buryatia).

## 3 Results

To achieve this goal, we first considered the regulatory framework. Article 6 of the Criminal Procedure Code of the Russian Federation "Appointment of criminal proceedings" postulates the most important principle, according to which criminal proceedings have their purpose of protecting the rights and legitimate interests of people and organizations who suffered from crimes, as well as to protect the individual from illegal and unfounded accusations, convictions, and restrictions of rights and freedoms (part 1). The legislator explains that criminal prosecution and the imposition of fair punishment on the guilty are as much in line with the purpose of criminal proceedings as the refusal to prosecute innocent people, release them from punishment, and rehabilitate everyone subjected to criminal prosecution unreasonably (part 2).

It would seem that forensic science, as the primary applied legal science in Russia, should be the first of the anti-criminal sciences to promote this principle. Unfortunately, this is far from reality. It seems strange that forensic scientists rarely pay attention to this problem. Noteworthy is the position of A. N. Khalikov, who notes that in some textbooks and training courses on criminalistics, a line is drawn in full to develop only the first part of the purpose of criminal proceedings. The author states and, using generalized examples of publications, proves that scientific and educational methodical works are missing the second purpose aimed at protecting human rights and interests. "However, the second appointment works on par with the first!"—exclaims the author (Khalikov 2019).

Meanwhile, the accusatory bias phenomenon has been scrutinized by scholars in the criminal procedure for many years. Thus, A. D. Nazarov believes that the accusatory bias consists of ignoring the representatives of the prosecution (investigator, prosecutor, etc.) and the judges of the evidence in favor of the accused, in the unwillingness to check and take into account the arguments of the defense. The author notes that the accusatory bias leads to the fact that investigative and judicial errors are committed in criminal cases, which distorts the true essence of the administration of justice and conflicts with the appointments and principles of criminal proceedings (Kolokolov 2015; Nazarov 2015).

The accusatory bias becomes dominant when investigating specific types of crimes, in which the personality of the offender has high special knowledge and skills of their use in criminal activities, i.e., crimes in the field of computer information (Polyakov 2019). In our opinion, this happens naturally due to the complexity of the investigation of such crimes and hence the increased prevalence of various kinds of investigative, operational-search, and prosecutorial errors and omissions (Rodriguez and Molina 2017). A clear accusatory bias replaces shortcomings in the evidence base, the proof of complex, and issues related to special knowledge not clear to law enforcement officials, i.e., technical aspects included in how these crimes were committed (Polyakov and Lapin 2018). Moreover, even the court de facto "turns a blind eye" to some shortcomings of the prosecution (including procedural) when subjectively convinced of the defendants' guilt.

V. Ya. Gorelik notes the "total triumph" of the accusatory bias in the activities of prosecution and the Russian courts. He also notes the important (in this regard) painful processes to which the Russian justice system and law enforcement, in general, were subjected (Gorelik 2015).

In the context of this study, it is essential to once again and in more detail dwell on the conclusions of A. N. Khalikov, as one of the few forensic scientists who paid attention to the problem. In his opinion, all scientific and applied research in the field of forensic science, its general provisions, and especially forensic tactics and forensic methods (a section of Russian forensic science as a science) should be simultaneously aimed at bringing the perpetrators and not bringing the innocent to justice. Forensic science should exist in an adversarial process, evaluate the arguments and evidence of the prosecution and defense parties on an equal footing, without giving preference to any of them. Thus, the author summarizes that there should be two equal vectors for the development of forensic science (Khalikov 2019).

Investigative actions are accompanied by the defense attorney (Muratova and Rakhmatullin 2015), which can vary depending on the current situation. O. Ya. Baev proposed to call such situations "protective and investigative" (Baev, 411). At the same time, most researchers believe that such situations are not forensic, and a lawyer cannot be considered a subject of forensic situations, even though this would contribute to justice and reduce the accusatory bias in situations where there is fertile ground for him.

The stated scientific approach, which can be conditionally designated as "adversarial," can be illustrated by the example of creating a forensic methodology for investigating corruption crimes. For example, the study of investigative and judicial practice in criminal cases of criminal corruption in higher education (Gulina 2019) identified many cases where there is a clear negative trend of accusatory bias on the part of representatives of the prosecution against university employees and other participants in legal relations in higher education, especially against ordinary representatives of faculty.

This is indirectly indicated, among other things, by the following statistics. During the study, the authors obtained official data on the prosecution of people working in higher education in several Russian regions (the Republic of Buryatia, Irkutsk and Novosibirsk regions, and Zabaikalsky and Primorsky Krais). According to the information centers of the departments (ministries) of internal affairs of the indicated Russian subjects (Gulina 2019), in 2013–2018, 189 people were convicted under articles of the Criminal Code of the Russian Federation providing for liability for corruption-related crimes:

- Part 3 of art. 159 of the Criminal Code of the Russian Federation—20 people;
- Part 3 of art. 160 of the Criminal Code of the Russian Federation—32 people;

- Art. 285, 285.1, and 285.2 of the Criminal Code of the Russian Federation—3 people;
- Art. 286 of the Criminal Code of the Russian Federation—1 person;
- Art. 286, 290, 291, 291.1, 291.2 of the Criminal Code of the Russian Federation—70 people;
- Art. 292 of the Criminal Code of the Russian Federation—63 people.

Simultaneously, only 8 people were acquitted in the same regions (one defendant accused of committing a crime under part 3 of art. 160 of the Criminal Code of the Russian Federation, in the Irkutsk Region, and seven defendants in the Novosibirsk region, two of them were accused under art. 159 of the Criminal Code of the Russian Federation and five—under art. 160 of the Criminal Code of the Russian Federation). In the Primorsky Krai, there was not a single acquittal for the indicated period.

Let us give an example of the manifestation of an accusatory bias in a situation typical of Russian anti-corruption law enforcement practice of the illegal prosecution of university employees and other participants in legal relations in higher education. This situation can be designated as an accusation of university workers of embezzlement of wage funds in connection with the allegedly not performed educational load at the primary or another place of work.

Based on the materials of one of the criminal cases, in November 2014 in the Chelyabinsk region, the acquittal of the district court came into force against the university professor M. She was acquitted on charges of fraud (part 3 of Art. 159 of the Criminal Code of the Russian Federation), connected with the alleged illegal receipt of the wage in the amount of 83,000 rubles. According to the investigation, in 2011–2012, Professor M. flew to scientific conferences abroad but did not warn the university administration about the trips. The timesheet was marked with the presence of a professor at the workplace, conducting classes with students.

At the request of the defense, the court found that the professor was conducting classes online using the Skype program. It was also established that, according to the employment contract between the university and the professor, participation in international conferences, scientific research (including those carried out on a trip), as well as the use of modern technologies are laid down as the employee's obligations. The representatives of the prosecution, the former students of this university, could not fail to know that the professor's actions fully fit into the customs prevailing in all universities of the country. The acquitted did not have and could not have the intent to steal the university funds. While abroad, she properly performed her official duties (Kalininsky District Court of the city of Chelyabinsk 2014; Katina 2014; Chelyabinsk Regional Court 2014). To these arguments of the court, we can add the thesis about the seeming insignificance (part 2 of article 14 of the Criminal Code of the Russian Federation) of this particular and a considerable number of other similar acts occurring in practice.

During the study, we found that in many regions of the country, criminal cases were initiated and pre-investigation checks were carried out in relation to university teachers on the grounds of embezzlement, allegedly related to non-fulfillment of professional duties, failure to teach at a university in situations of the physical absence of at the workplace. Such cases were initiated especially against people not living in the corresponding region of the university's location, but working there (i.e., part-time) and rarely come to the region. The vast majority of such criminal cases are terminated. However, appropriate attempts by individual law enforcement officers are still being made. This fact is understandable, since, to identify such "crimes," minimal labor costs and the level of qualification of a law enforcement officer are required, compared with what is needed to combat real "top" or multi-episode criminal corruption. (Sazonov 2019).

## 4 Discussion

Sharing A. N. Khalikov's position on the adversarial approach to forensic research methodology and the presentation of their results, it is necessary to make two important clarifications before presenting the author's approach.

Firstly, we agree with A. A. Exarhopoulo, who noted that the interest of forensic science in crime, the offender, and his defender is due to the desire to create scientific means that ultimately contribute to the knowledge of the truth in criminal proceedings (Exarhopoulo 2004). That is, forensic science as a science and academic discipline can be neither "accusatory" nor "defensive." It aims to achieve the truth. Meanwhile, the defense lawyer and his client are not always interested in establishing the truth due to the requirements of Russian legislation [i.e., a lawyer "... is obliged to honestly, reasonably, and conscientiously defend the rights and legitimate interests of the client by all means not prohibited by the Russian legislation" (Russian Federation 1996)]. That is why A. A. Exarhopulo, N. P. Yablokov, and other forensic scholars spoke about the inadmissibility of including research into the subject area of forensic science within the framework of the so-called "forensic defense," "forensic advocacy," or "the theory of advocacy in criminal proceedings" (Baev 2002; Zashlyapin 2007), without denying the relevance of this scientific direction (Exarhopulo 2004).

Thus, scientists-developers of forensic methods and forensic tactics can and should develop applied recommendations of a protective nature, but only to establish the truth, i.e., to prevent the prosecution of innocent people. Other tasks of the defense, such as proving the innocence of a person known to the lawyer to have committed a crime, although legitimate from the point of view of his powers, cannot be the subject of scientific forensic research.

Secondly, we proceed from the premise that there is no need to confuse the lawful accusatory position with the notorious accusatory bias. The accusatory position is always based on the law, and the accusatory bias is a manifestation of improper performance of the prosecution's role. However, the accusatory position is not only permissible but also the responsibility of the prosecution's representatives.

## 5 Conclusion

Undoubtedly, such illegal and ethically incorrect actions of law enforcement officers in educational and other spheres, if allowed systematically as a stable pattern, destroy people's trust in the law enforcement system and discredit the state as a whole (Cruz 2015). To solve the problem of accusatory bias, complex measures of a legal, socio-economic, and organizational nature are needed. However, in the system of such measures, changes in scientific, theoretical, and didactic approaches are also needed. It is also necessary to adjust the methodology of scientific forensic research and approaches to training future and existing lawyers.

For this purpose, we offer a special principle for developing forensic techniques and other forensic recommendations. It is about "the principle of inadmissibility of an accusatory bias." The application of this principle by a scientist-developer consists in the development and implementation of the following scientific provisions and applied recommendations:

1. On common, widespread non-criminal acts in a particular field of activity, which, as shown by the analysis of practice, are often mistakenly assessed by law enforcers as a criminal;
2. On typical investigative situations and versions that ensure this position of the developed forensic investigation technique;
3. On such tactical operations as "Checking the version of the person's innocence (less guiltiness), on self-incrimination";
4. On the prevention and elimination of typical mistakes and violations of the law by law enforcement officers, including warnings about criminal and other legal liability for illegal methods of investigation (with examples of relevant criminal and civil cases, disciplinary proceedings);
5. On the means and methods of work of defense lawyers in connection with the identification of an accusatory bias against innocent people;
6. On the investigation of crimes related to an accusatory bias in the activities of representatives of the prosecution. Similar studies are being conducted in the USA (Stinson et al. 2018).

In conclusion, it should be emphasized that forensic science, like all other anti-criminal sciences, should be involved in the fight against the notorious accusatory bias in law enforcement. It seems that it is precisely the forensic science that potentially has the most effective tools to combat this negative phenomenon in Russian legal reality.

## References

Baev OYa (2002) Forensic advocacy as a subsystem of the science of forensic science. In: Professional activity of a lawyer as an object of forensic research. Charoid, Yekaterinburg, Russia, pp 301–318

Baev OYa (2011) Selected works on the problems of criminalistics and criminal procedure. Eksmo, Moscow, Russia

Barabanov PK (2007) Procedural means of overcoming the accusatory bias in court activities. Russ Judge 7:26–28

Chelyabinsk Regional Court (2014) Appellate determination (10–6058/2014). Chelyabinsk Regional Court, Chelyabinsk, Russia

Cruz JM (2015) Police misconduct and political legitimacy in central America. J Lat Am Stud 47(2):251–283

D'Souza A, Weitzer R, Brunson RK (2019) Federal investigations of police misconduct: a multi-city comparison. Crime Law Soc Chang 71:461–482

Exarhopulo A (2004) Subject and system of forensic science: Development problems at the turn of the 20–21 centuries. Publishing House of the Saint Petersburg State University, St. Petersburg, Russia

Garmaev YuP (2018) Fundamentals of methods for investigating corruption crimes. Retrieved from http://www.iuaj.net/node/2490

Gorelik VY (2015) Triumph of the accusatory bias. Ezh-Yurist 15:5

Gulina EV (2019) Features of the forensic methodology for investigating corruption crimes in higher education. Buryat State University named after Dorzhi Banzarov, Ulan-Ude, Russia, Dissertation of the Candidate of Legal Sciences

Hough M, May T, Hales G, Belur J (2018) Misconduct by police leaders in England and wales: an exploratory study. Polic Soc 28(5):541–552

Kalininsky District Court of the city of Chelyabinsk (2014) Sentence (No. 1–439/2013). Kalininsky District Court of the city of Chelyabinsk, Chelyabinsk, Russian Federation

Katina E (2014) The court upheld the acquittal of the Chelyabinsk professor who lectured on Skype. New day. Retrieved from https://newdaynews.ru/chel/517643.html

Khalikov AN (2019) Forensics: theory and practice, 7th edn. Publishing House of the Krasnodar University of the Ministry of Internal Affairs of Russia, Krasnodar, Russia

Kolokolov NA (2015) Accusatory bias. Ezh-Yurist 36:1–6

Koryakin EA (2008) Administrative proceedings in the Russian federation. Russ Judge 2:15–18

Kudryavtsev VL (2008) "Accusatory bias" in the activities of the investigator: a problem of only legislative regulation? Crim Just 2:26–29

Kutnjak Ivković S, Haberfeld M, Peacock R (2018) Decoding the code of silence. Crim Justice Policy Rev 29(2):172–189

Luneev VV (1995) Crime control: are the indicators reliable? State and Law 7:87–90

Muratova NG, Rakhmatullin RR (2015) Problems of implementing tactical techniques by the parties in the production of identification: Investigative errors and other gaps. Judicial Branch Crim Proced 4:130–136

Nazarov AD (2015) Accusatory bias in subjects conducting criminal proceedings as a factor contributing to the appearance of errors in criminal proceedings. Topical Probl Russ Law 9:149–154

Polyakov V (2019) Criminalistics specifics of methods of committing computer crimes and peculiarities of their prevention. Religacion. J Soc Sci Human 4(21):90–97

Polyakov VV, Lapin AS (2018) The architecture of the Honeypot System for studying targeted attacks. In: Proceedings from APEIE-2018: 14th international scientific-technical conference on actual problems of electronic instrument engineering, vol 1, Part 2. Novosibirsk State Technical University, Novosibirsk, Russia, pp 202–205

Rodriguez GD, Molina F (2017) The preservation of digital evidence and its admissibility in the court. Int J Electron Secur Digit Forensics 9(1):1–18

Rozema K, Schanzenbach M (2019) Good cop, bad cop: using civilian allegations to predict police misconduct. Am Econ J Econ Pol 11(2):225–268
Russian Federation (1996) Criminal code of the Russian federation (13 June 1996 No. 63-FZ). Moscow, Russia
Russian Federation (2002) Federal law "on advocacy and the legal profession in the Russian Federation" (31 May 2002 No. 63-FZ). Moscow, Russia
Sazonov A (2019) In Samara, the criminal case against the professor accused of truancy was dropped. Kommersant. Retrieved from https://www.kommersant.ru/doc/3968018
Stinson PM Sr, Liederbach J, Buerger M, Steven L, Brewer Jr (2018) To protect and collect: a nationwide study of profit-motivated police crime. Criminal Justice Studies. Crit J Crime Law Soc 31(3):310–331
Zashlyapin LA (2007) The main components of the theory of advocacy in criminal proceedings. Ural University Publishing House, Yekaterinburg, Russia

# Privacy Restriction in Communications Transmitted via Electronic Networks: Analysis of Criminal Procedural Legislation of the Russian Federation and the Republic of Kazakhstan

**Sergey I. Davidov, Sergey A. Polyakov, and Liliya V. Smeshkova**

**Abstract** This paper analyzes the norms of the legislation of the Russian Federation and the Republic of Kazakhstan in the field of secrecy of correspondence transmitted via electronic channels, including the internet. The choice of objects for comparative legal analysis is due to the qualitative changes in the institute of investigative actions over the past five years in the context of the Criminal Procedure Law of both countries. We focus on the grounds and procedure for conducting investigative actions related to obtaining information from electronic messages. We emphasize that the utmost clarity of normative prescriptions should distinguish the rules for using such methods of collecting evidentiary information since the right of individuals to the secrecy of correspondence, postal, and telegraph messages enshrined in constitutional norms. The priority of these norms is evident, both in the system of legislation of Russia and Kazakhstan. We proposed implementing the Russian Criminal Procedure Law of the norms governing the detection, seizure, and examination of electronic messages provided for by the legislation of Kazakhstan, taking into account the needs of investigative and judicial practice.

**Keywords** Restriction of rights · Message privacy · Investigation · Actions

## 1 Introduction

Crimes committed with the use of information and telecommunication technologies are a dynamically developing criminal industry that threatens the security of the country, society, and individuals. According to the Prosecutor General's Office of

S. I. Davidov
Altai State University, Barnaul, Russia

S. A. Polyakov · L. V. Smeshkova (✉)
Novosibirsk State Technical University, Novosibirsk, Russia
e-mail: smeshkovalv@mail.ru

L. V. Smeshkova
Novosibirsk State University of Economics and Management, Novosibirsk, Russia



S. G. Maximova et al. (eds.), *Advances in Natural, Human-Made, and Coupled Human-Natural Systems Research*, Lecture Notes in Networks and Systems 234,
https://doi.org/10.1007/978-3-030-75483-9_85

the Russian Federation, in 2018, 1,746,774 crimes committed using information and communication, or computer technologies were registered (43,362 of these criminal cases were previously investigated and sent to court, which is 24%) Statistics for 2019 also show consistently low rates of detection of such crimes: only 65,238 (22%) of the 294,409 registered criminal cases were investigated. Another disturbing factor is the 1.7-fold increase in the number of crimes committed using computer or communication technologies in 2019 (Prosecutor General's Office of the Russian Federation. Portal of Legal Statistics 2018, 2019, 2020). The Ministry of Internal Affairs of the Russian Federation provides similar data in statistical reports (The Ministry of Internal Affairs of the Russian Federation 2020).

In this situation, the improvement of legal support for the investigation of such crimes is an urgent problem for investigative practice. It is also necessary to optimize the procedure for conducting investigations related to the extraction and examination of messages transmitted through electronic communication networks, including the internet. The existing provisions on their conduct do not allow effectively obtaining the necessary information and using it to prove crimes committed using information and telecommunication technologies in criminal cases.

The study aims to conduct a comparative analysis of the legal regulation of operational investigative actions related to the authorized restriction of the rights to confidentiality of correspondence transmitted over electronic communication networks in Russia and Kazakhstan. It is necessary to identify the positive experience of legal regulation and substantiate proposals for improving the Russian Criminal Procedure Law.

## 2 Materials and Methods

We analyze the results of previous procedural studies on the examination and seizure of electronic communications, such as the work of Gavrilin (2009), Ilyushin (2008), Khomkolov (2004), Shevchenko (2016), Avetisyan and Ryasov (2018), Bikmiyev and Burganov (2018), Borisevich (2019), Vasyukov (2016), Dobrovlyanina. (2019), Orumbaeva (2015), Pastukhov and Losavio (2017), Lipsky and Popov (2017), Suprun and Cherkasov (2017).

The choice of scientific methods is determined by some issues arising from the content of research tasks:

(1) Determination of the content of legal prescriptions serving as a normative basis of manufacture for conducting investigative actions related to the authorized restriction of rights to the secrecy of electronic correspondence, in the legislation of Russia and Kazakhstan;
(2) Identification of the most effective rules defining the object, subject, grounds, and procedure for checking and withdrawing electronic communications based on the results of a comparative legal analysis of the Criminal Procedure Law of both countries;

(3) Preparation of proposals for improving the Criminal Procedure Code of the Russian Federation in terms of the algorithm for checking and withdrawing messages transmitted over electronic communication networks.

Accordingly, to solve the first problem, we use the formal-logical method of studying and defining legal concepts, determining their features, and interpreting their meaning (Kelsen 1949). This approach allows us to define the conceptual framework used to describe the procedure for reviewing and withdrawing electronic communications and correctly interpret the meaning of the relevant legal provisions.

The use of the comparative-legal method enables us to draw a parallel between the regulatory regulation of verification and retrieval of electronic messages in criminal proceedings in Russia and Kazakhstan and to identify the most promising standards that provide positive results during these investigative actions.

We used the modeling method to form conclusions and proposals on the implementation of certain norms of the Criminal Procedure Legislation of the Republic of Kazakhstan in the Criminal Procedure Code of the Russian Federation to optimize the procedure for conducting investigative actions related to the restriction of the rights of individuals to the secrecy of messages transmitted via electronic communication networks (Garner 1987).

We applied the statistical method to substantiate the requirement to seize postal and telegraph items and other correspondence in the law enforcement practice of the investigative bodies of the Russian Federation. The historical-legal method made it possible to identify the historical prerequisites that influenced the decision of the legislators to introduce Part 7 of Article 185 of the Code of Criminal Procedure of the Russian Federation.

Concrete-sociological method and method of a retrospective analysis of traditional jurisprudence are not relevant for this study since the novelty of the Institute of inspection and seizure of electronic communications in the Russian criminal proceedings does not give opportunities for a comprehensive study of examples of its use in investigative and judicial practice, and to obtain a representative survey of employees from preliminary investigation bodies.

## 3 Results

The rapid qualitative and quantitative development of cybercrime in Russia has led to the need to develop and implement effective additional tools to counter crimes committed using communication and internet technologies. In 2016, the Russian government adopted the Code of Norms on countering terrorism in the form of Federal Law No. 374-FZ of July 6, 2016 "On amendments to the Federal Law "On Countering Terrorism" and certain legislative acts of the Russian Federation regarding the establishment of additional measures to counter terrorism and ensure public security." This law significantly expanded the ability of preliminary investigation bodies to obtain information relevant to criminal cases (Russian Federation

2016b). In the history of the Russian legislative process, these amendments were entitled "Yarovaya Package."

In addition to the normative act mentioned above, Federal Law No. 375-FZ of July 6, 2016 was adopted. Due to this law, Article 185 of the Criminal Procedure Code of the Russian Federation "the seizure of postal and telegraph items and their inspection" was supplemented by Part 7, which states that "if there are sufficient grounds to believe that the information relevant to a criminal case may be contained in electronic messages or other messages transmitted over telecommunication networks, the investigator may, by a court decision, perform their inspection and seizure" (Russian Federation 2016a).

Despite the almost four-year existence, the practice of using this legislative provision remains relatively unpopular. Commenting on the situation in the execution of the orders of Part 7 of Article 185 of the Code of Criminal Procedure of the Russian Federation, preliminary investigation officers note that obtaining the information transmitted via electronic networks is faster through sending an inquiry to the bodies about the relevant operational search measures in the form of withdrawal of information from technical communication channels, and not through the opening and checking electronic means of communication. The possibility of sending orders is provided for by the Criminal Procedure Code of the Russian Federation (Russian Federation 2001).

The above situation is illustrated by statistical data on the work of courts of general jurisdiction when considering applications of preliminary investigation bodies for conducting investigative actions and measures related to the restriction of citizens' rights to the secrecy of correspondence.

In 2017, in compliance with the decisions of the courts of the Russian Federation adopted at the request of the preliminary investigation bodies, investigative actions in the form of seizure of correspondence were carried out 13,593 times. During the same reporting period, the investigative authorities sent 5,881,012 petitions to restrict the constitutional rights of citizens to privacy in correspondence, telephone conversations, postal, telegraph, and other messages transmitted via electronic and postal networks. The court satisfied 99.7% of these cases. Additionally, in 2017–2019, the number of appeals to the courts by bodies engaged in operational search activities did not change dramatically, while the number of appeals by investigators under Part 2 of Article 185 of the Criminal Procedure Code of the Russian Federation decreased by 28% in 2018 and by 22% in 2019 (Judicial Department at the Supreme Court of the Russian Federation 2018, 2019a, b).

The reasons for the ineffectiveness of the normative provisions of Part 7 of Article 185 of the Code of Criminal Procedure of the Russian Federation are the lack of reasonableness. According to some processualists, another reason is the absence of a legal mechanism for checking and extracting messages and other information transmitted via electronic communication channels (Pastukhov and Losavio 2017). The validity of the observations substantiated the textual analysis of Article 185 of the Code of Criminal Procedure, under which the procedure information contained in electronic messages is subject to the general requirements formulated for the seizure of postal and telegraph messages, their examination, and seizure (Bikmiyev and Burganov 2018).

The criminal procedure legislation of the Republic of Kazakhstan also contains several provisions providing for the possibility of obtaining information transmitted via electronic communication channels. However, in the Criminal Procedure Code of the Republic of Kazakhstan, this task is implemented in a qualitatively different way.

The Criminal Procedure Code of the Republic of Kazakhstan, adopted in 2014, fundamentally changed the principles of the republican bodies of the preliminary investigation, prosecutor's office, and court (Republic of Kazakhstan 2014). The institution of investigative actions has been significantly transformed. In this regard, the most controversial is the consolidation of so-called secret investigative actions that promote the integration of criminal proceedings and operational investigative activities, which corresponds to global trends in countering organized forms of criminal activity and corruption (Orumbaev 2015).

Undercover investigative actions are carried out when to clarify the circumstances of the criminal case, it is necessary to obtain information about the facts without informing the persons involved in the criminal proceedings whose interests they affect (Part 1 of Article 232 of the Criminal Procedure Code of the Republic of Kazakhstan). Planned investigations are carried out in criminal cases on crimes for which a penalty of imprisonment for a term of more than one year is provided, and on crimes prepared and committed by a group. The procedure for conducting secret investigations is determined by law enforcement and special government bodies with the consent of the Prosecutor General of the Republic of Kazakhstan.

Article 231 of the Criminal Procedure Code of the Republic of Kazakhstan provides for eight types of tacit investigative actions, including (1) tacit control, (2) interception, and (3) seizure of information transmitted via electric (telecommunication) communication networks. To a certain extent, it can be considered similar to the seizure and investigation of electronic or other messages transmitted via telecommunication networks provided for by the Criminal Procedure Code of the Russian Federation. Although it should be noted that, based on the terminology used to the greatest extent, the tacit investigative action is similar to operational-search measures in the form of wiretapping and withdrawal of information from technical communication channels, provided for in Article 6 of the Federal Law "On Operational and Investigation Activity" of August 12, 1995 No. 144-FZ (Russian Federation 1995).

According to Article 234 of the Code of Criminal Procedure of the Republic of Kazakhstan, the secret control of electronic communication networks is understood as the secret listening or recording of speech information using scientific and technical means or computer programs transmitted by phone or other devices, allowing the transmission of speech information produced, if necessary, by secret intrusion or inspection.

Interception and removal of information transmitted over electronic communication networks is the interception and removal of (1) signs, (2) signals, (3) speech information, (4) written text, (5) images, (6) video images, (7) sounds, and other information transmitted over wired, radio, optical, and other electromagnetic systems.

The Code of Criminal Procedure of the Republic of Kazakhstan provides for the following procedure for the indicated investigative action:

- Sending to the specialized division of investigative judges a reasoned decision on behalf of the person conducting the preliminary investigation with the production of the necessary secret investigative action on the attachment of materials justifying the need for it, within 24 h from the date of adoption of this procedural act;
- Approval of the secret investigative action by the investigating judge within 12 h from the date of receipt by the court of the relevant decision, and if necessary, consideration of the decision beyond the established period, but not more than 24 h;
- The investigator or the person conducting the initial inquiry shall send an order to the body conducting the initial inquiry based on the decision of the investigating judge;
- The secret investigation results are recorded on the appropriate material carrier, which is packed, sealed, and certified with the signature of an official of the authorized body conducting the secret investigation. The content of these results is brought to the attention of the authorized prosecutor within two days after the end of the investigation.

Textual analysis of these norms of the Code of Criminal Procedure reveals some gaps in the legal regulation of the newly created institution. However, the problem of the procedure for conducting investigative actions related to the restriction of the right to secrecy of messages transmitted via electronic communication channels has been solved more successfully than in the Code of Criminal Procedure of the Russian Federation:

- The object and subject of the investigative action are determined;
- Fixed independent production mechanism for the interception and removal of information transmitted via electronic communication networks (Article 243 of the Code of Criminal Procedure of the Republic of Kazakhstan), which is separated from the procedure for controlling postal and other messages (Article 246 of the Code of Criminal Procedure of the Republic of Kazakhstan).

It seems that Russian legislators should use the experience of the Republic of Kazakhstan on the legal regulation of investigative actions related to the restriction of the right to secret messages transmitted over electronic communication networks.

## 4 Discussion

Due to the innovations provided for in Article 185 of the Code of Criminal Procedure of the Russian Federation, a spirited discussion among specialists, scientists, and practitioners developed on the pages of scientific publications.

A generalized analysis of the critical points of view of Russian processualists reveals several problematic, unresolved issues affecting the inefficiency of using novelties of Criminal Procedure Law in terms of studying and retrieving messages transmitted via telecommunication networks:

- The object and subject of investigative actions related to obtaining information from messages transmitted via electronic communication channels are not defined (Suprun and Cherkasov 2017).
- The subject is not identified as the subject of the investigation, which should enable the investigator and the person conducting the initial inquiry to make and verify electronic communications, including those transmitted via the internet (Suprun and Cherkasov 2017);
- The form and procedure for conducting the procedure for the seizure and examination of electronic messages are not established (Lipsky and Popov 2017).

We cannot disagree with the expert assessment. Moreover, according to surveys, investigators are experiencing difficulties in implementing innovations.

We believe that the experience of the Republic of Kazakhstan in regulating this area is quite interesting and meets the needs of the investigative practice. The idea of legal regulation of recording, seizure, and verification of electronic messages in the form of independent investigative action is a promising direction. Such regulation may be based on the principle of interception and seizure of information transmitted via electric communication networks, provided for by the Criminal Procedure Code of the Republic of Kazakhstan, namely, a court order issued at the request of the investigator and the person conducting the initial inquiry is sent for execution to the bodies carrying out operational search activities. The timing and procedure for obtaining the relevant results are controlled by the official who initiated the investigation.

## 5 Conclusion

A systematic assessment of the conflicts identified in the Criminal Procedure Law of the Russian Federation allows us to conclude that the procedure for working with detained postal and telegraph items, provided for by the Criminal Procedure Code of the Russian Federation, is not limited to the inspection and seizure of messages transmitted over electronic communication networks. Meanwhile, the importance of this investigative action for solving the preliminary investigation tasks is undeniable, especially in terms of cybercrime. The search in the electronic information space and the seizure of information about the circumstances of the commission of crimes are often the primary evidence in criminal cases. The Russian legislators can use the proposals formulated based on a comparative analysis of the norms of the Criminal Procedure Code of the Russian Federation and the Republic of Kazakhstan to improve the legal basis for the seizure and verification of messages transmitted via electronic communication channels.

## References


Avetisyan AD, Ryasov AA (2018) Peculiarities of investigator's work with data storage devices in detection and investigation of crime with modern information and communication technologies. Legal Sci Pract: Bull Nizhny Novgorod Acad Ministry of Internal Aff Russia 3(43):109–112

Bikmiyev RG, Burganov RS (2018) Seizure and inspection of electronic devices. Criminal Law 1:125–131

Borisevich GY (2019) Concerning the development of interdisciplinary institution regulating electronic information technologies usage in criminal court proceedings of Russia. Perm Legal Almanac 2:610–617

Dobrovlyanina OV (2019) Some aspects of procedural seizure (copy) of electronic information carriers. Perm Legal Almanac 2:641–649

Garner BA (1987) A dictionary of modern legal usage. Oxford University Press, New York, NY

Gavrilin YuV (2009) Investigation of crimes encroaching on information security in the economic sphere: Theoretical, organizational, tactical, and methodological foundations (Dissertation of Doctor of Legal Sciences). Moscow University of the Ministry of Internal Affairs of Russia, Moscow, Russia

Ilyushin DA (2008) Features of the investigation of crimes committed in the field of providing internet services (Dissertation of Doctor of Legal Sciences). Volgograd, Russia: Volgograd Academy of the Ministry of Internal Affairs of Russia

Judicial Department at the Supreme Court of the Russian Federation (2018) Summary statistics on the activities of federal courts of general jurisdiction and Justices of the Peace of 2017. http://www.cdep.ru/index.php?id=79&item=4476

Judicial Department at the Supreme Court of the Russian Federation (2019a) Summary statistics on the activities of federal courts of general jurisdiction and Justices of the Peace for the 1st half of 2019a. http://www.cdep.ru/index.php?id=79&item=5081

Judicial Department at the Supreme Court of the Russian Federation (2019b) Summary statistics on the activities of federal courts of general jurisdiction and Justices of the Peace of 2018. http://www.cdep.ru/index.php?id=79&item=4891

Kelsen H (1949) General theory of law and state. Harvard University Press, Cambridge, MA

Khomkolov VP (2004) Organizational and legal aspects of the investigation and prevention of crimes in the field of computer information (Dissertation of Candidate of Legal Sciences). Baikal State University of Economics and Law, Irkutsk, Russia

Lipsky NA, Popov EM (2017) Problems of electronic correspondence in the criminal-procedural proof. Sci Notes of the St. Petersburg Branch of the Russ Customs Acad 3(63):82–86

Orumbaev NE (2015) Undercover investigative actions under the new Code of Criminal Procedure of Kazakhstan. Vestnik of the Omsk Law Acad 3(28):75–77

Pastukhov PS, Losavio M (2017) Use of information technology to ensure security of the individual, society and state. Perm U Herald Jurid Sci 36:231–236

Prosecutor General's Office of the Russian Federation. The portal of legal statistics (2018) Monthly compilation on the state of crime in Russia for 2017. http://crimestat.ru/analytics

Prosecutor General's Office of the Russian Federation. The portal of legal statistics (2019) Monthly compilation on the state of crime in Russia for 2018. http://crimestat.ru/analytics

Prosecutor General's Office of the Russian Federation. The portal of legal statistics (2020) Monthly compilation on the state of crime in Russia for 2019. http://crimestat.ru/analytics
Republic of Kazakhstan (2014) Code of Criminal Procedure of Kazakhstan (July 4, 2014 No. 231-V). Astana, Kazakhstan. http://online.zakon.kz/m/document?doc_id=31575852#sub_id=2310000
Russian Federation (1995) Federal law "On the operational-search activity" (August 12, 1995 No. 144-FZ). Moscow, Russia. http://www.consultant.ru/document/cons_doc_LAW_7519/
Russian Federation (2001) Code of Criminal Procedure of the Russian Federation (December 18, 2001 No. 174-FZ). Moscow, Russia. Retrieved from http://www.consultant.ru/document/cons_LAW_34481/
Russian Federation (2016a) Federal law "On amendments to the Criminal Code of the Russian Federation and the Code of Criminal Procedure of the Russian Federation regarding the establishment of additional counter-terrorism measures and ensuring public safety" (July 6, 2016a No. 375-FZ). Moscow, Russia. http://www.consultant.ru/document/cons_doc_LAW_201087/
Russian Federation (2016b) Federal law "On amendments to the Federal Law "On countering terrorism and certain legislative acts of the Russian Federation regarding the establishment of additional counter-terrorism measures and ensuring public safety" (July 6, 2016b No. 374-FZ). Moscow, Russia. http://www.consultant.ru/document/cons_doc_LAW_201078/
Shevchenko ES (2016) Tactics of investigative actions in the investigation of cybercrime (Dissertation of Candidate of Legal Sciences). Moscow. Russia: Kutafin Moscow State Law University
Suprun SV, Cherkasov VS (2017) On the opposing nature of novella in the legislative regulation of the investigatory action "Arrest of postal and telegraph." Siberian Law Rev 14(1):59–64
The Ministry of Internal Affairs of the Russian Federation (2020) A brief description of the state of crime in the Russian Federation for January–December 2019. MIA FR Statistics Publishing, Moscow, Russia
Vasyukov VF (2016) Several aspects of investigative actions aimed at detection, fixation, and caption of electronic messages transmitted via user mobile devices. Russ Invest 23:15–18

# Constitutional Law of the CIS Countries in the Context of Supranational Integration Processes and National Security

**Yury V. Kim and Evgeny S. Anichkin**

**Abstract** According to the fundamental documents, the Commonwealth of Independent States is a form of cooperation between equal independent States formed during the dissolution of the Union of SSR. It is a regional inter-State organization recognized by the international community, whose main purpose is to form a deeply integrated economic and political association of the States concerned, ensuring the effective development of each of its participants. Among the promising targets designed to create favorable conditions for the further development of national economies and security, to strengthen the well-being of citizens, to involve in the world political and economic processes, is the further convergence of national legislation of the Commonwealth member States in various areas of multilateral cooperation on the basis of universally recognized principles and norms of international law. Over the three decades of its establishment and development, the Commonwealth has travelled a difficult path, the results of which cannot be assessed unequivocally. There are positive achievements that cannot be ignored. Many problems remain, which constrain integration processes and are caused by many inter-State contradictions of political, economic, legal and other properties. As decisive factors impeding the dynamic development of supranational integration mechanisms, the authors of the study see insufficient constitutional and legal support for the goals and objectives of the CIS.

**Keywords** Commonwealth · Statehood · Constitution · Law · Convergence · Integration · Harmonization · Security

Y. V. Kim
Kemerovo State University, Kemerovo, Russia

E. S. Anichkin (✉)
Altai State University, Barnaul, Russia
e-mail: rrd231@rambler.ru



S. G. Maximova et al. (eds.), *Advances in Natural, Human-Made, and Coupled Human-Natural Systems Research*, Lecture Notes in Networks and Systems 234,
https://doi.org/10.1007/978-3-030-75483-9_86

## 1 Introduction

National security largely depends on geopolitical factors (geographical, geophysical, political, social, demographic, economic, environmental, etc.). The national security of Russia is a state of protection of the individual, society, and the country from internal and external threats, which ensures constitutional rights and freedoms of citizens of Russia, decent quality and standard of living, sovereignty, independence, national and territorial integrity, and sustainable socio-economic development of Russia. The achievement of strategic stability and equal partnership with an active foreign policy aimed at creating a sustainable system of international relations based on international law is the priority area that strengthens the national security of Russia. The development of bilateral and multilateral cooperation with the member states of the Commonwealth of Independent States [CIS], regional and sub-regional integration, focused on fruitful cooperation, are essential elements of international relations (Presidential Executive Office 2016).

December 2021 marks the 30th anniversary of the establishment of the CIS. This major political and legal event requires analyzing and summarizing the results and clarifying future development prospects. The scientific community should not pass this event. The formation and development of the CIS were complicated and did not have the desired dynamics. This situation has a systemic impact on the quality of integration processes; it hinders the development potential of the economy, social sector, and other relevant fields of the functioning of the CIS member states.

Initially, the CIS was mainly considered a mechanism for the *civilized separation* of the former Soviet republics and the redistribution of the USSR property, a smooth transition from a single union Soviet state to several independent sovereign countries. Vladimir Putin, at his press conference in Yerevan on March 25, 2005, explained with utmost frankness the fundamental differences between European integration and the CIS:

> In Europe, the EU countries worked together for unification, while the CIS was created for a civilized separation. Here is the difference <...> There was one goal: the CIS was created in such a way so that the USSR collapse occurred in the most civilized way. The CIS coped with this task; simultaneously, the CIS did not have any super-tasks either in the integration or in the economic sector (President of the Russian Federation 2005).

Only later, when the desire for state isolation was somehow rethought, the political and socio-economic costs of the *waves* of sovereignty were revealed, and the motives for greater convergence and integration in the political, economic, and other fields increased. The countries are doing their best to strengthen international cooperation. Simultaneously, the ideas about turning the CIS into a more cohesive entity (for example, the Eurasian Union) with a joint parliament, common citizenship, united armed forces, and close economic coordination do not find support among the CIS member states. We assume that the Commonwealth has not yet realized its integration potential. We think that this thesis is reasonable, although the problems are much more profound.

The conditions and factors hindering the convergence of the state-legal systems of the CIS countries and the clarification of their understanding and targets should be thoroughly studied using the theoretical and cognitive resources of the constitutional law.

## 2 Materials and Methods

The research methodology is based on traditional approaches used in law—both general scientific methods (analysis, synthesis, comparison, generalization, abstraction, induction, deduction, system-structural, etc.) and private scientific methods (historical-legal, comparative-legal, legal, legal hermeneutics, etc.). We used them in the theoretical and methodological analysis of international and Russian normative legal acts. In particular, we studied and compared the main documents of the CIS, the United Nations [UN], European Union [EU], the Association of Southeast Asian Nations [ASEAN], the League of Arab States [LAS], and other international organizations (CIS Executive Committee n.d.; EU n.d.; ASEAN n.d.; LAS n.d.). In addition to the monographic and periodical literature, we paid particular attention to the dissertations of Russian scientists and specialists from the CIS countries (Armenia, Azerbaijan, Belarus, Kazakhstan, Kyrgyzstan, Tajikistan, etc.) (references to the relevant literature sources are given in below). The methodological principles underlying this work were complexity, consistency, objectivity, adequacy, and efficiency.

## 3 Results

The research was conducted in accordance with the essential documents of the CIS (Agreement on the Establishment of the Commonwealth of Independent States of December 08, 1991; the Almaty Declaration of December 21, 1991; the Charter of the Commonwealth of Independent States of January 22, 1993, etc.) (the primary CIS documents posted on the official website of the CIS Executive Committee. http://www.cis.minsk.by/site/about-cis.). The Commonwealth is a form of cooperation of equal independent states, recognized by the international community as a regional international organization. The promising tasks of the CIS include (1) creating favorable conditions for the further development of national economies and security; (2) strengthening the well-being of citizens, (3) engaging in global political and economic processes, and (4) converging the national legislation of the CIS member states in various areas of multilateral cooperation based on the generally recognized principles and norms of international law.

The CIS formation and development occurred under challenging conditions formed under the complex influence of external and internal geopolitical factors. Many prerequisites have a systemic impact on the quality of integration processes, limiting the full disclosure of the development potential in the economy, social sector, and other relevant CIS functioning areas. Another precondition is the lack of organizational and legal support for the harmonization of constitutional and legal systems.

The international and Russian experience of state-building shows that the successful implementation of large-scale strategic tasks is impossible without the combination of constitutional and legal instruments. Foreign policy is an area with an active convergence of institutions and norms of public international law and national constitutional (state) law.

Another important condition is the depth of perception of the ideology of international integration as a key political and legal instrument ensuring the disclosure and unification of the potential for self-development of the participating states to achieve optimal synergy effects in crucial areas of sustainable development. Therefore, the synthetic integration of the regulatory capabilities of international legal and constitutional legal documents in the CIS space are designed to provide *adequate logistics* of integration configured as *streets with multi-directional traffic*. Many international associations, especially the EU, are following this path and achieving impressive results.

Meanwhile, throughout the CIS existence, despite the noticeable intensification of cooperation in the legal field in recent years, the issues of constitutional and legal support for the CIS functioning and development and its institutions were not resolved. The described sluggishness and inertia in the development of the CIS initially had a specific effect on the organizational, managerial, resource, legal, and other support for integration.

The CIS Summit in Dushanbe on October 5, 2007, which approved the Concept of Further Development of the CIS and the Plan of Main Measures for its Implementation (CIS Executive Committee n.d.), was an important milestone that somewhat intensified interstate cooperation. The CIS Economic Development Strategy for the period up to 2020 was soon adopted by the Council of CIS Heads of Government on November 14, 2008, in Chisinau (CIS Executive Committee n.d.).

Following the CIS Charter (CIS Executive Committee n.d.), the CIS has the following goals:

- Cooperation in political, economic, environmental, humanitarian, cultural, and other fields;
- Comprehensive and balanced socio-economic development of member states within the common economic space, international cooperation, and integration;
- Guarantee of human rights and fundamental freedoms under the universally recognized principles and norms of international law and OSCE documents;
- Cooperation among member states in ensuring international peace and security, implementing effective measures to reduce armaments and military expenditures, eliminating nuclear and other weapons of mass destruction, and achieving general and complete disarmament;
- Assistance to citizens of the member states in free communication, contacts, and movement on the territory of the CIS;

- Mutual legal assistance and cooperation in other areas of legal relations;
- Peaceful settlement of disputes and conflicts between CIS member states.

According to the provisions listed above, legal interaction issues were initially established in the most general form, without distinguishing the areas of regulatory support for integration processes. A particular reservation relates generally accepted principles and norms of international law, which are relatively traditional for international instruments.

In the Concept of Further Development of the CIS of October 5, 2007 (CIS Executive Committee n.d.), aspects of cooperation in the legal field received a slightly broader interpretation.

Therefore, the main development goals of the CIS include (1) the implementation of fundamental international principles and standards in the field of democracy and human rights; and (2) further convergence of national legislation of the CIS in various areas of multilateral cooperation based on universally recognized principles and norms of international law (p. 2.3).

The inter-parliamentary cooperation of the CIS member states aims to (1) deepen the integration, convergence, and harmonization of economic systems and national legislation; (2) protect the rights and interests of citizens; and (3) enhance the development of model laws by the Inter-Parliamentary Assembly of the CIS member states and their implementation in the legislation of the member States (p. 4.3). It is envisaged that the CIS Economic Court will be modernized by the countries concerned, with the understanding that the judicial dispute settlement mechanism is an essential element of the modern infrastructure of international economic relations (p. 5.6).

Based on the presented provisions, we defined the legal cooperation of the CIS countries as an independent direction, which received intensive development and appropriate institutional, organizational, and information support. The Advisory Committee of Heads of Legal Services of the Ministries of Foreign Affairs of the State Parties of the CIS, Council of Ministers of Justice of the State Parties of the CIS, Council of Chairmen of the Supreme Arbitration, Economic and Other Courts Allowing Affairs on Disputes in the Field of Economy have been formed. Moreover, the Unified Register of Legal Acts and Other Documents of the Commonwealth of Independent States has been established. The CIS Internet portal (https://e-cis.info/) has a particular section *Legal cooperation* (CIS Executive Committee n.d.) which contains (1) relevant information, (2) analytical materials, (3) Final Statistical Indicators on the Ratification of CIS Instruments, (4) Certificate of Ratification of the Documents adopted by the Council of Heads of State of the CIS, (5) the Council of Heads of Government of the CIS, (6) information on the ratification of instruments adopted within the CIS, (7) information on the implementation of internal procedures with the instruments, adopted by the Council of Heads of State of the CIS, the Council of Heads of Government of the CIS; (8) information on the implementation of internal procedures with the instruments adopted within the CIS; (9) final statistical indicators on the implementation of internal procedures under the documents adopted within the CIS, and many others.

Inter-parliamentary legal cooperation within the Inter-Parliamentary Assembly of the CIS member states has become closer and more productive. According to the Convention on Inter-parliamentary Assembly of the State Parties of the CIS of May 26, 1995 (CIS Executive Committee n.d.), the Inter-Parliamentary Assembly is an international body of the CIS (Article 1) established to achieve the following goals:

- Adopt recommendations for the approximation of the legislation of the participating states;
- Adopt model legislation and transmit it with appropriate recommendations to the parliaments of the states parties to this Convention;
- Adopt recommendations on the timing of ratification (approval) of treaties concluded by the parliaments of the CIS (agreements) concluded within the Commonwealth and of the Council of Heads of State or Heads of Government of the CIS and other international instruments in which the participation of the Member States of the Commonwealth is highly desirable to achieve their common goals, as set out in the Charter of the Commonwealth of Independent States;
- Adopt recommendations on bringing the legislation of the participating states into line with the provisions of international treaties concluded by these states within the CIS;
- Facilitate the exchange of legal information between member states (Article 4).

Since 2009, the activities of the Inter-Parliamentary Assembly on Model Lawmaking have been carried out on a planned basis in cooperation with the CIS Executive Committee. According to the Plan of Interaction of the CIS Executive Committee with the Secretariat of the Council of the CIS Inter-Parliamentary Assembly on Model Lawmaking for 2016–2020 (CIS Executive Committee n.d.), normative work is carried out in eight areas: (1) economy; (2) social policy; (3) humanitarian field; (4) agrarian policy, ecology, and environmental management; (5) military cooperation and security; (6) legal field; (7) science and education; (8) state-building and local self-government. The latter direction provides for the development of methodological recommendations on legal regulation and the organization of inter-municipal cooperation, Draft Laws *On Strategic Planning and Anti-Crisis Regulation* and *On Road Transport*, a comparative legal analysis of the housing legislation of the CIS member states, and monitoring law enforcement practices. It is planned to develop laws *On Regulatory Legal Acts* and *On Mortgage* in the legal field.

When we studied this plan, we noticed some unsystematic and structural inconsistency of its directions. The activities in certain areas, especially in state-building and local self-government, are deplorable and do not fully correspond to the existing ideas (at least in Russian law) about the law. In previous periods (2009–2010; 2011–2015), activities in this field were not envisaged.

If we turn to the results of the activities in this area of another institutional entity—the Council of Ministers of Justice of the CIS member states, established by the Decision of the Council of Heads of State of the CIS of August 26, 2005 (CIS Executive Committee n.d.) and the crucial task of forming a coherent legal policy, including the harmonization of legislation of the CIS countries, then there is also specific unilateral nature. According to the Report on its Activities dated January 31, 2019 (CIS Executive Committee n.d.), since its establishment, 13 meetings have been held on the following topics:

- Comparative-legal analysis of the penal enforcement legislation of the CIS member states;
- Legal regulation of relations in the field of family law and education; the status of legal reform and prospects for the development of the legal system in the CIS member states;
- Activities of the civil registry offices;
- The work of the Council on improving the implementation of international humanitarian law in the CIS member states;
- Experience in using the latest information transfer and storage technologies in legal services of the CIS member states;
- Experience of legislative regulation in public control over the activities of bodies and institutions implementing punishment and criminal liability measures in the CIS member states;

Ironically, the constitutional and legal aspects of the CIS functioning and development are often out of sight. In general, the analysis showed that involving the resources of constitutional law in numerous documents of the CIS (more than 6,000 units) and the activities of its bodies (about 70 institutions and entities) are not raised today. Moreover, we should note the depressing state of the relevant constitutional and legal scientific developments.

## 4 Discussion

In the modern Russian science of constitutional law, there are no relevant, comprehensive studies on the development of the constitutional law of the CIS countries in the context of supranational integration processes in terms of international and Russian law. However, comparative studies of individual institutions of constitutional law and state legal processes in the CIS countries were conducted quite intensively (Bobrakov 2008; Dodonov 1997; Dostiev 1999; Dzhangiryan 2004; Gadzhi-Zadeh 2005; Gafurov 2006; Kainov 1999; Klishas 2007; Magomedov 2004; Mirsalimov 2013; Pasternak 1999; Raimanov 2003; Reshetnikova 2005; Zmievsky 2009 etc.).

Many comprehensive studies partially discuss specific state and legal aspects of the integration processes in the CIS countries (Khabrieva 2014, 2015, 2016).

Besides, several theoretical works on the constitutional law of the CIS countries in the 1990s and the first decade of the twenty-first century, published in Russia, are worth mentioning (Khudoley 2009; Mikhaleva 1999; Sulfugaradze 2008; Vishnyakov 2006).

In the Soviet period, there was no comparative legal research of the republics (currently CIS states), since they were part of the unified USSR. The first studies emerged after the USSR collapse due to intensifying international relations and strengthening international cooperation within the CIS. At the same time, modern scientists are interested in this topic. A few scientific publications cover local situations in constitutional law in the CIS countries, which are peculiar due to the fragmentary development and discussion of the proposed provisions. Thus, the problems of human rights and freedoms grasp the attention of Russian scientists (Ivanov and Petrenko 2018; Nechkin 2016; Novikova 2018; Zaitsev 2018), the constitutional and legal status of central electoral bodies (Artemova 2017), and many other aspects.

Kazakhstan researchers focus on constitutional law sources (Mataeva 2006, 2019; Sidorova 2016).

In Tadjikistan, attention is focused on the problem of the Institute of human rights in a comparative legal context (Iskandarov 2015; Safarzoda 2019).

In Kyrgyzstan, much attention is paid to the constitutional control bodies (Mirsalimov 2013; Nurdinova 2012).

For legal scholars of the Republic of Armenia, it has become traditional to compare national state legal institutions with Russian ones, especially in executive power and the institution of the presidency (Arutyunyan 2013, 2006; Konjakulyan 2010).

The Republic of Belarus is relatively active in comparative research (Ivashkevich 2012; Kambiz 2012; Murashko 2016; Sokolova 2013).

Simultaneously, in some countries, there are practically no studies aimed at the convergence and harmonization of constitutional and legal systems in the CIS (Republic of Moldova, Turkmenistan, Ukraine).

Improving the effectiveness of the Commonwealth and its institutions is impossible without using the regulatory potential of constitutional law and constitutional law itself.

The review reveals a significant gap in the institutional, doctrinal, and legal support for the goals of convergence and integration of the CIS geopolitical space stated in the main documents of the CIS. However, we assume that the main political and legal prerequisites for achieving these goals should be created by combining and harmonizing the constitutional and legal systems of CIS member states.

International and Russian state-building experience shows that the successful implementation of such large-scale strategic tasks is impossible without adequate constitutional and legal support. Regardless of the specific field of international cooperation, it is always accompanied by intensive functioning and contacts of public authorities; it has social and legal content and needs regulatory support from national constitutions and current legislation. There is no other way to ensure the proper legitimacy of international legal acts and acts committed by the public authorities of the CIS member states. Logically, foreign policy is a field of active interaction and interpenetration of public international law and national constitutional law. The international interaction experience also proves that international contacts and agreements are strengthened if they find counter-support from national constitutional legislation. If international legal relations, security agreements, and implementation of principles and norms are in force on the goodwill of States parties and their leaders, in Russian law, this issue concerns legal liability, including constitutional law. In any case, many international associations, especially the EU, are now following this path using the synergy effect obtained from the harmonious combination of the regulatory potential of international public and national constitutional law.

None of the key areas of the CIS development will succeed without the appropriate institutional and legal reproduction of international legal initiatives and initiatives at the level of national, primarily constitutional, law. On the other hand, the convergence and harmonization of the legislation of the CIS member states is a well-known synchronization of national legal processes taking place both within certain countries and at the level of international institutions. These events and actions always have an industry-specific (intersectoral) legal design. Constitutional law plays a key role here. In all cases, international law implementation (the actual implementation of international obligations at the national level) is primarily a constitutional and legal event. Only then do the sectoral regimes of legal regulation (administrative, criminal, and civil law) come to the fore. These events in the CIS and other international associations happen all the time. In the strategic dimension, the effectiveness of the activities of international entities is determined by the effectiveness of the state and legal institutions of their participants, which largely depends on the national constitutional legislation.

Russia and other CIS member states have almost 70 years of experience in joint state-building within the USSR. The influence of Soviet Russia on the region was manifested in socio-cultural and unifying legal influence. The legal systems of Russia and the CIS countries and their systemically significant legal structures still have a lot in common. Meanwhile, the differentiation of legal systems in general and constitutional law in the countries consideration is increasing.

Russia strives to be consistent in its leadership mission in the integration processes in the post-soviet space. Supranational integration is a necessary condition for the further existence of Russia and all CIS member states. Due to the genetic links between the Russian statehood and the statehood of the CIS countries, the mentality of the peoples and political traditions, the constitutional law of the CIS countries can be represented as an integral state-legal phenomenon; it implicitly represents an integration factor (legal awareness), which at the same time manifests itself. For example, it generates similar models of the constitution, similar parameters of the country's mechanism, and repeated precedents of the state-building. The effect of this factor can be used as a connecting rather than separating moment.

In this regard, Russia could offer a productive version of supranational integration, which would not contradict the principle of state sovereignty, since it would take into account the "limit of concordance" in the unification of the constitutional law and the development of a balance of interests and powers in connection with the supranational level. It is necessary to use productive theoretical models for an adequate assessment of the potential of constitutional and legal institutions involved in the implementation of integration tasks. They have yet to be developed and tested.

Thus, urgent scientific problems include (1) the current state of constitutional law and constitutional statehood in the CIS countries; (2) the main patterns and trends that characterize the dynamics of the constitutional space and constitutional institutions of the CIS; (3) doctrinal thinking; and (4) constitutional and legal support (including the design of legal models and legal structures).

## 5 Conclusion

The development of possible approaches to the successful solution of the scientific issue under consideration is possible within an innovative scientific project *Modern Constitutional Law of the Commonwealth of Independent States in the Context of Supranational Integration Processes: Status, Development Trends, and Reserves for Improving Efficiency*, which provides for the development of five specialized sections, presented in the form of the following thematic clusters:

- Constitutional space of the CIS as geopolitical and state-legal reality: constitutional and legal aspects of institutional support for CIS development;
- Dynamics of the constitutional and legal space of the CIS: main patterns and development trends;
- Constitutionalism and constitutional statehood in the CIS area: a typology of national models of constitutions and correlation of systemically significant elements of the constitutional system;
- Statehood, supranationality, and conventionality: a comparative legal analysis of the role and potential of constitutional and legal means in the international integration;
- Targets and main directions of integration and harmonization of the CIS constitutional space: current problems of improving institutional and constitutional-legal support.

Based on these clusters, it is necessary to (1) to reveal and present the constitutional law of the CIS countries from the point of view of interaction with regional international law as an integral state-legal phenomenon; (2) to conduct the necessary fundamental studies that allow to present and substantiate theoretically and practically significant provisions, conclusions, and recommendations for improving the efficiency of the CIS and its institutions by constitutional and legal means.

The research results can be presented in the form of a holistic concept containing a comprehensive solution to a fundamental scientific problem related to the study and assessment of the current state of constitutional law and constitutional statehood in countries. Besides, this concept should comprise identification and description of the main patterns and trends characterizing the dynamics of the constitutional space and CIS institutions and the development of practically significant recommendations for improving the constitutional and legal support for integration processes in the CIS.

In particular, it is advisable to give a theoretical description in a comparative legal context of systemically significant elements of the constitutional system and the parameters of constitutional statehood, as well as find approaches to an objective assessment of the depth of perception and quality of reproduction of the beginning of constitutionalism in the state-building in Russia and the CIS countries.

In addition, one should classify constitutional models emerging in CIS space and analyze factors influencing their evolution. Also, scientists should design a legal model of constitutionalization of the CIS international law and internationalization of the CIS constitutional law. Furthermore, it is necessary to justify the positive synergetic effect of a harmonious combination of the regulatory potential of the international public and national constitutional law of the CIS countries.

Proposals to adjust the Russian constitutional legislation in the context of supranational integration processes should be studied appropriately. As a next step, one should examine theoretical, methodological, and practical recommendations optimizing selecting legal instruments aimed at preserving the national constitutional traditions and the identity of the political and legal institutions of the CIS countries in the context of the intensification of international integration.

We suggest scientists analyze and substantiate the best practices of constitutional reforms in the CIS countries to adapt the foreign constitutional experience to Russian legal reality and, conversely, the mechanisms of translation of Russian constitutional experience into the CIS legal space.

Further research should focus on the constitutional and legal support for priority directions of the CIS with model provisions of acts of international public law, national constitutional legislation, and relevant laws on amendments to the constitution (on the example of Russia and CIS member states). This step can stimulate and harmonize the integration in the state-legal field.

The significance of the research results is to supplement the Russian constitutional law with theoretical provisions that expand the understanding of the essence and features of constitutionalism, constitutional statehood, and the current dynamics of the constitutional and legal space of the CIS.

More than that, the scholarship will get a clear understanding of the laws and trends in the functioning and interaction of the constitutional and legal systems of the CIS member states and the problems of integration arising from them, as well as the constraints on the CIS development.

As a result, scientific-based recommendations of an applied nature will be developed, aimed at intensifying the harmonization and integration of the CIS member states, contributing to the practical solution of the problems of improving the effectiveness of institutions and, in general, the potential for sustainable development of the CIS by constitutional and legal means.

The implementation of such a project dedicated to the 30th anniversary of the CIS, under the leadership of the CIS Executive Committee, could make a considerable contribution to the development of constitutional law and the implementation of the creative tasks of the CIS.

## References

Artemova OE (2017) Constitutional and legal status of central electoral authorities of Central Asia—members of CIS. Jurisprudence 3:85–88

Arutyunyan AS (2006) Institute of the president in the post-Soviet republics: constitution and political reality (comparative legal analysis). Petakantsarayutyun, Yerevan, Armenia

Arutyunyan AR (2013) Institutional and functional characteristics of the governments of the Russian Federation and the Republic of Armenia (comparative legal research). Dissertation of the

candidate of legal sciences. Moscow State University of Economics, Statistics and Informatics, Moscow, Russia
ASEAN. (n.d.). Official website. https://asean.org/
Bobrakov N (2008) The institute of the President in the CIS countries: comparative legal analysis. Dissertation of the candidate of legal sciences. Krasnodar, Russia
CIS Executive Committee (n.d.) Official website. http://www.cis.minsk.by/site/about-cis
Dodonov VN (1997) Constitutional reforms in the States that emerged on the territory of the former USSR. General and special. Dissertation of the candidate of legal sciences. Moscow, Russia
Dostiev AS (1999) Constitution of the Republic of Tajikistan 1994: history of development, adoption and main provisions. Dissertation of the candidate of legal sciences. Institute of Philosophy and Law of the Academy of Sciences of the Republic of Tajikistan, Dushanbe, Tadjikistan
Dzhangiryan ZhD (2004) Separation of powers in the Republic of Armenia: constitutional model and practice. Dissertation of the candidate of legal sciences. Moscow State University, Moscow, Russia
EU (n.d.) Official website. https://europa.eu/
Gadzhi-Zadeh EA (2005) Constitutional models of the form of government and the Institute of presidency in the CIS countries. Dissertation of the candidate of legal sciences. Volgograd Academy of the Ministry of Internal Affairs of Russia, Volgograd, Russia
Gafurov HM (2006) Judicial system of the Republic of Tajikistan (comparative legal analysis). Dissertation of the candidate of legal sciences, Moscow, Russia
Iskandarov ShF (2015) Comparative analysis of national institutes of human rights in Central Asia. In: Pavlenko EM (ed) Human rights in international and national law: collection of scientific articles on the 10th anniversary of the Departament of International Law and Human Rights of the Law Institute of MGPU, pp 256–264. Human Rights, Moscow, Russia
Ivanov IP, Petrenko SY (2018) The Role of the Institute of presidency and political parties in the countries of Central Asia. Sci Without Bord 8(25):33–38
Ivashkevich EF (2012) Constitutional justice in the post-Soviet space. In: Bochkov AA (ed) State-legal regulation of integration processes in the post-Soviet space: materials of the international scientific and practical conference. Vitebsk State University, Vitebsk, Belarus, pp 109–112
Kainov VI (1999) Institute of presidency: Constitutional and legal status. Dissertation of the candidate of legal sciences. St. Petersburg State University, St. Petersburg, Russia
Kambiz B (2012) Consolidation of social and economic rights in the constitutions of Russia, Belarus, Ukraine (comparative analysis). Probl Manag 4(45):140–145
Khabrieva TY (ed) (2014) Globalization and integration processes in the Asia-Pacific region (legal and economic research). SISP, INFRA, Moscow, Russia
Khabrieva TY (ed) (2015) BRICS: contours of a multipolar world. Institute of legislation and comparative law under the government of the Russian Federation, Moscow, Russia
Khabrieva TY (2016) Constitutional reform in the modern world. Nauka RAS, Moscow
Khudoley KM (2009) Constitutional law of foreign countries of the CIS and Baltic States. Perm State University, Perm, Russia
Klishas AA (2007) Constitutional control and constitutional justice in foreign countries. Dissertation of the candidate of legal sciences. Moscow University of the Ministry of Internal Affairs of the Russian Federation, Moscow, Russia
Konjakulyan KM (2010) Executive power and the Institute of presidency: correlation issues (administrative and legal research on the example of the Russian Federation and the Republic of Armenia). Dissertation of the candidate of legal sciences. Russian Academy of Public Service, Moscow, Russia
LAS (n.d.) Official website. https://dppa.un.org/en/league-of-arab-states
Magomedov AG (2004) Constitutional and legal foundations of the system of state authorities in the CIS countries. Dissertation of the candidate of legal sciences. Makhachkala, Russia
Mataeva MH (2006) Some features of the procedure for making changes and additions to the Constitution of Kazakhstan and the countries of Central Asia. Tomsk State Univ J 292–1:74–78

Mataeva MH (2019) Internationalization of national constitutions of the Republic of Kazakhstan and Central Asian countries. Law and state (KazGyuU, Astana) 2(83):25–34
Mikhaleva NA (2006) Constitutional law of foreign CIS countries: studies. Benefit. Jurist, Moscow, Russia
Mirsalimov RM (2013) The Bodies of the constitutional justice in the countries of Central Asia: problems of theory and practice. Dissertation of the candidate of legal sciences. RUDN University, Moscow, Russia
Murashko LO (2016) The Role of international covenants on human rights in strengthening the internationalization of constitutional law. Educ Law 11:64–67
Nechkin AV (2016) To the question of forms of government in the CIS countries. Const Municipal Law 12:11–16
Novikova AE (2018) Constitutional consolidation of guarantees of human and civil rights and freedoms in the CIS countries. Sci Rev Belg State Univ Ser Philos Soc Law 43(2):302–307
Nurdinova ZhT (2012) Comparative legal analysis of the constitutional Council of the Republic of Kazakhstan and the constitutional chamber of the Kyrgyz Republic in the constitutional mechanism. Izvestiya VUZov 5:204–208
Pasternak SN (1999) Constitutional justice in Central and Eastern Europe (on the example of the Polish Republic) (Dis. ... Cand. of Legal Sci) Dushanbe, Tadjikistan
President of Russian Federation (2005) Press conference following the results of the Russian-Armenian talks. http://kremlin.ru/events/president/transcripts/22883
President of Russian Federation (2016) Decree of the President of the Russian Federation about the strategy of national security of the Russian Federation (December 31, 2015 No 683). 1(Part II). Art 212, Moscow, Russia
Raimanov DF (2003) President in the mechanism of protection of constitutional rights and freedoms of man and citizen: comparative legal analysis of the legislation of Russia and Kazakhstan. Dissertation of the candidate of legal sciences. St. Petersburg, Russia
Reshetnikova TA (2005) Legal status, functions and powers of the Prosecutor's office in the CIS countries (comparative legal study). Dissertation of the candidate of legal sciences. Russia, Moscow
Safarzoda BA (2019) Consolidation of basic human and civil rights and freedoms in the Constitution of the Republic of Tajikistan and their correlation with international standards. Legal Life 1:14–26
Sidorova NV (2016) Comparative analysis of the constitutional legislation of foreign countries on the place of the international Treaty in the national law. Pub Secur Leg Leg Order III Millennium 1-1:304–307
Sokolova AA (2013) The principle of the rule of law in the constitutional models of post-Soviet States: problems of theoretical reconstruction. J Foreign Legis Comp Law 6(43):1065–1070
Sulfurated TE (2008) The Constitutional right of the states-participants of the CIS: the textbook. Russian New University, Moscow, Russia
Vishnyakov VG (2006) Comparative analysis of the constitutions of the CIS member States. Russia, Moscow
Zaitsev IA (2018) The Right to freedom of conscience and religion under the legislation of Russia and some Asian countries. In: Radnaeva EL, Khamnuev YG (eds) Actual problems of modern legal science and practice, pp 163–167. Buryat State University, Ulan-Ude, Russia
Zmievsky DV (2009) Constitutional and legal transformations in the post-Soviet States. Dissertation of the Candidate of Legal Sciences. Mytishchi, Russia

# Comparative Legal Acts on the Regulation of Personal Data Protection in Cloud Technology Applications

**Ekaterina A. Remmikh, Anton A. Vasiliev, and Meruert Muratkhanova**

**Abstract** The article deals with the most pressing issues related to the protection of personal data when using cloud technologies. In particular, the problem of applicable law to these relations is analyzed, as well as issues related to the legal qualification of contracts on the provision of cloud services. In addition, the authors touch upon aspects related to the possibility of developing general principles of data processing in the cloud, based on which legal conflicts may be resolved. Particular importance in the work is given to studying the experience of the use of cloud computing by foreign countries, including European countries, which have made a significant contribution to the development of concepts and terms that allow for a step closer to standardization and unification of the application of these technologies. Based on the study of domestic and foreign literature, the classification of cloud computing, as well as a detailed analysis of the pros and cons of their use in various areas of society. The authors conclude that the legislative regulation does not densely cover the rapidly developing market for cloud services. The lack of uniform regulation of personal data processing in the cloud, given the cross-border nature of services provided, leads to legal uncertainty and makes it difficult for users to apply legal protection tools.

**Keywords** Cloud technology · Personal data · Information security · Digitalization · Data operator

E. A. Remmikh (✉) · A. A. Vasiliev
Altai State University, Barnaul, Russia
e-mail: remmih.Katya@mail.ru

M. Muratkhanova
L. N. Gumilyov, Eurasian National University, Nur-Sultan, Republic of Kazakhstan



S. G. Maximova et al. (eds.), *Advances in Natural, Human-Made, and Coupled Human-Natural Systems Research*, Lecture Notes in Networks and Systems 234,
https://doi.org/10.1007/978-3-030-75483-9_87

## 1 Introduction

The launch of the World Wide Web Project in 1989, created by Tim Berners-Lee (Wikipedia 2020), was the starting point for many technologies' integral to everyone's life. Increasingly, experts say that today's economy is experiencing the fourth industrial revolution (Industry 4.0). Its main feature is the widespread introduction of information technology into production. Industry 4.0 includes, among others, cloud technologies (Ministry of Economic Development of Russia 2020).

At the same time, cloud technologies are an important tool for developing the economy and other areas of society. For example, the Russian Government is considering the possibility of using them to reduce the cost of creating and operating similar information systems in education (Russian Government 2014 (ed. 2018), for the automation of judicial and general office work (Russian Government 2012 (ed. 2019). "Cloud" technologies are successfully applied in other CIS countries as well. In Kazakhstan, telemedicine actively develops based on these technologies (CIS Economic Council 2016). Within the European Union, a European Cloud Partnership has been established to provide overall cloud computing implementation guidance.

According to one of the largest research companies specializing in information technology markets, Gartner, cloud computing is one of the fastest-growing IT expense segments (Gartner 2020). Almost everyone has come across cloud technologies, using e-mail or such services as Yandex.Disk, iCloud, OneDrive, Google Drive, etc.

Despite the wide scope of application, legal regulation of relations concerns to cloud technologies remains underdeveloped. Thus, Russia still lacks a legal definition of cloud services. Questions about applicable law are increasingly raised, as cloud services are usually cross-border. One of the most controversial issues remains the legal nature of a contract mediating these relations, parties, and rights and obligations. Given that cloud technologies do not provide 100% security for information, the above issues become important in the legal capacity to protect personal user data.

## 2 Materials and Methods

We use the material accumulated in Russian and foreign scientific and legal literature. It is worth mentioning that few works are devoted to protecting personal data in the context of introduction "cloud" technologies. It is caused by the technology itself that has appeared recently, and scientific publications do not keep up with its development dynamics. That is why scientific and legal articles and analytical publications by leading IT specialists, such as David Tollen (Tollen 2018) or Peter Melle and Tim Grance (Mell and Grance 2011), become important.

Articles by Kartskhiya are of high importance for our study (Kartskhiya 2018, 2019). The author analyzes European, Russian, and the USA's legislation regulating the provision of cloud resources. Saveliev's work allows to understand the main approaches to the qualification of contracts for the provision of cloud services (Saveliev 2015).

Russian legislation and the study of foreign experience in regulating personal data protection in the era of digitalization are of great importance for the research. Of particular interest is the General Regulation on Personal Data Protection resulting from the unification of European legislation in this area (European Union 2016). Besides, we analyze the laws "On Personal Data" of some CIS countries, in particular, the Republic of Kazakhstan (Parliament of the Republic of Kazakhstan 2013) and the Republic of Armenia (National Assembly of the Republic of Armenia 2015). Also, we use "Specification on the Security of Personal Information" (Shi et al. 2019) published by China National Technical Committee for Information Security Standardization (TC260).

The methodological basis for the study is the principle of dialectics, which is reflected in the meaningful refraction to consider cloud technology as a system consisting of interrelated elements and in constant development. As part of this approach, we use several methods of legal science. Comparative legal method is one of the most important, which in combination with the application of analysis, allows:

- Studying the experience of different countries in the sphere of regulation of personal data protection,
- Collecting and processing with the help of cloud technologies,
- Developing possible ways of solving the problems arising in this sphere.

## 3 Results

The study and synthesis of the material allow forming a number of important issues from the theoretical and practical point of view.

It should be noted that most authors refer to the formulation used by the US National Institute of Standards and Technology (NIST) when defining "cloud" technologies (computing). NIST defines cloud computing as a model providing convenient network access at a user's request to a set of configurable computing resources (networks, servers, storage, applications, and services) that can be quickly provisioned and released with minimal effort to manage or interact with the service provider (Mell and Grance 2011). Some authors suppose that this definition is technical and does not reflect the specifics of the legal regulation of this technology.

However, the Decree of the President of the Russian Federation accepts the formulation, in which cloud computing is an information technology model of providing universal and convenient access using the Internet to a common set of configurable computing resources (cloud), data storage devices, applications and services that can be quickly provided and relieved of the load with minimal operating costs or virtually without participation of the provider (President of the Russian Federation 2017).

In 2014, the International Organization for Standardization (ISO), jointly with the International Electrotechnical Commission (IEC), published a document containing general provisions and a glossary of terms used for cloud technologies (ISO and IEC 2014).

GOST (State Standards) adopted by Armenia, Belarus, Kyrgyzstan, Russia, and Tajikistan in 2019, which fixes, in particular, the definition of cloud computing as "a paradigm for enabling network access to a scalable and elastic pool of sharable physical or virtual resources with self-service provisioning and administration on-demand" (Interstate Council for Standardization, Metrology, and Certification [IGC] 2019).

The main reason for accepting the above documents is the desire to unify the conceptual apparatus to protect subjects whose data are contained in the cloud. Often, the regulation of relations between the purchaser of cloud services and the supplier (provider) are subject to the jurisdiction of several countries, in which the legal regulation of these relations may differ significantly. This is due to the very nature of cloud computing.

Not every country has a data center linked to specific cloud service providers. Thus, the famous American company Google has a network of its own data centers. However, their geography is quite narrow. Most of them are located in Europe and North America (Google, n.d.). Fulfilling the requirement of the legislation on personal data of some countries, companies use the services of commercial data centers located in the territory of the relevant state. Law adopted in July 2014 No. 242 *On Amending Certain Legislative Acts of the Russian Federation to Specify the Procedure for Processing Personal Data in Information and Telecommunication Networks* (Russian Federation 2014) established the obligation to store personal data of Russians in Russia. This circumstance forced many cloud service providers to promptly move servers to Russian data centers. In 2015, a similar provision appeared in the Republic of Kazakhstan. Relevant amendments were made to the Law *On Personal Data and their Protection* (Parliament of the Republic of Kazakhstan 2013). China and a number of other countries have followed this path.

However, according to the Asian Cloud Computing Association experts, the adoption of laws to ensure data localization is quite controversial. Local storage makes data centers an easy target for attackers and limits advanced ways to protect personal data by hiding the patterns that cloud technologies create by regrouping them between data centers in different locations (TechNode 2019).

In addition, the application of such laws is significantly hampered by deficiencies in the conceptual apparatus. The blurred definition of "personal data," fixed in the majority of laws, leads to the fact that subjects of the considered relations do not know, what exactly the data should be localized, and what can be processed without restrictions (personal data is usually understood as those data which allow directly or indirectly to identify the subject of such data).

At the same time, the requirement to store citizens' personal data in which they live does not allow for a clear decision on the applicable law. This problem arose a little later than the technology itself. Analysis of the literature reveals four main variants of solving the problem, called by the authors.

First is the definition of the applicable law based on their physical location in the cloud. This option is suitable for those countries that have legislated against storing data in multiple data centers. At the same time, defining jurisdiction based on the actual location of the data is not consistent with the ability of a cloud service provider to back up data to any data center it owns (James and Szewczyk 2017).

Secondly, to determine the applicable law depending on the place where the offense against personal data was committed. This approach also deserves criticism, as the term "cloud" itself is a metaphor and, based on the concept, it is not possible to determine the exact location of the cloud. In addition, it is often difficult to determine where data is uploaded or downloaded, as users can upload and download information using special software to hide their location.

The third point of view is to apply the laws of the country in which the subject of this data lives to the relations of personal data processing. It seems that the said way of solving the problem should not be absolute, because otherwise the balance of interest between the provider of services and their purchaser in favor of the latter will be violated. It should be agreed that the application of the legislation of the user's country should be limited to certain limits.

For example, one of the most known regulations for the personal data protection—the General Data Protection Regulation (GDPR) provides that personal data can be moved outside of the EU, but only if the jurisdiction in which the recipient is located provides an adequate level of data protection. (European Parliament and Council of the European Union 2016). The regulations thus do not set out specific stringent mandatory measures that non-EU member states must comply with. At the same time, it ensures the protection of European personal data, with a certain impact on the processing requirements and level of protection of such data.

It should be noted that the application of the pure law of the provider in some cases may lead to unjustified restriction of the rights of subjects of personal data, as there are cases when the legislation of the provider of "cloud" services provides less guarantees of protection of personal data than the legislation of the buyer of such services.

The fourth option was to define the jurisdiction under which the parties themselves could settle disputes, which was acceptable under the principle of freedom of contract. However, conflicts could arise between the provisions of the treaty and the laws of the countries in which the parties lived. For example, some authors note that the Russian law on personal data in Article 12 contains a super-imperative rule according to which "transboundary transfer of personal data on the territory of foreign states… is carried out in accordance with this Federal Law". According to the literal interpretation of this provision, all the provisions of the Russian Law on Personal Data (Terekhova 2013) should apply to the Russian operator when transferring personal data abroad for processing, including in case it involves a foreign processor. Under such conditions, the definition of the applicable law by the terms of the contract will be contrary to the Russian law.

The view seems to be correct that the participation of an international organization is necessary for the resolution of an emerging issue. Thus, in scientific literature, it was suggested that such an organization could be the UN, which currently unites 193 states (James and Szewczyk 2017).

In our opinion, creation of a special memorandum defining the possibilities of personal data transfer to the operator, ways, and means of the legal protection of interests of the personal data subject, the circle of processing operations, the circle of duties of the operator, responsibility for violations of obligations of the provider of "cloud services" on the processing of personal data, would allow excluding the possibility of legal collisions. The primacy of international law norms recognized in most countries would have made it possible to bring legislation into line with those general guarantees established in the memorandum. In addition, the resources and capabilities that the United Nations would have to use rationally to create unified contract forms for various cloud services.

Today, there are several types of cloud services:

- Storage-as-a-Service (SaaS);
- Database-as-a-Service (DBaaS);
- Information-as-a-Service (IaaS)—information as a service enables remote use of any type of information;
- Process-as-a-Service (PaaS)—process management as a service, allowing to link several resources together to create a single business process;
- Application-as-a-Service/Software-as-a-Service (AaaS/SaaS)—positioned as "software on demand" (for example, Google Docs);
- Platform-as-a-Service (PaaS)—computer platform with installed operating system and some software;
- Integration-as-a-Service (IaaS);
- Security-as-a-Service (SaaS);
- Management/Governace-as-a-Service (MaaS/GaaS);
- Infrastructure-as-a-Service (IaaS)—user is provided with computer infrastructure, usually virtual platforms (computers) connected to the network;
- Testing-as-a-Service (TaaS).

In addition, the types of cloud computing themselves (Microsoft Azure, n.d.) are also highlighted. This is primarily a private cloud, providing SaaS, PaaS, IaaS, and DBaaS to a limited number of users, usually belonging to the same organization. There is also a public cloud that provides SaaS, PaaS, IaaS, and DBaaS to an unlimited number of users. Additionally, a third type of cloud, a hybrid cloud, combining elements of private and public clouds, is also highlighted.

It seems that the above-mentioned contract templates should contain clearly defined rights and obligations of the parties, taking into account peculiarities of certain types of cloud services. This will allow ensuring an appropriate level of personal data protection and distribute responsibility between the parties.

It is worth noting that at present, there are many views on the legal nature of the treaties in question.

Regarding SaaS, many authors are inclined to conclude that this contract should be considered as a type of lease contract, because it leases computing power (Saveliev 2015). However, as Saveliev rightly notes, the subject of the lease agreement should be an individually determined and not the consumed thing. However, the pool of computing resources provided as part of cloud services does not have such qualities.

The authors consider a license agreement as one of the qualification options. From this point of view, Tech Contracts Academy disagrees, which notes that SaaS does not imply the copying of software on the user's computer. At the same time, the existence of a license agreement in the contract will not be beneficial to the provider of this service.

The most correct is to qualify contracts in the field of cloud computing as service contracts. In this case, the user gets an opportunity to assess the quality of the service. Besides, the advantage of such a contract will be also the terms of payment. The contract of reimbursable provision of services implies that the services provided are subject to payment. In this case, the user will pay only for the computing capacity that they have used (Taran 2018).

Recently, the most frequently used are the so-called SLA-contracts (Service Level Agreement), which is an extended agreement on the quality of services provided. The essential conditions of SLAs usually include:

- Definition of service level parameters; monitoring methodology, which includes characteristics and methodology for measuring security parameters in real time;
- Independent testing;
- Ranges of parameters, which cause Ad-Hoc warnings, incident response or recovery;
- Regular reports on service level (mirroring) and their content;
- Response thresholds, defined according to the risk profile of the organization;
- Liability of the parties and sanctions (Elin 2017).

In our opinion, SLA contracts are the most adapted to the IT environment.

## 4 Discussion

The research allows forming several important conclusions that can be used to write scientific works devoted to protecting personal data and at working out of the normative acts improving the system of their protection. Thus, given the dynamics of technology development, it is necessary to consolidate at the international level the universal principles of data processing when using cloud computing. This measure would make it possible to fill regulatory gaps, if necessary, quickly.

In our opinion, when improving the system of personal data protection in the context of the introduction of cloud technologies, it is necessary to balance the interests of the supplier of such technologies and the consumer. It is necessary to be more cautious concerning to any restrictive measures. For example, the requirement to localize data can only be applied to government agencies whose activities are directly related to the security of the country.

Further research prospects are connected with a more detailed study of the legislation on personal data of different countries, as well as the development of its own definition of "personal data" and "cloud computing."

## 5 Conclusion

Digitalization is a factor that has a significant impact on the transformation of law. However, legislation often fails to keep pace with rapidly changing market conditions. New technologies are emerging whose features prevent the application of established legal mechanisms. A major challenge is the cross-border nature of relations related to cloud services, resulting in legal conflicts. At the same time, ensuring the protection of personal data is a priority for each state. That is why scientific developments in this area become so urgent.

The study of the experience of legal regulation of "cloud" services in different countries allows identifying the most pressing problems today. There is no uniform terminology considering the technical and legal peculiarities of the technologies under consideration.

No less important is the question of the applicable law, as for the most effective protection of personal data, their subject must understand what rights and obligations he has.

The development of a contractual law regulating the provision of cloud services will also contribute to protecting personal data. Undoubtedly, the most correct in this case would be to develop a special type of contract, mediating the relationship between the provider of "cloud" services and their user, considering the specific types of services provided. However, as a basis, it is reasonable to use the legal constructions of the service contract and SLA-contract.

To summarize, it should be noted that legislation in the field of protection and use of personal data, in the context of the emergence of "cloud" technologies is in the formation stage. It is required to develop a regulatory and legal framework, including at the level of international organizations, considering the interests of both personal data subjects and providers of "cloud" services, creating conditions for further digitalization of the economy of different countries.

## References


CIS Economic Council (2016) Decision on the implementation of the strategy for cooperation of the CIS member states in building and developing the information society and measures of the action plan for its implementation for the period up to 2015. 27 May 2016, Unified register of legal acts and other CIS documents, Moscow, Russia

Elin VM (2017) "Cloud" services and features of their legal regulation in the Russian Federation. Inf Law 4:28–33

European Parliament and Council of the European Union (2016) Regulation on the protection of natural persons with regard to the processing of personal data and on the free movement of such data, and repealing Directive 95/46/EC (Data Protection Directive). 14 April 2016. Accessed https://gdpr-info.eu/

Gartner (2020) 4 trends impacting cloud adoption in 2020. Accessed https://www.gartner.com/smarterwithgartner/4-trends-impacting-cloud-adoption-in-2020/

Google (n.d.) Discover our data center locations. Accessed https://www.google.com/about/datacenters/locations/

Government of the Russian Federation (2012) Resolution "On the federal target program Development of the judicial system of Russia for 2013–2020". 27 Dec 2012 No. 1406, ed. by 25.12.2019, Meeting of Legislation of the Russian Federation, Moscow, Russia

Government of the Russian Federation (2014) Order "On approval of the Concept of regional informatization". 29 Dec 2014 No. 2769-r, ed. from 18.10.2018, Collection of Legislation of the Russian Federation, Moscow, Russia

James M, Szewczyk P (2017) Jurisdictional issues in cloud forensics. In: Clarke NL, Furnell S (eds) Proceedings of the eleventh international symposium on human aspects of information security and assurance. Lulu Press, Adelaide, Australia. pp 225–235. Accessed https://www.cscan.org/openaccess/?id=362

Kartskhiya AA (2018) Cloud technologies: legal aspect. Russ Leg Mag 6:162–172

Kartskhiya AA (2019) Digital technology (Online) platforms: Russian and foreign regulation experience. Civil Law 3:25–28

Mell P, Grance T (2011) The NIST definition of cloud computing. Accessed https://csrc.nist.gov/publications/detail/sp/800-145/final/

Ministry of Economic Development of Russia (2020) Order "On Approval of Methodological Recommendations for Statistical Assessment of the Level of Technological Development of the Russian Federation Economy as a whole and its Individual Branches". 12 Feb 2020 No. 66, Moscow, Russia

National Assembly of the Republic of Armenia (2015) Law of the Republic of Armenia "On Protection of Personal Data". 1 July 2015 No. ZR-49-N, Republic of Armenia, Yerevan

Parliament of the Republic of Kazakhstan (2013) Law of the Republic of Kazakhstan "On Personal Data and their Protection". 21 May 2013 No. 94-V, as amended on 28 Dec 2017, Egemen Kazakstan, Nur-Sultan, Republic of Kazakhstan

Russian Federation (2014) Federal law "On amending certain legislative acts of the Russian federation to specify the procedure for processing personal data in information and telecommunication networks". 21 July 2014 No 242-FZ, ed. 31 Dec 2014, Moscow, Russia

Russian President (2017) Decree "On the strategy for the development of the information society in the Russian Federation for 2017–2030", 9 May 2017 No. 203. Meeting of Legislation of the Russian Federation, Moscow, Russia

Saveliev AI (2015) The legal nature of "cloud" services: freedom of contract, copyright and high technology. Civ Rights Bull 5:62–99

Shi M, Sacks S, Chen Q, Webster G (2019) Translation: China's personal information security specification. Accessed https://www.newamerica.org/cybersecurity-initiative/digichina/blog/translation-chinas-personal-information-security-specification/

TechNode (2019) Chinese data containment laws can damage cloud security. Accessed https://ru.technode.com/2019/06/16/chinas-data-localization-laws-have-an-adverse-effect-on-cloud-security-report/

Terekhova EV (2013) Collisional issues of cross-border processing of personal data. Inf Law 5:33–36

The title of the standard (2014) ISO/IEC 17788:2014. Accessed https://www.iso.org/standard/60544.html

Tollen D (2018) Don't use license agreements for software-as-a-service. Accessed https://www.techcontracts.com/2018/06/01/dont-use-licenses-saas-contracts/

Wikipedia (2020) World wide web. Accessed https://en.wikipedia.org/wiki/World_Wide_Web

# Forensic Characteristics of Crime: Implications for Further Development

**Oleg V. Bespechniy, Maria A. Neymark, and Alexander S. Sokolov**

**Abstract** We investigated one of the main scientific categories of criminology—the forensic characteristics of crimes. The current state of development of forensic characteristics of crimes is considered, the theoretical and practical problems connected with it are analyzed. We underlined the significance of forensic characteristics for the development of private forensic methods of crime investigation. The concept of forensic characteristics of crime is studied and grounded on the basic approaches analysis, which is reflected in the scientific literature. The basic provisions, defining the content of forensic characteristics of crimes, included the nature of the forensic characteristics, the forensic aspect of the information it reflects, the systemic nature of the data contained in the forensic characteristics, the specificity of the object of forensic characteristics—a crime, as an independent and separate type of activity, the purpose of forensic characteristics as an integral part of forensic methods of investigation of crime. We also analyzed the prospects for further development of forensic characteristics of crimes as a scientific category with modern information technologies, like artificial intelligence, and suggested the necessity and importance of continuing scientific research in this area.

**Keywords** Forensic science · Criminalistics · Characteristics of crimes

## 1 Introduction

The globalization of economic and social processes in the world creates an objective basis for more significant interaction between individual countries in various areas of society. The development of common foundations and rules is required by the legal sphere, which serves as a guide not only for the organization of domestic life but also for interstate cooperation. Cooperation in combating crime occupies a special place in the legal cooperation of countries. At the same time, the efficiency of this activity at the interstate level is mostly related to the formation of a common understanding

O. V. Bespechniy (✉) · M. A. Neymark · A. S. Sokolov
Altai State University, Barnaul, Russia
e-mail: 5432236527584@mail.ru



S. G. Maximova et al. (eds.), *Advances in Natural, Human-Made, and Coupled Human-Natural Systems Research*, Lecture Notes in Networks and Systems 234,
https://doi.org/10.1007/978-3-030-75483-9_88

of (1) the goals and objectives of the law enforcement system and (2) the forms and methods of their solution. One of the theoretical foundations for the formation of forensic methods to combat crime is the forensic characteristics of crime.

## 2 Materials and Methods

The current research is based on the concept of construction of private forensic methods of the investigation established in forensics, including forensic characteristics of certain types of crime as a basic structural element. The research aims to establish the degree of influence of forensic characteristics on detecting and investigating certain crime types at the current stage. The authors rely on the open data of their dissertation research on crime investigation methods published in the educational portal (www.dissercat.com). They analyzed the relationship between information on significant features of crimes of certain types and developed recommendations for their detection and investigation.

## 3 Results

The forensic characteristics of crime as a scientific category of criminology still have significant potential for effective private forensic investigation techniques. Full development of this potential should be facilitated by the formation of unified criteria for the inclusion of certain information on crime characteristics in forensic elements. There is no disagreement among scientists that one of the main criteria is the influence of information on the disclosure and investigation of certain crimes. At the same time, the significance of certain circumstances related to the crime is not always obvious for the investigation.

In this regard, the use of artificial intelligence that allows processing large amounts of heuristic analysis data that reveals the correlation between individual elements of a criminal act can contribute to more complete disclosure of crime forensic characteristics as a scientific category.

## 4 Discussion

Since its emergence in the scientific field, the category "forensic characteristics of crimes" generated a broad discussion. Questions are discussed about the concept itself, the meaning of forensic characterization of crime, and its structure (Gavlo 2002).

Currently, the degree of acuteness of such a discussion decreased, but many previous questions never received a definite and clear answer. In a number of recent dissertations, the authors' diminished interest in developing theoretical problems related to the forensic characterization of crime draws attention. Quite often, the forensic characteristics of certain crimes cited in the research are only a description of their typical manifestations. Including forensic characteristics in the structure of private forensic methods of crime investigation, researchers do not always see a connection between forensic characteristics and developed recommendations for investigation. Meanwhile, the forensic characteristics of crimes as a scientific category have deeper content, serve as a base for the construction of private forensic methods, and should not be reduced to a simple description of the external features of criminal acts.

The authors evidently have different views on the considered category. In the light of the significant divergence of the authors' positions on the content of forensic characteristics of crimes, it is necessary to analyze the following individual provisions and define the concept:

- Forensic characteristics of crimes is an information model of a criminal event, a system of information about it, based on the analysis of law enforcement practices in the prevention, detection, and investigation of crimes (A. N. Kolesnichenko, V. E. Konovalova, I. M. Luzgin, N. P. Yablokov) (Kolesnichenko and Konovalova 1985). Koldin (1987) suggests the concept of a typical forensic model of criminal activity as an information system based on the generalization of a representative array of criminal cases of a certain category, reflecting the natural links between elements of a program-target complex essential for disclosure and investigation; the way of action of a criminal, the mechanism of the investigated event, the crime situation, and the peculiarities of the personality of a criminal;
- Forensic characteristics of crimes is information about a crime from the point of a forensic approach, i.e., information contributing to its disclosure and investigation. It is an orientation of crime, the data that makes it possible to differentiate forensic characteristics from criminal law. Such information has a systemic nature. The information model constituting the content of a criminalistic characteristic is a system of information on criminally significant components of a criminal act. A systemic approach to the analysis of such information ensures the identification of interdependencies between individual elements of a criminal event. As R. S. Belkin notes, the forensic characteristic as a single complex is of practical importance only when correlated links between its elements are established, which is natural and expressed in quantitative indicators (Belkin 1997). Data on these dependencies can serve as a basis for building typical versions of specific cases. Identifying such links between forensic characterization elements is possible only when they are components of a single system. Above all, this is the practical significance of the forensic characteristics of crimes;
- The main task an investigator faces when investigating a crime is to fill in the missing information. The solution to this problem should contribute largely to the forensic characteristics. Considering the existing interdependencies between the elements of the characteristics, having information about some of them, it is possible to assume the existence of others, not previously known. The elements of a

forensic characteristic can have both unambiguous and probabilistic relationships of different types (Selivanov 1977). In the case of an unambiguous connection, the presence of one element allows one to conclude about the existence of another categorically. In the case of a probabilistic relationship, the establishment of one element of a characteristic is possible with a great degree of probability (Selivanov 1977);

- The structure of forensic characteristics should include elements relating to the crime as a relatively independent type of activity, different from other objects of forensic study (crime detection and investigation activities, etc.);
- Forensic characteristics of crimes is an integral part of the investigation methodology. The information, which forms a typical forensic characteristic, determines the specific use of methods, means, and procedures developed by forensic techniques and tactics for the investigation of crimes and underlies the planning of an investigation.

The authors position on the structure of forensic characteristics of crimes, the number of elements, and their content are remarkably diverse. As constituent elements, we can distinguish (1) the method of crime (the method of committing a crime), (2) the mechanism of crime, (3) the situation of the crime, (4) the identity of the offender, and (5) the identity of the victim.

Certainly, the number of elements and their types are related to the peculiarities of the crime itself. At the same time, the literature analysis shows that the content of the same elements of forensic characteristics of crimes similar in criminal law features within the framework of private investigation methods offered by different authors may differ significantly.

It should be noted that such differences are not always of evident practical significance, since many of the private investigation methodologies proposed by researchers do not sufficiently reflect the connection between the forensic characteristics of specific crimes and the proposed recommendations for their investigation, nor do they include an analysis of the correlation between the individual elements (or such analysis is rather have a general nature).

Such shortcomings of private forensic investigation techniques do not reflect the scientific dishonesty of their authors but are the result of the objective complexity and laboriousness of identifying all criminally significant elements of a criminal act and establishing correlations between them.

Research of forensic characterization can get a new impulse in connection with the use of new information technologies, including artificial intelligence (Polyakov et al. 2019). Machine processing of large arrays of information (criminal case materials, etc.) will make it possible to construct detailed information models of various crime types and reveal correlations between the individual components of such models. At the same time, the range of considered elements can be significantly expanded and include not only traditional elements but also ones which forensic significance is not obvious. This would provide a basis for more specific investigator versions at the initial stage of the investigation and optimize the overall investigative process.

## 5 Conclusion

Significant differences among scientists in the understanding of forensic characterization essence, its content, the complexity of the process of converging positions on these issues, as well as the divergence of a significant part of research with the need to practice, raise doubts among individual authors about the prospects for scientific research in this area. Meanwhile, the emergence of the "forensic characterization of crime" category in forensic science is essential for developing its theoretical foundations and increasing the effectiveness of developed practical recommendations. These circumstances make it possible to assert the necessity to continue research on the forensic characteristics of crimes, which should be systematic and based on the active use of current technologies with artificial intelligence.

## References


Belkin RS (1997) A course on criminology, vol 3. Jurist, Moscow, Russia

Gavlo VK (2002) The legacy of R. S. Belkin and controversial issues. In: Averyanova TV (ed) Role and significance of prof. R. S. Belkin's activity in formation and development of modern criminology. Academy of Management of the Ministry of Internal Affairs of the Russian Federation, Moscow, Russia, pp 207–212

Koldin VY (1987) Forensic significance of criminal activity: modeling the function. Sovetskoe Gosudarstvo i Pravo 2:63–69

Kolesnichenko AN, Konovalova VE (1985) Criminalistic characteristics of crimes. YUridicheskij Institute, Kharkiv, Ukrainian SSR

Polyakov VV, Bespechniy OV, Neymark MA (2019) Artificial intelligence as an object of forensic study: perspectives from a border region. In: Bogoviz AV (ed) Proceedings of the ICSDCBR 2019: international conference on sustainable development of cross-border regions: economic, social and security challenges. Atlantis Press, Barnaul, Russia, pp 637–640. https://doi.org/10.2991/icsdcbr-19.2019.129

Selivanov NA (1977) Forensic characteristics of crimes and investigative situations in the methods of investigation. Sotcialisticheskaja Zakonnost 2:56–59

# The Legal Consciousness Questionnaire (LCQ): Assessment of Construct and Convergent Validity

**Victoria V. Molotova, Alexander V. Molotov, Dmitry V. Kashirsky, Natalia V. Sabelnikova, and Elina V. Gubernatorova**

**Abstract** This paper presents the results of the development and preliminary assessment of the psychometric properties of the new questionnaire of legal consciousness of an individual. In addition, we give a theoretical justification for the questionnaire structure containing six subscales. One scale relates to the awareness of individuals of their constitutional rights and freedoms, enshrined in the Universal Declaration of Human Rights and the Constitution of the Russian Federation (personal, economic, political, cultural, and social rights). The other five scales relate to the importance of constitutional rights and freedoms for the subject. The content validity of the tool was confirmed by expert assessments of professional lawyers (specialists in constitutional law), as well as by the results obtained with semi-structured interviews (N = 30). The constructive validity of the method was established using exploratory factor analysis and tested by the confirmatory factor analysis for a student sample (N = 100). We assessed converged validity on a sample of students of the Law Institute of Altai State University (N = 60).

**Keywords** Legal awareness of a person · Degree of the significance of rights and freedoms · Legal consciousness questionnaire [LCQ] · Construct validity · Convergent validity

---

V. V. Molotova (✉) · N. V. Sabelnikova
Altai State Pedagogical University, Barnaul, Russia
e-mail: advocatemolotova@mail.ru

A. V. Molotov · D. V. Kashirsky · E. V. Gubernatorova
Altai State University, Barnaul, Russia
e-mail: Lawmaking@yandex.ru

D. V. Kashirsky
Russian State Social University, Moscow, Russia



S. G. Maximova et al. (eds.), *Advances in Natural, Human-Made, and Coupled Human-Natural Systems Research*, Lecture Notes in Networks and Systems 234,
https://doi.org/10.1007/978-3-030-75483-9_89

## 1 Introduction

Currently, the issue of security is especially relevant. This problem requires solving a wide range of issues, such as ensuring the observance of the rights and freedoms of each person. This goal cannot be achieved without the legal awareness of each person.

A significant number of studies conducted by philosophers, sociologists, and lawyers was devoted to legal awareness. Psychology also paid attention to the law representation in the individual consciousness. The specifics of the legal consciousness, its concepts, and research tools are at the center of attention of the modern psychologist. The law regulates social relations and cannot be separated from the subjects of social life. Therefore, psychological research can significantly contribute to legal consciousness.

A functional-structural concept of Ratinov (1981) claims that the ambient reality is reflected in the legal consciousness of a person in a specific way. Legal consciousness has a particular language, a system of concepts and categories, unique criteria, and assessment methods. A. R. Ratinov determines three functions of legal consciousness: (1) cognitive, (2) evaluative, and (3) regulatory. According to Ratinov, all forms of mental reflection of the legal life of society can be attributed to the legal consciousness (Ratinov 1986).

According to Enikeev (2005), legal consciousness can express itself in the motive of legally significant behavior as a manifestation of solidarity with law, general legal principles, guidelines, and values, or by the manifestation of legal negativism, and the denial of legal values formed at a certain stage of the societal development. Baranov (1997) underlined the following aspects of legal consciousness: (1) legal feelings, (2) legal attitudes, (3) legal experience, and (4) legal illusions. Pristanskaya and Yutskova (1990) introduced the structure of legal consciousness including (1) knowledge of the law, (2) ideas about law, (3) attitude to law (4) legal requirements, (5) attitude to legal regulations. The four-component model of legal consciousness was introduced by Shchegortsev (1981). The model contains a substantive component which includes the following components: (1) knowledge and ideas of people about legal phenomena; (2) an evaluative component (assessment of legal phenomena by people); (3) a behavioral component (behavior and activities of people in right-handed situations); and (4) an energy component (emotions and feelings of people). Gulevich (2009a, b) defined a sense of justice as a set of social attitudes to (1) crimes, (2) criminals, (3) law, (4) punishment, (5) law enforcement, (6) judicial and penal systems. Our research is based on the structure of legal consciousness, which consists of three main components: (1) cognitive (legal concepts, knowledge, cognitive assessment of legal phenomena), (2) affective (emotional assessment of legal phenomena), and (3) behavioral (intention to behave in a certain way in legal situations).

We consider the legal consciousness of individuals as a reflection of their legal norms and values. The psychological structure of legal consciousness includes the following elements: (1) awareness of the subject of basic rights and freedoms (objective legal norms); (2) significance of particular rights and freedoms for the individual; and (3) peculiarities of the subject's behavior and activities within the legal field, (compliance, violation or denial of legal norms).

The analysis of secondary literature on legal consciousness revealed a lack of tools for assessing the legal consciousness of individuals. Primarily, it is due to insufficiently developed theoretical and methodological grounds for the related research. The developed tools assess specific aspects of legal consciousness and its particular components: (1) cognitive (Kuznetsova 2008; Gulevich, 2009a, b; Sorokina et al. 2012) and (2) emotional and behavioral (Beznosov 2004; Khashchenko and Shpak 2013; Khashchenko et al. 2011). Nonetheless, there are no valid and reliable tools for assessing the level of consciousness of basic rights and freedoms and the degree of significance of the initial legal values, acting as guidelines for individuals' behavior and their actions in a legally significant situation.

In our opinion, the psychological analysis of the legal consciousness of individuals should be based on the following provisions: (1) the legal consciousness of individuals is more related to the essence of law than to individual legal norms and laws; (2) analyzing the structure of an individual's legal consciousness and evaluating its qualitative characteristics, the researcher should, first of all, take into account its subjective (appraisal) side expressed in the personal attitude of the subject to the basic legal principles; (3) various fields of legal consciousness of an individual are directly related to the objective legal system-branches of law and specific types of legal relations reflected in international acts and national constitutions. In fact, these legal acts are nothing more than the universal human values officially established by the legislator.

Therefore, the purpose of this study is to create a theoretical model and evaluate the psychometric properties of the questionnaire, which reveals the integral structure of the legal consciousness of individuals. First of all, the developed tool aims to measure the level of consciousness of the subject of constitutional human rights and freedoms. Secondly, it emphasizes the importance of fundamental rights and freedoms that underlie the formation of law and current legal regulation.

## 2 Materials and Methods

We took the following steps to develop the Legal Consciousness Questionnaire [LCQ]:

1. Theoretical analysis of relevant scholarship on legal issues and development of diagnostic tools on a given topic;
2. Analysis of the Universal Declaration of Human Rights and the Constitution of the Russian Federation;
3. Development and conduct of semi-structured interviews;
4. Development of the new questionnaire structure and expert assessment of items;

5. Empirical validation of the test, that is, the analysis of the questionnaire structure and internal consistency of its subscales and qualitative analysis of the data obtained;
6. Assessment of convergent validity of the tool.

The questionnaire is based on legal maxims related to values and norms objectively established by the international community in the Universal Declaration of Human Rights (General Assembly of the United Nations 1948) and subsequently reflected in the constitutional rights and freedoms of citizens of the Russian Federation (Russian Federation 1993). Such tools express the rights and freedoms of citizens existing at a particular stage of cultural-historical development of society as a kind of "ideal form" (Vygotsky 2001) to develop the individual's legal consciousness. Therefore, these tools can serve as an essential objective criterion for studying the nature of individuals' legal consciousness, resulting from the internalization of rights and freedoms accepted in society (Molotova 2018). Consequently, the first stage of our research is devoted to the theoretical analysis of the Universal Declaration of Human Rights and the Constitution of the Russian Federation, aimed at reconstructing the main areas of the legal regulation that control legal relations of people in modern Russian society.

Next, we created and conducted a semi-structured interview (N = 30) aimed at identifying the qualitative characteristics of legal awareness, manifested in (1) legal literacy (awareness), (2) the level of understanding of the law, and (3) the varying degree of significance of constitutional rights and freedoms for individuals. Thus, the law is assumed to be an empirical phenomenon that exists at the levels of individuals' consciousness and behavior.

The first version of the LCQ was developed based on further generalizations of interview responses. After that, lawyers (experts in the field of constitutional law) evaluated the draft questionnaire to confirm the adequacy of its wording to the concept under study.

Next, the LCQ was empirically validated using exploratory factor analysis and verified by confirmatory factor analysis on a student sample (N = 100).

The participants were 60 male and 40 female students aged 17–19 (M = 18.24, SD = 0.56). Students were invited to participate in the study, convinced that they could leave at any time. Participants were asked about the nature of the study after completing and returning the paper-and-pencil survey.

Finally, we checked the validity structure of the new tool. For this purpose, we used a standard procedure for analyzing the relationships of this new tool with currently existing tools that measure similar constructs. The most common assessment tools for measuring an individual's legal consciousness in Russian science are the following: (1) the measure of everyday justice by L. A. Yasyukova, (2) "The level of justice development" questionnaire by Gulevich, (3) the "Legal awareness of the penal system by employees" questionnaire by Evstafieva and Krasnik, and (4) the instrument for the professional study of concepts, developed by Gornostaev (as sited in Yevstafeeva 2015). The tools listed above measure the individual's legal consciousness at two different levels—basic and professional. Therefore, the "Legal awareness of the penal system by employees" a questionnaire by Evstafieva, Krasnik, and

Gornostaev's tool for the professional study of concepts were developed to measure professional legal consciousness of the legal system employers. In contrast, the tools of L. A. Yasyukova and O. A. Gulevich were aimed at the evaluation of the legal consciousness of the ordinary people. Since the LCQ was created to study the level of legal consciousness of an ordinary individual, in our study, we chose the tools of L. A. Yasyukova and O. A. Gulevich for comparison with the LCQ scales.

Participants of this stage of the study were students of the Law School (N = 60), aged 17–23 (M = 19.3, SD = 1.7). They filled out questionnaires posted on Google forms via the internet. At the beginning of the survey, they signed an informed consent form and were free to withdraw at any time.

## 3 Results

### *3.1 The Theoretical Basis for the Scale Structure*

The analysis of the Universal Declaration of Human Rights (General Assembly of the United Nations 1948) and the current Constitution of the Russian Federation (Russian Federation 1993), as well as the works of prominent experts in the field of the theory of constitutional law (Baglay 2007; Kozlova and Kutafin 2006), allowed us to identify five primary areas of legal regulation:

1. Personal Rights (the right to life and health, freedom of speech and thought, freedom of religion, freedom of conscience, the right to privacy, and the right to freedom of movement)
2. Economic Rights (the right to own, use and dispose of a property based on the right of ownership, the right to freely dispose of the ability to work, and the right to run business activities);
3. Political Rights (the right to elect and be elected to the authorities, the right to participate in rallies and processions, the right to join trade unions, freedom of peaceful assembly and association, and the right of access to public service)
4. Social Rights (the right to social security, free medical care, and education, pensions, guarantees, and benefits for the poor, disabled people, children without parental care, and minorities);
5. Cultural Rights (the right to freedom of artistic expression and creativity, the right to education, and access to cultural heritage).

Consequently, the analysis of legal regulation constituted a five-factor hypothetical model of the questionnaire based on the theoretical ideas about the principles and mechanisms of legal regulation in the basic areas of life and social interaction of individuals. The structure of the questionnaire allows us to perform a holistic assessment of the legal consciousness of individuals.

Next, we added a scale of consciousness of constitutional rights and freedoms related to the cognitive component of the personal sense of justice, measuring the level of legal literacy of the subjects. Adding this scale to our questionnaire, we made it possible to reveal the coherence of the emotional and cognitive components of legal consciousness by comparing the scores that referred to one's knowledge of certain rights and freedoms and the scores showing their significance for that person.

## 3.2 Development of the Item Pool and the Extension of the Scale Structure

Precise points for measuring the a priori five-factor definition were also developed based on the analysis of the universal Declaration of Human Rights, the current Constitution of the Russian Federation, and publications of the leading experts in constitutional law. These questions were included in a semi-structured interview (N = 30) aimed at assessing legal literacy (awareness), the level of understanding of the law, and the significance of constitutional rights and freedoms for individuals. The respondents were asked to express their attitude to the existing constitutional norms and indicate how necessary and significant they are in social interaction and whether it is acceptable to cancel or restrict their rights and freedoms established by law.

Analysis of the semi-structured interview results allowed us to introduce a general idea of law as an empirical phenomenon that reflects the subjective reality of legal consciousness. We used the results obtained for the wording of the questionnaire items. Then, experts in constitutional law assessed the adequacy of the test questions. As a result of this work, we developed a 56-point questionnaire.

## 3.3 The Measure

The designed LCQ contains 56 items, representing six subscales:

1. Awareness of Rights and Freedoms;
2. Personal Significance of Personal Rights and Freedoms;
3. Personal Significance of Economic Rights and Freedoms;
4. Personal Significance of Political Rights and Freedoms;
5. Personal Significance of Social Rights and Freedoms;
6. Personal Significance of Cultural Rights and Freedoms.

According to the instructions, respondents were asked to read these statements and express their agreement with them on a seven-point Likert scale. Below, we gave examples of elements related to the LCQ scales.

Subscale 1 "Awareness of Rights and Freedoms" (ten items) assesses the respondents' awareness of their rights and freedoms (whether a person hears about them for the first time or, on the contrary, encountered these phrases before and fully understands them). The statements related to this scale are divided into two types (two statements for each of the five legal regulation areas).

The first type of statements is formulated in a general form and does not disclose the content of the legal regulation area for assessment, allowing a researcher an opportunity to reveal the general level of understanding of the relevant categories. For instance, "I am well aware of my personal rights" (item one, the sphere of personal rights regulation).

The second type of statements relates directly to the Universal Declaration of Human Rights and the Constitution of the Russian Federation. These statements specify the content of each of the five areas of legal regulation. For instance, "I am well aware that every person has the right to own, use and dispose of property on the basis of property rights; the right to freely dispose of one's ability to work, the right to conduct entrepreneurial activities" (item 18, the sphere of regulation of economic rights), etc.

Subscales 2–5 evaluate the subject's attitude toward the content of specific rights related to various areas of legal regulation and social interaction. Each of these scales contains five types of statements accounting for the following:

1. Significance for the subject of a particular group of rights and freedoms (e.g., "it is vital for me that no one can interfere in my private and family life, violate the privacy of correspondence, telephone conversations");
2. Willingness to restrict the right in order to serve the public good (e.g., to restrict the personal right to freedom of thought and speech, and permit to censor the internet to protect children from harmful information);
3. Willingness to restrict their rights in exchange for material remuneration or social guarantees (e.g., to waive the right to elect representatives to government bodies for material remuneration, or abandon the ownership of the real estate in favor of a contract of social renting real estate, decent wages, free medical care, and job security);
4. Willingness to waive or restrict rights in favor of a more significant category of rights (e.g., to waive the right to free medical care or the right to receive a pension to guarantee smooth-running business);
5. Inadmissibility of restriction of rights and freedoms or the denial of certain rights and freedoms (subjects under no circumstances consider the possibility of restricting or waiving their rights).

Subscale 2 "Personal Significance of Personal Rights and Freedoms" (14 items) assesses the subject's attitude to the content of personal rights and freedoms. For instance, "It is important for me that people are born with equal rights and freedoms, regardless of the family and the conditions in which they were born" (item three).

Subscale 3 "Personal Significance of Economic Rights and Freedoms," containing four items, evaluates the subject's attitude to the content of economic rights and freedoms (e.g., item 21, "Under no circumstances am I willing to give up or limit my ownership").

Subscale 4 "Personal Significance of Political Rights and Freedoms" (eight items), assesses the subject's attitude to the content of political rights and freedoms. For instance, "I am ready to give up the right to elect the authorities for remuneration" (item 29).

Subscale 5 "Personal Significance of Social Rights and Freedoms," consisting of six items, evaluates the subject's attitude to the content of social rights and freedoms (e.g., item 43, "Under no circumstances am I ready to give up my cultural rights").

Subscale 6 "Personal Significance of Cultural Rights and Freedoms," composed of 11 items, evaluates the subject's attitude to the content of cultural rights and freedoms. For instance, "It is very important for me that I have the right to free education and health insurance" (item 46).

The developed measure was further factorized and evaluated for internal consistency.

## *3.4 Construct Validity*

The Kaiser-Meier-Olkin [KMO] Measure of Sampling Adequacy and the Bartlett's test of Sphericity indicated the suitability of the data for the structure detection (KMO = 0.705, $\chi 2$ = 5,639.672, df = 1,540, $p \leq 0.0001$). We carried out an Exploratory Factor Analysis [EFA] (principal component method, Varimax axis rotation) for empirical confirmation of the questionnaire structure. In general, it confirmed six factors that had a saturation of 61.76% of the total variance.

However, some items of the initial version of the questionnaire had low loads on the extracted factors. These items were dropped, and the measure was reduced to 34 statements which formed six subscales:

1. Awareness of Rights and Freedoms (ten items);
2. Personal Significance of Personal Rights and Freedoms Subscale (five items);
3. Personal Significance of Economic Rights and Freedoms Subscale (five items);
4. Personal Significance of Political Rights and Freedoms Subscale (five items);
5. Personal Significance of Social Rights and Freedoms (four items);
6. Personal Significance of Cultural Rights and Freedoms (five items).

Next, we evaluated the questionnaire factor structure (34 items) using Confirmatory Factor Analysis [CFA] (N = 100). Subsequently, we assessed the factor structure of the newly developed questionnaire using CFA (n = 100). Statistical indicators showed an acceptable model fit: $\chi 2$ = 748.89, df = 455, $p \leq 0.0001$; CFI = 0.90, RMSEA = 0.08, LO 90 = 0.07, HI 90 = 0.09, TLI = 0.87. Therefore, the factor solution proved the appropriateness of the developed tool for assessing the significance of rights and freedoms for the subject and their awareness of the rights and freedoms regulated by the laws established in society. A more detailed description of the LCQ construct validity testing is given in the paper by Molotova et al. (2020).

## 3.5 Convergent Validity

Convergent validity of the LCQ was assessed with a sample of students of the Law Institute of Altai State University (N = 60). The students' scores on the LCQ subscales were correlated with their scores on the two tools aimed to diagnose justice—the tools of Yasyukova (2000) and Gulevich (2009a, b).

The questionnaire of L. A. Yasyukova contains 13 items. The respondents have to choose one of the three possible answers. The tool contains the integral scale measuring the general level of legal consciousness as well as the level of legal consciousness in three specific areas: (1) mode of life, (2) business, and (3) civil sphere. Besides, the questionnaire provides an opportunity to assess the level of legal knowledge of the respondents.

The tool of O. A. Gulevich consists of four unfinished sentences and offers a choice of four completion options, one of which is open: respondents can write their ideas. As a result of the assessment, we can indicate one of the three possible levels of an individual's legal consciousness: (1) law obedience, (2) law enforcement, (3) law-making (Tapp and Kohlberg 1971).

We assume that the awareness of rights and freedoms reflects the well-developed self-identity of a person. Therefore, we expected that the LCQ scales aimed at revealing a subject's awareness of personal, economic, political, social, and cultural rights would correlate with the general indicator of legal consciousness scores using the tool of L. A. Yasyukova and the level of development of justice measured with the tool of Gulevich. Moreover, given that these tools are based on theoretical constructions with different values, levels of legal consciousness (subjective morality and objective societal, legal regulation), we expected that the LCQ scales would negatively correlate with the former and positively correlate with the latter. We also anticipated correlations between the scores on scales to reveal the significance of personal, economic, political, social, and cultural rights and freedoms measured by the LCQ and the scores on self-awareness in the lifestyle, business, and civil scales of Yasyukova's tool.

*Yasyukova's scales*

The data obtained with the Yasyukova's scales showed that the respondents' scores on legal consciousness ($M = 15.57$, $SD = 2.63$) referred to the second (medium) level (Fig. 1).

Figure 1 shows that, on average, students have a medium level of legal awareness in lifestyle and business, a low level of legal awareness in the civil sphere, and a low level of legal knowledge.

The correlational analysis results presented in Tables 1 and 2 indicate the negative correlation between the scores on the integral scale of the Yasyukova's questionnaire and scores on the significance of social rights and freedoms measured by the LCQ ($r = -0.236$, $p \leq 0.06$).

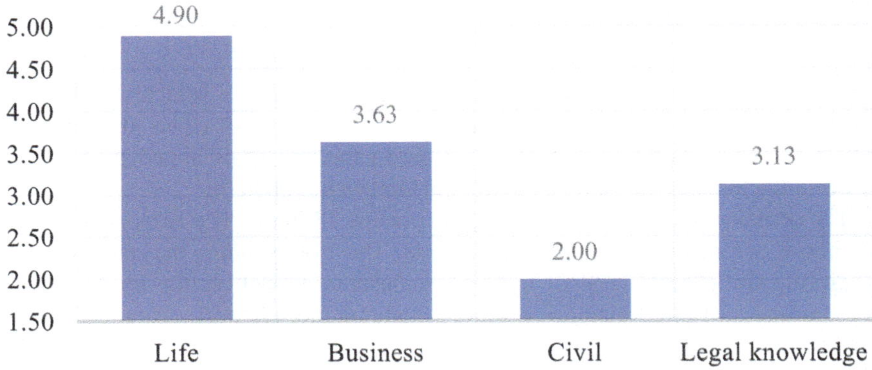


**Fig. 1** The students' scores on the specific areas of legal consciousness (Yasyukova's scales) *Source* Compiled by the authors

**Table 1** Table of correlations between the Awareness of the LCQ Rights and Freedoms subscales and Yasyukova's scales

| Yasyukova's scales | Awareness of Rights and Freedoms (LCQ) | | | | | |
|---|---|---|---|---|---|---|
| | Personal | Economic | Political | Cultural | Social | Overall score |
| Legal consciousness | 0.050 | 0.001 | 0.008 | 0.176 | –0.115 | 0.028 |
| Lifestyle | –0.044 | 0.040 | –0.113 | –0.052 | –0.074 | –0.057 |
| Business | 0.045 | 0.158 | 0.145 | 0.147 | 0.026 | 0.118 |
| Civil sphere | 0.205 | 0.082 | 0.085 | 0.049 | –0.100 | 0.058 |
| Legal knowledge | 0.091 | 0.032 | 0.121 | 0.012 | –0.101 | 0.024 |

*Source* Compiled by the authors

**Table 2** Table of correlations between the LCQ Significance of Rights and Freedoms subscales and the Yasyukova's scale

| Yasyukova's scales | Significance of Rights and Freedoms (LCQ) | | | | | |
|---|---|---|---|---|---|---|
| | Personal | Economic | Political | Cultural | Social | Overall score |
| Legal consciousness | –0.111 | –0.078 | 0.132 | 0.161 | –0.236* | –0.037 |
| Lifestyle | 0.013 | 0.102 | 0.148 | 0.154 | –0.171 | 0.076 |
| Business | –0.045 | –0.079 | 0.135 | –0.081 | –0.017 | –0.017 |
| Civil sphere | 0.164 | 0.147 | 0.063 | 0.144 | –0.060 | 0.135 |
| Legal knowledge | –0.210 | –0.037 | –0.015 | 0.092 | –0.207 | –0.118 |

*Note* $^*p \le 0.05$
*Source* Compiled by the authors

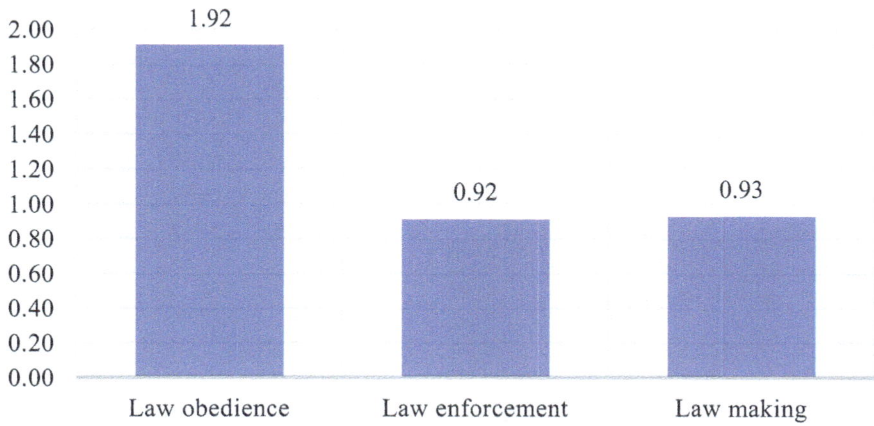


**Fig. 2** The students' scores on Gulevich's questionnaire. *Source* Compiled by the authors

*Gulevich's questionnaire*

The average scores on the scales of the questionnaire of O. A. Gulevich show that the participants had demonstrated low levels of perceived law obedience (Fig. 2).

The correlational analysis results, shown in Tables 3 and 4, indicate the negative correlation between the level of law obedience scores on the Gulevich's tool and the significance of cultural rights and freedoms measured by the LCQ ($r = -0.288$, $p \leq 0.026$).

Table 5 presents the results of the correlation analysis of Yasyukova's scales and Gulevich's questionnaire. One can see that there were no significant correlations between them, except for one case. The legal knowledge scores on the Yasyukova's scales negatively correlate with legal awareness scores on the Gulevich's questionnaire.

**Table 3** Table of correlations between the LCQ Awareness of Rights and Freedoms subscales and the subscales of the Gulevich's questionnaire

| The Gulevich's questionnaire scales | Awareness of Rights and Freedoms (LCQ) | | | | | |
|---|---|---|---|---|---|---|
| | Personal | Economic | Political | Cultural | Social | Overall score |
| Law obedience | –0.138 | –0.007 | 0.039 | –0.082 | 0.012 | –0.037 |
| Law enforcement | 0.071 | 0.046 | 0.074 | 0.049 | –0.072 | 0.032 |
| Law-making | 0.081 | –0.050 | –0.215 | 0.010 | 0.071 | –0.018 |

*Source* Compiled by the authors

**Table 4** Table of correlations between the LCQ Significance of Rights and Freedoms subscales and the subscales of the Gulevich's questionnaire

| The Gulevich's questionnaire scales | Significance of Rights and Freedoms (LCQ) | | | | | |
|---|---|---|---|---|---|---|
| | Personal | Economic | Political | Cultural | Social | Overall score |
| Law obedience | –0.141 | –0.083 | –0.031 | –0.288* | 0.038 | –0.152 |
| Law enforcement | 0.049 | 0.205 | –0.109 | 0.152 | –0.047 | 0.065 |
| Law-making | 0.047 | –0.096 | 0.083 | 0.169 | 0.067 | 0.093 |

*Note* *$p \leq 0.05$
*Source* Compiled by the authors

**Table 5** Correlations between the Yasyukova's scales and the Gulevich's questionnaire scales

| The Yasyukova's scales | The Gulevich's questionnaire scales | | |
|---|---|---|---|
| | Law obedience | Law enforcement | Law-making |
| Legal consciousness | –0.089 | 0.026 | 0.048 |
| Lifestyle | 0.100 | –0.025 | –0.033 |
| Business | 0.001 | –0.076 | –0.038 |
| Civil sphere | –0.088 | –0.028 | 0.000 |
| Legal knowledge | 0.202 | 0.014 | –0.294* |

*Note* *$p \leq 0.05$
*Source* Compiled by the authors

## 4 Discussion

The purpose of the study was to develop a tool for measuring the individual's legal consciousness in the following areas: (1) awareness of the rights and freedoms adopted in society, (2) significance of basic rights and freedoms that are the basis of legal regulation accepted in the society for the individual. Moreover, we aimed to assess its psychometric properties. At the first step, we performed a preliminary analysis of the Universal Declaration of Human Rights and the current Constitution of the Russian Federation to highlight the basic rights and freedoms of the citizens, which represent an objective ("intrapsychic," according to L. S. Vygotsky) form of legal consciousness. Five fundamental constitutional rights and freedoms were identified as five hypothetical factors of the model. Together with the factor named Awareness of Constitutional Rights and Freedoms, those factors were accounted for a six-factor questionnaire model.

Next, the initial version of the questionnaire was developed and subjected to discussion in the form of a semi-structured interview. The interview revealed the difficulties ordinary people face in understanding professional lawyers' the wording of items taken directly from the constitution. Consequently, we improved the wording of the items by making them clear for an ordinary person without losing a "letter and spirit of the law" in the Universal Declaration of Human Rights and the current Constitution of the Russian Federation. Then, two professional lawyers—experts in constitutional law, assessed these items. As a result of this work, we designed a 56 item LCQ.

Subsequently, the questionnaire was empirically validated. The proposed factor structure was generally confirmed using EFA and CFA. We omitted items with low load or double loads. The CFA confirmed the construct validity of the questionnaire.

Evaluation of the convergent validity of the developed questionnaire showed that some LCQ scales negatively correlated with the scales of the other two tools also aimed at the study of legal consciousness (Yasyukova's tool and Gulevich's tool). We assume the following explanation of the results obtained. The individual's consciousness is a complex systemic formation, and the techniques presented in our study reflect its different aspects. One can see that we dealt with two different constructs: the first, reflected in the concepts of L. A. Yasyukova and O. A. Gulevich, referred to the understanding of the law or the availability of legal knowledge; the second one concerned with the awareness of legal issues and evaluation of different legal situations (LCQ methodology). Both of them could be included in the structure of the individual's legal consciousness. Nonetheless, the connection between these two components is complex and possibly non-linear. Therefore, this phenomenon requires further comprehensive research.

The analysis of the tools of L. A. Yasyukova and O. A. Gulevich showed that they were based on different concepts of justice, determining opposite ways of assessing the levels of an individual's legal consciousness. Consequently, L. A. Yasyukova determines legal consciousness as a system of representation based on the understanding of the role of law and the comprehension of legal norms in the regulation of social relations (including those between the individual and the government) influencing one's attitudes to these norms (Yasyukova 2000). The concept of L. A. Yasynkova considers the orientation towards moral principles as a low level of legal consciousness named "legal nihilism." On the contrary, the concept of O. M. Gulevich is based on L. Kolberg and J. Tapp's moral theory. According to this concept, the highest level of legal consciousness is marked by compliance with one's moral and ethical standards in situations when legal regulation is considered unfair. This fact explains the negative correlation between legal knowledge (measured the Yasyukova's scales) and legal awareness (measured by Gulevich's questionnaire).

Despite the common usage of both tools in many Russian studies, they were not validated before. However, the traditional procedures for assessing validity are hardly suitable for these methods due to their basis on a qualitative methodology.

On the other hand, the absence of expected relationships between the LCQ and the two other questionnaires employed in the study also might be explained. The results obtained with Yasyukova's and Gulevich's measures showed a low level of legal consciousness of the study participants. Thus, the subjects might not cope with the LCQ methodology, requiring a certain level of development of a common legal culture and legal knowledge due to their low level of legal awareness. Then, further LCQ validation in a broader variety of samples is needed.

## 5 Conclusion

To summarize, the study introduced a theoretical model of the LCQ and described its preliminary empirical validation. The findings can be considered the first step in the development and validation of an individual's legal consciousness.

Currently, the research on the improvement of the psychometric characteristics of the LCQ measure is in progress. We plan to continue re-wording the dropped items by splitting them into smaller units to make them more comprehensible to respondents. Another direction of further research is a validation of the questionnaire on a wider variety of samples. Also, we plan to continue the evaluation of the convergent and discriminant validity of the LCQ.

## References


Baglay MV (2007) The constitutional law of the Russian Federation, 6th edn. Norma Press, Moscow, Russia

Baranov PP (1997) System theory and systems analysis of legal awareness of personnel of internal affairs agencies. Rostov Higher School of the Ministry of Internal Affairs of the Russian Federation Press, Rostov-on-Don, Russia

Beznosov DS (2004) Attitude to law as a manifestation of legal consciousness of cadets of universities of Ministry of Internal Affairs of the Russian Federation. Dissertation of candidate of psychology. St. Petersburg University of the Ministry of Internal Affairs of the Russian Federation, St. Petersburg, Russia

Enikeev MI (2005) Legal psychology. Norma INFRA-M Press, Moscow, Russia

General Assembly of the United Nations (1948) Universal declaration of human rights. https://www.un.org/en/universal-declaration-human-rights/index.html

Gulevich OA (2009a) Psychological aspects of jurisprudence. MPSU Press, Moscow, Russia

Gulevich OA (2009b) The structure of legal conscience and the behavior in legal area. Psikhologicheskie Issledovaniya 5(7):5. http://psystudy.ru

Khashchenko TG, Shpak MM (2013) Development of a questionnaire for the diagnosis of legitimate professional orientation of personality. http://psyedu.ru/journal/2013/4/Haschenko_Shpak.phtml

Khashchenko TG, Khashchenko AV, Shpak MM (2011) Psychology of lawful behavior. Questionnaire "legal behavior". UGSHA Press, Ulyanovsk, Russia

Kozlova EI, Kutafin OE (2006) The constitutional law of Russia, 4th edn. Prospect Press, Moscow, Russia

Kuznetsova OV (2008) The peculiarities of the formation of the legal consciousness of students of a technical university. Izvestia: Herzen Univ J Humanit Sci 63(2):119–122

Molotova VV (2018) The significance of constitutional rights and freedoms: the psychological aspect of the study of the legal consciousness. In: Anichkin ES (ed) The constitution of the Russian Federation: declarations and reality. ASU Press, Barnaul, Russia, pp 137–148

Molotova VV, Molotov AV, Kashirsky DV, Sabelnikova NV (2020) Survey for assessment of a person's legal consciousness: development and preliminary validation. Behav Sci 10(5):87

Pristanskaya OV, Yutskova EM (1990) Methods of criminalistic study of justice and public opinion about crime. All-Union scientific research Institute of Problems of Strengthening of Legality and the Rule of Law, Moscow, Russia

Ratinov AR (1981) The structure of legal consciousness and some methods of its research. In: Methodology and methods of social psychology, pp 201–214. Yuridicheskaya Literature, Moscow, Russia

Ratinov AR (1986) Legal conscience and illegal behavior. Methodology issues. In: Criminological problems of justice and public opinion about crime, pp 54–75. All-Union scientific research Institute, Moscow, Russia; Prague, Czech Republic

Russian Federation. (1993) The constitution of the Russian Federation. http://www.constitution.ru/en/10003000-01.htm

Shchegortsev VA (1981) Sociology of legal awareness. Mysl Press, Moscow, Russia

Sorokina AI, Popova MV, Esikova TV (2012) Components of professional mobility of young specialists of the Russian Federation and Scandinavian countries. Sci Opin 11:102–106

Tapp JL, Kohlberg L (1971) Developing senses of law and legal justice. J Soc Issues 27(2):65–91

Vygotskiy LS (2001) Lectures on pedology. Publishing House "Udmurt University", Izhevsk, Russia

Yasyukova LA (2000) Legal awareness: diagnostics and regularities of development. Prikladnaya Psikhologiya 4:1–13

Yevstafeyeva YA (2015) Legal awareness of employees of the penal system with different personal characteristics (Dis. of Cand. of Psychological Sci.). Chelyabinsk State University, Chelyabinsk, Russia

The manufacturer's authorised representative in the EU is Springer Nature Customer Service Centre GmbH, Europaplatz 3, 69115 Heidelberg, Germany. If you have any concerns regarding our products, please contact ProductSafety@springernature.com

Printed and bound by CPI Group (UK) Ltd, Croydon, CR0 4YY

30/09/2026

02225255-0001